Upconverting Nanoparticles

Upconverting Nanoparticles

From Fundamentals to Applications

Edited by Vineet K. Rai

WILEY-VCH

Editor

Prof. Dr. Vineet K. Rai
IIT (ISM) Dhanbad
Department of Physics
Police Line Road
Hirapur, Sardar Patel Nagar
826004 Dhanbad
India

Cover Image:
© BAIVECTOR/Shutterstock

Library of Congress Card No.: applied for

British Library Cataloguing-in-Publication Data
A catalogue record for this book is available from the British Library.

Bibliographic information published by the Deutsche Nationalbibliothek
The Deutsche Nationalbibliothek lists this publication in the Deutsche Nationalbibliografie; detailed bibliographic data are available on the Internet at <http://dnb.d-nb.de>.

Print ISBN: 978-3-527-34965-4
ePDF ISBN: 978-3-527-83486-0
ePub ISBN: 978-3-527-83487-7
oBook ISBN: 978-3-527-83488-4

Typesetting Straive, Chennai, India
Printing and binding CPI Group (UK) Ltd, Croydon, CR0 4YY

Printed on acid-free paper

C116837_060422

Contents

Preface

The conversion of low-energy photons into high-energy photons, known as "frequency upconversion," using advanced optical materials has become an emerging research field with wide consequence and impact in various scientific areas ranging from healthcare to energy and security. The materials showing frequency upconversion properties are known as upconversion (UC) materials. UC materials reveal variety of applications in different fields, viz. color display, two-photon imaging in confocal microscopy, WLEDs, high-density optical data storage, upconvertors, under sea communications, solid-state lighting, sensors, photovoltaics, photocatalysis, food industry, indicators, anti-counterfeiting, bioimaging, cancer therapy and other biological fields. It is known that in comparison to ultraviolet (UV) and visible light the near-infrared (NIR) light is abundant and non-destructive in nature. It has deep penetration in the organisms and less harmful quality. UC luminescent materials in nanosize range are known as UC nanomaterials or UC nanoparticles (UCNPs). UCNPs excited with non-destructive NIR light are a better choice than the conventional downconversion nanoparticles because they are free from autofluorescence, have low light penetration, and cause less severe photo-damage to living organisms. It is notable to mention that the low efficiency of UC materials definitely becomes a major barrier for their application in a wide range. For researchers, it is a top priority to overcome this problem. Several engineered UCNPs, e.g. organic, inorganic, hybrids, and thin films, have been explored widely to obtain highly efficient UC luminescent materials. Usually, organic luminescent materials suffer poor stability under harsh conditions and have poor long-term reliability, but have a greater ductility than inorganic materials. The inorganic luminescent materials are more durable and possess high thermal stability. So, the hybrid materials consisting of both inorganic and organic components, namely, metal organic frameworks (MOFs), have attracted researchers with enhanced luminescence properties as compared to the bare organic and inorganic materials. To enhance the upconversion efficiency, spherical metal nanoparticles showing plasmon resonance in close proximity of the UCNPs are utilized. The plasmonic nanostructures are widely used to evolve the UCNPs with improved electronic, metallic, and optical properties. When the surface plasmon resonance wavelength of the metallic nanostructure matches with the excitation wavelength of upconversion mechanism, signal enhancement occurs. Usually, the coating of gold (Au) and silver (Ag) nanoparticles is used to tune the luminescence properties of UCNPs, though the nanoparticles exhibit plasmon absorption in 400–600 nm range.

The upconversion emission efficiency can be enhanced by several ways, including doping with sensitizer, non-lanthanides, and coating with inorganic shell. The non-lanthanide co-doping in UCNPs has also been used frequently in order to get enhanced luminescence intensity along with the use of sensitizer ion. The co-doping of activator and sensitizer ions with proper concentration in an appropriate host matrix is essential to achieve highly efficient UC emission as the concentration quenching has a prejudicial effect on the luminescence intensity. The phonon frequency, stability, cost effectiveness, non-hygroscopic, and non-toxic nature of the UC materials are of utmost importance. The security of any important data, currency, etc. has become very crucial to prevent counterfeiting. UCNPs with high luminescence intensity can be validated in anti-counterfeiting applications. These materials are also utilized for visual exposure of fungicides, thiram, etc., which can be broadly applied in soybeans, apples, wine farming, etc., to avoid crop diseases and excessive use of pesticides. Rare-earth-ions-based UC emission has tremendous advantages in terms of long excited lifetime, sharp emission bandwidth, low autofluorescence, high photostability, high resolution, low toxicity, etc. Rare-earth ions are found to be very sensitive to even small changes in chemical surroundings. Therefore, it becomes essential to get information about the symmetry, bonding of the probe ion, and how they change their optical properties with chemical composition of the host materials. For getting the high quantum efficiency, concentration of the dopants should be high, but it may cause concentration quenching due to the interaction between the excited and unexcited neighbors. Therefore, the nano-structured materials containing metallic nanoparticles are of particular interest because the large local field around the rare-earth ions positioned near the nanoparticles may increase the luminescence efficiency. Among several strategies, the coating of upconversion nanoparticles with inorganic materials shell is an effective method to get enhanced UC luminescence. The core@shell approach offers shielding to the surface particles and thus reduces the surface defects and possibility of quenching. This core@shell architecture is very much beneficial in biomolecule conjugation and thus suitable for many biological applications. Different coating strategies have been employed according to the required application purposes. UCNPs probes can function as multiple contrast agents for concurrent use in altered medicinal imaging modalities by providing corresponding diagnostic information (i.e. MRI and CT). Bio-conjugation on the surface of the UCNPs shows a much enhanced imaging performance in comparison to the clinically used fluorescent dyes. Innovative bio-imaging methods are being established by combining the conventional medical imaging modalities using core-shell structured UCNPs.

The book entitled *Upconverting Nanoparticles: From Fundamentals to Applications* is completely different from the previously published books in all respects, including the basics, scientific and technological demands. It is divided into eighteen chapters. Chapter 1, authored by Mondal and Rai, introduces the basic concepts of upconversion, and upconversion of nano-particles. The introduction to frequency upconversion and its various mechanisms, excitation and de-excitation processes in hosts containing rare-earth ions along with the spectroscopic properties of rare-earth ions/transition metals are described in this chapter. The rate equations relevant to excited-state absorption and energy transfer processes with an overview of the UCNPs have been introduced. Chapter 2, authored by Mukhopadhyay and Rai, describes the synthesis protocol of upconversion nanoparticles. In this chapter introduction to host materials and synthesis strategies of UC nanomaterials like solid-state reaction, co-precipitation, sol–gel, hydrothermal, combustion, thermolysis, microwave-assisted synthesis, core@shell synthesis techniques, etc. have been described. Chapters 3 and 4, authored by Jain et al.; Ojha and Ojha, refer to characterization techniques and analysis; Raman and FTIR spectroscopic techniques and their applications, respectively. Various structural and optical techniques for the characterization of UCNPs, viz. X-ray diffraction (XRD), X-ray photoelectron spectroscopy (XPS), field emission scanning electron microscopy (FESEM), transmission electron microscopy (TEM), energy-dispersive X-ray spectroscopy (EDS), thermogravimetric analysis (TGA), ultraviolet–visible–near infrared (UV–Vis–NIR) absorption spectroscopy, dynamic light scattering (DLS), photoluminescence, Fourier transform infrared (FTIR), have been reported. Chapters 5, 6 and 7, authored by Ranjan et al.; Prasad and Rai; and Pattnaik and Rai, summarize the fundamental aspects of UCNPs based on their properties, frequency upconversion in UCNPs containing transition metal ions, and frequency upconversion in UCNPs containing rare-earth ions, respectively . Along with introduction the dynamics of UCNPs on the basis of fluorescence decay times, quantum yield measurement of UCNPs, frequency upconversion and its various mechanisms have also been interpreted. The various routes to enhance the upconversion luminescence along with the technological applications of UCNPs have been described.

Chapters 8, 9, and 10, authored by Singh; Dwivedi; and Ningthoujam et al., are devoted to the smart and new type of upconverting nanoparticles; surface modification and (bio) functionalization of upconverting nanoparticles, and frequency upconversion in core@shell nanoparticles, respectively. These chapters outline the upconverting core@shell nanostructures, hybrid upconverting nanoparticles, magnetic-upconverting nanoparticles, UC-based metal–organic frameworks, surface modification, bio-functionalization of upconverting materials, synthesis of core@shell and core@shell@shell UCNPs, and use of UCNPs for security, biological, and sensing applications. Chapters 11, 12, 13, 14, and 15, authored by Kumar, Mishra and Shwetabh; Singh et al.; Dey; Mahata, De and Lee; and Shahi and Rai, deal with the UCNPs in solar, forensic, security ink, and anti-counterfeiting applications; application of upconversion in photocatalysis and photodetectors; UCNPs in lighting and displays; upconversion nanoparticles in pH-sensing applications and upconversion nanoparticles in temperature-sensing and optical heating applications, respectively. Chapters 16, authored by Wang et al., throws the light on UCNPs applications in degradation of organic and inorganic pollutants along with the photocatalytic hydrogen generation. The visual detection of fungicides and plant viruses along with the future challenges have been explained by Kesarwani and

Rai in Chapter 17. Chapter 18, authored by Mukherjee and Sahu, involves the application of UCNPs in bio-imaging, drug delivery, photodynamic therapy, and photothermal therapy.

The present book is outcome of the untiring efforts of all the contributing authors. It will be very much helpful to the researchers as well as the undergraduate and post-graduate students studying physics, chemistry, materials science, biology, engineering, etc. in gaining a proper understanding about the upconversion luminescence. It was possible to complete this book only due to the great affection and blessings of Gurudev Pt. Shri Ram Sharma Acharya and Gurumataji Mata Bhagawati Devi Sharma. Special thanks to all my family members and research scholars for their motivation and kind support. I would also like to thank the Wiley team involved from the beginning till the completion of the book proposal. As a large number of topics related to the UCNPs and their applications have been covered in this book, there could be the possibility that some of the minute glitches have been missed out. Therefore, genuine suggestions and comments from the readers are welcome. Overall, the research developments on UCNPs and their uses in different fields starting from very basics to advanced level make the present book unique.

Department of Physics
Indian Institute of Technology
(Indian School of Mines),
Dhanbad, India

Professor (Dr.) Vineet K. Rai

1

Introduction to Upconversion and Upconverting Nanoparticles

Manisha Mondal[1,2] *and Vineet Kumar Rai*[1]

[1] *Indian Institute of Technology (Indian School of Mines), Department of Physics, Laser and Spectroscopy Laboratory, Dhanbad 826004, Jharkhand, India*
[2] *Tezpur University (Central University), Department of Physics, Napaam, Tezpur, Sonitpur 784028, Assam, India*

1.1 Introduction

Spectroscopy almost deals with the interaction of light and matter. It provides information about splitting of electromagnetic radiation into its constituent wavelengths. The beginning of spectroscopy lies since the observation of light dispersion through prism by Sir Isaac Newton. Among different spectroscopy techniques, optical spectroscopy delivers an exceptional tool by which one can find detailed information regarding the absorbing and emitting atoms, ions, molecules, defects, their local surroundings, etc. In a term, optical spectroscopy allows light to penetrate inside materials. Optical spectroscopy can be characterized into four parts: absorption, luminescence, reflection, and scattering. A marvelous dimension of research carried out in finding novel luminescent materials plays an important role in optical communication, lighting, medical diagnosis, etc. (Berthou and Jörgensen 1990; Cheng et al. 2013; Jiang et al. 2016; Lin et al. 2016; You et al. 2016; Dey and Rai 2017; Mehra et al. 2020). When an atomic system after absorbing the photons of appropriate frequency transits upward to a higher state and then by the spontaneous emission process, it may return to the ground state. This de-excitation route is familiar as the luminescence process. The occurrence of luminescence due to excitation of light is known as photoluminescence. On the other hand, luminescence due to excitation of an electron beam is termed as cathodoluminescence, which helps to identify impurities, lattice defects, and crystal distortions. Radioluminescence occurs due to excitation through the highly energetic electromagnetic radiations (i.e. α rays, β rays, and γ rays). The thermoluminescence phenomena are used in radiation dosimetry, dating of minerals and old ceramics, materials characterization, biology, forensic, etc. It occurs when a material radiates light as a consequence of release of energy kept in traps by thermal heating. Electroluminescence occurs due to the passage of electric current over a material. The emission of light due to mechanical disturbance originates triboluminescence. Conferring to the diverse positions of

Upconverting Nanoparticles: From Fundamentals to Applications, First Edition. Edited by Vineet K. Rai.

the excitation and emission bands, the luminescent materials can be categorized into Stokes- and anti-Stokes-type luminescent materials. These processes are typically exemplified by the Jablonski diagram (Jablonski 1935; Jablonski 1993). The luminescent materials are commonly known as phosphors, which means "light bearer," that consist of host and dopants. In these constituents, lanthanide materials are mainly introduced into the host matrix. Lanthanides have the most complicated electronic structures because of their large number of incomplete 4f energy levels. The present chapter presents a brief outlook on understanding the frequency conversion mechanisms, electronic energy levels of rare-earth (RE) ions, transition metal ions, theoretical description of the optical characteristics of RE ions, and Upconverting nanoparticles (UCNPs).

1.2 Frequency Conversion and Its Various Processes

The photoluminescent materials are able to display visible emissions via suitable ultraviolet (UV) or near-infrared (NIR) excitations. In the majority of cases, excitation energy is greater than emitted photon energy; this emission is called as Stokes emission, and the corresponding energy loss is known as Stokes shift. In certain circumstances, emitted energy is higher than absorbed energy; this is known as anti-Stokes emission.

1.2.1 Stokes Emission

The Stokes-type emission process possesses two types of features such as downconversion and quantum cutting (Huang et al. 2013; Loo et al. 2019). In quantum cutting process, two or more lower energy photons are emitted for each incident high-energy photon absorption. In this process, two, three or four low-energy photons are emitted because of the absorption of one NIR, visible, or ultraviolet photon. In this process, the conversion efficiency is more than 100%. In current years, quantum cutting has acknowledged considerable devotion as a budding method to improve the photovoltaic conversion efficiency of solar cells. On the other hand, in the downconversion process, emission of one lower energy photon takes place because of the absorption of one higher energy photon; thus, the conversion efficiency will not go beyond 100%.

1.2.2 Anti-Stokes Emission

The anti-Stokes emission process occur via three processes: two-photon absorption (TPA), second harmonic generation (SHG), and upconversion (UC) (Figure 1.1) (Pollnau et al. 2000; Gamelin and Gudel 2000; Suijver 2008; Grzybowski and Pietrzak 2013; Chen et al. 2015; Nadort et al. 2016). TPA is a type of nonlinear absorption process that can be defined as the simultaneous absorption of two photons of same or different frequencies by an atom, ion, or molecule. In this process, the electron is promoted from low energy level (i.e. ground state) to excited

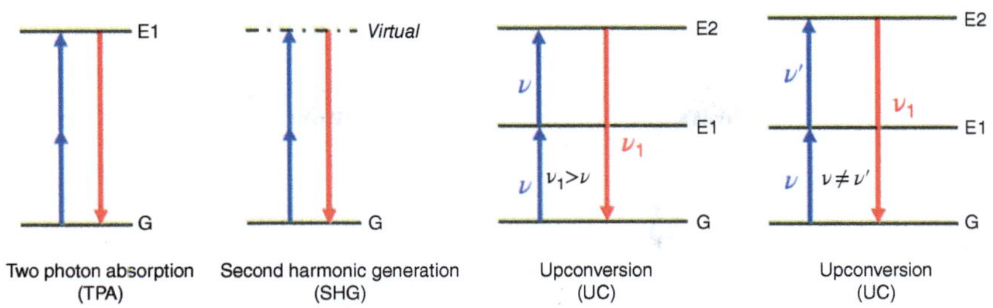

Figure 1.1 Basic energy-level diagrams depicting typical anti-Stokes processes.

level, and the energy of the emission transition is equal to the sum of two-photon energies. As this is a third-order nonlinear process, it is effective at precise high intensities. TPA was initially anticipated by Maria Goeppert-Mayar in the year 1931. This was experimentally verified by the laser after its discovery. A number of techniques are used to measure TPA, such as two-photon excited fluorescence, z-scan, nonlinear transmission, etc. On the other hand, SHG, "an optical nonlinear process," occurs from a virtual state in a medium having second-order nonlinear susceptibility. This was revealed and experimentally verified by Franken et al. (1961). They detected the second harmonic light when an intense beam of 6943 Å from the ruby laser was passed through the quartz crystal. In this process, two photons of the same frequency interact with a nonlinear material (i.e. medium) and give rise to a new photon of double the frequency or energy of the incident photons. Furthermore, UC is also an anti-Stokes process that converts the lower energy photons into high-energy photons, e.g. infrared to visible or UV light (Figure 1.1). It is a stepwise absorption process involving intermediate states (Auzel 1966; Ovsyakin and Feofilov 1966). Basically, among these three processes of converting lower energy photons into higher energy photons, TPA and SHG need a coherent beam as well as a very high excitation beam intensity. In the UC process, coherent pumping and high intensity of the excitation beam are not necessarily required. It occurs even at low intensity of the excitation beam because of the presence of real intermediate states (generally, of metastable nature).

The materials that exhibit the UC properties are known as upconverting materials. In recent years, these upconverting materials are extensively used in sensing, infrared counters, solid-state lasers, solar cells, fingerprint detection, security ink, upconverters, biological fields, etc. (Digonnet 1993; Wade et al. 2003; Rai 2007; Wang and Liu 2009; Gu et al. 2013; Li et al. 2013; Wang and Zhang 2014; Chen et al. 2014; Mondal and Rai 2020). Generally, the UC phenomenon observed in these materials is not as simple as depicted in Figure 1.1. Several processes accountable for UC mechanisms are as follows.

1.2.2.1 Ground/Excited-State Absorption (GSA/ESA)

Ground-state absorption (GSA) is one of the simplest routes for UC mechanism (Auzel 1973, 2004; Garlick 1976; Rai et al. 2013; Reddy et al. 2018). The process in which the ground-state ions (i.e. electrons) after absorbing the requisite energy from the pump photons are promoted to the first intermediate level is known as the GSA

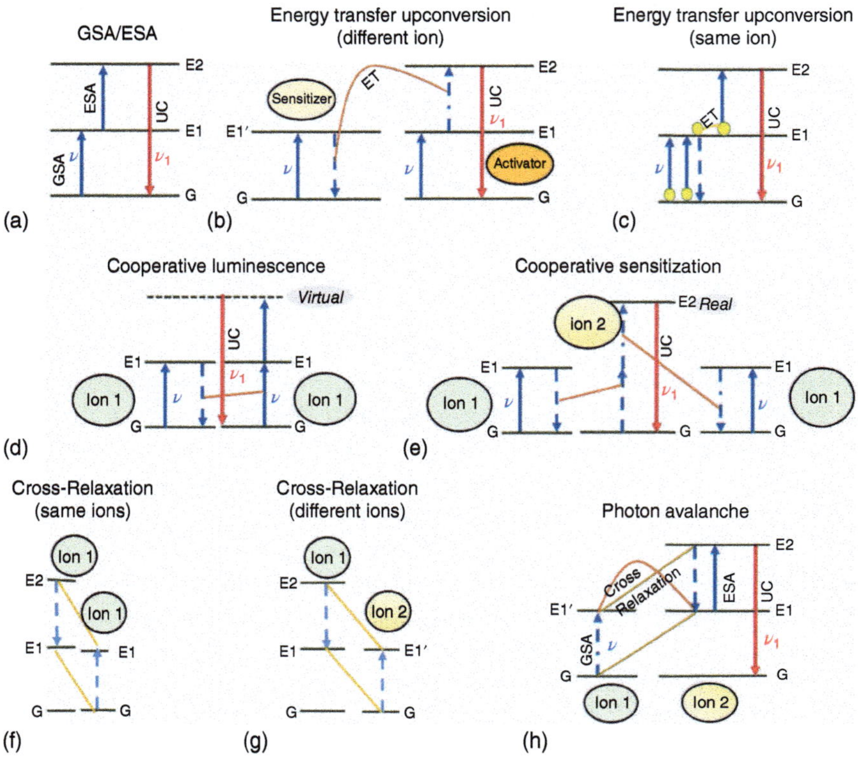

Figure 1.2 Schematic representation of possible UC mechanisms: (a) GSA/ESA, (b and c) ETU, (d) cooperative luminescence, (e) cooperative sensitization, (f and g) CR, and (h) PA processes.

process. Conversely, sequential absorption of two light quanta by a particular ion is known as ESA process (Auzel 1973, 2004; Garlick 1976; Rai et al. 2013). In the case of ESA process, the ion present in the intermediate state absorbs the second photon and transits upward to the next higher state. For example, the energy-level diagrams for GSA and ESA mechanisms are presented in Figure 1.2a. Here at first, an ion absorbs the pump photon of energy (=$h\nu$, where "h" is Planck's constant and "ν" is the frequency of the incident photon) and reaches to the intermediate state E1 (exhibit long lifetime) from the ground state G via the GSA process and then a second pump photon (of the same energy) excites the ion from E1 state to the next higher state E2. A radiative decay of the ion from the excited state (E2) to the ground state (G) results in UC emission. Thus, a single ion is involved in the whole ESA process. For getting proficient UC emission through the ESA process, a ladder-like energy-level arrangement in ions is essential.

1.2.2.2 Energy Transfer Upconversion (ETU)

Like the ESA process, the energy transfer upconversion (ETU) process also involves successive absorption of two energy quanta by the ions to occupy the intermediate (i.e. metastable) state (Figure 1.2). As in the ESA process there is an involvement of single ion, however, ETU operates within two (similar or different) ions. In this mechanism, the involved two dopant ions are termed as sensitizer and activator (Heer et al. 2003; Boyer et al. 2007; Shan et al. 2007; Soni et al. 2015; Mukhopadhyay and Rai 2020; Pattnaik and Rai 2020). At first, both the (different) ions absorb the pump photons from the ground state and then moves to their respective metastable states (E1$'$ and E1, where E1$' \cong$ E1) through the GSA process (Figure 1.2b). After that, the sensitizer ion (present in E1$'$ state) handovers its excitation energy to the neighboring activator ion (present in E1 state) and relaxes back to the ground state. The activator ion after gaining this excitation energy from the sensitizer reaches to the next higher energy state (E2).

When the two involved dopant ions are similar, these two ions are initially excited to the intermediate state (E1) after receiving the energy from pump photons (Figure 1.2c). The two ions present in the E1 state exchange their energy in such a way that one ion (i.e. donor), after transferring its excitation energy to the other excited ion (i.e. acceptor), decays nonradiatively to the lower energy level (G). The other ion (i.e. acceptor) after getting excitation energy from the first one (i.e. donor) is promoted to the next higher energy state (E2). A radiative transition from state E2 to the ground state (G) generates a photon of energy ($=h\nu_1$), which is higher than the incident photon energy ($=h\nu$) (Figure 1.2). This ETU process is the most efficient UC emission process (Auzel 2004; Rai et al. 2007, 2008). In this process, the dopant ion concentration (which regulates the average distance concerning adjacent dopant ions) plays a key role in the UC emission intensity.

1.2.2.3 Cooperative Luminescence and Cooperative Sensitization Upconversion (CSU)

UC emission by a cooperative energy transfer process involves two ions (one acts as a donor and the other ion as an acceptor). In the cooperative luminescence process, two ions absorb the pump photons successively and reach the higher (intermediate) state E1 (Figure 1.2d). In this intermediate level, these two ions transfer their energy in such a way that one ion (donor) transfers its excitation energy to the other one (acceptor) and the donor returns to the ground state (G). The acceptor, after gaining the excitation energy from the donor, transits upward to a higher energy state, "which is a virtual state." This virtual state is also known as the cooperative energy state (Lee et al. 1984; Maciel et al. 2000; Diaz-Torres et al. 2005). From this virtual state, it relaxes radiatively to the ground state (G) via emitting a photon of energy larger than the incident photon energy (Figure 1.2d). On the other hand, in the cooperative sensitization process, when the energy of the two excited ions are transferred to a third ion (ion 2), then it goes from the ground state to an excited state having energy equal to the sum of the energies of the two individual ions (Martín et al. 2001; Salley et al. 2001, 2003). In Figure 1.2e, the excitation energy of the two excited ions (ion 1) present in the state E1 is transferred to a third ion (ion 2). The third ion (ion 2) present in the ground state (G), after absorbing the excitation energy corresponding to the two excited ions (ion 1), moves to its higher state (E2). After that, the third ion from the excited state (E2) relaxes radiatively to the lower levels (say ground state) via emitting the photons of energy higher than that of the incident photon. This process is known as cooperative sensitization, and the emitting state (E2) in this process is a real state (Figure 1.2e). Thus, the cooperative sensitization is more effective than cooperative luminescence because it may compensate the low UC emission efficiency (Dwivedi et al. 2007; Liang et al. 2009).

1.2.2.4 Cross-relaxation (CR) and Photon Avalanche (PA)

The cross-relaxation (CR) process occurs due to ion–ion interaction (ions may be similar or different) (Chen et al. 2014; Pattnaik and Rai 2020) (Figure 1.2f,g. The cross-relaxation between two identical ions/molecules is responsible for self-quenching (Figure 1.2f). In the self-quenching process, the intermediate states of both the ions (ion 1) have the same energy (E1). When the cross-relaxation occurs between two different ions (Figure 1.2g), the first ion shares a part of its excitation energy to the second ion by the process E2 (ion 1) + G (ion 2) → E1 (ion 1) + E1′ (ion 2) (Figure 1.2g). In this process, the first ion (ion 1) initially present in the excited state (E2) interchanges a part of its excitation energy to the second ion (ion 2) that is initially available in the ground state (G). By this way, the decrease in the energy of the first ion (ion 1) is equal to the increase in the energy of the second ion. This results in both the ions/molecules changing simultaneously to the excited state (E1 and E1′). Among the other UC processes, the most exciting process is photon avalanche (PA), which was first experimentally observed in Pr^{3+}-doped infrared quantum counters (Chivian et al. 1979). Generally, this PA process occurs when the excitation energy exceeds its threshold limit. When the excitation energy is lower than the threshold energy, the emitted intensity is very poor, but as it exceeds the limit, the emitted intensity becomes enormously greater (Joubert 1999; Singh et al. 2011; Zhu et al. 2012; Mondal et al. 2016). For occurrence of PA process, at first, the intermediate level and the upper excited level are populated by the GSA, ESA, and ETU processes. By the CR process between these upper excited level and the ground state of a neighboring ion, two ions are generated in the intermediate level E1 (Figure 1.2h). Now, two ions are available in the intermediate state for the ESA process. Thus, with the feedback looping of ESA and CR processes simultaneously, the number of ions in the intermediate level increases, which give rise to strong UC emission.

The PA process is an unusual pumping process because it may lead to strong UC emission from the upper excited state E2 without any resonant GSA from the ground state (G) to the intermediate state (E1) of ion 2 (Figure 1.2h). The frequency of incident photon is in resonant with state E1′ of ion 1 and the upper excited state E2 of ion 2. An efficient CR process, i.e. E2 (ion 2) + G (ion 1) → E1 (ion 2) + E1 (ion 1), occurs between ion 1 and ion 2. This results in both the ions to occupy the intermediate state E1. These two ions readily populate the level E2 through ESA to further initiate the cross-relaxation. With the feedback looping of these efficient cross-relaxation and ESA processes, the number of ions in the intermediate state E1 increases rapidly, which results further an enormous increase in the population of level E2. Thus, in the PA process, a strong UC emission from state E2 to the ground state G (of ion 2) has been observed.

1.3 Transition Metals and Their Properties

The optical centers are necessary for the perfect crystals to exhibit the optical spectra. Depending on the absorption and emission bands of the optical centers present in the pure crystals, they are pertinent for diverse applications, such as optical amplifiers, solid-state lasers, color displays, absorbers, improving in luminescence brightness, fibers, optical switches, etc. Any element in the periodic table may act as a foreign element in the crystal. However, essentially, a few number of elements can be ionized, which can generate energy levels and thus yield optical features. For industrial applications, the two extremely important elements are transition metals and REs in the periodic table. Transition metal ions are especially used as optically active dopants in tunable solid-state lasers (Solé et al. 2005). These ions belong to the fourth period of the periodic table with electronic configuration $1s^2 2s^2 2p^6 3s^2 3p^6 3d^n$, where "$n$ (varies from 1 to 10)" is the number of 3d electrons present in the transition metal ions. Generally, valence electrons are responsible for optical transitions; hence, in the case of transition metals, 3d electrons are accountable. Because of the large radius of transition metal ions as compared to lanthanides and no shielding of valence electrons, strong field effect occurs; hence, they exhibit the broad bands.

The Sugano–Tanabe diagram explains the energy-level diagram for the transition metal ions (Figure 1.3) (Tanabe and Sugano 1954a,b). The spectroscopic terms for the free ion states of the transition metal ions due to the L-S interaction are described

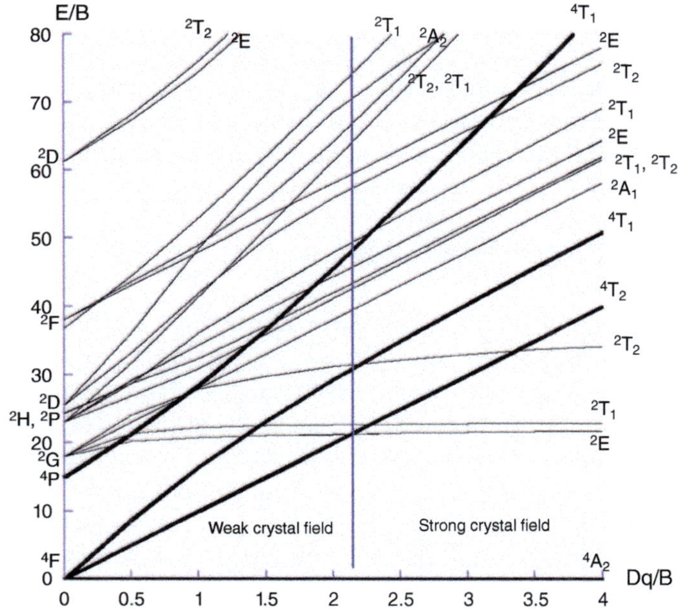

Figure 1.3 Tanabe–Sugano diagram for the d^3 electron configuration in the octahedral crystal field. Source: Brik et al. (2016). Reprinted with permission of The Electrochemical Society.

as $^{2S+1}L_J$, where, L, S, and J denote the total orbital angular momentum, total spin angular momentum, and total angular momentum, respectively. The energy separation among the ^{2S+1}L states, i.e. the strength of the electron–electron interaction, can be calculated with the help of Racah parameters (A, B, and C) (Solé et al. 2005). On the basis of octahedral crystal lattice, Sugano and Tanabe explained the occurrence of energy levels in the case of transition metal ions, but by using this diagram, one can also interpret the optical spectra arising from the transition metal ions in different types of host lattices.

This diagram explains the splitting of ^{2S+1}L free ion energy states with the ratio between the strength of the crystal field and the electron–electron interaction strength (symbolized as Dq/B) versus the free ion energy levels (E/B units). In this diagram, the *y*-axis is in terms of energy "E" scaled by B (one of the Racah parameters). The splitted terms for ^{2S+1}L energy states are termed as A, T, and E levels. This Sugano–Tanabe diagram also explains the nature of the optical bands for transition metal ions. In the case of strong crystal field approximation, the crystal field effect dominates over the electron–electron interaction among 3d ions. Accordingly, there are three single-electron orbitals for each orbital. Furthermore, according to the Sugano–Tanabe diagram, for low crystal field strength, the emission band is shifted toward the lower energy side. For this specific nature, the emission wavelength in the transition metal ions depends on a particular host material. Thus, doping of transition metal ions in different host materials directed to the advancement of countless varieties of tunable solid-state lasers. Most of the transition metal ions are incorporated in the octahedral crystal host matrix, so their energy level can be explained on the basis of Sugano–Tanabe diagram (Tanabe and Sugano 1956). However, in some cases, such as Ni^{2+}, Co^{2+}, and Cr^{2+} ions, these transition metal ions are incorporated in the tetrahedral crystal lattice for different applications; therefore, the Konig and Kremer diagram (Konig and Kremer 1997) is applicable in explaining the energy levels of transition metal ions other than the octahedral one.

1.4 Rare Earths and Their Properties

Most of the lasers, phosphors, amplifiers, etc., comprise RE elements. Surprisingly, the global applications of RE-based materials are increasing from industry applications to medical applications. There are 15 lanthanide elements along with two more elements i.e. scandium (Sc) and yttrium (Y). These 15 lanthanide elements are commonly named as lanthanum (La), cerium (Ce), praseodymium (Pr), neodymium (Nd), promethium (Pm), samarium (Sm), europium (Eu), gadolinium (Gd), terbium (Tb), dysprosium (Dy), holmium (Ho), erbium (Er), thulium (Tm), ytterbium (Yb), and lutetium (Lu). Most of the RE elements are entitled as per the name of the inventors or the name of their revealed places. These RE elements are incorporated in different host materials in their ionized (either divalent or trivalent) form. The divalent RE ions {Eu (+2), Yb (+2), and Sm (+2)} possess one more electron compared to the trivalent ions and thus exhibit different optical features and treat differently.

1.4.1 Trivalent Rare-Earth Ions

The outer most electronic configurations of divalent and trivalent RE ions are 5d $4f^n$ $5s^2$ $5p^6$ and $4f^n$ $5s^2 5p^6$, respectively, where n (varies from $n = 0$ to 14) specifies the number of electrons in the unfilled 4f shell. These $4f^n$ electrons are the valence electrons that are accountable for the spectroscopic transitions.

1.4.1.1 Electronic Structure

The presence of valence electrons in the 4f shell makes the RE ions as luminescent centers of any phosphor material. The group of 15 elements comprising atomic number starting from 57 to 71 in the sixth period of the periodic table together with scandium (Sc) and yttrium (Y) are known as RE elements. When these RE elements are introduced into the hosts, they easily convert into their either doubly or triply ionized states to acquire their stable electronic configurations. The outer most electronic configurations of lanthanum (La, atomic number Z = 57) and the last element lutetium (Lu, Z = 71) in their triply ionized state are $4f^0$ $5s^2$ $5p^6$ and $4f^{14}$ $5s^2$ $5p^6$, respectively. There are fifteen possibilities for filling these 4f orbitals as the f orbital contains seven suborbitals. Actually, these unfilled 4f valence electrons are in control for optical transitions. Table 1.1 presents the electronic arrangements and ground states of each triply ionized RE element. The actual electronic configuration of the 15 RE elements (i.e. from La to Lu) is [Xe] $5d^1$ $6s^2 4f^n$ ($n = 0$ to 14). However, in

Table 1.1 Electronic configuration of trivalent ionic states of RE elements (Shionoya et al. 1998).

Ion	Atomic number	Number of 4f electrons (n) and electronic configuration	$S = \Sigma s$	$L = \Sigma l$	$J = L - S$ ($n < 7$) $J = L + S$ ($n \geq 7$)	Ground state
La^{3+}	57	0 and [Xe]$4f^0$	0	0	0	$^1S_0$
Ce^{3+}	58	1 and [Xe]$4f^1$	1/2	3	5/2	$^2F_{5/2}$
Pr^{3+}	59	2 and [Xe]$4f^2$	1	5	4	$^3H_4$
Nd^{3+}	60	3 and [Xe]$4f^3$	3/2	6	9/2	$^4I_{9/2}$
Pm^{3+}	61	4 and [Xe]$4f^4$	2	6	4	$^5I_4$
Sm^{3+}	62	5 and [Xe]$4f^5$	5/2	5	5/2	$^6H_{5/2}$
Eu^{3+}	63	6 and [Xe]$4f^6$	3	3	0	$^7F_0$
Gd^{3+}	64	7 and [Xe]$4f^7$	7/2	0	7/2	$^8S_{7/2}$
Tb^{3+}	65	8 and [Xe]$4f^8$	3	3	6	$^7F_6$
Dy^{3+}	66	9 and [Xe]$4f^9$	5/2	5	15/2	$^6H_{15/2}$
Ho^{3+}	67	10 and [Xe]$4f^{10}$	2	6	8	$^5I_8$
Er^{3+}	68	11 and [Xe]$4f^{11}$	3/2	6	15/2	$^4I_{15/2}$
Tm^{3+}	69	12 and [Xe]$4f^{12}$	1	5	6	$^3H_6$
Yb^{3+}	70	13 and [Xe]$4f^{13}$	1/2	3	7/2	$^2F_{7/2}$
Lu^{3+}	71	14 and [Xe]$4f^{14}$	0	0	0	$^1S_0$

the triply ionized state, these elements lose their 5d and 6s orbital electrons; hence, the outer most electronic configuration of trivalent RE ions becomes $4f^n\ 5s^2 5p^6$ (Table 1.1). Therefore, due to the larger radii of 5s and 5p orbitals compared to the 4f orbital, these 4f electrons are shielded by 5s and 5p orbitals. Unlike transition metals when doped into solid materials, the outer 3d electrons are strongly affected by the crystal field effect; in the case of RE ions due to this shielding effect when they are incorporated into a solid host, the 4f electrons are weakly perturbed. Because of this shielding of 4f electrons by the completely filled outer electronic shells ($5s^2\ 5p^6$), the relative positions of 4f energy levels in the RE ions do not vary very much from one host to the other. Owing to this exceptional property of the triply ionized RE ions, they exhibit sharp absorption and emission spectra and thus have longer lifetime.

1.4.1.2 Interaction of Rare-Earth Ions

The 4f–4f transitions are parity forbidden according to the Laporte selection rule, but when incorporated into a host matrix, the electronic structure of the REs are perturbed, and because of the intermixing of $4f^n$ and $4f^{n-1}$ 5d orbitals, the optical transitions become allowed. Because of this intermixing, the parity of levels is changed and the transitions become allowed. These transitions are known as electric dipole allowed transitions. By using Schrodinger's equation, the energy levels of the RE ions responsible in optical transitions can be calculated as (Auzel 2004; Solé et al. 2005)

$$H\Psi = E\Psi \tag{1.1}$$

where "Ψ" is the eigenfunctions of the optical center and "H" denotes the Hamiltonian because of the diverse interactions of the 4f orbital electrons. Generally, the crystal field theory and the molecular orbital theory (MOT) have been used to describe the interaction between the RE elements and the host matrices.

Crystal Field Theory The crystal field theory was first described by Hans Bethe and John Hasbrouck van Vleck in the year 1930. In the case of transition metal ions when incorporated in a host material, the outer electrons are greatly affected by the surrounding host environment. However, in the case of RE ions when doped into a host material, the energy levels are only slightly perturbed because of the shielding by the $5s^2\ 5p^6$ orbitals. This breaking of degeneracy of the d and f electron orbitals can be described on the basis of crystal field theory (Wybourne 1965; Wybourne and Meggers 1965; Carnall et al. 1989; Auzel 2004; Liu 2005, 2015; Solé et al. 2005). The filled orbitals in the case of RE ions are 4d, 5s, and 5p, whereas 4f, 5d, and 6s are valence electron shells in their triply ionized state. Because of the larger radii of 5s and 5p orbitals, the 4f electrons of the RE ions are protected from the surrounding perturbation. The energy levels of the RE elements can be represented by some elementary quantum mechanical terms, total orbital angular momentum L (sum of total quantum number l), total spin angular momentum S (sum of total quantum number s), and the total angular momentum J ($=L+S$). With the help of Schrodinger's equation, the interaction between the RE ions and the host element can be realized. The overall Hamiltonian (H) of the RE ions subjected to the host surrounding can be defined as

$$H = H_F + H_C \tag{1.2}$$

where H_F is the Hamiltonian of the free ion (i.e. isolated) and H_C is the Hamiltonian arises due to the crystal field, which deals with the interaction between the 4f electrons of the RE ions and the crystal field of the host material.

In the case of isolated, i.e. free RE ions, the Hamiltonian (using the perturbation theory) can be described as

$$H_F = H_0 + H_{ee} + H_{SO} \tag{1.3}$$

where H_0, termed as the central field Hamiltonian, describes the interaction energy arises in the valence electrons owing to nucleus and electrons of triply ionized REs, H_{ee} describes the interaction energy due to the interaction among valence electrons, and H_{SO} is the spin–orbit interaction energy of the electrons.

The strongest interaction among three interactions is the Coulomb interaction, which signifies the repulsion among 4f valence electrons in the RE ions. The individual orbital angular momentum (l_i) and the spin angular momentum (s_i) of the individual electrons couple each other to form a total orbital angular momentum (L) and total spin angular momentum (S) due to the Coulombic interaction. Because of this interaction, the energy levels of any RE ions are divided into ^{2S+1}L folds. The degeneracy of the separated energy levels turn out to be $(2S+1)(2L+1)$ fold, and the energy separation concerning these levels are of the order of $\sim 10^4$ cm^{-1} (Figure 1.4).

The most significant interaction between the host material and the doped RE ions is the spin–orbit interaction. This interaction helps for calculating the free ion multiplet terms. Conferring to the Russell-Saunders scheme, the spin and orbital angular momentum couples with each other, and as a result, the total angular momentum term (J) generates. The value of J varies according to the variation for the filling of orbitals. For less than half-filled $J = L - S$ and for more than

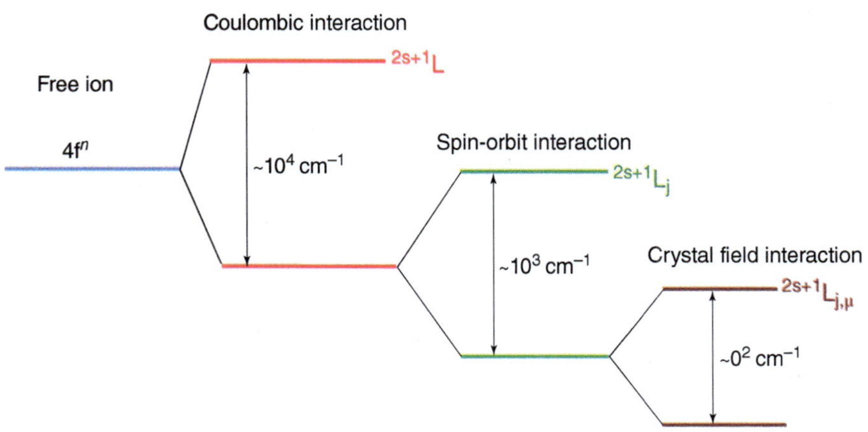

Figure 1.4 Splitting of 4f energy levels under different perturbations (drawn for visualization only).

half-filled shells, the value of $J = L + S$. Because of the spin–orbit interaction, the splitted levels are represented as $^{2S+1}L_J$ levels. In this interaction, the degeneracy of the splitted levels is $(2L+1)(2S+1)$ fold, and the order of the separated levels is $\sim 10^3$ cm^{-1} (Figure 1.4). The arrangement of energy levels of these multiplets is decided according to Hund's rule.

The shielding effect of the RE ions causes valence electrons weakly perturbed in the presence of host crystal field. Because of this, the spin–orbit interaction dominates over the crystal field interaction, and this results $(2L+1)(2S+1)$ fold degenerate energy levels of the RE ions. Because of this, the optical spectra of the RE ions incorporated into host materials are similar to the free ion spectra. However, while doping the RE ions in host materials, sometimes, a strong crystal field arises (generally because of the more asymmetric nature), and hence, the $^{2S+1}L_J$ levels of the RE ions again show splitting, known as Stark splitting. Degeneracy of Stark splitting depends on the value of electrons present in the outer orbitals. For even values of the 4f electrons, the degeneracy of the $^{2S+1}L_J$ level becomes $(2J+1)$ fold; however, it becomes $(J + 1/2)$ fold degenerate in the case of odd values of the valence electrons. Therefore, the general representation of the energy level of the dopant ions for Stark splitting is $^{2S+1}L_{J,\mu}$ (μ indicates the numbering of Stark sublevels) and the order of this splitting is $\sim 10^2$ cm^{-1} (Figure 1.4). In the case of transition metal ion-incorporated hosts, the d orbitals are the valence electrons, but they remain unshielded; hence, the crystal field interaction totally changes the energy levels of the transition metals. For example, if the outer arrangement is d^1, the electron–electron interaction energy term, i.e. H_{ee}, turns out to be zero because of the presence of a single electron in the valence state. Accordingly, in this circumstance, no variance occur between the weak (i.e. H_{SO} governs) and strong crystalline fields (i.e. H_{CF} leads). Then, again for d^n configuration, where $n > 1$, the electron–electron interaction term arises; therefore, the crystal field splitting turns into significantly complicated. Generally, in the case of RE-doped host materials (known as phosphors), because of the weak crystal field effect, the spin orbit interaction dominates (creates electronic states) and hence generates the ladder-like energy levels of the $4f^n$ valance shell configuration.

Molecular Orbital Theory (MOT) For the interpretation of the molecular electronic structure, MOT is the most important theory discovered by Friedrich Hund, Robert Mulliken, John C. Slater, and John-Lennard Jones (Mulliken 1967). This theory is based on quantum mechanical concepts. For understanding the structure of the energy levels of dopant ions at the quantum level, this theory works as the semi-empirical mode. Besides the GSA, ESA, ETU, and PA processes that are responsible for the generation of UC emission phenomena, the charge transfer mechanism is also an important process. To understand the charge transfer spectra, MOT is useful. In this process, valence electrons are equally shared by the ligand

and the dopant ion (Zhu et al. 2016). Therefore, the eigenfunctions of the molecular orbitals can be written as a direct combination of the eigenfunction of the ligand ion (Ψ_D) plus the eigenfunction of the dopant ion (Ψ_H), i.e.

$$\Psi = C(\Psi_D + \lambda\Psi_H) \tag{1.4}$$

where C is the normalization constant and λ is the mixing coefficient (Solé et al. 2005).

1.4.1.3 Dieke Diagram

The experimental structure of the electronic arrangement of trivalent RE ions is revealed by G. H. Dieke. The triply ionized RE ions can generate spectra (sharp energy levels) while incorporated in any crystal lattice. However, Dieke proved that the transition energies are independent of the host environment. In 1960, Dieke measured the characteristic energy levels ($^{2S+1}L_J$) of trivalent RE ions doped into $LaCl_3$ crystal, which is known as the Dieke diagram (Dieke and Crosswhite 1963; Dieke 1968; Dieke and Satten 1970). Since then, this Dieke diagram has become a reference for explaining the energy levels of any trivalent ions doped in any host material.

The extended version of the Dieke diagram has been given by Ogasawara et al. (2004). In Figure 1.5, the width of each and every level indicates the magnitude of the crystal field splitting.

1.4.2 Divalent Rare-Earth Ions

Similar to the trivalent RE ions, divalent RE ions possess a $4f^n$ outer electronic configuration. In the case of divalent RE ions, the $4f^{(n-1)}$ 5d configuration is not distant from the fundamental $4f^n$ configuration. The optical transitions take place from parity-allowed $4f^n \rightarrow 4f^{(n-1)}$ 5d in the case of divalent RE ions. As the transitions are parity allowed, the resultant emission spectra lead to an intense and broad band.

For instance, one of the most important examples of divalent RE ions is Eu^{2+}. Aguilar et al. reported the incorporation of Eu^{2+} ions in NaCl crystal at room temperature (Aguilar et al. 1982). Although Eu^{2+} ions are doped in NaCl crystal, they replace the Na^+ ions, and the spectrum of Eu^{2+}:NaCl shows two broad bands analogous to the transition state from the ground state of the $4f^7$ electronic configuration to the excited $4f^6$5d transition state. The crystal field yields a splitting of the 5d orbital into two components. With the help of X-ray irradiation upon the Eu^{2+}:NaCl crystal, Aguilar observed that some Eu^{2+} ions converted into Eu^{3+} ions; hence, additional peaks are observed in the optical spectra. The optical features of the bands show broad and intense band owing to Eu^{2+} ions (parity-allowed transition), while narrow and weak bands are due to the Eu^{3+} ions (parity-forbidden transition). Furthermore, unlike trivalent ions, divalent ions change their optical properties in different host crystal environments.

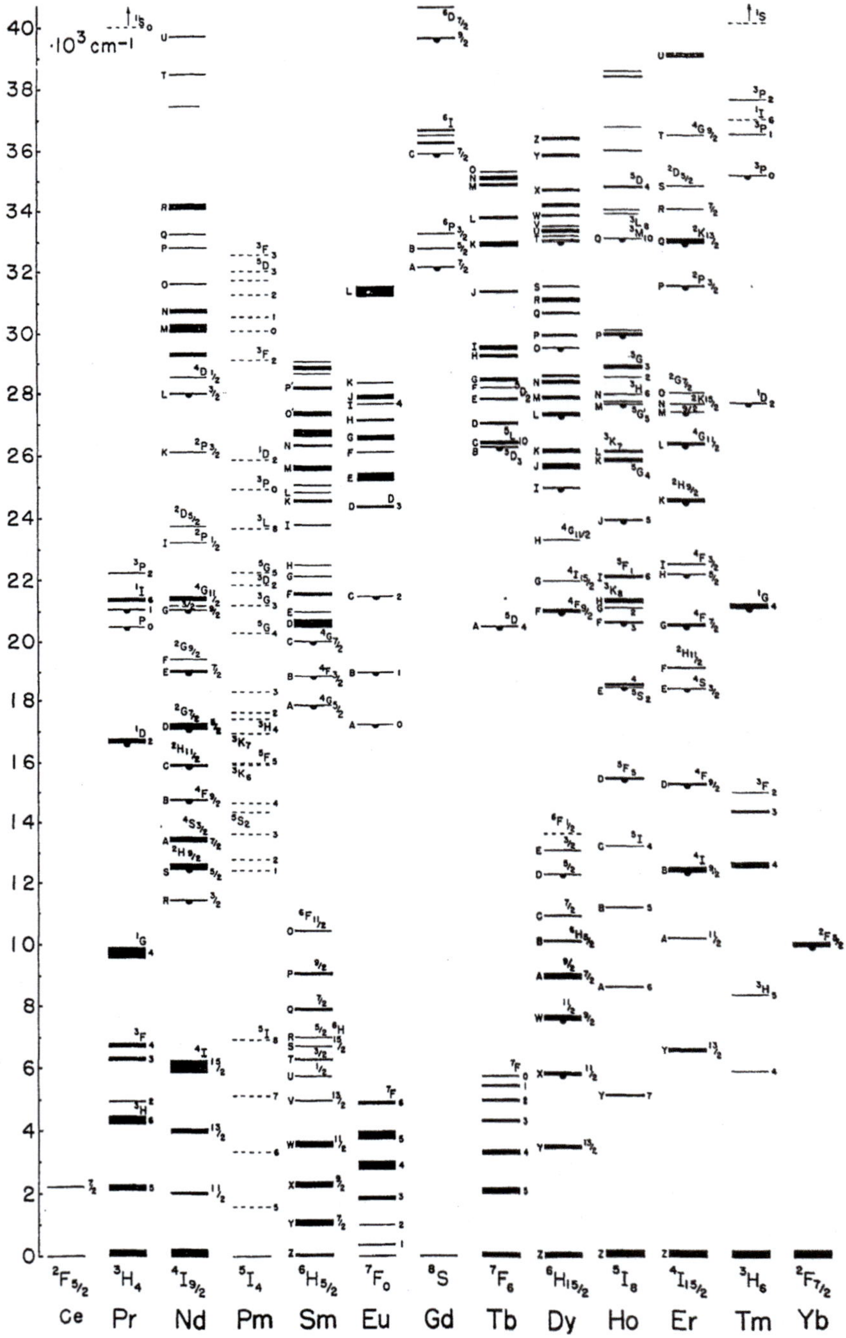

Figure 1.5 Energy levels of the trivalent RE ions. Reprinted with permission from ref. Source: Dieke and Crosswhite (1963). Reprinted with permission of The Optical Society.

1.5 Excitation and De-excitation Processes of Rare Earths in Solid Materials

1.5.1 Excitation Processes

Depending on the suitable excitation process, the divalent and trivalent RE ions are excited through various excitation routes (Shionoya et al. 1998). Specifically, three types of excitation processes occur.

1.5.1.1 f–f Transition

The trivalent RE ions with an outer electronic configuration $5s^25p^64f^n$ (n = 1 to 14) possess f–f transitions (Shionoya et al. 1998). Therefore, the valence electrons present in 4f shells are mainly accountable for optical transitions. According to the Laporte parity selection rule, 4f–4f transitions are forbidden, and they become allowed only in some host environment; hence, the absorption spectra lead to narrow cross sections.

1.5.1.2 f–d Transition

The f–d transition takes place within the 4f and 5d orbitals. In this process, the electrons present in the 4f orbitals get excited in the direction of higher energy orbitals, i.e. 5d orbitals, and the absorption process is represented as $4f^n \rightarrow 4f^{n-1}5d$. In this case, 4f–5d transitions are parity allowed according to the parity selection rule; hence, the absorption cross section leads to strong and broad type.

1.5.1.3 Charge Transfer Transition

Apart from f–f and f–d transitions, the charge transfer transition is an important transition process to populate the higher energy state of the RE ions (Shionoya et al. 1998). This process is based on the quantum mechanical concept, and it is related with one neighboring ion and one RE ion. In charge transfer transition, the mixing of two eigenfunction results in the transfer of electrons from neighboring ions to the 4f orbitals of RE ions. For instance, the transfer of 2p electrons of the neighboring oxygen ion (O^{-2}) to a 4f orbital of RE ions causes the charge transfer transition to arise. Considering the parity selection rules, these types of transitions are parity-allowed and thus generate broad and intense absorption spectra (Chen et al. 2000, 2004). These three (f–f, f–d, and charge transfer transition) processes are mainly accountable for the excitation process of the RE ions.

1.5.2 Emission Processes

After exciting the RE ions into higher energy state by suitable electromagnetic radiation, they relax to the ground state or low lying levels via some processes. These processes may be radiative or nonradiative in nature.

1.5.2.1 Emission via Radiative Transitions

The sources of radiations in any electromagnetic radiation process are oscillating electric dipole, oscillating magnetic dipole, oscillating electric quadrupole, and so on. Among these radiations, the electric dipole radiation and the magnetic dipole radiation contribute more. RE ions may be excited via electromagnetic (UV, visible, infrared, etc.) radiation. Thus, the electric dipole radiation and magnetic dipole radiation are responsible for any optical transition or emission. These transitions depend on parity of the initial and final states of the RE ions. The electric dipole transition takes place only when the initial and final states are of opposite parity; conversely, magnetic dipole transition originates if both the states have same parity. In the case of RE-doped phosphor materials, 4f–4f transitions are parity forbidden and become allowed because of the mixing of the wave functions of 4f orbitals and 5d orbitals in the presence of crystal field. These transitions are known as forced electric dipole transitions. The basic theory of 4f–4f transition properties, radiative emission rates, and lifetime of the states has been explained on the basis of Judd–Ofelt theory (Judd 1962; Ofelt 1962).

1.5.2.2 Emission via Nonradiative Transitions

The RE ions have ladder-like energy levels in the presence of host environment. When these ions are excited by the suitable electromagnetic radiation, some transitions lead to radiative, whereas some of them are nonradiative. When an electron relaxes from higher energy state to lower energy state without emitting a photon, then the relaxation process is known as nonradiative transition. Several mechanisms such as multiphonon relaxation, concentration quenching, etc., are involved for nonradiative transition process.

The multiphonon relaxation occurs when the ions in the excited state transfer their excitation energy to the lattice and no light (i.e. photon) emission takes place. The phonons of highest energy and the maximum density of state participate in this process. For possible emission via the multiphonon relaxation process, the maximum phonon energy should be five times the host vibrational energy. Therefore, the phonon energy should be minimum for higher UC emission intensity.

One more emission process via nonradiative transitions is the concentration quenching process. This process occurs when the dopant ion concentration becomes very high. When the concentration of the dopant ions become very high, the distance between the dopant ions seems to be very small. The ion–ion interaction takes place and cross-relaxation occurs. Thus, the excitation energy is lost via nonradiative relaxation. This process generally occurs in a similar type of dopant ions.

1.5.2.3 Energy Transfer Processes

The energy transfer (ET) process occurs in both similar and different ions (Förster 1948; Dexter 1953; Inokuti and Hirayama 1965). The energy transfer process is mainly of two types, radiative and nonradiative. The ET process occurs between the donor–donor or donor–acceptor system. Radiative energy transfer process arises when an excited sensitizer ion radiates a real photon, and this photon is reabsorbed by the activator ion. This type of ET process takes place through the dipole–dipole interaction process in which the energy transfer probability is proportional to $1/R^2$ (Auzel 1980).

The nonradiative energy transfer process occurs when the excited sensitizer ion transfers its excitation energy to an activator ion without emitting a photon.

The theoretical description can be understood by considering the energy transfer between a donor and an acceptor ion. Let donor ions, i.e. sensitizer, are termed as S and acceptor ions, i.e. activator, are termed as A. Under suitable excitation, the sensitizer ion is excited to the upper energy state S* by absorbing a photon, and when it returns to the ground state, it transfers its energy to the activator ion without photon emission and relaxes to the ground state; then, the activator ion is excited to the higher energy state by the following process: $S^* + A \rightarrow S + A^*$ (where "*" denotes the excited state).

The energy transfer probability between the sensitizer and the activator is given by (Förster 1948)

$$P = \frac{2\pi}{\hbar} |\langle S^*, A|H_{SA}|S, A^*\rangle|^2 \int g_S(\nu)g_A(\nu)d\nu \tag{1.5}$$

where H_{SA} is the interaction Hamiltonian, $\hbar$ is the Planck constant, $g_S(\nu)$ is the emission spectrum of the sensitizer, and $g_A(\nu)$ is the absorption spectrum of the activator.

Energy transfer interaction takes place via the exchange interaction or the electrostatic multipolar interaction. The exchange interaction arises as the distance between the donor (i.e. sensitizer) and acceptor (i.e. activator) ions are exactly adjacent such that charge distribution can exchange; hence, the orbitals of electrons overlap. In this case, S and A lie at a very small distance (≤6Å) to each other. On the other hand, the electrostatic interaction occurs via induced dipole oscillation in which no overlap occurs between the sensitizer and activator ions. The distance concerning the sensitizer and activator ions is enough (~20Å); therefore, no physical interaction take place among them. The energy transfer probability of electrostatic interaction can be described as (Förster 1948)

$$P = \frac{2\pi}{\hbar} |\langle S^*, A|H_{SA}^{el}|S, A^*\rangle|^2 \int g_S(\nu)g_A(\nu)d\nu \tag{1.6}$$

where H_{SA}^{el} signifies the electrostatic interaction Hamiltonian.

By considering the dipole–dipole interaction, this probability of energy transfer can be calculated as (Förster 1948)

$$P = \frac{1}{4\pi\varepsilon_0} \frac{3\pi\hbar e^4}{n^4 m^2 \nu^2 R^6} f_S f_A \int g_S(\nu)g_A(\nu)d\nu \tag{1.7}$$

where ε_0 is the permittivity in vacuum, n is the refractive index, e and m are the charge and mass of electrons, respectively, ν is the wavenumber of the coming transition, R is the distance between the donor and acceptor ions, and f_S and f_A are the oscillator strengths related to sensitizer emission and activator absorption, respectively. The energy transfer probability according to Förster can be written as (Förster 1948)

$$P = \frac{1}{\tau}\left(\frac{R_0}{R}\right)^6 \tag{1.8}$$

where τ is the excited-state lifetime of the sensitizer ion and R_0 is the Forster radius at which half (50%) of the energy of the sensitizer ion is transferred to the activator ion. Therefore, according to this theory, nonradiative energy transfer probability (followed by dipole–dipole interaction) varies as $1/R^6$.

Conversely, according to the Dexter theory, the energy transfer probability can be generalized as (Dexter 1953)

$$P = \frac{1}{\tau}\left(\frac{R_0}{R}\right)^a \tag{1.9}$$

where "a" is the positive integer, and for different types of interactions, it has different values. For dipole–dipole interaction, $a = 6$; for dipole–quadrupole interaction, $a = 8$; and for quadrupole–quadrupole interaction, $a = 10$.

1.6 Rate Equations Relevant to UC Mechanism

The rate equations are usually nontrivial for UC mechanisms because it involves different dopant ions for different host matrices. Depending on the dopant ions, the rate equations of UC mechanism differs. The rate equation corresponding to a particular UC phenomenon depends on various parameters such as different involved processes, pump-induced transition routes, energy transfer mechanisms, radiative and nonradiative decays, cross-relaxation processes, etc.

1.6.1 Rate Equations in a Basic Three-Level System

Let us consider a simplified three-level energy system in which GSA, ESA, and ETU processes are involved (Pollnau et al. 2000; Liu 2015; Mahata et al. 2015; Lu et al. 2016; Marciniak et al. 2016; Mondal et al. 2017; Pattnaik and Rai 2021). The population densities for the ground state G, the intermediate state E1, and the upconverting emitting state E2 are N_0, N_1, and N_2 respectively (Figure 1.6). In this case, the assumptions are taken as no pump depletion, the population density of the ground state is constant, and the scheme is constantly pumped by the GSA process.

If σ_i is the absorption cross section from some state $|Ei\rangle$, W_i is the relaxation rate for the ETU process, A_i is the rate of spontaneous emission, and ϕ is the pump rate (which is proportional to the incident pump power P), then the rate equation for a basic three-level system can be written as follows (Etchart 2010):

$$\frac{dN_1}{dt} = \phi\sigma_0 N_0 - \phi\sigma_1 N_1 - 2W_1 N_1^2 - A_1 N_1 \tag{1.10}$$

$$\frac{dN_2}{dt} = \phi\sigma_1 N_1 + W_1 N_1^2 - A_2 N_2 \tag{1.11}$$

where $\phi\sigma_0 N_0$ is the increase in population density for state E1 via the GSA process; on the other hand, the $\phi\sigma_1 N_1, 2W_1 N_1^2$, and $A_1 N_1$ terms indicate the reduction in population density of state E1 via ESA, ETU, and spontaneous emission process, respectively. For state E2, the $\phi\sigma_1 N_1$ and $W_1 N_1^2$ terms refer to the increase in population density through the GSA and ETU processes, respectively. Alternatively, the third term $A_2 N_2$ shows the decrease in population from state E2 via spontaneous emission.

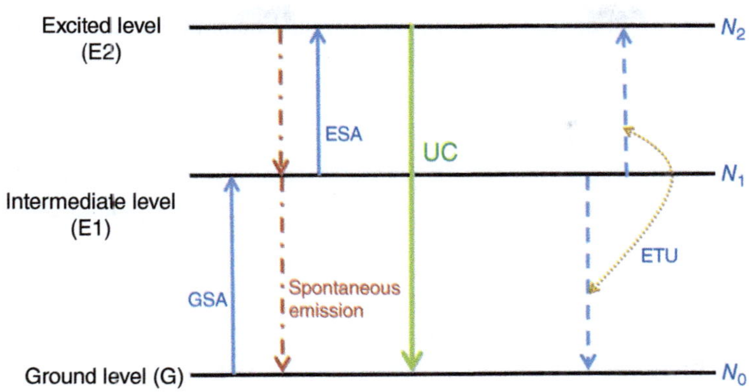

Figure 1.6 Schematic representation of a three-level energy system. The upconversion (UC) mechanism involves GSA, ESA, and ETU processes.

1.6.2 Rate Equation Related to Pump Power-Dependent UC Emission

On the basis of rate equations, the pump power-dependent UC phenomena in both cases, (i) where the ETU mechanism dominates over ESA and (ii) in the case of ESA dominated the UC emission process, can be explained.

Case (i): In the case of ETU-dominating UC emission process from Eqs. (1.10) and (1.11), the ESA, i.e. $\phi\sigma_1 N_{1,}$ term can be ignored. The rate equation can be written as

$$\frac{dN_1}{dt} = \phi\sigma_0 N_0 - 2W_1 N_1^2 - A_1 N_1 \tag{1.12}$$

$$\frac{dN_2}{dt} = W_1 N_1^2 - A_2 N_2 \tag{1.13}$$

Assuming the steady states, Eqs. (1.12) and (1.13) can be equated to zero. Therefore, $\frac{dN_1}{dt} = \phi\sigma_0 N_0 - 2W_1 N_1^2 - A_1 N_1 = 0$, or $\phi\sigma_0 N_0 = 2W_1 N_1^2 + A_1 N_1$, and $\frac{dN_2}{dt} = W_1 N_1^2 - A_2 N_2 = 0$. For instance, ETU is higher compared to the spontaneous relaxation rate (i.e. $A_1 N_1 \ll 2W_1 N_1^2$), then $\phi\sigma_0 N_0 = 2W_1 N_1^2$ or $N_1^2 \propto P_{\text{pump}}$. Also, $W_1 N_1^2 = A_2 N_2$, then $N_2 \propto N_1^2 \propto P_{\text{pump}}$. Thus, the population density of the upconverting emitting state is proportional to the excitation pump power (P_{pump}) (Dutta 2020). Conversely, when spontaneous relaxation takes over ETU (i.e. $A_1 N_1 \gg 2W_1 N_1^2$), then $N_1 \propto P_{\text{pump}}$ or $N_2 \propto N_1^2 \propto P_{\text{pump}}{}^2$. Therefore, in this case, the population density of the emitting state varies as the quadratic power of excitation pump power (P_{pump}).

Case (ii): In the case of ESA-dominating UC emission process, from Eqs. (1.10) and (1.11), the $W_1 N_1^2$ term can be ignored. Thus, the rate equations can be written as

$$\frac{dN_1}{dt} = \phi\sigma_0 N_0 - \phi\sigma_1 N_1 - A_1 N_1 \tag{1.14}$$

$$\frac{dN_2}{dt} = \phi\sigma_1 N_1 - A_2 N_2 \tag{1.15}$$

Under steady-state conditions, Eqs. (1.14) and (1.15) can be equated to zero. As the ESA process dominates over the spontaneous emission rate, i.e. $A_1N_1 \ll \phi\sigma_1 N_1$, then $N_1 \propto N_2 \propto P_{\text{pump}}$. On the other hand, if the spontaneous rate governs, then $A_1N_1 \gg \phi\sigma_1 N_1$, thus, $N_1 \propto P_{\text{pump}}$, so $N_2 \propto N_1 . P_{\text{pump}} \propto P_{\text{pump}}{}^2$.

Thus, while increasing the pump power, when the spontaneous emission rate dominates over ESA, the population density of the emitting state varies quadratically with the excitation pump power.

1.7 Theoretical Description of Optical Characteristics of Rare-Earth Ions

For instance, the electrons when de-excited from state E2 to state E1 give radiative emission; then, the rate of change of population density of N_2 is given by (Etchart 2010)

$$\frac{\mathrm{d}N_2}{\mathrm{d}t} = -A_T N_2(t) = -(A_r + A_{nr})N_2(t) \tag{1.16}$$

where A_T is the overall decay rate of radiative (A_r) and nonradiative (A_{nr}) terms.

The lifetime (experimental) of the upconverting emitting state is given by

$$\tau_{\exp} = \frac{1}{A_r + A_{nr}} \tag{1.17}$$

Conversely, the quantum efficiency (radiative) can be calculated as

$$\eta_r = \frac{A_r}{A_T} = \frac{A_r}{A_r + A_{nr}} = \frac{\tau_{\exp}}{\tau_r} \tag{1.18}$$

where τ_r is known as the radiative lifetime.

The optical transition probability can be calculated by using the term $M = \langle \Psi_i | \Gamma_D | \Psi_f \rangle$ where Ψ_i and Ψ_f are the wave functions related to the two initial and final states and Γ_D is the dipole momentum operator. Conferring to the group theory, the transition can be only allowed if $M \neq 0$. As the emitting wave is light wave, i.e. electromagnetic wave, it consists of electric dipole and magnetic dipole transitions. According to the Laporte rule, Ψ_i and Ψ_f must have opposite parities. The conditions for the allowed electric dipole transitions are as follows:

$$\Delta S = 0, \Delta L = \pm 1, \text{ and } \Delta J \le \pm 6 \text{ (but } J = 0 \to J' = 0 \text{ are forbidden)}$$

$$J = 0, \Delta L = \pm 1, \text{ and } \Delta J = \pm 2, \pm 4, \pm 6$$

On the other hand, the magnetic dipole operator is an even operator; hence, magnetic dipole transitions will be allowed only if Ψ_i and Ψ_f have the same parities. The conditions for magnetic dipole transitions are

$$\Delta S = 0, \Delta L = 0, \text{ and } \Delta J = 0, \pm 1 \text{ (but } J = 0 \to J' = 0 \text{ are forbidden)}$$

$$J = 0, \Delta L = 0, \text{ and } \Delta J = \pm 1$$

In rare-earth doped materials, the electric dipole transitions occur because of the forced electric dipole transitions.

1.7.1 Judd–Ofelt (J–O) Theory and Calculation of Radiative Parameters

The theoretical calculation on the intensities of the RE ions has been described by Judd and Ofelt in 1962. This theory gives the information about the transition probability, radiative lifetime, and oscillator strengths as well as the quantum efficiency. Conferring to the Judd–Ofelt theory, the oscillator strength (f) used for the absorption bands analogous to the electronic transition from initial state $^{2S+1}L_J$ to the final state $^{2S'+1}L_J{}'$ can be written as (Rai and Rai 2004; Liu 2005; Mahato et al. 2005; Rai et al. 2006; Etchart 2010; Azam and Rai 2017)

$$f = f_{\mathrm{ED}} + f_{\mathrm{MD}} = \frac{8\pi^2 mcv}{3he^2n^2(2J+1)}(\chi_{\mathrm{ED}}\, S_{\mathrm{ED}} + \chi_{\mathrm{MD}} S_{\mathrm{MD}}) \tag{1.19}$$

where m and e denote the mass and charge of electron, respectively, v is the wavenumber related to the observed transition, c is the speed of light, n is the refractive index of the host, S_{ED} and S_{MD} are the electric dipole line strength and magnetic dipole line strength of the electronic transition, respectively, and χ is the local field correction due to the crystal field effect of the host.

The local field correction factor and the line strength for electric dipole transition can be calculated as (Rai and Rai 2004; Mahato et al. 2005; Rai et al. 2006; Azam and Rai 2017)

$$\chi_{\mathrm{ED}} = n\left(\frac{n^2+2}{3}\right)^2 \tag{1.20}$$

$$S_{\mathrm{ED}} = e^2 \sum_{t=2,4,6} \Omega_t |\langle (S,L)J||U^{(t)}||(S',L')J'\rangle|^2 \tag{1.21}$$

Equally, the parameter local field correction and line strength for magnetic dipole transition can be written as

$$\chi_{\mathrm{MD}} = n^3 \tag{1.22}$$

$$S_{\mathrm{MD}} = \mu_{\mathrm{B}}^2 |\langle (S,L)J||L+2S||(S',L')J'\rangle|^2 \tag{1.23}$$

where μ_{B} denotes the Bohr magneton.

The Judd–Ofelt intensity parameters are Ω_2, Ω_4, and Ω_6. Carnall and Kaminskii (Carnall et al. 1968; Kaminskii 1990) described the value of reducible matrix element $|\langle (S,L)J||U^{(t)}||(S',L')J'\rangle|$, which depends on the crystal field effect of the host for different RE ions.

The spontaneous decay rate can be determined as (Liu 2005; Etchart 2010)

$$A_{\mathrm{r}} = A_{\mathrm{ED}} + A_{\mathrm{MD}} = \frac{64\pi^4 v^3}{3h(2J+1)}(\chi_{\mathrm{ED}} S_{\mathrm{ED}} + \chi_{\mathrm{MD}}\, S_{\mathrm{MD}}) \tag{1.24}$$

This radiative term strongly depends on the energy gap of the two involved electronic states, the higher the energy gap, the higher the radiative emission rate.

The radiative lifetime can be calculated as

$$\tau_{\mathrm{r}} = \frac{1}{\sum_J A_{\mathrm{r}}(J' \to J)} \tag{1.25}$$

1.7.2 Nephelauxetic Effect

In the inorganic materials, a metal ion is encircled by ligands, which modifies the electron surroundings of the metal ion. When the ligands bind to a metal ion, the metal ion orbitals spread out because of the delocalization of electrons. The spreading out of the electron cloud is known as the nephelauxetic effect (Jørgensen 1962, Luo et al. 2010; Som et al. 2015, Mukhopadhyay and Rai 2018; Azam and Rai 2019). As a consequence, the energy needed for pairing of electrons in metal ions is smaller than the energy required in the case of free ions. For metal ions, Racah parameters can be governed by the nature of the ligand ions attached to the metal ion. In the case of lanthanide ion-doped materials, the nature of bonding between the RE ions and their surrounding oxygen atoms may be studied from the absorption spectra. The slight variations in the peak positions of the absorption bands in all the RE ion-doped materials arise because of the variation in the environment around the RE ions on complexion. Also, the deviation in the positions of absorption bands arises because of the expansion of the partially filled 4f shells.

The nephelauxetic ratio (β^-) can be mathematically expressed by using the following relation (Luo et al. 2010; Som et al. 2015; Mukhopadhyay and Rai 2018; Azam and Rai 2019):

$$\beta^- = \frac{1}{n}\sum \frac{\nu_{\text{complex}}}{\nu_{\text{free ion}}} \tag{1.26}$$

where ν is the wavenumber of absorption band transition in the corresponding material and "n" is the number of the observed absorption transitions.

The covalency (δ) and the bonding parameters ($b^{1/2}$) can be calculated by using the following expressions (Mukhopadhyay and Rai 2018; Azam and Rai 2019):

$$\delta = \left[\frac{1-\beta^-}{\beta^-}\right] \tag{1.27}$$

$$\text{and } b^{1/2} = \left[\frac{1-\beta^-}{2}\right]^{1/2} \tag{1.28}$$

The positive or negative value of "δ" explains the covalent or ionic nature of bonding between the RE ions and their surrounding oxygen atoms.

1.8 An Introduction to Upconverting Nanoparticles

Generally, UCNPs with a size less than or equal to 100 nm consist of a host and an optically active center where dopants (e.g. RE ions) act as the optical center. Despite promising results achieved from luminescent materials such as organic dyes and quantum dots, their weak photostability and broad absorption and emission bands with moderate quantum yield restrict their various practical applications (Cheng et al. 2011). UCNPs play an effective role in modern days because of their nonblinking property, long lifetime, high chemical stability, sharp emission bands, large anti-Stokes shift, etc. (Wang et al. 2011; Chen et al. 2014; Mondal et al. 2017; Shao et al. 2018; Soni et al. 2019). The emission characteristics of these UCNPs differ with the selection of different dopants, hosts, excitation wavelengths, etc. In the UCNPs

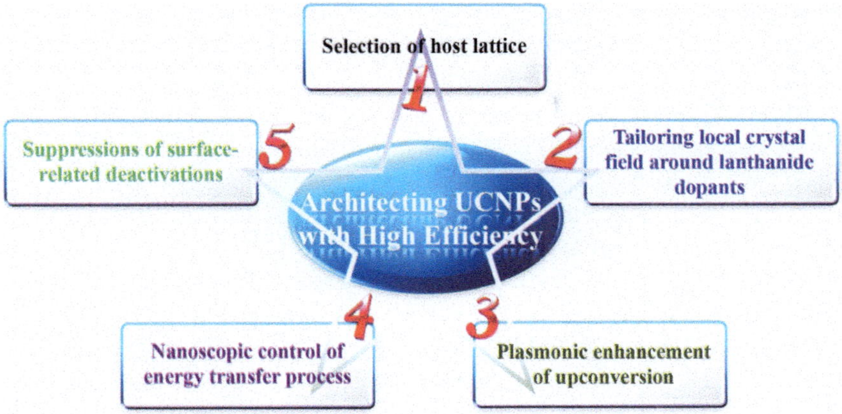

Figure 1.7 General strategies to achieve the high efficiency of UCNPs. Source: Chen et al. (2014). Reprinted with permission of American Chemical Society.

containing RE ions, the emission features, because of the shielded 4f–4f orbital electronic transitions, are not influenced greatly by the host NPs or bulk materials. However, the host with nanometer size plays an important and unique role in enhancing the UC emission intensity and emission profile. Figure 1.7 describes the general strategies to achieve the high efficiency of UCNPs. Primarily, a host material with low vibrational energy is essential to minimize the energy loss because of the nonradiation effect and hence improves the radiative UC emission property.

For nanoparticles, owing to their high surface-to-volume ratio, the dopant ions present near the surface of nanoparticles lose their emission properties by surface quenching centers and hence decrease the UC efficiency. For this, the different surface engineering approaches in UCNPs have been developed (Figure 1.7), such as homogeneous core@shell structure, heterogeneous core@shell structure, active core/active shell structure, plasmonic UCNPs, etc., to enhance the conversion efficiency (Chen et al. 2014; Fan et al. 2014; Mondal et al. 2017; Shao et al. 2018; Soni et al. 2019; Sui et al. 2019).

Acknowledgments

Authors are grateful to the Council of Scientific and Industrial Research (CSIR), New Delhi, India, project no. 03(1354)/16/EMR-II, for providing financial assistance in the form of a research project.

References

Aguilar, M., García Solé, G.J., H. Murrieta, S. et al. (1982). Trivalent europium in x-irradiated NaCl: Eu. *Physical Review B* 26: 4507. https://doi.org/10.1103/physrevb.26.4507.

Auzel, F. (1966). Quantum counter by energy transfer between two rare earth ions in a mixed tungstate and in a glass. *Comptes Rendus de l'Académie des Sciences (Paris)* 262: 1016.

Auzel, F.E. (1973). Materials and devices using double-pumped-phosphors with energy transfer. *Proceedings of the IEEE* 6: 758–786. https://doi.org/10.1109/proc.1973.9155.

Auzel, F. (1980). Multiphonon processes, cross-relaxation and up-conversion in ion-activated solids, exemplified by minilaser materials. In: *Radiationless Processes*, 213–286. https://doi.org/10.1007/978-1-4613-3174-2_6.

Auzel, F. (2004). Upconversion and anti-Stokes processes with f and d ions in solids. *Chemical Reviews* 104: 139–174. https://doi.org/10.1021/cr020357g.

Azam, M. and Rai, V.K. (2017). Ho^{3+}-Yb^{3+} codoped tellurite based glasses in visible lasers and optical devices: Judd–Ofelt analysis and frequency upconversion. *Solid State Sciences* 66: 7–15. https://doi.org/10.1016/j.solidstatesciences.2017.02.001.

Azam, M. and Rai, V.K. (2019). Effect of the addition of Pb_3O_4 and TiO_2 on the optical properties of Er^{3+}/Yb^{3+}:TeO_2-WO_3 glasses. *ACS Omega* 4: 16280–16291. https://doi.org/10.1021/acsomega.9b00609.

Berthou, H. and Jörgensen, C.K. (1990). Optical-fiber temperature sensor based on upconversion-excited fluorescence. *Optics Letters*. https://doi.org/10.1364/ol.15.001100.

Boyer, J.C., Cuccia, L.A., and Capobianco, J.A. (2007). Synthesis of colloidal upconverting $NaYF_4$:Er^{3+}/Yb^{3+} and Tm^{3+}/Yb^{3+} monodisperse nanocrystals. *Nano Letters* 7: 847–852. https://doi.org/10.1021/nl070235+.

Brik, M.G., Camardello, S.J., Srivastava, A.M. et al. (2016). Spin-forbidden transitions in the spectra of transition metal ions and nephelauxetic effect. *ECS Journal of Solid State Science and Technology* 5: R3067–R3077. https://doi.org/10.1149/2.0091601jss.

Carnall, W.T., Fields, P.R., and Rajnak, K. (1968). Spectral intensities of the trivalent lanthanides and actinides in solution. II. Pm^{3+}, Sm^{3+}, Eu^{3+}, Gd^{3+}, Tb^{3+}, Dy^{3+}, and Ho^{3+}. *The Journal of Chemical Physics* 49: 4412–4423. https://doi.org/10.1063/1.1669892.

Carnall, W.T., Goodman, G.L., Rajnak, K. et al. (1989). A systematic analysis of the spectra of the lanthanides doped into single crystal LaF_3. *The Journal of Chemical Physics* 90: 3443–3457. https://doi.org/10.1063/1.455853.

Chen, W., Malm, J.O., Zwiller, V. et al. (2000). Energy structure and fluorescence of Eu^{2+} in ZnS:Eu nanoparticles. *Physical Review B* 61: 11021–11024. https://doi.org/10.1103/physrevb.61.11021.

Chen, W., Joly, A.G., Malm, J.O. et al. (2004). Upconversion luminescence of Eu^{3+} and Mn^{2+} in ZnS:Mn^{2+}, Eu^{3+} codoped nanoparticles. *Journal of Applied Physics* 95: 667–672. https://doi.org/10.1063/1.1633345.

Chen, G.Y., Qju, H.L., Prasad, P.N. et al. (2014). Upconversion nanoparticles: design, nanochemistry, and applications in theranostics. *Chemical Reviews* 114: 5161–5214. https://doi.org/10.1021/cr400425h.

Chen, X., Peng, D., Ju, Q. et al. (2015). Photon upconversion in core–shell nanoparticles. *Chemical Society Reviews* 44: 1318. https://doi.org/10.1039/c4cs00151f.

Cheng, L., Yang, K., Li, Y. et al. (2011). Facile preparation of multifunctional upconversion nanoprobes for multimodal imaging and dual-targeted photothermal therapy. *Angewandte Chemie International Edition* 50: 7385–7390. https://doi.org/10.1002/anie.201101447.

Cheng, L., Wang, C., and Liu, Z. (2013). Upconversion nanoparticles and their composite nanostructures for biomedical imaging and cancer therapy. *Nanoscale* 5: 23–37. https://doi.org/10.1039/C2NR32311G.

Chivian, J.S., Case, W.E., and Eden, D.D. (1979). The photon avalanche: a new phenomenon in Pr^{3+} based infrared quantum counters. *Applied Physics Letters* 35: 124. https://doi.org/10.1063/1.91044.

Dexter, D.L. (1953). A theory of sensitized luminescence in solids. *The Journal of Chemical Physics* 21: 836–850. https://doi.org/10.1063/1.1699044.

Dey, R. and Rai, V.K. (2017). Er^{3+}-Tm^{3+}-Yb^{3+}:$CaMoO_4$ phosphor as an outstanding upconversion-based optical temperature sensor and optical heater. *Methods and Applications in Fluorescence* https://doi.org/10.1088/2050-6120/aa5e31.

Diaz-Torres, L.A., De la Rosa, E., Salas, P. et al. (2005). Enhanced cooperative absorption and upconversion in Yb^{3+} doped YAG nanophosphors. *Optical Materials* 27: 1305–1310. https://doi.org/10.1016/j.optmat.2004.10.020.

Dieke, G.H. (1968). *Spectra and Energy Levels of Rare-Earth Ions in Crystals*, vol. 261, 263. New York: Wiley-Interscience.

Dieke, G.H. and Crosswhite, H.M. (1963). The spectra of the doubly and triply ionized rare earths. *Applied Optics* 2: 675–686. https://doi.org/10.1364/ao.2.000675.

Dieke, G.H. and Satten, R.A. (1970). Spectra and energy levels of rare earth ions in crystals. *American Journal of Physics* 38: 399–400. https://doi.org/10.1119/1.1976350.

Digonnet, M.J.F. (1993). *Rare Earth Doped Fiber Lasers and Amplifier*. New York: Dekker. https://doi.org/10.1201/9780203904657.

Dutta, J. (2020). Preparation and characterization of lanthanides doped Yttrium based materials for multifunctional applications. Doctoral thesis. IIT(SM) Dhanbad.

Dwivedi, Y., Thakur, S.N., and Rai, S.B. (2007). Study of frequency upconversion in Yb^{3+}/Eu^{3+} by cooperative energy transfer in oxyfluoroborate glass matrix. *Applied Physics B* 89: 45–51. https://doi.org/10.1007/s00340-007-2747-y.

Etchart, I. (2010). Metal oxides for efficient infrared to visible upconversion. Doctoral thesis. University of Cambridge.

Fan, W., Shen, B., Bu, W. et al. (2014). A smart upconversion-based mesoporous silica nanotheranostic system for synergetic chemo-/radio-/photodynamic therapy and simultaneous MR/UCL imaging. *Biomaterials* 35: 8992. https://doi.org/10.1016/j.biomaterials.2014.07.024.

Förster, T. (1948). Intermolecular energy migration and fluorescence. *Annalen der Physik* 437: 55–75. https://doi.org/10.1002/andp.19484370105.

Franken, P.A., Weinreich, G., Peters, C.W. et al. (1961). Generation of optical harmonics. *Physical Review Letters* 7: 118–119. https://doi.org/10.1103/physrevlett.7.118.

Gamelin, D.R. and Gudel, H.U. (2000). Design of luminescent inorganic materials: new photophysical processes studied by optical spectroscopy. *Accounts of Chemical Research* 33: 235–242. https://doi.org/10.1021/ar990102y.

Garlick, G.F.J. (1976). Infrared to visible light conversion. *Contemporary Physics* 17: 127–144. https://doi.org/10.1080/00107517608210848.

Grzybowski, A. and Pietrzak, K. (2013). Maria Goeppert-Mayer (1906–1972): two-photon effect on dermatology. *Clinics in Dermatology* 31: 221–225. https://doi.org/10.1016/j.clindermatol.2012.06.002.

Gu, Z.J., Yan, L., Tian, G. et al. (2013). Recent advances in design and fabrication of upconversion nanoparticles and their safe theranostic applications. *Advanced Materials* 25: 3758–3779. https://doi.org/10.1002/adma.201301197.

Heer, S., Lehmann, O., Haase, M. et al. (2003). Blue, green, and red upconversion emission from lanthanide-doped $LuPO_4$ and $YbPO_4$ nanocrystals in a transparent colloidal solution. *Angewandte Chemie International Edition* 42: 3179–3182. https://doi.org/10.1002/anie.200351091.

Huang, X., Sanyang Han, S., Huang, W. et al. (2013). Enhancing solar cell efficiency: the search for luminescent materials as spectral converters. *Chemical Society Reviews* 42: 173–201. https://doi.org/10.1039/C2CS35288E.

Inokuti, M. and Hirayama, F. (1965). Influence of energy transfer by the exchange mechanism on donor luminescence. *The Journal of Chemical Physics* 43: 1978–1989. https://doi.org/10.1063/1.1697063.

Jablonski, A. (1935). About the mechanism of photo-luminescence of dye phosphors. *Zeitschrift für Angewandte Physik* 94: 38–46. https://doi.org/10.1007/bf01330795.

Jablonski, A. (1993). Efficiency of anti-Stokes fluorescence in dyes. *Nature* 131: 839–840. https://doi.org/10.1038/131839b0.

Jiang, G., Zhou, S., Wei, X. et al. (2016). 794 nm excited core–shell upconversion nanoparticles for optical temperature sensing. *RSC Advances* 6: 11795–11801. https://doi.org/10.1039/C5RA27203C.

Jørgensen, C.K. (1962). The nephelauxetic series. In: *Progress in Inorganic Chemistry*, vol. 4, 73–124. Wiley https://doi.org/10.1002/9780470166055.ch2.

Joubert, M.F. (1999). Photon avalanche upconversion in rare earth laser materials. *Optical Materials* 11: 181–203. https://doi.org/10.1016/s0925-3467(98)00043-3.

Judd, B.R. (1962). Optical absorption intensities of rare-earth ions. *Physical Review* 127: 750–761. https://doi.org/10.1103/physrev.127.750.

Kaminskii, A.A. (1990). *Laser Crystals, Their Physics and Properties*. New York/Berlin/Heidelberg: Springer. https://doi.org/10.1007/978-3-540-70749-3.

Konig, E. and Kremer, S. (1997). *Ligand Field Energy Diagrams*. New York: Plenum Press.

Lee, L.S., Rand, S.C., and Schawlow, A.L. (1984). Cooperative energy transfer among Pr^{3+} ions in LaF_3. *Physical Review B* 29: 6901–6906. https://doi.org/10.1103/physrevb.29.6901.

Li, X., Wang, R., Zhang, F. et al. (2013). Nd^{3+} sensitized up/down converting dual-mode nanomaterials for efficient in-vitro and in-vivo bioimaging excited at 800 nm. *Scientific Reports* 3: 3536. https://doi.org/10.1038/srep03536.

Liang, H.J., Chen, G.Y., Li, L. et al. (2009). Upconversion luminescence in Yb^{3+}/Tb^{3+}-codoped monodisperse $NaYF_4$ nanocrystals. *Optics Communications* 282: 3028–3031. https://doi.org/10.1016/j.optcom.2009.04.006.

Lin, C.C., Meijerink, A., and Liu, R.S. (2016). Critical red components for next-generation white LEDs. *The Journal of Physical Chemistry Letters* 7: 495–503. https://doi.org/10.1021/acs.jpclett.5b02433.

Liu, G. (2005). *Electronic Energy Level Structure*, Springer Series in Materials Science, 1–94. Springer. https://doi.org/10.1007/3-540-28209-2_1.

Liu, G. (2015). Advances in the theoretical understanding of photon upconversion in rare-earth activated nanophosphors. *Chemical Society Reviews* 44: 1635. https://doi.org/10.1039/c4cs00187g.

Loo, J.F.C., Chien, Y.H., Yin, F. et al. (2019). Upconversion and downconversion nanoparticles for biophotonics and nanomedicine. *Coordination Chemistry Reviews* 400: 213042. https://doi.org/10.1016/j.ccr.2019.213042.

Lu, H., Meng, R., Hao, H. et al. (2016). Stark sublevels of Er^{3+}–Yb^{3+} codoped $Gd_2(WO_4)_3$ phosphor for enhancing the sensitivity of a luminescent thermometer. *RSC Advances* 6: 57667. https://doi.org/10.1039/c6ra10138k.

Luo, W., Liao, J., Lia, R. et al. (2010). Determination of Judd–Ofelt intensity parameters from the excitation spectra for rare-earth doped luminescent materials. *Physical Chemistry Chemical Physics* 12: 3276. https://doi.org/10.1039/b921581f.

Maciel, G.S., Biswas, A., Kapoor, R., and Prasad, P.N. (2000). Blue cooperative upconversion in Yb^{3+}-doped multicomponent sol-gel-processed silica glass for three-dimensional display. *Applied Physics Letters* 76: 1978–1980. https://doi.org/10.1063/1.126228.

Mahata, M.K., Kumar, K., and Rai, V.K. (2015). Er^{3+}–Yb^{3+} doped vanadate nanocrystals: a highly sensitive thermographic phosphor and its optical nanoheater behavior. *Sensors and Actuators B* 209: 775–780. https://doi.org/10.1016/j.snb.2014.12.039.

Mahato, K.K., Rai, A., and Rai, S.B. (2005). Optical properties of Dy^{3+} doped in oxyfluoroborate glass. *Spectrochimica Acta Part A* 61: 431–436. https://doi.org/10.1016/j.saa.2004.02.038.

Marciniak, L., Waszniewska, K., Bednarkiewicz, A. et al. (2016). Sensitivity of a nanocrystalline luminescent thermometer in high and low excitation density regimes. *Journal of Physical Chemistry C* 120: 8877–8882. https://doi.org/10.1021/acs.jpcc.6b01636.

Martı´ n, I.R., Yanes, A.C., Méndez-Ramos, J. et al. (2001). Cooperative energy transfer in Yb^{3+}–Tb^{3+} codoped silica sol-gel glasses. *Journal of Applied Physics* 89: 2520–2524. https://doi.org/10.1063/1.1344216.

Mehra, S., Bishnoi, S., Jaiswal, A. et al. (2020). A review on spectral converting nanomaterials as a photoanode layer in dye-sensitized solar cells with implementation in energy storage devices. *Energy Storage* 2: 1–28. https://doi.org/10.1002/est2.120.

Mondal, M. and Rai, V.K. (2020). Optical thermometry using stark sublevels in charge compensated transition metal molybdate upconverting phosphors. *Journal of Optics and Laser Technology* 130: 106341. https://doi.org/10.1016/j.optlastec.2020.106341.

Mondal, M., Rai, V.K., Srivastava, C. et al. (2016). Enhanced frequency upconversion in $Ho^{3+}/Yb^{3+}/Li^{+}$:$YMoO_4$ nanophosphors for photonic and security ink applications. *Journal of Applied Physics* 120: 233101. https://doi.org/10.1063/1.4971966.

Mondal, M., Rai, V.K., and Srivastava, C. (2017). Influence of silica surface coating on optical properties of Er^{3+}-Yb^{3+}:$YMoO_4$ upconverting nanoparticles. *Chemical Engineering Journal* 327: 838–848. https://doi.org/10.1016/j.cej.2017.06.166.

Mukhopadhyay, L. and Rai, V.K. (2018). Investigation of photoluminescence properties, Judd–Ofelt analysis, luminescence nanothermometry and optical heating behaviour of $Er^{3+}/Eu^{3+}/Yb^{3+}:NaZnPO_4$ nanophosphors. *New Journal of Chemistry* 42: 13122–13134. https://doi.org/10.1039/C8NJ02320D.

Mukhopadhyay, L. and Rai, V.K. (2020). Thermally stable red emitting xenotime phosphate nanophosphors for displays. *Materials Research Bulletin* 121: 110628. https://doi.org/10.1016/j.materresbull.2019.110628.

Mulliken, R.S. (1967). Spectroscopy, molecular orbitals, and chemical bonding. *Science* 157: 13–24. https://doi.org/10.1126/science.157.3784.13.

Nadort, A., Zhao, J., and Goldys, E.M. (2016). Lanthanide upconversion luminescence at the nanoscale: fundamentals and optical properties. *Nanoscale* 8: 13099–13130. https://doi.org/10.1039/C5NR08477F.

Ofelt, G.S. (1962). Intensities of crystal spectra of rare-earth ions. *The Journal of Chemical Physics* 37: 511–520. https://doi.org/10.1063/1.1701366.

Ogasawara, K., Watanabe, S., Sakai, Y. et al. (2004). Calculations of complete $4f^n$ and $4f^{n-1}$ $5d^1$ energy level schemes of free trivalent rare-earth ions. *Japanese Journal of Applied Physics* 43: L611–L613. Available at: https://doi.org/10.1143/jjap.43.l611.

Ovsyakin, V.V. and Feofilov, P.P. (1966). Cooperative sensitization of luminescence in crystals activated with rare earth ions. *Soviet Journal of Experimental and Theoretical Physics Letters* 4: 317.

Pattnaik, S. and Rai, V.K. (2020). Impact of charge compensation on optical and thermometric behaviour of titanate phosphors. *Materials Research Bulletin* 125: 110761. https://doi.org/10.1016/j.materresbull.2019.110761.

Pattnaik, S. and Rai, V.K. (2021). Insight into the spectroscopic and thermometric properties of titanate phosphors via a novel co-excited laser system. *Materials Science and Engineering B* 272: 115318. https://doi.org/10.1016/j.mseb.2021.115318.

Pollnau, M., Gamelin, D.R., Lüthi, S.R. et al. (2000). Power dependence of upconversion luminescence in lanthanide and transition-metal-ion systems. *Physical Review B* 61: 3337–3346. https://doi.org/10.1103/physrevb.61.3337.

Rai, V.K. (2007). Temperature sensors and optical sensors. *Applied Physics B* 88: 297–303. https://doi.org/10.1007/s00340-007-2717-4.

Rai, V.K., de Menezes, L.S., and de Araújo, C.B. (2007). Stokes luminescence and frequency upconversion in Pr^{3+} doped TeO_2–PbO glass. *Journal of Applied Physics* 101: 123514. https://doi.org/10.1063/1.2745314.

Rai, V.K. and Rai, S.B. (2004). Optical transitions of Dy^{3+} in tellurite glass: observation of upconversion. *Solid State Communications* 132: 647–652. https://doi.org/10.1016/j.ssc.2004.07.067.

Rai, V.K., Rai, S.B., and Rai, D.K. (2006). Optical studies of Dy^{3+} doped tellurite glass: Observation of yellow-green upconversion. *Optics Communications* 257: 112–119. https://doi.org/10.1016/j.optcom.2005.07.022.

Rai, V.K., de Araújo, C.B., Ledemi, Y. et al. (2008). Frequency upconversion in a Pr^{3+} doped chalcogenide glass containing silver nanoparticles. *Journal of Applied Physics* 103: 103526. https://doi.org/10.1063/1.2927402.

Rai, V.K., Pandey, A., and Dey, R. (2013). Photoluminescence study of $Y_2O_3:Er^{3+}$-Eu^{3+}-Yb^{3+} phosphor for lighting and sensing applications. *Journal of*

Applied Physics 113: 083104. http://link.aip.org/link/doi/10.1063/1.4793265?ver=pdfcov.

Reddy, K.L., Balaji, R., Kumar, A. et al. (2018). Lanthanide doped near infrared active upconversion nanophosphors: Fundamental concepts, synthesis strategies, and technological applications. *Small* 14: 1801304. http://dx.doi.org/10.1002/smll.201801304.

Salley, G.M., Valiente, R., and Guedel, H.U. (2001). Luminescence upconversion mechanisms in Yb^{3+}–Tb^{3+} systems. *Journal of Luminescence* 94: 305–309. https://doi.org/10.1016/s0022-2313(01)00310-6.

Salley, G.M., Valiente, R., and Güdel, H.U. (2003). Cooperative Yb^{3+}–Tb^{3+} dimer excitations and upconversion in $Cs_3Tb_2Br_9$: Yb^{3+}. *Physical Review B* 67: https://doi.org/10.1103/physrevb.67.134111.

Shan, J., Qin, X., Yao, N. et al. (2007). Synthesis of monodisperse hexagonal $NaYF_4$:Yb, Ln (Ln = Er, Ho and Tm) upconversion nanocrystals in TOPO. *Nanotechnology* 18: 445607. https://doi.org/10.1088/0957-4484/18/44/445607.

Shao, Q., Yang, Z., Zhang, G. et al. (2018). Multifunctional lanthanide-doped core/shell nanoparticles: integration of upconversion luminescence, temperature sensing, and photothermal conversion properties. *ACS Omega* 3: 188–197. https://doi.org/10.1021/acsomega.7b01581.

Shionoya, S., and Yen, W.M. (eds.) 1998. *Phosphor Handbook*, CRC Press, Boca Raton, Florida.

Singh, A.K., Kumar, K., Pandey, A.C. et al. (2011). Photon avalanche upconversion and pump power studies in LaF_3:Er^{3+}/Yb^{3+} phosphor. *Applied Physics B* 104: 1035–1041. https://doi.org/10.1007/s00340-011-4673-2.

Solé, J.G., Bausá, L.E., and Jaque, D. (2005). Applications: rare earth and transition metal ions, and color centers. In: *An Introduction to the Optical Spectroscopy of Inorganic Solids*, 199–234. Wiley https://doi.org/10.1002/0470016043.

Som, S., Das, S., Dutta, S. et al. (2015). Synthesis of strong red emitting Y_2O_3: Eu^{3+} phosphor by potential chemical routes: comparative investigations on the structural evolutions, photometric properties and Judd–Ofelt analysis. *RSC Advances* 5: 70887. https://doi.org/10.1039/c5ra13247a.

Soni, A.K., Dey, R., and Rai, V.K. (2015). Stark sublevels in Tm^{3+}–Yb^{3+} codoped $Na_2Y_2B_2O_7$ nanophosphor for multifunctional applications. *RSC Advances* 5: 34999–35009. https://doi.org/10.1039/C4RA15891A.

Soni, A.K., Joshi, R., Singh, B.P. et al. (2019). Near-infrared- and magnetic-field-responsive $NaYF_4$:Er^{3+}/Yb^{3+}@SiO_2@AuNP@Fe_3O_4 nanocomposites for hyperthermia applications induced by fluorescence resonance energy transfer and surface plasmon absorption. *ACS Applied Nano Materials* 2: 7350–7361. https://doi.org/10.1021/acsanm.9b01867.

Sui, J., Chen, Z., Liu, G. et al. (2019). Multifunctional Ag@$NaGdF_4$:Yb^{3+}, Er^{3+} core-shell nanocomposites for dual-mode imaging and photothermal therapy. *Journal of Luminescence* 209: 357–364. https://doi.org/10.1016/j.jlumin.2019.01.046.

Suijver, J.F. (2008). *Luminescence: From Theory to Applications*, ch. 6 (ed. C. Ronda), 134. Weinheim: Wiley-VCH Verlag GmbH and Co. KGaA.

Tanabe, Y. and Sugano, S. (1954a). On the absorption spectra of complex ions I. *Journal of the Physical Society of Japan* 9: 753–766. https://doi.org/10.1143/jpsj.9.753.

Tanabe, Y. and Sugano, S. (1954b). On the absorption spectra of complex ions II. *Journal of the Physical Society of Japan* 9: 766–779. https://doi.org/10.1143/jpsj.9.766.

Tanabe, Y. and Sugano, S. (1956). On the absorption spectra of complex ions, III. The calculation of the crystalline field strength. *Journal of the Physical Society of Japan* 11: 864–877. https://doi.org/10.1143/jpsj.11.864.

Wade, S.A., Collins, S.F., and Baxter, G.W. (2003). Fluorescence intensity ratio technique for optical fiber point temperature sensing. *Journal of Applied Physics* 94: 4743–4756. https://doi.org/10.1063/1.1606526.

Wang, F. and Liu, X.G. (2009). Recent advances in the chemistry of lanthanide-doped upconversion nanocrystals. *Chemical Society Reviews* 38: 976–989. https://doi.org/10.1039/b809132n.

Wang, R. and Zhang, F. (2014). NIR luminescent nanomaterials for biomedical imaging. *Journal of Materials Chemistry B* 2: 2422–2443. https://doi.org/10.1039/c3tb21447h.

Wang, M., Abbineni, G., Clevenger, A. et al. (2011). Upconversion nanoparticles: synthesis, surface modification and biological applications. *Nanomedicine: Nanotechnology, Biology and Medicine* 7: 710–729. https://doi.org/10.1016/j.nano.2011.02.013.

Wybourne, B.G. (1965). *Spectroscopic Properties of Rare Earths*. New York: Wiley.

Wybourne, B.G. and Meggers, W.F. (1965). Spectroscopic properties of rare earths. *Physics Today* 18: 70–72. https://doi.org/10.1063/1.3047727.

You, M., Lin, M., Wang, S. et al. (2016). Three-dimensional quick response code based on inkjet printing of upconversion fluorescent nanoparticles for drug anti-counterfeiting. *Nanoscale* 8: 10096. https://doi.org/10.1039/c6nr01353h.

Zhu, Y., Xu, W., Li, C. et al. (2012). Broad white light and Infrared emission bands in YVO_4:Yb^{3+}, Ln^{3+} (Ln^{3+}=Er^{3+}, Tm^{3+} or Ho^{3+}). *Applied Physics Express* 5: 092701. https://doi.org/10.1143/APEX.5.092701.

Zhu, S., Wang, C., Li, Z. et al. (2016). High-efficiency broadband anti-Stokes emission from Yb^{3+}-doped bulk crystals. *Optics Letters* 41: 2141–2144. https://doi.org/10.1364/ol.41.002141.

2

Synthesis Protocol of Upconversion Nanoparticles

Lakshmi Mukhopadhyay[1,2] *and Vineet Kumar Rai*[1]

[1] *Indian Institute of Technology (Indian School of Mines), Department of Physics, Laser and Spectroscopy Laboratory, Dhanbad, Jharkhand 826004, India*
[2] *DRDO, Terminal Ballistics Research Laboratory, Chandigarh 160030, India*

2.1 Introduction

In the past few decades, the upconversion (UC) phenomenon becomes an active research field in the recent scientific era, viz, displays, non-contact thermometry, optical communication amplifiers, anticounterfeiting, photovoltaics, biomedical applications, etc. The UC luminescent materials are fetching significant research attention because of their pioneering characteristics including high resistance to photobleaching, less light scattering, non-blinking, and allowing for deep tissue penetration with excitation in the near-infrared (NIR) region (Kim et al. 2000; Huang et al. 2013; Song et al. 2016; Wang et al. 2016; Prasad and Rai 2020; Pattnaik and Rai 2021). The properties governing their suitability in different applications strongly depend on the particle size, morphology, crystallinity, pH, compositional phase, and surface area of the nanomaterials. To this end, a remarkable amount of fundamental research has been focused in searching new UC nanomaterials having desired photophysical effects. The synthesis strategy plays an important role in preparing well-defined, uniform, and monodisperse nanostructures and also effectively tailoring the properties of developed nanomaterials for some specific applications. The selection of a particular synthesis technique is not only important for optimizing the properties of particles but it should be simple enough to be reproducible, easily scalable, green, clean, cheap, and not lead to any hazardous and waste production (Cushing et al. 2004; Rao et al. 2007; Mao et al. 2007; Gupta and Mao 2020).

Moreover, the main limiting factor of UC nanoparticles (UCNPs) is low emission efficiency, which encountered difficulties in their widespread applications. The spectroscopic investigation reveals that the detrimental effect of the concentration quenching phenomenon is one of the major reasons in achieving low emission efficiency from these UCNPs. The quenching process associated with the energy migration of dopant ions can be largely diminished by leveraging the size and structure features of the nanomaterials. In this regard, the core@shell

Upconverting Nanoparticles: From Fundamentals to Applications, First Edition. Edited by Vineet K. Rai.

nanostructures are studied in a broad sense and become very popular to minimize the surface-related quenching problem. Beside this, the proper choice of the host matrix is also very crucial to get significant emission efficiency from the synthesized UCNPs. The coordination environment of dopant ions, i.e. the site symmetry and the interionic distance, largely depends on host crystallinity. In addition, these factors determine the intraparticle energy transfer and thus the efficiency of the photon UC process (Qiu et al. 2013; Han et al. 2014; Mukhopadhyay and Rai 2021).

Herein, the properties of a suitable host matrix and various aspects of different synthesis techniques adopted to prepare the UCNPs have been discussed with their respective pros and cons.

2.2 Host Matrix

For designing novel UCNPs, selection of an appropriate host material is of paramount importance. The UC emission efficiency has significant dependency on the crystal field environment of the host lattice and the interaction between the host and dopant ions. Hence, close lattice matching of the host matrix with dopant ions is essential. An appropriate host lattice is capable of determining the distance among dopants, anion types encircled by dopant ions, and their coordination number (Zhang 2015, p. 10). The vibrational properties of the host lattice have considerable impact on the UC efficiency by influencing the non-radiating de-excitation rate from the excited level of dopant ions. The non-radiative radiation is basically a multiphonon-assisted relaxation process in which the energy gap between the emitting level and the next low lying level is bridged by a number of phonons. The involved phonon number in this process can be known by the cutoff phonon frequency of the host lattice (Chen et al. 2013; Speghini et al. 2016, p. 39). In general, the larger is the phonon number required to convert excitation energy into phonon energy, the higher is the radiative emission efficiency. Thus, in order to achieve high luminescence efficiency by enhancing the radiative emission, the rare earths should be embedded in a host matrix with low phonon energy. Fluorides are known to have very low phonon frequency (e.g. $NaYF_4$ ~350 cm^{-1}) and therefore exhibits highest UC efficiency by minimizing non-radiative losses in the emitting and intermediate states (Ivaturi et al. 2013; Chen et al. 2014). Also, some heavy halides, viz, chlorides, bromides, and iodides, possess low phonon frequency (~300 cm^{-1}). However, these fluorides and halides are very hygroscopic and toxic in nature, which limit their relevance practically. In this instance, some other host materials such as oxides (e.g. Y_2O_3 ~550 cm^{-1} and ZrO_2 ~500 cm^{-1}), vanadates (e.g. YVO_4 ~890 cm^{-1}), niobates ($YNbO_4$ ~660 cm^{-1}), etc., are very much promising for UC emissions (Vetrone et al. 2003; Patra et al. 2003a; Mialon et al. 2010; Wang et al. 2018).

The crystal structure of the host material has profound influence on the UC emission efficiency as the optical properties depend on the local crystal symmetry (Peacock 1975). Typically, a less symmetric crystal phase is favorable for UC emission because of the presence of more uneven components around the lanthanides (Krämer et al. 2004; Schäfer et al. 2009; Lin et al. 2014). As a result, the intermixing of the 4f states with higher electronic configurations of lanthanide ions can be more manifested and subsequently enhances the f–f transition probabilities. For instance, the hexagonal phase of $NaYF_4$:Yb/Er exhibits ∼10 times higher intensity than the cubic phase counterpart (Heer et al. 2004).

The inorganic host materials based on trivalent lanthanide ions are ideal for UC because of the similarity of ionic radius and other chemical properties of the lanthanides used as dopants. The decrease in the cation size (or unit-cell volume) of the host lattice results in an increase in the crystal field strength around the dopants, which further improves the UC efficiency (Wang and Liu 2009). Moreover, host matrices consisting of some alkali cations (Na^+, Ca^{2+}, Sr^{2+}, and Ba^{2+}) and transition metal ions (Zr^{4+} and Ti^{4+}) with ionic radii close to that of lanthanides are frequently used in the UC process. The lanthanide doping in these materials generates less lattice stress and crystal defects such as interstitial anions and cation vacancies to maintain charge neutrality (Bol et al. 2002). Thus, the crystal with lower symmetry, low phonon energy, and less lattice mismatch with lanthanides possesses significant UC efficiency. For example, $NaYF_4$:Yb^{3+},Er^{3+} gives ∼20 times higher UC efficiency compared to La_2O_3:Yb^{3+},Er^{3+} and ∼6 times higher than that of $La_2(MoO_4)_3$:Yb^{3+},Er^{3+} (Haase and Schäfer 2011). Furthermore, an ideal host matrix should have some more important properties including large transparency in the spectral range of interest, high refractive index, non-hygroscopic nature, low toxicity, high chemical and thermal stability, and high optical damage threshold (Rai 2012, p. 210).

2.3 Synthetic Strategy of UC Nanomaterials

The optimization of synthetic strategies for efficient luminescent nanomaterials with defined crystal size, morphology, chemical composition, and phase is very critical. The first cornerstone in nanotechnology is to produce nanostructured materials, paving the way for understanding their several characteristics and applications as most of their properties are tremendously based on the size and dimensions (Zhang 2015, p. 21; Speghini et al. 2016, p. 40). This chapter covers several facile synthesis techniques to obtain UCNPs having excellent chemical and optical features and thus desirable for commercialization in diverse fields. Generally, there are two paradigms that are employed to prepare nanoparticles (NPs), namely, top–down and bottom–up method. The top–down (physical) method is a breakdown method in which a solid breaks up into smaller particles systematically bit-after-bit upon application of some external force. The conventional top–down approaches are electron beam lithography, photolithography, milling techniques, anodization, and ion and plasma etching. The benefit of the physical methods is large-scale production, but it is very difficult to control the size and uniformity of the prepared nanocrystals. These strategies face severe problem of noticeable crystallographic damage to the processed patterns, which may introduce further defects. Also, to carry out these techniques, a clean room or a vacuum environment and large machinery are also needed. However, bottom–up (wet-chemistry) is a buildup method that fabricates NPs by coalescence or assembling of atoms of gas or liquids based on atomic transformations or molecular condensations. The bottom–up methods are mainly based on chemistry and driven by reduction of Gibbs free energy, where the as-formed nanostructures are in a state closer to the thermodynamic equilibrium state (Zhang 2015, p. 22; Dhand et al. 2015; Gupta and Mao 2020). Thus, a variety of nanostructures can be produced with less defects, more homogeneity, and better short- and long-range ordering by using these techniques at relatively low cost, with high throughput and the potential for large-scale industrial production. Some examples of bottom–up strategy are chemical or electrochemical nanostructural precipitation, sol–gel processing, hydro(solvo)thermal, combustion, thermolysis, self-assembly of monomer/polymer molecules, laser pyrolysis, plasma or flame spraying synthesis, and bio-assisted synthesis (Gai et al. 2014; Dey et al. 2015; Zhang 2015, p. 22; Dhand et al. 2015; Das et al. 2020, p. 375). A brief discussion about the synthesis procedures and recent progress in the preparation of nanostructured materials has been given in this chapter.

2.3.1 Solid-State Reaction Technique

The solid-state reaction method is the easiest route for the preparation of upconverting phosphor materials, which involves direct reaction of starting reagents. This technique is generally used to synthesize the single crystals and polycrystalline powder materials. In this method, the precursors are weighted as desired quantities and then mixed homogeneously in an agate mortar with the help of acetone. This thorough grinding process is continued for some stipulated time. Then, the obtained powder is placed in an alumina crucible and heated. Sometimes, the heat treatment is accomplished by two steps, viz, preheated the grinded powder between 350 and 400 °C for few hours, which lets decomposition of the initial precursors and removes the volatile products such as NH_3, NO_2, CO_2, and H_2O, only the oxides remain. After that, the mixture is again ground to make it more homogeneous, minimize the grain size, and then again heated at elevated temperature. For the appropriate phase formation of phosphors, it is necessary to treat the starting reagents at very high pressure and temperature to increase the diffusion of ions across the boundaries and thus enhancing the rate of formation of product (Pan et al. 2012; Liu et al. 2014a; Mukhopadhyay et al. 2017). The step-by-step procedure involved in the synthesis technique has been schematically given in Figure 2.1 (Mukhopadhyay 2021).

A few numbers of parameters such as pressure through grinding, diffusion rate while heating, etc., are acting for governing the solid-state reaction process, but it is very difficult to have control over them. As a result, large agglomeration occurs among the particles. The shortcomings of this approach are that it is very slow and requires a lot of energy. The reaction takes place at very high temperatures (500–2000 °C) for several hours and sometimes for several days. Moreover, because of the use of high temperature, large-sized particles are obtained, and sometimes, they decompose so that the desired phase of the particles cannot be obtained. A large number of samples have been prepared by several researchers via this method to obtain highly crystalline materials with bright luminescence intensity, mainly suitable in several display applications (Soni and Rai 2014; Wang et al. 2018; Prasad and Rai 2020).

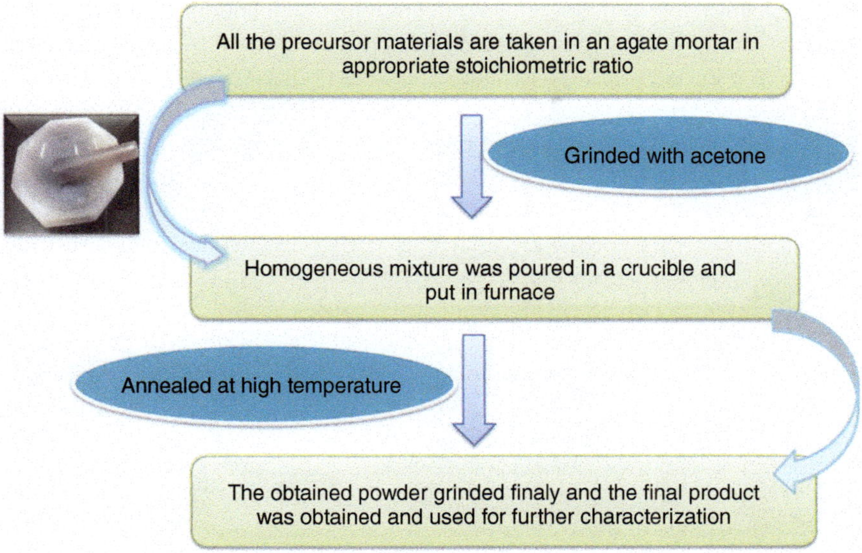

Figure 2.1 Flow chart representation of solid-state reaction method. Source: PhD thesis of Lakshmi Mukhopadhyay (2021): Structural and spectroscopic studies of rare earth doped phosphate phosphors for emerging applications.

2.3.2 Coprecipitation Technique

Coprecipitation is possibly the most convenient and simplest way to prepare UCNPs as it involves short processing time and facile and cheap reaction protocols. Additionally, by using this procedure, one can get homogeneous mixing, narrow size distribution, and good stoichiometric control among the particles. In this method, raw materials are mixed in the form of nitrates, acetates, and precipitates in the form of carbonates, oxalates, citrates, or hydroxides. The resultant precipitates are washed in distilled water and ethanol several times, and consequently, heat treatment is applied to the precipitate to obtain the desired phase of the material. Annealing at higher temperature of the obtained product is frequently performed to get increased particle size and enhanced luminescence intensity. In order to get proper precipitation, the solubility of the used reagents should be very close to each other. There are some important parameters that need to be regulated including stirring speed, pH of the precursor solution, concentration, and temperature of the mixture (Yi et al. 2004; Wang and Liu 2009; Dey et al. 2015; Speghini et al. 2016, p. 40). The step-by-step procedure involved in the synthesis technique has been viewed in flow chart format (Figure 2.2) (Mukhopadhyay 2021).

van Veggel and Stouwdam have done one of the earliest works on UCNPs based on the LaF_3 host doped with various lanthanide ions ($Ln^{3+} = Er^{3+}$, Eu^{3+}, Nd^{3+}, and Ho^{3+} ions) (Stouwdam and van Veggel 2002). This process was expanded and refined later by synthesizing UC LaF_3 nanocrystals with smaller particle size (~5 nm) and narrower size distribution from simple water-soluble inorganic precursors (Yi and Chow 2005). Additionally, the researchers have developed $NaYF_4$:Yb/Er(Tm), $LuPO_4$:Yb/Tm, and $YbPO_4$:Er nanocrystals via the

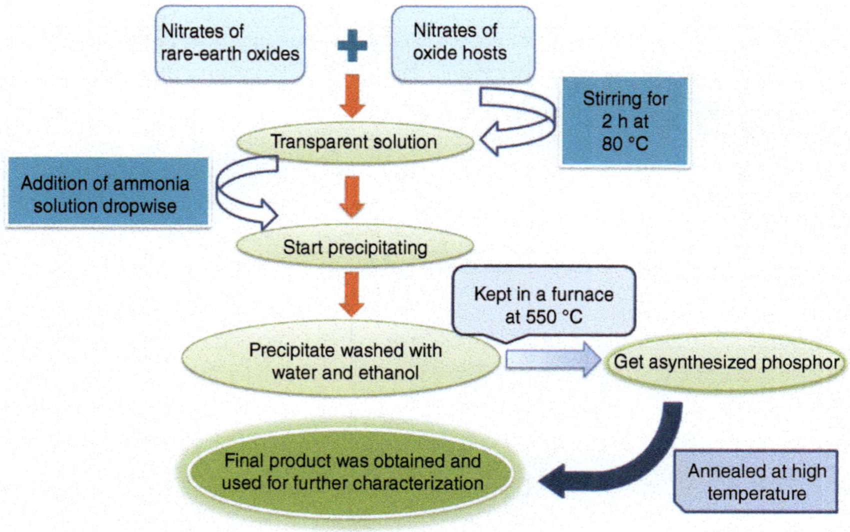

Figure 2.2 Flow chart representation of coprecipitation technique. Source: PhD thesis of Lakshmi Mukhopadhyay (2021): Structural and spectroscopic studies of rare earth doped phosphate phosphors for emerging applications.

coprecipitation method with further annealing process for obtaining the improved UC efficiency (Heer et al. 2003, 2004; Guo et al. 2004; Zeng et al. 2005). Some perovskite synthesis ($LaMn_{1-x}Fe_xO_3$ [$x = 0, 0.1$, and 0.2]) has also been reported by the coprecipitation technique. The advantages of this approach are stringent reaction conditions, low temperature, energy efficiency, no need for costly equipment, no use of organic solvents, rapid preparation, less time consumption and control over particle size, composition, and morphology. The disadvantages of this method are the high hydrophobicity of the product, low monodispersity, and the trace impurities may also get precipitated with the product. Also, this method fails for the reactants that have very different precipitation rates (Dey et al. 2015; Speghini et al. 2016, p. 41).

2.3.3 Sol–Gel Technique

Sol–gel processing is a very fascinating approach over other types of synthetic methods to fabricate NPs perhaps because of the ability to produce materials with controlled shape and size. The key features of this technique that have attracted the researchers are low temperature chemistry, cost effectiveness, less energy consumption, reproducibility, purity, and high surface-to-volume ratio of the obtained products. It is a very simple process with the contribution of two different phases, viz, "sol," which is a formation process of colloidal suspension of solid particles in liquid, and "gel," which are polymers in liquid. The gel is formed by a three-dimensional skeleton adjacent reticular pore containing a liquid phase (Reiser et al. 2019; Kaya and Deveci 2020). The sol–gel reaction is sensitive to numerous parameters, namely, the pH of the solution, the properties of the precursors, the nature of the solvent, the types and concentrations of the surfactants and catalysts, the temperature, the aging time, and the ratio of water-to-metal alkoxides. In the polymerization process, solvent plays a key role as it must be competent to maintain the NPs dissolved in the liquid and remain connected with each other. The hydrolysis and polycondensation reaction rate depend on the coordination number, metal ion radius, and the electronegativity of the precursors used. As the number of partial positive charges on the metal atom enhances, the coordination number increases subsequently, which further improves the reaction rate. Some metal alkoxide solutions (mostly tetraethyl orthosilicate [TEOS], citric acid, and ethylene glycol) and catalysts (hydrochloric acid and ammonia) are used to stimulate the rate of hydrolysis process. Moreover, the reaction rate and reaction time depend on the temperature. As the reaction temperature increases, the rate of hydrolysis increases with consuming low reaction time. The ratio of water to precursors (R) is an important parameter that affects the reaction rate and measured by the amount of water. If the amount of water is less than the stoichiometric amount of water (N), then with increasing R, the viscosity increases and thus shortened the gelation time as the amount of water maximizes the degree of crosslinking and accelerates the degree of polymerization of polycondensate. However, if there is excess water, then the concentration of polycondensate diminishes and enhances the drying time (Livage and Sanchez 1992, Buckley and Greenblatt 1994). The basic reaction steps of the sol–gel method are as follows (Hench and West 1990; Buckley and Greenblatt 1994; Periyasamy et al. 2020):

(1) *Solvation*: The metal cation M^{z+} attracts water molecules to form the solvent unit $M(H_2O)_n{}^{z+}$. In order to maintain its coordination number, it has a strong tendency to release H^+.

$$M(H_2O)_n{}^{Z+} \rightarrow M(H_2O)_{n-1}(OH)_{z-1} + H^+$$

(2) *Hydrolysis reaction*: Non-ionizing molecular precursors, such as metal alkoxide $M(OR)_n$, react with water.

$$M(OR)_n + xH_2O \rightarrow M(OR)_{n-x}(OH)_x$$

$$M(OR)_{n-x}(OH)_x + xROH \rightarrow M(OH)_n$$

(3) *Polycondensation reaction*: According to the type of molecules removed, it can be divided into two categories;
   (a) Dehydration polycondensation: $-M-OH + HO-M- \rightarrow -M-O-M- + H_2O$
   (b) Dealcoholization polycondensation: $-M-OR+HO-M- \rightarrow -M-O-M- + ROH$

The van der Waals forces or hydrogen bonds mainly dominate among the sol's particles, whereas covalent interactions govern the gel systems, and so gelation process is usually irreversible. This gelation reaction involved in the sol–gel process makes it superior as it allows mixing in atomic level to generate small particles with homogeneous distribution and can be easily sintered. Thus, the sol–gel process involves hydrolysis and condensation reaction of metal alkoxide precursors and subsequently the formation of thick gel by polycondensation process, followed by aging, solvent extraction, and ultimately drying the gel as per the requirement of the final product. Depending on the steps of heat treatment, the formation of crystalline (NPs and thin films) and nanocrystalline materials (ceramics and glasses) will take place. When the gel is dried by evaporation, densification occurs, i.e. the capillary forces will result in shrinkage, the pores removed from the gel, and a xerogel is formed. On the other hand, under supercritical conditions, the aerogel is formed with high surface area, large pore volume, low density, and the original network structure retained. If drying is done by freezing the solvents, then the cryogel forms, and in this case, the gel network shrinkage is relatively lower than the xerogel. The improved crystallinity of a sample is directly associated with its UC emission efficiency, so usually, the dried gel is calcined at higher temperature. During this calcination process, several reactions take place such as the release of the adsorbed solvent and water from the walls of micropores, decomposition of organic groups, collapse of small and larger pores, and continued polycondensation throughout the whole temperature range (Periyasamy et al. 2020; Parashar et al. 2020). The mechanisms involved in the synthesis technique and different synthesized form of nanostructures have been schematically shown in Figure 2.3.

A variety of materials can be fabricated using sol–gel synthesis techniques having different morphologies including monoliths, films, fibers, and powders. The sol–gel technique is very much appropriate for preparing metal oxides owing to the presence of metal–oxygen bonds corresponding to the alkoxide precursor. Several researchers have reported a number of UC nanomaterials based on metal oxide host materials using the sol–gel synthesis technique including TiO_2:Er, $BaTiO_3$:Er, $Lu_3Ga_5O_{12}$:Er, ZrO_2:Er, and YVO_4:Yb/Er nanocrystals (Patra et al. 2002, 2003b; Kuisheng et al. 2006; Venkatramu et al. 2008). The UC emission properties in

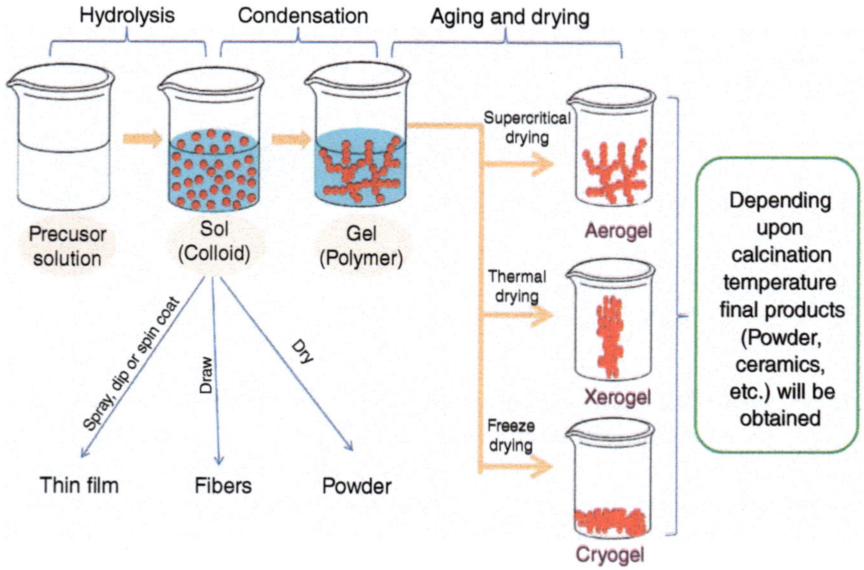

Figure 2.3 Schematic representation of sol–gel technique.

the $Tm^{3+}/Er^{3+}/Yb^{3+}:Gd_2O_3$ NPs synthesized by the sol–gel technique have been studied (Guo et al. 2004). Nowadays, the synthesis of various ceramic materials such as γ-Al_2O_3, γ-Fe_2O_3, TiO_2, etc., via the sol–gel method is also reported.

2.3.4 Hydro(solvo)thermal Technique

The hydro(solvo)thermal technique is one of the most significant solution-based chemical synthesis approaches for the preparation of complex oxides, ceramics, fluorides, luminescent phosphors, microporous crystals, conducting solids, magnetic materials, etc. The term "hydrothermal" was first used by the British Geologist Sir Roderick Murchison for describing the formation of minerals by hot water solutions rising from cooling magma. The word "hydrothermal" originates from the Greek words "hydros" meaning water and "thermos" meaning heat (Zhang 2015, p. 39). As the term indicates, the synthesis process is referred to as hydrothermal process when water is used as a solvent, whereas it is referred to as solvothermal process when organics are used as solvents. Some organic solvents usually used in the solvothermal route are methanol, 1,4-butanediol, toluene, amines, etc. The hydro(solvo)thermal technique is defined as to perform chemical reactions in the presence of solvents in a sealed environment at high temperature and autogenous pressure that facilitate the interaction between precursors during synthesis (Wang et al. 2007; Feng and Li 2017, p. 73). The pressure and temperature should be quite high, i.e. above the critical point of the solvent used so that the solubility and reactivity of the materials is enhanced as compared to the ordinary conditions. In this technique, the required precursors, solvents, and surfactants are mixed and heated in a sealed vessel (Li et al. 2015, p. 2). Commonly, ethylenediaminetetraacetic acid (EDTA), oleic acid (OA), cetyltrimethylammonium bromide (CTAB), and polyethylenimine (PEI) are used as surfactants, which offer chelating ability with cationic ions to regulate their reacting concentration (Qiu et al. 2011a; Wang and Liu 2008; Wang et al. 2006). This method is very useful to prepare monodisperse and homogeneous NPs as well as nanohybrids and nanocomposites. The main aspect of the hydrothermal scheme is to accelerate the reaction rate among the complex ions by intensifying the hydrolysis reaction and by considerably varying the redox potential of the reactants. The advantages of this technique is the capability to prepare a large amount of NPs with high crystallinity, homogeneity, controlled size, morphology, and other surface chemistry at comparatively lower temperature by adjusting some parameters including solvent types, surfactants, reaction time, and temperature. Furthermore, the nanomaterials with high vapor pressure and that are not stable at elevated temperatures can be synthesized through hydrothermal synthesis with minimum loss of materials. Also, with the help of this technique, the products of intermediate state and specific phase can be easily synthesized. In contrast to the hydrothermal process, the products obtained in solvothermal process are having relatively smaller and narrower size distribution at a lower temperature and pressure. Most importantly, the solvothermal technique is very much promising for handling the precursors that are sensitive to water (Wang et al. 2007; Byrappa and Adschiri 2007).

The synthesis of $NaYF_4$ prismatic micro-tubes based on multicomponent fluorides by hydrothermal method has been demonstrated (Zhuang et al. 2007). It was observed that with the addition of EDTA, the UC emission intensity decreases, but it contributes to the formation of perfect hexagonal. The synthesis of complex fluorides ABF_4 (A = Li, Na, and K; B = Y, Eu, Ho, and Er) using hydrothermal technique at 140–240 °C have been reported (Zhao et al. 1997; Xun et al. 1998). The doping of Eu^{3+} ions in $LiYF_4$ and KYF_4 has been carried out by directly adding Eu_2O_3 into the reaction mixtures (Zhao et al. 1997; Xun et al. 1998). Zhang et al. have reported an OA-mediated hydrothermal method to synthesize uniform $NaYF_4$ nanorods, nanotubes, and flower-patterned nanodisks (Zhang et al. 2007). A novel way to dope Gd^{3+} in $NaYF_4$ UCNPs via hydro(solvo)thermal synthesis with controlled crystal phase, size, and optical properties depending on temperature has been constructed (Wang et al. 2010a). Senthilselvan et al. have synthesized Er^{3+}/Yb^{3+}:SrF_2 NPs by hydrothermal method using an EDTA surfactant and reported the temperature sensing and biological applications (Senthilselvan et al. 2019). The schematic diagram describing the procedure involved in this work has been shown in Figure 2.4.

Nevertheless, this synthesis technique faces some serious drawbacks that need to be solved. Most of the reactions require long processing time and high reaction temperature, which make the experiments to be a certain risk. To accomplish the experiment, there is the need of a thick and steel-walled cylindrical vessel having airtight sealing known as an autoclave that can withstand high temperature and pressure safely for a prolonged period of time, and thus, it is impossible to observe NPs as they grow (Wang and Liu 2009). Moreover, to protect the interior cavity of the autoclave from the solvent corrosion, the coating of silver, gold, titanium, platinum, and mainly Teflon is applied based on the solvent and temperature used in the reaction.

2.3.5 Combustion Technique

Combustion synthesis, also known as self-propagating high-temperature synthesis, is an effective, low-cost method for the development of new catalysts, nanocarriers, various industrially useful materials with better properties. Depending on the nature of the initial reaction medium, the combustion synthesis procedure is mainly of three types, namely, condensed phase combustion, solution combustion, and gas-phase combustion synthesis. Among different types of combustion processes, solution combustion synthesis is a versatile, simple, and rapid process for nanoscale materials synthesis. The solution combustion synthesis can be carried out in an aqueous solution of the oxidizer and fuel that should be sufficiently exothermic with temperatures ranging from 500 to 3000 °C occurring in the form of a combustion wave that propagates through the reaction to retain a self-sustained chemical reaction (Wang and Liu 2009). Generally, hydrated metal nitrates are used as oxidizers, while fuels signify a wide range of compounds such as urea, glycine, citric acid, etc. It is worth noting that the role of the fuel is not only to provide adequate heat to the system but also to make sure the formation of stable complexes with the metal ions to increase their solubility and prevent selective precipitation

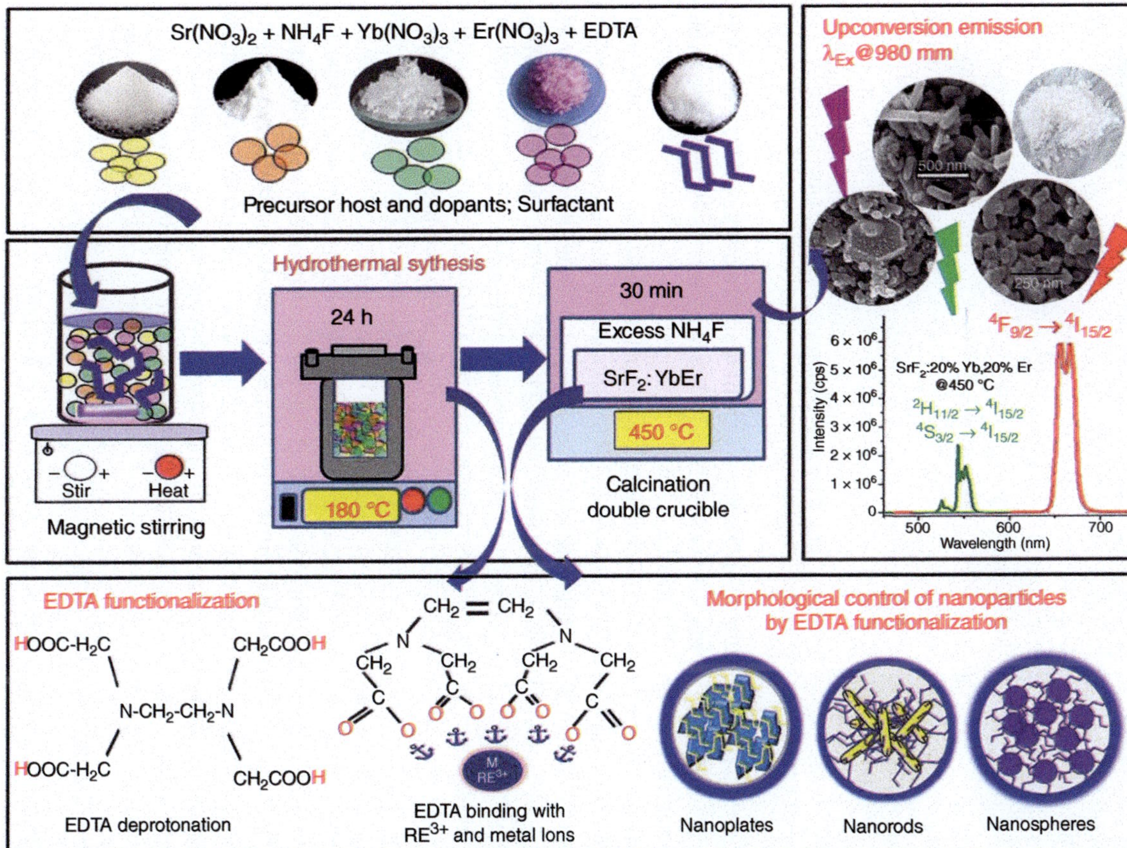

Figure 2.4 Schematic of hydrothermal synthesis and EDTA functionalization of the Yb- and Er-co-doped SrF_2 UCNPs. Source: Senthilselvan et al. (2019)/Reprinted with permission of Springer.

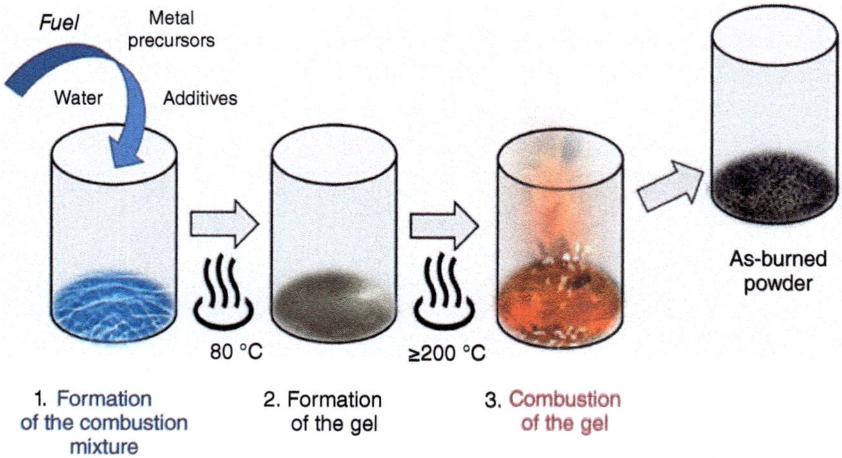

Figure 2.5 Schematic description of the three main steps in solution combustion synthesis. Source: Deganello and Tyagi (2018). Reprinted with permission of Elsevier.

of the metal ions during water removal (Aruna and Mukasyan 2008; Deganello and Tyagi 2018). The main steps involved in the synthesis technique have been schematically shown in Figure 2.5.

Based on the precursor types and conditions used, the solution combustion synthesis may happen as volume or layer-by-layer propagating combustion modes. With the help of this method, preparation of oxide-based UCNPs as well as homogeneous doping of rare earth (REs) can be performed (Zhou et al. 2014; Varma et al. 2016). The Er^{3+}-doped Y_2O_3 upconverting NPs have been prepared by using the combustion technique in which glycine was used as a fuel (Capobianco et al. 2000). This paper describes the UC emission properties of the sample upon 815 nm laser excitation and the dynamics of UC emission by decay time analysis. The optical thermometric properties of the Er^{3+}, Yb^{3+}-doped Gd_2O_3 upconverting NPs synthesized by combustion synthesis using urea as a fuel have been reported (Singh et al. 2009a,b). The spectroscopic analysis of blue and ultraviolet emission from 1% each Tm^{3+}- and Yb^{3+}-codoped $Gd_3Ga_5O_{12}$ nanostructure has been studied (Pandozzi et al. 2005). The main advantage of the combustion method is that it is a controlled explosion in which the product can be obtained within a few minutes, but in this process, it is very difficult to control the morphology of the developed materials.

2.3.6 Thermolysis Technique

The thermolysis approach is used to prepare high-quality nanocrystals with controlled dimension, monodispersity, and good photoluminescence properties. In this strategy, organometallic compounds are generally used as precursors (e.g. acetate or trifluoroacetate salts) that decompose in a non-polar organic solvent with high boiling point assisted by surfactants at an elevated temperature. The commonly used solvents are namely oleylamine (OM), trioctylphosphine oxide (TOPO), or 1-octadecene (ODE). OA is used as a surfactant with capping groups to regulate the size of NPs, coordinate the metallic elements, and reduce aggregation due to long hydrocarbon chains. By controlling some factors, viz, concentration of metal precursors, solvent nature, capping agents, reaction time, and temperature, highly monodispersed NPs with very good crystallinity can be produced. In spite of having all these features, this synthesis method faces some major hindrance including expensive procedures, high temperature, handling of air-sensitive starting materials, formation of highly toxic fluorinated by-products, etc. (Nguyen 2013; Tiwari et al. 2016). On the basis of solvents used during the synthesis, the thermolysis process is of three types:

2.3.6.1 Thermolysis in OA-Based Mixed Solvents

The LaF_3 NPs by using the thermolysis method has been developed by Yan group (Zhang et al. 2005). This approach has been further extended to synthesize the cubic Yb^{3+}/Er^{3+}- or Yb^{3+}/Tm^{3+}-codoped $NaYF_4$ NPs via thermal decomposition of metal trifluoroacetate precursors in OA/ODE (Boyer et al. 2006, 2007). In this investigation, ODE having high boiling point (315 °C) acts as the primary solvent, and OA performs the role of a solvent as well as a passivating ligand, which prevents the NPs from agglomeration. Murray et al. used this same procedure and reported the preparation of highly uniform hexagonal Yb^{3+}/Er^{3+}:$NaYF_4$ NPs with controlled sizes and morphologies (spheres, nanorods, and hexagonal prisms) (Ye et al. 2010). Moreover, this decomposition method has been employed to develop other RE fluorides, oxides, and oxyfluorides such as $LiREF_4$ (Mahalingam et al. 2009; Wang et al. 2010b; Chen et al. 2011), KRE_3F_{10} (Peng et al. 2012), $NaScF_4$ (Teng et al. 2012), YF_3 (Sun et al. 2007; Chen et al. 2012a), MF_2 (M = Mg, Ca, and Sr) (Du et al. 2009a), $BaREF_5$ (Vetrone et al. 2009), CeO_2 (Zhou et al. 2008; Qiu et al. 2011b), and LnOF (Ln = Er, Tm, Yb, Lu, Tb, Dy, Ho, and Y) (Sun et al. 2007).

2.3.6.2 Thermolysis in OM-Based Mixed Solvents

Yan et al. have reported the synthesis of Ln_2O_3 (Ln = Tb, Dy, Ho, Er, Tm, Yb, Lu, and Y) (Mai et al. 2007) and LnOF (Ln = Ce, Pr, Nd, Sm, Tb, Dy, Ho, and Y) NPs in the presence of OA/OM and $NaYF_4$ and $NaLnF_4$ (Ln = Pr, Nd, Sm, Eu, Tb, Dy, Ho, Er, Tm, Yb, Lu, and Gd) NPs in OA/OM/ODE solvents (Mai et al. 2006, 2007; Du et al. 2008; Zhou et al. 2009; Wang et al. 2012). Numerous tremendous works on the synthesis of monodispersed lanthanide-doped Eu_2O_2S (Zhao et al. 2006), La_2O_2S (Ding et al. 2011), Gd_2O_3 (Si et al. 2007; Cao 2004), LnOCl (Du et al. 2009b), $LiYF_4$ (Zhou et al. 2009), and $NaLnS_2$ (Ln = La, Sm, Tb, Ce) (Ding et al. 2012) NPs have been reported by the thermolysis approach using OA/OM, OA/OM/ODE, or OM solvents. The solvent OM has the capability to form ultrasmall UCNPs, which is superior in bioimaging applications. Also, the feeble coordination bond among the amine group and the lanthanides present on the surface of UCNPs facilitate the utilization of a popular ligand exchange procedure for phase transfer.

2.3.6.3 Thermolysis in TOPO-Based Mixed Solvents

Shan et al. reported the use of TOPO as a coordinating ligand for synthesizing ultrasmall, monodispersed, and pure hexagonal $NaYF_4$ UCNPs (Shan and Ju 2007;

Shan et al. 2007, 2010). Compared to OA and OAM solvents, the free energy barrier for the phase transition from cubic to hexagonal was notably decreased in TOPO. However, in some solvents including cyclohexane and chloroform, the TOPO-capped NPs possess low solubility, and this issue can be resolved by using the combination of trioctylphosphine (TOP)/OA. Instead of using only TOPO, the use of TOP/OA leads to preparation of hexagonal $NaYF_4$ NPs at lower temperature by reducing the energy barrier of α/β phase transition.

2.3.7 Microwave-Assisted Synthesis Technique

Most recently, the microwave heating strategy has been on the increasing demand starting from biochemical processes to nanotechnology as it can address the difficulty of inhomogeneous heating. Compared to the other conventional heating methods, microwave irradiation provides a clean, cheap, and convenient method because of its high efficiency in energy utilization and thus produces NPs with higher yield in shorter reaction time. The microwave radiation region is located between the infrared radiation and radiowaves in the electromagnetic spectrum. Telecommunication and microwave radar apparatus have many band frequencies in this region. To avoid interference with these, generally, a fixed frequency of operation of ~2.45 GHz is used in microwaves. Microwave-assisted synthesis methods endow with superior control over reaction mixing, separation of the nucleation and growth stages, endure high temperatures, and pressures, good reproducibility, etc. This technique has the ability to selectively activate and heat the solvent or the precursor materials, which is significant for scalability (Horikoshi 2013; Panda et al. 2007). Previously, a lot of works have been carried out by using domestic microwave ovens. However, it exhibits some disadvantages such as poor control over reaction, inability to measure temperature and microwave power precisely, and unavoidable hazardous situation in the case of flammable solvents and reagents. These shortcomings can be conquered by using scientific microwave systems (Horikoshi 2013). The temperature, pressure, reaction time, and power can be adjusted and measured accurately with the help of the predetermined program. Therefore, a number of parallel reactions can be performed in one run, and the parameters obtained in each reaction can be recorded easily. Researchers have reported the synthesis of ZnO crystals with various structures including nanorods, nanocandles, nanoneedles, nanodisks, and nanonuts, by adjusting precursor chemicals, capping agents, and aging times with the help of microwave irradiation method (Cho et al. 2008). Zhu and Cao have developed α-Fe_2O_3 hierarchically nanostructured hollow spheres through a microwave-assisted solvothermal route combined with subsequent thermal decomposition (Cao and Zhu 2008). He et al. have synthesized water-dispersed and highly luminescent CdTe–CdS core@shell nanocrystals by microwave-assisted method with ~75% photoluminescence quantum yield (He et al. 2006). Tsuji et al. have demonstrated the formation of Au–Ag core–shell nanocrystals with different shapes by utilizing microwave-polyol technique (Tsuji et al. 2010). The drawback of microwave technology is the constrained penetration depth of microwave irradiation into absorbing materials, i.e. the solvents or reagents in the vessel are heated by convection and not by direct "in core" microwave dielectric heating. Moreover, there is a lack of knowledge about the interaction of the electromagnetic radiation with matter, and it is very difficult to determine the existing microwave effects (thermal and non-thermal) so that a predictive model can be developed. This phenomenon has become a handicap to make new chemistry in microwave-assisted synthesis techniques (Priecel and Lopez-Sanchez 2019).

2.4 Synthesis Techniques for Fabricating Core@shell Architectures

Lanthanide-doped UCNPs, i.e. the "Core" material, prepared via different synthetic strategies, suffer from severe detrimental effect of surface defects and the concentration quenching phenomenon. Furthermore, lanthanides located on the surface of dispersed NPs experience an enhanced non-radiative multiphonon relaxation, owing to the solvent phonons, in comparison to that present in the interior of the NPs. Thus, to get rid of this problem and strengthen the emission efficiency, core@shell designing becomes a very promising strategy in which the core NPs are protected by an additional shell. These core@shell NPs are widely applicable in several biological applications including bio-labeling, photodynamic therapy (PDT), drug targeting, and delivery applications (Ghosh Chaudhuri and Paria 2012; Chen et al. 2015). Typically, core@shell NPs are synthesized in two steps: at first, preparation of core NPs and then deposition of shell layer over it. In accordance with the several principles of formations, the synthesis procedure of core@shell NPs can be of four types, viz, solid-phase reaction, liquid-phase reaction, gas-phase reaction, and mechanical mixing (Chen et al. 2020).

2.4.1 Solid-Phase Reaction

In this method, the core materials and shell precursors are mixed through dispersion, mechanical stirring, grinding, ball-milling, and decomposition process and calcined to get superfine coated particles. Generally, this method is used to synthesize the materials that need phase transformation to gain specific properties by calcination under high temperature, which cannot be obtained by wet chemistry strategy. Bian et al. have studied the magnetic properties of monodisperse single-crystal Co(Fe)Pt@C core–shell NPs with an L10 structure by one-step solid-phase synthesis (Bian et al. 2015). Nalluri et al. have reported cost-effective and quantifiable synthesis of catalytically active Fe_3O_4@M (M = Au, Ag, and Au–Ag alloy) core–shell nanostructures by a solid-state approach (Nalluri et al. 2019). Though the solid-phase reaction is based on directly mixing the raw materials and then calcined at high temperature, some limits cannot be overlooked such as insufficient contact between the reagents and inadequate reactions. Moreover, in the nucleation process, large amount of energy and high temperature are required so that the different lattice structures and atomic arrangements of the precursors could be greatly adjusted.

2.4.2 Liquid-Phase Reaction

The liquid-phase reaction method refers to the chemical reactions under a wet environment to deposit modifiers or films on the surface of preformed particles. Compared with other methods, the liquid-phase reaction method has the advantages of simple equipment and low reaction temperatures, and it is widely used in the laboratory and industry to prepare core@shell NPs. The most popular surface modification technique is to coat the core NPs with an inorganic shell layer via surface silanization owing to the highly biocompatible nature of silica. Additionally, the amorphous and inert SiO_2 shell protects the core NPs against the influence of physiological conditions and outside environments (Ansari et al. 2016; Mondal et al. 2017). Based on the polarity of the capping ligands on the NP surface to coat silica, there are two routes to coat silica, namely, Stöber and reverse microemulsion (water-in-oil [W/O]) method. Both these two techniques can be considered as the part of sol–gel method.

2.4.2.1 Stöber Technique

In 1968, Stöber et al. developed a proficient sol–gel technique, familiar to Stöber method, to prepare colloidal silica particles (Stöber et al. 1968). This method is like the ammonia-catalyzed sol–gel reaction in which hydrolysis of TEOS occurs in the presence of water and low molecular weight alcohols (ethanol, methanol), followed by subsequent condensation. The undergoing reaction mechanisms involved in the formation of silica NPs have been described elsewhere (Ghimire and Jaroniec 2021; Van Blaaderen and Kentgens 1992).

Hydrolysis The hydrolysis process takes place in the presence of water and an efficient catalysis that may be an acid or a base. Usually, in the Stöber synthesis of silica, TEOS is the most common precursor, and ammonia plays the role of a catalyst by accelerating the rate of hydrolysis and condensation. The nucleophilic substitution of ethoxy group (Si–OR) by silanol groups (–Si–OH) via a pentacomplexation transition state occurs in the hydrolysis process. The OH^- ions in ammonia solution are more efficient nucleophiles in contrast to that present in H_2O molecules. Based on the number of ethoxy groups being converted to hydroxyl groups, the degree of hydrolysis can be 1, 2, 3, or 4. This conversion leads to an increase in the rate of hydrolysis by increasing the positive charge of silicon atoms and reducing steric hindrance (Aelion et al. 1950).

Condensation The consequent process of hydrolysis is condensation in which silanol groups (–Si–OH) of silica monomers or oligomers condense to form the siloxane (Si—O—Si) bond by the nucleophilic substitution reaction. The condensation rate is much faster than the hydrolysis rate as the silanol groups are deprotonated more easily than water molecules to increase the positive charge density on the silicon atom to make favorable nucleophilic attack. This is also the reason why silanol monomers prefer to attach to larger siloxane network clusters rather than to other monomers or small oligomers (Ghimire and Jaroniec 2021). Figure 2.6 shows the step-by-step procedure involved in the synthesis technique (Mukhopadhyay 2021).

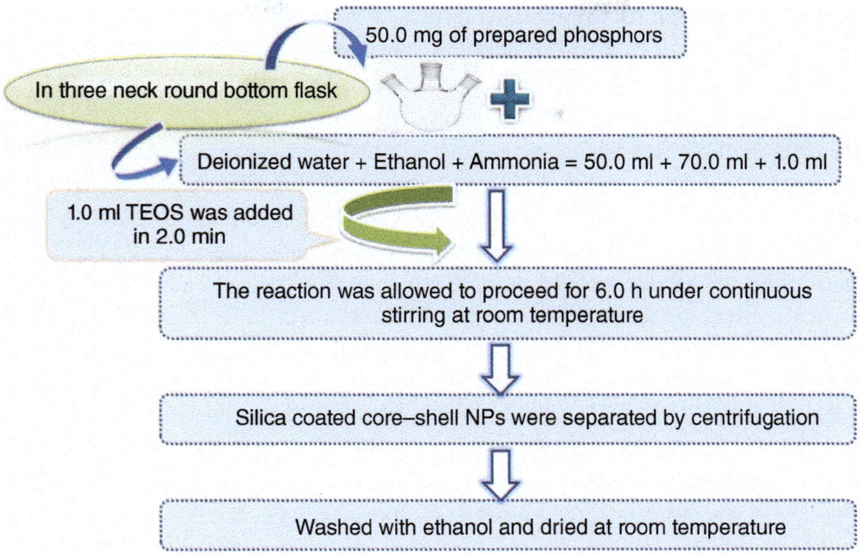

Figure 2.6 Flow chart of Stöber technique. Source: PhD thesis of Lakshmi Mukhopadhyay (2021): Structural and spectroscopic studies of rare earth doped phosphate phosphors for emerging applications.

With the help of this technique, well-defined, monodispersed, uniform silica coating can be achieved. The size of silica spheres can be controlled by adjusting the reaction rate, which depends on several parameters including the concentration of NH_3, TEOS, H_2O, cosolvent, pH, and temperature (Ciriminna et al. 2013; Lofgreen and Ozin 2014). The ammonia concentration is very crucial because it does not only control the rate of reaction but also control the aggregation between the particles and produces OH^- ions in aqueous solution. The degree of polymerization increases with increase in ammonia concentration, which in turn enhances the particle size. Similarly, the increase in the concentration of TEOS increases the particle size, but the monodispersity of silica particles can be maintained up to an optimum concentration. High water concentration raises the rate of hydrolysis reaction and also increases the separation between monomers, thus affecting the growth rate. It has been reported that the final particle size of silica increases with increasing molecular weight of the alcohol by following the order 1-butanol > 2-propanol > 1-propanol > ethanol > methanol (Sadasivan et al. 1998). The effect of cosolvent on hydrolysis and condensation rate follows the trend 1-butanol > methanol > 1-propanol > ethanol > 2-propanol. The different trend is due to the combination of two opposite forces, namely, steric effect and H-bonding, that governs the hydrolysis rate. The steric effect increases with increasing molecular weight, and if it is the only factor affecting the hydrolysis reaction, then the order should be methanol > ethanol > 1-propanol > 2-propanol > 1-butanol. A weaker H-bond between the solvent and nucleophile increases the mobility as well as the rate of hydrolysis, but H-bonding decreases with increasing molecular weight. If H-bonding is the single reason, then the order should be

1-butanol > 2-propanol > 1-propanol > ethanol > methanol (Sadasivan et al. 1998). The temperature and pH of the solution have also great impact on the rate of hydrolysis and thus on the growth of monodispersed and controlled size particles.

The renaissance of the Stöber technique in recent years has benefited the numerous field of nanoscience and material chemistry including ceramics, chromatography, catalysis, paint industries, pharmaceutical industries, high-tech computer applications, optical imaging, etc. The strong UC fluorescence has been reported in silica-coated polyvinylpyrrolidone (PVP)-stabilized $NaYF_4$:Yb,Er/Tm NPs through the Stöber method with 10 nm shell thickness (Li and Zhang 2006). They have observed that by adjusting the initial amount of TEOS, the shell thickness was controlled within 1–3 nm. Flores et al. obtained Ag-deposited silica spheres by modified Stöber synthesis on adding $AgNO_3$ as the silver precursor and PVP as the reducing agent (Flores et al. 2008). Moreover, the NIR luminescent $NaYF_4:Ln^{3+}$ (Ln = Nd, Er, Pr, or Ho) particles coated with SiO_2 has been developed, in which a layer of silica was built to improve the chemical stability and photostability (Salgueiriño-Maceira et al. 2006; Wang et al. 2009; Zhang et al. 2008). The presence of silica not only improves the colloidal stability but also enhances the luminescence intensity because of the increased band gap as a result of carbon impurities and oxygen-related defects in the silica matrix.

Sometimes, the organic template method is used to prepare core@shell NPs in which surfactants such as CTAB and hexadecyl trimethyl ammonium chloride (cetyltrimethylammonium chloride [CTAC]) are employed as a template. In this process, first of all, a hydrophilic complex of CTAB or CTAC with UCNPs is constructed and then processed with the Stöber method to produce a template-patterned silica shell. The patterned mesoporous silica shell is obtained by extracting CTAB or CTAC using ion exchange. Lin et al. reported a set of works on multifunctional mesoporous core@shell nanostructures in which CTAB is used as a template such as $Gd_2O_3:Er^{3+}@SiO_2@mSiO_2$, β-$NaYF_4:Yb^{3+}/Er^{3+}$@β-$NaGdF_4:Yb^{3+}/Er^{3+}@mSiO_2$, $NaYF_4:Yb^{3+}/Er^{3+}@SiO_2@mSiO_2$, and $Fe_3O_4@SiO_2@mSiO_2@NaYF_4:Yb^{3+}/Er^{3+}$ nanocomposites (Gai et al. 2010; Xu et al. 2011a; Kang et al. 2011; Li et al. 2013). Liu et al. studied the NIR light-triggered anticancer drug release based on azobenzene-modified mesoporous $NaYF_4:Yb^{3+}/Tm^{3+}@mSiO_2$ core/shell UCNPs by utilizing a CTAC template. Moreover, the same group, with the help of etching method based on water-phase Stöber method, has produced the $NaYF_4:Yb^{3+}/Er^{3+}@NaGdF_4@mSiO_2$, $NaYF_4$:Yb/Tm@$NaGdF_4@hmSiO_2$, and $NaYF_4$:Yb/Tm@$NaYF_4@hmSiO_2$ yolk nanostructures for applications in synergetic chemo/radiotherapy, dual-modal imaging modality, real-time in vivo quantitative monitoring of drug release, and hypoxia imaging (Liu et al. 2013; Liu et al. 2014b,c).

2.4.2.2 Microemulsion Technique

Microemulsion method is also a popular core@shell synthesis technique that enables the control of size, geometry, morphology, homogeneity, and surface area of the NPs. This scheme is usually used for the synthesis of several inorganic NPs in which the chemical reaction continues within the aqueous phase. Microemulsions are isotropic, macroscopically homogeneous, and thermodynamically stable solutions with a minimum of three components, viz, a polar phase (usually water), a non-polar phase (usually oil), and a surfactant. In the formation of microemulsion, the surfactants used can be either ionic or non-ionic with a cosurfactant and oil as the continuous phase. There are mainly three categories of microemulsions, namely, W/O, oil-in-water (O/W), and water-in-sc-CO_2 (w/sc-CO_2). According to the percentage of different components and the hydrophilic–lipophilic balance (HLB) value of the surfactant used, the formation of microdroplets can be in the form of oil-swollen micelles dispersed in water as O/W microemulsion (normal micelle) or water-swollen micelles dispersed in oil as for W/O microemulsion (reverse micelle) (Fan et al. 2013; Malik et al. 2012; Lu and Yin 2012) (Figure 2.7).

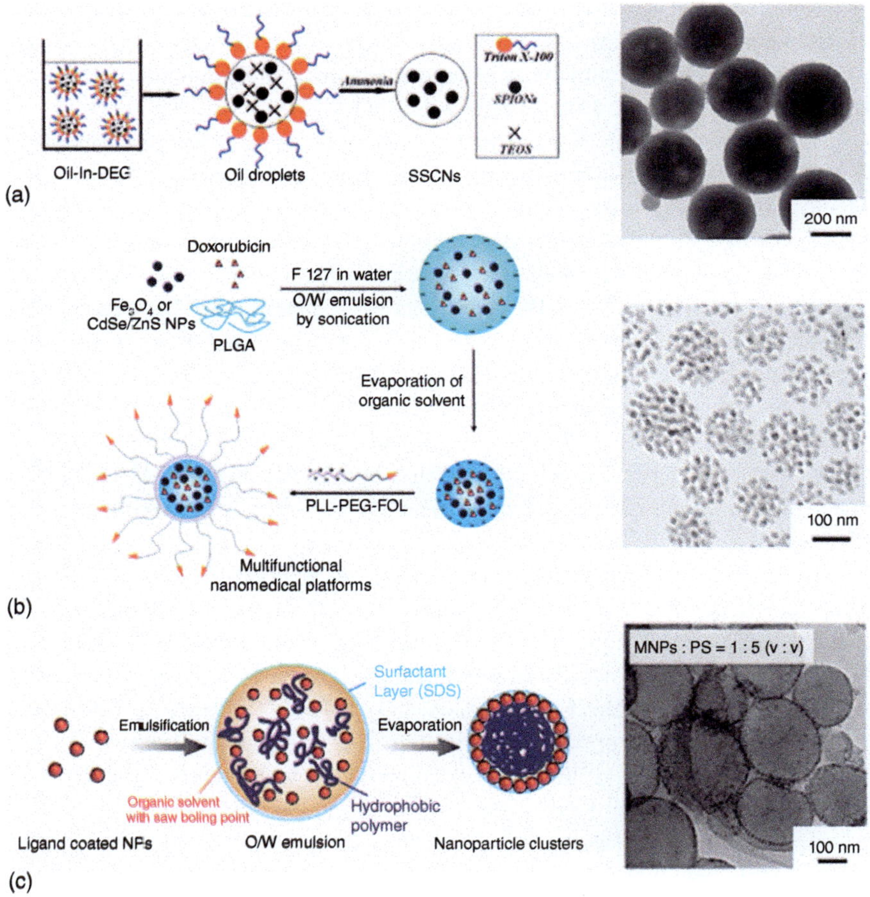

Figure 2.7 Schematic illustration of three emulsion evaporation-based preparation methods for colloidal nanoparticle clusters (CNC) structures and the corresponding transmission electron microscopy (TEM) images of the products: (a) TEOS-assisted clustering, (b) polymer-assisted clustering, and (c) polymer-assisted clustering, followed by phase segregation. DEG, diethylene glycol; SSCN, superparamagnetic silica composite nanospheres; SPION, superparamagnetic iron oxide nanoparticles; PLGA, poly D,L-lactic-co-glycolic acid; PLL, poly L-lysine; PEG, poly ethylene glycol; FOL, folate; MNP, magnetic nanoparticles; PS, polystyrene Source: Lu and Yin (2012). Reprinted with permission of Royal Society of Chemistry.

Recently, the synthesis of NPs using W/O microemulsion, also known as reverse microemulsion, created huge interest among researchers. The thermodynamically driven surfactant molecules dissolved in organic solvents form spheroidal aggregates (minimum surface energy) are called as reverse micelles. Surfactants form reverse micelles, and cosurfactant molecules assist in the micellization process by decreasing the electrostatic repulsive forces between the charged head groups of the surfactant molecules and leads to particle formation reaction. These micelles can act as "nanoreactors" giving proper environment for controlled nucleation and epitaxial growth of the NPs. With the variation of the concentration of surfactant and dispersed phase, different microemulsions can be synthesized that provide micelles with different droplet sizes. Also, the droplet size depends on surfactant HLB value, chain length of the surfactant, and water-to-surfactant ratio. The size and morphology of particles prepared by microemulsion techniques mainly depend on the size of the droplets (Malik et al. 2012; Li et al. 2010).

A few numbers of organic NPs can be synthesized by using O/W microemulsion called as emulsion polymerization in which the droplets of oil mixture (core materials and monomers) are emulsified in the water with the help of surfactants by mechanical stirring or ultrasound. After that, the monomers will polymerize under the action of initiators, and the phase separation will occur inside the droplets to acquire core@shell NPs. This phase separation is the major issue that causes hindrance of using this technique. The droplets that form dispersed phase of an O/W microemulsion can be exploited as carriers of lipophilic solutes across an aqueous environment. Using this polymerization reaction in O/W microemulsion, hydrophobic NPs (dispersed in water) with 10–40 nm size can be prepared. This method is benefited in terms of fast polymerization rates and high molar mass of polymers, but the need of high weight ratio of surfactant to polymer is the main drawback (Li et al. 2010; Wang et al. 2011).

Osseo-Asare and Arriagada first invented the novel W/O reverse-microemulsion method for the synthesis of silica NPs with more uniform size (Osseo-Asare and Arriagada 1990). From then, several researchers have reported the synthesis of different core@shell NPs by microemulsion method. Chen et al. have synthesized methylene blue based $NaYF_4$:Er/Yb/Gd@SiO_2 NPs via reverse microemulsion technique for PDT, optical/magnetic bimodal imaging applications (Chen et al. 2012b). Qian et al. utilized the microemulsion technique to prepare the $NaYF_4$:Yb^{3+}/Er^{3+}@SiO_2@mSiO_2 core@shell NPs to incorporate a photosensitizer or a genetic material into the mesoporous silica for UC PDT (Qian et al. 2009). Liu et al. have demonstrated the preparation of monodisperse core/shell NPs by a one-pot reverse microemulsion technique together with both $NaYF_4$:Yb,Er UCNPs and Fe_3O_4 superparamagnetic nanocrystals that act as a core and were encapsulated in silica shells because of the hydrolysis of TEOS and aminopropyltriethoxysilane (APS) for different biomedical applications (Liu et al. 2008). Xu et al. have described the synthesis of hydrophobic UCNPs in combination with the iron oxide NPs through the microemulsion strategy by using an amphiphilic block copolymer, poly(styrene-*block*-allylalcohol) (PS16-*b*-PAA10) for multimodal imaging and drug delivery applications (Xu et al. 2011b). Zhang et al. constructed the hydrophilic multifunctional mesoporous UC luminescent and magnetic nanorattles, Fe_3O_4@SiO_2@α-$NaYF_4$/Yb,Er, using the reverse microemulsion process, followed by a facile ion exchange route in the presence of hydrogen fluoride (HF) and NaF solution (Zhang et al. 2012). These nanorattles exhibit very good superparamagnetic property and promising for a variety of biological applications. Xiao et al. reported multifunctional core/satellite nanocomposites by decorating ultrasmall CuS semiconductor NPs onto the surface of silica-coated UCNPs synthesized by the W/O reverse-microemulsion method (Xiao et al. 2013). Recently, Li et al. have revealed a novel surface modification technique together with the fluorescence resonance energy transfer (FRET) mechanism to produce multicolor silica-coated UCNPs (Li et al. 2008). They encapsulated some organic dyes and quantum dots with $NaYF_4$:Yb,Er and $NaYF_4$:Yb,Tm NPs, respectively, via Stober-based microemulsion method and described their potentialities in multiplexed bioassays. Therefore, the liquid-phase reaction method has numerous advantages and widely used to synthesize the core@shell NPs. Nevertheless, the lack of scalability, requirement of harmful organic solvents, and further purification limit its applicability in some cases and thus need to be improved in the near future.

2.4.3 Gas-Phase Reaction

This technique is utilized to deposit the shell in the gas-phase supersaturated system onto the target surface for the formation of core@shell NPs. The method is mainly of two types including physical vapor deposition (PVD) and chemical vapor deposition (CVD). In the PVD technique, the material sources are vaporized into gaseous atoms, molecules, or partially ionized into ions by means of physical methods under the vacuum conditions and transported by low-pressure gas or plasma to deposit on the surfaces of the core materials with certain functions. Generally, the PVD methods can be categorized as cathodic arc deposition, electron beam PVD, evaporative deposition, pulsed laser deposition, and sputter deposition (Chen et al. 2020). Langlois et al. have described the preparation of Fe@Au core–shell NPs by sequential deposition from two elemental targets in an ultrahigh vacuum sputtering device (Langlois et al. 2015). The core@shell NPs prepared by this method possess higher impact strength, temperature, and abrasion resistance as PVD coatings are harder and more corrosion resistant. However, the need of high temperature and vacuum in some PVD methods restricts their applications in laboratories and industries. On the other hand, in CVD method, the chemical reactions of gases or steams on the surfaces of core materials occur to fabricate the core@shell NPs. The CVD technique can be of several types, namely, atomic layer CVD, combustion CVD, hot filament CVD, hybrid physical CVD, rapid thermal CVD, and photo-initiated CVD. Cao et al. have reported thermal CVD and thermal annealing processes to synthesize the SiC@SiO_2 core@shell NPs for blue light optoelectronic devices (Cao et al. 2010). However, the CVD method is used to develop thin films, core@shell NPs, and their characteristics can be easily controlled by adjusting the reactants, but the low deposition rate, high equipment requirements, and damages to the environment because of the flammable, explosive, or toxic residual gases limit their practical applications.

2.4.4 Mechanical Mixing

In this technique, two or more particles having good adhesion or adsorption properties at a definite temperature can be evenly adsorbed on the surfaces of modified particles with the help of convection and diffusion. Subsequently, the particles are made closely combined to cover the surface of the target particles by means of some external forces such as aerodynamic force, gravity, and mechanical forces. He et al. reported the synthesis of core@shell-structured paclitaxel (PTX)-loaded gel nanocarriers together with 12-hydroxystearic acid as an organic gelator by mechanical mixing and high-pressure homogenization method for sustained drug release and enhanced antitumor effect (He et al. 2015). It is very difficult to coat the particle surface uniformly, and thus, the mechanical mixing method of core@shell NP synthesis is very rarely used presently.

2.5 Other Synthesis Strategies to Develop Lanthanide-Doped UCNPs

There are some other synthesis methods to prepare UCNPs including molten salt synthesis, ionic liquid-based synthesis, polyol-mediated synthesis, flaming synthesis, electrospinning, melt and quenching process, Ostwald ripening strategy, polymer complex solution, etc. (Wang et al. 2005, 2011; Qin et al. 2007; Dong et al. 2008; Speghini et al. 2016, p. 43; Fiévet et al. 2018; Gupta and Mao 2020). The molten salt synthesis is one of the most convenient, effective, and environmentally friendly approaches to prepare UCNPs at relatively lower temperature and shorter reaction time compared to the conventional solid-state reaction method (Gupta and Mao 2020). The flaming synthesis comprised five steps including precursor reaction, nucleation, growth, polymerization, and ion deposition, mainly used to prepare oxide phosphor powders (Qin et al. 2007).

The ionic liquid-based synthesis is a green synthetic approach as it requires low reaction temperature, short reaction time, and does not need any organic solvents. Nanomaterials that are synthesized in ionic liquid media are called green solvents because of their high chemical stability, low vapor pressure, and non-flammability but with broader size distribution, lower monodispersity, and less uniformity (Wang et al. 2005; Speghini et al. 2016, p. 52).

The polyol-mediated synthesis is also a very popular technique to prepare uniform monodisperse lanthanide-doped UCNPs in large scale and simple manner with less waste products. Polyols with high boiling points, high viscosity, and good water solubility are used not only as a solvent but also as a reducing agent to synthesize highly crystalline UCNPs having controlled structure and morphology (Wang et al. 2011). The polyols are mainly a 1,2-diol, such as ethylene glycol and its derivatives up to poly(ethylene glycol); its different isomers, viz, propanediol, butanediol, pentanediol, etc.; and compounds with more than two hydroxyl groups, namely, glycerol, pentaerythritol, and carbohydrates. Because the presence of large number of OH group polyols exhibit excellent coordinating properties; they can adsorb on the particle surface and provide good colloidal stabilization (Wei et al. 2008; Fiévet et al. 2018).

2.6 Conclusion

Despite having pioneering characteristics, the UCNPs still face some limitations in terms of their practical application because of low emission yield. This mainly attributes to their non-trivial nature and, sometimes, poorly reproducible synthetic procedures. The quest of highly efficient UCNPs with controlled aesthetic size and shape endows an important milestone in the current research era. The morphology of the synthesized UCNPs is not only important to analyze the physicochemical properties but also to know their suitability in different scientific and technological applications. This chapter provides a brief account on the various synthesis strategies that are normally used to prepare the UCNPs. The step-by-step procedure, mechanisms involved, recent progresses, and with emphasis on the feasibility as well as drawback of each synthetic protocol have been discussed. From the above discussions, it can be summarized that till date, thermolysis process is very classic and is an advance approach to regulate the size and shape of the NPs. The sol–gel process is very much useful to construct different nanostructures. Moreover, surface modification of the NPs by means of coating of multishell is a fundamental step from application point of view. The microemulsion process, especially reverse micelle, allows us to form uniform UCNPs but requires a large amount of solvent and produces small quantities of products. However, the synthesis of uniform, monodisperse highly efficient UCNPs with controlled size and morphology is still in demand. Hopefully, the present chapter will be a powerful paradigm for the material scientists' interest in developing new UCNPs.

References

Aelion, R., Loebel, A., and Eirich, F. (1950). Hydrolysis of ethyl silicate. *Journal of the American Chemical Society* 72 (12): 5705–5712. https://doi.org/10.1021/ja01168a090.

Ansari, A.A., Yadav, R., and Rai, S.B. (2016). Enhanced luminescence efficiency of aqueous dispersible $NaYF_4$:Yb/Er nanoparticles and the effect of surface coating. *RSC Advances* 6 (26): 22074–22082. https://doi.org/10.1039/c6ra00265j.

Aruna, S.T. and Mukasyan, A.S. (2008). Combustion synthesis and nanomaterials. *Current Opinion in Solid State and Materials Science* 12 (3, 4): 44–50. https://doi.org/10.1016/j.cossms.2008.12.002.

Bian, B.R., He, J.H., Du, J. et al. (2015). Growth mechanism and magnetic properties of monodisperse $L1_0$-Co(Fe)Pt@C core–shell nanoparticles by one-step solid-phase synthesis. *Nanoscale* 7 (3): 975–980. https://doi.org/10.1039/c4nr04986a.

Bol, A.A., van Beek, R., and Meijerink, A. (2002). On the incorporation of trivalent rare earth ions in II–VI semiconductor nanocrystals. *Chemistry of Materials* 14 (3): 1121–1126. https://doi.org/10.1021/cm011195s.

Boyer, J.C., Vetrone, F., Cuccia, L.A., and Capobianco, J.A. (2006). Synthesis of colloidal upconverting $NaYF_4$ nanocrystals doped with Er^{3+}, Yb^{3+} and Tm^{3+}, Yb^{3+} via thermal decomposition of lanthanide trifluoroacetate precursors. *Journal of the American Chemical Society* 128 (23): 7444–7445. https://doi.org/10.1021/ja061848b.s001.

Boyer, J.C., Cuccia, L.A., and Capobianco, J.A. (2007). Synthesis of colloidal upconverting $NaYF_4$:Er^{3+}/Yb^{3+} and Tm^{3+}/Yb^{3+} monodisperse nanocrystals. *Nano Letters* 7 (3): 847–852. https://doi.org/10.1021/nl070235.

Buckley, A.M. and Greenblatt, M. (1994). The sol–gel preparation of silica gels. *Journal of Chemical Education* 71 (7): 599–602. https://doi.org/10.1021/ed071p599.

Byrappa, K. and Adschiri, T. (2007). Hydrothermal technology for nanotechnology. *Progress in Crystal Growth and Characterization of Materials* 53 (2): 117–166. https://doi.org/10.1016/j.pcrysgrow.2007.04.001.

Cao, Y.C. (2004). Synthesis of square gadolinium-oxide nanoplates. *Journal of the American Chemical Society* 126 (24): 7456–7457. https://doi.org/10.1021/ja0481676.

Cao, S.W. and Zhu, Y.J. (2008). Hierarchically nanostructured α-Fe_2O_3 hollow spheres: preparation, growth mechanism, photocatalytic property, and application in water treatment. *The Journal of Physical Chemistry C* 112 (16): 6253–6257. https://doi.org/10.1021/jp8000465.

Cao, L., Jiang, H., Song, H. et al. (2010). Thermal CVD synthesis and photoluminescence of SiC/SiO_2 core–shell structure nanoparticles. *Journal of Alloys and Compounds* 489 (2): 562–565. https://doi.org/10.1016/j.jallcom.2009.09.109.

Capobianco, J.A., Vetrone, F., D'alesio, T. et al. (2000). Optical spectroscopy of nanocrystalline cubic Y_2O_3:Er^{3+} obtained by combustion synthesis. *Physical Chemistry Chemical Physics* 2 (14): 3203–3207. https://doi.org/10.1039/b003031g.

Chen, G., Ohulchanskyy, T.Y., Kachynski, A. et al. (2011). Intense visible and near-infrared upconversion photoluminescence in colloidal $LiYF_4$:Er^{3+} nanocrystals under excitation at 1490 nm. *ACS Nano* 5 (6): 4981–4986. https://doi.org/10.1021/nn201083j.

Chen, G., Qiu, H., Fan, R. et al. (2012a). Lanthanide-doped ultrasmall yttrium fluoride nanoparticles with enhanced multicolor upconversion photoluminescence. *Journal of Materials Chemistry* 22 (38): 20190–20196. https://doi.org/10.1039/c2jm32298f.

Chen, F., Zhang, S., Bu, W. et al. (2012b). A uniform sub-50 nm-sized magnetic/upconversion fluorescent bimodal imaging agent capable of generating singlet oxygen by using a 980 nm laser. *Chemistry A European Journal* 18 (23): 7082–7090. https://doi.org/10.1002/chem.201103611.

Chen, G., Yang, C., and Prasad, P.N. (2013). Nanophotonics and nanochemistry: controlling the excitation dynamics for frequency up- and down-conversion in lanthanide-doped nanoparticles. *Accounts of Chemical Research* 46 (7): 1474–1486. https://doi.org/10.1021/ar300270y.

Chen, G., Qiu, H., Prasad, P.N., and Chen, X. (2014). Upconversion nanoparticles: design, nanochemistry, and applications in theranostics. *Chemical Reviews* 114 (10): 5161–5214. https://doi.org/10.1021/cr400425h.

Chen, G., Ågren, H., Ohulchanskyy, T.Y., and Prasad, P.N. (2015). Light upconverting core–shell nanostructures: nanophotonic control for emerging applications. *Chemical Society Reviews* 44 (6): 1680–1713. https://doi.org/10.1039/c4cs00170b.

Chen, H., Zhang, L., Li, M., and Xie, G. (2020). Synthesis of core–shell micro/nanoparticles and their tribological application: a review. *Materials* 13 (20): 4590. https://doi.org/10.3390/ma13204590.

Cho, S., Jung, S.H., and Lee, K.H. (2008). Morphology-controlled growth of ZnO nanostructures using microwave irradiation: from basic to complex structures. *The Journal of Physical Chemistry C* 112 (33): 12769–12776. https://doi.org/10.1021/jp803783s.

Ciriminna, R., Fidalgo, A., Pandarus, V. et al. (2013). The sol–gel route to advanced silica-based materials and recent applications. *Chemical Reviews* 113 (8): 6592–6620. https://doi.org/10.1021/cr300399c.

Cushing, B.L., Kolesnichenko, V.L., and O'connor, C.J. (2004). Recent advances in the liquid-phase syntheses of inorganic nanoparticles. *Chemical Reviews* 104 (9): 3893–3946. https://doi.org/10.1021/cr030027b.

Das, S., Mukherjee, A., Sengupta, G., and Singh, V.K. (2020). Overview of nanomaterials synthesis methods, characterization techniques and effect on seed germination. In: *Nano-Materials as Photocatalysts for Degradation of Environmental Pollutants* (ed. P. Singh, A. Borthakur, P.K. Mishra and D. Tiwary), 371–401. Elsevier.

Deganello, F. and Tyagi, A.K. (2018). Solution combustion synthesis, energy and environment: best parameters for better materials. *Progress in Crystal Growth and Characterization of Materials* 64 (2): 23–61. https://doi.org/10.1016/j.pcrysgrow.2018.03.001.

Dey, R., Kumari, A., Soni, A.K., and Rai, V.K. (2015). $CaMoO_4:Ho^{3+}–Yb^{3+}–Mg^{2+}$ upconverting phosphor for application in lighting devices and optical temperature sensing. *Sensors and Actuators B: Chemical* 210: 581–588. https://doi.org/10.1016/j.snb.2015.01.007.

Dhand, C., Dwivedi, N., Loh, X.J. et al. (2015). Methods and strategies for the synthesis of diverse nanoparticles and their applications: a comprehensive overview. *RSC Advances* 5 (127): 105003–105037. https://doi.org/10.1039/c5ra19388e.

Ding, Y., Gu, J., Ke, J. et al. (2011). Sodium doping controlled synthesis of monodisperse lanthanide oxysulfide ultrathin nanoplates guided by density functional calculations. *Angewandte Chemie International Edition* 50 (51): 12330–12334. https://doi.org/10.1002/ange.201105025.

Ding, Y., Gu, J., Zhang, T. et al. (2012). Chemoaffinity-mediated synthesis of $NaRES_2$-based nanocrystals as versatile nano-building blocks and durable nano-pigments. *Journal of the American Chemical Society* 134 (6): 3255–3264. https://doi.org/10.1021/ja211103b.

Dong, B., Song, H., Yu, H. et al. (2008). Upconversion properties of Ln^{3+} doped $NaYF_4$ polymer composite fibers prepared by electrospinning. *The Journal of Physical Chemistry C* 112 (5): 1435–1440. https://doi.org/10.1021/jp076958z.

Du, Y.P., Zhang, Y.W., Sun, L.D., and Yan, C.H. (2008). Luminescent monodisperse nanocrystals of lanthanide oxyfluorides synthesized from trifluoroacetate precursors in high-boiling solvents. *The Journal of Physical Chemistry C* 112 (2): 405–415. https://doi.org/10.1021/jp076717r.

Du, Y.P., Sun, X., Zhang, Y.W. et al. (2009a). Uniform alkaline earth fluoride nanocrystals with diverse shapes grown from thermolysis of metal trifluoroacetates in hot surfactant solutions. *Crystal Growth and Design* 9 (4): 2013–2019. https://doi.org/10.1021/cg801371r.

Du, Y.P., Zhang, Y.W., Sun, L.D., and Yan, C.H. (2009b). Atomically efficient synthesis of self-assembled monodisperse and ultrathin lanthanide oxychloride nanoplates. *Journal of the American Chemical Society* 131 (9): 3162–3163. https://doi.org/10.1021/ja8095416.

Fan, W., Shen, B., Bu, W. et al. (2013). Rattle-structured multifunctional nanotheranostics for synergetic chemo-/radiotherapy and simultaneous magnetic/luminescent dual-mode imaging. *Journal of the American Chemical Society* 135 (17): 6494–6503. https://doi.org/10.1021/ja312225b.

Feng, S.H. and Li, G.H. (2017). Hydrothermal and solvothermal syntheses. In: *Modern Inorganic Synthetic Chemistry* (ed. R. Xu and Y. Xu), 73–104. Elsevier.

Fiévet, F., Ammar-Merah, S., Brayner, R. et al. (2018). The polyol process: a unique method for easy access to metal nanoparticles with tailored sizes, shapes and compositions. *Chemical Society Reviews* 47 (14): 5187–5233. https://doi.org/10.1039/c7cs00777a.

Flores, J.C., Torres, V., Popa, M. et al. (2008). Preparation of core–shell nanospheres of silica–silver: SiO_2@Ag. *Journal of Non-Crystalline Solids* 354 (52–54): 5435–5439. https://doi.org/10.1016/j.jnoncrysol.2008.09.014.

Gai, S., Yang, P., Li, C. et al. (2010). Synthesis of magnetic, up-conversion luminescent, and mesoporous core–shell-structured nanocomposites as drug carriers. *Advanced Functional Materials* 20 (7): 1166–1172. https://doi.org/10.1002/adfm.200902274.

Gai, S., Li, C., Yang, P., and Lin, J. (2014). Recent progress in rare earth micro/nanocrystals: soft chemical synthesis, luminescent properties, and biomedical applications. *Chemical Reviews* 114 (4): 2343–2389. https://doi.org/10.1021/cr4001594.

Ghimire, P.P. and Jaroniec, M. (2021). Renaissance of Stöber method for synthesis of colloidal particles: new developments and opportunities. *Journal of Colloid and Interface Science* 584: 838–865. https://doi.org/10.1016/j.jcis.2020.10.014.

Ghosh Chaudhuri, R. and Paria, S. (2012). Core/shell nanoparticles: classes, properties, synthesis mechanisms, characterization, and applications. *Chemical Reviews* 112 (4): 2373–2433. https://doi.org/10.1021/cr100449n.

Guo, H., Dong, N., Yin, M. et al. (2004). Visible upconversion in rare earth ion-doped Gd_2O_3 nanocrystals. *The Journal of Physical Chemistry B* 108 (50): 19205–19209. https://doi.org/10.1021/jp048072q.

Gupta, S.K. and Mao, Y. (2020). Molten salt synthesis of metal oxide nanomaterials: status, opportunity, and challenge. *Progress in Materials Science* 100734. https://doi.org/10.1016/j.pmatsci.2020.100734.

Haase, M. and Schäfer, H. (2011). Upconverting nanoparticles. *Angewandte Chemie International Edition* 50 (26): 5808–5829. https://doi.org/10.1002/anie.201005159.

Han, S., Deng, R., Xie, X., and Liu, X. (2014). Enhancing luminescence in lanthanide-doped upconversion nanoparticles. *Angewandte Chemie International Edition* 53 (44): 11702–11715. https://doi.org/10.1002/anie.201403408.

He, Y., Lu, H.T., Sai, L.M. et al. (2006). Microwave-assisted growth and characterization of water-dispersed CdTe/CdS core–shell nanocrystals with high photoluminescence. *The Journal of Physical Chemistry B* 110 (27): 13370–13374. https://doi.org/10.1021/jp057498h.

He, W., Lv, Y., Zhao, Y. et al. (2015). Core–shell structured gel-nanocarriers for sustained drug release and enhanced antitumor effect. *International Journal of Pharmaceutics* 484 (1, 2): 163–171. https://doi.org/10.1016/j.ijpharm.2015.02.053.

Heer, S., Lehmann, O., Haase, M., and Gudel, H.U. (2003). Blue, green, and red upconversion emission from lanthanide-doped $LuPO_4$ and $YbPO_4$ nanocrystals in a transparent colloidal solution. *Angewandte Chemie International Edition* 42 (27): 3179–3182. https://doi.org/10.1002/anie.200351091.

Heer, S., Kömpe, K., Güdel, H.U., and Haase, M. (2004). Highly efficient multicolour upconversion emission in transparent colloids of lanthanide-doped $NaYF_4$ nanocrystals. *Advanced Materials* 16 (23, 24): 2102–2105. https://doi.org/10.1002/adma.200400772.

Hench, L.L. and West, J.K. (1990). The sol–gel process. *Chemical Reviews* 90 (1): 33–72. https://doi.org/10.1021/cr00099a003.

Horikoshi, S. (2013). Nanoparticle synthesis through microwave heating. In: *Microwaves in Nanoparticle Synthesis: Fundamental and Applications* (ed. S. Horikoshi and N. Serpone), 75–105. Wiley https://doi.org/10.1002/9783527648122.

Huang, X., Han, S., Huang, W., and Liu, X. (2013). Enhancing solar cell efficiency: the search for luminescent materials as spectral converters. *Chemical Society Reviews* 42 (1): 173–201. https://doi.org/10.1039/c2cs35288e.

Ivaturi, A., MacDougall, S.K., Martín-Rodríguez, R. et al. (2013). Optimizing infrared to near infrared upconversion quantum yield of β-$NaYF_4:Er^{3+}$ in fluoropolymer matrix for photovoltaic devices. *Journal of Applied Physics* 114 (1): 0135.05. https://doi.org/10.1063/1.4812578.

Kang, X., Cheng, Z., Li, C. et al. (2011). Core–shell structured up-conversion luminescent and mesoporous $NaYF_4:Yb^{3+}/Er^{3+}$@ $nSiO_2$@$mSiO_2$ nanospheres as carriers for drug delivery. *The Journal of Physical Chemistry C* 115 (32): 15801–15811. https://doi.org/10.1021/jp203039t.

Kaya, G.G. and Deveci, H. (2020). Synergistic effects of silica aerogels/xerogels on properties of polymer composites: a review. *Journal of Industrial and Engineering Chemistry* 89: 13–27. https://doi.org/10.1016/j.jiec.2020.05.019.

Kim, C.H., Kwon, I.E., Park, C.H. et al. (2000). Phosphors for plasma display panels. *Journal of Alloys and Compounds* 311 (1): 33–39. https://doi.org/10.1016/s0925-8388(00)00856-2.

Krämer, K.W., Biner, D., Frei, G. et al. (2004). Hexagonal sodium yttrium fluoride based green and blue emitting upconversion phosphors. *Chemistry of Materials* 16 (7): 1244–1251. https://doi.org/10.1021/cm031124o.

Kuisheng, Y., Fang, Z., Rina, W. et al. (2006). Upconversion luminescent properties of $YVO_4:Yb^{3+},Er^{3+}$ nano-powder by sol–gel method. *Journal of Rare Earths* 24 (1): 162–166. https://doi.org/10.1016/s1002-0721(07)60350-0.

Langlois, C., Benzo, P., Arenal, R. et al. (2015). Fully crystalline faceted Fe–Au core–shell nanoparticles. *Nano Letters* 15 (8): 5075–5080. https://doi.org/10.1021/acs.nanolett.5b02273.

Li, Z. and Zhang, Y. (2006). Monodisperse silica-coated polyvinylpyrrolidone/$NaYF_4$ nanocrystals with multicolor upconversion fluorescence emission. *Angewandte*

Chemie International Edition 118 (46): 7896–7899. https://doi.org/10.1002/anie.200602975.

Li, Z., Zhang, Y., and Jiang, S. (2008). Multicolor core/shell-structured upconversion fluorescent nanoparticles. *Advanced Materials* 20 (24): 4765–4769. https://doi.org/10.1002/adma.200801056.

Li, Z., Guo, H., Qian, H., and Hu, Y. (2010). Facile microemulsion route to coat carbonized glucose on upconversion nanocrystals as high luminescence and biocompatible cell-imaging probes. *Nanotechnology* 21 (31): 315105. https://doi.org/10.1088/0957-4484/21/31/315105.

Li, C., Yang, D., Ma, P.A. et al. (2013). Multifunctional upconversion mesoporous silica nanostructures for dual modal imaging and in vivo drug delivery. *Small* 9 (24): 4150–4159. https://doi.org/10.1002/smll.201301093.

Li, J., Wu, Q., and Wu, J. (2015). Synthesis of nanoparticles via solvothermal and hydrothermal methods, Chapter. 12. In: *Handbook of Nanoparticles*, 1–28. Switzerland: Springer International Publishing https://doi.org/10.1007/978-3-319-13188-7_17-1.

Lin, M., Zhao, Y., Liu, M. et al. (2014). Synthesis of upconversion $NaYF_4$:Yb^{3+},Er^{3+} particles with enhanced luminescent intensity through control of morphology and phase. *Journal of Materials Chemistry C* 2 (19): 3671–3676. https://doi.org/10.1039/c4tc00129j.

Liu, Z., Yi, G., Zhang, H. et al. (2008). Monodisperse silica nanoparticles encapsulating upconversion fluorescent and superparamagnetic nanocrystals. *Chemical Communications* 6: 694–696. https://doi.org/10.1039/b715402j.

Liu, J., Bu, W., Pan, L., and Shi, J. (2013). NIR-triggered anticancer drug delivery by upconverting nanoparticles with integrated azobenzene-modified mesoporous silica. *Angewandte Chemie International Edition* 125 (16): 4471–4475. https://doi.org/10.1002/ange.201300183.

Liu, F., Liang, Y., and Pan, Z. (2014a). Detection of up-converted persistent luminescence in the near infrared emitted by the $Zn_3Ga_2GeO_8$:Cr^{3+},Yb^{3+},Er^{3+} phosphor. *Physical Review Letters* 113 (17): 177401. https://doi.org/10.1103/physrevlett.113.177401.

Liu, J., Bu, J., Bu, W. et al. (2014b). Real-time in vivo quantitative monitoring of drug release by dual-mode magnetic resonance and upconverted luminescence imaging. *Angewandte Chemie International Edition* 126 (18): 4639–4643. https://doi.org/10.1002/ange.201400900.

Liu, J., Liu, Y., Bu, W. et al. (2014c). Ultrasensitive nanosensors based on upconversion nanoparticles for selective hypoxia imaging in vivo upon near-infrared excitation. *Journal of the American Chemical Society* 136 (27): 9701–9709. https://doi.org/10.1021/ja5042989.

Livage, J. and Sanchez, C. (1992). Sol–gel chemistry. *Journal of Non-Crystalline Solids* 145: 11–19. https://doi.org/10.1016/s0022-3093(05)80422-3.

Lofgreen, J.E. and Ozin, G.A. (2014). Controlling morphology and porosity to improve performance of molecularly imprinted sol–gel silica. *Chemical Society Reviews* 43 (3): 911–933. https://doi.org/10.1039/c3cs60276a.

Lu, Z. and Yin, Y. (2012). Colloidal nanoparticle clusters: functional materials by design. *Chemical Society Reviews* 41 (21): 6874–6887. https://doi.org/10.1039/c2cs35197h.

Mahalingam, V., Vetrone, F., Naccache, R. et al. (2009). Colloidal Tm^{3+}/Yb^{3+}-doped $LiYF_4$ nanocrystals: multiple luminescence spanning the UV to NIR regions via low-energy excitation. *Advanced Materials* 21 (40): 4025–4028. https://doi.org/10.1002/adma.200901174.

Mai, H.X., Zhang, Y.W., Si, R. et al. (2006). High-quality sodium rare-earth fluoride nanocrystals: controlled synthesis and optical properties. *Journal of the American Chemical Society* 128 (19): 6426–6436. https://doi.org/10.1021/ja060212h.

Mai, H.X., Zhang, Y.W., Sun, L.D., and Yan, C.H. (2007). Highly efficient multicolor up-conversion emissions and their mechanisms of monodisperse $NaYF_4$:Yb,Er core and core/shell-structured nanocrystals. *The Journal of Physical Chemistry C* 111 (37): 13721–13729. https://doi.org/10.1021/jp073920d.

Malik, M.A., Wani, M.Y., and Hashim, M.A. (2012). Microemulsion method: a novel route to synthesize organic and inorganic nanomaterials: 1st nano update. *Arabian Journal of Chemistry* 5 (4): 397–417. https://doi.org/10.1016/j.arabjc.2010.09.027.

Mao, Y., Park, T.J., Zhang, F. et al. (2007). Environmentally friendly methodologies of nanostructure synthesis. *Small* 3 (7): 1122–1139. https://doi.org/10.1002/smll.200700048.

Mialon, G., Turkcan, S., Dantelle, G. et al. (2010). High up-conversion efficiency of YVO_4:Yb,Er nanoparticles in water down to the single-particle level. *The Journal of Physical Chemistry C* 114 (51): 22449–22454. https://doi.org/10.1021/jp107900z.

Mondal, M., Rai, V.K., and Srivastava, C. (2017). Influence of silica surface coating on optical properties of Er^{3+}–Yb^{3+}:$YMoO_4$ upconverting nanoparticles. *Chemical Engineering Journal* 327: 838–848. https://doi.org/10.1016/j.cej.2017.06.166.

Mukhopadhyay, L. (2021). Structural and spectroscopic studies of rare earth doped phosphate phosphors for emerging applications. PhD thesis. IIT (ISM) Dhanbad, India.

Mukhopadhyay, L. and Rai, V.K. (2021). Colloidal stability and optical thermometry in mesoporous silica coated phosphate based upconverting nanoparticles. *Journal of Alloys and Compounds* 878: 160351. https://doi.org/10.1016/j.jallcom.2021.160351.

Mukhopadhyay, L., Rai, V.K., Bokolia, R., and Sreenivas, K. (2017). 980 nm excited $Er^{3+}/Yb^{3+}/Li^{+}/Ba^{2+}$:$NaZnPO_4$ upconverting phosphors in optical thermometry. *Journal of Luminescence* 187: 368–377. https://doi.org/10.1016/j.jlumin.2017.03.035.

Nalluri, S.R., Nagarjuna, R., Patra, D. et al. (2019). Large scale solid-state synthesis of catalytically active Fe_3O_4@M (M = Au, Ag and Au–Ag alloy) core–shell nanostructures. *Scientific Reports* 9 (1): 1–11. https://doi.org/10.1038/s41598-019-43116-7.

Nguyen, T.D. (2013). From formation mechanisms to synthetic methods toward shape-controlled oxide nanoparticles. *Nanoscale* 5 (20): 9455–9482. https://doi.org/10.1039/c3nr01810e.

Osseo-Asare, K. and Arriagada, F.J. (1990). Preparation of SiO_2 nanoparticles in a non-ionic reverse micellar system. *Colloids and Surfaces* 50: 321–339. https://doi.org/10.1016/0166-6622(90)80273-7.

Pan, Z., Lu, Y.Y., and Liu, F. (2012). Sunlight-activated long-persistent luminescence in the near-infrared from Cr^{3+} doped zinc gallogermanates. *Nature Materials* 11 (1): 58–63. https://doi.org/10.1038/nmat3173.

Panda, A.B., Glaspell, G., and El-Shall, M.S. (2007). Microwave synthesis and optical properties of uniform nanorods and nanoplates of rare earth oxides. *The Journal of Physical Chemistry C* 111 (5): 1861–1864. https://doi.org/10.1021/jp0670283.

Pandozzi, F., Vetrone, F., Boyer, J.C. et al. (2005). A spectroscopic analysis of blue and ultraviolet upconverted emissions from $Gd_3Ga_5O_{12}$:Tm^{3+},Yb^{3+} nanocrystals. *The Journal of Physical Chemistry B* 109 (37): 17400–17405. https://doi.org/10.1021/jp052192w.

Parashar, M., Shukla, V.K., and Singh, R. (2020). Metal oxides nanoparticles via sol–gel method: a review on synthesis, characterization and applications. *Journal of Materials Science: Materials in Electronics* 31 (5): 3729–3749. https://doi.org/10.1007/s10854-020-02994-8.

Patra, A., Friend, C.S., Kapoor, R., and Prasad, P.N. (2002). Upconversion in Er^{3+}:ZrO_2 nanocrystals. *The Journal of Physical Chemistry B* 106 (8): 1909–1912. https://doi.org/10.1021/jp013576z.

Patra, A., Friend, C.S., Kapoor, R., and Prasad, P.N. (2003a). Effect of crystal nature on upconversion luminescence in Er^{3+}:ZrO_2 nanocrystals. *Applied Physics Letters* 83 (2): 284–286. https://doi.org/10.1063/1.1592891.

Patra, A., Friend, C.S., Kapoor, R., and Prasad, P.N. (2003b). Fluorescence upconversion properties of Er^{3+}-doped TiO_2 and $BaTiO_3$ nanocrystallites. *Chemistry of Materials* 15 (19): 3650–3655. https://doi.org/10.1021/cm020897u.

Pattnaik, S. and Rai, V.K. (2021). Insight into the spectroscopic and thermometric properties of titanate phosphors via a novel co-excited laser system. *Materials Science and Engineering B* 272: 115318. https://doi.org/10.1016/j.mseb.2021.115318.

Peacock, R.D. (1975). The intensities of lanthanide f–f transitions. *Structure and Bonding* 22: 83–122. http://doi.org/10.1007/BFb0116556.

Peng, Y., Xing, M.M., Luo, X.X., and Wang, L.Q. (2012). Synthesis and characteristic research of nanoparticles KY_3F_{10}:Yb, RE (RE = Er, Ho, Tm) by thermal decomposition. *Acta Physica Sinica* 61: 137201. https://doi.org/10.7498/aps.61.137201.

Periyasamy, A.P., Venkataraman, M., Kremenakova, D. et al. (2020). Progress in sol–gel technology for the coatings of fabrics. *Materials* 13 (8): 1838. https://doi.org/10.3390/ma13081838.

Prasad, M. and Rai, V.K. (2020). Influence of transition metal ions: broad band upconversion emission in thermally stable phosphors. *Journal of Alloys and Compounds* 837: 155289. https://doi.org/10.1016/j.jallcom.2020.155289.

Priecel, P. and Lopez-Sanchez, J.A. (2019). Advantages and limitations of microwave reactors: from chemical synthesis to the catalytic valorization of biobased chemicals. *ACS Sustainable Chemistry & Engineering* 7 (1): 3–21. https://doi.org/10.1021/acssuschemeng.8b03286.

Qian, H.S., Guo, H.C., Ho, P.C.L. et al. (2009). Mesoporous-silica-coated up-conversion fluorescent nanoparticles for photodynamic therapy. *Small* 5 (20): 2285–2290. https://doi.org/10.1002/smll.200900692.

Qin, X., Yokomori, T., and Ju, Y. (2007). Flame synthesis and characterization of rare-earth (Er^{3+}, Ho^{3+}, and Tm^{3+}) doped upconversion nanophosphors. *Applied Physics Letters* 90 (7): 073104. https://doi.org/10.1063/1.2561079.

Qiu, H., Chen, G., Sun, L. et al. (2011a). Ethylenediaminetetraacetic acid (EDTA)-controlled synthesis of multicolor lanthanide doped $BaYF_5$ upconversion nanocrystals. *Journal of Materials Chemistry* 21 (43): 17202–17208. https://doi.org/10.1039/c1jm12950c.

Qiu, H., Chen, G., Fan, R. et al. (2011b). Tuning the size and shape of colloidal cerium oxide nanocrystals through lanthanide doping. *Chemical Communications* 47 (34): 9648–9650. https://doi.org/10.1039/c1cc13707g.

Qiu, P., Zhou, N., Chen, H. et al. (2013). Recent advances in lanthanide-doped upconversion nanomaterials: synthesis, nanostructures and surface modification. *Nanoscale* 5 (23): 11512–11525. https://doi.org/10.1039/c3nr03642a.

Rai, V.K. (2012). Frequency upconversion in rare earth ions. In: *Solid State Laser* (ed. A.H. Al-Khursan), 209–224. IntechOpen.

Rao, C.N.R., Vivekchand, S.R.C., Biswas, K., and Govindaraj, A. (2007). Synthesis of inorganic nanomaterials. *Dalton Transactions* 34: 3728–3749. https://doi.org/10.1039/b708342d.

Reiser, J.T., Ryan, J.V., and Wall, N.A. (2019). Sol–gel synthesis and characterization of gels with compositions relevant to hydrated glass alteration layers. *ACS Omega* 4 (15): 16257–16269. https://doi.org/10.1021/acsomega.9b00491.

Sadasivan, S., Dubey, A.K., Li, Y., and Rasmussen, D.H. (1998). Alcoholic solvent effect on silica synthesis – NMR and DLS investigation. *Journal of Sol-Gel Science and Technology* 12 (1): 5–14. https://doi.org/10.1023/A:1008659708390.

Salgueiriño-Maceira, V., Correa-Duarte, M.A., Spasova, M. et al. (2006). Composite silica spheres with magnetic and luminescent functionalities. *Advanced Functional Materials* 16 (4): 509–514. https://doi.org/10.1002/adfm.200500565.

Schäfer, H., Ptacek, P., Eickmeier, H., and Haase, M. (2009). Synthesis of hexagonal Yb^{3+}, Er^{3+}-doped $NaYF_4$ nanocrystals at low temperature. *Advanced Functional Materials* 19 (19): 3091–3097. https://doi.org/10.1002/adfm.200900642.

Senthilselvan, J., Thomas, S., Anbharasi, L. et al. (2019). EDTA functionalization of SrF_2:Yb,Er nanoparticles by hydrothermal synthesis: intense red upconversion, NIR-to-NIR emission and luminescence nanothermometry characteristics. *Journal of Materials Science: Materials in Electronics* 30 (23): 20376–20392. https://doi.org/10.1007/s10854-019-02311-y.

Shan, J. and Ju, Y. (2007). Controlled synthesis of lanthanide-doped $NaYF_4$ upconversion nanocrystals via ligand induced crystal phase transition and silica coating. *Applied Physics Letters* 91 (12): 123103. https://doi.org/10.1063/1.2783476.

Shan, J., Qin, X., Yao, N., and Ju, Y. (2007). Synthesis of monodisperse hexagonal $NaYF_4$:Yb,Ln (Ln = Er, Ho and Tm) upconversion nanocrystals in TOPO. *Nanotechnology* 18 (44): 445607. https://doi.org/10.1088/0957-4484/18/44/445607.

Shan, J., Kong, W., Wei, R. et al. (2010). An investigation of the thermal sensitivity and stability of the β-$NaYF_4$:Yb,Er upconversion nanophosphors. *Journal of Applied Physics* 107 (5): 054901. https://doi.org/10.1063/1.3298905.

Si, R., Zhang, Y.W., Zhou, H.P. et al. (2007). Controlled-synthesis, self-assembly behavior, and surface-dependent optical properties of high-quality rare-earth oxide nanocrystals. *Chemistry of Materials* 19 (1): 18–27. https://doi.org/10.1021/cm0618392.

Singh, S.K., Kumar, K., and Rai, S.B. (2009a). Multifunctional Er^{3+}–Yb^{3+} codoped Gd_2O_3 nanocrystalline phosphor synthesized through optimized combustion route. *Applied Physics B* 94 (1): 165–173. https://doi.org/10.1007/s00340-008-3261-6.

Singh, S.K., Kumar, K., and Rai, S.B. (2009b). Er^{3+}/Yb^{3+} codoped Gd_2O_3 nano-phosphor for optical thermometry. *Sensors and Actuators A: Physical* 149 (1): 16–20. https://doi.org/10.1016/j.sna.2008.09.019.

Song, Z., Lin, T., Lin, L. et al. (2016). Invisible security ink based on water-soluble graphitic carbon nitride quantum dots. *Angewandte Chemie International Edition* 128 (8): 2823–2827. https://doi.org/10.1002/ange.201510945.

Soni, A.K. and Rai, V.K. (2014). Intrinsic optical bistability and frequency upconversion in Tm^{3+}–Yb^{3+}-codoped Y_2WO_6 phosphor. *Dalton Transactions* 43 (36): 13563–13570. https://doi.org/10.1039/c4dt01266f.

Speghini, A., Pedroni, M., Zaccheroni, N., and Rampazzo, E. (2016). Synthesis of upconverting nanomaterials: designing the composition and nanostructure. In: *Upconverting Nanomaterials Perspectives, Synthesis, and Applications* (ed. C. Altavilla), 61–92. CRC Press.

Stöber, W., Fink, A., and Bohn, E. (1968). Controlled growth of monodisperse silica spheres in the micron size range. *Journal of Colloid and Interface Science* 26 (1): 62–69. https://doi.org/10.1016/0021-9797(68)90272-5.

Stouwdam, J.W. and van Veggel, F.C.J.M. (2002). Near-infrared emission of redispersible Er^{3+}, Nd^{3+}, and Ho^{3+} doped LaF_3 nanoparticles. *Nano Letters* 2 (7): 733–737. https://doi.org/10.1021/nl025562q.

Sun, X., Zhang, Y.W., Du, Y.P. et al. (2007). From trifluoroacetate complex precursors to monodisperse rare-earth fluoride and oxyfluoride nanocrystals with diverse shapes through controlled fluorination in solution phase. *Chemistry A European Journal* 13 (8): 2320–2332. https://doi.org/10.1002/chem.200601072.

Teng, X., Zhu, Y., Wei, W. et al. (2012). Lanthanide-doped Na_xScF_{3+x} nanocrystals: crystal structure evolution and multicolor tuning. *Journal of the American Chemical Society* 134 (20): 8340–8343. https://doi.org/10.1021/ja3016236.

Tiwari, S.P., Kumar, A., and Kumar, K. (2016). Upconversion phosphor materials for beginners: synthesis and applications. *Research Frontiers in Sciences* 23–42. https://doi.org/10.1166/asl.2015.6427.

Tsuji, M., Yamaguchi, D., Matsunaga, M., and Alam, M.J. (2010). Epitaxial growth of Au@Cu core–shell nanocrystals prepared using the PVP-assisted polyol reduction method. *Crystal Growth and Design* 10 (12): 5129–5135. https://doi.org/10.1021/cg100860d.

Van Blaaderen, A. and Kentgens, A.P.M. (1992). Particle morphology and chemical microstructure of colloidal silica spheres made from alkoxysilanes. *Journal of Non-Crystalline Solids* 149 (3): 161–178. https://doi.org/10.1016/0022-3093(92)90064-q.

Varma, A., Mukasyan, A.S., Rogachev, A.S., and Manukyan, K.V. (2016). Solution combustion synthesis of nanoscale materials. *Chemical Reviews* 116 (23): 14493–14586. https://doi.org/10.1021/acs.chemrev.6b00279.

Venkatramu, V., Falcomer, D., Speghini, A. et al. (2008). Synthesis and luminescence properties of Er^{3+}-doped $Lu_3Ga_5O_{12}$ nanocrystals. *Journal of Luminescence* 128 (5, 6): 811–813. https://doi.org/10.1016/j.jlumin.2007.11.015.

Vetrone, F., Boyer, J.C., Capobianco, J.A. et al. (2003). A spectroscopic investigation of trivalent lanthanide doped Y_2O_3 nanocrystals. *Nanotechnology* 15 (1): 75. https://doi.org/10.1088/0957-4484/15/1/015.

Vetrone, F., Mahalingam, V., and Capobianco, J.A. (2009). Near-infrared-to-blue upconversion in colloidal $BaYF_5$:Tm^{3+},Yb^{3+} nanocrystals. *Chemistry of Materials* 21 (9): 1847–1851. https://doi.org/10.1021/cm900313s.

Wang, F. and Liu, X. (2008). Upconversion multicolor fine-tuning: visible to near-infrared emission from lanthanide-doped $NaYF_4$ nanoparticles. *Journal of the American Chemical Society* 130 (17): 5642–5643. https://doi.org/10.1021/ja800868a.

Wang, F. and Liu, X. (2009). Recent advances in the chemistry of lanthanide-doped upconversion nanocrystals. *Chemical Society Reviews* 38 (4): 976–989. https://doi.org/10.1039/b809132n.

Wang, X., Zhuang, J., Peng, Q., and Li, Y. (2005). A general strategy for nanocrystal synthesis. *Nature* 437 (7055): 121–124. https://doi.org/10.1038/nature03968.

Wang, F., Chatterjee, D.K., Li, Z. et al. (2006). Synthesis of polyethylenimine/$NaYF_4$ nanoparticles with upconversion fluorescence. *Nanotechnology* 17 (23): 5786. https://doi.org/10.1088/0957-4484/17/23/013.

Wang, L., Li, P., and Li, Y. (2007). Down-and up-conversion luminescent nanorods. *Advanced Materials* 19 (20): 3304–3307. https://doi.org/10.1002/adma.200700144.

Wang, M., Mi, C.C., Wang, W.X. et al. (2009). Immunolabeling and NIR-excited fluorescent imaging of HeLa cells by using $NaYF_4$:Yb,Er upconversion nanoparicles. *ACS Nano* 3 (6): 1580–1586. https://doi.org/10.1021/nn900491j.

Wang, F., Han, Y., Lim, C.S. et al. (2010a). Simultaneous phase and size control of upconversion nanocrystals through lanthanide doping. *Nature* 463 (7284): 1061–1065. https://doi.org/10.1038/nature08777.

Wang, J., Wang, F., Xu, J. et al. (2010b). Lanthanide-doped $LiYF_4$ nanoparticles: synthesis and multicolor upconversion tuning. *Comptes Rendus Chimie* 13 (6, 7): 731–736. https://doi.org/10.1016/j.crci.2010.03.021.

Wang, M., Abbineni, G., Clevenger, A. et al. (2011). Upconversion nanoparicles: synthesis, surface modification and biological applications. *Nanomedicine: Nanotechnology, Biology and Medicine* 7 (6): 710–729. https://doi.org/10.1016/j.nano.2011.02.013.

Wang, Y.F., Sun, L.D., Xiao, J.W. et al. (2012). Rare-earth nanoparticles with enhanced upconversion emission and suppressed rare-earth-ion leakage. *Chemistry A European Journal* 18 (18): 5558–5564. https://doi.org/10.1002/chem.201103485.

Wang, H., Han, R.L., Yang, L.M. et al. (2016). Design and synthesis of core–shell–shell upconversion nanoparticles for NIR-induced drug release, photodynamic therapy, and cell imaging. *ACS Applied Materials & Interfaces* 8 (7): 4416–4423. https://doi.org/10.1021/acsami.5b11197.

Wang, X., Li, X., Zhong, H. et al. (2018). Up-conversion luminescence, temperature sensing properties and laser-induced heating effect of Er^{3+}/Yb^{3+} co-doped $YNbO_4$ phosphors under 1550 nm excitation. *Scientific Reports* 8 (1): 1–13. https://doi.org/10.1038/s41598-018-23981-4.

Wei, Y., Lu, F., Zhang, X., and Chen, D. (2008). Polyol-mediated synthesis and luminescence of lanthanide-doped $NaYF_4$ nanocrystal upconversion phosphors. *Journal of Alloys and Compounds* 455 (1, 2): 376–384. https://doi.org/10.1016/j.jallcom.2007.01.107.

Xiao, Q., Zheng, X., Bu, W. et al. (2013). A core/satellite multifunctional nanotheranostic for in vivo imaging and tumor eradication by radiation/photothermal synergistic therapy. *Journal of the American Chemical Society* 135 (35): 13041–13048. https://doi.org/10.1021/ja404985w.

Xu, Z., Li, C., Hou, Z. et al. (2011a). Facile synthesis of an up-conversion luminescent and mesoporous Gd_2O_3:Er^{3+}@nSiO_2@mSiO_2 nanocomposite as a drug carrier. *Nanoscale* 3 (2): 661–667. https://doi.org/10.1039/c0nr00695e.

Xu, H., Cheng, L., Wang, C. et al. (2011b). Polymer encapsulated upconversion nanoparticle/iron oxide nanocomposites for multimodal imaging and magnetic targeted drug delivery. *Biomaterials* 32 (35): 9364–9373. https://doi.org/10.1016/j.biomaterials.2011.08.053.

Xun, X., Feng, S., and Xu, R. (1998). Hydrothermal synthesis of complex fluorides $LiHoF_4$ and $LiErF_4$ with scheelite structures under mild conditions. *Materials Research Bulletin* 33 (3): 369–375. https://doi.org/10.1016/s0025-5408(97)00240-7.

Ye, X., Collins, J.E., Kang, Y. et al. (2010). Morphologically controlled synthesis of colloidal upconversion nanophosphors and their shape-directed self-assembly. *Proceedings of the National Academy of Sciences of the United States of America* 107 (52): 22430–22435. https://doi.org/10.1073/pnas.1008958107.

Yi, G.S. and Chow, G.M. (2005). Colloidal LaF_3:Yb,Er, LaF_3:Yb,Ho and LaF_3:Yb,Tm nanocrystals with multicolor upconversion fluorescence. *Journal of Materials Chemistry* 15 (41): 4460–4464. https://doi.org/10.1039/b508240d.

Yi, G., Lu, H., Zhao, S. et al. (2004). Synthesis, characterization, and biological application of size-controlled nanocrystalline $NaYF_4$:Yb,Er infrared-to-visible up-conversion phosphors. *Nano Letters* 4 (11): 2191–2196. https://doi.org/10.1021/nl048680h.

Zeng, J.H., Su, J., Li, Z.H. et al. (2005). Synthesis and upconversion luminescence of hexagonal-phase $NaYF_4$:Yb^{3+},Er^{3+} phosphors of controlled size and morphology. *Advanced Materials* 17 (17): 2119–2123. https://doi.org/10.1002/adma.200402046.

Zhang, F. (2015). *Photon Upconversion Nanomaterials*, vol. 416. Berlin: Springer.

Zhang, Y.W., Sun, X., Si, R. et al. (2005). Single-crystalline and monodisperse LaF_3 triangular nanoplates from a single-source precursor. *Journal of the American Chemical Society* 127 (10): 3260–3261. https://doi.org/10.1021/ja042801y.

Zhang, F., Wan, Y., Yu, T. et al. (2007). Uniform nanostructured arrays of sodium rare-earth fluorides for highly efficient multicolor upconversion luminescence. *Angewandte Chemie International Edition* 119 (42): 8122–8125. https://doi.org/10.1002/ange.200702519.

Zhang, M., Shi, S., Meng, J. et al. (2008). Preparation and characterization of near-infrared luminescent bifunctional core/shell nanocomposites. *The Journal of Physical Chemistry C* 112 (8): 2825–2830. https://doi.org/10.1021/jp076079c.

Zhang, F., Braun, G.B., Pallaoro, A. et al. (2012). Mesoporous multifunctional upconversion luminescent and magnetic "nanorattle" materials for targeted chemotherapy. *Nano Letters* 12 (1): 61–67. https://doi.org/10.1021/nl202949y.

Zhao, C., Feng, S., Xu, R. et al. (1997). Hydrothermal synthesis and lanthanide doping of complex fluorides, $LiYF_4$, KYF_4 and $BaBeF_4$ under mild conditions. *Chemical Communications* 10: 945–946. https://doi.org/10.1039/a607066c.

Zhao, F., Yuan, M., Zhang, W., and Gao, S. (2006). Monodisperse lanthanide oxysulfide nanocrystals. *Journal of the American Chemical Society* 128 (36): 11758–11759. https://doi.org/10.1021/ja0638410.

Zhou, H.P., Zhang, Y.W., Mai, H.X. et al. (2008). Spontaneous organization of uniform CeO_2 nanoflowers by 3D oriented attachment in hot surfactant solutions monitored with an in situ electrical conductance technique. *Chemistry A European Journal* 14 (11): 3380–3390. https://doi.org/10.1002/chem.200701348.

Zhou, H.P., Zhang, C., and Yan, C.H. (2009). Controllable assembly of diverse rare-earth nanocrystals via the Langmuir–Blodgett technique and the underlying size-and symmetry-dependent assembly kinetics. *Langmuir* 25 (22): 12914–12925. https://doi.org/10.1021/la9018986.

Zhou, Q., Mou, Y., Ma, X. et al. (2014). Effect of fuel-to-oxidizer ratios on combustion mode and microstructure of Li_2TiO_3 nanoscale powders. *Journal of the European Ceramic Society* 34 (3): 801–807. https://doi.org/10.1016/j.jeurceramsoc.2013.10.004.

Zhuang, J., Liang, L., Sung, H.H. et al. (2007). Controlled hydrothermal growth and up-conversion emission of $NaLnF_4$ (Ln = Y, Dy–Yb). *Inorganic Chemistry* 46 (13): 5404–5410. https://doi.org/10.1021/ic070220e.

3

Characterization Techniques and Analysis

Neha Jain[1], Prince K. Jain[1,2], Rajan K. Singh[1,3], Amit Srivastava[4], and Jai Singh[5]

[1] *Dr. Harisingh Gour Central University, Department of Physics, University road, Sagar 470003, Madhya Pradesh, India*
[2] *Government Arts and Commerce College, Department of Physics, Forest Department, Tili Rd, Housing Board Colony, Sagar, Madhya Pradesh, India*
[3] *National Taiwan University, Department of Chemical Engineering, Roosevelt Rd, Da'an District, Taipei 10617, Taiwan, ROC*
[4] *VBS Purvanchal University, TDPG College, Department of Physics, Husainabad, Jaunpur 222001, India*
[5] *Guru Ghasidas Vishwavidyalaya (A Central University), Department of Pure and Applied Physics, Koni, Bilaspur, 495009, Chhattisgarh, India*

3.1 Introduction

The materials revolution has always been qualitative as well as quantitative. It breeds the attitude of purposeful development of different new materials and the evolution of the existing ones with new functionalities to fulfill the demands in various fields of different aspects. Researchers/scientists are eager to study the materials having functional modalities with their physical, mechanical, thermal, electrical, magnetic, and optical properties. The imminent advent of technologies and information access in the research realm intends to accelerate the research on ubiquitous functional materials for a variety of applications. From the functional and innovative perspective, these days, upconversion nanoparticles (UCNPs) have apparently shown great significance in myriad fields of interest for mankind.

The unique optical properties and a high degree of tunability inherent in UCNPs make them promising new generation candidates in photonics, bioimaging, drug delivery, and energy harvesting (DaCosta et al. 2014; Demina et al. 2020). Notably, the unique properties of lanthanide ions arise from the 4f–4f orbital electronic transitions, and to achieve high upconversion efficiency, it is essential to codope sensitizer ions alongside activator ions that have closely matched intermediated excited states (DaCosta et al. 2014; Wen et al. 2018). The proper management of doping in a host modulates its inherent properties and decides the leveraging of the energy transfer process and in turn the luminescence performance. However, practical applications of UCNPs embark on the requirements on optical properties, size and shape, surface chemistry, photoluminescence (PL) downconversion and upconversion emission, and their ability to incorporate several desired modalities. Moreover, to develop

Upconverting Nanoparticles: From Fundamentals to Applications, First Edition. Edited by Vineet K. Rai.

cutting-edge technologies, synthesis, investigations to design these novel functional UCNP materials and finally integrating into devices are required through the steady advancing state of art instruments. Several powerful invasive tools have been used by researchers to visualize and study the recent progress in diverse aspects of UCNP materials.

Primarily, the present chapter extensively illustrates the basic descriptions of the fundamental principle, design, operation, observations, data acquisition, and analysis approach involved in these techniques. It involves techniques dealing with the information on the structure, elemental composition, optical properties, and other common and more specific physical properties of UCNPs such as X-ray diffraction (XRD), X-ray photoelectron spectroscopy (XPS), field emission scanning electron microscopy (FESEM), transmission electron microscopy (TEM), energy-dispersive X-ray spectroscopy (EDS), thermogravimetric analysis (TGA), UV–VIS–NIR absorption spectroscopy, dynamic light scattering (DLS), photoluminescence (PL), pump power-dependent UC, recognition of emission color and calorimetric theory, etc. These techniques are sometimes exclusive for the study of a particular property, while in other cases, they are combined. Table 3.1 illustrates a brief description of characterization techniques to be selected to analyze the different features/properties of the envisaged UCNPs. A number of developed UCNPs are listed in Table 3.2.

Table 3.1 List of upconversion nanomaterial's properties and name of the techniques used to measure these properties.

S.No.	Entity/property characterized	Characterization techniques to be employed
1	Size (structural properties)	XRD, TEM, SEM, AFM, UV–VIS
2	Shape	TEM, HRTEM, AFM, EPLS, FMR
3	Elemental–chemical composition	XRD, XPS, EDS
4	Crystal structure	XRD, HRTEM, electron diffraction, STEM
5	Size distribution	DCS, DLS, SAXS, DTA, TRPS, SEM
6	Chemical state–oxidation state	XPS, EELS, Mössbauer
7	Surface area, specific surface area	BET
8	Surface charge	Zeta potential
9	Agglomeration state	Zeta potential, DLS, DCS, UV–vis, SEM, Cryo-TEM, TEM
10	Ligand binding/surface composition	XPS, FTIR
11	Concentration	UV–vis, PTA, DCS, TRPS
12	3D visualization	3D tomography, AFM, SEM
13	Optical properties	UV–vis–NIR, PL, EELS-STEM
14	Magnetic properties	SQUID, VSM, Mössbauer, MFM, FMR, XMCD, magnetic susceptibility

Table 3.2 List of upconversion nanoparticles (UCNPs) and their phase structure.

Category	Name	Phase structure	References
Oxide	Bi_xO_y:Er,Y	Tetragonal	Back et al. (2017)
	ZrO_2–Al_2O_3:Er^{3+}	Tetragonal	Cong et al. (2016)
	$Bi_4Ti_3O_{12}$:Er^{3+},Yb^{3+},W^{6+}	Orthorhombic	Bokolia et al. (2017)
	Gd_2O_3:Ho^{3+},Yb^{3+}	Cubic	Antić et al. (2014)
	Yb_2O_3:Er^{3+},Yb^{3+}	Cubic	Lojpur et al. (2013)
	$La_2Ti_2O_7$:Eu^{3+},Er^{3+}	Phase mixture of monoclinic phase and cubic phase	Szczepanski et al. (2020)
	$Na_{0.5}Er_{0.5}Bi_4Ti_4O_{15}$	Orthorhombic	Hui et al. (2014)
	Pr-doped CeO_2 and ZrO_2	Tetragonal and monoclinic	Tiseanu et al. (2014)
Fluoride	$Re_{10}Pb_{25}F_{65}$ (Re-Er^{3+}, Yb^{3+}, Eu^{3+}, Dy^{3+}, Ho^{3+}, Tm^{3+})	Face-centered cubic	Tikhomirov et al. (2008)
	α-$NaMF_4$ (M = Dy, Ho, Er, Tm, Y, Yb, and Lu)	Cubic	Zeng et al. (2006)
	β-$NaMF_4$ (M = Pr, Nd, Sm, Eu, Gd, Tb, Dy, Ho, Er, Tm, Y, Yb, and Lu)	Hexagonal	Zhang et al. (2009), Zhou et al. (2008)
	ReF_3 (Re-La, Pr, Nd)	Hexagonal	Quimby et al. (1987)
	Pr, Yb-codoped KY_3F_{10}	Cubic	Kim et al. (2007)
Nitride	Graphitic carbon nitride nanoparticle-decorated g-C_3N_4-TiO_2	Tetragonal anatase	Raja and Ali (2021)
Sulfide	GaLaS	Glass	Voronov et al. (2019)
	Ga_2S_3:La_2O_3	Monoclinic cubic	Felix et al. (2000)
	SrS	Cubic	Shuchun and Fengmei (1996)

3.2 X-Ray Diffraction (XRD)

XRD is one of the most extensively used techniques to provide information regarding the crystalline structure, nature of the phase, lattice parameters, and crystalline grain size. This technique is based on the principle of diffraction of electromagnetic waves and Bragg's equation for diffraction (Cullity 1956). Among electromagnetic radiation, X-ray is used to study the crystal structure and its composition of atoms/molecules. Its wavelength lies in the range of 0.01–10 nm and is categorically divided into hard and soft X-rays. For crystal structure examination, a soft X-ray is usually used as the spacing between the lattice planes in the crystal (0.2–0.3 nm) is comparable to the X-ray wavelength. When the X-ray beam incidents on a target atom and interacts with the atom, X-ray photons are scattered in diverse directions. The scattering of photons is elastic and inelastic depending on the energy difference between the incident and scattered photons, and it may be zero or some positive value. The scattered photons lose their energy in inelastic scattering. The scattering of photons may cause phase differences, which lead to the formation of constructive and destructive interference. Generally, the diffracted radiation gets detected through a photomultiplier tube (PMT) employed. The diffraction pattern gives the structural information such as phase, crystallinity, strain, lattice parameters, etc. (Warren 1990; Whittig and Allardice 1986). There are two approaches for performing diffraction measurements; in the first approach, both the detector and the source move together, whereas in the other approach, only the detector moves (Bunaciu et al. 2015). The diffraction angle (2θ) is the angle between the transmitted X-rays and the detector as shown in Figure 3.1a. The XRD spectra consist of diffraction angle and intensity (number of diffracted photons)

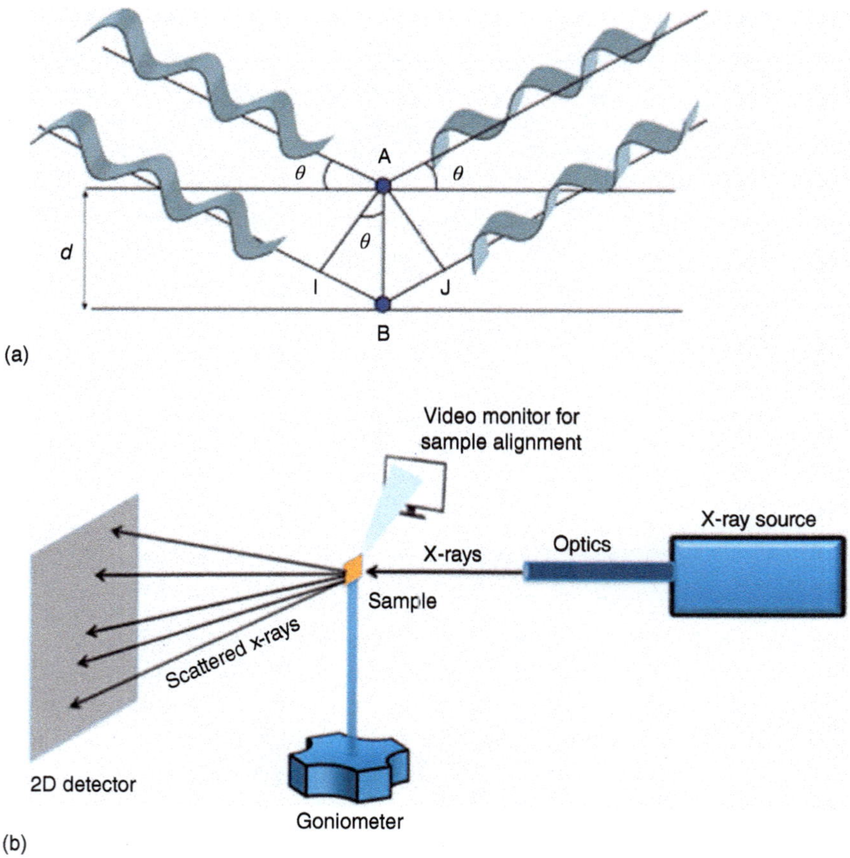

Figure 3.1 Representation of diffraction from atomic plane: (a) Bragg's law (imaginary planes represent two-dimensional crystal lattice). Source: Di Gianfrancesco (2016) and Lindon et al. (2016). Reprinted with permission of Elsevier. (b) Instrumentation of XRD. Source: Di Gianfrancesco (2016) and Lindon et al. (2016). Reprinted with permission of Elsevier.

(Cullity and Stock 2001). The diffracted beams are assumed to follow Bragg's law, which can be represented as (Bragg and Bragg 1913)

$$2d \sin \theta = n\lambda \tag{3.1}$$

Here, λ represents the X-ray wavelength, d is the interplanar distance, θ represents the angle of diffraction, and n is the order of diffraction.

The XRD instrument consists of three main components: an X-ray source, a sample holder, and a detector (Figure 3.1b). Generally, the Cu-Kα source is used for X-ray generation, and the wavelength generated to this source is 1.5405 Å. X-rays are generated by the application of high voltage and current, thereby high electron bombarding on a target material to eject X-ray photons. These photons incident on the sample and interact with the sample, resulting in diffraction of photons. Both the sample holder and the detector are rotated when the X-ray falls on the sample and get diffracted from the sample. These diffracted X-rays are then recorded by the detector. The detector converts the recorded data into spectra for spectral analysis of a material (Cullity 1956).

The composition of UCNPs can be accessed by comparing the position and intensity of the peaks with the reference patterns available from the International Centre for Diffraction Data (ICDD, previously known as Joint Committee on Powder Diffraction Standards, JCPDS) database. However, it is not suitable for amorphous materials, and the XRD peaks are too broad for particles with size below 3 nm. The crystalline grain size parameter can be estimated by employing Scherrer's equation using the broadening of the most intense peak of an XRD measurement for a specific sample. The structure of UCNPs highly influences the emission properties; therefore, XRD is the basic technique to study the structure. Ding et al. studied β-$NaYF_4$ by modifying the structure and morphology through citric acid. Also, they have reported the phase transition from cubic to hexagonal after increasing the reaction time from 3 to 12 hours (Figure 3.2) (Ding et al. 2015).

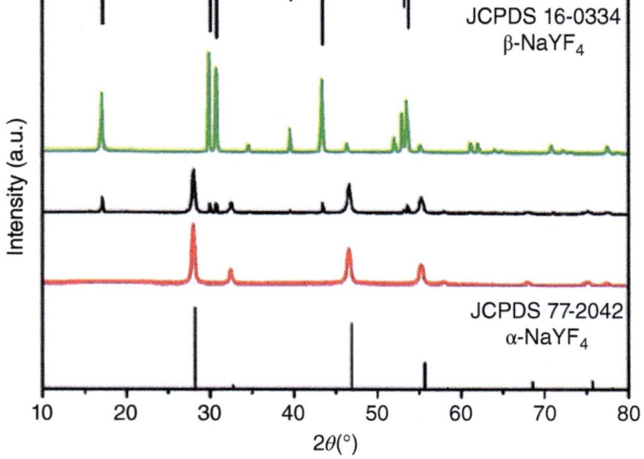

Figure 3.2 XRD patterns for cubic α-$NaYF_4$ phase (JCPDS no. 77-2042), mixed α, β-phase $NaYF_4$, and hexagonal β-$NaYF_4$ (JCPDS no. 28-1192). Source: Ding et al. (2015). Reprinted with permission of Elsevier.

The phase transition has been identified through XRD spectra by comparing the spectra with the standard JCPDS card.

3.3 X-ray Photoelectron Spectroscopy (XPS)

XPS is the most widely used analytical technique for surface chemical analysis and is also employed for the characterization of UCNPs. Its underlying physical principle is the photoelectric effect. XPS is a powerful quantitative technique useful to elucidate the electronic structure, elemental composition, and oxidation states of elements in a material. XPS spectra are obtained by irradiating a solid surface with a beam of X-ray. The properties of the material are derived by measuring the kinetic energy and the number of electrons that are emitted from the top (1–10 nm) of the material. A photoelectron spectrum is recorded by counting the ejected electrons over a range of kinetic energy of the electrons. The basic principle involved is based on the photoelectric effect, a phenomenon in which electrons are released from or within a material when photons of electromagnetic radiation having sufficient energy fall on it (Figure 3.3). These emitted electrons are known as photoelectrons. The kinetic energy of the emitted photoelectrons depends on the frequency of the incident radiation, and the number of photoelectrons depends on the intensity of the incident radiation (Van der Heide 2011).

The XPS instrument consists of the following basic components: an X-ray source, an energy analyzer, and a detector (Figure 3.4). Commonly, A1-Kα1 emissions have been used to produce a narrow X-ray beam. The energy filters used in XPS are concentric hemispherical analyzers or spherical mirror analyzers operated in the constant analyzer energy mode and with the energy resolution in the range of 0.3–0.5 eV. Furthermore, detectors are employed to measure the intensity of electrons in the form of current. In XPS, usually, the current of a single electron has been observed to be of the order of 1.6×10^{-19} Å. To measure this current, a highly sensitive detector is required, and therefore, an electron multiplier is used as a detector in ultrahigh vacuum conditions. The electronic state and oxidation number of an

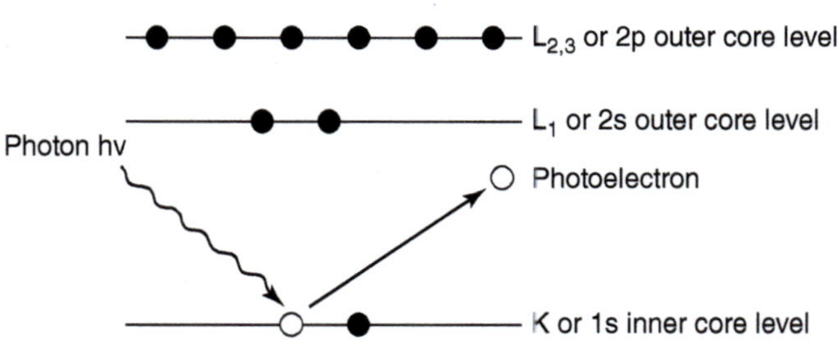

Figure 3.3 Schematic representation of ejection of an electron from an atom by X-ray radiations. Source: Based on Mather (2009).

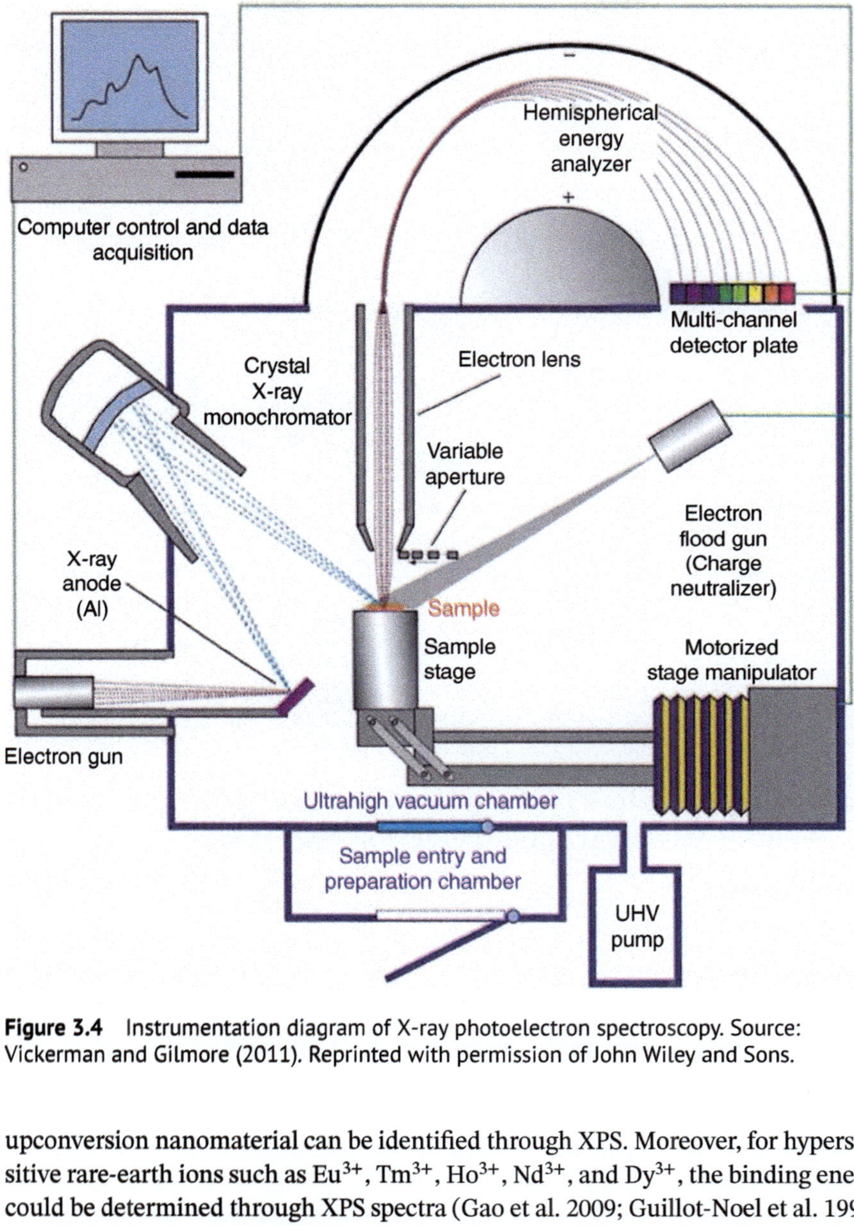

Figure 3.4 Instrumentation diagram of X-ray photoelectron spectroscopy. Source: Vickerman and Gilmore (2011). Reprinted with permission of John Wiley and Sons.

upconversion nanomaterial can be identified through XPS. Moreover, for hypersensitive rare-earth ions such as Eu^{3+}, Tm^{3+}, Ho^{3+}, Nd^{3+}, and Dy^{3+}, the binding energy could be determined through XPS spectra (Gao et al. 2009; Guillot-Noel et al. 1999). The emission wavelength of rare-earth ions (Eu^{3+}/Eu^{2+} or Dy^{3+}/Dy^{2+}) depends on the oxidation state, which can be identified by XPS spectra. Figure 3.5 represents the XPS spectra of $NaYF_4$@SnO_2 CSNPs (red line) and $NaYF_4$@SnO_2@Ag NPs. XPS spectra gives the binding energy of each element present in the sample (Du et al. 2018).

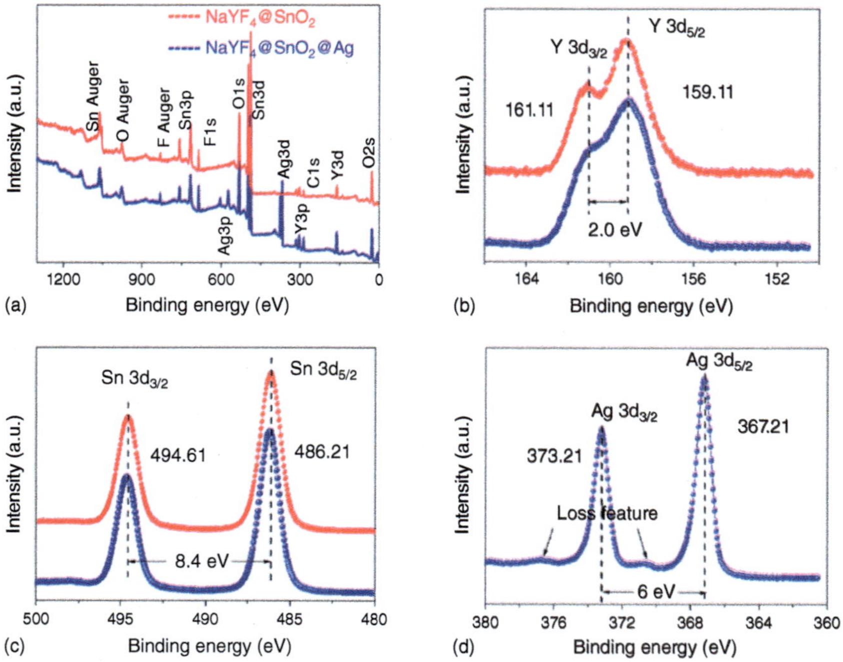

Figure 3.5 XPS spectra of $NaYF_4@SnO_2$ CSNPs (red line) and $NaYF_4@SnO_2@Ag$ NPs (blue line): (a) full-scale and high-resolution spectra of (b) Y 3d, (c) Sn 3d, and (d) Ag 3d. Source: Tian et al. (2017). Reprinted with permission of American Chemical Society.

3.4 Field Emission Scanning Electron Microscopy (FESEM)

Scanning electron microscopy (SEM) is a thoroughly accepted technique to access high-resolution imaging of surfaces that can be employed to characterize the UCNPs. It basically uses electrons for imaging, much as a light microscope uses visible light. Electron microscopy imaging removes resolving limitations of a light microscope, i.e. the resolution of the light microscope is about 200 nm; however, the resolution of an electron microscope is ~0.5 nm. The magnification power of an electron microscope can be increased to ×500000. These benefits of electron microscopy reveal that it allows applications of small objects at the subcellular, molecular, and atomic levels. According to the de-Broglie hypothesis, an electron has dual nature of particle as well as wave. The minimum wavelength as calculated by the de-Broglie principle is 12.3 Å. The basic principle of SEM is focused on the interaction of an electron with the sample and the loss in the energy of electrons through the scattering or absorption process. In addition, several processes have been involved during the interaction of an electron with the sample (Figure 3.6) as backscattered electrons (BSEs), secondary electrons (SEs), Auger electrons, characteristic X-ray, cathodoluminescence, and continuum X-ray (Bremsstrahlung). The BSEs are elastically scattered and reflected in the direction of the electron

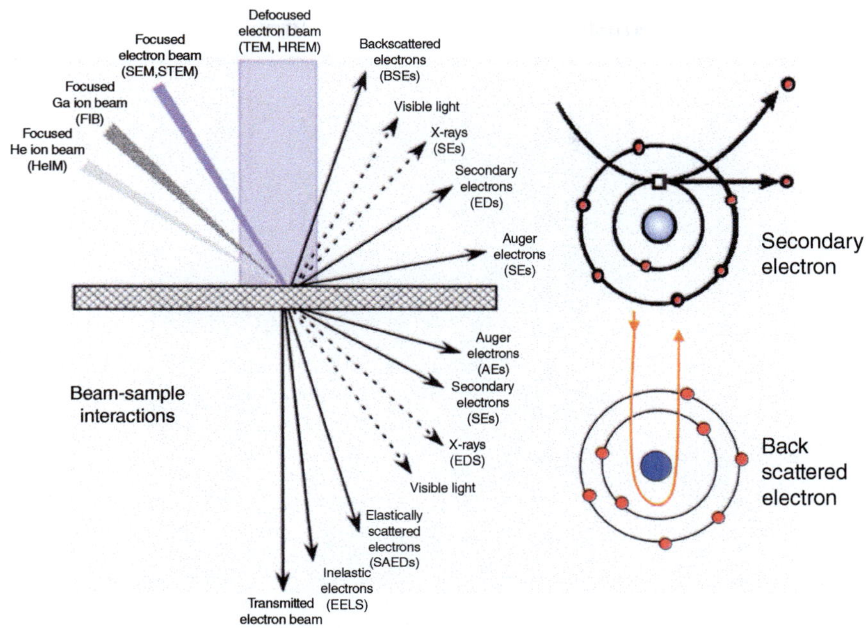

Figure 3.6 Schematic of the interactions of an electron beam with a sample and the types of signals that can be generated, secondary and backscattered electron emission. Source: Williams and Carter (1996). Reprinted with permission of Springer-Nature.

beam. The secondary electrons are generated through an inelastic interaction of the primary electron beam and the sample. The secondary electrons have lower energy than the BSEs. The energetic electrons incident on the sample and remove the electron from the shell, and the incident electrons also go back from the surface of the sample. These electrons are collected and recorded by a secondary electron detector. It gives information about the composition of the sample.

Figure 3.8 displays the SEM instrumentation that consists of various parts such as an electron gun, electromagnetic condenser lenses, an aperture, a scanning system, a specimen chamber, and an electron detector (Goldstein 2012). An electron gun produces the electron beam through the thermionic emission process. Generally, an electron gun is fabricated by using a tungsten filament or lanthanum hexaboride (LaB_6). This electron beam passes through the condenser lenses so that the electron beam converges and passes through a focal point of electromagnetic lenses. The focused electron beam further goes through an aperture, and irrelevant electrons are excluded. It is also used to determine the spot size of the beam. The electron beam interacts with the specimen and an image is formed. Here, the electron beam should be circular in shape and controlled by a stigmator. The secondary electrons emitted from the specimen are collected through a detector by applying a positive charge. The detector collects the electrical signal and converts it into a digital image to be displayed on a computer screen. The SEM image will be further used to examine the morphology of the upconversion nanomaterial, its particle size, and distribution of particle size (Vernon-Parry 2000). During emission measurement, the morphology

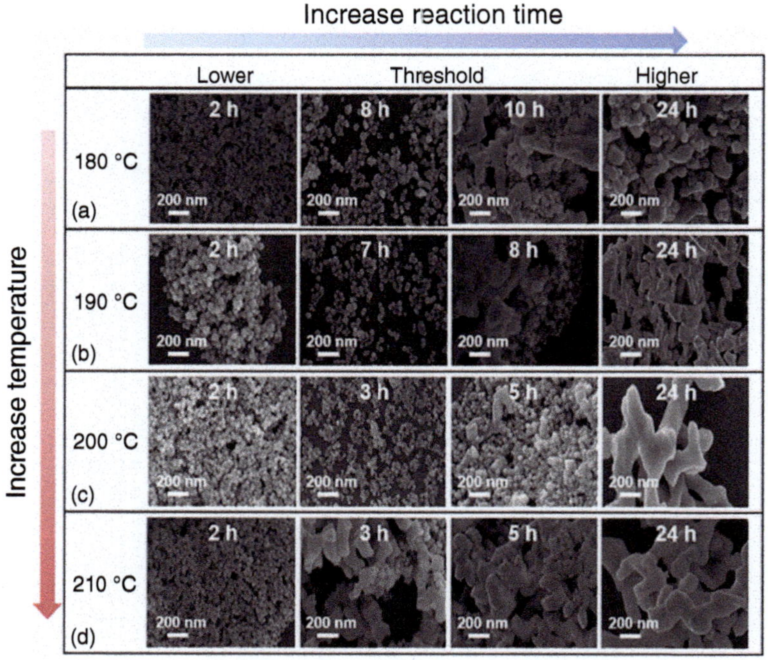

Figure 3.7 SEM images of the $NaYF_4:Yb^{3+}/Er^{3+}$ UCNPs synthesized using different HR temperatures and times. (a) 2, 8, 10, and 24 hours at 180 °C. (b) 2, 7, 8, and 24 hours at 190 °C. (c) 2, 3, 5, and 24 hours at 200 °C. (d) 2, 3, 5, and 24 hours at 210 °C. Source: Rafique et al. (2018). Reprinted with permission of Springer-Nature.

and size of the particle is controlled by using certain surfactants such as polyethylene glycol, polyvinylpyrrolidone, cetyltrimethylammonium bromide, etc. There are several factors, viz, pH, temperature, reaction time, surfactant, etc., which affect the shape and size of the UCNPs. Thus, SEM is used to optimize the size and shape of the UCNPs (Rafique et al. 2018; Yao et al. 2019). Figure 3.7 depicts the SEM images of the $NaYF_4:Yb^{3+}/Er^{3+}$ UCNPs synthesized at 180, 190, 200, and 210 °C temperatures for different reaction times 2, 8, 10, and 24 hours. The size and shape of UCNPs analyzed through SEM vary with reaction temperature and time (Rafique et al. 2018).

3.5 Transmission Electron Microscopy (TEM)

To get an insight into the microstructure and analyze the size and shape of nanoparticles, TEM is the most common technique to be used, as it provides not only the direct images of the sample but also the most accurate estimation of the nanoparticle's homogeneity. It exploits the interaction between a uniform current density beam of electrons (the energy is usually within a range of 60–150 keV) and a thin sample. When the electron beam reaches the sample, part of the electrons are transmitted, while the rest are elastically or inelastically scattered. The magnitude of the interaction depends on several factors, such as size, sample density, and elemental

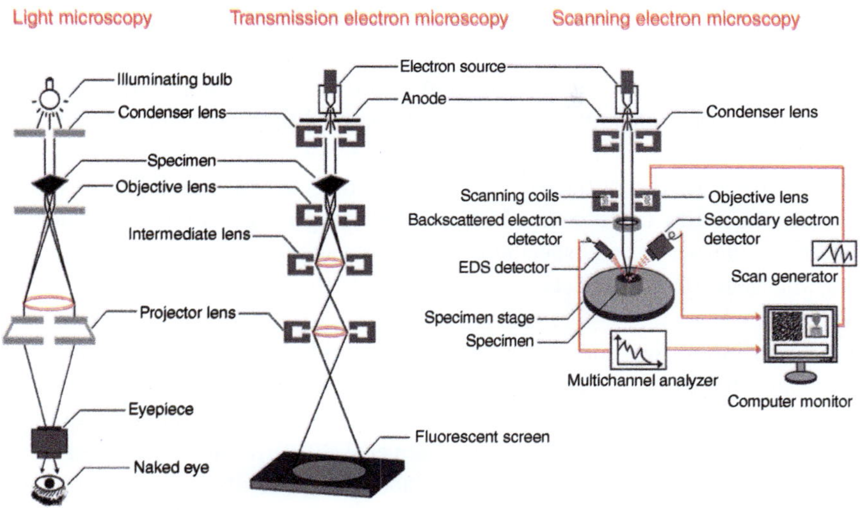

Figure 3.8 Comparative illustration of different components of a simple microscope, a transmission electron microscope, and a scanning electron microscope. Source: Ul-Hamid (2018). Reprinted with permission of Springer-Nature.

composition. The final image is built with the information acquired from the transmitted electrons.

In this technique, an electron beam is transmitted through a specimen, and the respective image is formed through diffraction and get displayed on a fluorescent screen/photographic film or recorded by a charge-coupled device (CCD) camera (Figure 3.8). It requires high voltage (in kilovolt) to emit an electron beam by the filament (cathode) (Hayat 2012). Generally, a tungsten filament or LaB_6 has been used as an electron emitter, and then, the electron beam is focused by electrostatic and electromagnetic lenses. The electron beam interacts with a specimen, and BSEs, secondary electrons (SEs), and diffracted electrons get released. Figure 3.6 shows the mechanism of electron interaction with the specimen. BSEs are reflected after elastic interactions between the beam and the sample. Higher atomic number atoms seem to have a brighter appearance in the image by this technique (Reimer 2013).

Elastically scattered electrons also follow Bragg's law and are used to get the information about the crystal structure using selected area electron diffraction (SEAD). Secondary electrons are generated through ionization of the sample surface and deliver the information about the surface of the sample. They originate as a result of inelastic interactions between a beam of electrons and the sample and provide information about the composition of the sample. The energy of secondary electrons emitted from the surface is less than that of BSEs (Michael et al. 2011). BSE originates from the deeper regions of the specimen, whereas SE generates from the surface of the specimen. SE imaging can provide more detailed surface information. The diffracted electrons are those that are spread through the crystal planes (Williams and Carter 1996).

However, the transmitted or direct electrons are advantageous for TEM image formation. TEM microscopy can be divided into two classes based on the electron recording by detector, i.e. bright-field and dark-field microscopy. In bright-field microscopy, the specimen seems darker in the bright background. In this technique, it would be possible to see the natural color of the specimen. It appears bright or natural color as the transmitted electrons are recorded through an objective lens. Moreover, dark-field microscopy displays bright specimens in the dark background. In this microscopy, the diffracted electrons are detected by the objective lens. Figure 3.8 illustrates the construction of TEM comprising several components (filament, condenser lenses, specimen holder, objective lenses, and fluorescent screen) and clearly the path of electron beam focused through the lenses (Ul-Hamid 2018).

TEM is an innovative microscopy technique by which we can determine the morphology and particle size of upconversion nanomaterials. High-resolution transmission electron microscopy (HRTEM) gives the information about the atomic plane spacing (*d*-spacing) and (*hkl*) planes (Jain et al. 2019). Moreover, selected area diffraction pattern (SADP) gives information about single-crystal or polycrystalline formation of upconversion nanomaterials. Also, the *d*-spacing and crystal plane can also be examined through the SADP pattern. Jain et al. have reported TEM, HRTEM, SADP, and particle size distribution plot of the $SrSnO_3$:Ho^{3+},Yb^{3+} UCNPs (shown in Figure 3.9) (Jain et al. 2018). Zhou et al. studied semiconductor plasmon-induced

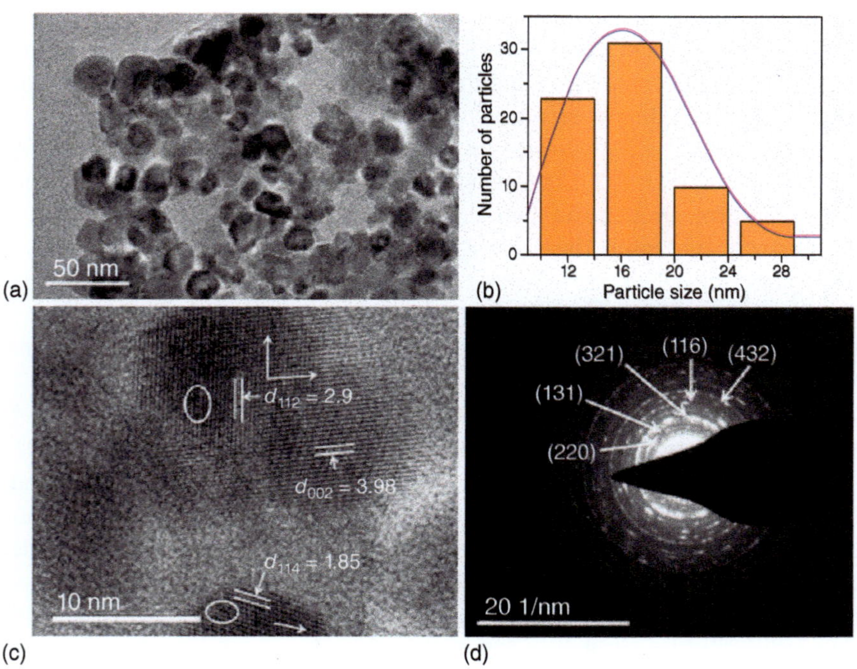

Figure 3.9 (a) TEM image of the $SrSnO_3$:1 at.% Ho^{3+} and 3 at.% Yb^{3+}; (b) particle distribution curve; and (c) HRTEM and (d) SADP patterns of UCNPs. Source: Jain et al. (2018) / Reprinted with permission of Springer-Nature.

upconversion properties of $mCu_{2-x}S@SiO_2@Y_2O_3:Yb^{3+}/Er^{3+}$ core–shell. They have analyzed the core–shell formation by using the TEM micrograph (Zhou et al. 2017). Thus, there are several reports based on the application of TEM for the determination of a variety of structure and size of UCNPS as well as the core–shell morphology of UCNPS (Grzyb et al. 2021; Santelli et al. 2021).

3.6 Energy-Dispersive X-ray Spectroscopy (EDS)

EDS is an elemental analysis technique based on recording of X-ray emission from the surface of the sample (Shindo and Oikawa 2002). An EDS detector is often attached with a transmission electron microscope and a scanning electron microscope. The interaction of an electron beam with the specimen involves several processes, and X-ray emission is one of them (Figure 3.6). In this process, an electron beam of high energy interacts with the sample, and an electron from the inner shell of an atom is ejected. This ejection of an electron ionizes the specimen, and it can be neutralized when an electron from the outermost shell falls into an electron's vacant position. This transition of an electron from the outer shell to the inner shell of the electron generates energy. The energy emitted in this process is observed in the order of X-ray energy (Kannan 2018; Williams and Carter 1996). As X-rays are observed from distinct transitions of each atom, the emission of X-ray is discrete and represented by discrete peaks. Each element has a specific energy of its electronic state, so the transitions between the states give a unique energy value. The X-ray generated by elements has a unique value so the element can be identified by comparing the energy value with its characteristic X-ray emission. Therefore, EDS spectra are plotted between the emitted X-ray energy and the number of photons emitted (Ngo 1999).

The construction of the EDS instrument is similar to SEM, and an X-ray detector is attached to record EDS spectra as shown in Figure 3.10. Here, the main role played by the detector is to record the X-ray and convert it into voltage signals. These signals pass through the pulse processor and then from the analyzer. Then, the analyzer detects the data signal and displays on a computer screen. At present, the Si(Li) detector is equipped with SEM or TEM techniques. These detectors are cooled to cryogenic temperatures with liquid nitrogen or a Peltier cooling system (Yefimov 2009). Here, cooling is very important because it reduces noise from spectra, narrows the X-ray emission peaks, and obstructs thermal current generation, which might be generated through the electron–hole pair. The spectra showed several emission peaks corresponding to the emission of X-rays through each element. The element can be analyzed by comparing these spectra with the characteristic emission spectra of that element (Shindo and Oikawa 2002). The distribution of elements can be seen from the spectra, and it may be concluded that UCNPs may exhibit uniform emission after device fabrication (Jain et al. 2019; Manurung et al. 2018). Figure 3.11 shows energy-dispersive X-ray spectra of the $NaYF_4$:Yb,Tm UCNPs and percentage of the elements.

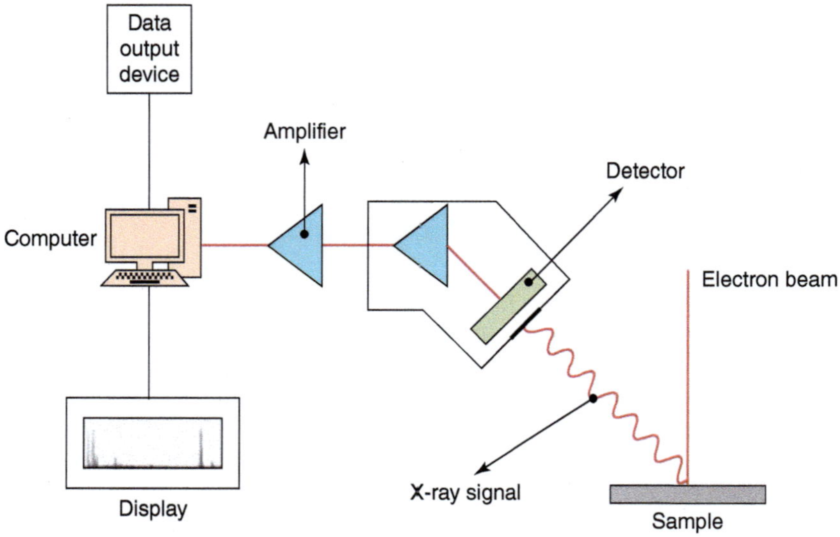

Figure 3.10 Instrumentation of energy-dispersive X-ray spectroscopy. Source: Dincer (2018). Reprinted with permission of Elsevier.

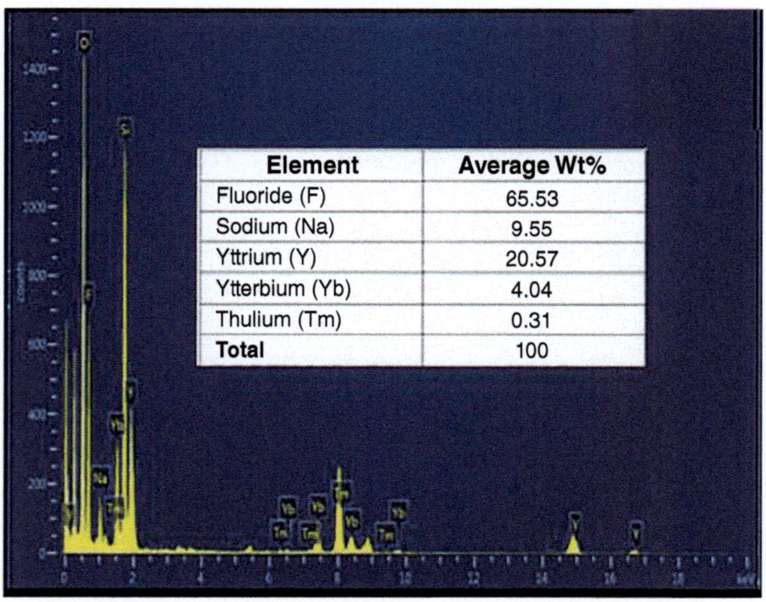

Element	Average Wt%
Fluoride (F)	65.53
Sodium (Na)	9.55
Yttrium (Y)	20.57
Ytterbium (Yb)	4.04
Thulium (Tm)	0.31
Total	100

Figure 3.11 Energy-dispersive spectra (EDS) of the $NaYF_4$:Yb,Tm UCNPs. Source: Manurung et al. (2018). Reprinted under Creative Commons License.

3.7 Thermogravimetric Analysis (TGA)

TGA techniques are generally employed to determine the thermal stability and oxidative stability of the material and the composition of multicomponent systems (Gabbott 2008) and to observe the change in the weight of the substance with temperature (Lothenbach et al. 2016). In this technique, a sample is heated up to a certain higher temperature, and the corresponding thermal parameters are noted. During this process, several changes are observed in the sample so that the melting point, the purity of the material, and transition temperature could be found. In the present approach, a sample and a reference material are used to compare the corresponding changes in a sample in the defined temperature range. If a thermal fluctuation, such as chemical or physical variation causes emission or absorption of heat, occurs in the sample, thermal energy is added into the sample or the reference container. As the energy absorption or emission occurs in the transition, it is recorded as heat flow for further analysis (Höhne et al. 2013).

Figure 3.12 describes the basics of the TGA system. The sample holders in TGA are made up of platinum, aluminum, and aluminum oxide. It consists of a platinum unit for heat flow at high temperature and performs the high-temperature operation. Each holder contains a resistance heater and a temperature sensor. To raise the temperature with the selected rate, a current is supplied to the heater. A constant flow of nitrogen gas is also maintained across the chamber (sample and reference chamber) to create a reproducible and inert atmosphere. The gas atmosphere reduces the possibility of air oxidation of the samples at high temperatures. The positive and negative enthalpy can be recorded throughout the heating of a sample from room temperature to its decomposition temperature. The peaks indicate specific processes such as crystallization or melting or phase change of the sample at a

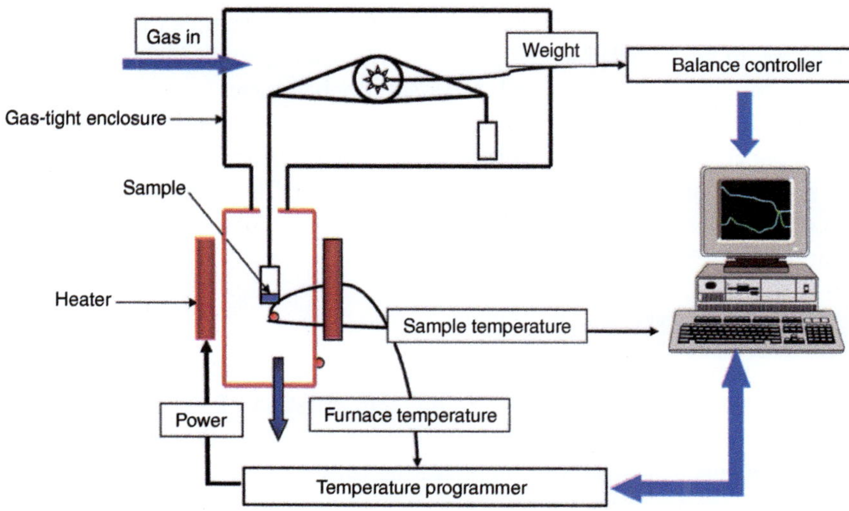

Figure 3.12 Schematic diagram of a thermobalance and a control unit. Source: Akash and Rehman (2020). Reprinted with permission of Springer-Nature.

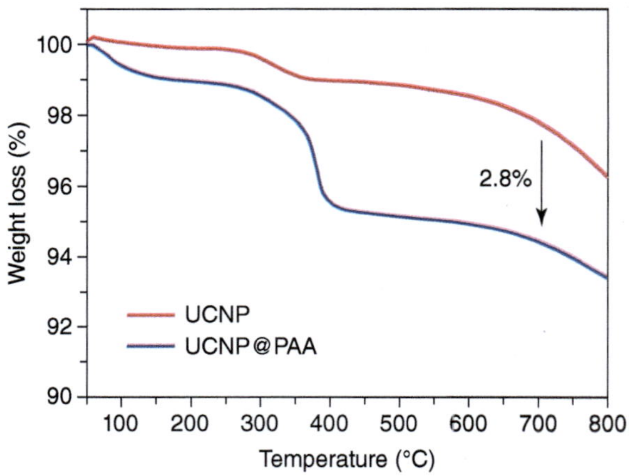

Figure 3.13 TGA of the $NaYF_4:Yb^{3+}/Er^{3+}$ UCNPs after coating with polyacrylic acid (PAA). Source: Rafique et al. (2018). Reprinted with permission of Springer-Nature.

definite temperature (Lothenbach et al. 2016). Wang et al. have studied the thermal stability of UCNPs@COP materials through TGA (Gao et al. 2009). Ji et al. have performed TGA to quantitatively confirm the substitution amount of polymers on the surface of UCNPs (Duong et al. 2018). Figure 3.13 demonstrates the TGA image of the $NaYF_4:Yb^{3+}/Er^{3+}$ UCNPs, which shows that the weight percentage is slightly decreased on heating the sample up to 800 °C (Rafique et al. 2018).

3.8 Ultraviolet–Visible–Near-Infrared (UV–Vis–NIR) Absorption Spectroscopy

UV-visible–NIR absorption spectroscopy is a versatile method to measure the optical bandgap of the material. When a UV-visible–NIR light beam is incident on the specimen, part of the incident photons get absorbed by the sample (Schoonheydt 2010). By recording several absorbed photons corresponding to the incident photon energy, the bandgap can be evaluated. A UV spectrophotometer is designed so that it follows the Beer–Lambert law (Mäntele and Deniz 2017). According to this law, when a monochromatic light beam is passed through a fixed concentration solution of any material, the decreasing rate of the photons of the incident light in the direction of path length is proportional to the concentration of the solution and the incident radiation. It can be expressed as

$$A = \log \frac{I_0}{I} = \varepsilon cl \tag{3.2}$$

Here, A represents the absorbance, I_0 and I describe the intensity of incident and transmitted light, respectively, c is the concentration of the solution, l is the length of the sample cell (i.e. optical path length), and ε denotes the molar absorptivity. Figure 3.14 depicts a simplified schematic diagram of the construction of a UV-visible spectrophotometer.

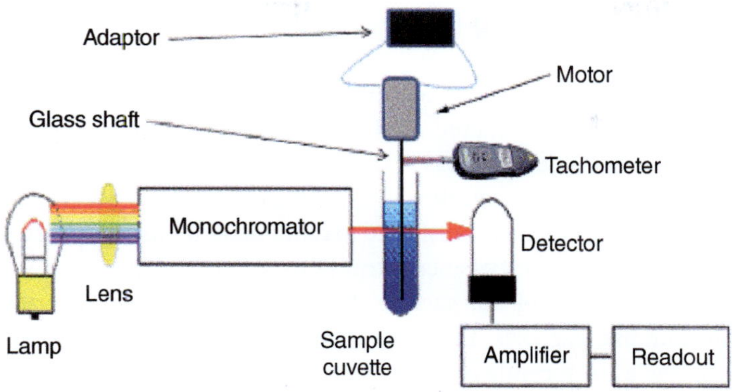

Figure 3.14 Schematic diagram of a UV-visible spectrophotometer. Source: Fahool and Sabet (2015). Reprinted with permission of Springer-Nature.

The spectra are plotted in between the transmittance/absorbance and wavelength of the incident light. The detectors used are generally CCD cameras, photodiodes, or PMTs. The bandgap of the material can be calculated by the Wood and Tauc method. In this method, a graph is plotted between the incident photon energy $h\upsilon$ and the absorbed energy $(\alpha h\upsilon)^2$, where α is the absorption coefficient of the material. The spectrophotometer can also be used for photocatalytic degradation measurement. In this procedure, absorption of dye (degrading material) is measured in a definite time interval. UV–visible absorption spectroscopy has been used to determine the bandgap and absorbance of the sample (Lee et al. 2020; Que et al. 2001). Moreover, UV–visible spectra can also be used to measure the photocatalytic activity of the sample. Fischer et al. have reported the photocatalytic activity of the SrYbF:1% Tm UCNPs up to three hours (shown in Figure 3.15) (Fischer et al. 2020).

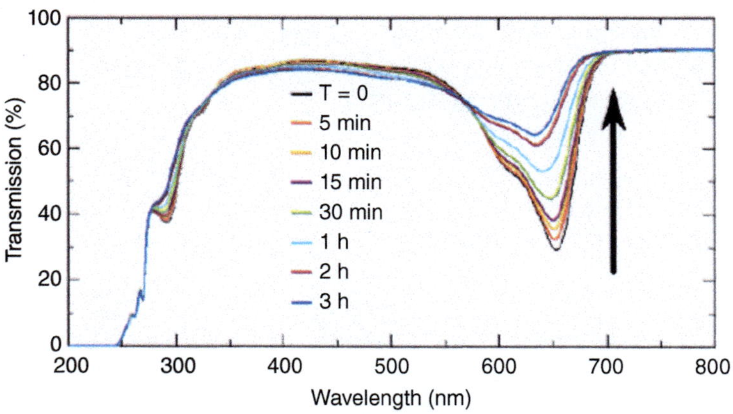

Figure 3.15 UV–vis spectra of UCNPs SrYbF:1% Tm dispersed in chloroform with methylene blue dye. Source: Fischer et al. (2020). Reprinted with permission of John Wiley and Sons.

3.9 Dynamic Light Scattering (DLS)

The DLS instrument is used to measure the particle size of the nanomaterials. The principle of this instrument is centered on the Brownian motion of nanoparticles dispersed in a solvent (Goldburg 1999). The dispersed particle moves and collides with other particles in the solution. Such collisions induce a definite amount of energy; this energy is further transferred to other particles for movement in solution (Pecora 2013). The amount of energy transferred and its ability to move the particles depends on the size of the particle. If the collision-induced energy is transferred to smaller size particles, these particles move at a faster speed as compared to larger particles. A formula is given by Stoke–Einstein to calculate the speed of the particle that involves diffusion coefficient (D) (Shaw 2014) as

$$D = \frac{k_B T}{6\pi\eta R_H}$$

Here, k_B is the Boltzmann constant, T represents the temperature, η denotes the viscosity, and R_H defines the hydrodynamic radius. However, particles that are sediment in solution do not exhibit Brownian motion. Therefore, their particle size cannot be measured by employing this instrument. In this instrument, upper and lower sizes are limited by the signal-to-noise ratio.

Figure 3.16 displays the basic schematic construction of the DLS instrument. A monochromatic light beam passes through a sample filled in a cuvette. The particles that are dispersed in a solution scatter incident photons in all the possible directions. A detector located at a certain angle records the scattered photons over time. Such obtained data are further used to calculate the diffusion coefficient and subsequently the particle size by using the Stokes-Einstein equation. Moreover, the incident monochromatic light is typically attenuated by a gray filter to get sufficient and processable signals through the detector (Leong et al. 2018). The DLS instrument detects the signal from the sample by two measurement positions, side scattering (90°) and backscattering (175°). If the sample is aggregated, then a forward angle (15°) is used to observe aggregation.

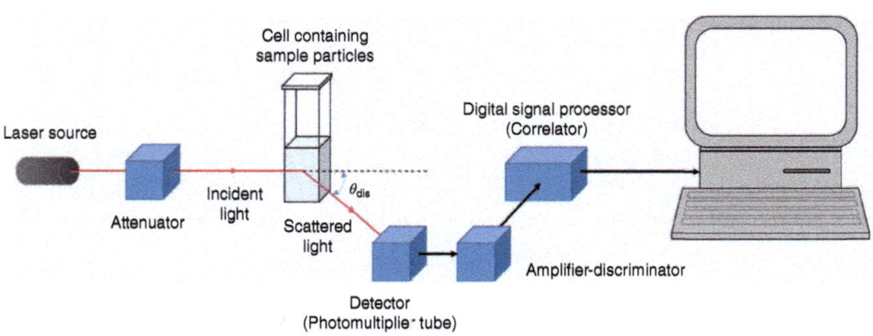

Figure 3.16 Basic setup of a DLS measurement system. The sample is contained in a cuvette. The scattered light of the incident laser beam can be detected at different angles. Source: Leong et al. (2018). Reprinted with permission of Springer-Nature.

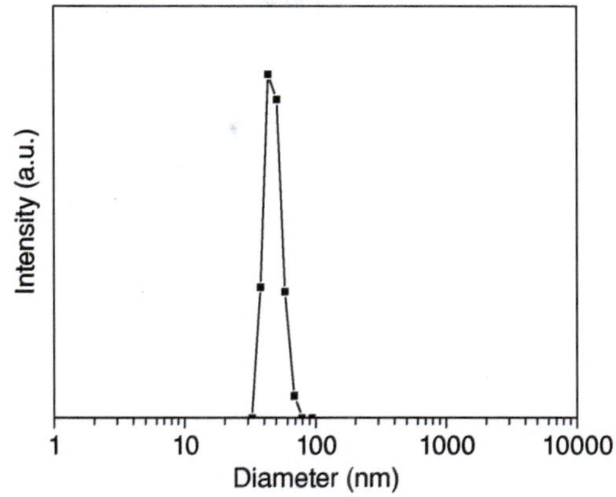

Figure 3.17 DLS size distribution plot of oleic acid capped hydrophobic $NaYF_4$:Yb,Er UCNPs. Source: Wu et al. (2012). Reprinted with permission of Royal Society of Chemistry.

There are three different DLS data analysis techniques, CUMULANT, CONTIN, and CORENN (Stetefeld et al. 2016). The first method CUMULANT can be used to estimate the average particle size for noisy data. In this method, particle size distribution was calculated by Gaussian distribution. The second method of analysis is CONTIN, used for bimodal particle size distribution; however, this method is very sensitive for noisy data. Moreover, CORENN based on the algorithm is processed by Latent Semantic Indexing. This method is fitted by the nonlinear or least-square fitting method and provides precise results even for particle scattering noisy data. A number of reports have been published on UCNPs size by using the DLS technique (Jiang et al. 2018; Sarmanova et al. 2020; Wu et al. 2012). Figure 3.17 illustrates the particle size distribution plot of oleic acid capped hydrophobic $NaYF_4$:Yb,Er UCNPs, and the size of the nanoparticles estimated through the DLS spectra is 45 nm (Wu et al. 2012).

3.10 Photoluminescence (PL) Study

A photoluminescence spectrophotometer is used to measure the photoluminescence emission from the material. It is a fast, simple, and inexpensive technique to determine the photoluminescence properties of a material. In this spectroscopy, a beam with an excitation wavelength ranging from 200 to 1000 nm passes through a solution kept in a cuvette or a solid sample (Gfroerer 2006). Further, photoluminescence is measured at an angle that is emitted by the sample. When the incident photons are absorbed by a fluorophore, it is excited and its valence electrons transit to higher excited state. After few moments, the electron relaxes from higher energy state to ground state, resulting in radiative or nonradiative emissions. The radiative emission can be recorded by a photoluminescence spectrophotometer. In fluorescence spectrometry, both the excitation and emission spectra can be recorded (Lakowicz 2013). Excitation spectra are those in which light is absorbed by the sample, and emission spectra show the emitted photons by the sample. Several parameters influence the emission intensity. The emission intensity depends on the excitation wavelength, solvent concentration, path length of the cuvette, and absorption coefficient.

In this spectroscopic technique, excitation sources such as a Xe lamp, a diode laser, and He-Cd, Kr, and Ar lasers are used. Upconversion emission spectra have been observed because of the multiphoton absorption from the sample, which can be only possible if a laser is used as an excitation source. Here, a beam passes through the excitation monochromator to acquire a desired excitation wavelength. These radiations are incident on the material, and the emitted photons from the sample pass through the emission monochromator. Further, the emitted radiation is recorded by the detector such as PMT (Gregersen 2014). Figure 3.18 shows a schematic diagram for the construction of a fluorescence spectrophotometer.

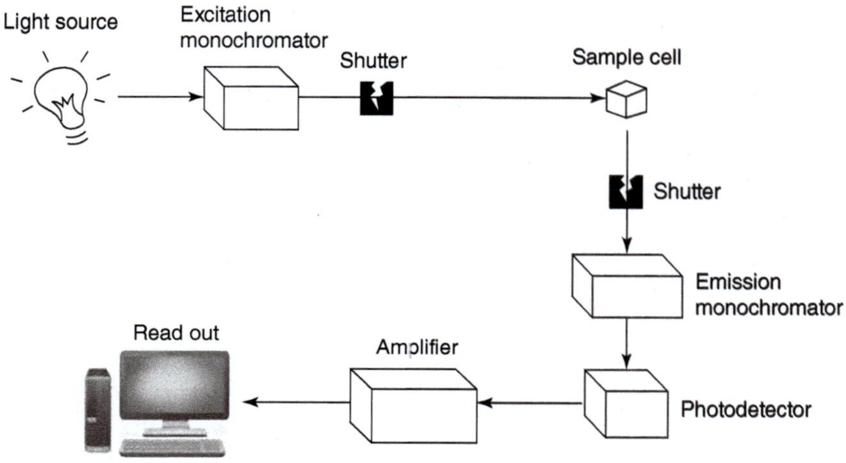

Figure 3.18 Schematic representation of a fluorescence spectrophotometer. Source: Munjanja and Sanganyado (2015). Reprinted with permission of Springer.

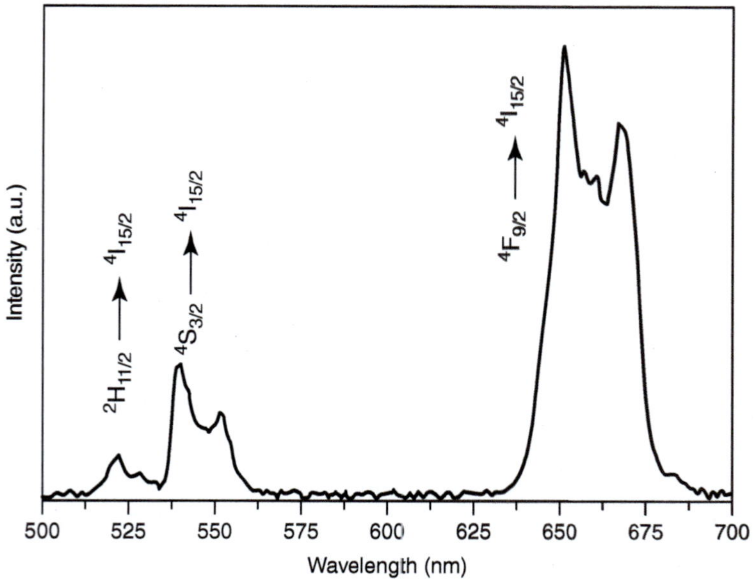

Figure 3.19 Upconversion luminescence spectrum of the $NaYF_4$:Yb/Er UCNPs excited at 980 nm. Source: Reddy et al. (2017). Reprinted with permission of Springer-Nature.

A lot of research papers have been published on upconversion emissions in the lanthanide-doped oxides, fluorides, sulfides, and aluminates. The upconversion emission is observed by doping of some lanthanides or dopants that have ladder-like structure such as Er^{3+}, Tm^{3+}, Ho^{3+}, Nd^{3+}, etc. (Kumar et al. 2017; Pu et al. 2018; Rajeswari et al. 2020; Righini et al. 2019). Therefore, a photoluminescence spectrophotometer is used to record the emission spectra of these fluorophores (Figure 3.19) (Reddy et al. 2017).

3.11 Pump Power-Dependent UC

Recently, upconversion materials have gained enormous attention because of their applications in sensors, biomedical field, displays, high-density optical data storage, etc. Upconversion emission spectra are observed because of two-photon or multiphoton absorption by a fluorophore. The multiphoton absorption could be possible by using the long wavelength lasers as the excitation source (Lakowicz 2013). Here, the excitation source may be continuous (CW) or pulsed, and the excitation wavelength could lie in the visible as well as the near-infrared region. In the pump power-dependent study, the UC emission spectra are recorded as a function of the laser pump power as the UC emission intensity (I) is related to the pump power (P) of the incident radiation by the relation, $I \propto P^n$, where "n" is number of pump photons involved in the UC emission process (Han et al. 2019). Furthermore, the UC emission spectra are recorded by using a spectrometer with the help of a PMT or a spectrograph (Figure 3.20) (Do Nascimento et al. 2016; Guo et al. 2004). The rare-earth ions Er^{3+}, Tm^{3+}, Ho^{3+}, Nd^{3+}, etc., doped in the suitable hosts absorb two or more low-energy photons and emit a single photon of higher energy in the upconversion process (Jain et al. 2018). The pump power-dependent UC emission spectra of the $NaYF_4$:Gd^{3+},Yb^{3+},Tm^{3+} UCNPs are shown in Figure 3.21a. The number of pump photons involved in the UC emission process has been calculated by using the logarithmic plot of the pump power versus the UC emission intensity (Figure 3.21b).

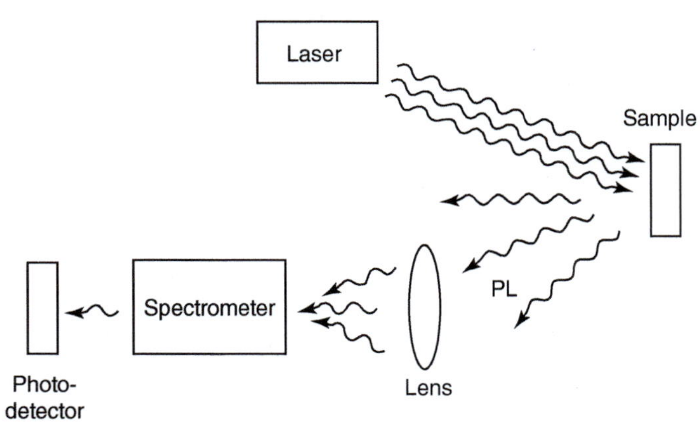

Figure 3.20 Experimental setup used to observe the UC emission by using the laser source. Source: Gfroerer (2006). Reprinted with permission of John Wiley and Sons.

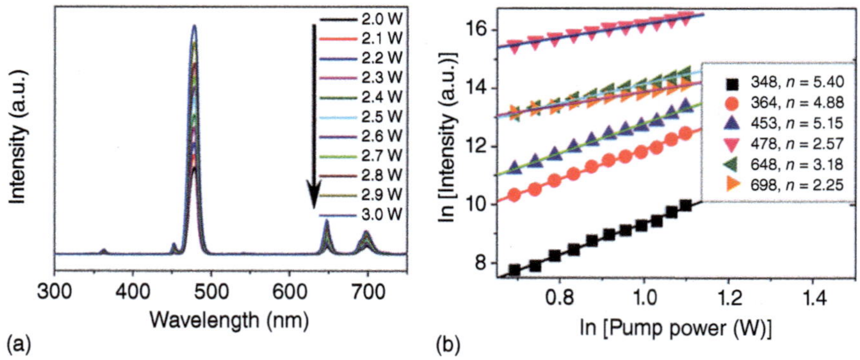

Figure 3.21 Pump power-dependent upconversion luminescence spectra of the $NaYF_4$:Gd^{3+}, Yb^{3+}, and Tm^{3+} UCNPs under irradiation at 980 nm with different pump powers from (a) 2.0 to 3.0 W and (b) its logarithmic plot. Source: Duo et al. (2018). Reprinted with permission of Springer-Nature.

3.12 Recognition of Emission Color and Colorimetric Theory

Colorimeter is a device used to identify the color of a solution or dispersed particles (Asheim et al. 2014). It primarily measures the concentration of the solute in the solution based on Beer–Lambert's law (Burnham 1952; Gordon and Harman 2002). When a light beam passes through a sample, a particular wavelength is absorbed by the sample; other wavelengths are transmitted and some parts of the wavelength get reflected. The concentration of the solute in the solution can be calculated by following Beer's law as

$$\log_{10} I_o/I_t = a_s c$$

Here, "I_o" and "I_t" represent the intensities of the incident and transmitted beams, respectively, "a_s" represents the absorbency index, and "c" is the concentration of the solute in the solution. According to Lambert's law, the amount of absorbance depends on the thickness or length of the cuvette in which the solution is filled. Thus,

$$A = \log_{10} I_o/I_t = a_s b$$

Here, "A" is the absorbance of the test sample. After combining these two laws, Beer–Lambert's law can be expressed as

$$A = \varepsilon c l$$

Here, "ε" represents the molar absorptivity and "l" denotes the length of the solution through which light is passed. Figure 3.22 represents the basic instrumentation of colorimetry, which comprises a filament lamp, an aperture, a colored filter, a cuvette holder, a photodetector, and a digital converter. It works in the spectral wavelength range of ~ 400–700 nm (Ly et al. 2020; Fang et al. 2016). Cui et al. designed a

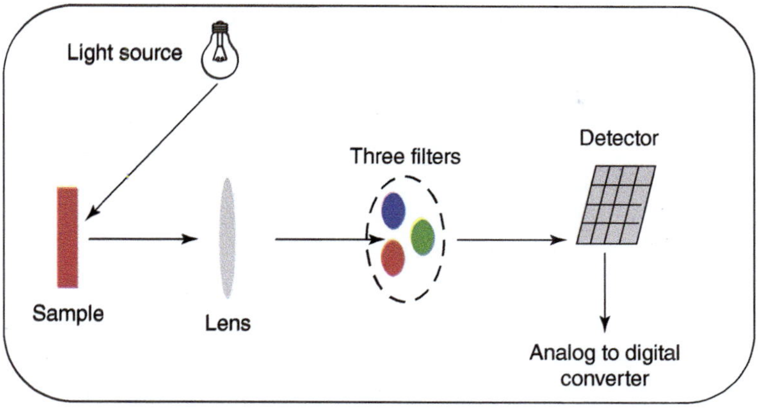

Figure 3.22 Instrumentation of colorimetry. Source: Ly et al. (2020). Reprinted with permission of Elsevier.

colorimetric upconversion nanothermometer by using the Er^{3+}-Mn^{2+}:$KMnF_3$ luminescent material, which works in the 303–343 K temperature range with ultrahigh sensitivity for biological detections (Cui et al. 2018). In addition, Fu et al. used the colorimetric detection method for the recognition of a cellulose paper substrate functionalized with telomerase (Fu et al. 2018).

Acknowledgment

The authors would like to acknowledge Prof. R.A. Singh (Dr. Harisingh Gour Central University, Sagar, India) and Dr. Kausal Kumar (IIT-ISM Dhanbad) for fruitful discussions and DST-CRG/2020/005709 PROJECT, Govt. of India, for financial support.

References

Akash, M.S.H. and Rehman, K. (2020). Thermo gravimetric analysis. In: *Essentials of Pharmaceutical Analysis*, 215–222. Springer https://doi.org/10.1007/978-981-15-1547-7_19.

Antić, Ž., Lojpur, V., Nikolić, M.G. et al. (2014). Strong emission via up-conversion of Gd_2O_3:Yb^{3+},Ho^{3+} nanopowders co-doped with alkali metals ions. *Journal of Luminescence* 145: 466–472. https://doi.org/10.1016/j.jlumin.2013.08.007.

Asheim, J., Kvittingen, E.V., Kvittingen, L., and Verley, R. (2014). A simple, small-scale lego colorimeter with a light-emitting diode (LED) used as detector. *Journal of Chemical Education* 91 (7): 1037–1039. https://doi.org/10.1021/ed400838n.

Back, M., Trave, E., Mazzucco, N. et al. (2017). Tuning the upconversion light emission by bandgap engineering in bismuth oxide-based upconverting nanoparticles. *Nanoscale* 9 (19): 6353–6361. https://doi.org/10.1039/C6NR09350G.

Bokolia, R., Mondal, M., Rai, V., and Sreenivas, K. (2017). Enhanced infrared-to-visible up-conversion emission and temperature sensitivity in (Er^{3+}, Yb^{3+}, and W^{6+}) tri-doped $Bi_4Ti_3O_{12}$ ferroelectric oxide. *Journal of Applied Physics* 121 (8): 084101. https://doi.org/10.1063/1.4977006.

Bragg, W.H. and Bragg, W.L. (1913). The reflection of X-rays by crystals. *Proceedings of the Royal Society of London Series A* 88 (605): 428–438. https://doi.org/10.1098/rspa.1913.0082.

Bunaciu, A.A., UdriŞTioiu, E.G., and Aboul-Enein, H.Y. (2015). X-ray diffraction: instrumentation and applications. *Critical Reviews in Analytical Chemistry* 45 (4): 289–299. https://doi.org/10.1080/10408347.2014.949616.

Burnham, R.W. (1952). A colorimeter for research in color perception. *The American Journal of Psychology* 65 (4): 603–608. https://doi.org/10.2307/1418043.

Cong, Y., Dong, B., Yu, N. et al. (2016). Enhanced upconversion emission in ZrO_2–Al_2 O_3 composite oxide. *Dalton Transactions* 45 (15): 6627–6633. https://doi.org/10.1039/C5DT04551G.

Cui, X., Cheng, Y., Lin, H. et al. (2018). Towards ultra-high sensitive colorimetric nanothermometry: constructing thermal coupling channel for electronically independent levels. *Sensors and Actuators B: Chemical* 256: 498–503. https://doi.org/10.1016/j.snb.2017.10.131.

Cullity, B.D. (1956). *Elements of X-ray Diffraction*. Addison-Wesley Publishing.

Cullity, B.D. and Stock, S.R. (2001). *Elements of X-ray Diffraction*, vol. 3. New Jersey: Prentice Hall.

DaCosta, M.V., Doughan, S., Han, Y., and Krull, U.J. (2014). Lanthanide upconversion nanoparticles and applications in bioassays and bioimaging: a review. *Analytica Chimica Acta* 832: 1–33. https://doi.org/10.1016/j.aca.2014.04.030.

Demina, P.A., Sholina, N.V., Akasov, R.A. et al. (2020). A versatile platform for bioimaging based on colominic acid-decorated upconversion nanoparticles. *Biomaterials Science* 8 (16): 4570–4580. https://doi.org/10.1039/D0BM00876A.

Di Gianfrancesco, A. (2016). *Materials for Ultra-Supercritical and Advanced Ultra-Supercritical Power Plants*. Woodhead Publishing https://doi.org/10.1016/C2014-0-04826-5.

Dincer, I. (2018). *Comprehensive Energy Systems*. Elsevier. ISBN: 978-0-12-814925-6.

Ding, M., Yin, S., Ni, Y. et al. (2015). Controlled synthesis of β-$NaYF_4$:Yb^{3+}/Er^{3+} microstructures with morphology-and size-dependent upconversion luminescence. *Ceramics International* 41 (6): 7411–7420. https://doi.org/10.1016/j.ceramint.2015.02.054.

Do Nascimento, J., Sales, A., Sousa, D. et al. (2016). Temperature-, power-, and concentration-dependent two and three photon upconversion in Er^{3+}/Yb^{3+} co-doped lanthanum ortho-niobate phosphors. *RSC Advances* 6 (72): 68160–68169. https://doi.org/10.1039/C6RA12941B.

Du, K., Xu, X., Yao, S. et al. (2018). Enhanced upconversion luminescence and controllable phase/shape of $NaYF_4$:Yb/Er crystals through Cu^{2+} ion doping. *CrystEngComm* 20 (14): 1945–1953.

Duo, S., Zhong, C., Zhang, J. et al. (2018). Preparation of $NaYF_4$:Gd^{3+},Yb^{3+},Tm^{3+}@TiO_2 and $NaYF_4$:Gd^{3+},Yb^{3+},Tm^{3+}@TiO_2@Au nanocomposites and their upconversion and photocatalytic properties under simulated solar light with or without an UV filter. *Journal of Materials Science: Materials in Electronics* 29 (4): 2974–2987. https://doi.org/10.1007/s10854-017-8228-0.

Duong, H.T., Chen, Y., Tawfik, S.A. et al. (2018). Systematic investigation of functional ligands for colloidal stable upconversion nanoparticles. *RSC Advances* 8 (9): 4842–4849. https://doi.org/10.1039/C7RA13765F.

Fahool, M. and Sabet, A.R. (2015). UV-visible assessment of hydrocluster formation and rheological behaviour in bimodal and mono-disperse shear thickening fluids. *Rheologica Acta* 54 (1): 77–83. https://doi.org/10.1007/s00397-014-0821-z.

Fang, A., Wu, Q., Lu, Q. et al. (2016). Upconversion ratiometric fluorescence and colorimetric dual-readout assay for uric acid. *Biosensors and Bioelectronics* 86: 664–670. https://doi.org/10.1016/j.bios.2016.07.055.

Felix, S., Gouveia, E., de Araujo, M. et al. (2000). Up-conversion pumped light amplification with temperature tunable gain in Er^{3+}/Yb^{3+}-codoped chalcogenide glasses. *Journal of Luminescence* 87: 1020–1022. https://doi.org/10.1016/S0022-2313(99)00515-3.

Fischer, S., Siefe, C., Swearer, D.F. et al. (2020). Bright infrared-to-ultraviolet/visible upconversion in small alkaline earth-based nanoparticles with biocompatible CaF_2 shells. *Angewandte Chemie* 132 (48): 21787–21796. https://doi.org/10.1002/anie.202007683.

Fu, A.C., Hu, Y., Zhao, Z.-H. et al. (2018). Functionalized paper microzone plate for colorimetry and up-conversion fluorescence dual-mode detection of telomerase based on elongation and capturing amplification. *Sensors and Actuators B: Chemical* 259: 642–649. https://doi.org/10.1016/j.snb.2017.12.124.

Gabbott, P. (2008). *Principles and Applications of Thermal Analysis*. Wiley https://doi.org/10.1002/9780470697702.

Gao, G., Wang, G., Yu, C. et al. (2009). Investigation of 2.0 μm emission in Tm^{3+} and Ho^{3+} co-doped oxyfluoride tellurite glass. *Journal of Luminescence* 129 (9): 1042–1047. https://doi.org/10.1016/J.JLUMIN.2009.04.024.

Gfroerer, T.H. (2006). Photoluminescence in analysis of surfaces and interfaces. In: *Encyclopedia of Analytical Chemistry: Applications, Theory and Instrumentation*. https://doi.org/10.1002/9780470027318.a2510.

Goldburg, W.I. (1999). Dynamic light scattering. *American Journal of Physics* 67 (12): 1152–1160. https://doi.org/10.1119/1.19101.

Goldstein, J. (2012). *Practical Scanning Electron Microscopy: Electron and Ion Microprobe Analysis*. Springer Science & Business Media https://doi.org/10.1007/978-1-4613-4422-3.

Gordon, J. and Harman, S. (2002). A graduated cylinder colorimeter: an investigation of path length and the Beer–Lambert law. *Journal of Chemical Education* 79 (5): 611. https://doi.org/10.1021/ed079p611.

Gregersen, S. (2014). Fluorescent Peptide-Stabilized Silver-Nanoclusters: A Solid-Phase Approach for High-Throughput Ligand Discovery. Corpus ID: 208751143.

Grzyb, T., Kamiński, P., Przybylska, D. et al. (2021). Manipulation of up-conversion emission in $NaYF_4$ core@shell nanoparticles doped by Er^{3+}, Tm^{3+}, or Yb^{3+} ions by excitation wavelength—three ions—plenty of possibilities. *Nanoscale* 13 (15): 7322–7333. https://doi.org/10.1039/D0NR07136F.

Guillot-Noel, O., Bellamy, B., Viana, B., and Gourier, D. (1999). Correlation between rare-earth oscillator strengths and rare-earth–valence-band interactions in neodymium-doped YMO_4(M=V, P, As), $Y_3Al_5O_{12}$, and $LiYF_4$ matrices. *Physical Review B* 60 (3): 1668. https://doi.org/10.1103/PhysRevB.60.1668.

Guo, H., Dong, N., Yin, M. et al. (2004). Visible upconversion in rare earth ion-doped Gd_2O_3 nanocrystals. *The Journal of Physical Chemistry B* 108 (50): 19205–19209. https://doi.org/10.1021/jp048072q.

Han, X., Song, E., Zhou, B., and Zhang, Q. (2019). Color tunable upconversion luminescent perovskite fluoride with long-/short-lived emissions toward multiple anti-counterfeiting. *Journal of Materials Chemistry C* 7 (27): 8226–8235. https://doi.org/10.1039/C9TC02171J.

Hayat, M.E. (2012). *Basic Techniques for Transmission Electron Microscopy*. Elsevier https://doi.org/10.1016/B978-0-12-333926-3.X5001-5.

Höhne, G., Hemminger, W.F., and Flammersheim, H.-J. (2013). *Differential Scanning Calorimetry*. Springer Science & Business Media https://doi.org/10.1007/978-3-662-06710-9.

Hui, X., Peng, D., Zou, H. et al. (2014). A new multifunctional aurivillius oxide $Na_{0.5}Er_{0.5}Bi_4Ti_4O_{15}$: up-conversion luminescent, dielectric, and piezoelectric properties. *Ceramics International* 40 (8): 12477–12483. https://doi.org/10.1016/j.ceramint.2014.04.102.

Jain, N., Singh, R.K., Sinha, S. et al. (2018). Color tunable emission through energy transfer from Yb^{3+} co-doped $SrSnO_3$: Ho^{3+} perovskite nano-phosphor. *Applied Nanoscience* 8 (6): 1267–1278. https://doi.org/10.1007/s13204-018-0778-9.

Jain, N., Paroha, R., Singh, R.K. et al. (2019). Synthesis and rational design of europium and lithium doped sodium zinc molybdate with red emission for optical imaging. *Scientific Reports* 9 (1): 1–14. https://doi.org/10.1038/s41598-019-38787-1.

Jiang, T., Tian, Y., Xing, M. et al. (2018). Research on the photoluminescence and up-conversion luminescence properties of $Y_2Mo_4O_{15}$:Yb,Ho under 454 and 980 nm excitation. *Materials Research Bulletin* 98: 328–334. https://doi.org/10.1016/j.materresbull.2017.09.025.

Kannan, M. (2018). Scanning electron microscopy: principle, components and applications. In: *Handbook of Materials Characterization*. Cham: Springer https://doi.org/10.1007/978-3-319-92955-2_4.

Kim, K.J., Jouini, A., Yoshikawa, A. et al. (2007). Growth and optical properties of Pr, Yb-codoped KY3F10 fluoride single crystals for up-conversion visible luminescence. *Journal of Crystal Growth* 299 (1): 171–177. https://doi.org/10.1016/j.jcrysgro.2006.10.220.

Kumar, V., Ntwaeaborwa, O., Soga, T. et al. (2017). Rare earth doped zinc oxide nanophosphor powder: a future material for solid state lighting and solar cells. *ACS Photonics* 4 (11): 2613–2637. https://doi.org/10.1021/acsphotonics.7b00777.

Lakowicz, J.R. (2013). *Principles of Fluorescence Spectroscopy*. Springer Science & Business Media https://doi.org/10.1007/978-0-387-46312-4.

Lee, J.-W., Cho, K.-H., Yoon, J.-S., and Sung, Y.-M. (2020). Enhanced IR-driven photoelectrochemical responses of CdSe/ZnO heterostructures by up-conversion UV/visible light irradiation. *Nanoscale* 12 (15): 8525–8535. https://doi.org/10.1039/D0NR00477D.

Leong, S.S., Ng, W.M., Lim, J., and Yeap, S.P. (2018). Dynamic light scattering: effective sizing technique for characterization of magnetic nanoparticles. In: *Handbook of Materials Characterization*, 77–111. Springer https://doi.org/10.1007/978-3-319-92955-2_3.

Lindon, J.C., Tranter, G.E., and Koppenaal, D. (2016). *Encyclopedia of Spectroscopy and Spectrometry*. Academic Press https://doi.org/10.1016/B978-0-12-409547-2.12138-9.

Lojpur, V.M., Ahrenkiel, P.S., and Dramićanin, M.D. (2013). Color-tunable up-conversion emission in Y_2O_3:Yb^{3+},Er^{3+} nanoparticles prepared by polymer complex solution method. *Nanoscale Research Letters* 8 (1): 1–6. https://doi.org/10.1186/1556-276X-8-131.

Lothenbach, B., Durdzinski, P., and De Weerdt, K. (2016). Thermogravimetric analysis. In: *A Practical Guide to Microstructural Analysis of Cementitious Materials*, vol. 1. CRC Press https://doi.org/10.1201/b19074.

Ly, B.C.K., Dyer, E.B., Feig, J.L. et al. (2020). Research techniques made simple: cutaneous colorimetry: a reliable technique for objective skin color measurement. *Journal of Investigative Dermatology* 140 (1): 3-12.e11. https://doi.org/10.1016/j.jid.2019.11.003.

Mäntele, W. and Deniz, E. (2017). *UV–Vis Absorption Spectroscopy: Lambert-Beer Reloaded*. Elsevier https://doi.org/10.1016/j.saa.2016.09.037.

Manurung, R., Wiranto, G., & Hermida, I. (2018). Synthesis and characterization of monodisperse core–shell lanthanide upconversion nanoparticles NaYF4:Yb,Tm/SiO_2. Paper presented at the IOP Conference Series: Materials Science and Engineering. CAMST 2018 Novotel, Semarang, Indonesia. https://doi.org/10.1088/1757-899X/367/1/012043.

Mather, R.R. (2009). Surface modification of textiles by plasma treatments. In: *Surface Modification of Textiles*, 296–317. Elsevier https://doi.org/10.1533/9781845696689.296.

Michael, J.R., Brewer, L.N., and Kotula, P.G. (2011). Electron beam-based methods for bioforensic investigations. In: *Microbial Forensics*, 421–729. Elsevier https://doi.org/10.1016/B978-0-12-382006-8.00025-6.

Munjanja, B. and Sanganyado, E. (2015). UV-visible absorption, fluorescence, and chemiluminescence spectroscopy. In: *Handbook of Food Analysis* (ed. L.M. Nollet and T. Fidel), 572–583. https://doi.org/10.1201/b18668.

Ngo, P.D. (1999). Energy dispersive spectroscopy. In: *Failure Analysis of Integrated Circuits*, 205–215. Springer https://doi.org/10.1007/978-1-4615-4919-2_12.

Pecora, R. (2013). *Dynamic Light Scattering: Applications of Photon Correlation Spectroscopy*. Springer Science & Business Media https://doi.org/10.1007/978-1-4613-2389-1.

Pu, Y., Leng, J., Wang, D. et al. (2018). Recent progress in the green synthesis of rare-earth doped upconversion nanophosphors for optical bioimaging from cells to animals. *Chinese Journal of Chemical Engineering* 26 (10): 2206–2218. https://doi.org/10.1016/j.cjche.2018.03.005.

Que, W., Zhou, Y., Lam, Y. et al. (2001). Up-conversion luminescence of erbium(III) oxalate nanoparticles/titania/ÿ-glycidoxypropyltrimethoxysilane composite sol-gel thin films. *Journal of Electronic Materials* 30 (1): 7–10. https://doi.org/10.1007/S11664-001-0207-8.

Quimby, R., Drexhage, M., and Suscavage, M. (1987). Efficient frequency up-conversion via energy transfer in fluoride glasses. *Electronics Letters* 23 (1): 32–34. https://doi.org/10.1049/el:19870024.

Rafique, R., Baek, S.H., Park, C.Y. et al. (2018). Morphological evolution of upconversion nanoparticles and their biomedical signal generation. *Scientific Reports* 8 (1): 1–11. https://doi.org/10.1038/s41598-018-35513-1.

Raja, V. and Ali, B.J. (2021). Synergy of photon up-conversion and Z-scheme mechanism in graphitic carbon nitride nanoparticles decorated g-C_3N_4-TiO_2. *Colloids and Surfaces A: Physicochemical and Engineering Aspects* 611: 125862. https://doi.org/10.1016/j.colsurfa.2020.125862.

Rajeswari, R., Islavath, N., Raghavender, M., and Giribabu, L. (2020). Recent progress and emerging applications of rare earth doped phosphor materials for dye-sensitized and perovskite solar cells: a review. *The Chemical Record* 20 (2): 65–88. https://doi.org/10.1002/tcr.201900008.

Reddy, K.L., Prabhakar, N., Arppe, R. et al. (2017). Microwave-assisted one-step synthesis of acetate-capped $NaYF_4$:Yb/Er upconversion nanocrystals and their application in bioimaging. *Journal of Materials Science* 52 (10): 5738–5750. https://doi.org/10.1007/s10853-017-0809-z.

Reimer, L. (2013). *Transmission Electron Microscopy: Physics of Image Formation and Microanalysis*, vol. 36. Springer https://doi.org/10.1007/978-0-387-40093-8.

Righini, G., Enrichi, F., Zur, L. et al. (2019). Rare-earth doped glasses and light managing in solar cells. Paper presented at the Journal of Physics: Conference Series. https://doi.org/10.1088/1742-6596/1221/1/012028.

Santelli, J., Lepoix, C., Lechevallier, S. et al. (2021). Custom NIR imaging of new up-conversion multimodal gadolinium oxysulfide nanoparticles. *Particle & Particle Systems Characterization* 38 (4): 2000216. https://doi.org/10.1002/ppsc.202000216.

Sarmanova, O.E., Burikov, S.A., Laptinskiy, K.A. et al. (2020). In vitro temperature sensing with up-conversion $NaYF_4$:Yb^{3+}/Tm^{3+}-based nanocomposites: peculiarities and pitfalls. *Spectrochimica Acta Part A: Molecular and Biomolecular Spectroscopy* 241: 118627. https://doi.org/10.1016/j.saa.2020.118627.

Schoonheydt, R.A. (2010). UV-vis-NIR spectroscopy and microscopy of heterogeneous catalysts. *Chemical Society Reviews* 39 (12): 5051–5066. https://doi.org/10.1039/C0CS00080A.

Shaw, R. (2014). *Dynamic Light Scattering Training*. Malvern Instruments Ltd.

Shindo, D. and Oikawa, T. (2002). Energy dispersive X-ray spectroscopy. In: *Analytical Electron Microscopy for Materials Science*, 81–102. Springer https://doi.org/10.1007/978-4-431-66988-3_4.

Shuchun, C. and Fengmei, D. (1996). Infrared stimulated up-conversion luminescence and optical storage properties in SrS (Cu, Sm). *Journal of Inorganic Materials* 04.

Stetefeld, J., McKenna, S.A., and Patel, T.R. (2016). Dynamic light scattering: a practical guide and applications in biomedical sciences. *Biophysical Reviews* 8 (4): 409–427. https://doi.org/10.1007/s12551-016-0218-6.

Szczepanski, F., Bayart, A., Katelnikovas, A. et al. (2020). Luminescence and up-conversion properties in $La_2Ti_2O_7$:Eu^{3+},Er^{3+} oxides under UV and NIR radiations towards a two-color sensor. *Journal of Alloys and Compounds* 826: 154157. https://doi.org/10.1016/j.jallcom.2020.154157.

Tian, Q., Yao, W., Wu, W. et al. (2017). Efficient UV–vis-NIR responsive upconversion and plasmonic-enhanced photocatalyst based on lanthanide-doped $NaYF_4/SnO_2/Ag$. *ACS Sustainable Chemistry & Engineering* 5 (11): 10889–10899. https://doi.org/10.1021/acssuschemeng.7b02806.

Tikhomirov, V., Mortier, M., Gredin, P. et al. (2008). Preparation and up-conversion luminescence of 8 nm rare-earth doped fluoride nanoparticles. *Optics Express* 16 (19): 14544–14549. https://doi.org/10.1364/OE.16.014544.

Tiseanu, C., Parvulescu, V., Avram, D. et al. (2014). Structural, down-and phase selective up-conversion emission properties of mixed valent Pr doped into oxides with tetravalent cations. *Physical Chemistry Chemical Physics* 16 (12): 5793–5802. https://doi.org/10.1039/C3CP54899F.

Ul-Hamid, A. (2018). *A Beginners' Guide to Scanning Electron Microscopy*, vol. 1. Springer https://doi.org/10.1007/978-3-319-98482-7_2.

Van der Heide, P. (2011). *X-Ray Photoelectron Spectroscopy. An Introduction to Principles and Practices*. Wiley https://doi.org/10.1002/9781118162897.

Vernon-Parry, K. (2000). Scanning electron microscopy: an introduction. *III-Vs Review* 13 (4): 40–44. https://doi.org/10.1016/S0961-1290(00)80006-X.

Vickerman, J.C. and Gilmore, I.S. (2011). *Surface Analysis: The Principal Techniques*. Wiley https://doi.org/10.1002/9780470721582.

Voronov, M., Pevtsov, A., Skvortsov, A. et al. (2019). Intense anti-Stokes emission of erbium ions in gallium lanthanum sulphide-oxide glass GaLaS(O) in visible spectral range. *Physica Status Solidi (a)* https://doi.org/10.1002/pssa.202000448.

Warren, B.E. (1990). *X-ray Diffraction*. Courier Corporation.

Wen, S., Zhou, J., Zheng, K. et al. (2018). Advances in highly doped upconversion nanoparticles. *Nature Communications* 9 (1): 1–12. https://doi.org/10.1038/s41467-018-04813-5.

Whittig, L. and Allardice, W. (1986). X-ray diffraction techniques. Methods of soil analysis: Part 1 physical and mineralogical. *Methods* 5: 331–362. https://doi.org/10.2136/sssabookser5.1.2ed.c12.

Williams, D.B. and Carter, C.B. (1996). The transmission electron microscope. In: *Transmission Electron Microscopy*, 3–17. Springer https://doi.org/10.1007/978-0-387-76501-3.

Wu, Z., Guo, C., Liang, S. et al. (2012). A pluronic F127 coating strategy to produce stable up-conversion $NaYF_4$:Yb,Er(Tm) nanoparticles in culture media for bioimaging. *Journal of Materials Chemistry* 22 (35): 18596–18602. https://doi.org/10.1039/C2JM33626J.

Yao, L., Xu, D., Li, Y. et al. (2019). Regulation of morphologies and luminescence of β-$NaGdF_4$:Ybc^{+}, Er^{3+} upconversion nanoparticles by hydrothermal method and their dual-mode thermometric properties. *Applied Surface Science* 466: 320–327. https://doi.org/10.1016/J.APSUSC.2018.09.209.

Yefimov, N. (2009). *Handbook of Non-Ferrous Metal Powders: Technologies and Applications*. Elsevier https://doi.org/10.1016/C2014-0-03938-X.

Zeng, J.H., Li, Z.H., Su, J. et al. (2006). Synthesis of complex rare earth fluoride nanocrystal phosphors. *Nanotechnology* 17 (14): 3549. https://doi.org/10.1088/0957-4484/17/14/032.

Zhang, F., Li, J., Shan, J. et al. (2009). Shape, size, and phase-controlled rare-earth fluoride nanocrystals with optical up-conversion properties. *Chemistry A European Journal* 15 (41): 11010–11019. https://doi.org/10.1002/chem.200900861.

Zhou, X., Li, S., Chen, X. et al. (2008). Formation of nanocrystalline matrix of sodium rare earth fluoride up-conversion phosphors. *Journal of Nanoscience and Nanotechnology* 8 (3): 1392–1397. https://doi.org/10.1166/jnn.2008.18202.

Zhou, D., Li, D., Zhou, X. et al. (2017). Semiconductor plasmon induced up-conversion enhancement in $mCu_{2-x}S@SiO_2@Y_2O_3:Yb^{3+}/Er^{3+}$ core–shell nanocomposites. *ACS Applied Materials & Interfaces* 9 (40): 35226–35233. https://doi.org/10.1021/acsami.7b09850.

4

Raman and FTIR Spectroscopic Techniques and Their Applications

Saurav K. Ojha and Animesh K. Ojha

Department of Physics, Motilal Nehru National Institute of Technology, Allahabad, Prayagraj 211004, India

4.1 Raman Spectroscopy

Raman spectroscopy is started in 1928 after the discovery of the Raman effect by a great Indian physicist Chandrashekhara Venkata Raman (Cotton 1971). Since the discovery, the Raman effect has proved to be a powerful analytical tool for the investigation of structural properties of simple and complex systems. The Raman effect is described as an inelastic scattering of light where the energy difference of the incident and scattered photons corresponds to the separation between the energy level of different states of the scatterer molecule. The energy difference between the incident and scattered photons depends on the chemical structure of the scatterer molecules. In a Raman scattering experiment, light of a known frequency and polarization is scattered from a sample. The scattered light is subsequently analyzed in terms of frequency and state of polarization. Raman scattered light is frequency shifted with respect to the excitation frequency, but the magnitude of the shift is independent of the excitation frequency. This shift, known as the "Raman Shift," is therefore characteristic of the sample under investigation. A schematic presentation of Stokes Raman, anti-Stokes Raman, and Rayleigh lines is shown in Figure 4.1.

In general, only some excitations ($\nu = 0$ to $\nu = 1$) of a given system are Raman active, which essentially means that the wave numbers of only some normal modes of a molecule may be obtained. The vibrational modes are sensitive to changes in the molecular environment. The peak positions, line widths, line shapes, and intensities of spectral lines provide useful information about the environment around a molecule and its association. The information obtained by analyzing the Raman peaks is shown in Figure 4.2.

The analytical capability of Raman spectroscopy is improved when the exciting radiation is in resonance with an electronic state. Thus, an overall enhancement of the Raman signal and also selectivity can be achieved by tuning the frequency of the exciting laser line to specific electronic transitions. In the so-called Raman excitation

Upconverting Nanoparticles: From Fundamentals to Applications, First Edition. Edited by Vineet K. Rai.

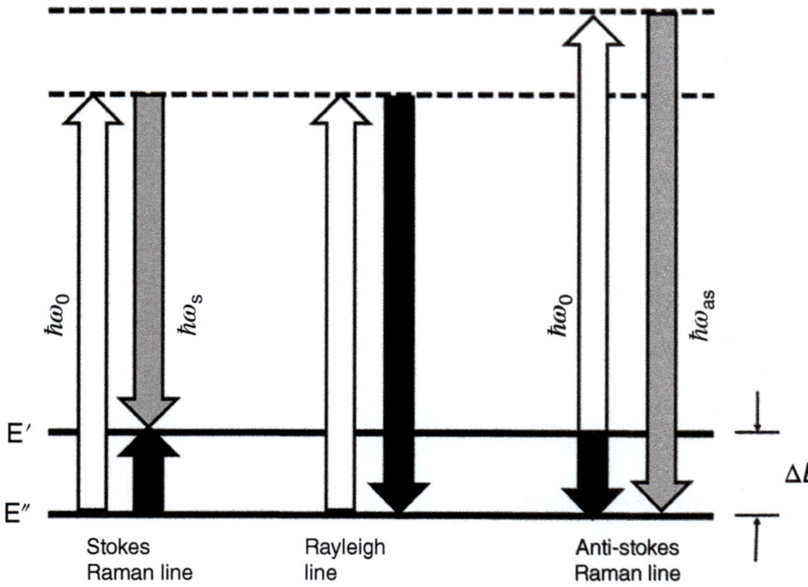

Figure 4.1 A schematic presentation of Stokes Raman, anti-Stokes Raman, and Rayleigh lines.

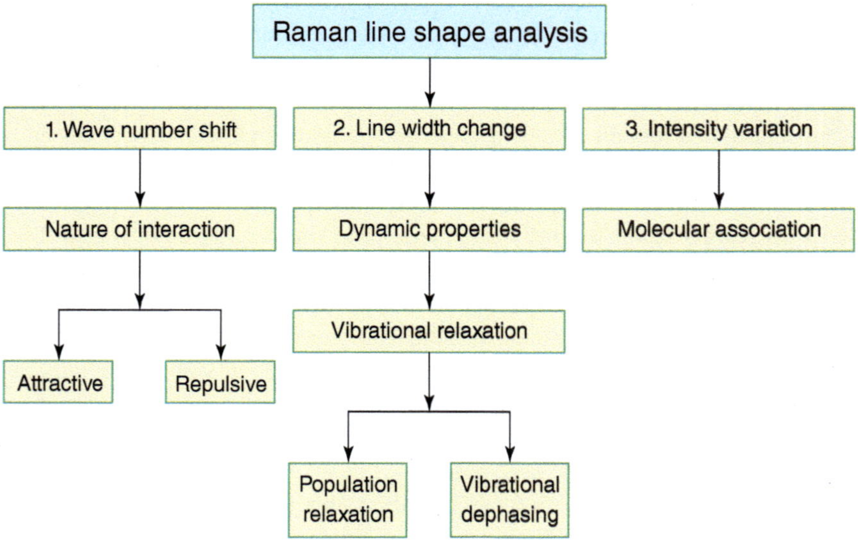

Figure 4.2 Information obtained by the analysis of Raman peak.

spectroscopy, Raman line positions, intensities, and line shapes are observed as a function of the frequency of the incident excitation line. In this way, the exciting radiation can be tuned to one or more of the electronic resonances. The Raman spectroscopy technique is widely used to explore various processes that occur in solid, liquid, and gaseous sates at the molecular level.

4.2 Fourier Transform Infrared (FTIR) Spectroscopy

Infrared (IR) spectroscopy was the first spectroscopic technique used to obtain structural information of the sample. It is an analytical method used to characterize the type of bonding between atoms of matter, on the basis of their interaction with the IR radiation, and measures the frequencies at which the substance absorbs the IR radiation, which lead to the production of vibrations in molecules. It provides a fast, non-destructive, and real-time measurement technique for the identification and characterization of chemical structures of a wide range of samples, from biological to composite materials and from liquids to gaseous state of the samples (Son et al. 2001; Vaghri et al. 2012). A Fourier transform infrared (FTIR) spectrometer uses the combination of an interferometer and the mathematical principle of Fourier transform to probe the samples. The FTIR spectrum of a sample is a graphical representation of the transmittance, in percent (T%) or absorbance, versus the IR frequency in terms of wave number (cm^{-1}). In an IR spectrum, the absorption bands are characterized by a wave number at which absorption occurs. These absorption bands correspond to the chemical bonds, and the intensity of absorption is proportional to the number of chemical bonds present in the solution.

Based on the spectral regions, the IR spectrum is divided into three categories: (i) near infrared (NIR), near the visible region (13 000 to 4000 cm^{-1} or 770–2500 nm), which mainly excites overtone or higher harmonic vibrations, (ii) mid infrared (MIR) (4000 to 400 cm^{-1} or 2500–25 000 nm), which mainly excites the fundamental vibrations, and (iii) far infrared (FIR) (400 to 10 cm^{-1} or 25 000–1 000 000 nm), which excites lattice vibrations, and below 300 cm^{-1}, FIR can be used for rotational spectroscopy. FTIR spectroscopy is a non-destructive and real-time measurement characterization technique that is used for identifying the unknown materials and their concentrations. FTIR can be used for the investigation of organic and inorganic samples in solid, liquid, or gaseous states. In IR spectroscopy, each chemical bond has its characteristic vibrational frequency corresponding to the energy difference in its vibrational energy level. The vibrational mode of a molecule represents a variation in the inter-atomic bond length, called stretching, or a vibration in the bond angle between two bonds, called bending or deformation (Stuart 2004; Reichenbacher and Popp 2012; Singh et al. 2012). A non-linear molecule with N number of atoms usually has $(3N - 6)$ fundamental vibrational modes, with $(N - 1)$ stretching vibrational modes and $(2N - 5)$ bending vibrational modes, whereas a linear molecule has $(3N - 5)$ fundamental vibrational modes. For example, N_2 (symmetrical) is IR inactive and does not have any vibrational modes, whereas CO (asymmetrical) has one bond ($n = 1$) and thus have only one vibrational mode lying in the spectral range, 2300 to 2000 cm^{-1}. The water molecule with $N = 3$ has two stretching vibrational modes, symmetric and asymmetric, lying in the range of 3450 to 3200 cm^{-1} and one bending vibrational mode lying in the region 1640 to 1630 cm^{-1} (Rakic et al. 2003). When the IR beam falls on a sample, it can be either reflected or scattered off from the surface of the particle or be transmitted through the particle. Based on the interaction of IR beam with the sample and methods to detect these beams after interaction, various IR measurement techniques have been developed.

4.2.1 FTIR in Transmission Mode

It is the most commonly applied method of FTIR spectroscopy for measuring the IR spectrum of a sample. In this mode, the IR radiation is passed through the sample, and the amount of radiation transmitted through the sample is measured as a function of IR wave number. The spectra thus obtained will represent whole of the volume of sample. However, in this technique, along with the sample preparation, such as the manufacturing of potassium bromide (KBr) discs or Nujol mull preparation, "localized" (e.g. surface) properties are also lost. It is better only for liquid samples or thin (<10 μm) samples.

4.2.2 Attenuated Total Reflectance (ATR)

In this technique, IR radiation is passed through an internal reflection element (IRE) crystal material whose surface is in contact with the sample (in liquid or solid state) to be investigated. As the IR light passes through the IRE crystal, it interacts with the sample at the crystal/sample interface and then gets reflected toward the detector. For many attenuated total reflectance (ATR) measurements, the IR light interacts with the sample at a 45° incident angle to the crystal/sample interface. This technique requires no sample preparation. Further ATR can be used for the investigation of very thin films and surface coatings that cannot be accessible through any standard chemical methods. It is an analytical tool employed for the investigation of membranes and polymers.

4.2.3 Diffuse Reflectance Infrared Fourier Transform Spectroscopy (DRIFTS)

It is a technique that records the IR energy scattered from the sample's rough surface. It is used to measure the IR-scattered signal from the fine particles, powders, and rough surface where traditional techniques failed. In this technique, the sampling is fast and easy compared to other techniques. In this case, the IR energy reflecting off the surface is typically lost and, therefore, the surface studies of the samples cannot be done using this technique.

4.3 Applications of Raman Spectroscopy

4.3.1 Raman Study of Molecular Association in Hydrogen-Bonded Systems

The phenomenon of hydrogen bonding is widely present in nature. Hydrogen bonds, in general, for instance, stabilize the secondary and tertiary structures of proteins, and they are thought to play a major role in molecular recognition. The hydrogen bond formed between solute and solvent molecules in the mixture causing wave number shift/line width changes in Raman bands corresponding to some vibrational modes of the solute molecule with the change of its concentration in the solution.

When a reference system (solute) having electronegative atoms, e.g. N, O, Cl, Br, and F, is mixed with hydrogen donor solvents, e.g. CH_3OH, C_2H_5OH, and H_2O, the molecules of the solute and solvent are no longer independent entities (Asthana et al. 2000; Raman and Krishnan 1928). A net attractive or repulsive interaction between the solute and solvent molecules may cause a red shift or blue shift, respectively, in the wave number position of the reference mode of the solute molecule. Such molecular associations do not form through a permanent chemical bond. In the solution, the molecular association takes place via the hydrogen bond interaction between two species in a structural arrangement where the hydrogen atom is placed in between two species that form a hydrogen bond. The effect of hydrogen bond interaction on molecular motions can, therefore, be used to understand the vibrational relaxation processes, e.g. vibrational dephasing in the liquid mixtures. Vibrational dephasing is an important probe for studying molecular structure and dynamics in a condensed phase. The vibrational wave function of an active isolated oscillator can be expressed in terms of a quantum state and a phase. When the oscillator is perturbed through the surrounding environment, its wave function can be modified by both the ways. When the quantum state changes, the relaxation is called vibrational energy relaxation, and when the phase of the oscillator is modified, it leads to relaxation through phase, called dephasing. The dephasing phenomena in liquids take place on a picosecond time scale, governing the line width of a Raman band. A schematic presentation of the relaxation processes is given in Figure 4.3. A number of studies (Srivastava et al. 2003, 2005; Ojha et al. 2004, 2005, 2006; Ojha 2007; Singh et al. 2007; Mishra et al. 2009; Singh et al. 2008a; Singh et al. 2008b; Vyas and Ojha 2010, 2011) have been carried out to explore the effect of hydrogen bonding on relaxation processes, e.g. vibrational dephasing using Raman spectroscopy.

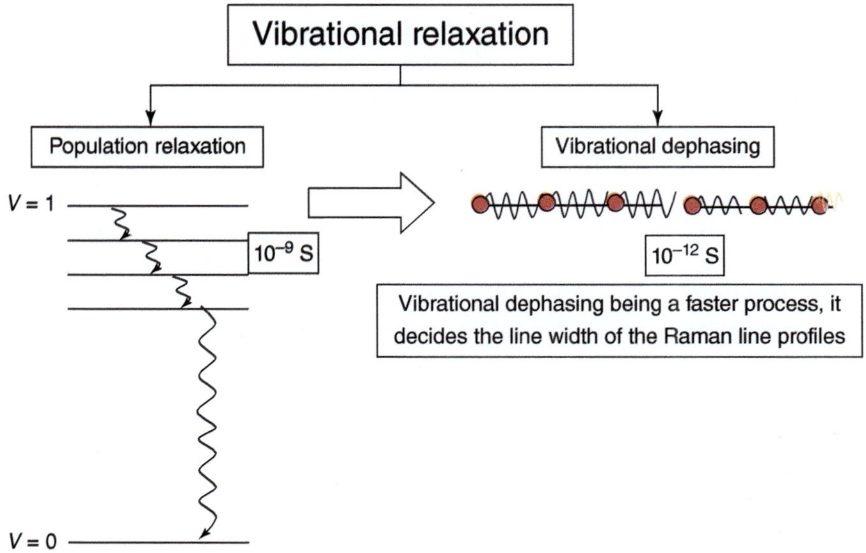

Figure 4.3 Schematic presentation of relaxation processes in molecules.

Raman spectroscopy is the most suited technique for studying a hydrogen-bonded system because the hydrogen bond can easily be estimated from the position of the symmetric/antisymmetric hydrogen bond bridging mode or studying the wave number shift of higher vibrational modes. In this regard, Srivastava et al. had studied the Raman non-coincidence effect (NCE) by recording the Raman spectra of the polar liquids, e.g. $(CH_3)_2C{=}O + (CD_3)_2C{=}O$ (Srivastava et al. 2003).

A wave number shift of ~2 cm^{-1} in C–H stretching mode was observed in diluted acetone (0.1 mol fraction), which confirmed the existence of a local short-range order as required for the existence of NCE in polar liquids. The change in the line width of Γ(C–H) mode with the dilution of acetone is also investigated and compared with the Γ(C–H) mode. Ojha et al. studied the formation of a self-associated dimer and hydrogen-bonded complex in binary mixture of formamide ($HCONH_2$) and methanol (CH_3OH) with varying concentration (Ojha et al. 2004). The effect of varying concentration was studied on C=O stretching, –N–H bending, –C–H bending, and C–N stretching modes, and it was found that the Raman peaks corresponding to the NH_2 bending and C=O stretching modes shift toward a higher wave number on decreasing the concentration of formamide because of the repulsive interaction between hydrogen-bonded complexes in the solution, whereas peaks corresponding to –C–N (1312 cm^{-1}) stretching and C–H (1393 cm^{-1}) bending modes are not influenced by the change in the concentration of solution, thereby indicating that the C—H bond length is not altered by the variation in the concentration of solution. In another study, Ojha et al. used Raman spectroscopy to analyze the concentration-dependent Raman spectra of tetrahydrofuran (THF) (C_4H_8O) dissolved in water, as THF is the simplest model compound for understanding the vibrations of furanose ring in RNA and DNA (Ojha et al. 2005). In this study, the concentration-dependent shift in the peak position and the variation in the line width of C–O (910 cm^{-1}) and C–C (914 cm^{-1}) stretching modes were investigated. It has been found that the decrease in the concentration of THF in THF + H_2O solution causes an increase in line width and a downshift in the C–O stretching mode, whereas the Raman peak corresponding to the C–C stretching mode shows an increase in line width and an upshift in peak position. The experimental results are well supported by the theoretical calculations (Ojha et al. 2005). In a study, Srivastava et al. studied the binary mixture of propylamine ($C_3H_7NH_2$) and methanol (CH_3OH) with varying concentration by Raman spectroscopy (Srivastava et al. 2005). They found that upon dilution, the O atom of methanol forms a hydrogen bond with a H atom of propylamine that weakens the N—H bond, which was supported by the downshift in Raman peak position corresponding to the symmetric N–H stretching mode (3305 cm^{-1}) and antisymmetric N–H stretching mode (3326 cm^{-1}) (Srivastava et al. 2005). Ojha et al. studied the effect of variation in pH on the charge transfer characteristic of the colloidal solution of phenylalanine (Phe) in silver (Ag) with surface-enhanced Raman scattering (SERS) (Ojha et al. 2006). They found that increasing the pH of the solution leads to an increase in the charge transfer between metal particles (Ag) and the reference molecule (Phe). The increased charge transfer increases the vibrational force constant, and as a result, the peak position of different vibrational modes (at

~1005, 1380, and 1582 cm^{-1}) shows an upshift with an increase in the pH of the solution (Ojha et al. 2006). Similarly, in another study, Ojha et al. performed SERS to study the tyrosine in colloidal Ag solution at different pHs and found similar results as reported in a previous study (Ojha et al. 2006). Singh et al. theoretically calculated the strength of the hydrogen bond and the corresponding wave number shift for pyridine, pyridazine, pyrimidine, pyrazine, and their complex with water molecules (Singh et al. 2007). Out of the three different types of hydrogen bonds (N...H—O, O...H—C, and O...H—C) studied in this article, the hydrogen bond in the case of O...H—C is strongest followed by N...H—C and then C...H—C. The strength of hydrogen bonds is also found to increase with an increase in the concentration of water (Singh et al. 2007). Similarly, Mishra et al. studied the Raman spectra of benzonitrile (Bz) in silver colloidal solution at different millimolar concentrations of benzonitrile by using SERS (Mishra et al. 2009). Comparison of SERS data with the theoretically calculated results shows that the redistribution of the electronic charge and the change in bond length in conjugation with the Ag atom is the main cause for wave number shift in the Raman peaks. Singh et al. studied the Raman spectra of thiophene (THP) and its highly diluted solution in dimethyl sulfoxide (DMSO) and methanol with varying concentration of THP to understand the influence of molecular interaction on spectral feature (Singh et al. 2008a). The variation of concentration of THP into the solvents, e.g. DMSO and methanol, has minimal effect on the experimentally observed vibrational spectra, which is also well supported by theoretical calculations. In another study, Singh et al. (2008b) studied the dependence of vibrational spectra (peak position, line width, and intensity) of pyrrole in aqueous solution and found that the variation in pH does not show any significant effect on the position of ring breathing mode, but Raman peak position corresponding to symmetric –N–H stretching mode shows an increase for lower pH value (pH 7.2) and a decrease for higher pH values (pH > 7.2) (Singh et al. 2008b). To understand the spectra–structure correlation, the Raman spectra of pure acetonitrile and its mixture with methanol at different mole fractions along with their self-associated and hydrogen-bonded complexes are given in Figure 4.4.

Vyas and Ojha found that the water molecule forms a stronger hydrogen bond with the –COOH group of alanine (Ala) compared to the $-NH_2$ group (Vyas and Ojha 2010). They also found that the stepwise addition of the water molecule into the Ala–$(H_2O)_n$ complex increases the binding energy of the confirmer. In another study, Vyas and Ojha examined the effect of coordination of metal cation and water molecule on the gas-phase structural stability of D-alanine (Ala) and zwitterionic alanine (ZAla) (Vyas and Ojha 2011). They found that addition of five water molecules in the –OO-coordinated mode of ZAla–M^+ complexes provides more stability in the gas phase. The Raman spectra of pure $CH_3C{\equiv}N$ and $C_2H_5C{\equiv}N$ and their binary mixtures ($CH_3C{\equiv}N + CH_3OH$) and ($C_2H_5C{\equiv}N + CH_3OH$) were recorded by varying the mole fractions (0.9–0.1) of the reference molecule, $CH_3C{\equiv}N/C_2H_5C{\equiv}N$. The structures of self-associated $CH_3C{\equiv}N$, $C_2H_5C{\equiv}N$, and their hydrogen-bonded complexes were optimized using density functional theory (DFT) for better spectra–structure correlation (Srivastava et al. 2006).

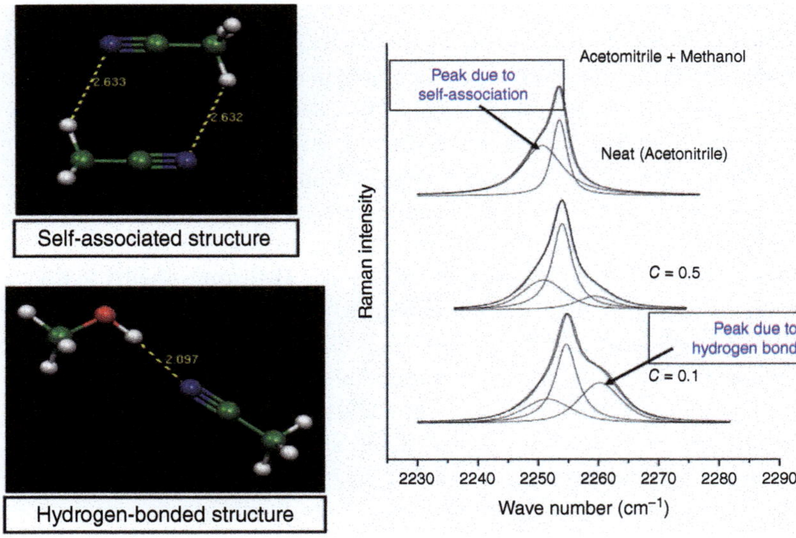

Figure 4.4 Raman spectra of pure $CH_3C{\equiv}N$ and ($CH_3C{\equiv}N + CH_3OH$) at different mole fractions of the reference molecule and their self-associated as well as hydrogen-bonded complexes. Source: Srivastava et al. 2006. Reprinted with the permission of John Wiley and Sons, Ltd.

4.3.2 Surface-Enhanced Raman Spectroscopy (SERS)

SERS is one of the most sensitive tools for enhancing the intensity of the Raman peaks using metal surface. In this technique, analyte molecules are brought close to the metal nanostructure (rough surface) of different shapes and sizes and the Raman spectra are record, wherein the Raman peaks are enhanced because of the presence of surface plasmons. In this technique, the plasmons present on the surface of some metals oscillate perpendicular to the surface of the metal and cause enhancement in the intensity of Raman scattering by a factor of 10^4–10^6. This drastic increase in the Raman intensity is called surface-enhanced Raman scattering (SERS), which allows the detection of even a very low concentration of analytes.

Researchers have observed intense Raman scattering from pyridine molecules adsorbed onto a roughened silver electrode surface in aqueous solution, but they did not recognize that it was a new phenomenon (SERS) (Fleischmann et al. 1974; Jeanmaire and van Duyne 1977). Several mechanisms were proposed to account for this tremendous intensity enhancement. These were accepted as viable mechanisms, and they are based on electromagnetic and chemical enhancement. In the electromagnetic enhancement, the optical field is enhanced because of the electronic resonance of surface plasmons, whereas chemical enhancement results from the electronic resonance via a charge transfer intermediate state.

The enhancement in the Raman peak arises from enhanced optical fields because of the presence of metallic structures, which is characterized by an enhancement factor (Kneipp et al. 1999)

$$A_{\rm em}^{\rm SERS}(r_{\rm m}, \nu) = \frac{\{E(r_{\rm m}, \nu)\}^4}{\{E_{\rm in}(r_{\rm m}, \nu)\}^4}$$

$E(r_m, \nu)$ is the total electric field at r_m distance from the molecule and $E(r_m, \nu)$ is the field of incident light at frequency ν (Kneipp et al.1999). In a study, a large cross-sectional area of the order of 10^{-17}–10^{-16} cm^2/molecule was reported using the SERS experiment (Kneipp et al. 1997). This provided the hope to see the Raman fingerprint of a single molecule. Kneipp et al. (1997) recorded single-molecule Raman scattering of single crystal violet (CV) molecule in an aqueous colloidal silver solution.

Additionally, it was observed that the maximum enhancement could be obtained at the tip of the spheroidal metal structures and increased with the aspect ratio. This effect is called the lightning rod effect, in which the electric field is largest near the sharpest surface of the rod. The enhancement in the intensity depends on the size and shape of the substrate structures, which can be metal clusters or parts of a rough surface. A maximum enhancement can be obtained for particle diameters between 10 and 150 nm. However, these mechanisms are not yet fully understood as the inhomogeneous enhancement effects as well as the structural changes of the molecules make it complex to further make a quantitative explanation of the SERS results.

Campion and Kambhampati presented a nice review on SERS, which reviews the basic experimental results and essential features of SERS mechanisms (Campion and Kambhampati 1998). The fundamental aspects of the phenomenon are still to be understood. The ability to isolate and characterize single particles and small clusters has provided a powerful new approach to study the electromagnetic enhancement. Loo et al. made a SERS study of imidazole molecules adsorbed on Cu and Ag electrodes and on Cu, Ag, and Au colloids at a pH ranging from 5.6 to 11.0. In this study, it is reported that the imidazole molecule is coordinated through its unshared electron pair on its pyridine nitrogen to the metal surface (Loo et al. 2006). Materny (Sackmann and Materny 2006) presented concentration-dependent SERS spectra of adenine in silver solution. A reproducible concentration dependence of the SERS intensity, in the concentration range between 1×10^{-7} and 1×10^{-6} molar, has been discussed. As another variant, Rasmussen and Deckert (Rasmussen and Deckert 2006) have studied tip-enhanced Raman scattering (TERS) spectra of pyrimidine bases of DNA and compared them with the results obtained from standard SERS. Surprisingly, the TERS spectra showed much more similarity to the Raman spectra of bulk samples. Pagliai et al. (Pagliai et al. 2006) have studied the Raman spectra of pyridine adsorbed onto silver colloids, and the results are compared with the theoretical results obtained using DFT and molar perturbation theory (MP2) calculations.

Kneipp et al. (Kneipp et al. 1997) have discussed the applications of SERS for a very sensitive chemical analysis. Xu et al. examined single-molecule sensitivity of SERS on the basis of electromagnetic theory by considering the shape and size of the Ag or Au nanostructures and found an enhancement of the order of 10^{11} (Xu et al. 2000). This enhancement factor was obtained at a special condition, namely, at interstitial sites between particles. Enhancement factors greater than 10^{14}–10^{15} are likely to involve additional enhancement mechanisms such as chemisorption-induced resonance Raman effects. In Xu et al. (1999), molecular vibrations in a single hemoglobin molecule have been recorded using SERS measurements by attaching the hemoglobin molecule with silver nanostructures. The theoretical calculations showed that the SERS spectrum of a single hemoglobin molecule was from the molecule lying in between the two silver nanoparticles. McFarland and Duyne described the localized surface plasmon resonance response of individual Ag nanostructures using dark-field optical microscopy (McFarland and Duyne 2003). It was reported that the adsorption of a certain amount of 1-hexadecanethiol molecules on single Ag nanoparticles causes to shift a localized surface plasmon resonance by 40.7 nm. SERS can be applied to detect even a smaller amount (femtomolar concentration) of hazardous materials in a highly controlled environment (Das and Agrawal 2011). SERS has also been employed to detect the presence of other contaminants such as TNT (2,4,6-trinitrotoluene), fullerenes, and uranium in environmental samples (Halvorson and Vivesland 2010). Using a new SERS substrate, based on (aminomethyl)phosphonic acid (APA)-modified gold nanostructures, a very small amount (8×10^{-7} M) of uranium can also be detected in groundwater (Halvorson and Vivesland 2010). The same technique has been employed for the detection of small traces (up to 10–100 g/l) of perchlorate (ClO_4^-) in ground and surface water (Gu et al. 2004). Optical detection methods in combination with Raman spectroscopy, fluorescence spectroscopy, FTIR spectroscopy, and digital imaging have been reported to detect the trace levels of harmful biological agents, e.g. *Bacillus anthracis*, *Yersinia pestis*, and *Francisellatularensis tularensis* (Kalasinsky et al. 2007).

4.3.3 Resonance Raman Spectroscopy (RRS)

Resonance Raman spectroscopy (RRS) is a special class of scattering in which the energy of incident light (or exciting laser beam) matches approximately with the energy required for an electronic transition in the analyte molecules. This technique is also used to enhance the intensity of normal Raman scattering. This technique is used in the analysis of various chromophoric biological samples, e.g. enzymes, various parts of biomolecules, and protective pigments of photosynthetic organisms. The RRS spectra are used to perform a deeper analysis of different experimental approaches to quantify the layer thickness (Zhang et al. 2017). Ultraviolet resonance Raman (UVRR) spectroscopy was used to study molecular recognition of peptides or proteins with guanidiniocarbonyl-pyrroles (GCP). In this study, 244 nm laser excitation is employed to characterize GCP as well as guanidiniocarbonyl indole (GCI) (Holtum et al. 2020). In another study, the UVRR spectroscopy was utilized to identify optimum conditions for other supramolecular ligands for peptide/protein recognition (Kumar et al. 2021). In this study, the RRS is used to record the non-zone center phonon modes in solids, in which weak Raman signals appear in a wide frequency region. By changing the excitation laser energy, phonon dispersion relations of a single-wall carbon nanotubes (SWCNTs) were derived (Saito et al. 2003). RRS is employed to record the Raman spectra of rhodamine 6G adsorbed on graphene with a relatively suppressed background. The fluorescence is reduced to nearly 10^3 times as compared to pure rhodamine 6G. This study infers that graphene can be used as a substrate to suppress fluorescence (Xie et al. 2009). The resonant Raman spectroscopy can be employed to monitor very small traces (below 200 ppb) of hazardous materials (NO_2^-, NO_3^-) in wastewater (Lanoul et al. 2002).

4.3.4 Raman Spectroscopy of Semiconducting, Superconducting, and Perovskite Materials

Raman spectroscopy is known as an effective and non-destructive technique for structural characterization of materials (Wood et al. 2017). The Raman spectrum of semiconductors is sensitive to the size, shape, and chemical structure of the nanostructured materials (Wood et al. 2017; Nanda et al. 1999) and is used to investigate the shape, size, crystallinity, and chemical composition of semiconducting materials (Wood et al. 2017; Sims et al. 2007). Recently, Mou et al. reported that the cadmium sulphide (CdS) nanostructures have two Raman bands (1OL and 2OL) (Mou et al. 2021). The peak parameters, Raman shift, line width, and relative intensity of these peaks depend on the size and shape of the nanostructures (Yilmaz et al. 2017). Because of the polarization dependence of Raman scattering, it can also be used to probe the orientation of nanowires and nanocones such as semiconducting materials (Wood et al. 2017). This technique has been applied by several groups to study the effect of different deposition conditions on orientation of the organic molecules (Seto and Furukawa 2012; Mino et al. 2012). Raman spectroscopy has been used to study stress, crystal lattice disorder, and phase separation of supersaturated solid solutions and homogeneity of semiconducting materials (Nesheva 2005). Some materials that show exceptional electrical conductivity (resistivity approaches to zero) at low temperature are called superconductors. In a superconducting state, electrical resistance vanishes and magnetic flux expelled from the material (Fan and Malozovsky 2013). Because of its very high conductivity, superconductors find many applications. The superconductivity was explained on the basis of Bardeen–Cooper–Schrieffer (BCS) theory, but this theory was unable to explain the phenomenon of superconductivity at higher temperature (Bardeen et al. 1957). Raman spectroscopy has been used to study the structural change arising due to doping in the superconducting materials at the molecular level. In superconducting materials, the strong electron phonon interaction gives rise to E_{2g} and A_{1g} Raman modes, which are characteristics of the superconducting state (Parisiades et al. 2009; Mu et al. 2021; Monteiro et al. 2017).

The perovskites are a group of materials with the crystal structure similar to the calcium titanium oxide mineral. The general formula for perovskites is ABX_3. Hybrid organic–inorganic type perovskites (HOIPs) with an organic cation ($CH_3NH_3^+$) at the A site, inorganic cations (Pb^{2+} and Sn^{2+}) at the B site, and halogens (I^-, Cl^-, and Br^-) at the X site are important for their applications in the field of solar cells (SCs). In HOIPs, the organic cations are inside the tetrahedron formed by the inorganic cation and the halogens, and the only interaction between them is hydrogen bonding between the hydrogen of the organic cation and the halogen of the inorganic network (Ojha et al. 2020a). This hydrogen bonding is crucial for the stability of HOIP structure and is affected by orientation of organic cations inside the inorganic tetrahedron. Along with the orientation, the vibration of inorganic cations inside the tetrahedron also affects the electric and optical properties of HOIPs. Raman and FTIR spectroscopy are useful characterization techniques for analyzing the effect of orientation of organic cations on the electrical and optical properties and stability of HOIPs.

Raman spectroscopy of perovskites is sensitive to orientation and vibrations of organic cations in octahedron cavity and hydrogen bonding between the organic cation and the inorganic network. The Raman spectra of HOIPs contain different peaks in the spectral region starting from 50 to 300 cm^{-1}. Among these peaks, the peaks in spectral region 100–250 cm^{-1} are assigned to the vibrational and torsional vibrational modes of methylamine (MA). It has been reported that the peaks at ~120, ~150, and ~250 cm^{-1} are sensitive to the interaction of the organic cation with the inorganic cage (mainly with iodine through hydrogen bonding) and represent an important marker for orientation disorder in the material, and a shift in these Raman peaks represents a change in the orientation of organic cations relative to the inorganic network. The peak at ~150 cm^{-1} is associated with hydrogen bonding between the organic cation with the inorganic cage, and an increase in the intensity of this peak indicates strengthening of the hydrogen bond between the hydrogen of the MA cation and the iodine of the inorganic cage (Ojha et al. 2020a). Raman spectroscopy has been employed by many researchers to investigate the effect of doping and synthesis techniques on the strength of hydrogen bonding between organic cations and inorganic network. On the basis of Raman spectra of undoped and doped samples, Ojha et al. reported that partial doping of $CoCl_2$ in methylammonium lead iodide ($MAPbI_3$) increases the strength of hydrogen bonding, leading to enhanced stability of $MAPbI_3$ (Ojha et al. 2020b). Likewise, Park also used Raman spectroscopy and photoluminescence to study the effect of stronger hydrogen bonding on the optical properties of $MAPbI_3$ (Park 2015).

4.4 Applications of FTIR Spectroscopy

4.4.1 FTIR Spectroscopy of Semiconductor, Superconductor, Hazardous, and Perovskite Materials

FTIR is most popular for studying the chemical bonding and other properties of semiconducting materials. FTIR spectroscopy has been reported to be used for examining the shift in different peak positions and their relative intensity in semiconducting polymers (Bhargava et al. 2003; Ludvigsson et al. 2000). Along with these applications, FTIR combined with a focal plane array (FPA) detector and a step-scan interferometer (Lewis et al. 1995) can also be used for imaging of samples. FTIR imaging has been employed for studying the defects in the materials. The crystallinity and local orientation of the crystals in semiconductors (Bhargava et al. 2003) are also determined using FTIR measurements. The effect of atomic vibrations and deformations in bond (such as bond stretching or tilting) on carrier density and Fermi level can also be described by measuring their FTIR spectra (Khan et al. 2004; Yusuf et al. 2011; Sedky and Salah 2020). FTIR can also be employed for the detection of harmful biological agents such as bacteria in groundwater. Based on the observation that different bacteria display different IR signatures, FTIR can be used for characterization of different types of bacteria (Schmitt and Flemming 1998).

FTIR spectroscopy is used to study the different bonding in the perovskite and interaction of perovskite with other chemicals. Hou et al. used FTIR to study the interaction between perovskite and dopant terephthalic acid (TPA), which leads to the formation of stable perovskite SCs (Hou et al. 2017). Likewise, the FTIR spectra confirm the addition of carbon quantum dots over the $MAPbI_3$ thin film, which leads to the fabrication of stable SCs (Yavari et al. 2018).

4.5 Raman and FTIR Spectroscopy of Upconverting Nanoparticles

Upconversion is a type of anti-Stokes emission in which the emitted photon has more energy compared to the incident photon. Unlike anti-Stokes emission wherein the energy difference between emission and incident photons is of the order of few kT (k is the Boltzmann constant), the energy of unconverted photon is increased by a factor of 10–100 times kT (Auzel 2004). This process is done with the help of nanoparticles, called upconverting nanoparticles (UCNPs). The nanoparticles of lanthanide-doped inorganic materials convert lower energy NIR radiation into visible radiation (Wang et al. 2010). Because of ladder-like arrangement of energy levels in lanthanides (Wang et al. 2010) UCNPs can be super-excited by a process called excited-state absorption in which an excited electron is raised to even higher energy level by absorbing a second photon (Auzel 2004). During de-excitation, the super-excited atom emits photon with energy more than the energy of photons absorbed during excitation (Figure 4.5) (Wang et al. 2010).

Photon upconversion was first reported in 1966 by François Auzel in ytterbium–erbium and ytterbium–thulium systems (Auzel 1966). With the development of IR light amplification by stimulated emission of radiation (LASER) and advancement in the fiber optic, the research in the field of UCNPs is also increased appreciably (Auzel 2004). Because of its unique luminescent properties (high penetration depth into tissues, low background signals, sharp emission bands, and high resistance to photobleaching), UCNPs find applications as a fluorescent probe in imaging and biodetection assays (Wang et al. 2011). Other applications of UCNPs include fluorescent microscopy, security labeling optogenetics, and volumetric display (Wen et al. 2018). However, the decrease in luminescence behavior due to thermal and concentration quenching restricts the development of bright UCNPs with a large number of dopants. Different efforts have been made to overcome quenching in nanoparticles

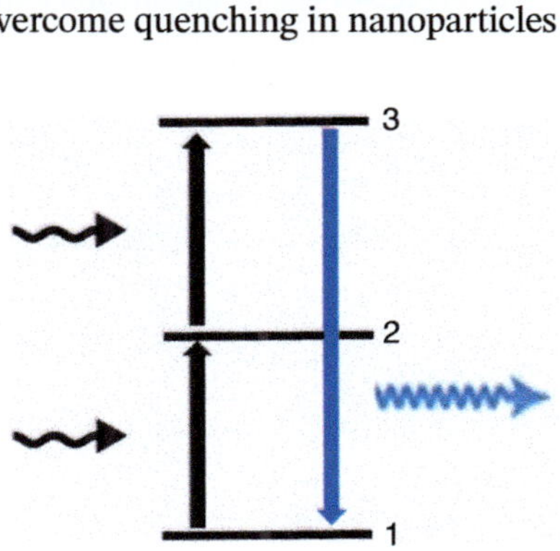

Figure 4.5 A schematic energy level diagram of upconversion nanoparticles. Source: Modified from Wang et al. (2010).

by adding core–shell structure and improving doping uniformity in the host nanomaterials (Wen et al. 2018). Raman spectroscopy and FTIR spectroscopy are used to confirm the formation of UCNPs and characterize the phase of UCNPs along with different bonding and chemical groups present on the surface of UCNPs (Afanaseva et al. 2021; Lin et al. 2021; Zhou et al. 2018; Rong et al. 2021). Rong et al. and Zhou et al. used FTIR for confirming the coating of oleic acid on UCNPs (Rong et al. 2021). Raman spectroscopy is employed to study the phonon bands in UCNPs (Zhou et al. 2018). On the basis of FTIR and Raman spectroscopy measurements, Zhou et al. found a kind of heat-favorable phonons existing at the surface of lanthanide-doped UCNPs to combat thermal quenching (Zhou et al. 2018). These phonons favor energy transfer from a sensitizer to an activator to enhance the upconversion process and minimize the luminescence loss due to thermal quenching (Zhou et al. 2018). Additionally, upconversion fluorescence coupled with SERS is used for the detection of harmful and toxic materials. Lin et al. developed an upconversion luminescence and SERS dual-mode aptasensor for the detection of small concentration of toxic mycotoxin ochratoxin A (OTA) (Lin et al. 2021). Similarly, by coupling SERS and upconversion fluorescence, Rong at al. developed a nanosensor for detecting phthalate acid esters (PAEs) in food samples (Rong et al. 2021).

References

Afanaseva, E.V., Vaishlia, E.I., Lähderanta, E., and Kolesnikov, I.E. (2021). Synthesis and study of upconversion $Lu_2(WO_4)_3$:Yb^{3+},Tm^{3+} nanoparticles synthesized by modified Pechini method. *Optical Materials* 117: 111179. https://doi.org/10.1016/j.optmat.2021.111179.

Asthana, B.P., Deckert, V., Shukla, M.K., and Kiefer, W. (2000). Isotopic dilution study of self association in ($CH_3CN + CD_3CN$) mixture by scanning multichannel Raman difference technique and ab-initio calculations. *Chemical Physics Letters* 326: 123–128. https://doi.org/10.1016/S0009-2614(00)00728-4.

Auzel, F. (1966). Compteur quantique par transfert d'energie entre deux ions de terres rares dans un tungstate mixte et dans un verre. *Comptes Rendus de l'Academie des Sciences* 262: 1016.

Auzel, F. (2004). Upconversion and anti-Stock process with f and d ions in solids. *Chemical Reviews* 104: 139–173. https://doi.org/10.1021/cr020357g.

Bardeen, J., Cooper, L.N., and Schrieffer, J.R. (1957). Theory of superconductivity. *Physics Review* 108: 1175. https://doi.org/10.1103/PhysRev.108.1175.

Bhargava, R., Wang, S.Q., and Koenig, J.L. (2003). FTIR microscopy of polymeric systems. *Advances in Polymer Science* 163: 137–191. https://doi.org/10.1007/b11052.

Campion, A. and Kambhampati, P. (1998). Surface-enhanced Raman scattering. *Chemical Society Reviews* 27, 241 and reference cited therein. https://doi.org/10.1039/A827241Z.

Cotton, F.A. (1971). *Chemical Applications of Group Theory*, 3e. New York: Wiley ISBN: 978-0-471-51094-9.

Das, R.S. and Agrawal, Y.K. (2011). Raman spectroscopy: recent advancements, techniques and applications. *Vibrational Spectroscopy* 57: 163–176. https://doi.org/10.1016/j.vibspec.2011.08.003.

Fan, J.D. and Malozovsky, Y.M. (2013). The origin of superconductivity. *Physica C: Superconductivity* 493: 12–14. https://doi.org/10.1016/j.physc.2013.03.008.

Fleischmann, M., Hendra, P.J., and Mcquillan, A.J. (1974). Raman spectra of pyridine adsorbed at a silver electrode. *Chemical Physics Letters* 26: 163–166. https://doi.org/10.1016/0009-2614(74)85388-1.

Gu, B., Tio, J., Wang, W. et al. (2004). Raman spectroscopic detection for perchlorate at low concentration. *Spectroscopy* 58: 741–744. https://doi.org/10.1366/000370204872890.

Halvorson, R.A. and Vivesland, P.J. (2010). Surface-enhanced Raman spectroscopy (SERS) for environmental analyses. *Journal of Environmental Science and Technology* 44: 7749–7755. https://doi.org/10.1021/es101228z.

Holtum, T., Kumar, V., Sebena, D. et al. (2020). UV resonance Raman spectroscopy of the supramolecular ligand guanidiniocarbonyl indole (GCI) with 244 nm laser excitation. *Beilstein Journal of Organic Chemistry* 16: 2911–2919. https://doi.org/10.3762/bjoc.16.240.

Hou, X., Huang, S., Ou-Yang, W. et al. (2017). Constructing efficient and stable perovskite solar cells via interconnecting perovskite grains. *ACS Applied Materials & Interfaces* 9: 40. https://doi.org/10.1021/acsami.7b08488.

Jeanmaire, D.L. and van Duyne, R.P. (1977). Heterocyclic, aromatic and aliphatic amines adsorbed on the anodized silver electrode. *Journal of Electroanalytical Chemistry* 84: 1–20. https://doi.org/10.1016/S0022-0728(77)80224-6.

Kalasinsky, K.S., Hadfield, T., Shea, A.A. et al. (2007). Raman chemical imaging spectroscopy reagent less detection and identification of pathogens: signature development and evaluation. *Analytical Chemistry* 79: 2658–2673. https://doi.org/10.1021/ac0700575.

Khan, N. A. Mumtaz A. Sabeeh K. et al. (2004). The study of phonon modes of $Cu_{1-x}Tl_xBa_2Ca_3Cu_4O_{12-y}$ superconductor thin films by FTIR absorption spectroscopy. *Physica C: Superconductivity*, 407, 103–114. https://doi.org/10.1016/j.physc.2004.04.028.

Kneipp, K., Wang, Y., Kneipp, H. et al. (1997). Single molecule detection using surface-enhanced Raman scattering (SERS). *Physical Review Letters* 78: 1667. https://doi.org/10.1103/PhysRevLett.78.1667.

Kneipp, K., Kniepp, H., Itzkan, I. et al. (1999). Ultrasensitive chemical analysis by Raman spectroscopy. *Chemical Reviews* 99: 2957–2975. and reference cited there in. https://doi.org/10.1021/cr980133r.

Kumar, V., Holtum, T., Sebena, D. et al. (2021). Ultraviolet resonance Raman spectroscopy with a continuously tunable picosecond laser: application to the supramolecular ligand guanidiniocarbonyl pyrrole (GCP). *Spectrochimica Acta Part A: Molecular and Biomolecular Spectroscopy* 250: 119359. https://doi.org/10.1016/j.saa.2020.119359.

Lanoul, A., Coleman, T., and Asher, S.A. (2002). UV resonance Raman spectroscopic detection of nitrate and nitrite in wastewater treatment processes. *Analytical Chemistry* 74: 1458–1461. https://doi.org/10.1021/ac010863q.

Lewis, E.N., Treado, P.J., Reeder, R.C. et al. (1995). Fourier transform spectroscopic imaging using an infrared focal-plane array detector. *Analytical Chemistry* 67: 3377–3381. https://doi.org/10.1021/ac00115a003.

Lin, X., Li, C., He, C. et al. (2021). Upconversion nanoparticles assembled with gold nanourchins as luminescence and surface-enhanced Raman scattering dual-mode aptasensors for detection of ochratoxin A. *ACS Applied Nano Materials* 4 (8): 8231–8240. https://doi.org/10.1021/acsanm.1c01421.

Loo, B.H., Tse, Y., Parsons, K. et al. (2006). Surface-enhanced Raman spectroscopy of imidazole absorbed on electrode and colloidal surface of Cu, Ag, and Au. *Journal of Raman Spectroscopy* 37: 299–304. https://doi.org/10.1012/jrs.1478.

Ludvigsson, M., Lindgren, J., and Tegenfeldt, J. (2000). Crystallinity of cast nafion. *Journal of the Electrochemical Society* 147: 1303–1305. https://doi.org/10.1149/1.1393354.

McFarland, A.D. and Duyne, R.P.V. (2003). Single silver nanoparticles as real-time optical sensors with zeptomole sensitivity. *Nano Letters* 3: 1057. https://doi.org/10.1021/nl034372s.

Mino, T., Saito, Y., Yoshida, H. et al. (2012). Molecular orientation analysis of organic thin films by z-polarization Raman microscope. *Journal of Raman Spectroscopy* 43: 2029–2034. https://doi.org/10.1002/jrs.4118.

Mishra, S., Singh, R.K., and Ojha, A.K. (2009). Investigation on bonding interaction of benzonitrile with silver nano particles probed by surface enhanced Raman scattering and quantum chemical calculations. *Chemical Physics* 355: 14–20. https://doi.org/10.1016/j.chemphys..2008.10.006.

Monteiro, J.F.H.L., Jurelo, A.R., and Siqueira, E.C. (2017). Raman spectroscopy of the superconductor $CuCrO_2$ delafossite oxide. *Solid State Communications* 252: 64–67. https://doi.org/10.1016/j.ssc.2017.01.014.

Mou, Q., Guo, Z., Chai, Y. et al. (2021). Visible-light assisted photoreduction of CO_2 using CdS-decorated $Bi_{24}O_{31}Br_{10}$. *Materials Science in Semiconductor Processing* 134: 106011. https://doi.org/10.1016/j.mssp.2021.106011.

Mu, Q.G., Fan, F.R., Borrmann, H. et al. (2021). Pressure-induced superconductivity and modification of Fermi surface in type-II Weyl semimetal $NbIrTe_4$. *Quantum Materials* 6: 55. https://doi.org/10.1038/s41535-021-00357-y.

Nanda, K.K., Sarangi, S.N., Sahu, S.N. et al. (1999). Raman spectroscopy of CdS nanocrystalline semiconductors. *Physica B: Condensed Matter* 262: 31–39. https://doi.org/10.1016/S0921-4526(98)00474-8.

Nesheva, D. (2005). Raman scattering from semiconductor nanoparticles and superlattices. *Journal of Optoelectronics and Advanced Materials* 7: 185–192.

Ojha, A.K. (2007). pH dependent SERS and solvation studies of tyrosine adsorbed on silver colloidal nano particles combined with DFT calculations. *Chemical Physics* 340: 69–78. https://doi.org/10.1016/j.chemphys..2007.07.042.

Ojha, A.K., Srivastava, S.K., Koster, J. et al. (2004). Investigation of hydrogen bonding and self-association in neat $HCONH_2$ and the binary mixture ($HCONH_2 + CH_3OH$) by concentration dependent Raman study and ab initio calculations. *Journal of Molecular Structure* 689: 127–135. https://doi.org/10.1016/j.molstruc.2003.10.030.

Ojha, A.K., Srivastava, S.K., Pieca, J. et al. (2005). Concentration dependent wave number shift and line width changes of some prominent vibrational mode of C_4H_8O in binary system ($C_4H_8O + H_2O$) by polarized Raman study and ab initio calculations. *Journal of Molecular Structure* 735, 736: 127–135. https://doi.org/10.1016/j.molstruc.2004.11.010.

Ojha, A.K., Singha, A., Dasgupta, S. et al. (2006). pH dependent surface enhanced Raman study of Phe + Ag complex and DFT calculation for spectral analysis. *Chemical Physics Letters* 431: 121–126. https://doi.org/10.1016/j.cplett.2006.09.038.

Ojha, S.K., Singh, A., and Ojha, A.K. (2020a). Modifications in structural morphology of $CH_3NH_3PbI_3$ perovskite using nitrilotriacetic acid and glycine as habit modifiers. *Materials Chemistry and Physics* 240: 122149. https://doi.org/10.1016/j.matchemphys.2019.122149.

Ojha, S.K., Kumar, A., Singh, A., and Ojha, A.K. (2020b). Improved environmental stability of cobalt incorporated methylammonium lead iodide perovskite for resistive switching applications. *Chemical Physics* 538: 100900. https://doi.org/10.1016/j.chemphys.2020.110900.

Pagliai, M., Bellucci, L., Muniz-Miranda, M. et al. (2006). A combined Raman, DFT and MD study of the solvation dynamics and the adsorption process of pyridine in silver hydrosols. *Physical Chemistry Chemical Physics* 8: 171–178. https://doi.org/10.1039/b509976e.

Parisiades, P., Liarokapis, E., Zhigadlo, N.D. et al. (2009). Raman investigation of C-, Li- and Mn-doped MgB_2. *Journal of Superconductivity and Novel Magnetism* 22: 169. http://doi.org/10.1007/s10948-008-0402-4.

Park, N.G. (2015). Perovskite solar cells: an emerging photovoltaic technology. *Materials Today* 18: 65–72. https://doi.org/10.1016/j.mattod.2014.07.007.

Rakic, V., Dondur, V., and Hercigonja, R. (2003). FTIR study of carbon monoxide adsorption on ion-exchanged X, Y and mordenite type zeolites. *Journal of the Serbian Chemical Society* 68 (4, 5): 409–416. https://doi.org/10.1.1.430.2368.

Raman, C.V. and Krishnan, K.S. (1928). A new class of spectra due to secondary radiation. *Nature* 121: 501. https://www.nature.com/articles/121501c0.

Rasmussen, A. and Deckert, V. (2006). Surface- and tip-enhanced Raman scattering of DNA components. *Journal of Raman Spectroscopy* 37: 311–317. https://doi.org/10.1012/jrs.1480.

Reichenbacher, M. and Popp, J. (2012). Challenges in molecular structure determination, Chapter 2. In: *Vibrational Spectroscopy*. https://doi.org/10.1007/978-3-642-24390-5_2 (ed. M. Reichenbächer and J. Popp), 21–482. Berlin, Heidelberg: Springer.

Rong, Y., Ali, S., Ouyang, Q. et al. (2021). Development of a bimodal sensor based on upconversion nanoparticles and surface-enhanced Raman for the sensitive determination of dibutyl phthalate in food. *Journal of Food Composition and Analysis* 100: 103929. https://doi.org/10.1016/j.jfca.2021.103929.

Sackmann, M. and Materny, A. (2006). Surface-enhanced Raman scattering (SERS) – a quantitative analytical tool? *Journal of Raman Spectroscopy* 37: 305–310. https://doi.org/10.1012/jrs.1443.

Saito, R., Grüneis, A., Samsonidze, G.G. et al. (2003). Double resonance Raman spectroscopy of single-wall carbon nanotubes. *New Journal of Physics* 5: 157. https://doi.org/.1088/1367-2630/5/1/157.

Schmitt, J. and Flemming, H.C. (1998). FTIR- spectroscopy in microbial and material analysis. *International Biodeterioration & Biodegradation* 41: 1–11. https://doi.org/10.1016/S0964-8305(98)80002-4.

Sedky, A. and Salah, A. (2020). Excess conductivity, diamagnetic transition and FTIR spectra of Ca substituted by La in (Bi,Pb):2212 superconductor. *Journal of Low Temperature Physics* 201: 294–310. https://doi.org/. doi: 10.1007/s10909-020-02492-5.

Seto, K. and Furukawa, Y. (2012). Study on solid structure of pentacene thin films using Raman imaging. *Journal of Raman Spectroscopy* 43: 2015–2019. https://doi.org/10.1002/jrs.4090.

Sims, M., Tuladhar, S.M., Nelson, J. et al. (2007). Correlation between microstructure and charge transport in poly(2,5-dimethoxy-pphenylenevinylene) thin films. *Physical Review B* 76: 195206. https://doi.org/10.1103/PhysRevB.76.195206.

Singh, D., Srivastava, S.K., Ojha, A.K. et al. (2007). DFT study of hydrogen bond bridging mode of pyridine and diazenes in water environment. *Journal of Molecular Structure* 819: 88–94. https://doi.org/10.1016/j.theochem.2007.05.030.

Singh, D.K., Srivastava, S.K., Ojha, A.K., and Asthana, B.P. (2008a). Vibrational study of thiophene and its solvation in two polar solvents, DMSO and methanol by Raman spectroscopy combined with ab initio and DFT calculation. *Journal of Molecular Structure* 892: 384–391. https://doi.org/10.1016/j.molstruc.2008.06.008.

Singh, D.K., Srivastava, S.K., Ojha, A.K., and Asthana, B.P. (2008b). pH dependent Raman study of pyrrole and its vibrational analysis using DFT calculation. *Spectrochimica Acta, Part A: Molecular and Biomolecular Spectroscopy* 71: 823–829. https://doi.org/10.1016/j.saa.2008.02.032.

Singh, K., Ohlan, A., and Dhawan, S.K. (2012). Polymer-graphene nanocomposites: preparation, characterization, properties, and applications, Chapter 3. In: *Nanocomposites – New Trends and Developments*. https://doi.org/10.5772/50408 (ed. F. Ebrahimi), 37–71. IntechOpen.

Son, Y.H., Jung, W.C., and Jeong, J.I. (2001). FTIR characteristics of hydrogenated amorphous carbon films prepared by ECR-PECVD. *Journal of the Korean Physical Society* 39 (4): 713–717.

Srivastava, S.K., Ojha, A.K., Koster, J. et al. (2003). Isotopic dilution, self-association, and Raman non-coincidence in the binary system $(CH_3)_2C{=}O + (CD_3)_2C{=}O$ reinvestigated by polarized Raman measurement and ab initio calculations. *Journal of Molecular Structure* 661, 662: 11–21. https://doi.org/10.1016/j.molstruc.2003.07.004.

Srivastava, S.K., Ojha, A.K., Kiefer, W., and Asthana, B.P. (2005). Investigation of ν(N–H) and ν(C–N) stretching mode of propylamine ($C_3H_7NH_2$) in a binary system $C_3H_7NH_2 + CH_3OH$ via concentration dependent Raman study and ab initio calculations. *Spectrochimica Acta, Part A: Molecular and Biomolecular Spectroscopy* 61: 2832–2839. https://doi.org/10.1016/j.saa.2004.10.031.

Srivastava, S.K., Ojha, A.K., Raghuvansh, P. et al. (2006). Influence of self-association and inter-molecular hydrogen bonding on the ν(C≡N) stretching mode of $CH_3C{\equiv}N$ and $C_2H_5C{\equiv}N$ in binary mixtures with CH_3OH – A comparative study via concentration-dependent polarized Raman study and *ab initio* calculation. *Journal of Raman Spectroscopy* 37: 1287–1295. https://doi.org/10.1002/jrs.1550.

Stuart, B.H. (2004). *Infrared Spectroscopy: Fundamentals and Applications*. New York: Wiley ISBN: 978-0-470-85428-0.

Vaghri, E., Khalaj, Z., and Ghoranneviss, M. (2012). Preparation and characterization of diamond-like carbon films on various substrates by PECVD system. *Studia Universitatis Babeş-Bolyai. Chemia* 3: 143–150. https://doi.org/10.1117/12.2041194.

Vyas, N. and Ojha, A.K. (2010). Interaction of alanine with small water clusters; Ala-$(H_2O)_n$ ($n = 1$, 2 and 3): a density function study. *Journal of Molecular Structure* 940: 95–102. https://doi.org/10.1016/j.theochem.2009.10.015.

Vyas, N. and Ojha, A.K. (2011). Effect of regular hydration on gas phase structural stability of [zwitterionic alanine+M^+] ($M^+ = Li^+$, Na^+, K^+,) complexes: a quantum chemical study. *Chemical Physics* 382: 5–14. https://doi.org/10.1016chemphys.2010.11.013.

Wang, F., Banerjee, D., Liu, Y. et al. (2010). Upconversion nanoparticles in biological labeling, imaging, and therapy. *Analyst* 135: 1839–1854. https://doi.org/10.1039/C0AN00144A.

Wang, M., Abbineni, G., Clevenger, A. et al. (2011). Upconversion nanoparticles: synthesis, surface modification and biological applications. *Nanomedicine* 7: 710–729. https://doi.org/10.1016/j.nano.2011.02.013.

Wen, S., Zhou, J., Zheng, K. et al. (2018). Advances in highly doped upconversion nanoparticles. *Nature Communications* 9: 2415. https://doi.org/10.1038/s41467-018-04813-5.

Wood, S., Hollis, J.R., and Kin, J.S. (2017). Raman spectroscopy as an advanced structural nanoprobe for conjugated molecular semiconductors. *Journal of Physics D: Applied Physics* 50: 073001. https://doi.org/10.1088/1361-6463/50/7/073001.

Xie, L., Ling, X., Fang, Y. et al. (2009). Graphene as a substrate to suppress fluorescence in resonance Raman spectroscopy. *Journal of the American Chemical Society* 131 (29): 9890–9891. https://doi.org/10.1021/ja9037593.

Xu, H., Bjerneld, E.J., Kall, M., and Borjesson, L. (1999). Spectroscopy of single hemoglobin molecules by surface enhanced Raman scattering. *Physical Review E* 83: 4357. https://doi.org/10.1103/physrevlett.83.4357.

Xu, H., Aizpurai, J., Kall, M., and Apell, P. (2000). Electromagnetic contributions to single-molecule sensitivity in surface-enhanced Raman scattering. *Physical Review E* 62: 4318. https://doi.org/10.1103/PhysRevE.62.4318.

Yavari, M., Mazloum-Ardakani, M., Gholipour, S. et al. (2018). Carbon nanoparticles in high-performance perovskite solar cells. *Advanced Energy Materials* 8 (12): 170719. https://doi.org/10.1002/aenm.201702719.

Yilmaz, M., Babur, E., Ozdemir, M. et al. (2017). Nanostructured organic semiconductor films for molecular detection with surface-enhanced Raman spectroscopy. *Nature Materials* 16: 918. https://doi.org/10.1038/NMAT4957.

Yusuf, A.A., Yahya, A.K., Khan, N.A. et al. (2011). Effect of Ge^{4+} and Mg^{2+} doping on superconductivity, fluctuation induced conductivity and interplaner coupling of $TlSr_2CaCu_2O_{7-\delta}$ superconductor. *Physica C: Superconductivity* 471: 363–372. https://doi.org/10.1016/j.physc.2011.03.007.

Zhang, C., Griciuc, A., Hudry, E. et al. (2017). Cromolyn reduced levels of the Alzheimer's disease-associated amyloid β-protein by promoting microglial phagocytosis. *Scientific Reports* 7: 1144. https://doi.org/10.1038/s41598-018-19641-2.

Zhou, J., Wen, S., Liao, J. et al. (2018). Activation of the surface dark-layer to enhance upconversion in a thermal field. *Nature Photonics* 12: 154–158. https://doi.org/10.1038/s41566-018-0108-5.

5

Fundamental Aspects of Upconverting Nanoparticles (UCNPs) Based on Their Properties

Sushil K. Ranjan, Sasank Pattnaik, Vishab Kesarwani, and Vineet Kumar Rai

Indian Institute of Technology (Indian School of Mines), Department of Physics, Laser and Spectroscopy Laboratory, Dhanbad, Jharkhand 826004, India

5.1 Introduction

The particles of diameter in the range of 1–100 nm showing the property of photon upconversion are called upconverting nanoparticles (UCNPs). Generally, in UCNPs, the emission occurs in the visible or ultraviolet (UV) region followed by the absorption in the infrared region of electromagnetic spectrum. These materials are generally composed of dopants and hosts. They have wide applications in non-contact thermometry, security, upconverters, bioimaging, biosensing, nanomedicines, optical nanoheater, photovoltaics, light emitting diode (LEDs), etc. Lanthanides (or rare-earth elements) are mostly stable in their triply ionized state, and the emission from these ions mainly appears because of 4f–4f or 5d–4f transitions. The frequency upconversion (UC) and the thermal heating nature in the materials containing rare-earth (RE^{3+}) ions depend on the nature of hosts, dopants, suitable excitation wavelength, etc. There are various ways to identify the processes involved in the UC phenomena. Among these, the fluorescence decay time measurement is found to be one of the essential assets in the field of luminescence to deliberate various characteristics of UCNPs. In spite of being a useful non-linear optical phenomenon, upconversion luminescence (UCL) had a very limited number of applications during the early days of its discovery (Midwinter and Warner 1967; Auzel 1973; Piguet 2014). However, with the advent of nanotechnology, the research in the field of upconversion gets accelerated, as highly intense UCL has been reported in UCNPs (Heer et al. 2004; Yi et al. 2004). Their properties have a crucial impact on a certain number of fundamental aspects. Two such aspects are dynamics and quantum yield of UCNPs.

The luminescence decay kinetics has been widely studied to evaluate many intermolecular processes such as rotational diffusion, resonance energy transfer, dynamic quenching, etc., which cannot be obtained from steady-state fluorescence strategy (Omary and Patterson 2017; Díaz García and Badía-Laíño 2005, p. 129). Over the years, many researchers have proposed several methodologies for measuring and

Upconverting Nanoparticles: From Fundamentals to Applications, First Edition. Edited by Vineet K. Rai.

analyzing lifetime data whose main purpose is to get time-dependent luminescence that are helpful in many branches of multidisciplinary fields such as biochemistry, biomedical, industrial, security, etc. (Chubb and Wolford 2000; Rai 2007; Pandey and Rai 2012; Shanmugam et al. 2014; Ranjan et al. 2016; Nadort et al. 2016; Soni and Rai 2017; Mondal and Rai 2018; Loo et al. 2019; Hany et al. 2019; Pattnaik and Rai 2021). For a single-component system, the decay of the emitting state can be represented by a single exponential curve. The mathematical formulation for the above process is a first-order differential equation defined as

$$-\frac{dN}{dt} = k_f N \tag{5.1}$$

where N represents the number density of the emitting state and k_f is the fluorescence constant.

The decrease in the number of excited atoms/molecules can be denoted as

$$N(t) = -N_0 e^{(-k_f t)} \tag{5.2}$$

where N_0 defines the initial number of excited atoms/molecules before the start of the decay process. The time at which the initially excited atoms/molecules are reduced to N_0/e represents the fluorescence lifetime τ, and the radiative fluorescence lifetime is the average time for which the atom/molecule remains in the excited state before emission occurs. It is denoted by τ_r, which is the reciprocal of a radiative lifetime. Hence, it is represented as

$$\tau_r = \frac{1}{k_f} \tag{5.3}$$

Based on the deactivation of the excited state, the change in intensity corresponding to the emission line can be written as

$$I(t) = I_0 e^{(-k_f t)} \tag{5.4}$$

where $I(t)$ indicates the intensity of the emission line at time t and I_0 implies the initial luminescence signal. In reality, the depopulation involves both radiative and non-radiative processes, so the natural lifetime, τ_f, can be represented as

$$\tau_f = \frac{1}{k_f + k_{NR}} \tag{5.5}$$

Mathematically, τ_f can be understood as the time when the luminescence intensity reduces to $1/e$ times its initial intensity (Figure 5.1a). It should be noted that the luminescence lifetime is the average time for the atoms/molecules to be in the excited states, out of which few atoms/molecules may decay faster, and few molecules stay in the excited states for a longer period of time. The luminescence lifetime from the single exponential decay curve can be derived by talking natural logarithm on both sides of Eq. (5.4).

$$\ln I(t) = \ln I_0 - k_f t \tag{5.6}$$

The reciprocal of the negative slope returns the luminescence lifetime for the system (Figure 5.1b).

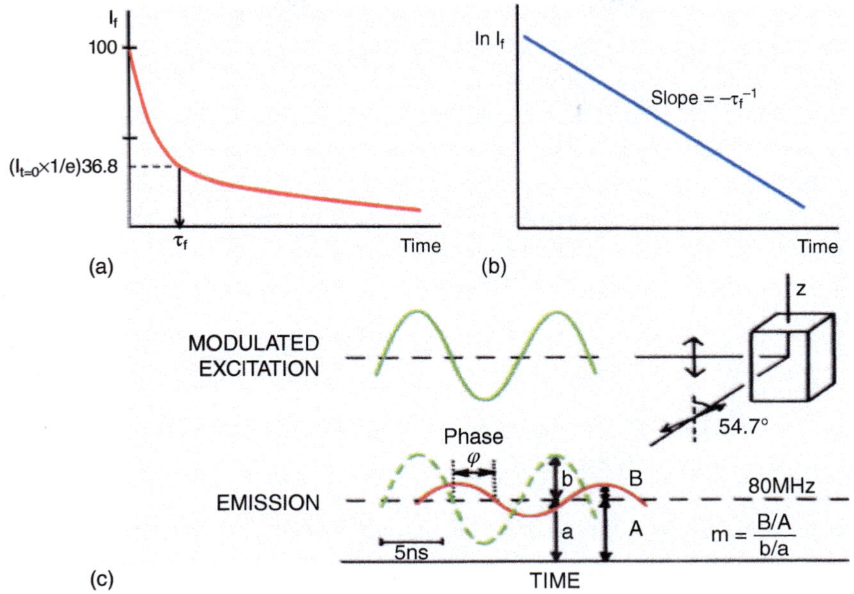

Figure 5.1 (a) Hypothetical decay of fluorescence from an electronically excited state. (b) Logarithmic slope method to determine the fluorescence lifetime. Source: Díaz García and Badía-Laíño (2005). Reprinted with the permission of Elsevier. (c) Phase modulation or frequency domain lifetime measurements. The ratios B/A and b/a represent the modulation of the emission and excitation, respectively. Source: Lakowicz (1999), p. 96. Reprinted with the permission of Springer, Boston, MA.

Although the procedure for finding the luminescence lifetime is easy, this is not always the case, as many researchers have observed biexponential and triexponential transient. For the multicomponent and non-exponential decay system, the intensity profile could be written as

$$I(t) = \sum_{i=1}^{n} A_i \exp\left(-\frac{t}{\tau_i}\right) \tag{5.7}$$

where A_i represents the pre-exponential factor and τ_i refers to the total lifetime for the system.

One of the major advantages of calculating a system's lifetime is distinguishing the luminescence phenomena, viz, fluorescence and phosphorescence. The lifetime corresponding to fluorescence phenomena (microseconds–picoseconds) falls shorter compared to phosphorescence (milliseconds-seconds) (Omary and Patterson 2017).

Basically, two methods have been excessively used to determine the time-resolved fluorescence, i.e. time domain and frequency domain methods (Lakowicz 1999, p. 96). The time domain method is otherwise called pulse fluorometry, in which the system is excited via a narrow pulse of light. The decay time could be calculated from the slope of the plot between log $I(t)$ versus t. However, in a frequency domain method, the sample is excited through intensity-modulated light, i.e. sine-wave modulation. The condition for the above method is such that when the sample is excited at a high frequency, the reciprocal frequency should be comparable to the reciprocal of the decay time. During this method, the delay in the emission from the fluorophore relative to the excitation can be expressed as a phase shift (φ) that can be used to calculate the decay time for the system (Figure 5.1c).

On the other hand, the quantum yield value of UCNPs also gets highly influenced by their properties as described later on in this chapter. In this chapter, after discussing the dynamics of upconversion on the basis of lifetime decay, a brief introduction of quantum yield of UCNPs has been presented followed by its measurement methods.

5.2 Elucidation of Dynamics of UCNPs on the Basis of Fluorescence Decay Times

5.2.1 General Understanding of Depopulation Processes and UC Decay

The four major upconversion mechanisms that involve the population and depopulation processes are ground-state absorption (GSA), excited-state absorption (ESA), energy transfer upconversion (ETU), and cross-relaxation (CR). Out of the four major mechanisms, the first two are quite simple as they follow the trend of populating the higher/intermediate energy states from the intermediate/ground energy state under suitable excitation condition (Figure 5.2a,b). However, for the ETU to occur, two ions should be present in their excited states. During the process, the first ion depopulates to the lower energy state and the second ion elevates to the higher energy state (Figure 5.2c). In contrary, the cross-relaxation process involves the excitation of the first ion to the intermediate state, whereas the second ion goes for the depopulation from the higher excited state, and thus, CR is often coined as the reverse mechanism of ETU (Figure 5.2d). The distinction between the ETU and CR process can be identified on the basis of the presence of ions in the highest excited energy state after the process. On the whole, if the ion is present at a higher energy state after the process, then it can be coined as ETU process; otherwise, it can be termed as CR process. Here, it is important to note that for both ETU and CR to occur, it is not necessary for ions to be identical (Suijver 2007).

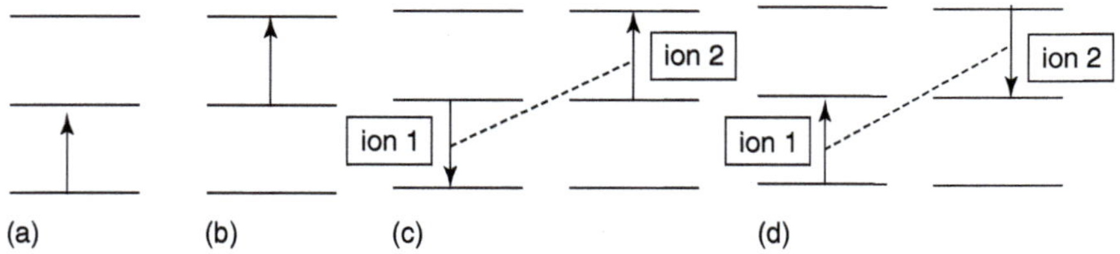

Figure 5.2 The four main energy transfer processes that are important for upconversion: (a) ground-state absorption (GSA), (b) excited-state absorption (ESA), (c) energy transfer upconversion (ETU), and (d) cross-relaxation. Source: Suijver (2007). Reprinted with the permission of Wiley.

All the UC processes can be summarized to obtain the desired radiative emission from the excited state, but the depopulation often encounters through radiative as well as undesired non-radiative pathways. The simplest way to understand the degree of non-radiative transitions between the levels in a RE^{3+}-doped material is by the non-radiative rate (k_{NR}), which can be expressed using energy gap law

$$k_{NR} \propto \exp\left[-\beta\frac{\Delta E}{\hbar\omega_{max}}\right] \tag{5.8}$$

where β implies a specific constant of the material, ΔE is the conjugative energy gap between the excited state and the subsequent intermediate state, and $\hbar\omega_{max}$ defines the highest vibrational energy of the system. This equation predicts that as the energy gap increases, the non-radiative rate decreases exponentially. Moreover, the radiative rates dominate only when the energy gap ΔE becomes five times greater than the maximum phonon energy available in the system. It is noteworthy to mention that other than f-electron system, the energy gap for the d-metals can be 10–20 times greater than the highest phonon energy available in the system.

The UC decay also narrates the energy diffusion mechanism in a crystal lattice. When the diffusion of excitation energy happens, the excitation energy will propagate through the sensitizer–activator pairs within the crystal lattice to reach a suitable activator. Generally, the sensitizer–activator pairs bearing shortest radial separation favors highest energy transfer rates, which are frequently observed. The sensitizer–activator ion pairs having different radial separation will end up with a non-exponential UCL decay curve because of different transfer rates between different ion pairs.

5.2.2 Differentiating the ESA and ETU Mechanism Based on the Decay Profile

The time evolution of the upconversion emission is often witnessed through two different kinds of UC mechanisms, namely, ETU and ESA. As the process of the population of the emitting state by the two mechanisms is different, the plots corresponding to the temporal evolution of UC emission should be unique. When ESA is the governing excitation process, the intermediate and excited states populate rapidly within the excitation time. If we consider a three-level system and represents the population corresponding to the state, then the luminescence decay follows the equation given by

$$n_3 \propto \exp\left(-{}^{t}/_{\tau_3}\right) \tag{5.9}$$

where τ_3 represents the decay time of the emitting state whose population is n_3.

Therefore, after completing the short excitation pulse, the depopulation of the excited state of the activator ions traces a single exponential curve (Figure 5.3a,b). In contrast to the ESA, the population of the activators in ETU goes through at least two subsequent energy transfer processes. Hence, the population takes a longer time than the excitation pulse in ESA. As shown in Figure 5.3c,d, after the excitation pulse is applied, the decay curve shows no initial intensity followed by a rise

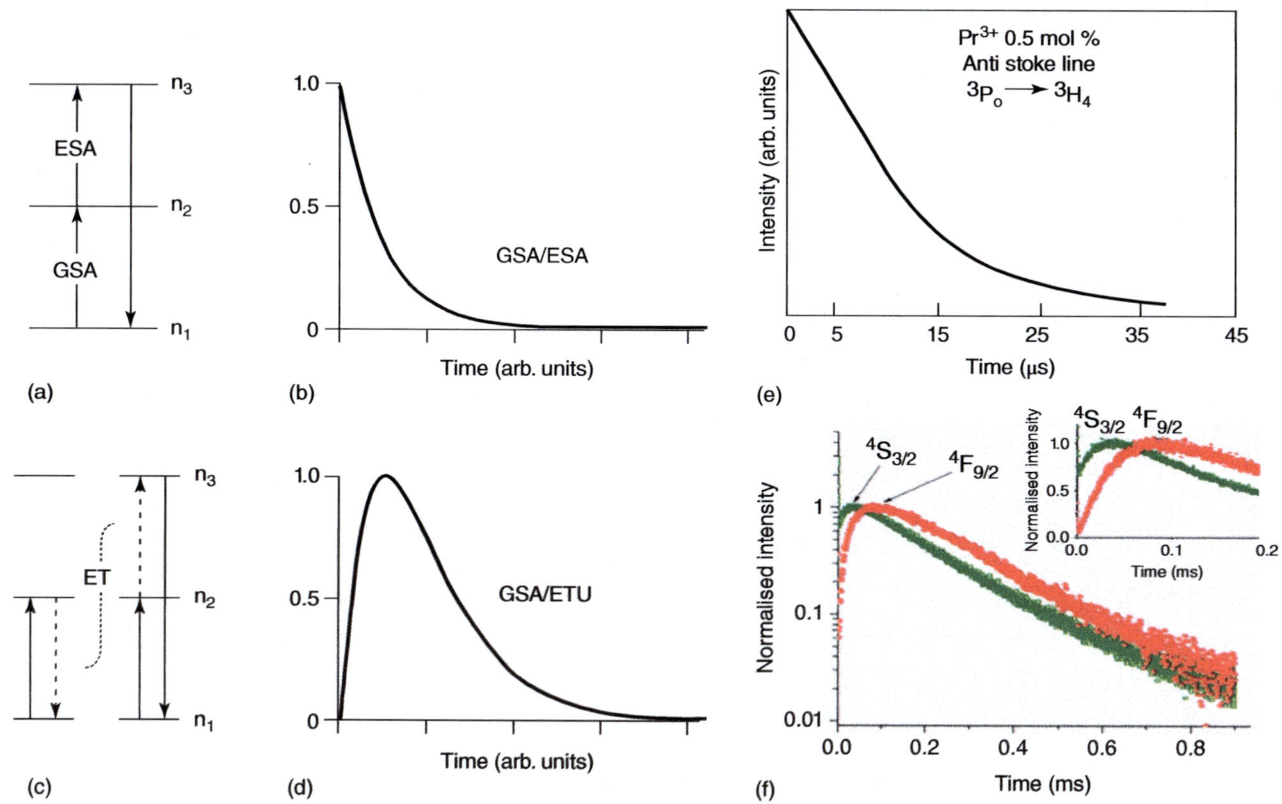

Figure 5.3 Schematic representation of the GSA/ESA (a) and GSA/ETU (c) processes. Dotted and straight arrows represent non-radiative and radiative ET, respectively. The graphs in (b) and (d) describe the UCL time evolution of the emission after a short excitation pulse, resulting from the two processes. Source: Suyver et al. (2005). Reprinted with the permission of Elsevier. (e) Decay curve for anti-Stokes emission corresponding to the $^3P_0 \rightarrow {}^3H_4$ transition. Source: Rai and Rai (2007). Reprinted with the permission of Elsevier. (f) Temporal evolution of the green $^4S_{3/2}$ and red $^4F_{9/2}$ level populations under pulsed 977 nm excitation on a log scale. The inset presents the intensity data at the short times on a linear scale. Source: Etchart et al. (2010). Reprinted with the permission of Royal Society of Chemistry.

in the intensity (resembles the rise in the population of the emitting states) with subsequent exponential decay at a longer time. Rai and Rai observed the characteristics of frequency upconversion of the prepared Pr^{3+}–Li_2O–TeO_2 binary glass under non-resonant excitation at ~890 nm radiation from the Ti–Sapphire laser. The possible population mechanisms for the emitting state of the activator could be assumed to be either ESA or ETU. The authors analyzed the decay curve for the emitting state of the Pr^{3+} ion corresponding to the $^3P_0 \rightarrow {}^3H_4$ transition and monitored a decay time of ~8.9 μs with no rise time in the curve (Figure 5.3e). Thus, the nature of the curve implies the dominance of ESA over ETU (Rai and Rai 2007). Later, Etchart et al. have studied the temporal evolution of the emitting states ($^4S_{3/2}$, $^2H_{11/2}$, and $^4F_{9/2}$) of the Er^{3+} ion in the Y_2BaZnO_5:Yb^{3+},Er^{3+} material under a 7 ns pulsed excitation at 977 nm. A typical rise followed by delayed decay observed from the $^4F_{9/2}$ emitting state suggests that the ETU process is responsible for the transient (Etchart et al. 2010). Nevertheless, in practice, both ESA and ETU processes simultaneously exist in the system (Suyver et al. 2005; Nadort et al. 2016). Therefore, a convoluted decay curve of both ESA and ETU has been observed from the $^4S_{3/2}$ emitting level, and the nature of the decay curve eventually reflects its biexponential characteristics (Figure 5.3f).

5.2.3 Theoretical and Experimental Approach of Understanding the Factors Affecting Upconversion Decay

Despite the advancement in the UC nanomaterials and their associated applications in the past few decades, the understanding of the UCL decay characteristics is not elucidated appropriately. Bergstrand et al. demonstrated the rate equation and matrix laboratory (MATLAB)-based simulation of UCL decay to obtain a clear idea of the dependence of external factors on the intrinsic decay time of the emitting state of activators (Bergstrand et al. 2019). According to the authors, the UCL decay profile depends on the contributions from the sensitizer's excited state, energy transfer, and cross-relaxation processes. A two-photon upconversion system has been taken into account with the Yb^{3+} ion as the sensitizer to analyze the behavior of the UCL decay profile (Figure 5.4a). ESA and ETU have been considered as two major processes involved to populate the emitting level. After the short pulse is applied, the time-dependent population of the energy states can be written in terms of rate equations as

$$\frac{dn_1}{dt} = -W_1 n_5 n_1 + \frac{n_2}{\tau_2} + \frac{n_3}{\tau_3} \tag{5.10}$$

$$\frac{dn_2}{dt} = W_1 n_5 n_1 - W_2 n_5 n_2 - \frac{n_2}{\tau_2} \tag{5.11}$$

$$\frac{dn_3}{dt} = W_2 n_5 n_2 - \frac{n_3}{\tau_3} \tag{5.12}$$

$$\frac{dn_4}{dt} = -\frac{\sigma\rho(t)}{h\nu} n_4 + \frac{n_5}{\tau_5} + W_1 n_5 n_1 + W_2 n_5 n_2 \tag{5.13}$$

$$\frac{dn_5}{dt} = \frac{\sigma\rho(t)}{h\nu} n_4 - \frac{n_5}{\tau_5} - W_1 n_5 n_1 - W_2 n_5 n_2 \tag{5.14}$$

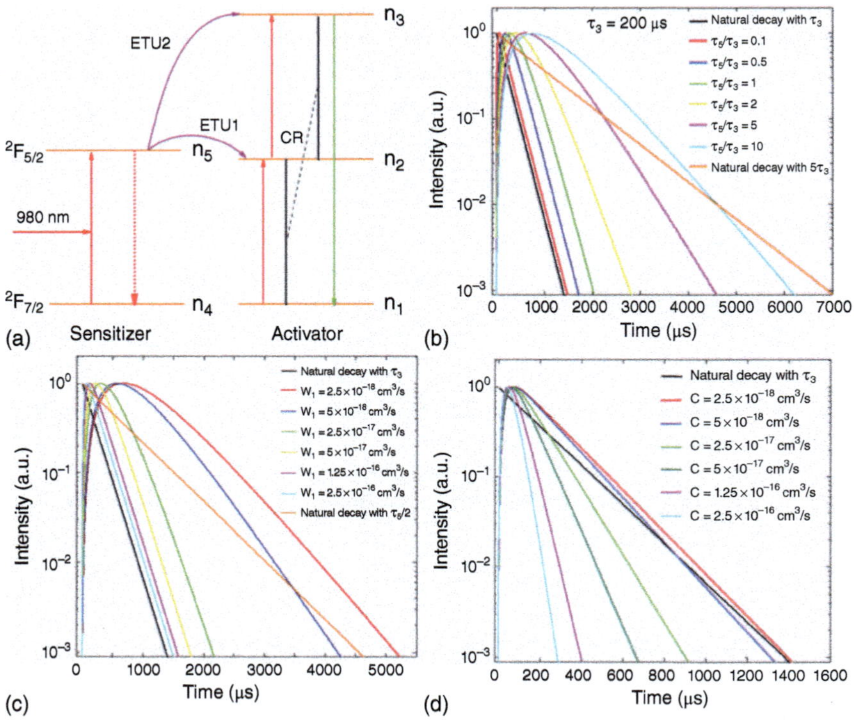

Figure 5.4 (a) Energy transfer upconversion (ETU) mechanism for two-photon upconversion luminescence with the presence of a cross-relaxation process, (b) simulated UCL (from state 3) decay profiles under short-pulse upconversion excitation with different τ_5/τ_3, (c) simulated UCL (from state 3) decay profiles under short pulse excitation with different W_1, and (d) decay profiles of UCL from state 3 under short pulse excitation with different cross-relaxation coefficients. Source: Bergstrand et al. (2019). Reprinted with the permission of Royal Society of Chemistry.

$$n_1 + n_2 + n_3 = n_A \tag{5.15}$$

$$n_4 + n_5 = n_S \tag{5.16}$$

where $n_i(i = 1 - 5)$ signifies the population of ions at different states, $n_{A(S)}$ is the number density of the activator (sensitizer), τ_j (j = 2, 3, 5) is the lifetime of jth excited state, W_1 and W_2 represent the energy transfer rates for ETU1 and ETU2, respectively, σ and $\rho(t)$ denote the absorption cross section of the sensitizer ions and time-dependent excitation intensity function, respectively.

Considering τ_3 as the desired intrinsic lifetime of the activator ion and estimating the parameters from the reported literature by Liu et al., Villanueva-Delgado et al., Jung et al., and Bansal et al., the MATLAB-based simulation for the decay time of the UCL has been carried out with varying conditions (Liu et al. 2013; Jung et al. 2015; Bansal et al. 2016; Villanueva-Delgado et al. 2017). Initially, while observing the modification in the lifetime of the emitting states, the CR mechanisms have been ignored. Keeping the pump power density of the excitation source at 1 W/cm^2, τ_3 at

Table 5.1 Summary of the constant parameters used in the simulations of studying the effect of sensitizer's excited-state lifetime (Bergstrand et al. 2019).

σ (cm^2)	n_S (cm^{-3})	n_A (cm^{-3})	τ_2 (ms)	τ_3 (ms)	W_1 (cm^3/s)	W_2 (cm^3/s)
1.5×10^{-20}	1.5×10^{21}	1.5×10^{20}	1.32	0.2	2.5×10^{-18}	5×10^{-17}

200 μs, and the parameters listed in Table 5.1, the value of sensitizer lifetime (τ_5) from 0.1–10 τ_3 is varied (Bergstrand et al. 2019). When τ_5 is insignificant compared to τ_3, the behavior of the decay curve resembles the intrinsic decay. While increasing the ratio of τ_5/τ_3 to 10, it has been observed that the rate constant reaches $\tau_5/2$, which eventually lengthens the lifetime (Figure 5.4b). Thus, the study reveals that the lifetime of the sensitizer's excited state under longer wavelength excitation affects the lifetime of the emitting state of the activator ions. In another simulation, the lifetime of the intermediate state τ_2 is varied from 0.1–10 τ_3, keeping the value of τ_5 to 0.02 ms and other parameters unchanged. Interestingly, no noticeable change in the lifetime has been observed, which further reclaims that the lifetime of the emitting state is more sensitive to the lifetime of sensitizer's excited state.

The effect of energy transfer rate on the lifetime of the emitting state has been understood with the above rate equations. It has come to notice that as the rate of energy transfer increases, the influence of the lifetime of sensitizer's excited state decreases. Moreover, authors have found that with the increase in energy transfer rates, the lifetime tends toward the intrinsic decay time (Figure 5.4c). The impact of concentration dependence on the lifetime of the emitting state cannot be ignored under anti-Stokes radiation conditions because the inter-ionic separation varies with the concentration of activators/sensitizers.

Further, the role of CR has been studied by following the modified rate equations as

$$\frac{dn_1}{dt} = -W_1 n_5 n_1 - Cn_1 n_3 + \frac{n_2}{\tau_2} + \frac{n_3}{\tau_3} \tag{5.17}$$

$$\frac{dn_2}{dt} = W_1 n_5 n_1 + 2Cn_1 n_3 - W_2 n_5 n_2 - \frac{n_2}{\tau_2} \tag{5.18}$$

$$\frac{dn_3}{dt} = W_2 n_5 n_2 - Cn_1 n_3 - \frac{n_3}{\tau_3} \tag{5.19}$$

where C represents the rate constant for the CR process and all other terms have their usual meanings. During the simulation, the energy transfer rate has been set to a higher value so as to eliminate the effect of τ_5. As the route of depopulation has increased because of the inclusion of the CR process, the short lifetime of the emitted state has been expected and the same has been depicted in Figure 5.4d.

Following the above results, Cheng and coworkers have studied the temporal response of β-NaYF$_4$:Yb/Tm microrods under 980 nm pulsed laser excitation. The study revealed that the lifetime of the emission states depends on the energy transfer coefficients, back energy transfer (BET) coefficients, cross-relaxation coefficients, and pump power of the laser excitation (Han et al. 2021). Under low pump power densities, no noticeable fluctuation has been noticed, but at high pump power, an increase in the lifetime of the $^1G_4$ energy state of the Tm^{3+} ion has been observed. The reason could be attributed to multiple energy exchange processes between the Yb^{3+} and Tm^{3+} ions. Authors have reported that the higher excited states could be populated by different energy transfer (ET) pathways such as ET1, ET2, ET3, and ET4, etc. (Figure 5.5a). While evaluating the influence of ET coefficients (W_i, $i = 1, 2, 3, 4$, etc.) on the lifetime of the emission states, it has been found that when W_1 increases significantly, the temporal response brings out a fast rise time and the decay time. However, the temporal behavior of the excited states remains almost unaltered because of W_2, W_3, W_4, etc. (Figure 5.5b–d). They have also investigated the effect of BET and CR on the lifetime of the emitted states. The study unveils the shortening of a lifetime for the particular excited states involved in the depopulation channels because of CR and BET, and for all other excited states, the lifetime remains unchanged.

Along with the theoretical understanding, various UC mechanisms can be understood with the help of luminescence and time-resolved spectroscopy. The Stokes

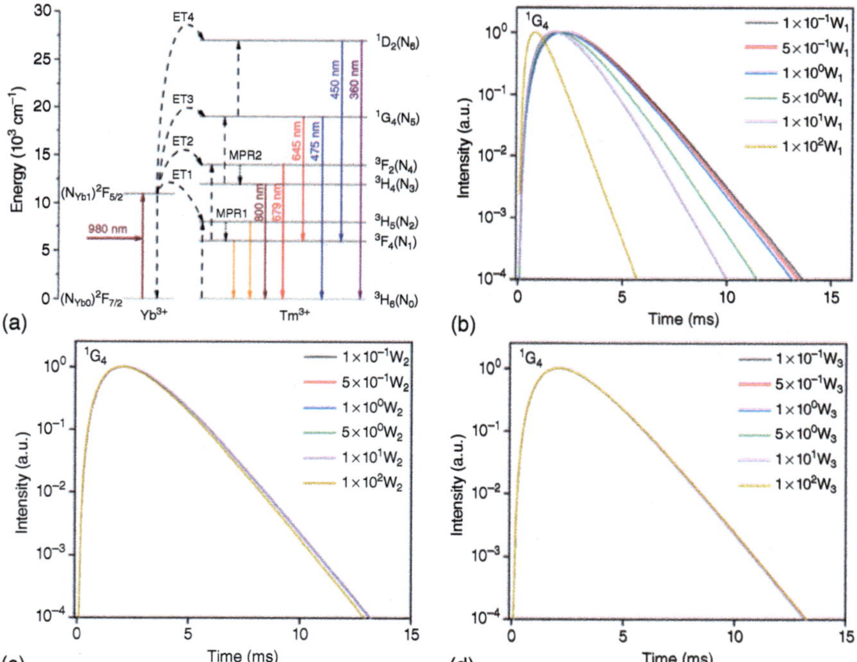

Figure 5.5 (a) Schematic diagram of Yb/Tm upconversion emission pathways. In the Yb/Tm upconversion diagram, different emission peaks from the ultraviolet to near-infrared wavelengths are obtained by successive optical transitions, including light absorption; four ET steps of ET1, ET2, ET3, and ET4; and multiphonon relaxation (MPR) steps of MPR1 and MPR2. Temporal responses with different ET coefficients. Normalized temporal responses from (b) the $^1G_4$ energy level with different W_1. (c) Normalized temporal responses of the $^1G_4$ with different W_2. (d) Normalized temporal responses of $^1G_4$ with different W_3. Source: Han et al. (2021). Reprinted with the permission of Elsevier.

luminescence and frequency upconversion in the Pr^{3+}-doped TeO_2–PbO glass has been performed by Rai et al. (2007a). It has been reported that the absorption transitions are inhomogeneously broadened because of the site-to-site variation of the crystal field strength. The absorption spectra observed for the samples at different concentrations of the dopants are similar, but the intensities of the bands are concentration dependent. The quadratic dependence of the UC emission intensity with the concentration of dopant ions and the laser intensity confirms the Pr^{3+} ions in pair and two laser photons responsible for the UC emission at ~484 nm. The concentration-dependent decay curves have been analyzed and observed that the experimental results fitted well with the biexponential fitting function as

$$I_{UC} \propto \left[\exp\left(-\frac{t}{\tau_R}\right) - \exp\left(-\frac{t}{\tau_D}\right)\right] \tag{5.20}$$

where τ_R and τ_D defines the rise time and decay time, respectively.

Moreover, the UC mechanisms has been deliberated by considering the dynamics of pair states, where $N_1(t)$ and $N_2(t)$ represents the population densities in the pair states $|1\rangle = |^1D_2, {}^1D_2\rangle$ and $|2\rangle = |^3P_0, {}^3H_4\rangle$, respectively. Considering the excitation of the state $|1\rangle$, the temporal evolution of the emitting state has been given as

$$N_2(t) = \frac{N_1(0)W_{12}}{\gamma_1 - \gamma_2 + W_{12}}\left[e^{-\gamma_2 t} - e^{-(\gamma_1+W_{12})t}\right] \tag{5.21}$$

where, γ_1 and γ_2 represent the relaxation rates from the states $|1\rangle$ and $|2\rangle$, respectively. Here, the relaxation rates from the states $|1\rangle$ to $|2\rangle$ due to ET have been ignored. W_{12} is the energy transfer rate from the state $|1\rangle$ to state $|2\rangle$. From Eqs. (5.20) and (5.21), it can be concluded that ${\tau_R}^{-1}$ and ${\tau_D}^{-1}$ can be assigned to γ_2 and $\gamma_1 + W_{12}$, respectively.

In the year 2009, Rai has reported the upconversion mechanism for the Pr^{3+}-doped TeO_2–Li_2O glasses under ~590 nm laser excitation (Rai 2009). The author studied the depopulation from the $^3P_0$ state using the rate equation model and predicts that the ETU between the Pr^{3+} ions pairs followed by the dipole–dipole interaction is responsible for the upconversion emission at ~480 nm. The temporal evolution of the $^3P_0$ level with varying concentrations of the Pr^{3+} ions has been depicted in Figure 5.6a–c. The decrease in the decay time on increasing the Pr^{3+} ion concentration establishes the ETU process for the observed UC emission. Earlier, a group of researchers has also studied the UC emission behavior of singly Er^{3+}-doped fluoroindate glasses as a function of concentration under 1.48 μm laser excitation (Maciel et al. 1997). They found that the emission in the green region is due to the ETU processes between the Er^{3+} ion pairs as the temporal evolution study detects the decrease in decay time upon increasing the concentration of the Er^{3+} ions from 1 to 3 mol%. Authors have also studied the temperature-dependent upconversion and found that the prepared glasses can be suitable for preparing the temperature sensors. Likewise, the influence of a sensitizer on the decay characteristics has been verified by the researchers in the Er^{3+}–Yb^{3+} co-doped La_2O_3 phosphors. The experimental results show that because of co-doping of the Yb^{3+} ions, the decay time decreases from ~839.5 to ~479.3 μs, which validates the ETU process (Dey and Rai 2014).

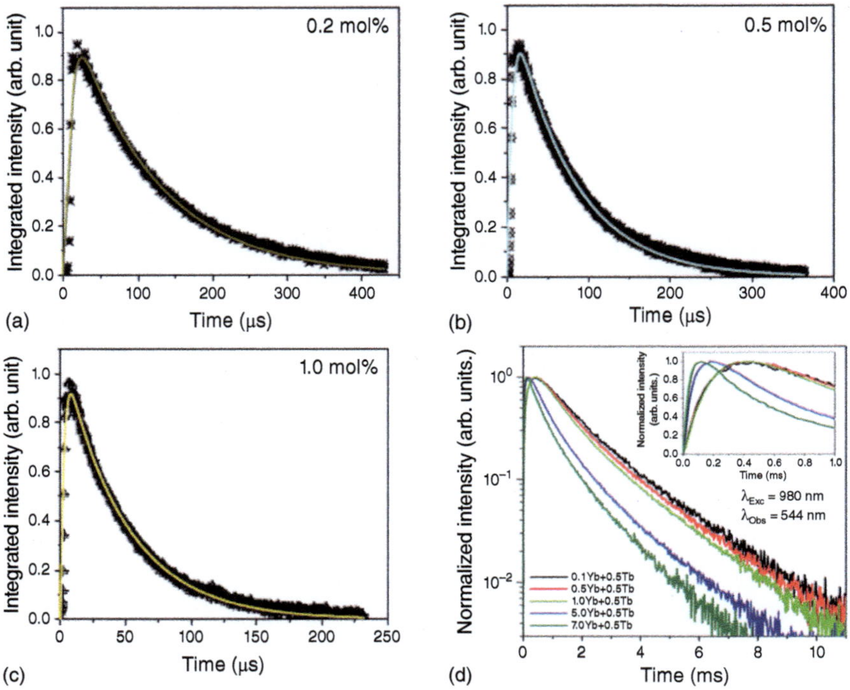

Figure 5.6 (a–c) Temporal evolution of upconversion emission at ~480 nm with different concentrations of the Pr^{3+} ions. Source: Modified from Rai (2009). Reprinted with the permission of Springer-Verlag, (d) Kinetic response of the Tb^{3+} UC luminescence for the samples xYb + 0.5 Tb with (x = 0.1, 0.5, 1.0, 5.0, and 7.0) upon excitation at λ_{Exc} = 980 nm monitored at λ_{Obs} = 544 nm (Tb^{3+}:$^5D_4 \rightarrow {}^7F_5$). The inset shows the responses at the first millisecond after the laser pulse.

Further, the researchers deliberated the UC processes in the Yb^{3+}–Tb^{3+} co-doped calcium lithium borate glasses. They confirmed that the cooperative energy transfer (CET), CR, and ESA processes are responsible for the UCL (Terra et al. 2020). The rising of fingerprint trend after the short pulse in the time-resolved measurement confirms the ETU process from the Yb^{3+} to Tb^{3+} ions under 980 nm, 5-ns pulsed laser excitation. They have examined the nature of the transient as a function of Yb^{3+} ion doping concentration. The rise time (τ_R) addresses the energy transfer probability, and the information about the deactivation processes could be obtained from the decay time (τ_D). The experimental values of τ_R and τ_D reveal that the energy transfer rates increase with the Yb^{3+} ion doping concentration. The non-exponential nature of the decay curve further indicates the involvement of non-radiative processes from the $^5D_4$ level, which can be designated to the CR processes between Yb^{3+} and Tb^{3+} ions. Thus, CR offers an additional decay pathway that enables shortening in the decay time with Yb^{3+} doping concentration, which is further confirmed by the CR efficiency values (Figure 5.6d). The dependence of concentration variation on the lifetime of the oxide-based Y_2BaZnO_5 phosphors doped with Yb^{3+} and Tm^{3+} ions has been investigated at room temperature (Etchart et al. 2011a). The temporal evolutions of the 479, 654, and 796 nm emission lines of the Tm^{3+} ion and the 1022 nm emission line of the Yb^{3+} ion have been monitored under the 7 ns excitation pulse of 974 nm wavelength. The transient behavior displays an initial rise followed by a long decay and confirms the involvement of the energy transfer process. A similar trend in the rise and decay time of 479 and 654 nm emission lines predicts that both the emission lines are arising from the same emitting level, i.e. $^1G_4$ of the Tm^{3+} ion. While examining the temporal evolution of the states associated with the 479, 654, 796, and 1022 nm emission, it has been found that the emitting states' lifetime decreases with the increase in the Tm^{3+} ion concentration. The decrease in the emitting states' lifetime of the Tm^{3+} ions could be attributed to the sensitizer's excited state ($^2F_{5/2}$ of Yb^{3+} ion) lifetime. As the lifetime of the sensitizer's (Yb^{3+}) excited state decreases with increasing the Tm^{3+} ion concentration, the probability of energy transfer rate from the Yb^{3+} to Tm^{3+} ions increases accordingly because of the reduction in the average radial distance between Yb^{3+}–Tm^{3+} ions. The spectroscopic investigation on the Nd^{3+}-doped $Ga_{10}Ge_{25}S_{65}$ glasses has been done with a non-resonant 1064 nm excitation to elucidate the frequency upconversion characteristics of the doped glasses (Rai et al. 2009). The experimental observation shows that the UC luminescence is due to the involvement of two-photon absorption by isolated Nd^{3+} ions and energy transfer among pairs of the Nd^{3+} ions (Figure 5.7). For a clear understanding of energy transfer UC mechanisms, the temporal evolution studies have been reported by several research groups (de Araújo et al. 1994; Etchart et al. 2011b; Qin et al. 2018).

Similarly, the increase in the Yb^{3+} ions' concentration substantially reduces the lifetime of the excited state associated with the 1022 nm emission line. In addition, the decrease in the excited-state lifetime of the sensitizer could be attributed to the increase in the energy transfer rate because of the short radial separation between the Yb^{3+} and Tm^{3+} ions. The results of concentration-dependent lifetime measurements have been depicted in Figure 5.8a.

Apart from the impact of UCL processes, several other parameters are also responsible for modulating the time-resolved spectroscopy measurements such as dopant concentration, nanocrystal size, external temperature, crystal field modifications by various dopants, etc. (Rai et al. 2007b; Etchart et al. 2011c; Zhao et al. 2013; Carrasco et al. 2019; Li et al. 2020). A group of researchers demonstrated the influence of the energy distributors (e.g. surface defects and the excess number of dopants) on the decay profile of sensitizers and emitters (Wang et al. 2016). They have also compared the relevance of pump power density with the spectral properties and the luminescence dynamics under 975 nm continuous wave (CW) excitation. In the developed $NaYF_4$:Yb/Er (20/2%) UCNPs, the lifetimes of the excited states responsible for the 540 nm emission of Er^{3+} and 985 nm emission of Yb^{3+} ions are determined to be 88 and 69 μs, respectively (Figure 5.8b). To observe the effect of luminescence decay on the number of energy distributors, they have implemented various approaches such as surface passivation by developing an inert shell around the UCNPs, size modulation, variation of the Yb^{3+} ion concentration in $NaYF_4$:Yb active shells, and Er^{3+} ion concentration variation. On increasing the inert shell thickness and increasing the size of UCNPs, a long decay time of the excited state of Yb^{3+} and the emitting states of Er^{3+} ions has been noticed. On the other hand, it

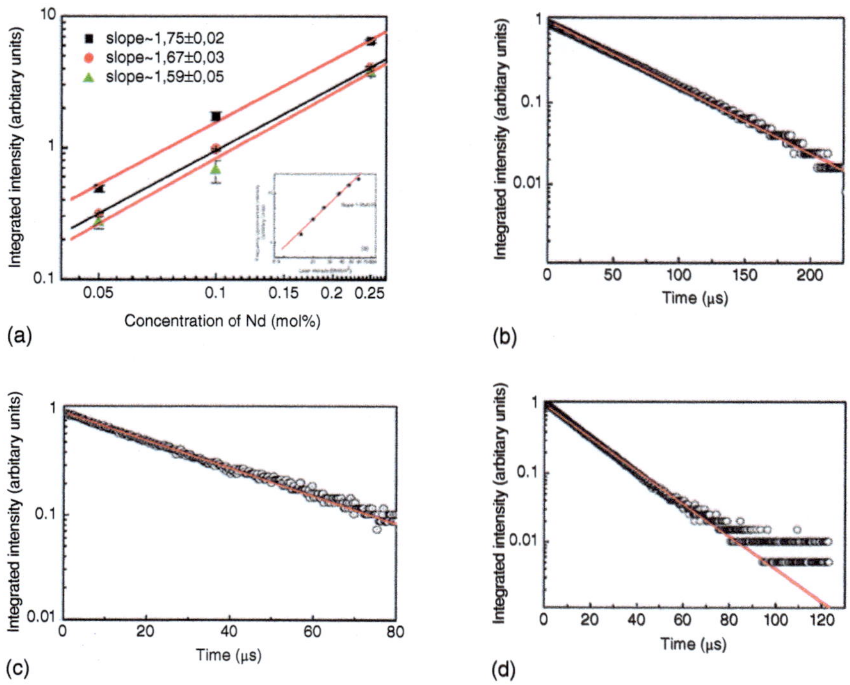

Figure 5.7 (a) Dependence of the frequency UC signal on the Nd^{3+} concentration. Transitions: ${}^4G_{7/2}{\rightarrow}{}^4I_{9/2}$ (squares), ${}^4G_{7/2}{\rightarrow}{}^4I_{11/2}$ (solid circles), and ${}^4G_{7/2}{\rightarrow}{}^4I_{13/2}$ (triangles). Inset of (a) shows the dependence of the frequency UC signal on the laser intensity corresponding to the ${}^4G_{7/2}{\rightarrow}{}^4I_{9/2}$ transition for 0.25 mol%. Temporal evolution of the UC signal because of the ${}^4G_{7/2}{\rightarrow}{}^4I_{9/2}$ transition. Samples with different values of the Nd^{3+} ion concentration: (b) 0.05 mol%, (c) 0.10 mol%, and (d) 0.25 mol%. Source: Rai et al. (2009). Reprinted with the permission of American Institute of Physics.

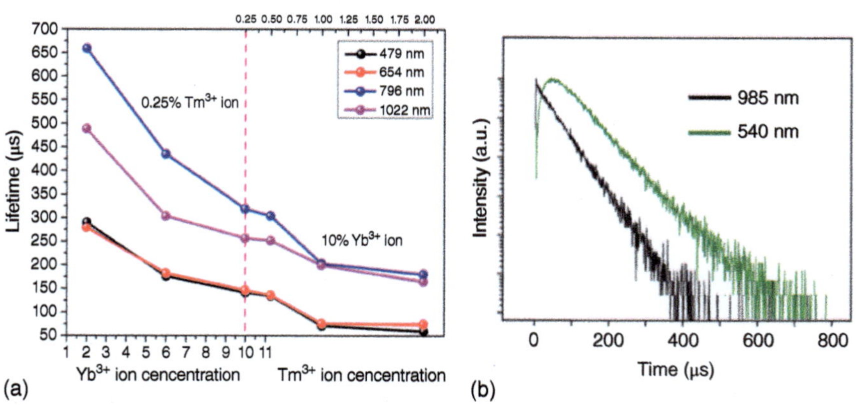

Figure 5.8 (a) Variation of the lifetime values of different emitting levels with the increasing concentrations of the Yb^{3+} and Tm^{3+} ions and (b) decay curves of 540 nm emissions from Er^{3+} as well as 985 nm from Yb^{3+} obtained upon excitation of the nanoparticles at 975 nm. Source: Modified from Wang et al. (2016). Reprinted with the permission of the Royal Society of Chemistry.

has been detected that the luminescence decay time of Er^{3+} and Yb^{3+} ions reduces remarkably with the increase in energy distributors. The number of pump photons required for the population decreases as the lifetime increases and vice versa.

5.3 Measurement of Quantum Yield of UCNPs

Luminescence refers to a phenomenon in which a material (commonly called luminescent material) emits photons after being excited by an excitation source. Depending on the type of excitation source, luminescence can be classified as chemiluminescence (excitation by chemical energy), electroluminescence (excitation by electrical energy), photoluminescence (excitation by photons), mechanoluminescence (excitation by mechanical energy), etc. In all these types, one point is common that the conversion of excitation energy into photons is not 100% because of various types of losses taking place through various paths. Because of this fact, two quantities came into existence, quantum yield (Φ) and quantum efficiency (η). In this section, first of all, a clear definition of both the quantities in context to fluorescence (type of photoluminescence) has been provided, and after that, the quantum yield of UCNPs has been discussed in detail.

The International Union of Pure and Applied Chemistry (IUPAC) has defined quantum yield (Φ) as (Wong et al. 2020)

$$\Phi(\lambda) = \frac{\text{number of events}}{\text{number of photons absorbed}} \tag{5.22}$$

In the case of light emission, the same is defined as

$$\Phi(\lambda) = \frac{\text{number of photons emitted}}{\text{number of photons absorbed}} \tag{5.23}$$

where λ denotes the wavelength of the photons absorbed. The quantum yield depends on many other external parameters, such as temperature, irradiance (particularly for UCNPs), emission wavelength range, experimental procedure of measurement, excitation power, etc. Hence, these factors have to be mentioned clearly while reporting the quantum yield value (Poelman and Smet 2010).

Unlike quantum yield, quantum efficiency (η) refers to the amount of output energy with respect to the amount of input energy and hence defined as (Wong et al. 2020)

$$\eta = \frac{\text{energy output}}{\text{energy input}} \tag{5.24}$$

Thus, from both the definitions, it is clear that the quantum efficiency of a luminescent material may or may not be equal to its quantum yield, depending on the energy output. Both the quantities are equal to each other only for a primary photophysical process such as resonant luminescence from a two-level system (see Figure 5.9). Also, from Eq. (5.23), it is clear that for a two-photon upconversion process, the quantum yield can have a maximum value of 50%, and for a three-photon upconversion process, it can have a maximum value of 33% only. In general, for

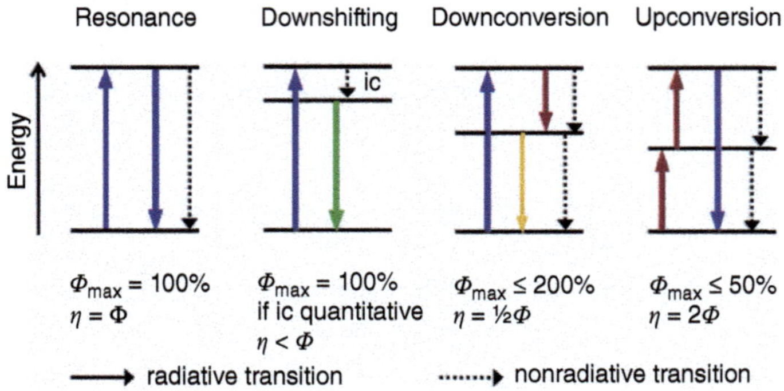

Figure 5.9 Comparative values of quantum yield (Φ) and quantum efficiency (η) for various types of photoluminescence (here "ic" stands for internal conversion). Source: Wong et al. (2020). Reprinted with the permission of Elsevier.

an n-photon upconversion process, the maximum value of quantum yield will be $100\%/n$.

5.3.1 Role of Quantum Yield in Upconversion

Owing to its unique optical properties, UCL has been widely studied over the decades by many researchers in context to various applications such as security ink, display devices, solar cells, optical sensing, etc. (Kumari et al. 2017, 2019; Dutta et al. 2019). However, one thing that badly affects the practicality of rare-earth-doped UCNPs is their low intensity or, more specifically, low brightness. In the field of luminescence, the term brightness (B) is defined as (Wong et al. 2020)

$$B = \Phi(\lambda) \times \varepsilon(\lambda) \tag{5.25}$$

where $\varepsilon(\lambda)$ denotes the molar absorption coefficient at excitation wavelength "λ."

Hence, to enumerate the brightness of UCNPs, one needs to quantify the following two quantities:

- The ability to absorb the excitation wavelength, i.e. the absorption coefficient.
- Converting the absorbed energy into photons of the desired wavelength, i.e. the quantum yield.

Thus, the emission intensity of UCNPs could either be enhanced by improving the absorption coefficient or by increasing the quantum yield. The former has been successfully achieved by co-doping of sensitizers while synthesizing the UCNPs (Huang et al. 1994; Strohhöfer and Polman 2003; Turri et al. 2009; Xie et al. 2013; Wang et al. 2013; Mukhopadhyay and Rai 2018), and the strategies of achieving the latter have been discussed by many researchers (Han et al. 2014; Wisser et al. 2016; Wilhelm 2017; Wen et al. 2019; Tessitore et al. 2019).

However, the triplet–triplet annihilation (TTA) systems that produce upconversion emission by diffusion-mediated annihilation of two triplet excitons have large absorption cross sections and high upconversion efficiencies. However, these systems suffer from a number of problems such as relatively small anti-Stokes shift (as it follows a two-photon upconversion process) and instability in oxygen and aqueous environment (Gray et al. 2018). On the other hand, rare-earth-doped UCNPs can undergo two-to-six photon upconversion processes producing large anti-Stokes shifts from near infrared (NIR) excitation into visible and UV emissions (Zhou et al. 2015).

5.3.2 Optical Methods of Measuring Quantum Yield of Upconverting Nanoparticles (UCNPs)

The concept of quantum yield is very valuable for deciding the applicability of UCNPs in various fields such as solar cells, solid-state lighting, security ink, optical sensing, etc. (Wong et al. 2020). As discussed earlier, the quantum yield of a luminescent material also depends on the experimental procedure applied for its measurement. Hence, a proper knowledge of the measurement method for getting accurate quantum yield is necessary. The optical methods of quantum yield measurement include the relative method, the absolute method, the lifetime method, etc. Some of the most widely used optical methods of determining the quantum yield of UCNPs are described below.

5.3.2.1 Relative Method of Measuring Quantum Yield

The relative method of measurement requires a standard fluorescent sample having same excitation and emission properties as that of the sample under investigation. This method is mostly applicable for powder samples because for such samples, a plenty of standard references are available. Some of the prominently used standards are as follows (Bril and de Jager-Veenis 1976),

- Standards issued by National Bureau of Standards, Washington, DC.
- Sodium salicylate, also suitable for excitations lying in the far ultraviolet region.
- "Ekta S10" proposed by Grum.
- "Lumogen T red GG", suitable for the 190–550 nm excitation range.

A detailed discussion of this method is provided elsewhere (Bril and de Jager-Veenis 1976; Wong et al. 2020).

5.3.2.2 Absolute Method of Measuring Quantum Yield

In this method, an experimental setup consisting of an integrating sphere is used to determine the quantum yield and is applicable for solutions, powders, as well as thin films (Greenham et al. 1995; Johnson et al. 2007; Ishida et al. 2010). The integrating sphere collects all the light emitted and scattered from the luminescent sample, thereby achieving an accuracy of about 10–15%, and it does not require any reference sample for measurement. The experimental setup of this method is shown in Figure 5.10. This method utilizes two types of measurement procedures:

- Three measurement procedures
- Two measurement procedures

A proper discussion of both the procedures using the integrating sphere theory has been discussed by Leyre et al. (2014).

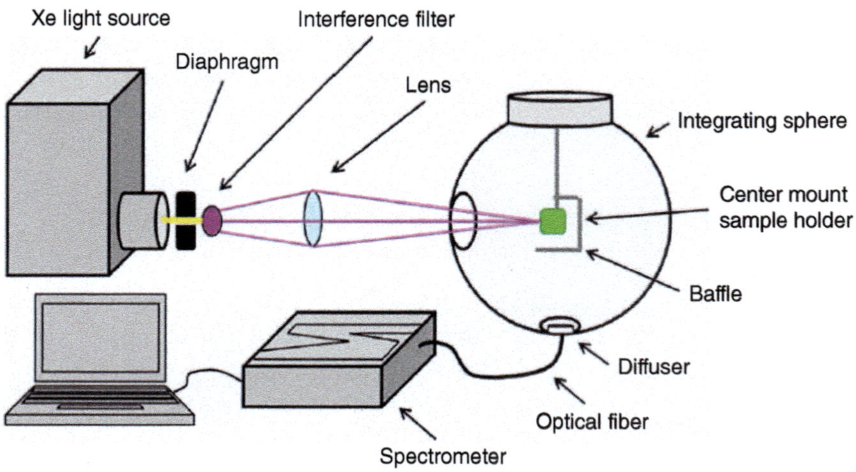

Figure 5.10 Experimental setup of the absolute method of quantum yield measurement. Source: Leyre et al. (2014). Reprinted with the permission of American Institute of Physics.

5.3.2.3 Measurement of Intrinsic Quantum Yield of Lanthanide-Based Materials Using Lifetimes

Intrinsic quantum yield (Φ_{int}) comes into picture when there is a direct excitation into the luminescent level of the activator and is defined as (Wong et al. 2020)

$$\Phi_{int} = \frac{\tau_{obs}}{\tau_r} \tag{5.26}$$

where τ_{obs} and τ_r represent the measured decay lifetime and radiative lifetime of the relevant state, respectively.

The radiative lifetime can be easily determined experimentally from the fluorescence decay curves or with the help of Judd–Ofelt theory.

5.3.3 Some Other Methods of Determining Quantum Yield

Besides the optical methods of quantum yield determination, several other methods such as photo-acoustic spectroscopy (PAS), thermal lensing (TL), and calorimetry have been used for determining the quantum yield. The former two have been briefly described below.

5.3.3.1 Photo-acoustic Spectroscopy (PAS)

In this method, the sample is illuminated using a pulsed radiation. Because of this, a periodic variation in the sample temperature is induced, and during this variation, sound is generated by the sample. The quantum yield is measured by comparing the sound intensities generated by the fluorescent sample (whose quantum yield is to be determined) and a non-fluorescent sample under identical conditions. The accuracy of this method is about 10%, and this method has been successfully applied in solutions, solid-state materials, gas-phase bearing aerosols, and turbid suspensions (Tomasini et al. 2009; Würth et al. 2012).

5.3.3.2 Thermal Lensing (TL) Method

The concept of temperature-dependent refractive index is applied in this method. In this method, a refractive index gradient is generated in the sample when the excitation beam (Gaussian in nature) is absorbed by it. The gradient causes the beam to converge or diverge, hence the name "thermal lens." Now, the generated gradient is used to evaluate the absorption losses, from which the quantum yield of the sample can be calculated (Brannon and Magde 1978).

References

de Araújo, L.E., Gomes, A.S.L., De Araújo, C.B. et al. (1994). Frequency upconversion of orange light into blue light in Pr^{3+}-doped fluoroindate glasses. *Physical Review B* 50 (22): 16219. https://doi.org/10.1103/PhysRevB.50.16219.

Auzel, F.E. (1973). Materials and devices using double-pumped-phosphors with energy transfer. *Proceedings of the IEEE* 61 (6): 758–786. https://doi.org/10.1109/PROC.1973.9155.

Bansal, A., Liu, H., Jayakumar, M.K.G. et al. (2016). Quasi-continuous wave near-infrared excitation of upconversion nanoparticles for optogenetic manipulation of *C. elegans*. *Small* 12 (13): 1732–1743. https://doi.org/10.1002/smll.201503792.

Bergstrand, J., Liu, Q., Huang, B. et al. (2019). On the decay time of upconversion luminescence. *Nanoscale* 11 (11): 4959–4969. https://doi.org/10.1039/c8nr10332a.

Brannon, J.H. and Magde, D. (1978). Absolute quantum yield determination by thermal blooming. Fluorescein. *The Journal of Physical Chemistry* 82 (6): 705–709. https://doi.org/10.1021/j100495a018.

Bril, A. and de Jager-Veenis, A.W. (1976). Some methods of luminescence efficiency measurements. *Journal of Research of the National Bureau of Standards, Section A, Physics and Chemistry* 80 (3): 401. https://doi.org/10.6028%2Fjres.080A.039.

Carrasco, I., Laversenne, L., Bigotta, S. et al. (2019). Super-quadratic upconversion luminescence among lanthanide ions. *Optics Express* 27 (23): 33217–33232. https://doi.org/10.1364/oe.27.033217.

Chubb, D.L. and Wolford, D.S. (2000). Rare earth optical temperature sensor. *Review of Scientific Instruments* 71 (5): 2233–2240. https://doi.org/10.1063/1.1150616.

Dey, R. and Rai, V.K. (2014). Yb^{3+} sensitized Er^{3+} doped La_2O_3 phosphor in temperature sensors and display devices. *Dalton Transactions* 43 (1): 111–118. https://doi.org/10.1039/c3dt51773j.

Díaz García, M.E. and Badía Laíño, R. (2005). Fluorescence | time-resolved fluorescence. In: *Encyclopedia of Analytical Science* (ed. P. Worsfold, A. Townshend and C. Poole), 128–138. Elsevier, ISBN: 978-0-12-369397-6. https://doi:10.1016/b0-12-369397-7/00173-4.

Dutta, J., Rai, V.K., Durai, M.M., and Thangavel, R. (2019). Development of Y_2O_3:Ho^{3+}/Yb^{3+} upconverting nanophosphors for enhancing solar cell efficiency of dye-sensitized solar cells. *IEEE Journal of Photovoltaics* 9 (4): 1040–1045. https://doi.org/10.1109/JPHOTOV.2019.2912719.

Etchart, I., Huignard, A., Bérard, M. et al. (2010). Oxide phosphors for efficient light upconversion: Yb^{3+} and Er^{3+} co-doped Ln_2BaZnO_5 (Ln = Y, Gd). *Journal of Materials Chemistry* 20 (19): 3989–3994. https://doi.org/10.1039/c000127a.

Etchart, I., Bérard, M., Laroche, M. et al. (2011a). Efficient white light emission by upconversion in Yb^{3+}-, Er^{3+}- and Tm^{3+}-doped Y_2BaZnO_5. *Chemical Communications* 47 (22): 6263–6265. https://doi.org/10.1039/c1cc11427a.

Etchart, I., Hernández, I., Huignard, A. et al. (2011b). Oxide phosphors for light upconversion; Yb^{3+} and Tm^{3+} co-doped Y_2BaZnO_5. *Journal of Applied Physics* 109 (6): 063104. https://doi.org/10.1063/1.3549634.

Etchart, I., Hernández, I., Huignard, A. et al. (2011c). Efficient oxide phosphors for light upconversion; green emission from Yb^{3+} and Ho^{3+} co-doped Ln_2BaZnO_5 (Ln = Y, Gd). *Journal of Materials Chemistry* 21 (5): 1387–1394. https://doi.org/10.1039/c0jm01652g.

Gray, V., Moth-Poulsen, K., Albinsson, B., and Abrahamsson, M. (2018). Towards efficient solid-state triplet–triplet annihilation based photon upconversion: supramolecular, macromolecular and self-assembled systems. *Coordination Chemistry Reviews* 362: 54–71. https://doi.org/10.1016/j.ccr.2018.02.011.

Greenham, N.C., Samuel, I.D.W., Hayes, G.R. et al. (1995). Measurement of absolute photoluminescence quantum efficiencies in conjugated polymers. *Chemical Physics Letters* 241 (1, 2): 89–96. https://doi.org/10.1016/0009-2614(95)00584-Q.

Han, S., Deng, R., Xie, X., and Liu, X. (2014). Enhancing luminescence in lanthanide-doped upconversion nanoparticles. *Angewandte Chemie International Edition* 53 (44): 11702–11715. https://doi.org/10.1002/anie.201403408.

Han, Y., Yang, J., Deng, Y. et al. (2021). Insight into the temporal response of Yb^{3+}/Tm^{3+} upconversion luminescence. *Optics Communications* 499: 127305. https://doi.org/10.1016/j.optcom.2021.127305.

Hany, R., Cremona, M., and Strassel, K. (2019). Recent advances with optical upconverters made from all-organic and hybrid materials. *Science and Technology of Advanced Materials* 20 (1): 497–510. https://doi.org/10.1080/14686996.2019.1610057.

Heer, S., Kömpe, K., Güdel, H.U., and Haase, M. (2004). Highly efficient multicolour upconversion emission in transparent colloids of lanthanide-doped $NaYF_4$ nanocrystals. *Advanced Materials* 16 (23, 24): 2102–2105. https://doi.org/10.1002/adma.200400772.

Huang, C.H., McCaughan, L., and Gill, D.M. (1994). Evaluation of absorption and emission cross sections of Er-doped $LiNbO_3$ for application to integrated optic amplifiers. *Journal of Lightwave Technology* 12 (5): 803–809. https://doi.org/10.1109/50.293972.

Ishida, H., Tobita, S., Hasegawa, Y. et al. (2010). Recent advances in instrumentation for absolute emission quantum yield measurements. *Coordination Chemistry Reviews* 254 (21, 22): 2449–2458. https://doi.org/10.1016/j.ccr.2010.04.006.

Johnson, A.R., Lee, S.J., Klein, J., and Kanicki, J. (2007). Absolute photoluminescence quantum efficiency measurement of light-emitting thin films. *Review of Scientific Instruments* 78 (9): 096101. https://doi.org/10.1063/1.2778614.

Jung, T., Jo, H.L., Nam, S.H. et al. (2015). The preferred upconversion pathway for the red emission of lanthanide-doped upconverting nanoparticles, $NaYF_4$:Yb^{3+},Er^{3+}.

Physical Chemistry Chemical Physics 17 (20): 13201–13205. https://doi.org/10.1039/c5cp01634g.

Kumari, A., Soni, A.K., and Rai, V.K. (2017). Near infrared to blue upconverting $Tm^{3+}/Yb^{3+}/Li^{+}:Gd_2(MoO_4)_3$ phosphors for light emitting display devices. *Infrared Physics and Technology* 81: 313–319. https://doi.org/10.1016/j.infrared.2017.01.024.

Kumari, A., Mukhopadhyay, L., and Rai, V.K. (2019). $Er^{3+}/Yb^{3+}/Li^{+}/Zn^{2+}:Gd_2(MoO_4)_3$ upconverting nanophosphors in optical thermometry. *Journal of Rare Earths* 37 (3): 242–247. https://doi.org/10.1016/j.jre.2018.05.021.

Lakowicz, J.R. (1999). Time-domain lifetime measurements. In: *Principles of Fluorescence Spectroscopy*, 95–140. Boston, MA: Springer. https://doi.org/10.1007/978-1-4757-3061-6_4.

Leyre, S., Coutino-Gonzalez, E., Joos, J.J. et al. (2014). Absolute determination of photoluminescence quantum efficiency using an integrating sphere setup. *Review of Scientific Instruments* 85 (12): 123115. https://doi.org/10.1063/1.4903852.

Li, Y., Liu, C., Zhang, P. et al. (2020). Doping lanthanide nanocrystals with non-lanthanide ions to simultaneously enhance up-and down-conversion luminescence. *Frontiers in Chemistry* 8: https://doi.org/10.3389/fchem.2020.00832.

Liu, H., Xu, C.T., Dumlupinar, G. et al. (2013). Deep tissue optical imaging of upconverting nanoparticles enabled by exploiting higher intrinsic quantum yield through use of millisecond single pulse excitation with high peak power. *Nanoscale* 5 (20): 10034–10040. https://doi.org/10.1039/c3nr01917a.

Loo, J.F.C., Chien, Y.H., Yin, F. et al. (2019). Upconversion and downconversion nanoparticles for biophotonics and nanomedicine. *Coordination Chemistry Reviews* 400: 213042. https://doi.org/10.1016/j.ccr.2019.213042.

Maciel, G.S., De Araújo, C.B., Messaddeq, Y., and Aegerter, M.A. (1997). Frequency upconversion in Er^{3+}-doped fluoroindate glasses pumped at 1.48 μm. *Physical Review B* 55 (10): 6335. https://doi.org/10.1103/PhysRevB.55.6335.

Midwinter, J.E. and Warner, J. (1967). Up-conversion of near infrared to visible radiation in lithium-meta-niobate. *Journal of Applied Physics* 38 (2): 519–523. https://doi.org/10.1063/1.1709367.

Mondal, M. and Rai, V.K. (2018). An effective way to enhance upconversion emission and temperature sensing via Zn^{2+} incorporation in Er^{3+}–$Yb^{3+}:YMoO_4$ nanophosphors. *Journal of Industrial and Engineering Chemistry* 60: 125–132. https://doi.org/10.1016/j.jiec.2017.10.043.

Mukhopadhyay, L. and Rai, V.K. (2018). Investigation of photoluminescence properties, Judd–Ofelt analysis, luminescence nanothermometry and optical heating behaviour of $Er^{3+}/Eu^{3+}/Yb^{3+}:NaZnPO_4$ nanophosphors. *New Journal of Chemistry* 42 (15): 13122–13134. https://doi.org/10.1039/C8NJ02320D.

Nadort, A., Zhao, J., and Goldys, E.M. (2016). Lanthanide upconversion luminescence at the nanoscale: fundamentals and optical properties. *Nanoscale* 8 (27): 13099–13130. https://doi.org/10.1039/C5NR08477F.

Omary, M.A. and Patterson, H.H. (2017). Luminescence, theory. In: *Encyclopedia of Spectroscopy and Spectrometry*, 3e (ed. J.C. Lindon, G.E. Tranter and D.W. Koppenaal), 636–653. Academic Press. ISBN 9780128032244, https://doi.org/10.1016/B978-0-12-803224-4.00193-X.

Pandey, A. and Rai, V.K. (2012). Colour emission tunability in Ho^{3+}–Tm^{3+}–Yb^{3+} co-doped Y_2O_3 upconverted phosphor. *Applied Physics B* 109 (4): 611–616. https://doi.org/10.1007/s00340-012-5224-1.

Pattnaik, S. and Rai, V.K. (2021). Insight into the spectroscopic and thermometric properties of titanate phosphors via a novel co-excited laser system. *Materials Science and Engineering B* 272: 115318. https://doi.org/10.1016/j.mseb.2021.115318.

Piguet, C. (2014). Extricating erbium. *Nature Chemistry* 6 (4): 370–370. https://doi.org/10.1038/nchem.1908.

Poelman, D. and Smet, P.F. (2010). Photometry in the dark: time dependent visibility of low intensity light sources. *Optics Express* 18 (25): 26293–26299. https://doi.org/10.1364/OE.18.026293.

Qin, H., Wu, D., Sathian, J. et al. (2018). Tuning the upconversion photoluminescence lifetimes of $NaYF_4$:Yb^{3+}, Er^{3+} through lanthanide Gd^{3+} doping. *Scientific Reports* 8 (1): 1–8. https://doi.org/10.1038/s41598-018-30983-9.

Rai, V.K. (2007). Temperature sensors and optical sensors. *Applied Physics B* 88 (2): 297–303. https://doi.org/10.1007/s00340-007-2717-4.

Rai, V.K. (2009). Upconversion due to energy transfer involving Pr^{3+} ions in pairs. *Applied Physics B* 95 (2): 329–333. https://doi.org/10.1007/s00340-008-3322-x.

Rai, V.K. and Rai, S.B. (2007). Frequency upconversion in Pr^{3+}–Li_2O–TeO_2 binary glass by decay curve analysis. *Spectrochimica Acta Part A: Molecular and Biomolecular Spectroscopy* 68 (3): 460–462. https://doi.org/10.1016/j.saa.2006.11.051.

Rai, V.K., Menezes, L.D.S., and de Araújo, C.B. (2007a). Spectroscopy, energy transfer, and frequency upconversion in Tm^{3+}-doped TeO_2–PbO glass. *Journal of Applied Physics* 102 (4): 043505. https://doi.org/10.1063/1.2769788.

Rai, V.K., Menezes, L.D.S., and de Araújo, C.B. (2007b). Stokes luminescence and frequency upconversion in Pr^{3+} doped TeO_2–PbO glass. *Journal of Applied Physics* 101 (12): 123514. https://doi.org/10.1063/1.2745314.

Rai, V.K., de Araújo, C.B., Ledemi, Y. et al. (2009). Optical spectroscopy and upconversion luminescence in Nd^{3+} doped $Ga_{10}Ge_{25}S_{65}$ glass. *Journal of Applied Physics* 106 (10): 103512. https://doi.org/10.1063/1.3259439.

Ranjan, S.K., Dey, R., Soni, A.K., and Rai, V.K. (2016). Er^{3+}–Tm^{3+}–Yb^{3+}:Gd_2O_3 upconverting phosphors for sensing and laser-induced heating applications. *IEEE Sensors Journal* 16 (23): 8494–8500. https://doi.org/10.1109/JSEN.2016.2609139.

Shanmugam, N., Cholan, S., Viruthagiri, G. et al. (2014). Synthesis and characterization of Ce^{3+}-doped flowerlike ZnS nanorods. *Applied Nanoscience* 4 (3): 359–365. https://doi.org/10.1007/s13204-013-0217-x.

Soni, A.K. and Rai, V.K. (2017). Thermal and pump power effect in $SrMoO_4$:Er^{3+}–Yb^{3+} phosphor for thermometry and optical heating. *Chemical Physics Letters* 667: 226–232. https://doi.org/10.1016/j.cplett.2016.12.002.

Strohhöfer, C. and Polman, A. (2003). Absorption and emission spectroscopy in Er^{3+}–Yb^{3+} doped aluminum oxide waveguides. *Optical Materials* 21 (4): 705–712. https://doi.org/10.1016/S0925-3467(02)00056-3.

Suijver, J.F. (2007). Upconversion Phosphors. In: *Luminescence: From Theory to Applications* (ed. C. Ronda), 133–177. Wiley Online Library. https://doi.org/10.1002/9783527621064.ch6.

Suyver, J.F., Aebischer, A., Biner, D. et al. (2005). Novel materials doped with trivalent lanthanides and transition metal ions showing near-infrared to visible photon upconversion. *Optical Materials* 27 (6): 1111–1130. https://doi.org/10.1016/j.optmat.2004.10.021.

Terra, I.A., Borrero-González, L.J., Almeida, J.M. et al. (2020). Judd–Ofelt analysis of Tb^{3+} and upconversion study in Yb^{3+}–Tb^{3+} co-doped calibo glasses. *Química Nova* 43: 188–193. https://doi.org/10.21577/0100-4042.20170465.

Tessitore, G., Mandl, G.A., Brik, M.G. et al. (2019). Recent insights into upconverting nanoparticles: spectroscopy, modeling, and routes to improved luminescence. *Nanoscale* 11 (25): 12015–12029. https://doi.org/10.1039/C9NR02291K.

Tomasini, E.P., San Román, E., and Braslavsky, S.E. (2009). Validation of fluorescence quantum yields for light-scattering powdered samples by laser-induced optoacoustic spectroscopy. *Langmuir* 25 (10): 5861–5868. https://doi.org/10.1021/la803492k.

Turri, G., Jenssen, H.P., Cornacchia, F. et al. (2009). Temperature-dependent stimulated emission cross section in Nd^{3+}:YVO_4 crystals. *Journal of the Optical Society of America B: Optical Physics* 26 (11): 2084–2088. https://doi.org/10.1364/JOSAB.26.002084.

Villanueva-Delgado, P., Biner, D., and Krämer, K.W. (2017). Judd–Ofelt analysis of β-$NaGdF_4$:Yb^{3+},Tm^{3+} and β-$NaGdF_4$:Er^{3+} single crystals. *Journal of Luminescence* 189: 84–90. https://doi.org/10.1016/j.jlumin.2016.04.023.

Wang, Y.F., Liu, G.Y., Sun, L.D. et al. (2013). Nd^{3+}-sensitized upconversion nanophosphors: efficient in vivo bioimaging probes with minimized heating effect. *ACS Nano* 7 (8): 7200–7206. https://doi.org/10.1021/nn402601d.

Wang, Y., Deng, R., Xie, X. et al. (2016). Nonlinear spectral and lifetime management in upconversion nanoparticles by controlling energy distribution. *Nanoscale* 8 (12): 6666–6673. https://doi.org/10.1039/c6nr00812g.

Wen, S., Zhou, J., Schuck, P.J. et al. (2019). Future and challenges for hybrid upconversion nanosystems. *Nature Photonics* 13 (12): 828–838. https://doi.org/10.1038/s41566-019-0528-x.

Wilhelm, S. (2017). Perspectives for upconverting nanoparticles. *ACS Nano* 11 (11): 10644–10653. https://doi.org/10.1021/acsnano.7b07120.

Wisser, M.D., Fischer, S., Maurer, P.C. et al. (2016). Enhancing quantum yield via local symmetry distortion in lanthanide-based upconverting nanoparticles. *ACS Photonics* 3 (8): 1523–1530. https://doi.org/10.1021/acsphotonics.6b00166.

Wong, K.L., Bünzli, J.C.G., and Tanner, P.A. (2020). Quantum yield and brightness. *Journal of Luminescence* 224: 117256. https://doi.org/10.1016/j.jlumin.2020.117256.

Würth, C., González, M.G., Niessner, R. et al. (2012). Determination of the absolute fluorescence quantum yield of rhodamine 6G with optical and photoacoustic methods–providing the basis for fluorescence quantum yield standards. *Talanta* 90: 30–37. https://doi.org/10.1016/j.talanta.2011.12.051.

Xie, X., Gao, N., Deng, R. et al. (2013). Mechanistic investigation of photon upconversion in Nd^{3+}-sensitized core–shell nanoparticles. *Journal of the American Chemical Society* 135 (34): 12608–12611. https://doi.org/10.1021/ja4075002.

Yi, G., Lu, H., Zhao, S. et al. (2004). Synthesis, characterization, and biological application of size-controlled nanocrystalline $NaYF_4$:Yb,Er infrared-to-visible up-conversion phosphors. *Nano Letters* 4 (11): 2191–2196. https://doi.org/10.1021/nl048680h.

Zhao, J., Lu, Z., Yin, Y. et al. (2013). Upconversion luminescence with tunable lifetime in $NaYF_4$:Yb,Er nanocrystals: role of nanocrystal size. *Nanoscale* 5 (3): 944–952. https://doi.org/10.1039/c2nr32482b.

Zhou, B., Shi, B., Jin, D., and Liu, X. (2015). Controlling upconversion nanocrystals for emerging applications. *Nature Nanotechnology* 10 (11): 924–936. https://doi.org/10.1038/nnano.2015.251.

6

Frequency Upconversion in UCNPs Containing Transition Metal Ions

Manisha Prasad and Vineet Kumar Rai

Indian Institute of Technology (Indian School of Mines), Department of Physics, Laser and Spectroscopy Laboratory, Dhanbad, Jharkhand 826004, India

6.1 Introduction

Transition metals (TMs) are the elements in fourth period of the periodic table having 3d orbitals with outermost electronic configuration $3d^{1-10}\ 4s^2$. Although all the elements are in the fourth period, the outermost electronic configuration is $3d^5$ $4s^1$ and $3d^{10}\ 4s^1$ for chromium (Cr) and copper (Cu), respectively. Because of this reason, they are also known as d-block elements (Table 6.1). When the transition metal ions react with species, organic or inorganic, capable of donating a pair of electrons, a coordinate bond is formed between the metal ion and the electron-donating species. This property is very much useful in preparing the upconverting materials. The choice of host and codopant combinations covers a broad range of possibilities and the ability to use well-defined chemical properties to design and synthesize upconversion materials displaying exciting new photophysical properties. Here in the present chapter, in addition to several current concerns in the area of frequency upconversion, the upconversion (UC) emissions in the upconverting nanoparticles (UCNPs) containing transition metal ions have been deliberated (Gamelin and Güdel 2001). The transition metal ions are expected to show a number of differences in the relevant parameters that govern the upconversion dynamics. In particular, the larger radial extension of the d-orbitals than that of the f-orbitals may lead to large electron–phonon (ion lattice) coupling strengths and a greater susceptibility to crystal field and exchange perturbations. As a result, the spectra of TM ions present both broad ($S > 0$) and sharp ($S \approx 0$) bands, opposite to the spectra of $(RE)^{3+}$ ions.

Several d^2, d^3, d^4, and d^8 transition metal ions also luminesce from more than one excited state. Moreover, there is a relatively low gap in energy between the possible oxidation states of TMs and therefore exhibit many oxidation states. Among all the TMs, manganese (Mn) has very large number of oxidation states; therefore, it has a variety of applications. The TM ions have been largely neglected as upconversion ions, but recent developments indicate that several such ions are also suitable for multiphoton excitation and sometimes show properties different from those of

Upconverting Nanoparticles: From Fundamentals to Applications, First Edition. Edited by Vineet K. Rai.

Table 6.1 Electronic configuration of transition elements (d-block elements).

Transition elements	Atomic number	Electronic configuration
Sc	21	[Ar] $3d^1$ $4s^2$
Ti	22	[Ar] $3d^2$ $4s^2$
V	23	[Ar] $3d^3$ $4s^2$
Cr	24	[Ar] $3d^5$ $4s^1$
Mn	25	[Ar] $3d^5$ $4s^2$
Fe	26	[Ar] $3d^6$ $4s^2$
Co	27	[Ar] $3d^7$ $4s^2$
Ni	28	[Ar] $3d^8$ $4s^2$
Cu	29	[Ar] $3d^{10}$ $4s^1$
Zn	30	[Ar] $3d^{10}$ $4s^2$
Y	39	[Kr] $4d^1$ $5s^2$
Zr	40	[Kr] $4d^2$ $5s^2$
Nb	41	[Kr] $4d^4$ $5s^1$
Mo	42	[Kr] $4d^5$ $5s^1$
Tc	43	[Kr] $4d^5$ $5s^2$
Ru	44	[Kr] $4d^7$ $5s^1$
Rh	45	[Kr] $4d^8$ $5s^1$
Pd	46	[Kr] $4d^{10}$
Ag	47	[Kr] $4d^{10}$ $5s^1$
Cd	48	[Kr] $4d^{10}$ $5s^2$
Hf	72	[Xe]$4f^{14}5d^26s^2$
Ta	73	[Xe]$4f^{14}5d^36s^2$
W	74	[Xe]$4f^{14}5d^46s^2$
Re	75	[Xe]$4f^{14}5d^56s^2$
Os	76	[Xe]$4f^{14}5d^66s^2$
Ir	77	[Xe]$4f^{14}5d^76s^2$
Pt	78	[Xe]$4f^{14}5d^96s^1$
Au	79	[Xe]$4f^{14}5d^{10}6s^1$
Hg	80	[Xe]$4f^{14}5d^{10}6s^2$

the rare-earth ions, including the increased likelihood of the encountering broad ligand field-dependent excited-state features and strict spin selection rules. UC in doped hosts has been observed by various researchers. The first consideration of TM ions in the frequency upconversion can be found in the work of Cresswell et al. in 1978, where in Cs_2NaYCl_6, they considered Re^{4+}, a $5d^3$ ion, as a replacement for Yb in the infrared (IR) to blue upconversion in Yb–Tm systems (Auzel 2004).

However, the use of transition metals as upconverting materials is not a new concept, e.g. yttrium aluminium garnet (YAG):Cr^{3+} ions. Daniel R. Gamelin and Hans U. Güdel presented the unusual upconversion properties because of the exchange interactions between metal ions. In their work, including rare-earth/rare earth, rare-earth/transition metal, and transition metal/transition metal ion pairs, exchange interactions are found to be particularly important when the ions are situated in close proximity to one another by the geometric constraints of the host material. UC in most of the TM ions has been performed by Markus Wermuth and Hans U. Güdel, which is appreciable and provides a new exposure to the current researchers in studying the properties of TM ions. Mostly, they are low-temperature UC phenomena. TM–lanthanide ion-codoped systems show room temperature UC luminescence since the nonradiative relaxation rate of TM ions is generally large, resulting in depopulation of the intermediate energy levels of TM ions involved in the UC process. In comparison to f–f transitions in rare earths, the d–d transitions in the TM ions have higher oscillator strengths and ligand field-dependent spin-allowed d–d excitations. Thus, they are potentially capable for chemical tuning by introducing new intermediate levels such as in Mn^{2+} and Fe^{3+}. However, there are some drawbacks of using transition metals as electron–phonon coupling leads to more multiphonon relaxations and luminescence usually occurred from lowest excited states. The 4d and 5d TM ions have greater ligand field than 3d ions. Therefore, the following two transitions are defined (Aebischer et al. 2002):

i. ligand field-dependent spin-allowed d–d transitions.
ii. ligand field-independent spin–flip transitions.

The spin–flip excitations lie midway of the spin-forbidden f–f transitions of rare-earth ions and spin-allowed d–d transitions of 3d transition metal ions (Aebischer et al. 2002).

6.2 Synthesis of Transition Metal Ion-Activated Luminescent Nanomaterials

Similar to the rare-earth-doped UC nanomaterials, the TM-incorporated UC nanomaterials can also be synthesized via the solid-state reaction method, coprecipitation method, solvothermal, hydrothermal method, Bridgman method, etc.

TMs can be used in two ways to synthesize the luminescence UCNPs or nanomaterials:

1. as dopant/codopant
2. as a component of the host material.

6.3 Structural and Optical Characterizations

TMs are incorporated as sensitizers or as activators in the upconverting nanomaterials. Thus, for the detection of TMs, elemental analysis via energy dispersive X-ray analysis (EDAX), X-ray photoemission spectroscopy (XPS), etc., is required. Also, the formation of UC nanomaterials or UCNPs with TMs can be confirmed through X-ray diffraction (XRD), transmission electron microscopy (TEM), scanning electron microscopy (SEM), field emission scanning electron microscopy (FESEM), etc., analysis. The traditional and very basic way to optically characterize the upconverting nanomaterials are Fourier transform infrared spectroscopy (FTIR), Raman, frequency upconversion, ultraviolet (UV)–visible, etc.

6.4 Frequency Upconversion and Its Various Mechanisms

There are several research findings on the enhancement of UC emission intensity. The codoping of TMs may improve the efficiency of the upconverting nanomaterials or UCNPs in many ways:

i. increases the luminescence intensity.
ii. changes the spectra from sharp to broad band.
iii. combination of rare earths with TMs provides very fascinating results.

The mechanism of TMs in the upconverting host lattice is somehow different from the rare earths. In this section, all possible dynamics and theoretical models will be discussed. Similar to the rare-earths, optical emissions from TMs are due to crystal field modification, and TMs are slightly more sensitive than rare earths (REs). Because of some specific properties, TMs behave as effective luminescent centers. The cross-relaxation and photon avalanche are the very basic optical phenomena for the desired emission and modified spectra. Also, many TMs consist of charge transfer bands such as WO_3, VO_4, MoO_4, etc., because of which broad absorption and emission spectra have been observed by different researchers. Another important phenomenon is surface plasmon resonance (SPR) usually observed in TMs such as Au and Ag. SPR is a way of photoluminescence enhancement in which light matter interaction occurs because of which phenomena such as light absorption, energy transfer, and radiative emissions increase. The mechanism of TMs in any host matrix can be very well understood with Sugano–Tanabe (T–S) diagram, which is based on the number of available electrons in the outer d orbital.

6.5 Applications

The improved UCNPs after co-doping with TMs can be used for display screens, optical communication amplifiers, lamps, solid-state lasers, optoelectronics, 3-D imaging materials, screen printing ink, anticounterfeiting, bioimaging, etc. The transition metal ions have emerging applications in plant growth and photocatalytic activity, plant virus and fungi site detection, in vivo imaging, and drug delivery. The transition metals lying in the fourth period are scandium (Sc), titanium (Ti), vanadium (V), chromium (Cr), manganese (Mn), iron (Fe), cobalt (Co), nickel (Ni), copper

(Cu), and zinc (Zn). However, the period 5 metals are yttrium (Y), zirconium (Zr), niobium (Nb), molybdenum (Mo), technetium (Tc), ruthenium (Ru), rhodium (Rh), palladium (Pd), silver (Ag), and cadmium (Cd). The period 6 transition metals are hafnium (Hf), tantalum (Ta), tungsten (W), rhenium (Re), osmium (Os), iridium (Ir), platinum (Pt), gold (Au), and mercury (Hg). Starting with the d-block, there are certain exception such as the very first element scandium has the electronic structure [Ar] $3d^1 4s^2$. When it forms ions, it always loses the three outer electrons and ends up with an argon structure. The Sc^{3+} ion has no d electrons, so does not meet the definition and hence not considered as a transition metal. Similarly, zinc does not count as a transition metal but plays an effective role in the photoluminescence from the upconverting nanomaterials. Application of some of these metals is reported in downconversion emissions but not in upconversion emissions such as Zr^{3+} and V^{5+}. Most of the TMs are not reported as dopants or codopants in upconverting nanomaterials and still need attention such as Nb, Tc, Ru, Rh, Pd, Cd, Hf, Ta, Ir, Pt, and Hg. The fourth period TMs are frequently used, and they have wide varieties of applications as they are very much dependent on the crystal field. Among the various TMs, some of them are most frequently used such as Mn^{2+}, Mn^{4+}, Cr^{3+}, Fe^{3+}, and so on.

6.6 Mechanism of Transition Metal Ions in Crystal Field

The behavior of TMs is different in the octahedral and tetragonal crystal field and also in the strong and weak crystal field. Following the equation of perturbation, the Hamiltonian can be explained as (Solé et al. 2005)

$$H = H_{FI} + H_{CF} \tag{6.1}$$

$$H_{FI} = H_0 + H_{ee} + H_{SO} \tag{6.2}$$

where H_{CF} is the crystal field of the material, H_{FI} is the Hamiltonian related to the free ion usually situated at the center of the ligands, H_0 is the central field Hamiltonian acting on the valance electrons due to nucleus, H_{ee} is the perturbation due to the Coulomb interaction among valence electrons, and H_{SO} is the sum of spin–orbit interaction due to all electrons.

The single electron ($3d^1$) system has the simplest mechanism. In this case, the strong and weak field is not differentiated, and the term H_{ee} becomes zero. However, it is complex in more than two electrons ($3d^n$), and the term H_{ee} has a special role because of more interaction among valance electrons and can be explained with the help of the T–S diagram (Solé et al. 2005). TMs have low values of atomic number (Z) compared to rare earths. As the spin–orbit term H_{SO} is dependent on the Z^4 value, its effect is negligible in the case of TMs. Therefore, H_{ee} and H_{CF} are dominant over H_{SO} (Solé et al. 2005). Because of the electronic interaction, the states generated in TMs are given by ^{2S+1}L, where "S" and "L" are the spin and angular quantum number, respectively. The separation between these states is given by Racah parameters given by A, B, and C. With the help of Racah parameters, the strength can be determined.

"A" can be neglected as it is the same for an atom, "B" represents the electronic splitting energy, and "C" is given by B/4. In the T–S diagram, the vertical y-axis represents the E/B ratio of the free ion's states and D_q/B (ratio of crystal field strength and interelectronic interaction) in the x-axis. In the T–S diagram at $D_q/B = 2.2$, there is a vertical line that distinguishes the low and high crystal field, where D_q is given by the expression (Grinberg et al. 2017)

$$D_q = \frac{Ze^2}{6R^5} \int r^4 \varphi_{3d}^2(r) d^3 r \tag{6.3}$$

where $\varphi_{3d}(r)$, e, Z, and R are the 3d wave function, charge of electron, charge of ligand ions represented by the point charge, and distance between ions and ligands, respectively. The crystal field strength parameter is represented by $10D_q$. The T–S diagram is the interpretation of the spectra of TM ions in the various crystal fields. It shows that the ^{2S+1}L levels split as the D_q/B increases. Here, the T–S diagram is explained for the $3d^3$ electron such as Mn^{4+} in the octahedral and tetrahedral crystal field as Mn has a varying valance state among which Mn^{2+} and Mn^{4+} are the most residing states in the material. Here, both the states of Mn and its effects have been discussed in terms of the structural and optical properties on incorporating in the crystal field. For the d electrons, wave functions are given by d_{xy}, d_{yz}, d_{zx}, $d_{x^2-y^2}$, and $d_{3z^2-r^2}$ (Grinberg et al. 2017). In the octahedral crystal field, the d state splits into sixfold "t" state and fourfold "e" states. The "t" state is superposition of d_{xy}, d_{yz}, and d_{zx} and "e" state is the superposition of $d_{x^2-y^2}$ and $d_{3z^2-r^2}$. The crystal field configurations t^3, t^2e, te^2, and e^3 are having crystal field energies 0, $10D_q$, $20D_q$, and $30D_q$, respectively. Further, in a 3d system, there are three "d" electrons, whose combination results in three orbitals. Each orbital is having t_{2g} and e_g orbitals. Again, its four possible combinations after mixing of orbitals are t_{2g}^3, $(t_{2g}^2 e_g^1)$, $(t_{2g} e_g^2)$, and e_g^3. The T–S diagram of the $3d^3$ system has been explained with Figure 6.1a (Grinberg et al. 2017).

6.6.1 UC Mechanisms in Mn-Based System

For Mn^{4+} ($3d^3$) in octahedral crystal field, the ground state is ^4F, and it splits into ground state $^4A_{2g}$ and two excited states $^4T_{2g}$ and $^4T_{1g}$ (Solé et al. 2005). The other excited states ^4P, ^2G, and ^2F also split into A, T, and E levels, respectively. All the energy levels are at different positions in the T–S diagram. The energy of levels ^2E_g and $^2T_{1g}$ are almost independent of crystal field, having a slope value close to zero, as shown in Figure 6.1a. The level $^2T_{2g}$ is constant over $D_q/B > 1$. Thus, the transition between these states and ground state $^4A_{2g}$ is independent of the crystal field strength. In contrast to this, some levels such as $^4T_{1g}$, $^4T_{2g}$, $^2A_{1g}$, and $^4A_{2g}$ have large slope, which indicates that these levels are strongly dependent on the crystal field (Solé et al. 2005). Therefore, it could be concluded that larger slope resembles greater dependence to the crystal field environment and stronger is the optical absorption and emission field. The transition $^4A_{2g} \rightarrow {}^2E_g$, $^2T_{1g}$ is spin forbidden and $^4A_{2g} \rightarrow {}^4T_{1g}$, $^4T_{2g}$ is spin allowed (Solé et al. 2005). Another state Mn^{2+} is an outstanding luminescent center with variable emitting color from green to yellow and then red. It has a $3d^5$ electronic system. Similar to the $3d^3$ system of electrons (Mn^{4+}), Mn^{2+} has $^6A_1$ as

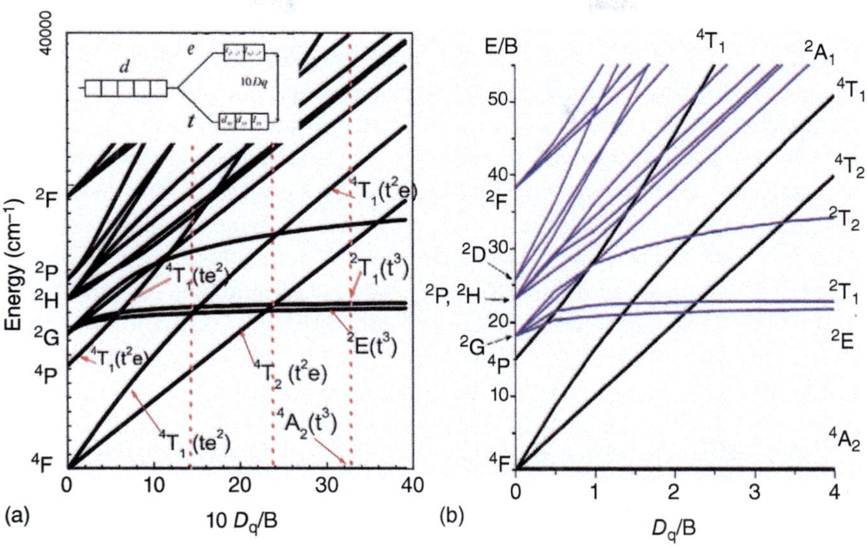

Figure 6.1 T–S diagram for (a) d^3 electron. Source: Grinberg et al. (2017). Reprinted with permission of Optical Materials. (b) d^5 Electron in octahedral crystal field. Source: Zhou et al. (2018). Reprinted with permission of the Royal Society of Chemistry.

the ground state and ${}^4T_1$, ${}^4T_2$, and ${}^4A_1$ as the excited states (Solé et al. 2005). The T–S diagram for $3d^5$ has been shown in Figure 6.1b (Zhou et al. 2018). In many works of photoluminescence, it has been reported that usually in octahedral crystal field, the transition from Mn^{2+} lies in the red region, and in tetrahedral field, the transitions occur in the green region. The energy of ${}^4T_1$ state of the sixfold octahedral Mn^{2+} is lower than that of the fourfold tetrahedral Mn^{2+}.

The luminescence from any material is a result of selection rules of radiative transitions and electron lattice interaction on spectral line shape as Mn is having five oxidation states +2, +3, +4, +5, and +6 out of which +2 and +4 are mostly occupied in the host materials. Mn can be incorporated in both ways, either as a dopant or itself as a host element. In both types of occupancies, its effects on the structural and optical properties have been discussed. The UC emissions in Mn-doped phosphors are generally giving red emission as reported by many authors. The mechanisms involved for TMs in any material can be explained with T–S diagram. The photoluminescence enhancement with cooperative sensitization has been analyzed with Mn^{2+} codoping for tuning both morphology and optical properties (Wang et al. 2015). In this work, Mn^{2+} is used to manipulate cooperative energy transfer and to adjust the luminescence spectrum. The population of photons in the case of Mn^{2+}–Yb^{3+} can take place in two ways – firstly, the cooperative sensitization, in which the two Yb^{3+} ions simultaneously transfer their excitation energy to the Mn^{2+} ions, takes place usually at more concentrations of Yb^{3+}. Secondly, the concept of dimers that is very general in the case of frequency UC arising from the Mn^{2+}-based phosphors doped with rare-earth ions occurs. It is a superexchange interaction model and takes place when there is no overlap between the excitation band of Yb^{3+} and emission band of

Mn^{2+} and the distance between the two ions is very less <5Å. In the superexchange interaction, Yb^{3+} and Mn^{2+} form new ground, intermediate, and emitting states as ($^2F_{7/2}$, $^6A_1$), ($^2F_{5/2}$, $^6A_1$), and ($^2F_{7/2}$, $^4T_1$), respectively. In the first excitation ground state absorption (GSA) step, the dimer is promoted from ground state to intermediate state, which is mainly localized in the Yb^{3+} ions. In the second excitation, the excited state absorption (ESA) step populates the emitting level localized at Mn^{2+} ions. From emitting levels, the excited ions relax to the ground state giving the visible photons.

The Yb^{3+} ions do not contain any f state above $^2F_{5/2}$ (~10 000 cm^{-1}), and Mn^{2+} ions do not have any *d* state below $^4T_1$ (~17 000 cm^{-1}). Therefore, UC emission in the Mn^{2+}–Yb^{3+}-codoped system under near-infrared (NIR) excitation takes place via cooperative sensitization or exchange-coupled Mn^{2+}–Yb^{3+} dimer model. The coupling strength may weaken with the increase in distance between them. Thus, the answer of "how do dimers work?" is the GSA/ESA excitation process of dimers. In another reported work, the transition metal silver cavity is used for enhancing light absorption, energy transfer, and radiative emission. $NaYF_4$:Er^{3+}/Yb^{3+}/Mn^{2+} phosphors have been prepared with a two-step anodization process and then spin-coated with AgNPs. The phosphors are excited with 960 nm excitation, and enhancement in red to green UC intensity is a joint effect of energy transfer from the $^4T_1$ level of Mn^{2+} and surface resonance energy transfer from AgNPs. Three UC emission peaks occur at 520, 540, and 655 nm corresponding to the transitions from the $^2H_{11/2}$, $^4S_{3/2}$, and $^4F_{9/2}$ excited states to the $^4I_{15/2}$ ground state, respectively. On codoping with Mn^{2+}, the Er^{3+} ions in the $^4S_{3/2}$ state transfer its excitation energy to the $^4T_1$ level and then this excitation energy from the $^4T_1$ level is transferred to the $^4F_{9/2}$ level because of which the red emission intensity increases, which further increases on codoping of the AgNPs (Wang et al. 2015). In another experimental work, $NaYF_4$:Er^{3+}/Yb^{3+}/Mn^{2+} phosphors have been prepared with the solvothermal method for the application of in vivo imaging and drug delivery. The phase transformation from hexagonal to cubic has been observed with the addition of Mn^{2+}, and the red/green intensity ratio increases because of the exchange energy transfer from Er^{3+} to Mn^{2+}. The poly ethylene glycol (PEG) phospholipid layer has been coated in the surface of UCNPs for in vivo imaging. For drug delivery, the drugs are loaded in these PEGylated UCNPs (Tian et al. 2012). Zeng et al. prepared a series of Er^{3+}/Yb^{3+}-doped $KMnF_3$ phosphors using the hydrothermal method. Mn^{2+} is having $^6A_1$ as the ground state, $^4T_1$ as the metastable excited state, and the two lower energy levels E_1 and E_2. Upon 980 nm excitation, if the Mn^{2+} ions are excited, it will go to the $^4T_1$ level. The excited Er^{3+} ions populate the $^4S_{3/2}$ level, which is just above the level E_1. By transferring its excitation energy to neighboring Mn^{2+} ions, the Er^{3+} ions relax. These Mn^{2+} ions again transfer energy to the lower levels of Er^{3+} and hence the red to green intensity ratio of the phosphors increases (Zeng et al. 2007). The upconverted luminescence has been obtained from the Yb^{3+}–Mn^{2+}-doped CaO phosphors (Wang et al. 2016). The phosphors have been synthesized by high-temperature solid-state reaction method. The phosphors have been structurally characterized by XRD and lattice contraction has been noticed with co-doping of the Mn^{2+} ions. Upon 980 nm excitation the UC emission occurred at 508 nm because of the $^4T_1 \rightarrow {}^6A_1$ transition, and the emission is further tunable from 560 to 725 nm wavelength by adjusting the concentration of the Mn^{2+} ions (Wang et al. 2016). In another work, broadband (530–710 nm) yellowish emitting $Sr_5(PO_4)_3Cl:Yb^{3+}/Mn^{2+}$ phosphors have been prepared by Xiao et al. by using the solid-state reaction method. The introduction of Mn^{2+} ions forms dimers combing with Yb^{3+} ions. Upon 976 nm excitation, the UC emission has been found at 585 nm because of the interaction of the Yb^{3+}–Mn^{2+} dimers (Xiao et al. 2020). Rodríguez et al. prepared the Yb^{3+}–Mn^{2+}-doped $LaMgAl_{11}O_{19}$ phosphors with the co-precipitation method. Green emission at 19 450 cm^{-1} has been observed because of the $^4T_1 \rightarrow {}^6A_1$ transition when excited with 10 250 cm^{-1}. The exchange-coupled Yb^{3+}–Mn^{2+} dimer mechanism is found to be responsible for this transition (Martín-Rodríguez et al. 2010). E.H. Song et al. obtained the tunable white luminescence from the Yb^{3+}–Tm^{3+}–Mn^{2+}:$KZnF_3$ phosphors prepared by the hydrothermal technique. Upon 976 nm excitation, a broadband peaking at 585 nm is originated from the Yb^{3+}–Mn^{2+} dimers. On codoping with the Tm^{3+} ions in this phosphors, white light is generated from the Yb^{3+}–Tm^{3+}–Mn^{2+}:$KZnF_3$ phosphors. The UC emission exhibited peaks at 480, 650, and 700 nm because of the $^1G_4 \rightarrow {}^3H_6$, $^1G_4 \rightarrow {}^3F_4$, and $^3F_{2,3} \rightarrow {}^3H_6$ transitions (Figure 6.2). On increasing the concentration of Mn^{2+} from 1 to 20 mol%, the UC emission intensity changes. The peaks at 480 and 650 nm almost diminished, and the peaks at 585 and 700 nm rise. At 5 mol% concentration of the Mn^{2+} ions, the emission at ~700 nm is maximum. The energy transfer from the Yb^{3+} ions and further from the Tm^{3+} ions can be realized (Song et al. 2014).

Also, Mn^{2+} ions can be used as a sensitizer in rare-earth-doped phosphors prepared by using solid-state reaction method upon 1550 nm excitation. An enhancement of about 2–3 times has been reported with codoping of Mn^{2+} in the Er^{3+}-doped Y_2O_2S phosphors because of the energy transfer from matching levels of Er^{3+} to Mn^{2+} ions (Yuan et al. 2014). The screen-printing ink has been prepared with $RbCaF_3$:Mn^{2+}–Yb^{3+} for anticounterfeiting application by E. Song et al. $RbCaF_3$:Mn^{2+}–Yb^{3+} is a afterglow phosphor (luminescence glow >300 ms) having an emission peak at 565 nm upon 980 nm excitation. On varying the temperature from 300 to 77 K, the luminescence appears from 62 to 138 ms. The superexchange mechanism is responsible for the UC emission. The codoping of Mn^{2+} in the phosphors reduces the lifetime of Yb^{3+}, which shows Mn^{2+}–Yb^{3+} energy transfer (Song et al. 2019). Qiang and Wang observed the enhancement in the UC emission by back energy transfer from the Ho^{3+} ions to Mn^{2+} ions in the $NaGdF_4$:$Ho^{3+}/Yb^{3+}/Mn^{2+}$ phosphors prepared with a hydrothermal synthesis technique. The back energy transfer initiates the $^3F_3$ and $^5F_4$, $^5S_2$ (Ho^{3+}) → $^4T_1$ (Mn^{2+}) → $^5F_5$ (Ho^{3+}) transition because of which stronger peaks at 630–680 nm occur. On codoping with Mn^{2+} in the phosphors, both the green and red emission intensities increase. With the increasing temperature, color tunability has been observed (Qiang and Wang 2019). Enhanced UC emission has also been observed by Yao et al. in the Mn^{2+}-codoped $NaYF_4$:Er^{3+}/Yb^{3+} phosphors prepared via a solvothermal method. With the incorporation of Mn^{2+}, a slight change in the rod-like structure to cubic shape has been observed. In the red UC emission band at 659 nm, the intensity enhancement has been observed because of the energy transfer from the Er^{3+} ions

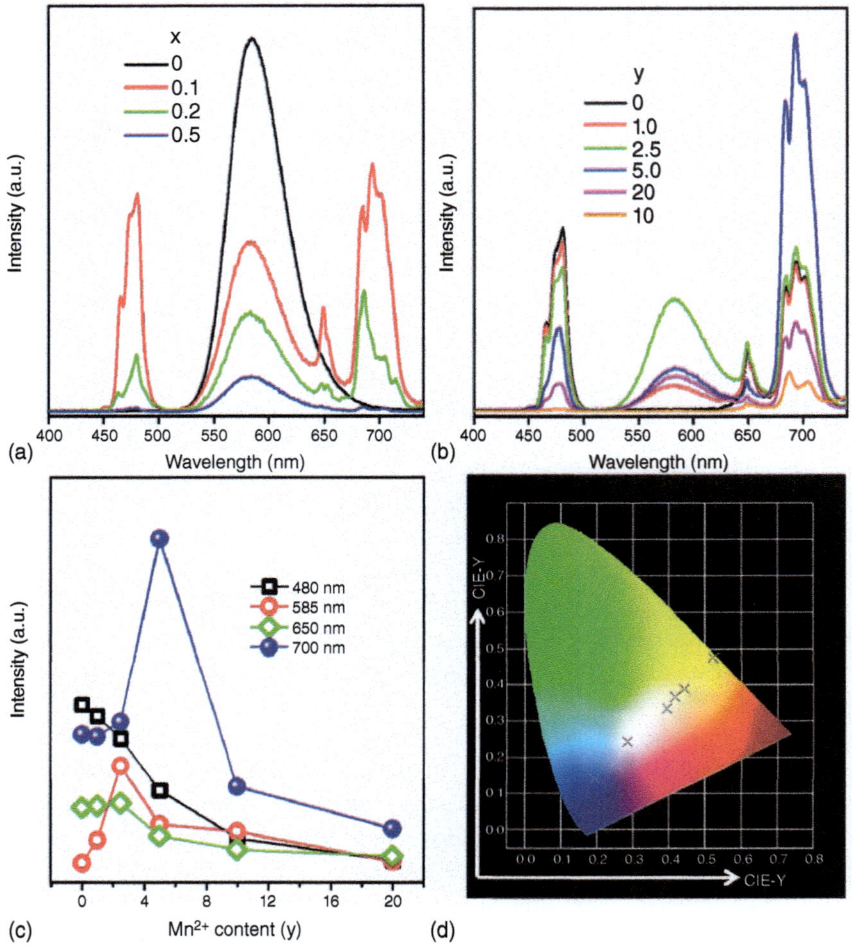

Figure 6.2 (a) The UC emission spectra of $KZnF_3$:1% Yb^{3+}, x%Tm^{3+}, 2.5% Mn^{2+}(x = 0, 0.1, 0.2, and 0.5), (b) $KZnF_3$:1% Yb^{3+}, 0.1% Tm^{3+}, and y% Mn^{2+} (y = 0, 1, 2.5, 5, 10, and 20). (c) UC emission intensity of $KZnF_3$:1%Yb^{3+}, 0.1%Tm^{3+}, and y%Mn^{2+} as a function of Mn^{2+} content and (d) Commission Internationale de l'Eclairage (CIE) chromaticity coordinates of $KZnF_3$:1%Yb^{3+}, x%Tm^{3+}, y%Mn^{2+} (1–5: for x = 0.1, y = 0, 1, 5.0, 2.5, and 10, respectively; 6: for x = 0 and y = 2.5). Source: Song et al. (2014). Reprinted with permission of The Optical Society of America.

and the sensing properties improved. The temperature sensitivity is enhanced from 0.0036 to 0.0057 K^{-1} from $NaYF_4$:Er^{3+}/Yb^{3+} to $NaYF_4$:Er^{3+}/Yb^{3+}/Mn^{2+} phosphors (Yao et al. 2019). The enhancement in the red UC emission band has been observed in the $NaYF_4$:Ho^{3+}/Yb^{3+} phosphors (Lingeshwar Reddy et al. 2017). The UC emission intensity enhancement through addition of Mn^{2+} has also been studied in the $CaZnF_4$ and $ZnWO_4$ phosphors (Reinhard et al. 2004; Ren and Yang 2018). A yellowish broad emission has been noticed by Xie et al. upon 980 nm excitation in the Mn^{2+}/Yb^{3+}:$Ca_3Y(GaO_3)(BO_3)_4$ phosphors prepared by using a solid-state reaction method. The UC emission covers the broad range from 450 to 750 nm peaking at 505 nm because of the transition from exchange-coupled Mn^{2+}–Yb^{3+}. The Mn^{2+}–Yb^{3+} pair creates new energy states because of the overlap of their wave functions (Xie et al. 2020). UC emission intensity enhancement has been observed in the $Mn^{4+}/Er^{3+}/Yb^{3+}$-doped glass ceramics prepared by the conventional melting method (Dan et al. 2014). The green UC emission intensity decreases, and hence, the intensity of red/green ratio increases after codoping of Mn. A broad band spectrum within 450–700 nm has been obtained because of Mn codoping upon 980 nm excitation (Dan et al. 2014). The red UC emission transition has been observed in the Mn^{2+}-codoped Eu^{3+}/Yb^{3+}:$KCdF_3$ phosphors prepared with the solvothermal method. Upon 980 nm excitation, the red emission increases with the incorporation of Mn because of energy transfer from exchange-coupled Mn^{2+}–Yb^{3+} dimer to Eu^{3+}. The enhancement in the lifetime of Eu^{3+} has been noticed after codoping with the Mn^{2+} ions (Ye et al. 2016).

6.6.2 UC Mechanisms in Mn^{4+}- and Ti^{2+}-Based Systems

$Yb^{3+}/Ho^{3+}/Mn^{4+}$:$SrLaAlO_4$ phosphors have been synthesized by Yan et al. using a solid-state reaction technique. Upon 980 nm excitation, the phosphors give broad red UC emission covering the 600–800 nm wavelength range because of the $^5F_5 \rightarrow {}^5I_8$ and $^5S_2/{}^5F_4 \rightarrow {}^5I_7$ transitions from the Ho^{3+} ions and $^2E_g \rightarrow {}^4A_{2g}$ transition from the Mn^{4+} ions. The luminescence from the Mn^{4+} ions is dependent on the concentration of the Ho^{3+} and Yb^{3+} ions, which ensures the energy transfer from rare-earths ions to Mn^{4+} ions. For the optimized phosphors, the energy transfer efficiency is ~7%. On increasing the Mn^{4+} ion concentration, the lifetime of the green level ($^5S_2/{}^5F_4$) decreases, and the energy transfer (ET) process from the Ho^{3+} to Mn^{4+} ions increases. Thus, the broad emitting material can be explored for lighting and optical communication applications (Yan et al. 2020). UC emission in the Mn^{4+}-codoped Ln^{3+}/Yb^{3+} $YAlO_3$ phosphors prepared with the solid-state reaction method has been observed by Chen et al. The spectral overlap between the Ln^{3+} (Er^{3+}, Tm^{3+}, and Ho^{3+}) and Mn^{4+} ions shows the energy transfer because of which the red UC emission occurred at 714 nm corresponding to Mn^{4+}:$^2E \rightarrow {}^4A_2$ transition. All the prepared phosphors have been analyzed for temperature sensing with enhanced sensitivity (Chen et al. 2017). The mutual energy transfer from the Er^{3+} to Mn^{4+} ions has been observed in the codoped Gd_2ZnTiO_6 phosphors prepared by Liao et al. The phosphors synthesized by using the solid-state reaction method give deep red emission upon 980 nm excitation. The energy transfer efficiency from the Er^{3+} to Mn^{4+} ions increases with increasing concentration of Mn^{4+}, which results in enhancement of 705 nm UC emission (Liao et al. 2017). The energy transfer from the Er^{3+} to Mn^{4+} ions has also been observed in the Er^{3+}/Mn^{4+}-doped $NaY(MoO_4)_2$ phosphors (Jiang et al. 2021). Also, the UC emission transition occurs at 712 nm upon 980 nm excitation in the $Ca_{14}Zn_6Ga_{10}O_{35}$:Mn^{4+},Ln^{3+} phosphors (Lv et al. 2019). Researchers have reported MnO_2 as a nanosheet shielded with UCNPs for pesticide detection based on the luminescence resonance energy transfer (Ouyang et al. 2021). This method is based on the quenching of the luminescence because of the strong overlap of the absorption spectrum of UCNPs and MnO_2. The UCNPs

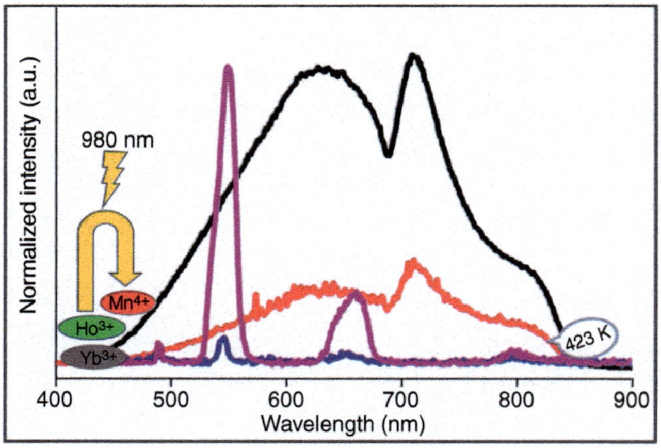

Figure 6.3 Room temperature UC emission spectra of the optimized Ho^{3+}, Ho^{3+}/Yb^{3+}, $Ho^{3+}/Yb^{3+}/Mn^{4+}$:SnO_2 phosphors and UC emission spectra of the optimized $Ho^{3+}/Yb^{3+}/Mn^{4+}$:SnO_2 phosphors at 423 K. Source: Modified from Prasad and Rai (2020). Reprinted with permission of Elsevier.

act as a donor and MnO_2 acts as an accepter (Ouyang et al. 2021). The influence of Mn^{4+} has been studied in the Ho^{3+}/Yb^{3+}-doped thermally stable SnO_2 phosphors synthesized with the solid-state reaction method (Prasad and Rai 2020). Codoping of Mn^{4+} has large effect in the bandgap of the phosphors. The temperature-dependent UC emission study has also been performed. With the incorporation of Mn^{4+}, the spectra of sharp green UC emitting Ho^{3+}/Yb^{3+}:SnO_2 phosphors has been modified. Upon 980 nm excitation, a broad band has been observed ranging from 400 to 900 nm wavelength range peaking at 640 and 710 nm as shown in Figure 6.3. The responsible mechanisms for the UC emissions have been discussed on the basis of energy transfer, cooperative energy transfer, and photon avalanche processes (Prasad and Rai 2020). The TMs Ti^{3+} and V^{4+} ions have the simplest $3d^1$ electronic configuration. In V^{4+}, only downconversion spectroscopic properties have been studied until now; however, some works on frequency UC have been reported, which will be discussed further.

Wenger et al. reported that external hydrostatic pressure can be used to tune the UC properties and efficiencies. They reported the NIR to visible photon upconversion properties of Ti^{2+}-doped NaCl at 15 K as a function of external hydrostatic pressure. When 0.8% Ti^{2+} ions, having $3d^2$ electronic configuration, are doped into NaCl, it substitutes Na^+ and is thus in octahedral Cl^- coordination. On excitation with ~9394 cm^{-1}, two broad bands centered at ~8200 and ~15 500 cm^{-1} are observed, which are assigned to the spin-allowed ${}^3T_{1g} \rightarrow {}^3T_{2g}$ and ${}^3T_{1g} \rightarrow {}^3T_{1g}$ transitions, respectively. The basic concept behind this phenomenon is with increasing pressure, the crystal field strength $10D_q$ increases. The responsible mechanism GSA/energy transfer upconversion (ETU) becomes 10 times more efficient than GSA/ESA at 34 kbar. The overall upconversion efficiency of 0.8% Ti^{2+}:NaCl at 15 K has been increased by a factor of 20 between ambient pressure and 34 kbar (Wenger et al. 2002a).

6.6.3 UC Mechanisms in Cr^{3+}-Based System

The effect of co-doping with the Cr^{3+} ions in the $ZnGa_2O_4$:Yb^{3+},Er^{3+},Cr^{3+} nanoparticles (NPs) synthesized via a hydrothermal method at 120 °C has been studied by Srivastava et al. The prepared NPs emit a bright single red emission under 980 nm excitation. This process is facilitated by the absorption of green emission via the $Er^{3+} \rightarrow Cr^{3+}$ ET, followed by back ET to the $^4F_{9/2}$ level of Er^{3+}. The red-to-green ratio around 23 times has been achieved from the developed phosphors at an annealing temperature of 800 °C. These NPs also directly emit NIR emission under 254 nm excitation. This luminescence ability can only be deciphered when dual NIR and UV excitations exist simultaneously, which makes illegal imitation very difficult to implement, thus improving the level of anticounterfeiting. The prepared phosphors may be suitable for biological imaging, security applications, and optical probes for magnetic resonance imaging (Srivastava et al. 2020). Upon infrared femtosecond laser excitation at ~800 nm, the Cr^{3+}-doped Al_2O_3 crystal shows red UC luminescence at 694 nm because of the $^2E \rightarrow {}^4A_2$ transition (Yang et al. 2005). The process responsible for this emission is the two-photon simultaneous absorption. Heer et al. have reported the frequency upconversion under NIR excitation in yttrium gallium garnet (YGG) and YAG codoped with the Cr^{3+} and Yb^{3+} ions. They observed the UC emission around 700 nm corresponding to the $^2E \rightarrow {}^4A_2$ (Cr^{3+}) transition followed by the cooperative energy transfer process from the Yb^{3+} to Cr^{3+} ions (Heer et al. 2001). Further, in many works, Cr^{3+} has been codoped with two rare-earth ions, Er^{3+} and Yb^{3+}, acting as an activator and a sensitizer. The Cr^{3+} ion in the codoped system influences the crystal field, which results the change in UC mechanism, and hence, an increment in the intensity has been observed. Upon 808 nm excitation, the UC phenomenon has been achieved by Maciejewska et al. in the Cr^{3+}–Nd^{2+}:GdP_5O_{14} nanocrystals. The novel GdP_5O_{14}:Cr^{3+},Nd^{3+} nanocrystals have been prepared using the two steps: (i) precipitation of hydroxycarbonates and (ii) annealing with an excess of ammonium dihydrogen phosphate. Upon λ_{exc} = 808 nm, the Stokes ($^4F_{3/2} \rightarrow {}^4I_{11/2}$) and anti-Stokes (i.e. UC) emissions ($^4G_{5/2}$, $^2G_{7/2}$) $\rightarrow {}^4I_{9/2}$ have been obtained. Additionally, it was found that the increased concentration of the Nd^{3+} ions have beneficial influence on the thermometric properties because of the fact that shortening of the average distance between ions facilitates the UC process (Maciejewska and Marciniak 2020). The strong broadband Cr^{3+}-sensitized UC emission in the $LiScSi_2O_6$:Cr^{3+}/Er^{3+} phosphors synthesized through the solid-state reaction method has been achieved by Duan et al. under 690 nm laser excitation. No upconversion phenomenon is observed in singly doped Cr^{3+}or Er^{3+} ions. The effect of codoping with the Er^{3+} and Yb^{3+} ions in the $LiScSi_2O_6$:Cr^{3+} phosphors has been investigated. Upon 690 nm excitation in the singly Cr^{3+}-doped $LiScSi_2O_6$ phosphors, the emission peak is observed at ~845 nm because of the spin-allowed $^4T_2 \rightarrow {}^4A_2$ transition of Cr^{3+} (Figure 6.4). Strong green emission from the $^2H_{112}$ and $^4S_{3/2}$ levels of the Er^{3+} ions at ~522, 536, 548, and 562 nm has been detected because of the codoping of the Er^{3+} ions, which shows that there is energy transfer process from $Cr^{3+} \rightarrow Er^{3+}$ (Duan et al. 2020). Huy et al. have reported ~20 times improvement in the UC emission intensity with codoping of Cr^{3+} in the $NaLuGdF_4$:Yb/Er phosphors

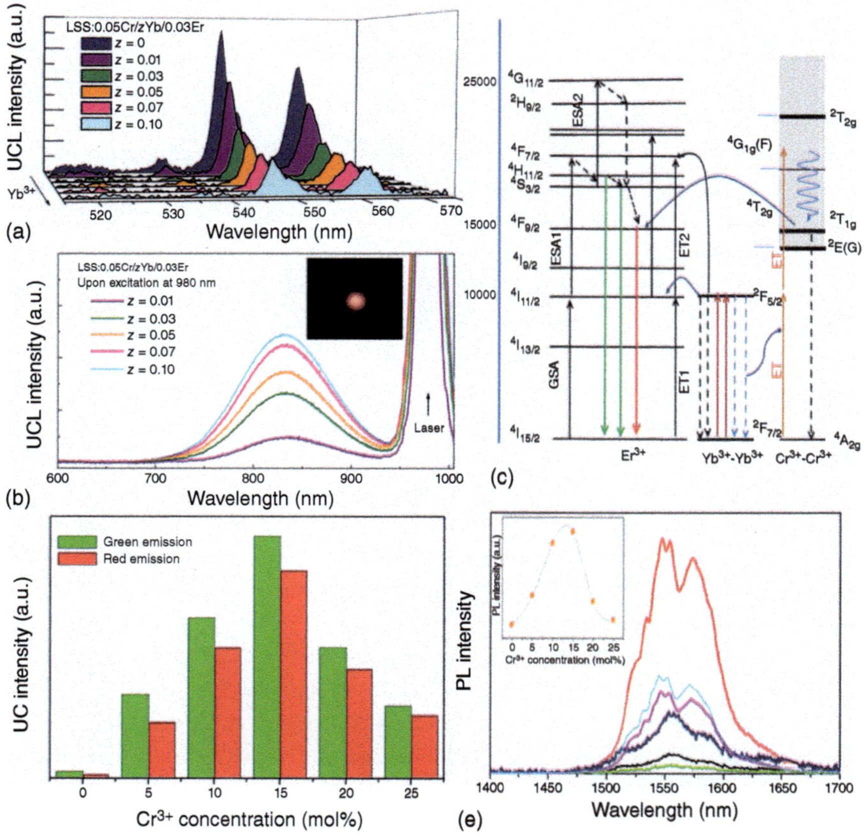

Figure 6.4 (a) Upconversion luminescence (UCL) spectra of $LiScSi_2O_6$ (LSS):$0.05Cr^{3+}/zYb^{3+}/0.03Er^{3+}$ with various Yb^{3+} ions concentrations under the 690 nm laser excitation and (b) UCL spectra of LSS:$0.05Cr^{3+}/zYb^{3+}/0.03Er^{3+}$ under the 980 nm laser excitation, and the inset presents the corresponding UCL picture taken by a visible camera. Source: Duan et al. (2020). Reprinted with permission of Elsevier. (c) Proposed energy transfer mechanism of Cr^{3+} -doped upconversion particle (UCP):Yb, Er. (d) Effect of Cr^{3+} concentration on green and red band emission under excitation at 980 nm wavelength, and (e) NIR emission spectra of UC-YbErCr samples under excitation at 980 nm wavelength (inset: photoluminescence (PL) intensity of NIR band versus Cr^{3+} concentrations). Source: Huy et al. (2017). Reprinted with permission of Elsevier.

prepared by the solvothermal method. The main reason behind the mechanism is the formation of Yb^{3+}–Cr^{3+} dimers, which are able to transfer energy to the Er^{3+} ions (Figure 6.4) (Huy et al. 2017). The broadband Cr^{3+}-sensitized UC luminescence in the $La_3Ga_5GeO_{14}$:$Er^{3+}/Yb^{3+}/Cr^{3+}$ phosphors have been analyzed upon 976 nm diode laser excitation by Ye et al. The UC emission $^4T_2 \rightarrow {}^4A_2$ transition of Cr^{3+} is observed from the developed phosphors, and the slope values of the emission peak intensities on pump are deviated from 2. This may be attributed to the large depopulation rate of intermediate energy levels in the UC process created by forward and backward energy transfer among the Yb^{3+}, Er^{3+}, and Cr^{3+} ions. The UC emission intensity of Er^{3+} decreases, while that of Cr^{3+} varies slightly, as the Cr^{3+} content slightly increases (Ye et al. 2014).

6.6.4 UC Mechanisms in the Fe^{3+}-Based System

UC in singly doped Fe^{3+} ions has not been reported till date, but the combination of Fe^{3+} with the Er^{3+}/Yb^{3+} ions has been reported by various researchers. The structural and optical properties of Fe^{3+} ion-incorporated $NaGdF_4:Er^{3+}/Yb^{3+}$ nanorods prepared by the coprecipitation method have been reported by Tang et al. The structural modification has been noticed because of the shift in the diffraction angle and shrinkage in the unit cell after codoping of Fe^{3+} ions. Upon 980 nm excitation, the UC emissions are obtained at 521, 541, and 657 nm from the Er^{3+} ions. The UC emission intensities have been sharply improved on adding the Fe^{3+} ions. The UC mechanism has been explained with two-photon energy transfer from the Yb^{3+}–Fe^{3+} dimers to Er^{3+}. The maximum absolute and relative sensor sensitivities of the $NaGdF_4:Er^{3+}/Yb^{3+}/0.005\ Fe^{3+}$ nanorods have been found to be ~0.0191 and ~0.567% K^{-1}, respectively, which are higher as compared to the $NaGdF_4:Er^{3+}/Yb^{3+}$ nanorods. This shows that the $Er^{3+}/Yb^{3+}/Fe^{3+}:NaGdF_4$ nanorods can be utilized as a potential upconverting compounds for ratiometric nanothermometers and optical heaters (Tang et al. 2020). The increase in UC emission has been reported in the $Er^{3+}/Yb^{3+}/Fe^{3+}:NaGdF_4$ phosphors prepared by the solvothermal process. About 10 times increment in the green band and 3 times in red band have been observed upon 980 nm excitation (Muraleedharan and Zhang 2020). In improving the red UC emission band, the concept of energy transfer from the Yb^{3+}–Fe^{3+} dimers to Er^{3+} in the $Er^{3+}/Yb^{3+}/Fe^{3+}:NaYF_4$ phosphors prepared using the hydrothermal method has been developed by Tang et al. On increasing the doping concentration of Fe^{3+} ions, the transformation from cubic to hexagonal structure, expansion in the host, increase in the unit cell volume, increase in the interplanar distance, and increase in particle size have been reported. The intensity of the red UC emission is increased by seven times with addition of the Fe^{3+} ions (Figure 6.5) (Tang et al. 2015). The frequency upconversion and temperature sensing in the $Er^{3+}/Yb^{3+}/Fe^{3+}:NaBiF_4$ nanoparticles prepared by using the coprecipitation method have been reported by Du et al. The shrinkage in the host lattice and the increase in the crystallite size have been noticed with the addition of Fe^{3+} ions. After codoping with Fe^{3+} ions, enhancement in the optical properties and intensity of both green and red UC emission bands have been noticed. The responsible mechanism is found to be ET transfer from the Yb^{3+}–Fe^{3+} dimers to the Er^{3+} ions. The maximum sensor sensitivity for the $Er^{3+}/Yb^{3+}/Fe^{3+}:NaBiF_4$ nanoparticles is obtained to be 0.0053 K^{-1} at 543 K. The increased ability to transform excitation pump power to heat makes the phosphors applicable for cancer therapy treatment (Du et al. 2019). Structural and photoluminescence studies of the $Yb^{3+}/Fe^{3+}:NaErF_4$ microcrystals synthesized via a hydrothermal technique have been carried out by Zhang et al. The structural phase transition from cubic to hexagonal has been reported for the prepared codoped microcrystals. Also, an increase in the particle size causes reduction in the surface quenching, which may further contribute to the

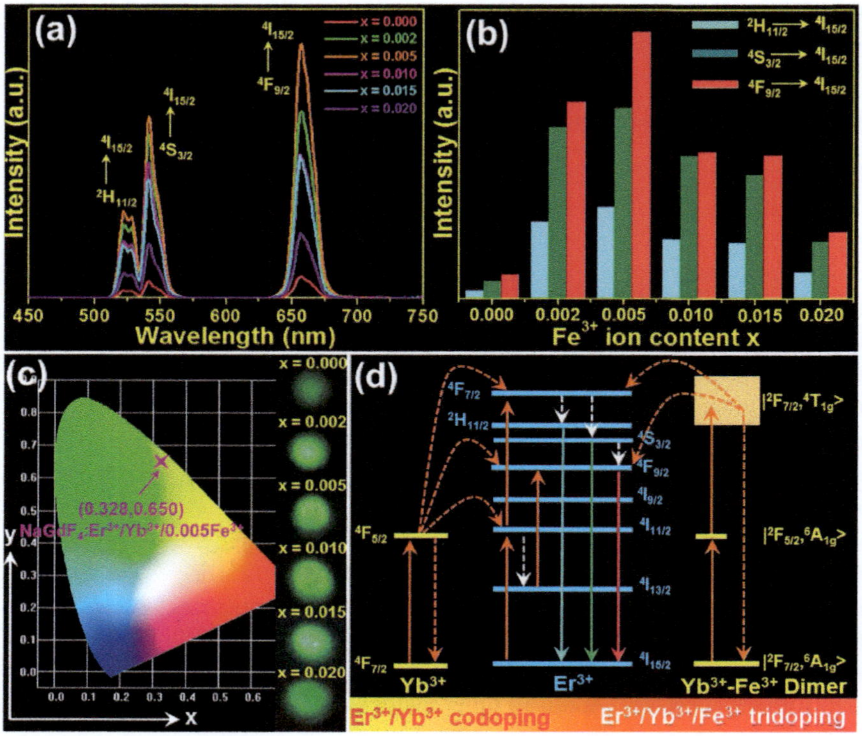

Figure 6.5 (a) UC emission spectra of the NaGdF$_4$:Er^{3+}/Yb^{3+}/xFe^{3+} nanorods. (b) Dependence of UC emission intensity on the Fe^{3+} ion content. (c) CIE chromaticity diagram of the NaGdF$_4$:Er^{3+}/Yb^{3+}/0.005Fe^{3+} nanorods. The inset shows the optical images of the resultant samples excited at 980 nm light. (d) Simplified energy-level diagram of the Yb^{3+} and Er^{3+} ions as well as the Yb^{3+}–Fe^{3+} dimer. Source: Modified from Tang et al. (2015). Reprinted with permission of Elsevier.

enhanced UC emission. Dual excitations at 980 and 1560 nm have been performed in the developed system. As the energy transfer mechanism from Yb^{3+} to Er^{3+} changes the population mechanism, thus, upon 980 nm excitation, the material exhibits red-green emission, and on 1560 nm excitation, it gives green–yellow emission. The UC emission intensity increases with the incorporation of the Fe^{3+} ions. The red to green band ratio varies with the variation of concentration of the Fe^{3+} ions and thus may be used as a double-color controllable material and anti-counterfeiting applications. The magnetic hysteresis has also been performed in the Yb^{3+}/Fe^{3+}:NaErF$_4$ microcrystals at room temperature. On increasing the concentration of the Fe^{3+} ions, the magnetization increases because of the intrinsic magnetic moment of Fe^{3+}, which broadens the application of the prepared microcrystals in magnetic resonance imaging (MRI), charge transfer (CT), and bioseparation (Zhang et al. 2019). The effect of addition of Fe^{3+} has been realized in the CeO$_2$ material by Bai et al. They prepared the Er^{3+}/Yb^{3+}/Fe^{3+}:CeO$_2$ nanomaterials by using a hydrothermal technique. The codoping of Fe^{3+} improved the broad-spectrum absorption and light scattering properties in dye-sensitized solar cells. No structural change has been observed; however, the UC emission intensity of green and red bands increases by seven and three times, respectively, under 980 nm excitation. When the sunlight is irradiated, the NIR photons can be converted into the visible photons by the $Er^{3+}/Yb^{3+}/Fe^{3+}$-codoped CeO_2 nanomaterials. Thus, the developed nanomaterials improve the light capturing capacity and photoconversion efficiency (Bai et al. 2020).

6.6.5 UC Mechanisms in Co^{3+}- and Ni^{2+}-Based System

The frequency UC in Co^{3+} has been reported by Manaa et al. in MgF_2 and $KZnF_3$ crystals (Manaa and Moncorge 1990). They explained the excited-state absorption spectra and absolute cross section in the Co^{3+}-doped MgF_2 and $KZnF_3$ crystals. The Racah parameters have been calculated for both the crystals. The band around 800 nm due to the $^4T_2$ (4F) → $^4T_1$ (4P) transition is more intense relative to the band occurred around 1300 corresponding to the $^4T_2$ (4F) → $^4A_2$ (4F) transition. Oliver et al. reported that the external hydrostatic pressure can be used to tune the UC properties and efficiencies. The NIR to visible photon upconversion properties of the Ni^{2+}-doped $CsCdCl_3$ crystal at 15 K have been studied as a function of external hydrostatic pressure. When Ni^{2+} is doped into $CsCdCl_3$, it enters into the site of trigonally distorted octahedral Cl_2 coordination. UC luminescence in the Ni^{2+}:$CsCdCl_3$ crystal occurs because of the $^1T_{2g} \rightarrow {}^3A_{2g}$ transition at 15 K. On increasing the pressure from 1 to 46 kbar with a rate of 19 cm^{-1}/kbar upon 810 nm excitation, the UC emission band shifted from 610 to 585 nm. Thus, because of the $^1T_{2g} \rightarrow {}^3A_{2g}$ transition of Ni^{2+} ions, the upconversion luminescence undergoes a color change from red to yellow (i.e. blue shift) in the 1–46 kbar pressure range (Figure 6.6) (Wenger et al. 2002b). The increase in the ligand field strength with increasing pressure results in reduction in the Racah parameters B and C produced by the increasing degree of covalency in the Ni^{2+}–Cl–interaction toward higher pressures. Thus, high pressure may be a useful tool to tune the UC properties, and it may potentially become an important technique to study the photophysical properties such as spectral overlaps, competition between radiative and nonradiative relaxation processes, and energy-level sequences in many UC systems (Wenger et al. 2002b). The Ni^{2+} is alone enough to give green UC luminescence as mentioned in the MgF_2:Ni^{2+} phosphors prepared by Tonucci et al. The luminescence band occurs because of the $^1T_2$ (^{1}D) → $^3T_1$ (^{3}F) transition, which potentially gives a second pathway for non-radiative excitation energy transfer by cross-relaxation in crystals of higher Ni^{2+} concentration (Tonucci et al. 1990).

In another way, Ni^{2+} can be introduced with the Er^{3+} ions in the hosts and used as solar cell materials. Luitel et al. reported the Ni^{2+}, Er^{3+}-codoped $CaZrO_3$ broadband-sensitive upconverters prepared by the solid-state reaction method, which ensures the absorption of those photons that are not received by c-Si itself or Er^{3+} ions. $CaZrO_3$:Ni^{2+},Er^{3+} absorbs photons in a broad wavelength range (1060–1600 nm) and emits photons at 980 nm, which needs to be efficiently absorbed by c-Si solar cells. Thus, the conversion efficiency of c-Si solar cells has been improved by 18%. For the single Ni^{2+}-doped $CaZrO_3$, a broad emission band

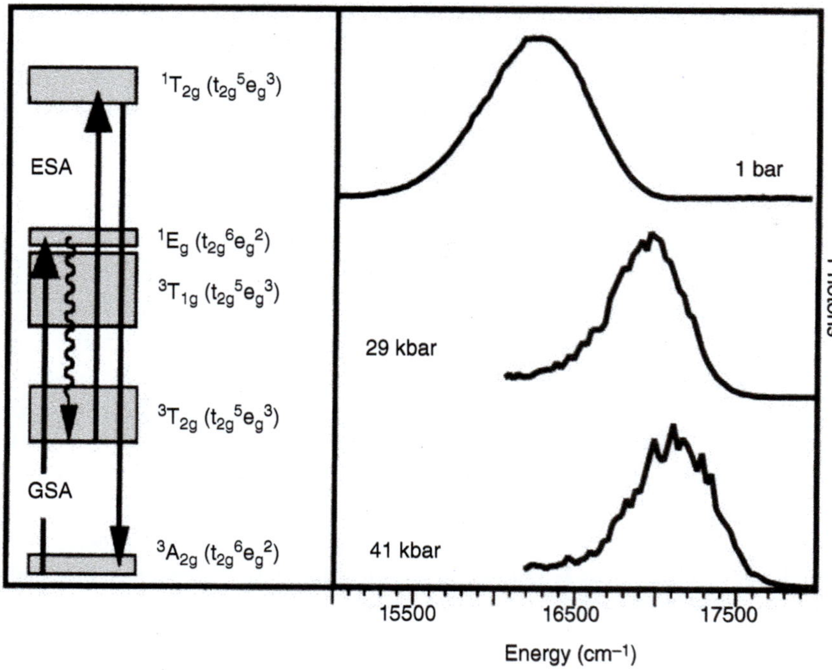

Figure 6.6 (a) Energy-level diagram for Ni^{2+} in octahedral Cl_2 coordination illustrating the Ni^{2+} UC mechanism. The solid upward arrows represent the ground-state and excited-state absorption, respectively, and the solid downward arrow represents the UC emission. The wavy arrow stands for nonradiative multiphonon relaxation. (b) $^1T_{2g} \rightarrow {}^3A_{2g}$ upconversion luminescence of Ni^{2+}:$CsCdCl_3$ at 15 K excited around 12 400 cm^2 at 1, 29, and 41 kbar. Source: Wenger et al. (2002b). Reprinted with permission of Taylor and Francis.

ranging from 1300 to 1600 nm originated from the $^3T_2$ (^3F) → $^3A_2$ (^3F) transition of Ni^{2+} is observed. However, the UC emission intensity was realized maximum at 15 mol% Er^{3+} doping concentration in $CaZrO_3$:Ni^{2+},Er^{3+} under both 1300 and 1490 nm excitations. On introducing the Er^{3+} (Li^+) codopants, the Ni^{2+} emission almost disappeared, and the Er^{3+} emission around 1550 nm ($^4I_{13/2} \rightarrow {}^4I_{15/2}$) and 980 nm ($^4I_{11/2} \rightarrow {}^4I_{15/2}$) appeared. This suggests that the energy absorbed by the Ni^{2+} ions is transferred to the Er^{3+} ions producing emissions from the Er^{3+} ions, i.e. the Ni^{2+} ions act as donors and the Er^{3+} ions as acceptors. Further, under Er^{3+} co-doping at ≥15 mol%, the emission arising from Ni^{2+} almost perfectly quenched, and only the Er^{3+} emission appeared, thereby showing the Ni → Er energy transfer efficiency close to unity. Based on the experimental observations, it is concluded that this design for broadband-sensitive upconversion may cover a new direction for the improvement in efficiency of the present c-Si solar cells to go beyond the limiting conversion efficiency of single-junction solar cells (Luitel et al. 2016).

6.6.6 UC Mechanisms in Cu^{2+}-, Zn^{2+}-, and Zr^{4+}-Based System

Cu^{2+} as a TM ion has not been reported for giving the UC emission directly but can be used in the UCNP-based probe for signal regulator in monitoring biothiols in living cells and tissues (Liang et al. 2016). In the work reported by Liang et al., seed-mediated growth approach has been used to prepare the sandwich-structured upconversion nanoparticles (SWUCNPs), i.e. $NaYF_4$@$NaYF_4$:Yb,Tm@$NaYF_4$, capped with oleate. The $NaYF_4$:Tm^{3+}/Yb^{3+} UCNPs have been prepared and coated with Cu^{2+} particles. In the present case, Cu^{2+} has been taken as a characteristic to explore the quenching effect of heavy metal ions to the UC emission. The researchers have observed that the heavy metal ions, Cu^{2+} and Fe^{3+}, may cause the quenching of UC luminescence to >95%. They concluded that the developed system with heavy metal ions may describe a new way to design the UCNP-based probe for biological applications (Liang et al. 2016).

Zinc has the electronic structure [Ar] $3d^{10}4s^2$. When it forms ions, it always loses the two 4s electrons to give a 2+ ion with the electronic structure [Ar] $3d^{10}$. Incorporation of non-lanthanide (Zn^{2+}) in the upconverting nanophosphors/UCNPs increases the luminescence intensity, arising from the rare-earth ions, many times. In almost all the research works reported by the researchers, intensity enhancement is noticed (Figure 6.7). The nanophosphors containing the rare-earth and non-rare-earth ions have been synthesized by different preparation techniques (Pandey and Rai 2013; Mahata et al. 2015; Yadav et al. 2018; Tiwari et al. 2015; Dey et al. 2012; Kumari et al. 2014, 2017).

The doping of lanthanide and non-lanthanide ions can simultaneously enhance the UC and downconversion luminescence intensity. The effect of codoping with the Zr^{4+} ions in the $NaGdF_4$:Er/Yb nanocrystals, prepared by the co-precipitation method, on its photoluminescence properties have been performed by L. Yingying et al. With the incorporation of Zr^{4+}, because of the change in the local crystal field around the rare-earth ions, a great change has been observed in the

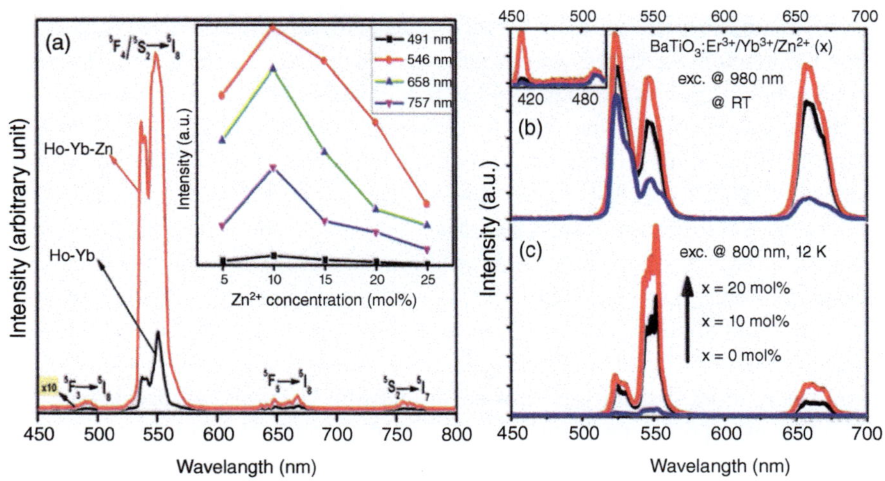

Figure 6.7 (a) Upconversion emission spectra of Y_2O_3:Ho^{3+}–Yb^{3+} and Y_2O_3:Ho^{3+}–Yb^{3+}–Zn^{2+} nanophosphors along with concentration dependence of Zn^{2+} ions. Source: Pandey and Rai (2013). Reprinted with permission of Dalton Transactions. UC emission spectra of $BaTiO_3$:Er^{3+}/Yb^{3+} and $BaTiO_3$:Er^{3+}/Yb^{3+}/Zn^{2+} phosphors annealed at 850 °C (b) excited by 980 nm at room temperature (295 K) and (c) excited by 800 nm at 12 K. Source: Mahata et al. (2015). Reprinted with permission of Royal Society of Chemistry.

structure, decrease in particle size, optical characteristics, and lifetime of the developed nanocrystals containing rare-earth ions. Upon 980 nm excitation, UC emissions in the green and red regions have been observed in the $NaGdF_4$:Er/Yb nanocrystals. The drastic change in the lifetime of 541 nm emission has been observed, from 82.1 μs for the un-doped Zr^{4+} nanocrystals to 114.1 μs for the Zr^{4+}-codoped nanocrystals containing rare-earth ions. On introducing the Zr^{4+} ions, the green UC emission intensity enhancement of about seven times to that of the $NaGdF_4$:Er^{3+}/Yb^{3+} phosphors have been observed. Based on the study, it is concluded that to manipulate the UC and downconversion luminescence coming from the rare-earth ions, as emitters, the local crystal field around the emitters could be an effective parameter (Li et al. 2020).

6.6.7 UC Mechanisms in Nb^{5+}-, Mo^{3+}-, Ru-, and Ag^+-Based System

It has been observed that TMs with *d* electrons play an important role in enhancing the optical properties and in varieties of applications. Zeng et al. have reported the bluish-green UC emission in the Nb^{5+} ion-doped silicate glasses with 800 nm femtosecond laser pulse excitation. The UC emission contains a broad peak from 350 to 650 nm peaking at 480 nm. The three-photon UC mechanism has been confirmed by the pump power study. The charge transfer from O^{2-} (2d electron) to Nb^{5+} ($4d^0$ electron) is another responsible reason for the appearance of UC emission (Zeng et al. 2008). These authors anticipated that the developed glasses may find applications in high-density optical storage, colorless transparent fluorescence material, and three-dimensional color displays.

Upconversion upon NIR excitation in Mo^{3+}, another $4d^3$ ion, has been studied in two different hosts, viz Cs_2NaYCl_6 and Cs_2NaYBr_6, from 10 to 150 K. These Mo^{3+} ions are excited from ground state $^4A_2$ to $^4T_2$ via the metastable $^2T_1/^2E$ levels because of the two-photon absorption process followed by the ESA and ETU mechanisms. The ESA-to-ETU ratio is found to be ~15% to 85% and ~35% to 65%, respectively, for the chloride- and bromide-based hosts, respectively (Auzel 2004).

The TM Ru in its complex (Ru polypyridyl) form has been used in the $NaYF_4$:Er^{3+}/Yb^{3+} UCNPs to affect the optical properties of UCNPs by Ruggiero et al. TM complexes are utilized as light-activatable drugs for biological applications such as photochemotherapy and as a tool for the controlled release of biologically active ligands. Upon 980 nm excitation, the complex dissociates, and release of pyridine and formation of aqua adduct take place, which can be easily inserted into the biological body (Ruggiero et al. 2014).

The singly ionized silver can be used to intensify UC emission because of the SPR effect as reported by Kumar et al. The Ag^+ ions are incorporated in the Ho^{3+}/Yb^{3+}-doped GdF_3 phosphors. Upon 980 nm excitation, the UC emissions are observed in the green and red regions from the Ho^{3+} ions. With 10 mol% of the Ag^+ ion incorporation, the UC emission intensity has been increased by 10-fold as compared to that of rare-earth-doped phosphors. Thus, using the mechanism of SPR, TMs can be used to enhance the UC emission intensity significantly (Figure 6.8) (Kumar et al. 2019).

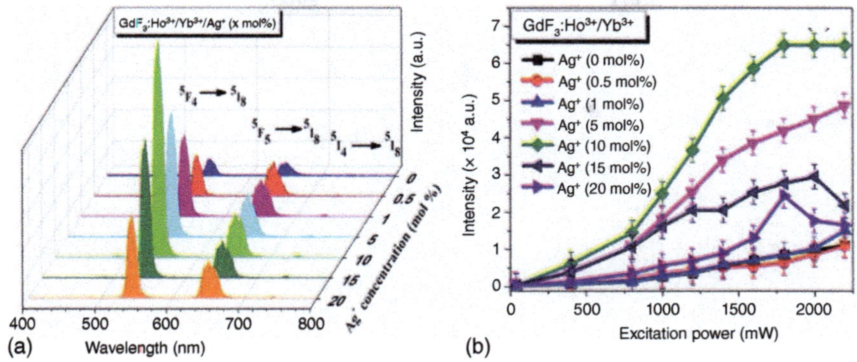

Figure 6.8 (a) Comparative UC emission spectra for different concentrations of Ag^+ ions in the GdF_3:Ho^{3+} (0.2 mol%)/Yb^{3+}(3 mol%) *phosphors* and (b) variation of UC *luminescence* intensity for all samples with pump power from 40 to 2200 mW. Source: Kumar et al. (2019). Reprinted with permission of Elsevier.

6.6.8 UC Mechanisms in W^{6+}- and Re^{4+}-Based System

Tungsten is the sixth period transition metal and most often found in the +6-oxidation state and a stable element. Bokolia et al. reported the frequency UC in the $Er^{3+}/Yb^{3+}/W^{+6}$:$Bi_4Ti_3O_{12}$ phosphors prepared with the solid-state reaction method. The structural modification has been noticed with the substitution of W^{+6} ions for Ti^{4+} in the $Bi_4Ti_3O_{12}$ phosphors. With the increasing concentration of W^{+6}, the transformation in the crystallographic structure from the orthorhombic to the tetragonal structure has been observed, which causes the change in the optical properties. With the introduction of non-luminescent W^{+6} ions in the Er^{3+}/Yb^{3+}:$Bi_4Ti_3O_{12}$ phosphors, an increase in the UC emission intensity upon 980 nm excitation is observed. The green UC emission intensity increases distinctly with increasing W^{+6} content (0.01–0.06 at.%) in the tri-doped $Bi_4Ti_3O_{12}$ system and begins to quench at a higher content of W^{+6} (0.1 at.%). The observed enhancement in the UC emission, because of W^{+6} ion incorporation up to 0.06 at.%, is expected because of the change in the local crystal field nearby the Er^{3+} ions and the increased probability of electronic–dipole (ED) transition. Also, this causes an enhancement in the temperature sensitivity of the developed phosphors (Figure 6.9) (Bokolia et al. 2017). Increase in intensity with the incorporation of W^{6+} has also been reported with glass matrix by Wen et al. The Er^{3+}/Yb^{3+}-codoped La_2O_3–TiO_2–WO_3 glasses have been prepared with the aerodynamic levitation method. The increase in UC emission intensity has been noticed with W^{6+} co-doping upon 980 nm excitation. Also, the effect of W^{6+} incorporation has been observed in thermal stability, Raman spectra, and florescence lifetime (Wen et al. 2019).

Re^{4+} is another $5d^3$ TM ion having three metastable states. Re^{4+} has been studied in two hosts, namely, Cs_2ZrBr_6 and Cs_2ZrCl_6. Aebischer et al. have reported the frequency upconversion in the Re^{4+}-doped Cs_2ZrBr_6 phosphors grown in the form of single crystal at 900 °C using the Bridgman technique upon NIR excitation. The emission transitions occur due to the decay of $5d^3$ electrons from the energy

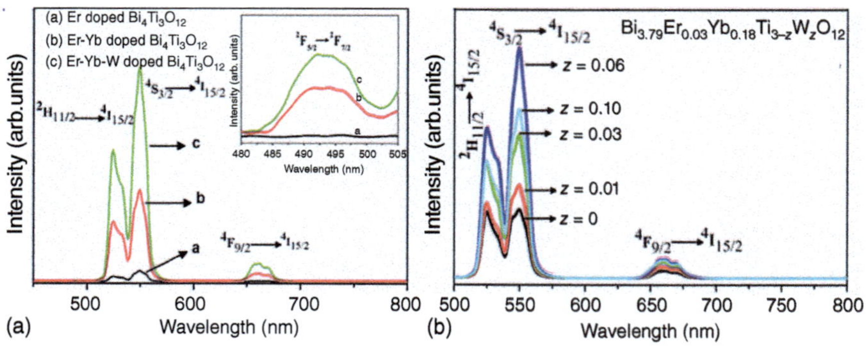

Figure 6.9 (a) Comparative study of upconversion emission spectra of $Bi_{4x}Er_xTi_3O_{12}$, $Bi_{4xy}Er_xYb_yTi_3O_{12}$, and $Bi_{4xy}Er_xYb_yTi_{3z}W_zO_{12}$ ceramic compositions with x¼ 0.03, y¼ 0.18, and z¼ 0.06 under 980 nm excitation. The inset shows the upconverted blue emission in the range of 480–505 nm. (b) UC emission spectra with varying W^{6+} ion content (z) in the $Bi_{3.79}Er_{0.03}Yb_{0.18}Ti_{3z}W_zO_{12}$ composition. Source: Bokolia et al. (2017). Reprinted with permission of American Institute of Physics.

levels Γ_8 ($^2T_{1g}$) in NIR, Γ_7 ($^2T_{2g}$), and Γ_8 ($^2T_{2g}$) in the visible region because of the large splitting of the ligand field terms. Upon NIR excitation at ~9563 cm^{-1}, the red UC emission at ~13 000 and ~14 800 cm^{-1} has been observed. The GSA/ETU processes are responsible for the transitions with significant cross-relaxations, which quenched the UC visible luminescence (Aebischer et al. 2002). Secondly, the time-dependent UC study has been performed in the Re^{4+}-doped Cs_2ZrCl_6 by Gamelin et al. Three emissions at ~7400, ~13 600, and ~6200/4800 cm^{-1} corresponding to the Γ_8 ($^2T_{1g}$) → Γ_8 ($^4A_{2g}$), Γ_7 ($^2T_{2g}$) → Γ_8 ($^4A_{2g}$), and Γ_7 ($^2T_{2g}$) → $^2T_{1g}$/^2E_g transitions, respectively, have been observed with different rate constants at two different temperatures 10 and 300 K, and the ESA/ETU process is found dominant for the frequency upconversion (Gamelin and Güdel 1999).

6.6.9 UC Mechanisms in Os^{4+}- and Au-Based System

The halide lattices doped with Os^{4+} or Er^{3+} are well known for its visible luminescence under excitation at ~10 000 cm^{-1}. Wermuth and Güdel observed the photon avalanche upconversion in the Cs_2ZrBr_6 crystal doped with Os^{4+} (i.e. a 5d system). The crystal doped with Os^{4+} has been prepared by using the Bridgman technique. The 5d^4 electrons of Os^{4+} give $^3T_{1g}$ in octahedral symmetry, which splits into Γ_1, Γ_3, Γ_4, and Γ_5 energy levels. The doped crystal having low vibrational energy of ~220 cm^{-1} is easily excited with an argon-ion laser-pumped Ti: sapphire laser. Upon excitation in the range of 10 000–14 000 cm^{-1}, the NIR–vis UC emission occurred. The two luminescence bands at 15 900 and 13 200 cm^{-1} correspond to the Γ_1 ($^1A_{1g}$) → Γ_1 ($^3T_{1g}$) and Γ_1 ($^1A_{1g}$) → Γ_4 ($^3T_{1g}$) transitions, respectively. The lifetimes of the excited states Γ_4 ($^3T_{1g}$), Γ_5 ($^3T_{1g}$), Γ_5 ($^1T_{2g}$), and Γ_1($^1A_{1g}$) at 15 K are 45 ms, 1.6 ms, 261 μs, and 8 μs, respectively. The photon avalanche is reported to be the main responsible mechanism in association with GSA/ESA. The whole dynamics has also been explained with rate equations. The UC emission has been observed only below 80 K (Wermuth and Güdel 1999).

The low-temperature NIR to visible upconversion has been observed in the Os^{4+}-doped cubic halide crystals, viz, Cs_2GeF_6, Cs_2ZrCl_6, and Cs_2ZrBr_6 by Wermuth and Güdel. No UC emission has been reported in the fluoride-based crystal, but both the chloride and the bromide-based crystals show visible luminescence upon excitation in the NIR below 80 K. At 15 K, the materials Cs_2ZrCl_6:Os^{4+} and Cs_2ZrBr_6:Os^{4+} are excited with ~11 226 and ~11 351 cm^{-1}, respectively. The mechanisms responsible for the visible frequency upconversion in the Cs_2ZrCl_6:Os^{4+} and Cs_2ZrBr_6:Os^{4+} crystals are found to be energy transfer and two excited-state absorption processes, respectively (Figure 6.10) (Wermuth and Güdel 2000). The low-temperature visible frequency UC in the Os^{4+}-codoped Er^{3+}:Cs_2NaYCl_6 crystal synthesized by Bridgman technique under excitation around 10 000 cm^{-1} has been analyzed at 15 K by Wermuth et al. This excitation energy around 10 000 cm^{-1} is transferred from the Γ_5 ($^1T_{2g}$)/Γ_3 (^1E_g) states of Os^{4+} to the $^4I_{11/2}$ and $^4I_{13/2}$ states of the Er^{3+} ions. The NIR to visible frequency upconversion observed corresponding to the $^4S_{3/2} \rightarrow {}^4I_{15/2}$ transition at ~18 154 cm^{-1} in the presently developed Os^{4+}–Er^{3+}-codoped crystal is due to the cooperative processes between the 5d transition metal ion Os^{4+} and the Er^{3+} ion. The UC efficiency upon excitation into the Γ_5 ($^1T_{2g}$)/Γ_3 (^1E_g) states of Os^{4+} is dependent on the doping concentration of the Er^{3+} ions. The maximum UC efficiency has been marked for the 0.1% Os^{4+} + 10% Er^{3+} combination, and it is limited by the back energy transfer process from the Er^{3+} to Os^{4+} and low concentration of the Os^{4+} ions (Wermuth et al. 2001).

The surface plasmon effect has been analyzed with the gold coating in the Er/Yb-doped $NaGdF_4$ upconverting phosphors by Liu et al. The SiO_2-coated UCNPs are modified with APTES (aminopropyltriethoxysilane) for further coating of Au

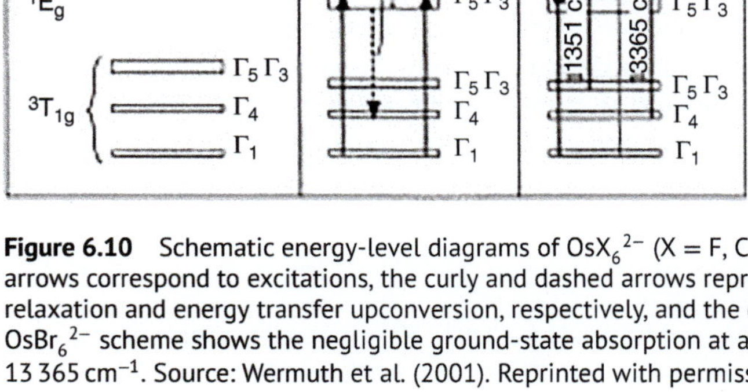

Figure 6.10 Schematic energy-level diagrams of OsX_6^{2-} (X = F, Cl, and Br). The solid arrows correspond to excitations, the curly and dashed arrows represent multiphonon relaxation and energy transfer upconversion, respectively, and the dotted arrow in the $OsBr_6^{2-}$ scheme shows the negligible ground-state absorption at a laser energy of 13 365 cm^{-1}. Source: Wermuth et al. (2001). Reprinted with permission of Elsevier.

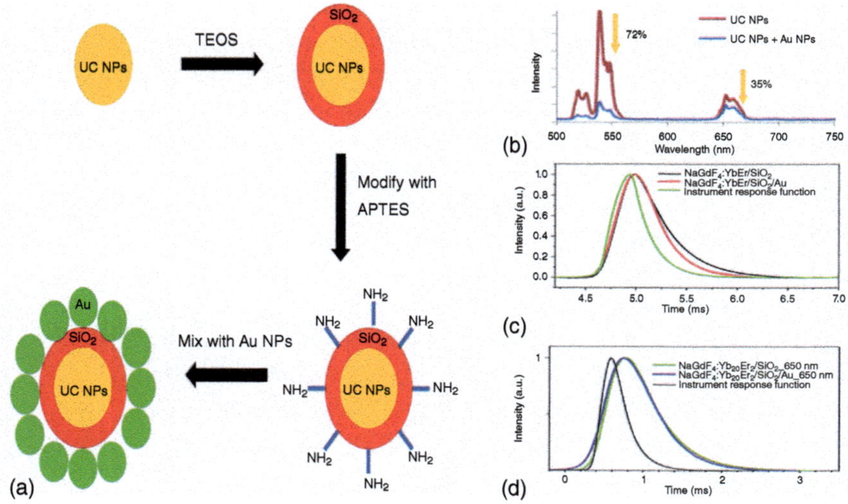

Figure 6.11 (a) Schematic illustration of synthesis procedures. (b) Photoluminescence spectra (c) and lifetime of green emission for UCNPs before and after decorating with Au NPs. (d) Decay of red emission for UCNPs before and after decorating with Au NPs. Source: Liu et al. (2013). Reprinted with permission of Springer.

particles. The prepared phosphors are excited with 975 nm fiber-coupled laser diode, and UC emission peaks are found in the 520–540 nm green region and at ~650 nm. The emission peaks are observed because of the Er^{3+} ions, and the UC emission intensity has been decreased because of the incorporation of Au NPs. The decrement in the green band by 72% and in the red band only by 35% due to plasmonic particles has been reported. The difference in attenuation of two bands can be used in UC color tuning and biosensors (Figure 6.11) (Liu et al. 2013). In another work, a significant UC enhancement has been noticed with the incorporation of Au NPs or Au shell in the $NaYF_4$:Yb^{3+}/Er^{3+}/Gd^{+3} nanorods prepared via hydrothermal synthesis technique. Upon 980 nm excitation, the UC emission peaks are observed at ~540 and ~660 nm wavelength. The nanorod with gold nanoparticles causes the increment of 3.8 and 4 times in the UC emission intensity of 540 and 660 nm peaks, respectively. In the case of a nanorod with a gold shell, it is by 3.2 and 9.6 times, respectively. The phosphors have been used as upconverters in solar cells and improved the external quantum efficiency by 0.14%. The I–V characteristics show that there is an improvement of 16–72-fold in the short-circuit current using the upconverter (Li et al. 2011).

References

Aebischer, A., Salley, G.M., and Güdel, H.U. (2002). Near infrared to visible photon upconversion in Re^{4+} doped Cs_2ZrBr_6. *The Journal of Chemical Physics* 117 (18): 8515–8522. https://doi.org/10.1063/1.1513455.

Auzel, F. (2004). Upconversion and anti-stokes processes with f and d ions in solids. *Chemical Reviews* 104 (1): 139–174. https://doi.org/10.1021/cr020357g.

Bai, J., Duan, P., Wang, X. et al. (2020). Upconversion luminescence enhancement by Fe^{3+} doping in CeO_2:Yb/Er nanomaterials and their application in dye-sensitized solar cells. *RSC Advances* 10 (32): 18868–18874. https://doi.org/10.1039/D0RA02308F.

Bokolia, R., Mondal, M., Rai, V.K., and Sreenivas, K. (2017). Enhanced infrared-to-visible up-conversion emission and temperature sensitivity in (Er^{3+}, Yb^{3+}, and W^{6+}) tri-doped $Bi_4Ti_3O_{12}$ ferroelectric oxide. *Journal of Applied Physics* 121 (8): 084101. https://doi.org/10.1063/1.4977006.

Chen, D., Xu, W., Yuan, S. et al. (2017). Ln^{3+}-sensitized Mn^{4+} near-infrared upconverting luminescence and dual-modal temperature sensing. *Journal of Materials Chemistry C* 5 (37): 9619–9628. https://doi.org/10.1039/C7TC02182H.

Dan, H.K., Zhou, D., Wang, R. et al. (2014). Effect of Mn^{2+} ions on the enhancement red upconversion emission of $Mn^{2+}/Er^{3+}/Yb^{3+}$ tri-doped in transparent glass-ceramics. *Optics & Laser Technology* 64: 264–268. https://doi.org/10.1016/j.optlastec.2014.05.002.

Dey, R., Rai, V.K., and Pandey, A. (2012). Green upconversion emission in Nd^{3+}–Yb^{3+}–Zn^{2+}:Y_2O_3 phosphor. *Spectrochimica Acta Part A: Molecular and Biomolecular Spectroscopy* 99: 288–291. https://doi.org/10.1016/j.saa.2012.09.001.

Du, P., Zhang, Q., Wang, X. et al. (2019). Upconversion luminescence, temperature sensing and internal heating behaviors of $Er^{3+}/Yb^{3+}/Fe^{3+}$-tridoped $NaBiF_4$ nanoparticles. *Journal of Alloys and Compounds* 805: 171–179. https://doi.org/10.1016/j.jallcom.2019.07.054.

Duan, Y., Liu, Y., Zhang, G. et al. (2020). Broadband Cr^{3+}-sensitized upconversion luminescence of $LiScSi_2O_6$:Cr^{3+}/Er^{3+}. *Journal of Rare Earths.* https://doi.org/10.1016/j.jre.2020.08.007.

Gamelin, D.R. and Güdel, H.U. (1999). Spectroscopy and dynamics of Re^{4+} near-IR-to-visible luminescence upconversion. *Inorganic Chemistry* 38 (22): 5154–5164. https://doi.org/10.1021/ic990556r.

Gamelin, D.R. and Gudel, H.U. (2001). Upconversion processes in transition metal and rare earth metal systems. In: *Transition Metal and Rare Earth Compounds,* Topics in Current Chemistry, vol. 214 (ed. H. Yersin), 1–56. Berlin, Heidelberg: Springer. https://doi.org/10.1007/3-540-44474-2_1.

Grinberg, M., Lesniewski, T., Mahlik, S., and Liu, R.S. (2017). $3d^3$ system–comparison of Mn^{4+} and Cr^{3+} in different lattices. *Optical Materials* 74: 93–100. https://doi.org/10.1016/j.optmat.2017.03.057.

Heer, S., Wermuth, M., Krämer, K. et al. (2001). Up-conversion excitation of sharp Cr^{3+} ^2E emission in YGG and YAG codoped with Cr^{3+} and Yb^{3+}. *Journal of Luminescence* 94: 337–341. https://doi.org/10.1016/S0022-2313(01)00395-7.

Huy, B.T., Gerelkhuu, Z., Chung, J.W. et al. (2017). Enhanced light harvesting with chromium in $NaLu_{0.70-x}Gd_{0.10}F_4$:$Yb_{0.18}Er_{0.02}Cr_x$ ($0 \leq x \leq 0.25$) upconversion system. *Materials Science and Engineering B* 223: 91–97. https://doi.org/10.1016/j.mseb.2017.06.001.

Jiang, X., Wang, X., Shi, X. et al. (2021). Effect of Mn^{4+} ions on the structure and luminescence properties of $NaY(MoO_4)_2:Yb^{3+}/Er^{3+}$ phosphor. *Optical Materials* 113: 110873. https://doi.org/10.1016/j.optmat.2021.110873.

Kumar, A., Tiwari, S.P., Kumar, K., and da Silva, J.C.E. (2019). Magnetic tuning in upconversion emission enhanced through Ag^+ ions co-doped in $GdF_3:Ho^{3+}/Yb^{3+}$ phosphor and a real-time temperature sensing demonstration. *Journal of Alloys and Compounds* 776: 207–214. https://doi.org/10.1016/j.jallcom.2018.10.156.

Kumari, A., Pandey, A., Dey, R., and Rai, V.K. (2014). Simultaneous influence of Zn^{2+}/Mg^{2+} on the luminescent behaviour of $La_2O_3:Tm^{3+}$–Yb^{3+} phosphors. *RSC Advances* 4 (42): 21844–21851. https://doi.org/10.1039/C4RA01400F.

Kumari, A., Mondal, M., Rai, V.K., and Singh, S.N. (2017). Photoluminescence study in $Ho^{3+}/Tm^{3+}/Yb^{3+}/Li^+:Gd_2(MoO_4)_3$ nanophosphors for near white light emitting diode and security ink applications. *Methods and Applications in Fluorescence* 6 (1): 015003. https://doi.org/10.1088/2050-6120/aa8c61.

Li, Z.Q., Li, X.D., Liu, Q.Q. et al. (2011). Core/shell structured $NaYF_4:Yb^{3+}/Er^{3+}/Gd^{+3}$ nanorods with Au nanoparticles or shells for flexible amorphous silicon solar cells. *Nanotechnology* 23 (2): 025402. https://doi.org/10.1088/0957-4484/23/2/025402.

Li, Y., Liu, C., Zhang, P. et al. (2020). Doping lanthanide nanocrystals with non-lanthanide ions to simultaneously enhance up-and down-conversion luminescence. *Frontiers in Chemistry* 8. https://doi.org/10.3389/fchem.2020.00832.

Liang, T., Li, Z., Song, D. et al. (2016). Modulating the luminescence of upconversion nanoparticles with heavy metal ions: a new strategy for probe design. *Analytical Chemistry* 88 (20): 9989–9995. https://doi.org/10.1021/acs.analchem.6b01963.

Liao, J., Wang, Q., Wen, H.R. et al. (2017). First observation of mutual energy transfer of Mn^{4+}–Er^{3+} via different excitation in $Gd_2ZnTiO_6:Mn^{4+}/Er^{3+}$ phosphors. *Journal of Materials Chemistry C* 5 (35): 9098–9105. https://doi.org/10.1039/C7TC03045B.

Lingeshwar Reddy, K., Srinivas, V., Shankar, K.R. et al. (2017). Enhancement of luminescence intensity in red emitting $NaYF_4$:Yb/Ho/Mn upconversion nanophosphors by variation of reaction parameters. *The Journal of Physical Chemistry C* 121 (21): 11783–11793. https://doi.org/10.1021/acs.jpcc.7b01334.

Liu, S., Chen, G., Ohulchanskyy, T.Y. et al. (2013). Facile synthesis and potential bioimaging applications of hybrid upconverting and plasmonic $NaGdF_4:Yb^{3+},Er^{3+}$/silica/gold nanoparticles. *Theranostics* 3 (4): 275. https://doi.org/10.7150/thno.4983.

Luitel, H.N., Mizuno, S., Tani, T., and Takeda, Y. (2016). Broadband-sensitive Ni^{2+}–Er^{3+} based upconverters for crystalline silicon solar cells. *RSC Advances* 6 (60): 55499–55506. https://doi.org/10.1039/C6RA10713C.

Lv, Y., Jin, Y., Sun, T. et al. (2019). Visible to NIR down-shifting and NIR to visible upconversion luminescence in $Ca_{14}Zn_6Ga_{10}O_{35}:Mn^{4+},Ln^{3+}$ (Ln = Nd, Yb, Er). *Dyes and Pigments* 161: 137–146. https://doi.org/10.1016/j.dyepig.2018.09.052.

Maciejewska, K. and Marciniak, L. (2020). Multimodal stokes and anti-stokes luminescent thermometers based on GdP_5O_{14} co-doped with Cr^{3+} and Nd^{3+} ions. *Chemical Engineering Journal* 402: 126197. https://doi.org/10.1016/j.cej.2020.126197.

Mahata, M.K., Koppe, T., Mondal, T. et al. (2015). Incorporation of Zn^{2+} ions into $BaTiO_3:Er^{3+}/Yb^{3+}$ nanophosphor: an effective way to enhance upconversion, defect luminescence and temperature sensing. *Physical Chemistry Chemical Physics* 17 (32): 20741–20753. https://doi.org/10.1039/C5CP01874A.

Manaa, H. and Moncorge, R. (1990). Excited-state absorption of Co^{2+} in MgF_2 and $KZnF_3$. *Optical and Quantum Electronics* 22 (1): S219–S226. https://doi.org/10.1007/BF02089011.

Martín-Rodríguez, R., Valiente, R., Rodríguez, F. et al. (2010). Temperature dependence and temporal dynamics of Mn^{2+} upconversion luminescence sensitized by Yb^{3+} in codoped $LaMgAl_{11}O_{19}$. *Physical Review B* 82 (7): 075117. https://doi.org/10.1103/PhysRevB.82.075117.

Muraleedharan, D. and Zhang, J. (2020). The effect of metallic ions on the enhanced upconversion emission of $NaGdF_4$ nanostructures. In: *4th International Conference on Theoretical and Applied Nanoscience and Nanotechnology (TANN'20). Paper No. 141*. Avestia Publishing. http://dx.doi.org/10.11159/tann20.141.

Ouyang, Q., Wang, L., Ahmad, W. et al. (2021). A highly sensitive detection of carbendazim pesticide in food based on the upconversion-MnO_2 luminescent resonance energy transfer biosensor. *Food Chemistry* 349: 129157. https://doi.org/10.1016/j.foodchem.2021.129157.

Pandey, A. and Rai, V.K. (2013). Improved luminescence and temperature sensing performance of Ho^{3+}–Yb^{3+}–Zn^{2+}:Y_2O_3 phosphor. *Dalton Transactions* 42 (30): 11005–11011. https://doi.org/10.1039/C3DT50592H.

Prasad, M. and Rai, V.K. (2020). Influence of transition metal ions: broad band upconversion emission in thermally stable phosphors. *Journal of Alloys and Compounds* 837: 155289. https://doi.org/10.1016/j.jallcom.2020.155289.

Qiang, Q. and Wang, Y. (2019). Enhanced optical temperature sensing and upconversion emissions based on the Mn^{2+} codoped $NaGdF_4$:Yb^{3+},Ho^{3+} nanophosphor. *New Journal of Chemistry* 43 (13): 5011–5019. https://doi.org/10.1039/C8NJ05079A.

Reinhard, C., Gerner, P., Rodrıguez, F. et al. (2004). Near-infrared to green photon upconversion in Mn^{2+} and Yb^{3+} doped lattices. *Chemical Physics Letters* 386 (1–3): 132–136. https://doi.org/10.1016/j.cplett.2003.12.127.

Ren, H. and Yang, F. (2018). Influence of Mn^{2+} on the up-conversion emission performance of Mn^{2+}, Yb^{3+}, Er^{3+}:$ZnWO_4$ green phosphors. *Journal of Materials Science: Materials in Electronics* 29 (18): 15396–15403. https://doi.org/10.1007/s10854-018-9053-9.

Ruggiero, E., Habtemariam, A., Yate, L. et al. (2014). Near infrared photolysis of a Ru polypyridyl complex by upconverting nanoparticles. *Chemical Communications* 50 (14): 1715–1718. https://doi.org/10.1039/C3CC47601D.

Solé, J., Bausa, L., and Jaque, D. (2005). *An Introduction to the Optical Spectroscopy of Inorganic Solids. Wiley*.

Song, E.H., Ding, S., Wu, M. et al. (2014). Tunable white upconversion luminescence from Yb^{3+}–Tm^{3+}–Mn^{2+} tri-doped perovskite nanocrystals. *Optical Materials Express* 4 (6): 1186–1196. https://doi.org/10.1364/OME.4.001186.

Song, E., Han, X., Zhou, Y. et al. (2019). Long-lived photon upconversion phosphorescence in $RbCaF_3$:Mn^{2+},Yb^{3+} and the dynamic color separation effect. *iScience* 19: 597–606. https://doi.org/10.1016/j.isci.2019.08.013.

Srivastava, B.B., Gupta, S.K., and Mao, Y. (2020). Single red emission from upconverting $ZnGa_2O_4$:Yb,Er nanoparticles co-doped by Cr^{3+}. *Journal of Materials Chemistry C* 8 (19): 6370–6379. https://doi.org/10.1039/D0TC00411A.

Tang, J., Chen, L., Li, J. et al. (2015). Selectively enhanced red upconversion luminescence and phase/size manipulation via Fe^{3+} doping in $NaYF_4$:Yb,Er nanocrystals. *Nanoscale* 7 (35): 14752–14759. https://doi.org/10.1039/C5NR04125B.

Tang, J., Du, P., Li, W., and Luo, L. (2020). Boosted thermometric performance in $NaGdF_4:Er^{3+}/Yb^{3+}$ upconverting nanorods by Fe^{3+} ions doping for contactless nanothermometer based on thermally and non-thermally coupled levels. *Journal of Luminescence* 224: 117296. https://doi.org/10.1016/j.jlumin.2020.117296.

Tian, G., Gu, Z., Zhou, L. et al. (2012). Dual-targeting upconversion nanoprobes across the blood-brain barrier for magnetic resonance/fluorescence imaging of intracranial glioblastoma. *Advanced Materials* 24: 1226. https://doi.org/10.1002/adma.201104741.

Tiwari, S.P., Mahata, M.K., Kumar, K., and Rai, V.K. (2015). Enhanced temperature sensing response of upconversion luminescence in $ZnO–CaTiO_3:Er^{3+}/Yb^{3+}$ nano-composite phosphor. *Spectrochimica Acta Part A: Molecular and Biomolecular Spectroscopy* 150: 623–630. https://doi.org/10.1016/j.saa.2015.05.081.

Tonucci, R.J., Jacobsen, S.M., and Yen, W.M. (1990). Dynamics of excited states in $Ni^{2+}:MgF_2$. *Journal of Luminescence* 46 (3): 155–161. https://doi.org/10.1016/0022-2313(90)90035-A.

Wang, Y., Nan, F., Cheng, Z. et al. (2015). Strong tunability of cooperative energy transfer in Mn^{2+}-doped (Yb^{3+}, Er^{3+})/$NaYF_4$ nanocrystals by coupling with silver nanorod array. *Nano Research* 8 (9): 2970–2977. https://doi.org/10.1007/s12274-015-0802-2.

Wang, J.L., Song, E.H., Wu, M. et al. (2016). Room-temperature green to orange color-tunable upconversion luminescence from Yb^{3+}/Mn^{2+} co-doped CaO. *Journal of Materials Chemistry C* 4 (42): 10154–10160. https://doi.org/10.1039/C6TC02773C.

Wen, H., Ai, F., Pan, X. et al. (2019). Thermal properties and upconversion luminescence of Er^{3+}/Yb^{3+} co-doped $La_2O_3–TiO_2–WO_3$ glasses prepared by containerless processing. *Materials Research Express* 6 (8): 085209. https://doi.org/10.1088/2053-1591/ab24a8.

Wenger, O.S., Salley, G.M., Valiente, R., and Güdel, H.U. (2002a). Luminescence upconversion under hydrostatic pressure in the 3d-metal systems Ti^{2+}:NaCl and $Ni^{2+}:CsCdCl_3$. *Physical Review B* 65 (21): 212108. https://doi.org/10.1103/PhysRevB.65.212108.

Wenger, O.S., Valiente, R., and Güdel, H.U. (2002b). Luminescence upconversion under high pressure in Ni^{2+} doped $CsCdCl_3$. *High Pressure Research* 22 (1): 57–62. https://doi.org/10.1080/08957950211364.

Wermuth, M. and Güdel, H.U. (1999). Photon avalanche in $Cs_2ZrBr_6:Os^{4+}$. *Journal of the American Chemical Society* 121 (43): 10102–10111. https://doi.org/10.1021/ja990847z.

Wermuth, M. and Güdel, H.U. (2000). NIR to VIS up-conversion in Os^{4+}-doped halide compounds. *Journal of Luminescence* 87: 1014–1016. https://doi.org/10.1016/S0022-2313(99)00510-4.

Wermuth, M., Schmitz, A., and Güdel, H.U. (2001). Cooperative processes and upconversion in Cs_2NaYCl_6 codoped with Os^{4+} and Er^{3+}. *Physical Review B* 63 (24): 245118. https://doi.org/10.1103/PhysRevB.63.245118.

Xiao, F., Xie, S., Yi, R. et al. (2020). Room temperature broadband upconversion luminescence in Yb^{3+} and Mn^{2+} codoped $Sr_5(PO_4)_3Cl$ phosphors. *Journal of Luminescence* 219: 116943. https://doi.org/10.1016/j.jlumin.2019.116943.

Xie, C., Xie, S., Yi, R. et al. (2020). Site-selective occupation for broadband upconversion luminescence in $Ca_3Y(GaO)_3(BO_3)_4$:Yb^{3+},Mn^{2+} phosphors. *The Journal of Physical Chemistry C* 124 (12): 6845–6852. https://doi.org/10.1021/acs.jpcc.0c00809.

Yadav, M., Mondal, M., Mukhopadhyay, L., and Rai, V.K. (2018). Intense blue upconversion emission and intrinsic optical bistability in Tm^{3+}/Yb^{3+}/Zn^{2+} tridoped YVO_4 phosphors. *Methods and Applications in Fluorescence* 6 (2): 025001. https://doi.org/10.1088/2050-6120/aa9e46.

Yan, Z., Huang, R., Yang, X., and Xiao, S. (2020). Broadband up-conversion in $SrLaAlO_4$:Yb^{3+}/Ho^{3+}/Mn^{4+} phosphor. *Optical Materials* 109: 110416. https://doi.org/10.1016/j.optmat.2020.110416.

Yang, L.Y., Dong, Y.J., Chen, D.P. et al. (2005). Upconversion luminescence from ^2E state of Cr^{3+} in A_2O_3 crystal by infrared femtosecond laser irradiation. *Optics Express* 13 (20): 7893–7898. https://doi.org/10.1364/OPEX.13.007893.

Yao, H., Shen, H., and Tang, Q. (2019). Enhanced upconversion luminescence and temperature sensitivity of $NaYF_4$:Er,Yb phosphors via Mn doping. *Materials Research Express* 6 (12): 125017. https://doi.org/10.1088/2053-1591/ab4009.

Ye, S., Song, E.H., Ma, E. et al. (2014). Broadband Cr^{3+}-sensitized upconversion luminescence in $La_3Ga_5GeO_{14}$:Cr^{3+},Yb^{3+},Er^{3+}. *Optical Materials Express* 4 (4): 638–648. https://doi.org/10.1364/OME.4.000638.

Ye, S., Song, E.H., and Zhang, Q.Y. (2016). Transition metal-involved photon upconversion. *Advanced Science* 3 (12): 1600302. https://doi.org/10.1002/advs.201600302.

Yuan, S., Zeng, H., Wu, X. et al. (2014). Significant enhancement of visible up-conversion emissions of Y_2O_2S:Er^{3+} phosphors by Mn^{2+} sensitizing under 1550 nm excitation. *RSC Advances* 4 (32): 16710–16715. https://doi.org/10.1039/C3RA47645F.

Zeng, J.H., Xie, T., Li, Z.H., and Li, Y. (2007). Monodispersed nanocrystalline fluoroperovskite up-conversion phosphors. *Crystal Growth and Design* 7 (12): 2774–2777. https://doi.org/10.1021/cg070477n.

Zeng, H., Song, J., Chen, D. et al. (2008). Three-photon-excited upconversion luminescence of niobium ions doped silicate glass by a femtosecond laser irradiation. *Optics Express* 16 (9): 6502–6506. https://doi.org/10.1364/OE.16.006502.

Zhang, X., Wang, X., Wang, L. et al. (2019). Double-color luminescence and magnetic characteristics in Fe^{3+} doped $NaErF_4$ microcrystals. *Optical Materials Express* 9 (8): 3379–3389. https://doi.org/10.1364/OME.9.003379.

Zhou, Q., Dolgov, L., Srivastava, A.M. et al. (2018). Mn^{2+} and Mn^{4+} red phosphors: synthesis, luminescence and applications in WLEDs. A review. *Journal of Materials Chemistry C* 6 (11): 2652–2671. https://doi.org/10.1039/c8tc00251g.

7

Frequency Upconversion in UCNPs Containing Rare-Earth Ions

Sasank Pattnaik and Vineet Kumar Rai

Indian Institute of Technology (Indian School of Mines), Laser and Spectroscopy Laboratory, Department of Physics, Dhanbad, Jharkhand 826004, India

7.1 Introduction

The ample use of lasers and photonic devices in almost every field of science and technology motivates the researchers worldwide to work in these fields (Savage 1987; Jackson 2003; Rai 2012, p. 209). Such an emerging and thought-provoking field is "luminescence," more specifically photoluminescence. Luminescent materials serve best in various technologies such as optical communication amplifiers, lamps, solid-state lasers, display screens, etc. (Zhang 2015, p. 1). Considering the exceptional photophysical properties and the flexibility offered in these research areas, many research groups have been devoted to develop strategies and characterize new luminescent materials. Recent advancements in luminescent materials mainly comprise organic dyes, quantum dots (QDs), upconverting nanoparticles, etc. The common phenomenon used to describe the emission mechanisms of the said materials obey Stokes' law, where the emission peak appears at a higher wavelength compared to the excitation peak, and the difference between these two peaks are often termed as "Stoke's shift." The luminescence associated with this shift is claimed as "downconversion luminescence" (Valeur 2003). The materials showing these properties are often exposed to ultraviolet (UV) or visible radiation. Conversely, cases are also found where shorter wavelength emissions appear from the relatively higher wavelength radiation sources. The difference in energy could be of a few $k_B T$ (k_B is the Boltzmann constant and T is the absolute temperature) or of the order of 10–100 $k_B T$ (Cesaria and Di Bartolo 2018, p. 28). The events associated with such energy differences are either termed as "anti-Stokes luminescence" or "upconversion luminescence (UCL)," respectively (Bloembergen 1959). The infrared (IR) radiations are commonly used as the excitation source to observe the UCL.

Upconverting Nanoparticles: From Fundamentals to Applications, First Edition. Edited by Vineet K. Rai.

The concept of UCL was first introduced by N. Bloembergen in 1959 with a device named "infrared quantum counter" and later explained by Porter in 1961 (Bloembergen 1959; Porter 1961). The UCL refers to a non-linear optical process in which two or more sequentially absorbed low-energy photons are converted into high-energy photons through the long-lived metastable states. With the successful explanation of UCL, another type of UC luminescence termed as "energy transfer upconversion (ETU)" has been proposed (Auzel 2004). Later, photon avalanche was also familiarized based on infrared quantum counters (Chivian et al. 1979). It should be noted that UCL is totally different from other non-linear phenomena (i.e. simultaneous two-photon absorption [STPA] and second harmonic generation [SHG]). The UCL phenomena could be understood with long-lived intermediate states (Auzel 2004; Suyver et al. 2005; Dong et al. 2013).

In recent decades, the rare-earth (RE)-based materials grasp attention owing to their inimitable optical characteristics. The appearance of new energy levels within the optical energy gap in a perfect crystal due to the influence of RE and transition metal ions has created strange opportunities in various fields such as solid-state lighting, temperature sensing, optical heaters, display devices, white light-emitting devices (wLEDs), security inks, etc. (Solé et al. 2005, p. 199; Rai et al. 2013; Dey and Rai 2014; Pandey et al. 2015; Soni et al. 2015; Dey et al. 2015; Mukhopadhyay and Rai 2017). One of the advantages of the RE ions over the transition metal ions is due to the abundance of energy levels present in RE ions because of the large quantum number ($n = 4$, $l = 3$). This results in many intra-configurational transitions. However, the transitions corresponding to the intra-configuration are initially restricted and governed by the Laporte parity forbidden rule. However, when the RE ions are ingrained into the inorganic lattice, the parity forbidden rule is broken partially because of the mixing of orbitals that ultimately lead to inter-configurational transitions. The photoluminescence phenomena are performed with upconverting nanoparticles, QDs, and organic dyes. The RE^{3+} ion-based upconverting nanoparticles are less affected by the microsurrounding because of the shielding effect from the outer most $5s^2 5p^6$ sub-shells. For example, the upconversion (UC) performance in semiconductor QDs depends highly on the particle size, whereas it is less affected for RE-doped upconverting nanoparticles (UCNPs) (Bünzli and Piguet 2005; Dong et al. 2013; Sun et al. 2015; Zhu et al. 2019). In addition to these, properties such as excellent photostability, a large anti-Stokes shift, long luminescence lifetime, and sharp band emissions are arising from the 4f orbitals (Bünzli and Piguet 2005; Binnemans 2009; Wang and Liu 2009; Sun et al. 2015; Gu and Zhang 2018). Along with RE ions, the infrared (IR) excitation sources play a big role in the luminescent properties of the materials. Apart from being gentle to the microbiological environment, the near-infrared (NIR) light possesses a low pump power density ($1–10^3$ W/cm^2) to initiate UC emission. The use of NIR radiation enables the UCNPs to exhibit high resistance to photo-bleaching and photochemical degradation under photoexcitation conditions and excellent penetration depth in bio-samples with weaker autofluorescence from backgrounds (Suyver et al. 2005; Shen et al. 2008; Haase and Schäfer 2011; Li et al. 2013; Sun et al. 2014; Gai et al. 2014; Idris et al. 2015).

In the early days of developing upconversion materials, bulk materials are considered the main focus in this field before introducing nanomaterials in the late 1990s. The tradition of miniaturization is becoming a prime asset to fulfill the growing demand of the society. With the advent of nanoparticles (NPs), this has been possible to manufacture miniaturized products on a large scale. Although the optical properties of bulk materials and nanomaterials show similar behavior in terms of peak positions because of the shielding effect of 4f orbitals, the relative intensity and the upconversion efficiency are way more improved in the case of nanomaterials. Moreover, considering the advantages of RE ions and NIR radiation along with the help of nanoscience and nanotechnology, new possibilities have opened up for UCNPs, which span many applications such as biosensing, photovoltaic devices, bioimaging, photochemical reactions, and many interdisciplinary areas (Wang et al. 2010a; Zhou et al. 2012; Huang et al. 2013; Chen et al. 2014). New strategies have been engineered for optimizing the upconversion emission, such as nanoengineering of UCNPs in 3D shapes, dielectric superlens-mediated approach, confinement of migration of excitation energy, dye sensitization strategy, super-resolution imaging, optogenetics, NIR image vision of mammals, etc. (Liu et al. 2016a; Chen et al. 2016a; Liu et al. 2017; Garfield et al. 2018; Wisser et al. 2018; Chen et al. 2018; Liang et al. 2019; Ma et al. 2019). Although many works related to luminescence have been reported before, significant modifications and efficient executions are still an open problem in the real world. Thus, excellent knowledge of the previous work is vital for exploring new applications of UPNPs.

7.2 Familiarization with the Spectroscopic Behavior of RE^{3+} Ion-Doped UCNPs

7.2.1 Physics of Trivalent Rare-Earth Ions

RE metals along with scandium (Sc) and yttrium (Y) are lanthanides in the periodic table, having similar physical and chemical characteristics. Although RE elements are found in +2, +3, and +4 oxidation states, +3 oxidation states are considered as the most stable one (Wang and Liu 2019. p. 362). The outer electronic configuration corresponding to the RE^{3+} ions is represented as $5s^2 5p^6 4f^n$ ($n = 0$–14), which reflects in a rich quantity of excited energy levels of 4f orbitals with $4f^n$ electrons as valance electrons responsible for the unique optical properties. Within the framework of Russell–Saunders notation, each energy level can be labeled $^{2S+1}L_J$, where S, L, and J imply the spin angular momentum, orbital angular momentum, and total angular momentum of the $4f^n$ electrons (Cesaria and Di Bartolo 2018, p. 33). Because of the shielding effect by the $5s^2$ and $5p^6$ orbitals, the electrons in $4f^n$ orbitals weakly interact with the surroundings of the crystal, which results in weak electron–phonon coupling responsible for sharp transition bands (Solé et al. 2005, pp. 200, 201; Sharma et al. 2017). Moreover, the low transition probabilities and long luminescence lifetime (~0.1 seconds) are the consequences of the Laporte-forbidden f–f transitions (Zhang 2015, p. 8). Although the spin–orbit interaction dominates over the crystalline field, RE^{3+} ions slightly perturbed when incorporated in the crystal, and the width of the states depends on the crystal field strength. The number of components associated with the $^{2S+1}L_J$ multiplate is proportional to $(J+1)$ for integer J and $(J+1/2)$ for half-integer J values. However, the splitting of energy levels with the reduction in symmetry around the active center decides the actual number of components of RE^{3+} ions.

Even with the peculiar intrinsic spectroscopic properties of RE ions, not all RE ions can be considered as dopant ions. Because the outermost orbitals (i.e. 4f) of Sc^{3+}, Y^{3+}, La^{3+}, and Lu^{3+} ions are filled, they cannot participate in intra-configurational transitions (Dong et al. 2013). Hence, these ions are often used as host matrices rather than dopants. In practice, the elements possessing ladder-like energy levels are less prone to the multiphonon relaxation between the excited levels and the next lower lying levels. Ions such as Er^{3+}, Nd^{3+}, Yb^{3+}, Tm^{3+}, Ho^{3+}, Eu^{3+}, and Tb^{3+} are often used as the luminescent centers because of having ladder-shaped energy levels (Dong et al. 2013; Zhang 2015, p. 8). Moreover, the energy-level separations in the Er^{3+}, Nd^{3+}, Ho^{3+}, Tm^{3+}, and Yb^{3+} ions are mostly resonant with the IR/NIR excitation. Thus, these ions are frequently considered as dopants for making the upconverting luminescent phosphors (Cesaria and Di Bartolo 2018 p. 34).

7.2.1.1 UC Mechanisms in Yb^{3+}- and Pr^{3+}-Based Systems

The immense reputation of ytterbium (Yb^{3+}) ions as a sensitizer has grasped the attention of researchers because of their simple energy-level scheme containing the ground ($^2F_{7/2}$) and excited ($^2F_{5/2}$) states separated by approximately 10^4 cm^{-1}. Despite playing the supporting role for the activators, Yb^{3+}-doped phosphors can emit in the green region under suitable NIR excitation. The cooperative luminescence in the Yb^{3+} doped $LiNbO_3$ matrix has been reported (Montoya et al. 2000). Under 920 nm excitation, four prominent UC emission bands at ~981, 1006, 1030, and 1060 nm corresponding to the transitions from the $^2F_{5/2}$ state (0) to the four different Stark levels of the $^2F_{7/2}$ state (0, 1, 2, and 3, respectively) of the Yb^{3+} ions have been noticed. These authors have claimed that along with the emission in the NIR region, the emission has been noticed in the green region, which is visible to the naked eyes. The energy corresponding to the peak in the green region due to the cooperative process was found to be about twice the energy of NIR excitation. Thus, it has been confirmed that the emission in the green region is a result of simultaneous annihilation of two excited states of the Yb^{3+} ions in a pair. Additionally, the decay time and power variation of the excitation measurement confirms the cooperative luminescence in the Yb^{3+}-doped $LiNbO_3$ phosphors. Among the RE ions, triply ionized praseodymium (Pr^{3+}) has been extensively studied in various systems because of its several metastable states that can provide the emissions in blue, green, orange, and IR regions upon suitable excitation. The influence of metallic (Ag) nanoparticles annealed for different time intervals on the optical behaviors of Pr^{3+}-doped TeO_2–ZnO glasses has been studied (Rai et al. 2008). The excitation wavelength at 590 nm from the dye laser, pumped by the second harmonic of a Nd:yttrium aluminum garnet (YAG) laser of nanosecond pulses, provides an excellent resonating condition for the $^3H_4 \rightarrow {}^1D_2$ absorption transition. The emission peaks at ~486 and ~692 nm ascribed to the $^3P_0 \rightarrow {}^3H_4$ and $^1D_2 \rightarrow {}^3H_5$ transitions were detected. The significant enhancement observed, for the UC emission band corresponding to the $^3P_0 \rightarrow {}^3H_4$ transition in the blue region, because the presence of metallic nanoparticles has been attributed to the proximity of the plasmon band with the (^3P_J ($J = 0, 1, 2$); $^1I_6$) manifold. The study on Pr^{3+}-doped glasses has been carried out by a group of researchers, in which they have claimed an enhancement of about 20 times in the blue region in the Pr^{3+}/Yb^{3+}-co-doped fluoroindate glasses under 1.064 μm excitation. This enormous enrichment in emission intensity at ~485 nm has been observed on varying the sample temperature from 20 to 260 °C. The investigation suggests that the temperature-dependent multiphonon-assisted anti-Stokes excitation in the Yb^{3+} ions and the excited-state absorption (ESA) of Pr^{3+} ions are the root cause of the UCL enhancement (Oliveira et al. 2000). The structural and spectroscopic behaviors of the Yb^{3+}/Pr^{3+}:Y_2O_3 nanophosphors have been reported (Yadav et al. 2015). These authors have observed an intense green UC emission at ~552 nm as a result of the $^3P_0 \rightarrow {}^3H_5$ transition under 976 nm laser irradiation. The prepared phosphors show an eightfold enhancement in the fluorescence intensity because of the removal of optical quenching centers after annealing at higher temperatures (Yadav et al. 2015).

7.2.1.2 UC Mechanisms in Er-Based Systems

The triply ionized erbium (Er^{3+}) ion offers high UC emission efficiency because of the resonant energy difference between ($^4I_{11/2}$ and $^4I_{15/2}$), ($^4F_{7/2}$ and $^4I_{11/2}$), and ($^4F_{9/2}$ and $^4I_{13/2}$) pairs under 980 nm excitation. As a result, the emissions in the green and red region from the Er^{3+}-doped UCNPs show absolute dominance over other lanthanides. For 980 nm excitation, the Yb^{3+} ion has been considered as a conventional and promising sensitizer because of high absorption cross section in that region.

The frequency upconversion mechanism in the Er^{3+}/Er^{3+}–Yb^{3+}-doped/co-doped La_2O_3 phosphors under 980 and 808 nm laser excitations has been reported (Dey and Rai 2014). Under 980 nm excitation, the UC emission bands around 409, 523, 548, and 660 nm have been monitored in both the Er^{3+}- and Er^{3+}–Yb^{3+}-based systems with different intensities (Figure 7.1a). The complete understanding of the UC mechanisms for the optimized Er^{3+}/Er^{3+}–Yb^{3+} system has been narrated using an energy-level diagram. The prominent green emission peaks at ~523 and ~548 nm have been assigned to the $^2H_{11/2} \rightarrow {}^4I_{15/2}$ and $^4S_{3/2} \rightarrow {}^4I_{15/2}$ transitions, respectively, while the $^4F_{9/2} \rightarrow {}^4I_{15/2}$ transition is referred to the red UC emission band centered at ~660 nm. In the Er^{3+}-doped La_2O_3 phosphors, the levels $^4I_{11/2}$, $^4F_{7/2}$, $^2H_{9/2}$, and $^4F_{9/2}$ have been populated via the ground-state absorption (GSA) and ESA processes. However, the enhancement in the UC emissions of up to ~8, ~4, and ~29 times in the green and red regions, respectively, has been witnessed because of co-doping with the Yb^{3+} ions. With the introduction of Yb^{3+} ions, the population at the $^4I_{11/2}$, $^4F_{7/2}$, $^2H_{9/2}$, and $^4F_{9/2}$ levels has been dominated by the energy transfer (ET) process (Figure 7.1b). Conversely, under 808 nm laser excitation, the UC emission intensities at green and red regions have been reduced by ~4, ~5, and ~1.3 times, respectively, because of the back energy transfer (BET) from the Er^{3+} to

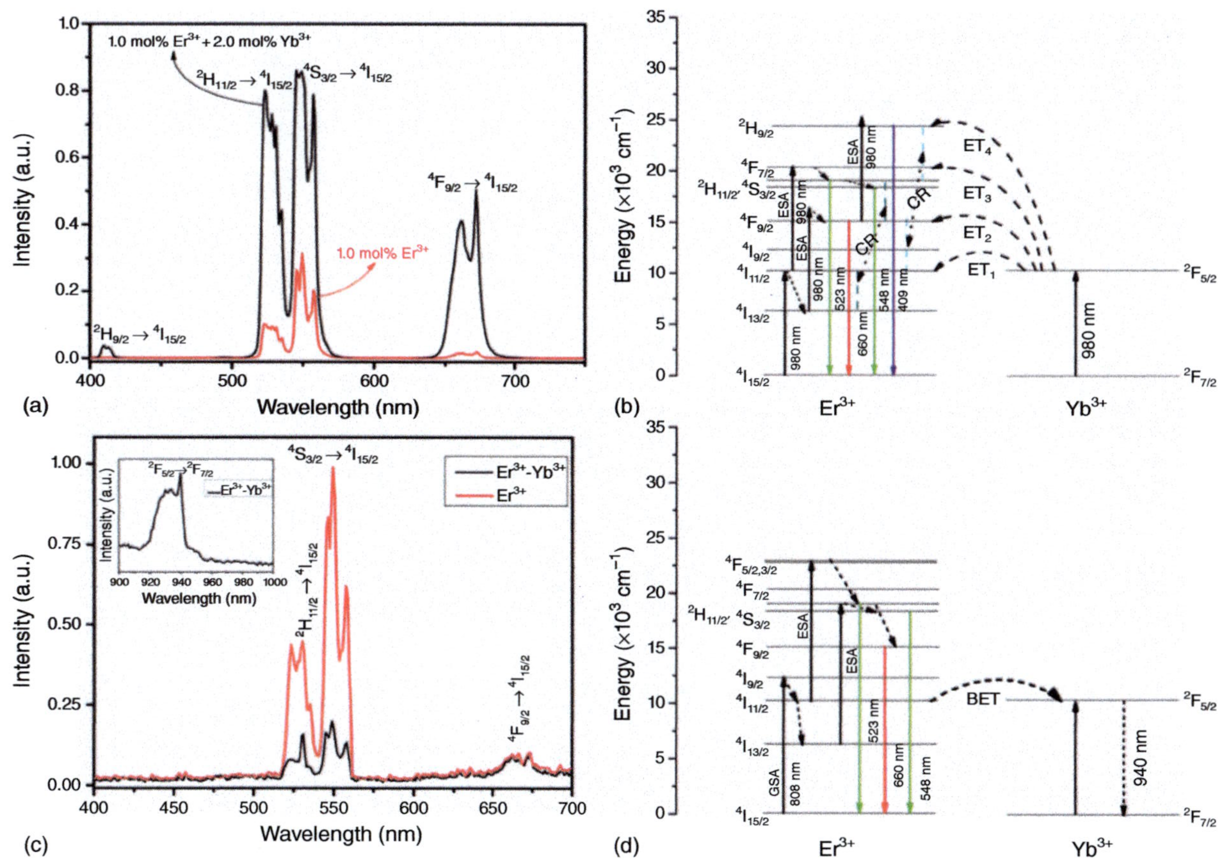

Figure 7.1 (a) Comparative study of upconversion emission spectra of 1.0 mol% Er^{3+}-doped La_2O_3 phosphor and 1.0 mol% Er^{3+} + 2.0 mol% Yb^{3+}-co-doped La_2O_3 phosphor powders heat treated at 800 °C temperature upon 980 nm pump excitation, (b) schematic energy-level diagram for the Er^{3+}–Yb^{3+} energy transfer system under 980 nm excitation, (c) upconversion luminescence spectra of 1.0 mol% Er^{3+} and 1.0 mol% Er^{3+} + 2.0 mol% Yb^{3+}-doped/co-doped La_2O_3 phosphor powders heat treated at 800 °C under 808 nm pump excitation, and (d) schematic energy-level diagram for Er^{3+}–Yb^{3+} energy transfer system under 808 nm excitation. Source: Dey and Rai (2014). Reprinted with permission of Royal Society of Chemistry.

Yb^{3+} ions (Figure 7.1c). The BET phenomenon has been verified with the emission resulted at ~940 nm because of the $^2F_{5/2} \rightarrow {}^2F_{7/2}$ transition of the Yb^{3+} ions (inset of Figure 7.1d). It has been reported that the $BaMoO_4$: 0.3 wt% Er^{3+} + 1.0 wt% Yb^{3+} phosphors generate high green UC emission under 980 nm excitation that could be seen by the naked eye even at very low pump power density (Soni et al. 2015). Initially, in Er^{3+}-doped phosphor, the ratio between green to NIR and green to red is ~105 and ~37, respectively. However, with the sensitization effect by the Yb^{3+} ions in the UC phosphors, the enrichment of ~2200 times in the green emission compared to NIR and ~45 times related to red has been monitored. The prime cause of increment in the UC system has been attributed to the ET processes from the Er^{3+} ions to Yb^{3+} ions in Er^{3+}–Yb^{3+}:$BaMoO_4$ nanophosphors. Although the sheer dominance of green light has been encouraged in Er^{3+}-doped phosphors, the domination of red emission has been reported by many researchers. The UC mechanisms in the $NaErF_4$ and Er^{3+}–Yb^{3+}:$NaYF_4$ phosphors under different laser excitations have been reported by the researchers (Lee et al. 2019). They described that the increase in the red emission compared to the green and NIR emissions in the $NaErF_4$ phosphors is resulted from the non-radiative relaxation (NRR) from the $^4F_{7/2}$, $^2H_{11/2}$, $^4S_{3/2}$ levels and cross-relaxation (CR) from the $^4F_{3/2}$ and $^4F_{5/2}$ levels, respectively, because of high Er^{3+} ion doping, whereas in the Er^{3+}–Yb^{3+}:$NaYF_4$ phosphors, the intense green light has been observed compared to red light. Similarly, the researchers have synthesized the $ZnGa_2O_4$:Yb^{3+}, Er^{3+}, and Cr^{3+} phosphors to obtain a single red emission (Srivastava et al. 2020). According to the authors, green light is undesirable for the tissues because it may cause cell damage and decrease the overall bioimaging sensitivity. Because the Cr^{3+} ion has a high absorption cross section at the green region, the ET from Er^{3+} to Cr^{3+} increases the red to green emission under the 980 nm excitation. Further, the BET to the $^4F_{9/2}$ levels of Er^{3+} ions intensifies the red emission for the prepared phosphors. As Er^{3+} is a more suitable activator ion for observing UC emission, various industrial and bio-related applications have been performed in numerous nanomaterials doped with Er^{3+} ions (Yin et al. 2012; Pandey et al. 2015; Berry and May 2015). Apart from the single excitation source, viz, 980, 808, and 1550 nm, etc., the UC mechanism has been understood with the help of co-excited systems. Frequency upconversion in the multifunctional Gd_2MoO_6:Er^{3+}/Yb^{3+} nanoparticles under 980 and 1550 nm dual-excitation wavelength mode has been monitored (Li et al. 2020). They reported the highly intense UC emission in the green region with high thermosensitivity. Likewise, to understand the spectroscopic behavior of the $CaTiO_3$:Er^{3+}–Nd^{3+}–Yb^{3+} phosphors, a novel co-excitation method has been chosen in which the UCL and thermometric behavior of the tri-doped system has been analyzed by 980, 808 nm, and co-excitation of both the laser radiations (Pattnaik and Rai 2021). According to the study, the co-excited system shows more excellent thermal stability, thus helpful for various applications.

7.2.1.3 UC Mechanisms in Ho-Based Systems

Resembling the Er^{3+} ion-doped systems, the Ho^{3+} ion-doped phosphors can also emit in the red and green regions depending on the local environment around the dopant ions and excitation situations. The thermally coupled $^5F_4$ and $^5S_2$ levels have been potentially used for optical temperature sensing applications. Ho^{3+} ions are also used to develop white LEDs. As the single Ho^{3+} ion-doped systems cannot potentially emit under 980 nm excitation, the Yb^{3+} ions have been introduced by the researchers (Barrera et al. 2014; Chai et al. 2017; Huang et al. 2020; Mahata et al. 2020). Ho^{3+}- and Yb^{3+}-co-doped lanthanum silicate and lanthanum zirconate nanoparticles (NPs) have been synthesized and frequency upconversion (UC) reported under 980 nm NIR excitation (Sangeetha and van Veggel 2009). The Ho^{3+}–Yb^{3+}-co-doped lanthanum silicate NPs exhibited red emission, whereas the green emission is predominantly observed in the case of Ho^{3+}–Yb^{3+}-co-doped lanthanum zirconate nanoparticles (Figure 7.2a,b). They observed no UC emission in either of the singly doped Ho^{3+} nanoparticles under 980 nm excitation, but the UC emissions at ~540, ~660, and ~750 nm corresponding to the $^5S_2 \rightarrow {}^5I_8$, $^5F_5 \rightarrow {}^5I_8$, and $^5S_2 \rightarrow {}^5I_7$ transitions, respectively, have been obtained for the Yb^{3+}-co-doped systems (Figure 7.2c). The dominance of red emission intensity compared to the green light in the lanthanum silicate matrices has been assigned to (i) the phonon

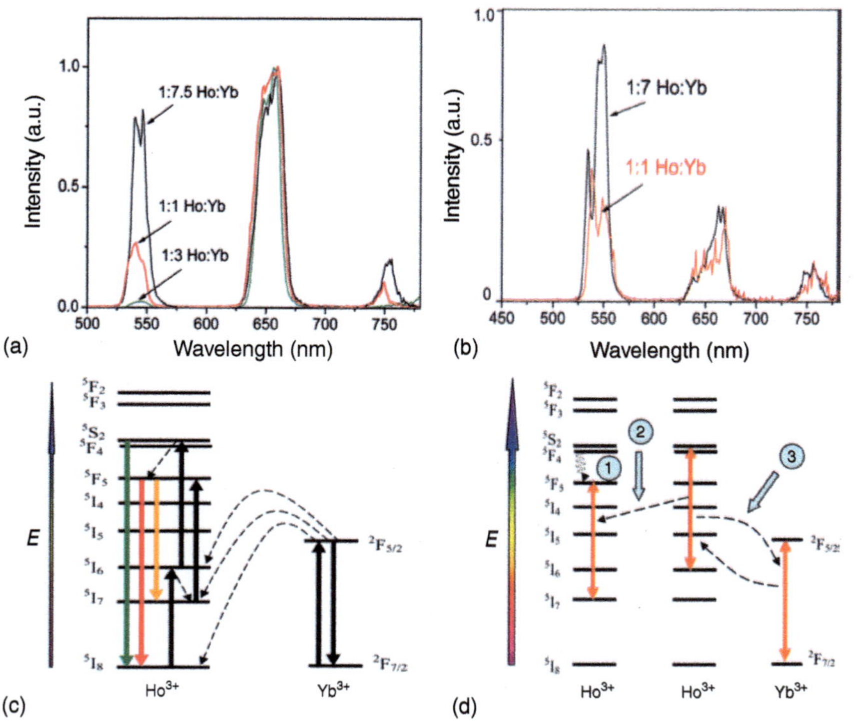

Figure 7.2 Upconversion luminescence from the Ho^{3+}-, Yb^{3+}-co-doped (a) lanthanum silicate and (b) lanthanum zirconate nanoparticles normalized to the red emission from Ho^{3+}, (c) schematic diagram for the upconversion mechanism in the Ho^{3+}–Yb^{3+} system, and (d) scheme depicting the non-radiative deactivation of Ho^{3+} ($^5S_2$). (1) Phonon relaxation of Ho^{3+} ($^5S_2$, $^5F_4$). (2) Cross-relaxation between two Ho^{3+} ions. (3) Energy back transfer from Ho^{3+} to Yb^{3+}. Source: Sangeetha and van Veggel (2009). Reprinted with permission of American Chemical Society.

relaxation of Ho^{3+} ($^5S_2$, $^5F_4$) to Ho^{3+} ($^5F_5$), (ii) cross-relaxation between two Ho^{3+} ions, i.e. $^5S_2$, $^5F_4$ (Ho^{3+}) + $^5I_7$ (Ho^{3+}) → $^5F_5$ (Ho^{3+}) + $^5I_6$ (Ho^{3+}), and (iii) energy back transfer (EBT) from Ho^{3+} to Yb^{3+}, i.e. $^5S_2$, $^5F_4$ (Ho^{3+}) + $^2F_{7/2}$ (Yb^{3+}) → $^5I_6$ (Ho^{3+}) + $^2F_{5/2}$ (Yb^{3+}) (Figure 7.2d). However, in the lanthanum zirconate-based system, the EBT is inefficient because of the mismatch of resonant energy separation between the Ho^{3+} and Yb^{3+} ions. Again, it has been observed that the multiphonon relaxation from the $^5I_6$ to $^5I_7$ levels in the said matrix is slower owing to the low-phonon energy of lanthanum zirconate (~510 cm^{-1}) compared to lanthanum silicate (~915 cm^{-1}). Therefore, the efficient ET from the Yb^{3+} ions to Ho^{3+} ions has been devised as the prime cause of dominant green emission in the lanthanum zirconate nanoparticles. The impact on the spectroscopic behavior of Y_2O_3:Ho^{3+}–Yb^{3+} phosphors on co-doping with Zn^{2+} ions under 980 nm excitation has been demonstrated (Pandey and Rai 2013). The authors have noticed a predominant green emission resulting from efficient ET from the Yb^{3+} ions to the Ho^{3+} ions under 980 nm excitation. The incorporation of Zn^{2+} ions in the nanophosphors increased the transition probability, followed by an increase in the green emission intensity. Similarly, the optical properties of non-lanthanide-incorporated UC phosphors have been studied (Dey et al. 2015). Authors have prepared $CaMoO_4$:Ho^{3+}–Yb^{3+}–Mg^{2+} phosphors by a chemical co-precipitation route. The UC emission spectra confirm three major peaks centered at ~545, 660, and 756 nm associated with the $^5S_2 \rightarrow {}^5I_8$, $^5F_5 \rightarrow {}^5I_8$, and $^5S_2 \rightarrow {}^5I_7$ transitions. Upon co-doping, with the Yb^{3+} ions, an enrichment of ~92, ~197, and ~92 times in the green, red, and NIR regions, respectively, has been observed. It is important to note that the red emission is more intense than the green emission because of the smaller lifetime of the $^5F_4$ and $^5S_2$ levels compared to the $^5F_5$ level. Thus, the transition probability for the red emission is greater than the green emission. Moreover, an enormous enhancement in the visible and NIR regions has been monitored with the Mg^{2+} co-doping owing to the local crystal field modification around the Ho^{3+} ions.

7.2.1.4 UC Mechanisms in Tm-Based Systems

The Tm^{3+} ion is one of the crucial dopant ions used for technological applications when placed in a suitable host matrix. It generally offers intense blue and NIR emissions under different excitations. The optical and thermometric properties of the Tm^{3+} ion-doped nanophosphors have been studied by various researchers and found that the developed nanophosphors are suitable for NIR to UV and visible upconverters, display devices, optical temperature sensors, and cancer cell detectors (Jin et al. 2017; Li et al. 2016; Kraft et al. 2019; Barrera et al. 2014). The UC mechanisms in the developed Tm^{3+}–Yb^{3+}-co-doped Y_2WO_6 phosphors under 980 nm laser excitation have been demonstrated (Soni and Rai 2014). The study revealed that in the Tm^{3+}:Y_2WO_6 phosphors, the UC emission in the visible and NIR regions is weaker because of the non-resonant GSA and ESA of the higher long-lived levels. With the incorporation of Yb^{3+} ions in the Tm^{3+}:Y_2WO_6 phosphors, the peak profile did not change, but the prominent peaks at ~485, ~649, and ~796 nm corresponding to the $^1G_4 \rightarrow {}^3H_6$, $^1G_4 \rightarrow {}^3F_4$, and $^3H_4 \rightarrow {}^3H_6$ transitions have been increased by ~260, ~140, and ~164 times, respectively (Figure 7.3a,b). It has been claimed that

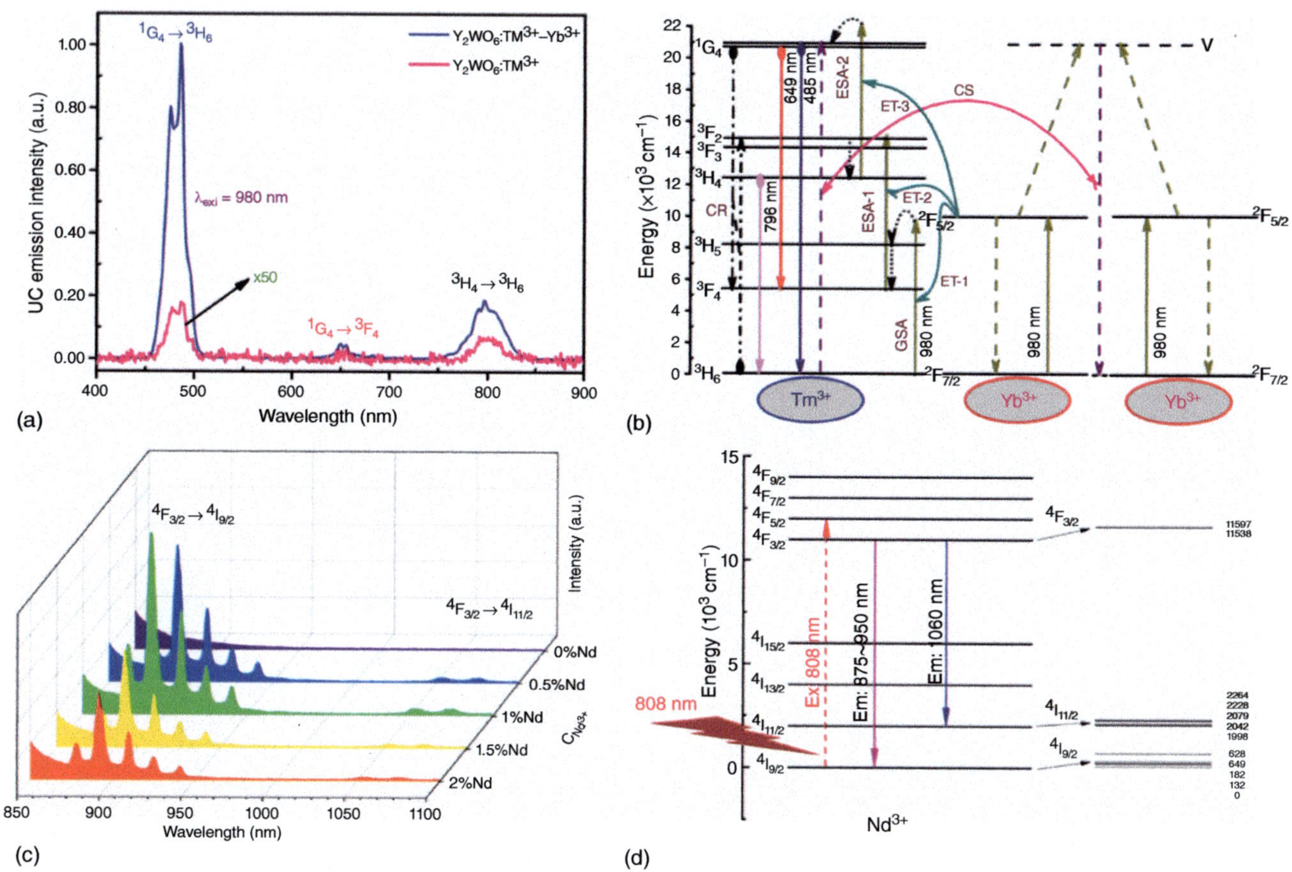

Figure 7.3 (a) UC emission spectra of the Y_2WO_6:Tm^{3+} and Y_2WO_6:Tm^{3+}–Yb^{3+} phosphors and (b) proposed energy-level diagram of the Tm^{3+}–Yb^{3+}-co-doped Y_2WO_6 phosphors. Source: Soni and Rai (2014). Reprinted with permission of Royal Society of Chemistry, (c) fluorescence spectra of Gd_2O_3:x mol% Nd^{3+} ($x = 0.0, 0.5, 1.0, 1.5$, and 2.0) under calcination at 800 °C and 808 nm laser excitation, and (d) energy transfer diagram of Nd^{3+} ions in Gd_2O_3 crystal. Source: Yang et al. (2021). Reprinted with permission of Elsevier.

the enhancement in the emission intensities could be attributed to ESAs, ETs, and cooperative energy transfer (CET). To prove the existence of CET from the Yb^{3+} ions to Tm^{3+} ions, the authors have conducted the pump power analysis. In a single Tm^{3+} phosphor, it has been observed that three photons are needed to populate the $^1G_4$ level, whereas the number of involved photons is ~2 for the Yb^{3+}-co-doped phosphors. The existence of the CET mechanism has been concluded as one of the important processes in the Yb^{3+}-co-doped phosphors. The researchers have reported that the Tm–Yb system dominates in the first therapeutic window in the NIR region along with blue emission (Kumari et al. 2014). They found no UC emission in a single Tm^{3+}-doped phosphor until the introduction of Yb^{3+} ions in the La_2O_3 matrix. This argument suggests that the emission in the blue region corresponding to the $^1G_4 \rightarrow {}^3H_6$ transition can be ascribed to the CET from a pair of Yb^{3+} ions to the Tm^{3+} ion. Apart from the CET process, two CR processes have been assigned in the energy-level diagram, which populates the $^3H_4$ and $^3F_2$ levels. Further, the effects of Zn^{2+} and Mg^{2+} ions on the spectral and thermal behavior of the phosphors have been elucidated.

7.2.1.5 UC Mechanisms in Nd-Based Systems

With the development of Nd^{3+}:$Y_3Al_5O_{12}$ laser, the demand of Nd^{3+} ions has shown specific attention in the solid-state laser industry. Although Nd^{3+} ions are emitting in the green and NIR regions under suitable excitation conditions, they can act as a suitable sensitizer in the 808 nm responsive systems. Besides, Nd^{3+}-doped materials have high demand in biomedicine, automotive sector, and military (Rai et al. 2009; Wang et al. 2013; Song et al. 2019; Yang et al. 2019). The UC emission mechanisms in the cubic Nd^{3+}-doped Gd_2O_3 phosphors has been illustrated under 808 nm excitation (Yang et al. 2021). They reported that the emission from the Nd^{3+} ions is mostly concentrated at the NIR regions. No emission peaks in UV and visible regions have been observed by these authors. The observed UC emissions at ~875–950 and ~1060 nm under 808 nm laser excitation are designated to the $^4F_{3/2} \rightarrow {}^4I_{9/2}$ and $^4F_{3/2} \rightarrow {}^4I_{11/2}$ electronic transitions of Nd^{3+} ions (Figure 7.3c,d). Several small peaks have been observed in the 875–950 nm range, which has been assigned to the Stokes splitting of the corresponding emission. The Nd^{3+}-doped nanophosphors can be realized as visible light emitting phosphors by introducing the Yb^{3+} ions under 980 nm laser excitation. The UC emissions in the Nd^{3+}–Yb^{3+}–Zn^{2+}:Y_2O_3 phosphors synthesized by a solution combustion method have been monitored and responsible processes have been explained via different routes with a suitable energy-level diagram (Dey et al. 2012). Two UC emission bands at ~550 and ~664 nm corresponding to the $^4G_{7/2} \rightarrow {}^4I_{9/2}$ and $^4G_{7/2} \rightarrow {}^4I_{13/2}$ transitions, respectively, are observed. The CET from the Yb^{3+} ion pair to the Nd^{3+} ions is considered to be responsible for the UC emission in the visible range.

7.2.1.6 Tri-Doped Systems

The triply doped upconverting phosphors have emerged as one of the promising candidates in the field of color tuning, wLEDs, and optoelectronic applications. The unique combination of the activators and sensitizers has led the phosphors to produce UC emissions ranging from UV to NIR regions. The UC mechanisms associated with the multicolor emission of tetragonal YPO_4:20% Yb^{3+}/1% Tm^{3+}/1% Er^{3+} nanoparticles have been described (Tymiński et al. 2020). According to these authors, the UC emission spectra can be modulated as a function of excitation sources and pump power variation ignoring the choice of suitable host matrices. Upon 975 nm laser excitation, a blend of different emission spectra resulting from Tm^{3+} and Er^{3+} ions has been seen. The emission spectra at 448, 471, 700, and 800 nm are associated with the $^1D_2 \rightarrow {}^3F_4$, $^1G_4 \rightarrow {}^3H_6$, $^3F_2 \rightarrow {}^3H_6$, and $^3H_4 \rightarrow {}^3H_6$ transitions of Tm^{3+} ions. Again, the emission bands at ~522 and 545 nm correspond to the $^2H_{11/2} \rightarrow {}^4I_{15/2}$ and $^4S_{3/2} \rightarrow {}^4I_{15/2}$ transitions of the Er^{3+} ions. Apart from these bands, it is observed that the emission at 650 nm of Tm^{3+} ions is partially overlapped with the band of the Er^{3+} ions (Figure 7.4a,b). They demonstrated the color tuning ability in the developed nanoparticles from bluish to greenish through white–blue in a Commission Internationale de I'Eclairage (CIE) chromaticity diagram by changing the pump power from 5 to 440 mW (the inset of Figure 7.4a). The analysis of UC spectra reveals that at low pump power, the absorption of three and four pump photons are favorable. In contrast, two-photon absorptions in the Er^{3+} ions are prominent under high power values. Besides the UC emissions by the 975 nm excitation source, the authors have also deliberated the UC emission mechanisms of the UC nanoparticles under 800, 1208, and 1532 nm excitations and a combination of laser source excitations, viz, 975 + 1208, 975 + 1532, and 1208 + 1532 nm. The white upconversion emission in the Y_2O_3:Er^{3+}–Tm^{3+}–Yb^{3+} phosphors has been reported under 980 nm excitation (Rai et al. 2013). Apart from Er^{3+}–Tm^{3+}–Yb^{3+}, Yb^{3+}–Tm^{3+}–Ho^{3+} is another combination of RE ions that can effectively be used for color modulation and other optical applications. The frequency of UC in the hexagonal Lu_2TeO_6:Yb^{3+}–Tm^{3+}–Ho^{3+} nanophosphors has been studied (Tang et al. 2015). Initially, the authors have conducted the UC emission study of Tm–Yb- and Ho–Yb-co-doped systems. Later, they performed the UC emission study of Lu_2TeO_6:Yb^{3+}–Tm^{3+}–Ho^{3+} tri-doped nanophosphors. Unlike the co-doped system, triply doped systems exhibit some extra peaks along with the same peaks as observed in the co-doped systems. As both Ho^{3+} and Tm^{3+}ions do not have the direct energy levels for 976 nm laser excitation, no UC emissions are monitored until the introduction of Yb^{3+} ions. Prominent UC emission peaks resulted from the excited Tm^{3+} ions at ~475, 650, and 700 nm have been assigned to the $^1G_4 \rightarrow {}^3H_6$, $^1G_4 \rightarrow {}^3F_4$, and $^3F_{2,3} \rightarrow {}^3H_6$ transitions, respectively. In contrast, the UC luminescence at the green and red regions results from $^5F_4$, $^5S_2 \rightarrow {}^5I_8$, and $^5F_5 \rightarrow {}^5I_8$, respectively (Figure 7.4c). Energy transfer processes have been considered as the principal cause of UC emission. Along with Yb^{3+}, the Ho^{3+} ion has experienced an additional sensitization from the Tm^{3+} ion through the energy transfer between Tm^{3+} and Ho^{3+} ($^1G_4 + {}^5I_8 \rightarrow {}^3H_6 + {}^5F_3$). All the UC mechanisms associated with the UC emission spectra could be discussed well through the energy-level diagram (Figure 7.4d). Although the Yb^{3+} ion-co-doped nanophosphors excited under 980 nm laser source have many advantages in various fields, in vivo biological application remains still challenging because of the overlapping absorption band of water with the NIR excitation range. In order to neutralize the risk of overheating

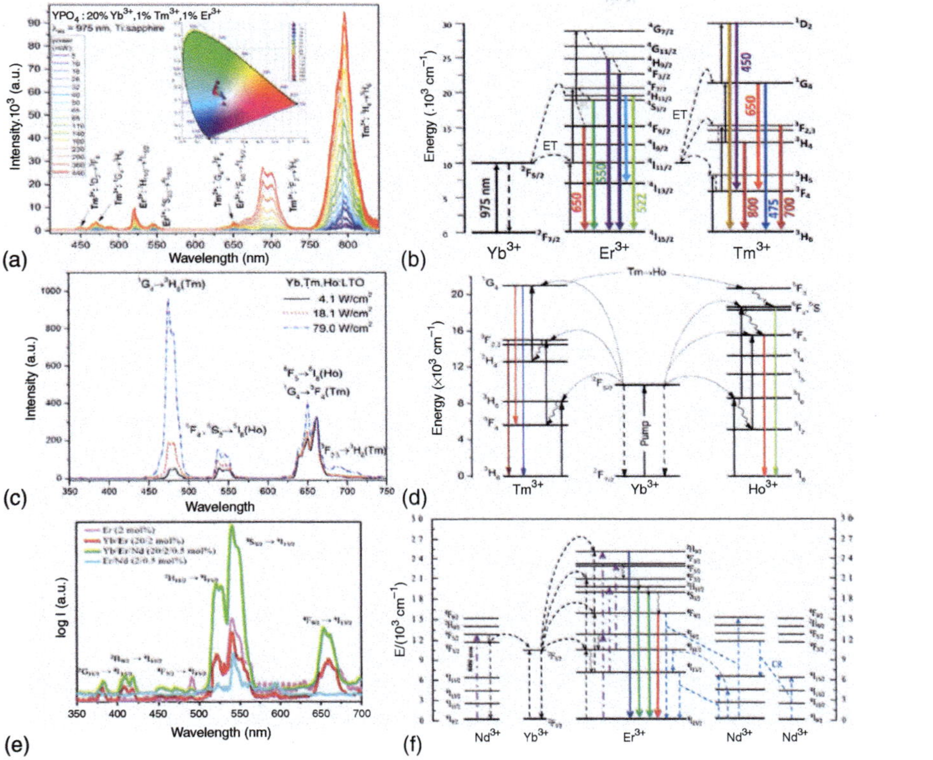

Figure 7.4 (a) The UC luminescence intensities of YPO_4:20% Yb^{3+}/1% Tm^{3+}/1% Er^{3+} samples under continuous excitation and different laser powers and (b) simplified diagram of energy levels and electronic transitions in the sample excited at 975 nm. Solid arrows represent radiative transitions and dashed ones represent non-radiative transitions. ET means the energy transfer process and the CR means the cross-relaxation process. The inset shows the CIE chromaticity coordinates of YPO_4 nanoparticle. Source: Tymiński et al. (2020). Reprinted with permission of John Wiley & Sons. (c) UC luminescence spectra of the Yb, Tm, and Ho:LTO under 976 nm laser excitation and (d) energy-level diagrams and schematic representation of the mechanisms responsible for the generation of UC luminescence for the Yb, Tm, and Yb, and Ho systems. Source: Tang et al. (2015). Reprinted with permission of Royal Society of Chemistry. (e) Upconversion emission spectra using a log scale for the emission intensity of $NaYF_4$:Er (2 mol%), $NaYF_4$:Yb/Er (20/2 mol%), $NaYF_4$:Yb/Er/Nd (20/2/0.5 mol%), and $NaYF_4$:Er/Nd (2/0.5 mol%) nanoparticles under 808 nm excitation and (f) proposed energy transfer mechanism under 808 nm excitation in $NaYF_4$:Yb/Er/Nd nanoparticles. The dashed–dotted, dashed, dotted, and full arrows represent photon excitation, energy transfer, multiphonon relaxation, and the emission process, respectively. For clarity, only dominant energy transfer processes are shown in the proposed mechanisms. Source: Tian et al. (2014). Reprinted with permission of MDPI.

and cell damage by the water molecules, the $Er^{3+}/Yb^{3+}/Nd^{3+}$:β-$NaYF_4$ nanocrystals have been synthesized, and the spectroscopic study of the developed nanocrystals was performed under 808 nm excitation (Tian et al. 2014). The UC emission spectra depicted in Figure 7.4e shows prominent green and red emission of Er^{3+} ions. Therefore, from this figure, it is clear that Nd^{3+} and Yb^{3+} ions act as sensitizers to transfer energy to the Er^{3+} ions. It has been observed that despite the efficient absorption of 808 nm photons by the Nd^{3+} ions, the UC emission intensities show weak emission peaks in the Nd^{3+}/Er^{3+} system compared to $Er^{3+}:NaYF_4$ phosphors because of the poor ET from the Nd^{3+} to Er^{3+} ions. The weak UC emission intensity in $Nd^{3+}/Er^{3+}:NaYF_4$ compared to the $Yb^{3+}/Er^{3+}:NaYF_4$ phosphors indicates strong EBT $Er^{3+} \rightarrow Nd^{3+}$ than $Er^{3+} \rightarrow Yb^{3+}$. However, the maximum luminescence intensity has been obtained for the $NaYF_4:Nd^{3+}/Yb^{3+}/Er^{3+}$ phosphors after co-doping with Yb^{3+} ions. This indicates that the Yb^{3+} ions play the key role of bridging ions between Nd^{3+} ions and Er^{3+} ions due to which the ET process has been enhanced significantly (Figure 7.4f).

The upconversion phenomena have been understood with the help of RE ions in association with non-lanthanide ions as well (Pandey and Rai 2013; Dwivedi et al. 2016; Prasad and Rai 2020; Srivastava et al. 2020). The role of different transition metal ions and other non-lanthanide ions with their corresponding UC mechanisms is demonstrated well later in this book.

7.2.2 Color Modulation in UCNPs

As mentioned in the previous section, a specific color can be attributed from certain RE^{3+} ion-based phosphors. As the green and red lights are emitted from the Er^{3+}- or Er^{3+}–Yb^{3+}-co-doped phosphors, blue and yellow emissions can be attained with the help of Ho^{3+}- and Tm^{3+}-based phosphors. Thus, fine-tuning of UC emission color is crucial for the varieties of applications such as light-emitting displays, multiplexed biological labeling, imaging, etc. (Dong et al. 2013; Zhang 2015). For example, the white light generation by phosphors is desirable for a liquid crystal display (LCD) back light, while well-resolved emission spectra with a narrow bandwidth is essential for multiplexed labeling. Various methods have been devoted to modulate the color of UCNPs by controlling the dopant ion concentration, selecting different host–dopant combinations, manipulating the phase, shape, and size of UCNPs, and through FRET (Wang and Liu 2009; Li et al. 2019). In this section, different techniques to tune the emission color and the routes to enhance the luminescence properties of UCNPs have been discussed.

7.2.2.1 Role of Dopant Concentration and Combination of RE^{3+} Ions in Color Modulation

The color variation in the UCNPs can be realized by varying the concentration and the combination of different RE^{3+} ions. By varying the dopant ion concentration, the distance between the RE^{3+} ions changes, which results in the change in the relative emission intensity of the UC nanophosphors. Several research papers have been published, addressing the influence of dopant ion concentration on the optical properties of UCNPs. The upconversion multicolor fine-tuning of $NaYF_4$ host co-doped with Yb^{3+}, Tm^{3+}, and Er^{3+} ions has been demonstrated (Wang and Liu 2008). The tuning of the output color from yellow to red has been realized in the $NaYF_4$:Er,Yb UCNPs with varying Yb ion concentrations from 25% to 60%. Because of the increase of Yb^{3+} ions in the upconverting system, the inter-ionic separation

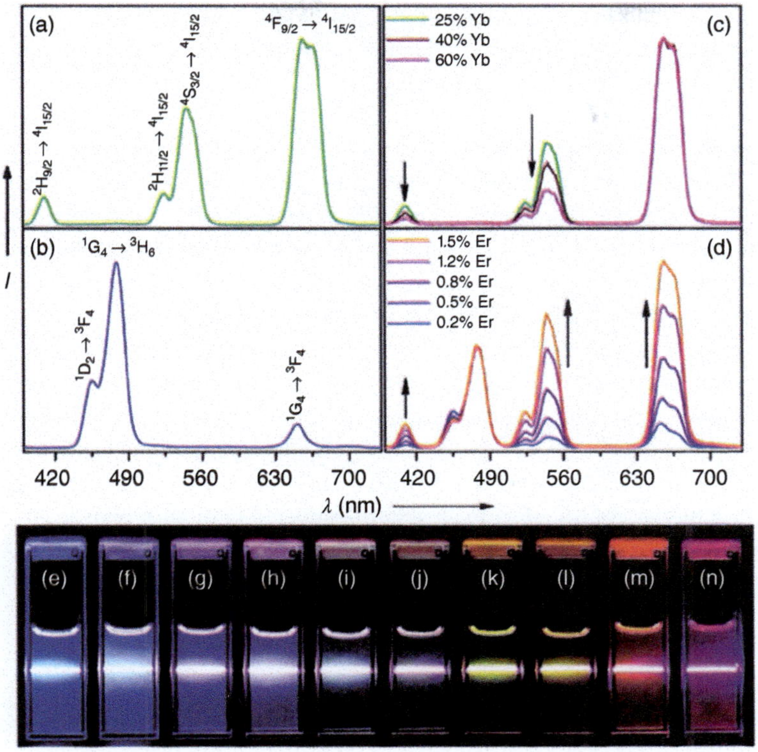

Figure 7.5 Room temperature UC emission spectra of (a) $NaYF_4$:Yb/Er (18/2 mol%), (b) $NaYF_4$:Yb/Tm (20/0.2 mol%), (c) $NaYF_4$:Yb/Er (25–60/2 mol%), and (d) $NaYF_4$:Yb/Tm/Er (20/0.2/0.2–1.5 mol%) particles in ethanol solutions (10 mM). The spectra in (c) and (d) have been normalized to Er^{3+} 660 nm and Tm^{3+} 480 nm emissions, respectively. Compiled luminescent photos showing the corresponding colloidal solutions of (e) $NaYF_4$:Yb/Tm (20/0.2 mol%), (f–j) $NaYF_4$:Yb/Tm/Er (20/0.2/0.2–1.5 mol%), and (k–n) $NaYF_4$:Yb/Er (18–60/2 mol%). Source: Wang and Liu (2008). Reprinted with permission of American Chemical Society.

between the Er^{3+} and Yb^{3+} ions decreases, which enables the EBT from the green emitting levels of Er^{3+} to Yb^{3+}. The depopulation of the $^2H_{9/2}$, $^2H_{11/2}$, and $^4S_{3/2}$ levels of Er^{3+} ions allows the color tuning from violet–yellow to red (Figure 7.5). They have also achieved the white light emission from the blue emission by varying the Er^{3+} ion concentration in $NaYF_4:Tm^{3+},Yb^{3+}$ UCNPs. The three-component ion-doped UCNP system ($NaYF_4$:Yb/Ho/Tm) has been established to observe the color tunability (Zhang et al. 2011). The authors stated that the blue emission is resulted from the $^1D_2 \rightarrow {}^3F_4$ and $^1G_4 \rightarrow {}^3H_6$ transition of the Tm^{3+} ions in $NaYF_4$:0.2% Tm/20% Yb UCNPs. It has been noticed that the relative emission spectra at the green and red regions change with co-doping of the second emitting ion (Ho^{3+}) precisely. For instance, by varying the concentration of Ho^{3+} ions from 0.2 to 2.0 mol%, the color of the UCNP changes from blue to green.

Similar results on concentration-dependent color tunability property have been published by many researchers with the different combination of Er-, Yb-, Tm-,

Ho-doped and co-doped systems (Vetrone et al. 2004; Wang et al. 2010b; Yin et al. 2010; Huang et al. 2011). These results indicate that the precise control of dopant ions is necessary to obtain a color tunable system.

7.2.2.2 Role of Host/Dopant Combination in Color Modulation

The upconversion mechanism is highly affected by the nature of host materials and the dopants chosen. The factors governed by the host lattice comprise the relative spatial position, coordination number, distance between the dopants, and the behaviors of the surrounding environment around the dopant ions, which affect the optical properties of RE^{3+}-doped UCNPs. The impact of host matrix on the spectroscopic properties of the UCNPs in a Er–Yb system has been explored (Soukka et al. 2005). It has been noticed that the oxides and oxychlorides exhibited larger red to green emission ratio than the oxysulfides and fluoride salts of Er^{3+} ions.

The effect of host matrix on the multicolor emission in the $NaREF_4$:Er–Yb system (RE = Sc^{3+} and Y^{3+}) has been investigated (Teng et al. 2012). As depicted in Figure 7.6a,b, the hexagonal $NaScF_4$:Er–Yb system emits red light with high red to green (R/G = 6 : 1) ratio, whereas the green light is dominant over the red light in the $NaYF_4$:Er–Yb system with the R/G ratio equals to 1 : 13. The unique dominant red light emission corresponding to the Sc^{3+}-based phosphors is resulted because of the difference in the ionic radii between the Sc^{3+} (0.83 Å) and Y^{3+} (1.06 Å) ions, which can help in the shortening of the Sc^{3+}–Sc^{3+} distance compared to the Y^{3+}–Y^{3+} distance. The pump power-dependent color tunability has been observed (Pandey and Rai 2012). The study reveals that the color tunability of the synthesized Ho^{3+}–Tm^{3+}–Yb^{3+}-co-doped Y_2O_3 phosphors can be viewed through naked eye under the pump power variation of 980 nm laser radiation. The clear picture of the color tunability has been depicted in Figure 7.6c–f.

7.2.2.3 Controlling the Emission Color Through Phonon Effects

The surface phonon effect can be manipulated by controlling the particle size of the UCNPs. An enhancement in the UC emission intensity of 20 nm Y_2O_3:Er^{3+}, Yb^{3+} UCNPs compared to the bulk materials has been reported (Vetrone et al. 2004). Likewise, the optical behavior of Y_2O_3:1% Er^{3+}/4% Yb^{3+} NPs ranging from 13 to 55 nm and their corresponding bulk materials under 978 nm excitation has been examined (Bai et al. 2007). They observed that the relative intensity of blue emission along with the red to green emission ratio increases with miniaturization of particle size. They observed that the number of photons required to populate the red emitting energy level vary between 2.0 and 1.0 with the increase in the particle size. A three-photon populating process has been noticed as the particle size reaches to 13 nm. It is necessary to understand that the size-dependent optical change is resulted from the surface effects rather the quantum confinement phenomena. In addition, phonon energy has a great impact on the multicolored UC emission as the population distribution between two closely situated energy levels depends on the phonon frequency of the phosphors. For the UCNPs, the UC emission intensities are affected by the surrounding environment because of the large surface-to-volume ratio. The size-dependent UC emission intensity of

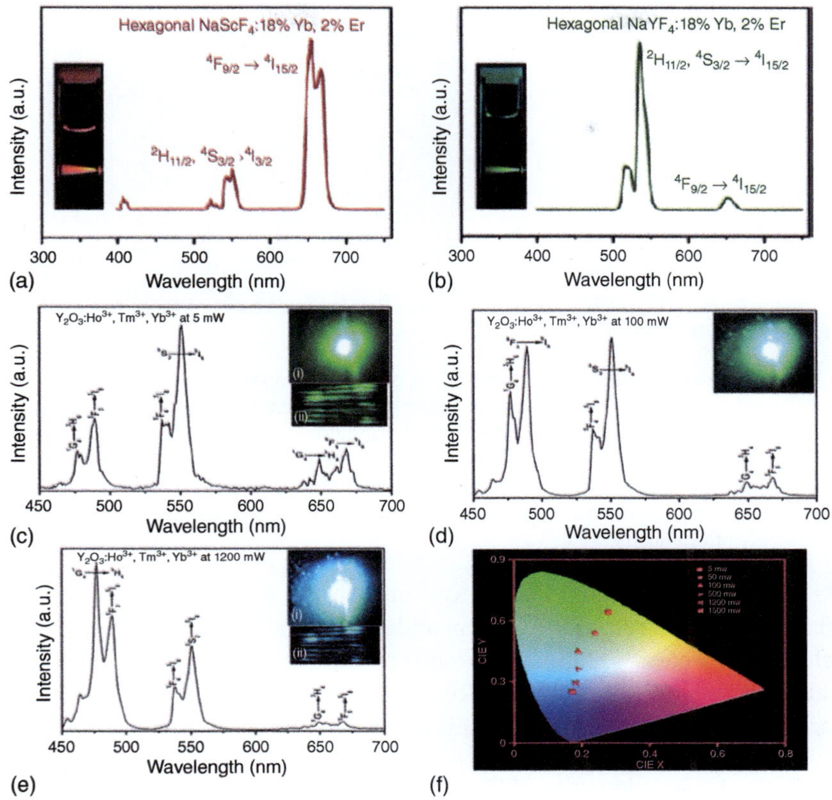

Figure 7.6 UC fluorescence spectra of hexagonal-phase (a) $NaScF_4$:Yb/Er and (b) $NaYF_4$:Yb/Er nanocrystals. The insets show the UC emission colors of the respective nanocrystals in cyclohexane solution. Source: Teng et al. (2012). Reprinted with permission of American Chemical Society. UC emission spectrum of Y_2O_3:Ho^{3+}–Tm^{3+}–Yb^{3+} phosphor on increasing power: (c) 5 mW, (d) 100 mW, and (e) 1200 mW. (f) CIE chromaticity diagram of Y_2O_3:Ho^{3+}–Tm^{3+}–Yb^{3+} phosphor at different pump powers. Source: Pandey and Rai (2012). Reproduced with permission of Springer-Verlag.

a-$NaYF_4$:Yb/Er UCNPs under 980 nm NIR irradiation has been demonstrated (Mai et al. 2007). They illustrated the variation in the emission color of the UCNPs at a very small change in the particle size from 5.1 to 8 nm. A facile method for tuning of the multicolor output of $NaYF_4$:20% Yb,2% Er UCNP has been reported (Niu et al. 2010). In the report, they have used octadecylamine (OM) and oleamide as the ligands attached to the surface of the UCNPs. By altering the ratio between the two coordinating ligands, the tunability of multicolour emission has been witnessed (Figure 7.7). With the success of the above experiment, authors have modified the pair of ligands attached and the dopant combination to observe the multicolor output of $NaYF_4$ phosphors. In another report, the fine-tuning in the UC emission of $NaYF_4$:Yb^{3+}/Er^{3+} UCNPs by the cooperative effect of carboxylic acid and amine group using amidation reaction has been described (Niu et al. 2011). Authors have reported that the formation of *N*-octadecyloleamide (OOA) at the surface of the

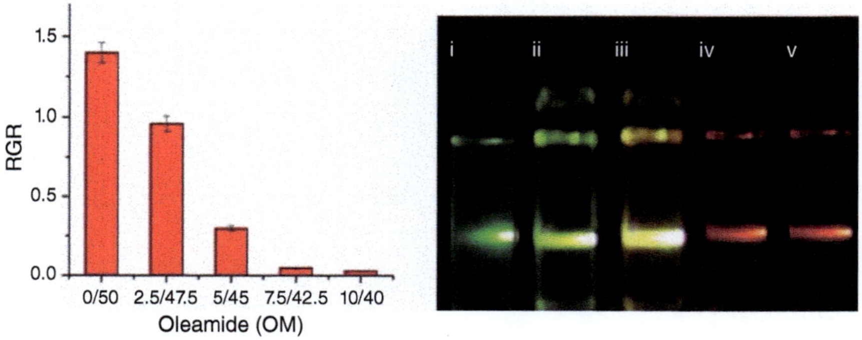

Figure 7.7 Ratio of green to red emissions (RGR), and photographs of $NaYF_4$:20% Yb,2% Er UCNPs prepared with various molar ratios of oleamide/OM in reaction solution (normalized to Er^{3+} 650 nm emission). Compiled luminescent photos showing corresponding chloroform solutions of particles excited at 980 nm diode laser, and the photographs were taken with exposure times of 3.2 seconds. (All the photographs were obtained under the same conditions without further illustration). Source: Niu et al. (2010). Reprinted with permission of Royal Society of Chemistry.

UCNPs during the amidation reaction can quench the green emitting levels of excited Er^{3+} ions. Apart from the color modulation by the surface-induced effects through various process, experimental results reveal that these surface effects can act as an obstacle in the UC luminescence. To overcome these difficulties, efficient approaches have been narrated later in this chapter.

7.2.2.4 Tuning UC Emission Using FRET

Förster or fluorescence resonance energy transfer (FRET) refers to a non-radiative energy transfer process from a donor fluorophore to an acceptor fluorophore via an electrostatic dipole–dipole interaction (Maier et al. 2003; Medintz et al. 2005; Ågren et al. 2017). The UCNPs have been considered as the donors, while the dyes and QDs are acting as the acceptors. Efficient multicolor emission by the FRET mechanism has been realized when the fluorophores such as dyes and QDs are attached to the surface of UCNPs (Li et al. 2019). They narrated the multicolor emission of hexagonal-silica/$NaYF_4$:Yb,Er/Tm UCNP-fluorophore-conjugated system by FRET mechanism (Li et al. 2008). The fluorescent dyes, such as fluorescein isothiocyanate (FITC) and tetramethylrhodamine isothiocyanate (TRITC), and QDs (QD605) have been considered as the acceptor fluorophores while the silica-coated $NaYF_4$:Yb,Er/Tm UCNPs have been treated as the donor. At first, the 980 nm NIR radiation is absorbed by the core UCNPs, and the fluorescence spectra in the visible region are detected. Then, the emitted light by the UCNPs is further absorbed by the organic dyes and QDs to get multicolor emission spectra. The authors have noticed that the emission spectra of undoped silica/$NaYF_4$:Yb,Tm UCNPs at 450 and 479 nm got quenched after introducing FITC and QD605. New emission peaks have formed at 536 and 605 nm corresponding to FITC and QD605, respectively. The above result indicates an efficient FRET between the UCNPs and organic fluorophores. Similar types of results have also been witnessed in the TRITC-doped silica/$NaYF_4$:Yb,Er UCNPs (Figure 7.8a–c).

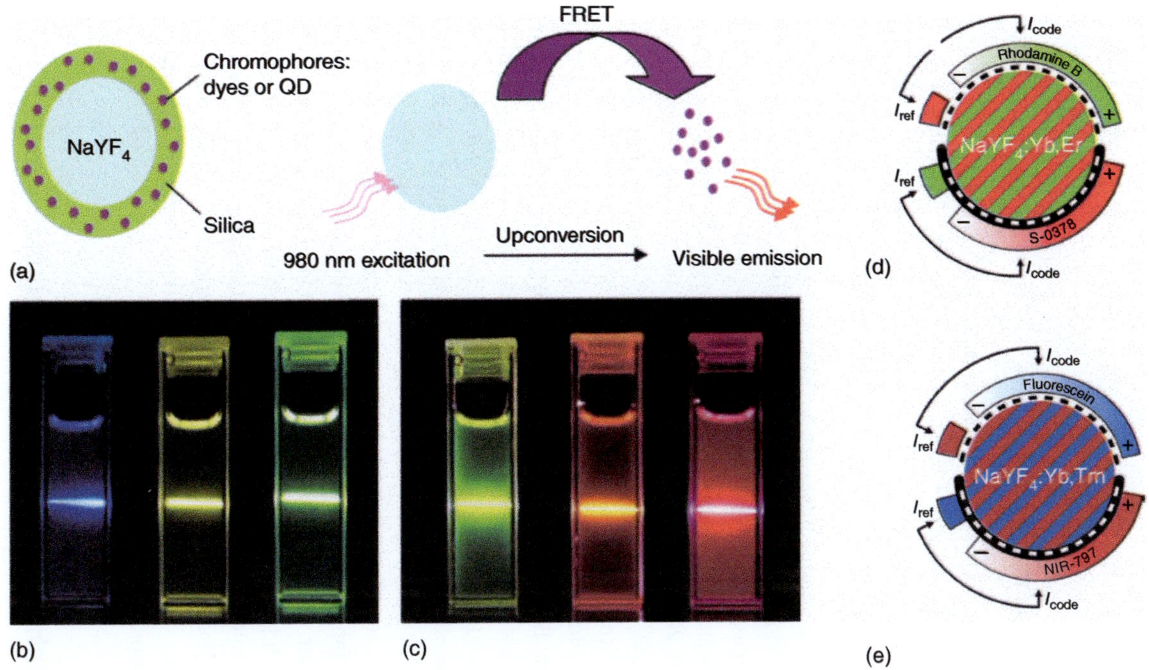

Figure 7.8 (a) Schematic drawing of FRET-based multicolor silica/$NaYF_4$ NIR-to-vis upconversion nanospheres and (b) photographs of silica/$NaYF_4$ nanospheres in hexane (1 wt%) under excitation of NIR laser (980 nm, power density = 50W/cm^2): total upconversion fluorescence of $NaYF_4$:Yb,Tm nanospheres (blue), total upconversion fluorescence of $NaYF_4$:Yb,Er nanospheres (yellow green), (c) photographs showing total fluorescence of FITC-doped silica/$NaYF_4$:Yb,Tm nanospheres (left), TRITC-doped silica/$NaYF_4$:Yb,Er nanospheres (middle), and QD605-doped silica/$NaYF_4$:Yb,Tm nanospheres (right). Source: Li et al. (2008). Reprinted with permission of John Wiley & Sons. (d) Different concentrations of either rhodamine B or the dye S-0378 on the surface of $NaYF_4$:Yb,Er UCNPs are used to screen off selectively either the green or the red emission band to various degrees. (e) Fluorescein or the dye NIR-797 screen off of the blue or the near-infrared emission band of $NaYF_4$:Yb,Tm UCNPs. In both cases, the second emission band, which is not absorbed, acts as an internal reference to yield ratiometric codes. Source: Gorris et al. 2011. Reprinted with permission of John Wiley & Sons.

In another report, the FRET-based color modulation in $NaYF_4$:Yb^{3+},Er^{3+} utilizing rhodamine B (RhB) and the dye S-0378 and in $NaYFF_4$:Yb,Tm UCNPs using the fluorescein and dye NIR-797 has been discussed (Gorris et al. 2011) (Figure 7.8d,e). Apart from using quenching dyes, plasmonic metals such as Au aced as energy acceptors by strong plasmonic absorption at 530 nm to quench the selective green light emission of Er^{3+} without upsetting the red emission at 650 nm in $NaYF_4$:Yb^{3+}/Er^{3+} UCNPs (Wang et al. 2005).

7.2.3 Quenching Mechanisms in UCNPs

Luminescence quenching is the non-radiative process that hinders the luminescence intensity and efficiency of the RE^{3+}-doped UCNPs. Therefore, it is off the center of attention to identify the detailed mechanisms responsible for the non-radiative losses. The prominent non-radiative mechanisms associated with the emitting states of RE ions responsible for the faster decaying of emission from the states are cross-relaxation, energy migration to the luminescent killing centers, and FRET to ligand or solvent molecules (Fischer et al. 2016; Rabouw et al. 2018). The quenching mechanism associated with the RE^{3+} ions depend highly on the system's stability and the resonant conditions with the vibrational modes of solvents and ligands (Tessitore et al. 2019). The UC emission intensity is affected by the size of the synthesized particles, the concentration of dopant ions, the power density of the excitation, and the environment around the dopant ions. With the miniaturization of particles, the surface-to-volume ratio increases rapidly with a decrease in particle size (Wang et al. 2010c; Boyer and Van Veggel 2010). Thus, the emitting RE^{3+} ions at the surface are exposed to the surface defects, ligands, and solvents with high vibrational modes (Escudero Belmonte et al. 2017). It is found that the ions present at the surface bear reactive dangling bonds, which are unsaturated, hence carry a partial charge that increases the surface energy. Because all the systems try to be at their minimum energy states, the dangling bonds tend to interact with the surrounding environment (e.g. solvents and ligands) (van Hest et al. 2017). The effect of miniaturization causes a decrease in UC emission efficiency because of high non-radiative energy transfer to the defect traps. Eventually, high pump power density is required to maximize the emission intensity as the particle size decreases. In view of the fact that the quenching mechanism is detrimental to the UC emission intensity and emission efficiency, several approaches have been deliberated to enhance the UC emission efficiency.

7.3 Routes to Enhance Upconversion Luminescence in Nanoparticles

Apart from the phenomenal characteristics and advantages of UCNPs over other luminescent materials, UCNPs experience low emission intensity and low quantum efficiency because of several aspects. The major factors that affect the luminescent efficiency are low absorption efficiency, concentration quenching, non-avoidable surface quenchers, and detrimental cross-relaxations (Zhu et al. 2019). Accordingly, numerous approaches such as confinement of energy migration, reduction of surface defects, attachment of organic dyes, etc., have been proposed (Liang et al. 2019; Wisser et al. 2018; Wang et al. 2010c; Huang et al. 2018; Vetrone et al. 2009; Zhan et al. 2015). In this section, different optimization routes and tailoring techniques to enhance UCL in nanoparticles have been mentioned briefly.

7.3.1 Dye Sensitization Techniques

One of the significant internal characteristics of low UC efficiency lies with low absorption cross section and narrow bandwidth in the NIR region of the dopant ions. Many dyes have been deployed in UCNPs to enhance the absorption cross section efficiency of UCNPs. The researchers have investigated that the molar extinction coefficient of IR-806 dye at 806 nm in $CHCl_3$ is $\sim 5 \times 106$ times greater than that of β-$NaYF_4$:Yb/Er UCNPs under 975 nm excitation (Zou et al. 2012). In the developed sensitized system, IR-806 effectively absorbs the 800 nm excitation photons and transfers the energy to Yb^{3+} ions by resonance energy transfer process. Because of the amplified absorption cross section and broad absorption bandwidth by dye-sensitized UCNPs, an increment of about 3300 times in the emission intensity has been reported compared to desensitized UCNPs. Further, the idea of energy-cascaded upconversion (ECU) has been explained in which the NIR-excited dye is attached to the epitaxial core/shell UCNPs (Chen et al. 2015a). The stated ECU technique follows the absorption of the excitation radiation via the NIR dye, followed by transfer of excited energy to sensitizer-1 in the core–shell and then to sensitizer-2 at the core and shell interface, which eventually transfers to the activators in the core to get UC emission. An enhancement of about 100 times has been reported in the ECU-sensitized UCNPs than the bare UCNPs. A similar unique hybrid organic–inorganic system has been proposed. In this system, an organic indocyanine green (ICG) dye is attached to the surface of UCNPs, which receives the NIR excitation (Shao et al. 2016). The energy migration pathway from the organic dye at the surface to the emitter at the core via sensitizers in the shell due to resonant energy levels helps in a four-time enrichment in the UC emission intensity (Figure 7.9a,b). Later, a facile concept to improve the quantum efficiency of UCNPs through the emission dye sensitization technique has been reported (Wisser et al. 2018). The study suggests that the commercial ATTO 542 dye attached to the surface of the UCNPs possess high radiative rates, and similar energy gaps between the lowest unoccupied molecular orbitals (LUMOs) of ATTO 542 dye and the $^2H_{11/2}$ state of Er^{3+} ions favor the efficient energy transfer from UCNPs to the dye. The extraordinary boost in the quantum yield and UC emission intensity reveals the importance of dye sensitization in UCNPs.

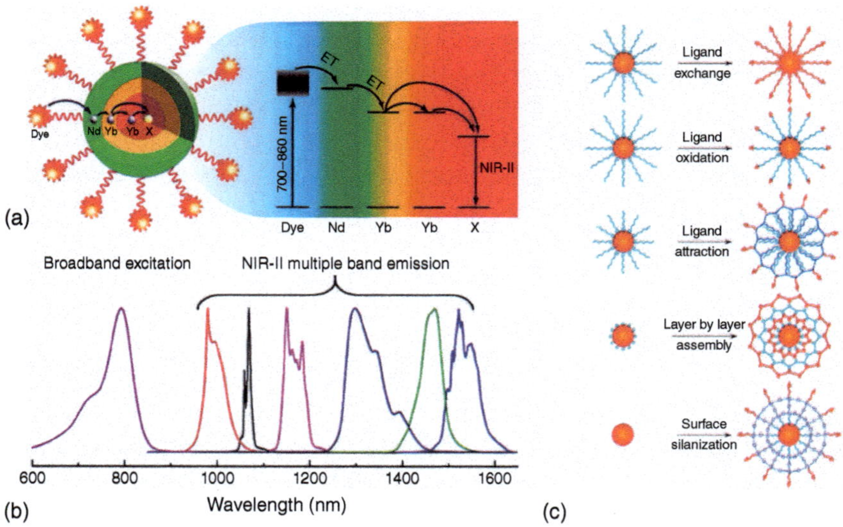

Figure 7.9 Schematic illustrations of (a) energy transfer pathway from ICG on the surface of $NaYF_4$:Yb^{3+}/X^{3+}@$NaYF_4$@$NaYF_4$:Nd^{3+} nanocrystal to the Nd^{3+} ions in the outer shell, then to the Yb^{3+} ions in the inner shell, and finally to the Yb^{3+}/X^{3+} ions (X = null, Er, Ho, Tm, or Pr) in the core, producing large Stokes-shifted NIR-II emissions; (b) the functional roles of ICG (providing excitation between 700 and 860 nm, absorption spectrum for 40 μg/ml in dimethylformamide [DMF]), the core/shell/shell structure (spatial isolation of the core from the surrounding quenching center and directing energy transfer to the core), and the activator of varying type (entailing defined narrow band emission in the NIR-II range). Source: Shao et al. (2016). Reprinted with permission of American Chemical Society. (c) Generic strategies for solubilization and functionalization of upconversion (UC) nanophosphors. Source: Wang and Liu (2009). Reprinted with permission of Royal Society of Chemistry.

7.3.2 Concentration Quenching Minimization

A typical UCNP contains thousands of photon sensitizers and hundreds of photon activators. Theoretically, the presence of higher number of activators and sensitizers in UCNPs could help in boosting the intensity and efficiency of UCNPs. However, after a certain threshold value of dopant concentration, the emission intensity started to decline with a further increase in dopant concentration. The process is named as concentration quenching. Presently, two major reasons have been investigated, which are responsible for concentration quenching; (i) enhanced energy migration to the surface killing centers and (ii) deleterious cross-relaxation between the dopant ions. Owing to the concentration quenching effect, the maximum concentration of the Yb^{3+} ions (sensitizer) is ~20 mol% and that for activator ions is around 0.5–5 mol%.

7.3.2.1 Suppression of Surface-Related Quenching

It has been reported by many researchers that the quenching centers such as defects, ligands, and solvents with high phonon frequency occur at the surface of the UCNPs (Kumar et al. 2009; Wang et al. 2010c; Boyer et al. 2010). Construction of inert core@shell around the UCNPs has been emerged as a promising approach to passivate the energy migration to the surfaces. A 450-fold enhancement in the UC emission intensity arising from the $NaGdF_4$:Yb/Tm@$NaGdF_4$ core@shell structure due to the attachment of an inert $NaGdF_4$ shell in the core $NaGdF_4$:Yb/Tm UCNPs has been observed (Wang et al. 2010c). Moreover, further research work claimed that the insertion of an inert $NaLuF_4$ shell around the Er^{3+}-doped $NaYF_4$ core can push the threshold concentration limit of Er^{3+} ions up to 100 mol% (Johnson et al. 2017). A wet chemical annealing technique to improve the UC emission efficiency of core UCNPs has been presented (Bian et al. 2018). They detected the amorphous nature at the surface of core $KLuF_7$ UCNPs via the aberration-corrected high-angle annular dark-field scanning transmission electron microscopy (HADDF-STEM). They observed an increase in crystallinity around the surface with the application of thermal annealing process, and the UC emission increases by an order of magnitude because of the recovery of surface defects.

Besides suppressing the quenching effect at the surface, it is essential to functionalize the surface of the RE-doped nanophosphors to be used in biological labeling applications. As most bulk materials follow a high-temperature synthesis route, the UC nanophosphors do not possess any intrinsic aqueous solubility. Therefore, for biological applications, surface functionalization of UCNPs with hydrophilic ligands is needed before the chemical attachment of biomolecules to become water compatible and dispersible (Sharma et al. 2017; Wang and Liu 2019). As depicted in the Figure 7.9c, different strategies such as ligand exchange, ligand oxidation, ligand attraction, layer-by-layer assembly, and surface silanization have been employed to observe the dual-surface properties of the UCNPs (Wang and Liu 2009).

7.3.2.2 Removal of Detrimental Cross-Relaxation

Cross-relaxation is one of the deleterious processes at high dopant concentrations, which causes reduction in luminescence intensity by transferring excited energy to the neighboring ions non-radiatively. Precisely, RE ions with many energy levels have higher chances of going through the destructive cross-relaxation process. At high dopant concentrations, the ion–ion distance is the order of a few nanometers, which helps in the detrimental non-radiative energy transfer between the neighboring ions. Keeping the concentration of the dopant ions into account, researchers have recommended an increase in the spatial distance between the RE^{3+} ions. To verify the suggestion, unique UCNPs consisting of an alternative illuminating shell ($NaYF_4$:Yb^{3+},Er^{3+}) and a separating shell ($NaYF_4$:Yb^{3+}) have been designed (Liu et al. 2011). They claimed an increase in the upper threshold of Er^{3+} ions from 2 to 5 mol%. An increase in the spatial distance between the emitters and minimization of energy transfer to the surface quenchers result in enhanced UC emission. A convenient core–shell has been constructed, in which the core–shell is filled with a high concentration of sensitizer ions, and the activator ions are located at the shell. Because of the high concentration of sensitizer ions, more excitation photons are harvested. The energy migration between the dopant ions and the surface quenchers is minimal, which subsequently gives rise to high UC emission (Huang et al. 2018).

7.3.3 Confinement of Energy Migration

As discussed earlier, quenching centers have been considered as the main hurdles in the path of efficient UC emission intensity and UC efficiency. The phenomena of spatial confinement of excited energy in the RE^{3+} ion-doped nanostructure could prevail detrimental cross-relaxation and energy migration to the surface defects, thereby increasing the UC emission intensity and efficiency. The researchers came out with the idea of improving the optical properties by tuning the energy confinement of the nanophosphors through manipulating the dimension of the crystal lattice (Chen et al. 2016a). They have fabricated hexagonal $NaYF_4$@$NaYF_4$:Tm@$NaYF_4$ core@shell@shell-constructed nanoparticles by a layer-by-layer epitaxial growth method and obtained the localization of excitation energy in the $NaYF_4$ inner shell (Figure 7.10a). Further, they monitored the prolongation of the decay time of Yb^{3+} ion as the size of the $NaYF_4$ shell thickness reduces from 17 to 1 nm, validating the suppression of energy migration to the host lattice in the shell, which results in an increase in the UC emission intensities (Figure 7.10b). The authors stated that the localization of energy transfer eventually ends up with efficient energy transfer to the nearby Tm^{3+} ions, responsible for five-photon UC emission of Tm^{3+} ions without concentration quenching. Additionally, they found an enrichment of lasing emission intensity at~311 nm because of high Gd^{3+} ion concentration under 980 nm excitation compared to Tm/Yb-based nanophosphors

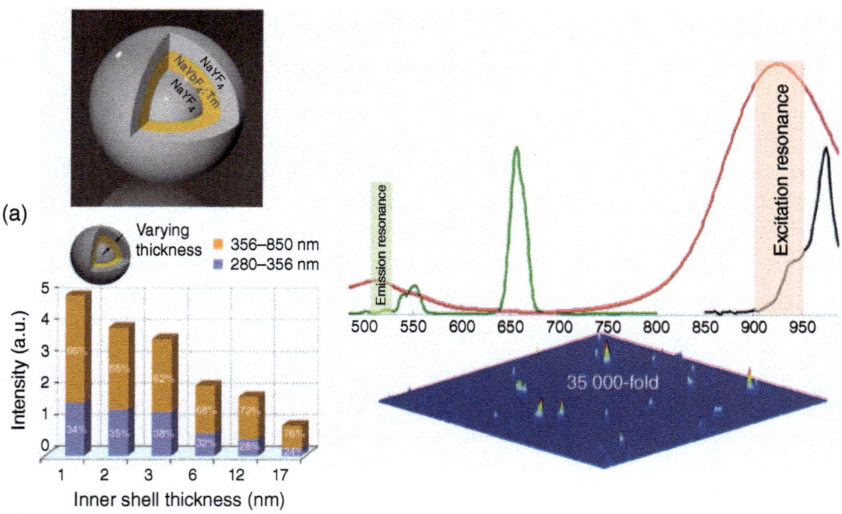

Figure 7.10 (a) Schematic design of a $NaYF_4$@$NaYF_4$:Tm@$NaYF_4$ core–shell–shell nanoparticle for confining the migration of excitation energy generated in the Yb^{3+} ions and (b) upconversion emission intensity versus inner shell thickness (1–17 nm). The emission intensities have been calculated by integrating the spectral intensity of the emission spectra that are normalized to the absorption of Yb^{3+} at 980 nm. Source: Chen et al. (2016a). Reprinted with permission of Springer Nature (c) Emission spectra of the ZrO_2 UCNPs (blue solid line) and absorption spectra of the GNRs (red dotted line) in a DMF solution. The orange line indicates the excitation wavelength at 920 nm. Source: Zhan et al. (2015). Reprinted with permission of John Wiley & Sons.

under 650 nm excitation. Later, the energy confinement of the excitation energy by introducing energy trapping centers such as Yb^{3+}, Tm^{3+}, Ho^{3+}, Eu^{3+}, and Pr^{3+} ions in $NaErF_4$@$NaYF_4$ core@shell nanoparticles has been studied (Shang et al. 2018). The investigation reveals that the addition of energy trapping centers reduces the energy migration losses to the quenching centers and the cross-relaxation between the ions. The detailed study shows that by varying the energy trapping centers, the UC emission intensity and color can be voluntarily tuned. Tm^{3+} and Ho^{3+} ions are helpful for populating the $^4F_{9/2}$ state of Er^{3+} ions (red emission), whereas for green color emission, Eu^{3+} and Pr^{3+} centers are beneficial, and Yb^{3+} ions are helpful for both green and red color emission. The authors have also noticed an improvement in the photoelectron conversion efficiency up to 15.25% after the incorporation of $NaErF_4$:10 at.% Yb^{3+}@$NaYF_4$ NPs into dye-sensitized solar cells (DSSCs).

7.3.4 Other Techniques to Enhance Upconversion Emission

7.3.4.1 Crystal-Phase Modification

The crystal phase of the host matrix is equally important to cultivate the spectroscopic properties in the UCNPs. Modulation in the luminescence properties has been examined by transforming the crystal phase using various synthesis techniques. For example, a stable hexagonal GdF_3:Er/Yb system by doping alkaline earth metal Sr^{2+} ions has been prepared (Zhao et al. 2013). The spectroscopic investigation suggests that the hexagonal GdF_3:Er^{3+}/Yb^{3+} UCNPs are more efficient than the orthorhombic system. Authors have found that doping of Sr^{2+} ions modifies the crystal field around the activator ions, which results in better UC emission efficiency. A similar report has been discussed, in which the hexagonal β-$NaYF_4$:Yb^{3+},Ln (Ln = Er^{3+} and Tm^{3+}) UCNPs show better UC efficiency compared to the cubic phase (Shan and Ju 2007). A 40-fold enhancement in the UC emission intensity has been reported when the as-prepared $NaYF_4$:Yb^{3+}, Er^{3+} UCNPs annealed to a temperature between 400 to 600 °C (Yi et al. 2004).

7.3.4.2 Constructing an Active Core/Active Shell Strategy

As discussed in the previous sections, the construction of an inert core–shell has improved the UC emission efficiency tremendously by minimizing the migration of excitation energy to surface quenchers. The shells primarily served as the protecting layer among the dopant ions as well as between the emitters and surface defects. Along with the traditional core@shell structure, the sensitizer-rich shell has proven beneficial because of the high absorbance rate of excitation energy by the sensitizers in the shell (Fischer et al. 2015; Ai et al. 2015; Hu et al. 2018; Reddy et al. 2018; Kostiv et al. 2020).

7.3.4.3 Conjugating Surface Plasmon Resonance Technique

The localized surface plasmon resonance (LSPR) is nothing but the collective oscillations of the conduction electrons triggered by the interaction between the incident electromagnetic radiation and noble metal nanoparticles (Unser et al. 2015; Petryayeva and Krull 2011). The resonance occurs when the electromagnetic fields match with the oscillating electrons at the surface of the nanoparticles. Noble plasmonic metals such as Ag, Au, and Pt are commonly used for improving the UC emission efficiency. Few probable methods have been predicted for enhancing the UC emission via LSPR properties of the plasmonic metals. One of which is the strengthening of the local incident electric field. Because the electric field is proportional to the flux of incident photons, stronger electric fields stimulate the flux of incident photons. The UCNPs eventually absorb these photons to generate enhanced UC emissions. Similarly, when the UC emission bands couple with the LSPR band, the photon density increases, promoting the radiative rates and hence improving the emission intensity of UCNPs. It is believed that if both the emission and excitation bands of UCNPs match well with the LSPR peak, LSPR-assisted increment of UC intensity will be more prominent. A plasmonic dual resonance tactic has been implemented, in which two distinct LSPR peaks of gold nanorods (GNRs) resonate with both the excitation and emission bands of ZrO_2 UCNPs (Zhan et al. 2015). According to this report, a tremendous enrichment of UC emission up to 35 000 times at 522 nm has been monitored in the ZrO_2 UCNPs because of the plasmonic effect of GNRs (Zhan et al. 2015) (Figure 7.10c). The researchers have synthesized $NaYF_4$:Tm,Yb UCNPs and incorporated gold nanoparticles as the plasmonic metal to the surface of the UCNPs and obtained an enrichment in the UC emission in the blue region at 452 and 476 nm by 150% and 50% enhancement in the red region at 647 nm (Zhang et al. 2010). A LSPR-assisted UC emission enhancement in UCNPs for the enhancement of UC emission in $NaYF_4$:Yb,Er UCNPs has been performed, where the Ag nanowire has contributed to the LSPR properties (Feng et al. 2009). The authors have found a 2.3- and 3.7-fold rise in the UC emission in the green and red regions, respectively.

7.3.4.4 Dielectric Superlensing-Mediated Strategy

High UC emission efficiency is desirable for cutting-edge applications such as photovoltaics and biophotonics, but obtaining high UC emission efficiency under low irradiance is challenging because of the low extinction coefficient of absorbing dopants. The dielectric superlensing technique is one of the latest approaches perceived by researchers to develop high UC emission efficiency. In this technique, dielectric microbeads are used as superlenses, which may confine the low power incident radiation to a sub-wavelength photonic hotspot of high local field intensity, thus achieving highly efficient UC emission. The giant intensification in the UC emission intensity by 5 orders of magnitude with modulating both excitation and emission wave fronts through dielectric superlensing technique has been reported (Liang et al. 2019). The authors have used transparent poly(ethylene glycol) diacrylate (PEG-DA) polymeric microbeads as the superlenses to enhance the UC emission. The dielectric superlensing mediated strategy for the UCNPs integrated with photonic platforms has been described (Ahn et al. 2020). In the study, they observed an enormous gain (i.e. by 3 orders of magnitude) in the UC emission when the dielectric microbeads are decorated with plasmonic metal nanoparticles under weak NIR excitation (Ahn et al. 2020). It has come to notice that the dielectric superlensing mediated strategy is helping in photon–photon interaction in UCNPs at sub-threshold excitation radiation.

7.4 Technological Applications

The success of pioneering synthesis techniques to develop nanoparticles along with promising optical tunability and efficient upconversion emission strategy offers the use of UCNPs in many applications. Few of them are briefly discussed in this section.

7.4.1 Photonic Applications

7.4.1.1 Light-Emitting Diodes (LEDs)

wLEDs is one of the most valued photonic applications produced till date. As white light is crucial for solid-state lighting applications, the production of high-quality LEDs is equally essential. So far, two approaches have been endorsed to obtain wLEDs. The first one is by combining three different blue, green, and red LEDs to a single device. However, this strategy failed to perform well in the market because of the lack in controlling constraints, complex manufacturing, and durability due to three different decay times of individual LEDs (Sharma et al. 2017). Similar to mixing three different LEDs, mixing three different color-emitting phosphors bears disadvantages over the directly white lighting yielding phosphors because of different degradation, re-absorption of blue light by other phosphors, etc. Thus, the second technique is based on a single phosphor that can emit white light, and this technique is coined as phosphor-converted light-emitting devices (pcLEDs). The phosphor-converted white light-emitting devices (pc-wLEDs) are the potential candidates for artificial lighting to replace other lighting sources such as incandescent light, fluorescent lamps, halogen lights, etc., because of their low energy consumption, high luminous efficiency, and environment-friendly nature (Pust et al. 2015; Pulli et al. 2015; Chen et al. 2017). Despite everything, the conventional pc-wLEDs could render an average color rendering index (CRI) ~ 80, which is not sufficient to define an object's true color such as under sunlight. However, the researchers have come up with a novel light-emitting device (LED) package (PKG) by considering two blue LEDs and a UV LED in association with green and red phosphors (Ahn et al. 2019). The authors have proclaimed to achieve a CRI point over 95 with luminous efficiency ~100 lm/W in pc-wLEDs. Halide perovskites are considered eminent members for the wLED preparation owing to their excellent photoluminescence quantum yield (PLQY) and UV excitation. The researchers have synthesized a gallium nitride (GaN) LED coated with $CsPbCl_3:Ce^{3+},Eu^{3+}$ nanocrystals dispersed in polymethyl methacrylate (PMMA) excited under UV radiation. The broadband emission ranging from the blue to green region due to Ce^{3+} ions and the red emission by the Eu^{3+} ions results in cold white light emission with 24 lm/W luminous efficiency (Pan et al. 2017). Similar to downconverting phosphors, white light can be achieved using upconverting phosphors. The photometric properties of the Ho^{3+}–Yb^{3+}–Tm^{3+}–$Li^{+}:Gd_2(MoO_4)_3$ nanophosphors synthesized by the solid-state reaction method under 980 nm excitation have been studied (Kumari et al. 2017). They obtained a white light emission because of co-doping with blue emitting (Tm^{3+}), green emitting, and red emitting (Ho^{3+}) ions (Kumari et al. 2017) (Figure 7.11a,b). Similarly, researchers have developed the Yb^{3+}–Ho^{3+}–Tm^{3+} triply

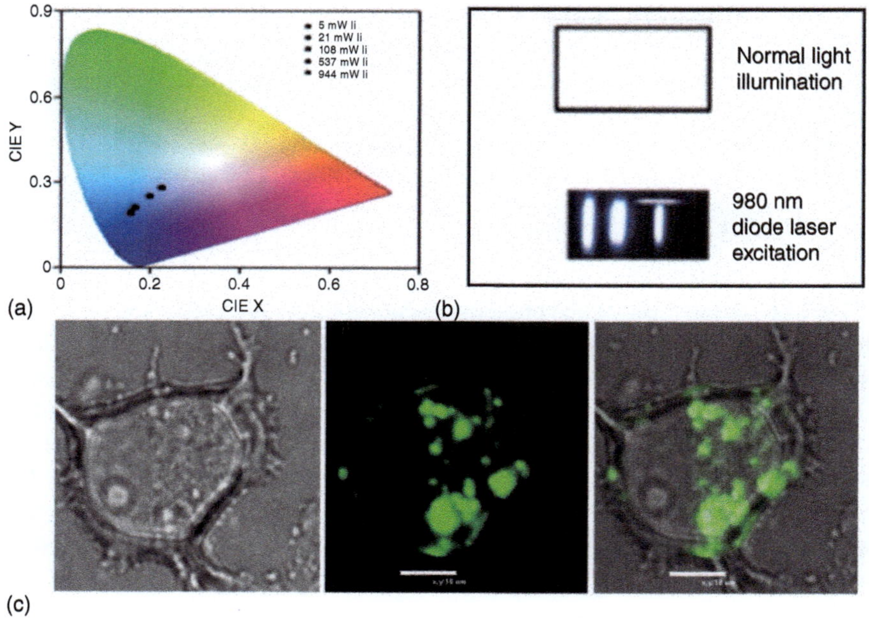

Figure 7.11 (a) Color coordinate diagram of 0.4 mol% Ho^{3+}–8.0 mol% Yb^{3+}–0.5 mol% Tm^{3+}–15.0 mol% Li^{+}:$Gd_2(MoO_4)_3$ nanophosphors at different pump power. (b) Written letters under normal light illumination and 980 nm diode laser excitation at 285 mW for 0.4 mol% Ho^{3+}–8.0 mol% Yb^{3+}–0.5 mol% Tm^{3+}–15.0 mol% Li^{+}:$Gd_2(MoO_4)_3$ nanophosphors. Source: Kumari et al. 2017. Reprinted with permission of IOP Publishing. (c) Confocal fluorescence imaging of MCF-7 cells using silica/$NaYF_4$:Yb,Er nanospheres. Bright-field (left), confocal fluorescence (middle), and superimposed (right) images of MCF-7 cells incubated with the nanospheres for 24 hours. Source: Li et al. (2008). Reprinted with permission of John Wiley & Sons.

doped hexagonal $NaYF_4$ nanorods via the hydrothermal method (Yang et al. 2009). While observing the optical phenomena of the prepared phosphors, the authors have found the white light emission as a combination of blue UC emission at 450 and 475 nm corresponding to the $^1D_2 \rightarrow {}^3F_4$ and $^1G_4 \rightarrow {}^3H_6$ transitions of Tm^{3+} ions, green emission at 545 nm corresponding to the $^5S_2/^5F_4 \rightarrow {}^5I_8$ transition of Ho^{3+} ions, and red emission at 695 nm corresponding to the $^3F_3 \rightarrow {}^3H_6$ transition of Tm^{3+} ions along with the emission at 650 nm that refers to both the $^5F_5 \rightarrow {}^5I_8$ transition of Ho^{3+} ions and the $^3F_2 \rightarrow {}^3H_6$ transition of Tm^{3+} ions. Although white light has been obtained using UC phosphors, still efficient and high CRI wLEDs are necessary for advanced applications.

7.4.1.2 Photovoltaic Applications

UCNPs have been considered as potential candidates in photovoltaics. Usually, conventional solar cells use a narrower band of the total solar radiation that falls in the UV region, i.e. about ~5%. Most of the low-energy radiations are not absorbed by the solar cells because of the energy mismatch. Because of the high bandgap of solar cells, the sub-bandgap photons could not generate electron–hole pairs, thus hinders the solar cells efficiency. For instance, the transmission loss is relatively low in the case of low-bandgap c-Si solar cells. The energy loss is still as high as ~20%. With the help of UCNPs, the energy loss could be minimized, followed by a high conversion of solar radiation to electricity. As UCNPs can absorb at a longer wavelength, i.e. NIR radiations, and can emit shorter wavelengths, i.e. in the visible or UV region, this can help to enhance the solar cell efficiency. First of all, upconverting phosphors or UCNPs sequentially absorb the NIR radiation of solar spectrum to the excited state, followed by emission at shorter wavelength. The emitted photons from the UCNPs are absorbed by the solar cells to create electron–hole pairs. An increase in the Shockley–Queisser efficiency limit from ~30% to 40.2% in a c-Si solar cell after UCNP incorporation under non-concentrated air mass (AM) 1.5 solar irradiation has been predicted by a theoretical study (Trupke et al. 2006). Although the theoretical results expect the solar cell efficiency to be more than 40%, such enhancement in solar cells is yet to be reported experimentally. Briggs et al. developed a thermodynamic model of an upconverter cell with a realistic narrow band. Despite a narrow bandwidth with better quantum yield of the upconverters, the researchers, after developing a thermodynamic model, reported an increase in solar cell efficiency from 28% to 34% (Briggs et al. 2013). The photovoltaic properties of substrate-free GaAs cells have been studied by the researchers (Gibart et al. 1996). These authors have fabricated a very thin layer of the cell over 100 mm-thick vitroceramic doped with Er^{3+} and Yb^{3+} ions. The optical spectra obtained under Ti–sapphire IR radiation at 1.391 eV fall in the green region, responsible for photo responses in GaAs cells. On attaching the Er^{3+}–Yb^{3+}-doped Vitro ceramic layer to the back of the solar cell, the efficiency of GaAs solar cells is noted to be 2.5%.

7.4.2 Bioimaging

The traditional optical imaging by using the UV and visible light radiations has been overlooked for bioimaging during the past few decades owing to some of the primary limitations offered by the excitation radiation and the processes involved in it. The bioimaging via high-energy radiations suffers from extensive autofluorescence from a biological background, less penetration depth, and photo damages to biosamples (O'connell et al. 2002; Michalet et al. 2005; Wang and Liu 2008). NIR-harvested UC fluorescence has been deliberated as the potential applicant for the in vivo and in vitro imaging because of low toxicity, low background autofluorescence, high penetration depth, high signal-to-noise ratio, etc. (Wang et al. 2009; Jin et al. 2011; Wang et al. 2012). REF_3 and $NaREF_4$ (RE = Y, Gd, La, and Lu) with silica coating are the common materials used for in vitro imaging. The pure hexagonal $NaYF_4$:Yb^{3+}–Er^{3+}–Tm^{3+} nanospheres coated with silica have been developed and used as a fluorescent probe for bioimaging (Li et al. 2008). They observed an efficient fluorescence signal from the MCF-7 cell incorporated with the nanoprobes under 980 nm excitation. They also claimed that the synthesized materials are advantageous for in vitro multiplexed bioassays and in vivo studies (Figure 7.11c). The polyethyleneimine (PEI)-coated $NaYF_4$:Er,Yb nanofluorophores have been synthesized and used for tissue imaging in small mammals under 980 nm excitation. They reported that the NIR-triggered PEI/$NaYF_4$:Yb,Er fluorophores show better fluorescence signal than the UV-excited QDs. The emission from the fluorophores can be detected from the depth of 10 mm after inserting the tissues inside the body of Wistar rats (Chatterjee et al. 2008). Researchers have prepared the β-$NaYF_4$:Yb,Er core–shell UCNPs with a thin layer of SiO_2 coating (Wang et al. 2009). After functionalizing the core–shell UCNPs with amino groups, they are further attached with rabbit anti-CEA8 covalently to obtain antibody–UCNP conjugates. The prepared conjugates have been considered as the fluorescent biolabels for detecting the cancer biomarker (i.e. carcinoembryonic antigen [CEA]) on the surface of HeLa cells. The UCNP-labeled cell started to glow under the 980 nm excitation, which confirms the use of the prepared UCNP conjugates beneficial for in vivo imaging, detection, and diagnosis of cancer. The in vivo imaging has also performed on live nematode *Caenorhabditis elegans* (Lim et al. 2006). These authors have prepared the Y_2O_3:Er/Yb nanoparticles and injected in the digestive system of the worm. The statistical distribution of the nanoparticles inside the intestine of the worm has been clearly observed under 980 nm excitation.

Apart from UCL imaging, several other imaging techniques are also wonderful in providing the physiological and pathological information in vivo. These imaging techniques that are capable of conveying the anatomical and molecular information are magnetic resonance imaging (MRI), computed tomography (CT), ultrasound, positron emission tomography (PET), and single-photon emission computed tomography (SPECT), etc. (Weissleder and Pittet 2008; Willmann et al. 2008). It has been noted that all the imaging techniques carry some advantages and disadvantages. To counter this problem, UCNP-based multimodal molecular imaging technique has been introduced where two or more imaging modalities combine to perform a single imaging technique to acquire enhanced diagnostic information (Jennings and Long 2009; Chen et al. 2015b). Five modal bioimaging capabilities by designing the β-$NaGdF_4$:Yb/Er@β-$NaGdF_4$:Yb@β-$NaNdF_4$:Yb@MS-Au25-PEG nanocomposites have been reported (He et al. 2015). The system containing Gd^{3+} ions is used for MRI, Yb^{3+} ions for CT, and Er^{3+} ions for UCL imaging, and Au_{25} has been considered for photoacoustic (PA) and photothermal (PT) modalities (He et al. 2015). Similarly, the $NaYF_4$:Yb/Er@$NaYF_4$:Yb@NaNdF4:Yb@$NaYF_4$@NaGdF4 UCNPs have been synthesized (Liu et al. 2016b). The ICG dye has been attached to the surface of the UCNPs to witness UCL imaging, PA, and MRI multimodalities (Liu et al. 2016b).

7.4.3 Photo-Induced Therapeutic Applications

Oncological phototherapy has been emerged as a promising remedy to treat superficial and internal tumors. The proficient therapeutic agents provide accurate cancer targeting, low invasiveness to diagnose cancer cells (Shi and Sadler 2020). The n discovery of new light sources and photo agents motivates future advancement in oncological treatment. In this section, various photo-induced therapeutic applications have been discussed briefly.

7.4.3.1 Photodynamic Therapy

Photodynamic therapy (PDT) is one of the non-invasive phototherapies used to cure cancer cells and other treatments using light-sensitive photosensitizers or photosensitizing agents. In the dark, photosensitizer (PS) is non-toxic, but when exposed to radiation of a specific wavelength comparable to the absorption of PS, it becomes cytotoxic and generates reactive oxygen species (ROS) (Prasad 2012, p. 398). The PDT system mainly depends on three major parameters, i.e. nature of the radiation, photosensitizers, and reacted oxygen species nearby the target sites. The applications associated with the conventional PDT using visible light are limited to cure surface tumors because of the low penetration depth. The UCNP-conjugated systems often using NIR radiation can thus treat cancer tissues in deep. The monodispersed PEI-coated $NaYF_4$:Er,Yb nanoparticles for imaging and treating cancer in deep tissues with zinc(II) phthalocyanine (ZnPc) as a photosensitizer, which can emit singlet oxygen under sufficient excitation, have been developed (Chatterjee and Yong 2008). Again, cancer cells contain folic acid (FA) receptors; thus, authors have coupled FA with ZnPc-UCNPs and injected it into the HT29 human colon cancer cells. Upon 980 nm excitation, they observed emissions in the green and red regions, which resonate with the excitation maximum of ZnPc at 670 nm. Thus, the excited ZnPc can generate singlet oxygen, which can be utilized to kill the cancer cells. A similar report has been reported by the researchers (Qian et al. 2009). They have synthesized the $NaYF_4$:Er/Yb@mesoporous SiO_2 UCNPs and ZnPc as photosensitizers. The mesoporous SiO_2 encapsulated with a photosensitizer helps in increasing the stability of UCNPs by avoiding the interaction of UCNPs with the harsh environment. Also, the large surface area of $NaYF_4$:Er/Yb@mesoporous SiO_2 UCNPs permits ample storage of ZnPc inside the cell. The authors have claimed that the PDT process can be triggered under 980 nm excitation, which can be effective for PDT treatments. A dual photosensitized tactic, in which two photosensitizers, namely, merocyanine 540 (MC540) and ZnPc, coupled to the $NaYF_4:Yb^{3+},Er^{3+}$@mesoporous SiO_2 composite, has been established (Idris et al. 2012). The in vivo study reveals that the growth of tumor has been constrained under 980 nm excited PDT treatment; thus, PDT treatments can be promising for future non-invasive cancer therapy.

7.4.3.2 Photothermal Therapy

Photothermal therapy (PTT) is the kind of phototherapy treatment that uses photoabsorbers to generate heat by absorbing light for the thermal ablation of cancer cells. UCNPs are often used as the imaging probe to monitor the thermal variation during the PTT process. Because the UCNPs do not directly convert heat from the excitation because of the low extinction coefficient of the RE^{3+} ions, they are coupled with photoabsorbers (Chen et al. 2014; Zhang 2015, pp. 329–330). Nanomaterials such as Ag, Au, CuS, and organic dyes are typical examples of photoabsorbers owning high photothermal conversion efficiency (Dong et al. 2011;

Xiao et al. 2013; Shan et al. 2013; Shen et al. 2014). A multifunctional approach of UCNPs integrated with a thin intermediate shell of iron oxide nanoparticles followed by a gold nanoparticle outer shell has been demonstrated (Cheng et al. 2011). The intermediate iron oxide nanoparticle cell helps to improve the system's magnetic properties, while Au nanoparticles increase the absorption cross section ranging from visible to NIR region. These authors have claimed that the prepared composites can be suitable for multimodal imaging and PTT-based destruction of cancer cells under NIR excitation. The biofunctional properties and versatility of $NaYF_4$:Yb^{3+}–Er^{3+}@Ag nanoparticles have been presented (Dong et al. 2011). The efficient fluorescence intensity ratio (FIR) change between the $^2H_{11/2} \rightarrow {}^4I_{15/2}$ and $^4S_{3/2} \rightarrow {}^4I_{15/2}$ transitions as a function of temperature under 980 nm excitation enables the UCNPs to perform as a temperature sensor. The study confirms that the Au core–shell helps to reduce the cytotoxicity of the developed composite NPs, and about 95% of mortality of the cancer cells has been reported at 1.5 W/cm^2 of pump density. A similar report based on the coupling of PDT and PTT to attain the synergetic effect has been narrated (Chan et al. 2018). They explained that the prepared $NaYF_4$:Yb^{3+}/Er^{3+}@$NaYF_4$:Yb^{3+}/Nd^{3+}@$NaYF_4$ UCNPs electrostatically combined with mesoporous silica shell-coated gold nanorods (AuNR@mSiO_2) can effectively kill the cancer cells via PDT with the help of merocyanine 540 (MC540) PS. While the UC emission from the UCNPs resonates with the LSPR-mediated Au nanorods, surface plasma peaks to witness the PTT effect.

7.4.3.3 Photoactivated Chemotherapy (PACT)

Conventional chemotherapy is one of the most preferred therapies, which employs one or more chemical drugs for cancer cell treatment. Despite the rigorous use in oncological treatment, conventional chemotherapy is constrained by the poor targeting efficiency, drug dependence, drug resistivity, etc. (Chabner and Roberts 2005; Holohan et al. 2013). The UCNP-based drug carrier has been extensively used for the efficient and safe drug delivery to boost the efficacy of conventional chemotherapy. Because of the exceptional physicochemical properties of UCNPs, the ability to improve the solubility, stability of chemotherapeutic drugs, and control over drug delivery, UCNPs have been broadly adopted in vivo and in vitro applications (Wang et al. 2011; Liu et al. 2012). The approaches describing the UCNP-based drug loadings include (i) attaching to the surface of the UCNPs by covalent bonding, (ii) fixing to the polymer-capped UCNPs, (iii) implanting in the UCNPs shielded with mesoporous SiO_2, and (iv) loading into the hollow mesoporous UCNP-coated spheres, etc. (Shen et al. 2013; Yang et al. 2015). A polyethylene glycol (PEG)-grafted amphiphilic polymer-functionalized UCNPs loaded with anticancer drug and doxorubicin (DOX) for intracellular drug delivery has been reported (Wang et al. 2011). The authors have observed the pH-regulated drug release of DOX at an acidic environment and found it advantageous of controlled drug release. Similarly, the pH-controlled anticancer drug delivery system for DOX-conjugated $NaYF_4$:Yb^{3+}/Tm^{3+} UCNPs have been studied (Dai et al. 2012). The authors have observed the optimum pH to be 5.0 for faster drug release. The UCNPs have been used to monitor the drug release following the efficient energy transfer from UCNPs to DOX.

7.4.4 Other Emerging Applications

7.4.4.1 Anticounterfeiting

Upconversion-based anticounterfeiting technology has been widely accepted across the globe to circumvent forgery or counterfeiting. Some popular fields in which anticounterfeiting is commonly used are paper currencies, identity cards, certificates, etc. (Yoon et al. 2013; Bae et al. 2015; Jiang et al. 2016). The trademark properties such as multicolour emission, remarkable optical tunability, high chemical stability, etc., have led the UCNPs in anticounterfeiting applications (Niu et al. 2013; Wang and Liu 2014; Liu et al. 2015). Various approaches have been reported in which the Yb-based UCNPs show remarkable optical tunability over the changing excitation power. For instance, an amplified power-sensitive optical tunability in the Ho^{3+}-doped $NaYF_4$ host has been reported (Chen et al. 2016b). They obtained the tunability of red to green emission intensity ratio from 0.37 to 5.19 while changing the excitation power from 1.2 to 46.25 W/cm^2, which is 5.6 times amplified than the Yb/Ho-co-doped $NaYF_4$ phosphors. Likewise, the Er^{3+}/Yb^{3+}:$BiYO_3$ nanophosphors synthesized by sol–gel method have been verified as the security ink under 980 nm continuous wave (CW) laser excitation (Dutta and Rai 2021). The prepared $SrBi_4Ti_4O_{15}$:Er^{3+} phosphors have defined the anticounterfeit abilities through multimode excitation (Wu et al. 2021) The study suggests that the effect of concentration variation of the Er^{3+} ions in the phosphors does not affect the green emission much under 980 nm laser excitation, but under the 1550 nm excitation, the emission color varies from orange to red by changing the concentration of the Er^{3+} ions from 0.5% to 3.0%. Therefore, by adjusting the concentration and excitations, different colors could be recognized with naked eye. The demonstration of different color emissions under sunlight and under 980 and 1550 nm laser excitations has been depicted in Figure 7.12a. As the application is a life savior in many aspects, starting from economic, security, and business-related problems, further study with efficient performance is vital.

7.4.4.2 Sensing and Detection

UCNPs-based sensing such as FRET-based biosensing and FIR-based temperature sensing has been a subject of attraction since the past few decades in nanoscience and nanotechnology. FRET-based sensing involves the energy transfer between the acceptors and donors, including a reduction in the fluorescence intensity of donors, followed by an increase in the intensity of acceptors (Su et al. 2017). Different biological and physicochemical processes such as temporal distribution of biological molecules, molecular interactions, protein–protein interactions, etc., have been understood with the help of FRET. Considering the advantages of FRET, UCNP integration as energy donors with FRET has led to detect various analytes such as pH, gas, temperature, metal ions, DNA, etc. The $NaYF_4$:Er,Yb UCNP-based pH sensor, in which the neutral red (NR) is chosen as a pH indicator, has been deigned (Meier et al. 2014). The emission peak of the UCNPs has been monitored in the green and red regions under 980 nm excitation. As the absorption peak of the pH indicator resonates at the green emission band, the energy transfer between the

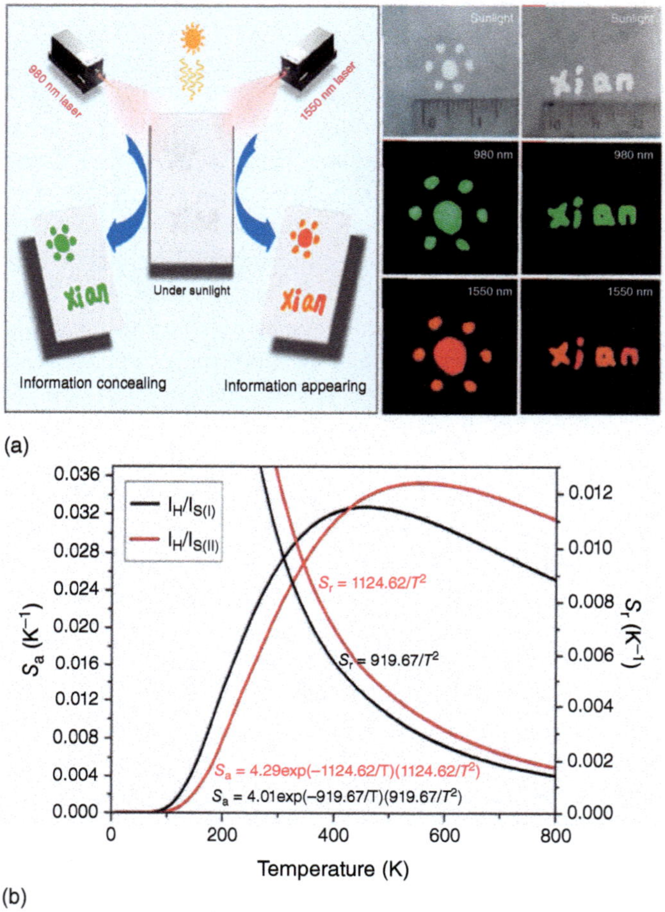

Figure 7.12 (a) Anticounterfeiting and information concealment schematic diagram and patterns written using transparent ink containing $SrBi_4Ti_4O_{15}$:xEr^{3+} (x = 0.5% and 3.0%) phosphors under sunlight, 980 nm laser, and 1550 nm laser excitations. Source: Wu et al. (2021). Reprinted with permission of Elsevier. (b) Plot of relative and absolute sensitivity versus temperature for Stark sub-levels. Source: Mondal and Rai (2020). Reprinted with permission of Elsevier.

UCNPs and the pH indicator results in a decrease in the UC emission intensity in the green region. In contrast, the intensity of red emission band remains unaffected and serves as the inert reference signal for ratiometric measurements. The authors have managed to get a resolution of about ±0.06 pH unit. Like pH sensors, gas sensors have also gathered considerable interest to monitor many physiological behaviors. The optical sensor based on the $NaYF_4$:Yb/Tm UCNPs to monitor the oxygen has been established (Achatz et al. 2011). Under the 980 nm excitation, the UCNPs emit at 455 and 475 nm, which itself acts as the excitation source for the oxygen probe [Ir(CS)2(acac)], pursuing the FRET mechanism. It has been noticed that the emission peak of the Ir-complex at 568 nm is quenched because of oxygen. A similar success has been reported with the FIR-based temperature sensing, where UCNPs have been considered as the temperature probe to monitor accurate results with great precision (Soni et al. 2017; Pattnaik and Rai 2020). The FIR-based temperature sensing, by considering the stark sub-levels of the $^2H_{11/2}$ and $^4S_{3/2}$ thermally coupled levels, has been performed (Mondal and Rai 2020). They found the maximum absolute sensitivity of $\sim 35.6 \times 10^{-3}$ K^{-1} and relative sensitivity of $\sim 12.5 \times 10^{-3}$ K^{-1} with better thermal stability (Figure 7.12b).

7.4.4.3 Optogenetic Stimulation

Optogenetics refers to the biological techniques to control the genetically modified light-sensitive neurons. Although the contribution of the visible light toward the neuromodification is enormous, effective stimulation of targeted neurons remained a challenging aspect because of low penetration depth. On the contrary, NIR light supports the high penetration depth to trigger the deeper tissues (All et al. 2019). As UCNPs display great assistantship toward the NIR light, UCNP-based optogenetics could be decisive in terms of minimal invasive optical stimulation with improved selectivity, sensitivity, and spatial resolution of neurons. The NIR-mediated optogenetic control of neural activity using UCNPs combined with the photo-responsive ion channels has been described (Shah et al. 2015). Authors have claimed that the UCNPs could activate the channelrhodopsin-2 (ChR2) ion channels under 980 nm excitation and perform efficient neuronal stimulation with comparable efficiency. A novel quasi-continuous wave (quasi-CW) excitation tactic has been implemented (Bansal et al. 2016). This unique NIR-triggered technique helps in improving the UC emission efficiency, followed by activating ChR2 in *C. elegans* with a considerably lower excitation power. Similarly, the non-invasive deep brain stimulation of mouse with the use of UCNPs as optogenetic actuators has been demonstrated (Chen et al. 2018).

7.4.4.4 NIR Image Vision of Mammals

The perceiving vision in the naturally invisible NIR region is still desirable and challenging. The sophisticated visual system of mammals including humans can detect the light ranging from 400 to 700 nm (Schnapf et al. 1988). This is due to the fact that the photoreceptors (rods and cones) comprising opsins and retinal in the mammal's eye cannot absorb light at a longer wavelength than visible. In an attempt to study the implementation and results, the researchers came up with an ocular injectable photoreceptor-binding UCNPs (pbUCNPs), rooted to the eyes of mouse that can activate the photoreceptors under NIR light irradiance (Ma et al. 2019). Authors have reported that the $NaYF_4$:Yb,Er@$NaYF_4$ UCNPs are capable of upconverting NIR light to green light under 980 nm excitation. Thus, the engineered UCNPs affixed to the eyes of mice can help the mice to observe the NIR radiation as a green light. To fully understand the effect of injected UCNPs, the authors have performed a pupillary light reflex test between pbUCNP-injected mice and non-injected mice. They found that the pbUCNP-injected mice show substantial constrictions, whereas the non-injected mice remain intact under 980 nm light illumination. In another experiment, the mice have been subjected to identifying the NIR patterns in a Y-shaped water maze. They observed that the pbUCNPs-injected trained mice have successfully differentiated the vertical or horizontal grating of NIR light. Hence, this technique can be applicable to witness the supervision in human eyes and helpful in curing visible spectrum-related deficit such as color blindness.

References

Achatz, D.E., Meier, R.J., Fischer, L.H., and Wolfbeis, O.S. (2011). Luminescent sensing of oxygen using a quenchable probe and upconverting nanoparticles. *Angewandte Chemie International Edition* 50 (1): 260–263. https://doi.org/10.1002/anie.201004902.

Ågren, H., Yang, C., and Chen, G. (2017). Dye-sensitized lanthanide-doped upconversion nanoparticles. *Chemical Society Reviews* 46 (14): 4150–4167. https://doi.org/10.1039/c7cs00053g.

Ahn, Y.N., Do Kim, K., Anoop, G. et al. (2019). Design of highly efficient phosphor-converted white light-emitting diodes with color rendering indices $(R_1 - R_{15}) \geq 95$ for artificial lighting. *Scientific Reports* 9 (1): 1–10. https://doi.org/10.1038/s41598-019-53269-0.

Ahn, I.H., Yeo, S.J., Jung, K. et al. (2020). A multi-functional highly efficient upconversion luminescent film with an array of dielectric microbeads decorated with metal nanoparticles. *Advanced Functional Materials* 30 (13): 1909445. https://doi.org/10.1002/adfm.201909445.

Ai, F., Ju, Q., Zhang, X. et al. (2015). A core-shell-shell nanoplatform upconverting near-infrared light at 808 nm for luminescence imaging and photodynamic therapy of cancer. *Scientific Reports* 5 (1): 1–11. https://doi.org/10.1038/srep10785.

All, A.H., Zeng, X., Teh, D.B.L. et al. (2019). Expanding the toolbox of upconversion nanoparticles for in vivo optogenetics and neuromodulation. *Advanced Materials* 31 (41): 1803474. https://doi.org/10.1002/adma.201803474.

Auzel, F. (2004). Upconversion and anti-stokes processes with f and d ions in solids. *Chemical Reviews* 104 (1): 139–174. https://doi.org/10.1021/cr020357g.

Bae, H.J., Bae, S., Park, C. et al. (2015). Biomimetics: biomimetic microfingerprints for anti-counterfeiting strategies. *Advanced Materials* 27 (12): 2123–2123. https://doi.org/10.1002/adma.201570084.

Bai, X., Song, H., Pan, G. et al. (2007). Size-dependent upconversion luminescence in Er^{3+}/Yb^{3+}-codoped nanocrystalline yttria: saturation and thermal effects. *The Journal of Physical Chemistry C* 111 (36): 13611–13617. https://doi.org/10.1021/jp070122e.

Bansal, A., Liu, H., Jayakumar, M.K.G. et al. (2016). Quasi-continuous wave near-infrared excitation of upconversion nanoparticles for optogenetic manipulation of *C. elegans*. *Small* 12 (13): 1732–1743. https://doi.org/10.1002/smll.201503792.

Barrera, E.W., Pujol, M.C., Carvajal, J.J. et al. (2014). White light upconversion in Yb-sensitized (Tm, Ho)-doped $KLu(WO_4)_2$ nanocrystals: the effect of Eu incorporation. *Physical Chemistry Chemical Physics* 16 (4): 1679–1686. https://doi.org/10.1039/c3cp53847h.

Berry, M.T. and May, P.S. (2015). Disputed mechanism for NIR-to-red upconversion luminescence in $NaYF_4:Yb^{3+},Er^{3+}$. *The Journal of Physical Chemistry A* 119 (38): 9805–9811. https://doi.org/10.1021/acs.jpca.5b08324.

Bian, W., Lin, Y., Wang, T. et al. (2018). Direct identification of surface defects and their influence on the optical characteristics of upconversion nanoparticles. *ACS Nano* 12 (4): 3623–3628. https://doi.org/10.1021/acsnano.8b00741.

Binnemans, K. (2009). Lanthanide-based luminescent hybrid materials. *Chemical Reviews* 109 (9): 4283–4374. https://doi.org/10.1021/cr8003983.

Bloembergen, N. (1959). Solid state infrared quantum counters. *Physical Review Letters* 2 (3): 84. https://doi.org/10.1103/physrevlett.2.84.

Boyer, J.C. and Van Veggel, F.C. (2010). Absolute quantum yield measurements of colloidal $NaYF_4:Er^{3+},Yb^{3+}$ upconverting nanoparticles. *Nanoscale* 2 (8): 1417–1419. https://doi.org/10.1039/c0nr00253d.

Boyer, J.C., Manseau, M.P., Murray, J.I., and Van Veggel, F.C. (2010). Surface modification of upconverting $NaYF_4$ nanoparticles with PEG – phosphate ligands for NIR (800 nm) biolabeling within the biological window. *Langmuir* 26 (2): 1157–1164. https://doi.org/10.1021/la902260j.

Briggs, J.A., Atre, A.C., and Dionne, J.A. (2013). Narrow-bandwidth solar upconversion: case studies of existing systems and generalized fundamental limits. *Journal of Applied Physics* 113 (12): 124509. https://doi.org/10.1063/1.4796092.

Bünzli, J.C.G. and Piguet, C. (2005). Taking advantage of luminescent lanthanide ions. *Chemical Society Reviews* 34 (12): 1048–1077. https://doi.org/10.1039/b406082m.

Cesaria, M. and Di Bartolo, B. (2018). Chapter 3: Nanophosphors: from rare earth activated multicolor-tuning to new efficient white light sources. In: *Quantum Nano-Photonics. NATO 2017*, NATO Science for Peace and Security Series B: Physics and Biophysics (ed. B. Di Bartolo, L. Silvestri, M. Cesaria and J. Collins), 27–77. Dordrecht: Springer. https://doi.org/10.1007/978-94-024-1544-5_3.

Chabner, B.A. and Roberts, T.G. (2005). Chemotherapy and the war on cancer. *Nature Reviews Cancer* 5 (1): 65–72. https://doi.org/10.1038/nrc1529.

Chai, X., Li, J., Wang, X. et al. (2017). Upconversion luminescence and temperature-sensing properties of Ho^{3+}/Yb^{3+}-codoped $ZnWO_4$ phosphors based on fluorescence intensity ratios. *RSC Advances* 7 (64): 40046–40052. https://doi.org/10.1039/c7ra05846b.

Chan, M.H., Chen, S.P., Chen, C.W. et al. (2018). Single 808 nm laser treatment comprising photothermal and photodynamic therapies by using gold nanorods hybrid upconversion particles. *The Journal of Physical Chemistry C* 122 (4): 2402–2412. https://doi.org/10.1021/acs.jpcc.7b10976.s001.

Chatterjee, D.K. and Yong, Z. (2008). Upconverting nanoparticles as nanotransducers for photodynamic therapy in cancer cells. *Nanomedicine* 3 (1): 73–82. https://doi.org/10.2217/17435889.3.1.73.

Chatterjee, D.K., Rufaihah, A.J., and Zhang, Y. (2008). Upconversion fluorescence imaging of cells and small animals using lanthanide doped nanocrystals. *Biomaterials* 29 (7): 937–943. https://doi.org/10.1016/j.biomaterials.2007.10.051.

Chen, G., Qiu, H., Prasad, P.N., and Chen, X. (2014). Upconversion nanoparticles: design, nanochemistry, and applications in theranostics. *Chemical Reviews* 114 (10): 5161–5214. https://doi.org/10.1021/cr400425h.

Chen, G., Damasco, J., Qiu, H. et al. (2015a). Energy-cascaded upconversion in an organic dye-sensitized core/shell fluoride nanocrystal. *Nano Letters* 15 (11): 7400–7407. https://doi.org/10.1021/acs.nanolett.5b02830.

Chen, G., Ågren, H., Ohulchanskyy, T.Y., and Prasad, P.N. (2015b). Light upconverting core–shell nanostructures: nanophotonic control for emerging applications. *Chemical Society Reviews* 44 (6): 1680–1713. https://doi.org/10.1039/c4cs00170b.

Chen, X., Jin, L., Kong, W. et al. (2016a). Confining energy migration in upconversion nanoparticles towards deep ultraviolet lasing. *Nature Communications* 7 (1): 1–6. https://doi.org/10.1038/ncomms10304.

Chen, B., Liu, Y., Xiao, Y. et al. (2016b). Amplifying excitation-power sensitivity of photon upconversion in a $NaYbF_4$:Ho nanostructure for direct visualization of electromagnetic hotspots. *The Journal of Physical Chemistry Letters* 7 (23): 4916–4921. https://doi.org/10.1021/acs.jpclett.6b02210.

Chen, H.W., Zhu, R.D., He, J. et al. (2017). Going beyond the limit of an LCD's color gamut. *Light: Science & Applications* 6 (9): 17043–17043. https://doi.org/10.1038/lsa.2017.43.

Chen, S., Weitemier, A.Z., Zeng, X. et al. (2018). Near-infrared deep brain stimulation via upconversion nanoparticle–mediated optogenetics. *Science* 359 (6376): 679–684. https://doi.org/10.1126/science.aaq1144.

Cheng, L., Yang, K., Li, Y. et al. (2011). Facile preparation of multifunctional upconversion nanoprobes for multimodal imaging and dual-targeted photothermal therapy. *Angewandte Chemie International Edition* 123 (32): 7523–7528. https://doi.org/10.1002/ange.201101447.

Chivian, J.S., Case, W.E., and Eden, D.D. (1979). The photon avalanche: a new phenomenon in Pr^{3+}-based infrared quantum counters. *Applied Physics Letters* 35 (2): 124–125. https://doi.org/10.1063/1.91044.

Dai, Y., Yang, D., Kang, X. et al. (2012). Doxorubicin conjugated $NaYF_4:Yb^{3+}/Tm^{3+}$ nanoparticles for therapy and sensing of drug delivery by luminescence resonance energy transfer. *Biomaterials* 33 (33): 8704–8713. https://doi.org/10.1016/j.biomaterials.2012.08.029.

Dey, R. and Rai, V.K. (2014). Yb^{3+} sensitized Er^{3+} doped La_2O_3 phosphor in temperature sensors and display devices. *Dalton Transactions* 43 (1): 111–118. https://doi.org/10.1039/c3dt51773j.

Dey, R., Rai, V.K., and Pandey, A. (2012). Green upconversion emission in Nd^{3+}–Yb^{3+}–Zn^{2+}:Y_2O_3 phosphor. *Spectrochimica Acta Part A: Molecular and Biomolecular Spectroscopy* 99: 288–291. https://doi.org/10.1016/j.saa.2012.09.001.

Dey, R., Kumari, A., Soni, A.K., and Rai, V.K. (2015). $CaMoO_4:Ho^{3+}$–Yb^{3+}–Mg^{2+} upconverting phosphor for application in lighting devices and optical temperature sensing. *Sensors and Actuators B: Chemical* 210: 581–588. https://doi.org/10.1016/j.snb.2015.01.007.

Dong, B., Xu, S., Sun, J. et al. (2011). Multifunctional $NaYF_4:Yb^{3+},Er^{3+}$@Ag core/shell nanocomposites: integration of upconversion imaging and photothermal therapy.

Journal of Materials Chemistry 21 (17): 6193–6200. https://doi.org/10.1039/c0jm04498a.

Dong, H., Sun, L.D., and Yan, C.H. (2013). Basic understanding of the lanthanide related upconversion emissions. *Nanoscale* 5 (13): 5703–5714. https://doi.org/10.1039/c3nr34069d.

Dutta, J. and Rai, V.K. (2021). Upconverting $BiYO_3$ nanophosphors in DSSCs applications. *Optics & Laser Technology* 140: 107087. https://doi.org/10.1016/j.optlastec.2021.107087.

Dwivedi, A., Mishra, K., and Rai, S.B. (2016). Role of Gd^{3+} ion on downshifting and upconversion emission properties of Pr^{3+}, Yb^{3+} co-doped $YNbO_4$ phosphor and sensitization effect of Bi^{3+} ion. *Journal of Applied Physics* 120 (4): 043102. https://doi.org/10.1063/1.4959156.

Escudero Belmonte, A., Becerro Nieto, A.I., Carrillo Carrión, C. et al. (2017). Rare earth based nanostructured materials: synthesis, functionalization, properties and bioimaging and biosensing applications. *Nano* 6 (5): 881–921. https://doi.org/10.1515/nanoph-2017-0007.

Feng, W., Sun, L.D., and Yan, C.H. (2009). Ag nanowires enhanced upconversion emission of $NaYF_4$:Yb,Er nanocrystals via a direct assembly method. *Chemical Communications* 29: 4393–4395. https://doi.org/10.1039/b909164e.

Fischer, S., Johnson, N.J., Pichaandi, J. et al. (2015). Upconverting core-shell nanocrystals with high quantum yield under low irradiance: on the role of isotropic and thick shells. *Journal of Applied Physics* 118 (19): 193105. https://doi.org/10.1063/1.4936119.

Fischer, S., Bronstein, N.D., Swabeck, J.K. et al. (2016). Precise tuning of surface quenching for luminescence enhancement in core–shell lanthanide-doped nanocrystals. *Nano Letters* 16 (11): 7241–7247. https://doi.org/10.1021/acs.nanolett.6b03683.

Gai, S., Li, C., Yang, P., and Lin, J. (2014). Recent progress in rare earth micro/nanocrystals: soft chemical synthesis, luminescent properties, and biomedical applications. *Chemical Reviews* 114 (4): 2343–2389. https://doi.org/10.1021/cr4001594.

Garfield, D.J., Borys, N.J., Hamed, S.M. et al. (2018). Enrichment of molecular antenna triplets amplifies upconverting nanoparticle emission. *Nature Photonics* 12 (7): 402–407. https://doi.org/10.1038/s41566-018-0156-x.

Gibart, P., Auzel, F., Guillaume, J.C., and Zahraman, K. (1996). Below band-gap IR response of substrate-free GaAs solar cells using two-photon up-conversion. *Japanese Journal of Applied Physics* 35 (8R): 4401. https://doi.org/10.1143/jjap.35.4401.

Gorris, H.H., Ali, R., Saleh, S.M., and Wolfbeis, O.S. (2011). Tuning the dual emission of photon-upconverting nanoparticles for ratiometric multiplexed encoding. *Advanced Materials* 23 (14): 1652–1655. https://doi.org/10.1002/adma.201004697.

Gu, B. and Zhang, Q. (2018). Recent advances on functionalized upconversion nanoparticles for detection of small molecules and ions in biosystems. *Advanced Science* 5 (3): 1700609. https://doi.org/10.1002/advs.201700609.

Haase, M. and Schäfer, H. (2011). Upconverting nanoparticles. *Angewandte Chemie International Edition* 50 (26): 5808–5829. https://doi.org/10.1002/anie.201005159.

He, F., Yang, G., Yang, P. et al. (2015). A new single 808 nm NIR light-induced imaging-guided multifunctional cancer therapy platform. *Advanced Functional Materials* 25 (25): 3966–3976. https://doi.org/10.1002/adfm.201500464.

van Hest, J.J., Blab, G.A., Gerritsen, H.C. et al. (2017). Probing the influence of disorder on lanthanide luminescence using Eu-doped $LaPO_4$ nanoparticles. *The Journal of Physical Chemistry C* 121 (35): 19373–19382. https://doi.org/10.1021/acs.jpcc.7b06549.

Holohan, C., Van Schaeybroeck, S., Longley, D.B., and Johnston, P.G. (2013). Cancer drug resistance: an evolving paradigm. *Nature Reviews Cancer* 13 (10): 714–726. https://doi.org/10.1038/nrc3599.

Hu, J., Zhan, S., Wu, X. et al. (2018). Core/shell upconversion nanoparticles with intense fluorescence for detecting doxorubicin in vivo. *RSC Advances* 8 (38): 21505–21512. https://doi.org/10.1039/c8ra02928h.

Huang, L., Wang, L., Xue, X. et al. (2011). Enhanced red upconversion luminescence in Er–Tm codoped $NaYF_4$ phosphor. *Journal of Nanoscience and Nanotechnology* 11 (11): 9498–9501. https://doi.org/10.1166/jnn.2011.5255.

Huang, X., Han, S., Huang, W., and Liu, X. (2013). Enhancing solar cell efficiency: the search for luminescent materials as spectral converters. *Chemical Society Reviews* 42 (1): 173–201. https://doi.org/10.1039/c2cs35288e.

Huang, K., Liu, H., Kraft, M. et al. (2018). A protected excitation-energy reservoir for efficient upconversion luminescence. *Nanoscale* 10 (1): 250–259. https://doi.org/10.1039/c7nr06900f.

Huang, Y., Bai, G., Zhao, Y. et al. (2020). Yb/Ho codoped layered perovskite bismuth titanate microcrystals with upconversion luminescence: fabrication, characterization, and application in optical fiber ratiometric thermometry. *Inorganic Chemistry* 59 (19): 14229–14235. https://doi.org/10.1021/acs.inorgchem.0c02015.

Idris, N.M., Gnanasammandhan, M.K., Zhang, J. et al. (2012). In vivo photodynamic therapy using upconversion nanoparticles as remote-controlled nanotransducers. *Nature Medicine* 18 (10): 1580. https://doi.org/10.1038/nm.2933.

Idris, N.M., Jayakumar, M.K.G., Bansal, A., and Zhang, Y. (2015). Upconversion nanoparticles as versatile light nanotransducers for photoactivation applications. *Chemical Society Reviews* 44 (6): 1449–1478. https://doi.org/10.1039/c4cs00158c.

Jackson, S.D. (2003). Continuous wave 2.9 μm dysprosium-doped fluoride fiber laser. *Applied Physics Letters* 83 (7): 1316–1318. https://doi.org/10.1063/1.1603353.

Jennings, L.E. and Long, N.J. (2009). 'Two is better than one' – probes for dual-modality molecular imaging. *Chemical Communications* 24: 3511–3524. https://doi.org/10.1039/b821903f.

Jiang, K., Zhang, L., Lu, J. et al. (2016). Triple-mode emission of carbon dots: applications for advanced anti-counterfeiting. *Angewandte Chemie International Edition* 128 (25): 7347–7351. https://doi.org/10.1002/ange.201602445.

Jin, J., Gu, Y.J., Man, C.W.Y. et al. (2011). Polymer-coated $NaYF_4:Yb^{3+},Er^{3+}$ upconversion nanoparticles for charge-dependent cellular imaging. *ACS Nano* 5 (10): 7838–7847. https://doi.org/10.1021/nn201896m.

Jin, L.M., Chen, X., Siu, C.K. et al. (2017). Enhancing multiphoton upconversion from $NaYF_4$:Yb/Tm@$NaYF_4$ core–shell nanoparticles via the use of laser cavity. *ACS Nano* 11 (1): 843–849. https://doi.org/10.1021/acsnano.6b07322.

Johnson, N.J., He, S., Diao, S. et al. (2017). Direct evidence for coupled surface and concentration quenching dynamics in lanthanide-doped nanocrystals. *Journal of the American Chemical Society* 139 (8): 3275–3282. https://doi.org/10.1021/jacs.7b00223.

Kostiv, U., Engstová, H., Krajnik, B. et al. (2020). Monodisperse core-shell $NaYF_4:Yb^{3+}/Er^{3+}$@ $NaYF_4:Nd^{3+}$-PEG-GGGRGDSGGGY-NH2 nanoparticles excitable at 808 and 980 nm: design, surface engineering, and application in life sciences. *Frontiers in Chemistry* 8: 497. https://doi.org/10.3389/fchem.2020.00497.

Kraft, M., Würth, C., Palo, E. et al. (2019). Colour-optimized quantum yields of Yb, Tm co-doped upconversion nanocrystals. *Methods and Applications in Fluorescence* 7 (2): 024001. https://doi.org/10.1088/2050-6120/ab023b.

Kumar, R., Nyk, M., Ohulchanskyy, T.Y. et al. (2009). Combined optical and MR bioimaging using rare earth ion doped $NaYF_4$ nanocrystals. *Advanced Functional Materials* 19 (6): 853–859. https://doi.org/10.1002/adfm.200800765.

Kumari, A., Pandey, A., Dey, R., and Rai, V.K. (2014). Simultaneous influence of Zn^{2+}/Mg^{2+} on the luminescent behaviour of $La_2O_3:Tm^{3+}–Yb^{3+}$ phosphors. *RSC Advances* 4 (42): 21844–21851. https://doi.org/10.1039/c4ra01400f.

Kumari, A., Mondal, M., Rai, V.K., and Singh, S.N. (2017). Photoluminescence study in $Ho^{3+}/Tm^{3+}/Yb^{3+}/Li^{+}:Gd_2(MoO_4)_3$ nanophosphors for near white light emitting diode and security ink applications. *Methods and Applications in Fluorescence* 6 (1): 015003. https://doi.org/10.1088/2050-6120/aa8c61.

Lee, C., Park, H., Kim, W., and Park, S. (2019). Origin of strong red emission in Er^{3+}-based upconversion materials: role of intermediate states and cross relaxation. *Physical Chemistry Chemical Physics* 21 (43): 24026–24033. https://doi.org/10.1039/c9cp04692e.

Li, Z., Zhang, Y., and Jiang, S. (2008). Multicolor core/shell-structured upconversion fluorescent nanoparticles. *Advanced Materials* 20 (24): 4765–4769. https://doi.org/10.1002/adma.200801056.

Li, X., Zhang, F., and Zhao, D. (2013). Highly efficient lanthanide upconverting nanomaterials: progresses and challenges. *Nano Today* 8 (6): 643–676. https://doi.org/10.1016/j.nantod.2013.11.003.

Li, M., Liu, X., Liu, L. et al. (2016). β-$NaYF_4$:Yb,Tm: upconversion properties by controlling the transition probabilities at the same energy level. *Inorganic Chemistry Frontiers* 3 (8): 1082–1090. https://doi.org/10.1039/c6qi00121a.

Li, H., Wang, X., Huang, D., and Chen, G. (2019). Recent advances of lanthanide-doped upconversion nanoparticles for biological applications. *Nanotechnology* 31 (7): 07200. https://doi.org/10.1088/1361-6528/ab4f36.

Li, Y., Li, Z., Guo, L. et al. (2020). Enhanced dual-wavelength upconversion luminescence, thermosensitivity and DMMP detection of multifunctional $Gd_2MoO_6:Er^{3+}/Yb^{3+}$ nanoparticles. *Journal of Alloys and Compounds* 156399. https://doi.org/10.1016/j.jallcom.2020.156399.

Liang, L., Teh, D.B., Dinh, N.D. et al. (2019). Upconversion amplification through dielectric superlensing modulation. *Nature Communications* 10 (1): 1–9. https://doi.org/10.1038/s41467-019-09345-0.

Lim, S.F., Riehn, R., Ryu, W.S. et al. (2006). In vivo and scanning electron microscopy imaging of upconverting nanophosphors in *Caenorhabditis elegans*. *Nano Letters* 6 (2): 169–174.

Liu, X., Kong, X., Zhang, Y. et al. (2011). Breakthrough in concentration quenching threshold of upconversion luminescence via spatial separation of the emitter doping area for bio-applications. *Chemical Communications* 47 (43): 11957–11959. https://doi.org/10.1039/c1cc14774a.

Liu, J.N., Bu, W., Pan, L.M. et al. (2012). Simultaneous nuclear imaging and intranuclear drug delivery by nuclear-targeted multifunctional upconversion nanoprobes. *Biomaterials* 33 (29): 7282–7290. https://doi.org/10.1016/j.biomaterials.2012.06.035.

Liu, X., Deng, R., Zhang, Y. et al. (2015). Probing the nature of upconversion nanocrystals: instrumentation matters. *Chemical Society Reviews* 44 (6): 1479–1508. https://doi.org/10.1039/c4cs00356j.

Liu, D., Xu, X., Du, Y. et al. (2016a). Three-dimensional controlled growth of monodisperse sub-50 nm heterogeneous nanocrystals. *Nature Communications* 7 (1): 1–8. https://doi.org/10.1038/ncomms10254.

Liu, Y., Kang, N., Lv, J. et al. (2016b). Deep photoacoustic/luminescence/magnetic resonance multimodal imaging in living subjects using high-efficiency upconversion nanocomposites. *Advanced Materials* 28 (30): 6411–6419. https://doi.org/10.1002/adma.201506460.

Liu, Y., Lu, Y., Yang, X. et al. (2017). Amplified stimulated emission in upconversion nanoparticles for super-resolution nanoscopy. *Nature* 543 (7644): 229–233. https://doi.org/10.1038/nature21366.

Ma, Y., Bao, J., Zhang, Y. et al. (2019). Mammalian near-infrared image vision through injectable and self-powered retinal nanoantennae. *Cell* 177 (2): 243–255. https://doi.org/10.1016/j.cell.2019.01.038.

Mahata, M.K., Koppe, T., Kumar, K. et al. (2020). Upconversion photoluminescence of Ho^{3+}–Yb^{3+} doped barium titanate nanocrystallites: optical tools for structural phase detection and temperature probing. *Scientific Reports* 10 (1): 1–12. https://doi.org/10.1038/s41598-020-65149-z.

Mai, H.X., Zhang, Y.W., Sun, L.D., and Yan, C.H. (2007). Highly efficient multicolor up-conversion emissions and their mechanisms of monodisperse $NaYF_4$:Yb,Er core and core/shell-structured nanocrystals. *The Journal of Physical Chemistry C* 111 (37): 13721–13729. https://doi.org/10.1021/jp073920d.s001.

Maier, S.A., Kik, P.G., Atwater, H.A. et al. (2003). Local detection of electromagnetic energy transport below the diffraction limit in metal nanoparticle plasmon waveguides. *Nature Materials* 2 (4): 229–232. https://doi.org/10.1038/nmat852.

Medintz, I.L., Uyeda, H.T., Goldman, E.R., and Mattoussi, H. (2005). Quantum dot bioconjugates for imaging, labelling and sensing. *Nature Materials* 4 (6): 435–446. https://doi.org/10.1038/nmat1390.

Meier, R.J., Simbürger, J.M., Soukka, T., and Schaferling, M. (2014). Background-free referenced luminescence sensing and imaging of pH using upconverting phosphors and color camera read-out. *Analytical Chemistry* 86 (11): 5535–5540. https://doi.org/10.1021/ac5009207.

Michalet, X., Pinaud, F.F., Bentolila, L.A. et al. (2005). Quantum dots for live cells, in vivo imaging, and diagnostics. *Science* 307 (5709): 538–544. https://doi.org/10.1126/science.1104274.

Mondal, M. and Rai, V.K. (2020). Optical thermometry using Stark sublevels in charge compensated transition metal molybdate upconverting phosphors. *Optics & Laser Technology* 130: 106341. https://doi.org/10.1016/j.optlastec.2020.106341.

Montoya, E., Espeso, O., and Bausa, L.E. (2000). Cooperative luminescence in Yb^{3+}:$LiNbO_3$. *Journal of Luminescence* 87: 1036–1038. https://doi.org/10.1016/s0022-2313(99)00531-1.

Mukhopadhyay, L. and Rai, V.K. (2017). Upconversion based near white light emission, intrinsic optical bistability and temperature sensing in Er^{3+}/Tm^{3+}/Yb^{3+}/Li^{+}:$NaZnPO_4$ phosphors. *New Journal of Chemistry* 41 (15): 7650–7661. https://doi.org/10.1039/c7nj01622k.

Niu, W., Wu, S., and Zhang, S. (2010). A facile and general approach for the multicolor tuning of lanthanide-ion doped $NaYF_4$ upconversion nanoparticles within a fixed composition. *Journal of Materials Chemistry* 20 (41): 9113–9117. https://doi.org/10.1039/c0jm01879a.

Niu, W., Wu, S., and Zhang, S. (2011). Utilizing the amidation reaction to address the "cooperative effect" of carboxylic acid/amine on the size, shape, and multicolor output of fluoride upconversion nanoparticles. *Journal of Materials Chemistry* 21 (29): 10894–10902. https://doi.org/10.1039/c1jm10985e.

Niu, W., Wu, S., Zhang, S. et al. (2013). Multicolor tunability and upconversion enhancement of fluoride nanoparticles by oxygen dopant. *Nanoscale* 5 (17): 8164–8171. https://doi.org/10.1039/c3nr01612a.

O'connell, M.J., Bachilo, S.M., Huffman, C.B. et al. (2002). Band gap fluorescence from individual single-walled carbon nanotubes. *Science* 297 (5581): 593–596. https://doi.org/10.1126/science.1072631.

Oliveira, A.S., Gouveia, E.A., De Araujo, M.T. et al. (2000). Twentyfold blue upconversion emission enhancement through thermal effects in Pr^{3+}/Yb^{3+}-codoped fluoroindate glasses excited at 1.064 μm. *Journal of Applied Physics* 87 (9): 4274–4278. https://doi.org/10.1063/1.373065.

Pan, G., Bai, X., Yang, D. et al. (2017). Doping lanthanide into perovskite nanocrystals: highly improved and expanded optical properties. *Nano Letters* 17 (12): 8005–8011. https://doi.org/10.1021/acs.nanolett.7b04575.

Pandey, A. and Rai, V.K. (2012). Colour emission tunability in Ho^{3+}–Tm^{3+}–Yb^{3+} co-doped Y_2O_3 upconverted phosphor. *Applied Physics B* 109 (4): 611–616. https://doi.org/10.1007/s00340-012-5224-1.

Pandey, A. and Rai, V.K. (2013). Improved luminescence and temperature sensing performance of Ho^{3+}–Yb^{3+}–Zn^{2+}:Y_2O_3 phosphor. *Dalton Transactions* 42 (30): 11005–11011. https://doi.org/10.1039/c3dt50592h.

Pandey, A., Rai, V.K., Kumar, V. et al. (2015). Upconversion based temperature sensing ability of Er^{3+}–Yb^{3+} codoped $SrWO_4$: an optical heating phosphor. *Sensors and Actuators B: Chemical* 209: 352–358. https://doi.org/10.1016/j.snb.2014.11.126.

Pattnaik, S. and Rai, V.K. (2020). Impact of charge compensation on optical and thermometric behaviour of titanate phosphors. *Materials Research Bulletin* 125: 110761. https://doi.org/10.1016/j.materresbull.2019.110761.

Pattnaik, S. and Rai, V.K. (2021). Insight into the spectroscopic and thermometric properties of titanate phosphors via a novel co-excited laser system. *Materials Science and Engineering: B* 272: 115318. https://doi.org/10.1016/j.mseb.2021.115318.

Petryayeva, E. and Krull, U.J. (2011). Localized surface plasmon resonance: nanostructures, bioassays and biosensing – a review. *Analytica Chimica Acta* 706 (1): 8–24. https://doi.org/10.1016/j.aca.2011.08.020.

Porter, J.F. Jr., (1961). Fluorescence excitation by the absorption of two consecutive photons. *Physical Review Letters* 7 (11): 414. https://doi.org/10.1103/physrevlett.7.414.

Prasad, P.N. (2012). *Introduction to Nanomedicine and Nanobioengineering*, vol. 7. Wiley.

Prasad, M. and Rai, V.K. (2020). Influence of transition metal ions: broad band upconversion emission in thermally stable phosphors. *Journal of Alloys and Compounds* 837: 155289. https://doi.org/10.1016/j.jallcom.2020.155289.

Pulli, T., Dönsberg, T., Poikonen, T. et al. (2015). Advantages of white LED lamps and new detector technology in photometry. *Light: Science & Applications* 4 (9): 332–332. https://doi.org/10.1038/lsa.2015.105.

Pust, P., Schmidt, P.J., and Schnick, W. (2015). A revolution in lighting. *Nature Materials* 14 (5): 454–458. https://doi.org/10.1038/nmat4270.

Qian, H.S., Guo, H.C., Ho, P.C.L. et al. (2009). Mesoporous-silica-coated up-conversion fluorescent nanoparticles for photodynamic therapy. *Small* 5 (20): 2285–2290. https://doi.org/10.1002/smll.200900692.

Rabouw, F.T., Prins, P.T., Villanueva-Delgado, P. et al. (2018). Quenching pathways in $NaYF_4:Er^{3+},Yb^{3+}$ upconversion nanocrystals. *ACS Nano* 12 (5): 4812–4823. https://doi.org/10.1021/acsnano.8b01545.

Rai, V.K. (2012). Frequency upconversion in rare earth ions. In: *Solid State Laser* (ed. A.H. Al-Khursan). In Tech https://doi.org/10.5772/36197.

Rai, V.K., Menezes, L.D.S., de Araújo, C.B. et al. (2008). Surface-plasmon-enhanced frequency upconversion in Pr^{3+} doped tellurium-oxide glasses containing silver nanoparticles. *Journal of Applied Physics* 103 (9): 093526. https://doi.org/10.1063/1.2919566.

Rai, V.K., de Araújo, C.B., Ledemi, Y. et al. (2009). Optical spectroscopy and upconversion luminescence in Nd^{3+} doped $Ga_{10}Ge_{25}S_{65}$ glass. *Journal of Applied Physics* 106 (10): 103512. https://doi.org/10.1063/1.3259439.

Rai, V.K., Pandey, A., and Dey, R. (2013). Photoluminescence study of $Y_2O_3:Er^{3+}$–Eu^{3+}–Yb^{3+} phosphor for lighting and sensing applications. *Journal of Applied Physics* 113 (8): 083104. https://doi.org/10.1063/1.4793265.

Rai, V.K., Dey, R., and Kumar, K. (2013). White upconversion emission in $Y_2O_3:Er^{3+}$–Tm^{3+}–Yb^{3+} phosphor. *Materials Research Bulletin* 48 (6): 2232–2236. https://doi.org/10.1016/j.materresbull.2013.02.064.

Reddy, K.L., Prabhakar, N., Rosenholm, J.M., and Krishnan, V. (2018). Core–shell structures of upconversion nanocrystals coated with silica for near infrared light enabled optical imaging of cancer cells. *Micromachines* 9 (8): 400. https://doi.org/10.3390/mi9080400.

Sangeetha, N.M. and van Veggel, F.C. (2009). Lanthanum silicate and lanthanum zirconate nanoparticles co-doped with Ho^{3+} and Yb^{3+}: matrix-dependent red and green upconversion emissions. *The Journal of Physical Chemistry C* 113 (33): 14702–14707. https://doi.org/10.1021/jp904516s.

Savage, J.A. (1987). Materials for infrared fibre optics. *Materials Science Reports* 2 (3): 99–137. https://doi.org/10.1016/s0920-2307(87)80001-4.

Schnapf, J.L., Kraft, T.W., Nunn, B.J., and Baylor, D.A. (1988). Spectral sensitivity of primate photoreceptors. *Visual Neuroscience* 1 (3): 255–261. https://doi.org/10.1017/s0952523800001917.

Shah, S., Liu, J.J., Pasquale, N. et al. (2015). Hybrid upconversion nanomaterials for optogenetic neuronal control. *Nanoscale* 7 (40): 16571–16577. https://doi.org/10.1039/c5nr03411f.

Shan, J. and Ju, Y. (2007). Controlled synthesis of lanthanide-doped $NaYF_4$ upconversion nanocrystals via ligand induced crystal phase transition and silica coating. *Applied Physics Letters* 91 (12): 123103. https://doi.org/10.1063/1.2783476.

Shan, G., Weissleder, R., and Hilderbrand, S.A. (2013). Upconverting organic dye doped core-shell nano-composites for dual-modality NIR imaging and photo-thermal therapy. *Theranostics* 3 (4): 267. https://doi.org/10.7150/thno.5226.

Shang, Y., Hao, S., Lv, W. et al. (2018). Confining excitation energy of Er^{3+}-sensitized upconversion nanoparticles through introducing various energy trapping centers. *Journal of Materials Chemistry C* 6 (15): 3869–3875. https://doi.org/10.1039/c7tc05742c.

Shao, W., Chen, G., Kuzmin, A. et al. (2016). Tunable narrow band emissions from dye-sensitized core/shell/shell nanocrystals in the second near-infrared biological window. *Journal of the American Chemical Society* 138 (50): 16192–16195. https://doi.org/10.1021/jacs.6b08973.

Sharma, R.K., Mudring, A.V., and Ghosh, P. (2017). Recent trends in binary and ternary rare-earth fluoride nanophosphors: how structural and physical properties influence optical behavior. *Journal of Luminescence* 189: 44–63. https://doi.org/10.1016/j.jlumin.2017.03.062.

Shen, J., Sun, L.D., and Yan, C.H. (2008). Luminescent rare earth nanomaterials for bioprobe applications. *Dalton Transactions* 42: 5687–5697. https://doi.org/10.1039/b805306e.

Shen, J., Zhao, L., and Han, G. (2013). Lanthanide-doped upconverting luminescent nanoparticle platforms for optical imaging-guided drug delivery and therapy. *Advanced Drug Delivery Reviews* 65 (5): 744–755. https://doi.org/10.1016/j.addr.2012.05.007.

Shen, J., Li, K., Cheng, L. et al. (2014). Specific detection and simultaneously localized photothermal treatment of cancer cells using layer-by-layer assembled multifunctional nanoparticles. *ACS Applied Materials & Interfaces* 6 (9): 6443–6452. https://doi.org/10.1021/am405924g.

Shi, H. and Sadler, P.J. (2020). How promising is phototherapy for cancer? *British Journal of Cancer* 123 (6): 871–873. https://doi.org/10.1038/s41416-020-0926-3.

Solé, J., Bausa, L., and Jaque, D. (2005). *An Introduction to the Optical Spectroscopy of Inorganic Solids*. Wiley https://doi.org/10.1002/0470016043.

Song, N., Zhou, B., Yan, L. et al. (2019). Understanding the role of Yb^{3+} in the Nd/Yb coupled 808-nm-responsive upconversion. *Frontiers in Chemistry* 6: 673. https://doi.org/10.3389/fchem.2018.00673.

Soni, A.K. and Rai, V.K. (2014). Intrinsic optical bistability and frequency upconversion in Tm^{3+}–Yb^{3+}-codoped Y_2WO_6 phosphor. *Dalton Transactions* 43 (36): 13563–13570. https://doi.org/10.1039/c4dt01266f.

Soni, A.K., Kumari, A., and Rai, V.K. (2015). Optical investigation in shuttle like $BaMoO_4$:Er^{3+}–Yb^{3+} phosphor in display and temperature sensing. *Sensors and Actuators B: Chemical* 216: 64–71. https://doi.org/10.1016/j.snb.2015.04.017.

Soni, A.K., Rai, V.K., and Mahata, M.K. (2017). Yb^{3+} sensitized $Na_2Y_2B_2O_7$:Er^{3+} phosphors in enhanced frequency upconversion, temperature sensing and field emission display. *Materials Research Bulletin* 89: 116–124. https://doi.org/10.1016/j.materresbull.2017.01.009.

Soukka, T., Kuningas, K., Rantanen, T. et al. (2005). Photochemical characterization of up-converting inorganic lanthanide phosphors as potential labels. *Journal of Fluorescence* 15 (4): 513–528. https://doi.org/10.1007/s10895-005-2825-7.

Srivastava, B.B., Gupta, S.K., and Mao, Y. (2020). Single red emission from upconverting $ZnGa_2O_4$:Yb,Er nanoparticles co-doped by Cr^{3+}. *Journal of Materials Chemistry C* 8 (19): 6370–6379. https://doi.org/10.1039/d0tc00411a.

Su, Q., Feng, W., Yang, D., and Li, F. (2017). Resonance energy transfer in upconversion nanoplatforms for selective biodetection. *Accounts of Chemical Research* 50 (1): 32–40. https://doi.org/10.1021/acs.accounts.6b00382.

Sun, L.D., Wang, Y.F., and Yan, C.H. (2014). Paradigms and challenges for bio-application of rare earth upconversion luminescent nanoparticles: small size and tunable emission/excitation spectra. *Accounts of Chemical Research* 47 (4): 1001–1009. https://doi.org/10.1021/ar400218t.

Sun, L.D., Dong, H., Zhang, P.Z., and Yan, C.H. (2015). Upconversion of rare earth nanomaterials. *Annual Review of Physical Chemistry* 66: 619–642. https://doi.org/10.1146/annurev-physchem-040214-121344.

Suyver, J.F., Aebischer, A., Biner, D. et al. (2005). Novel materials doped with trivalent lanthanides and transition metal ions showing near-infrared to visible photon upconversion. *Optical Materials* 27 (6): 1111–1130. https://doi.org/10.1016/j.optmat.2004.10.021.

Tang, J.F., Li, G.N., Yang, C. et al. (2015). Power driven tunable white upconversion luminescence from Lu_2TeO_6 tri-doped with Yb^{3+}, Tm^{3+} and Ho^{3+}. *CrystEngComm* 17 (47): 9048–9054. https://doi.org/10.1039/c5ce01734c.

Teng, X., Zhu, Y., Wei, W. et al. (2012). Lanthanide-doped Na_xScF_{3+x} nanocrystals: crystal structure evolution and multicolor tuning. *Journal of the American Chemical Society* 134 (20): 8340–8343. https://doi.org/10.1021/ja3016236.

Tessitore, G., Mandl, G.A., Brik, M.G. et al. (2019). Recent insights into upconverting nanoparticles: spectroscopy, modeling, and routes to improved luminescence. *Nanoscale* 11 (25): 12015–12029. https://doi.org/10.1039/c9nr02291k.

Tian, L., Xu, Z., Zhao, S. et al. (2014). The upconversion luminescence of $Er^{3+}/Yb^{3+}/Nd^{3+}$ triply-doped β-$NaYF_4$ nanocrystals under 808-nm excitation. *Materials* 7 (11): 7289–7303. https://doi.org/10.3390/ma7117289.

Trupke, T., Shalav, A., Richards, B.S. et al. (2006). Efficiency enhancement of solar cells by luminescent up-conversion of sunlight. *Solar Energy Materials and Solar Cells* 90 (18, 19): 3327–3338. https://doi.org/10.1016/j.solmat.2005.09.021.

Tymiński, A., Martín, I.R., and Grzyb, T. (2020). Upconversion in detail: multicolor emission of Yb/Er/Tm-doped nanoparticles under 800, 975, 1208, and 1532 nm excitation wavelengths. *Particle & Particle Systems Characterization* 37 (8): 2000068. https://doi.org/10.1002/ppsc.202000068.

Unser, S., Bruzas, I., He, J., and Sagle, L. (2015). Localized surface plasmon resonance biosensing: current challenges and approaches. *Sensors* 15 (7): 15684–15716. https://doi.org/10.3390/s150715684.

Valeur, B. (2003). Molecular fluorescence. *Digital Encyclopedia of Applied Physics* 477–531. https://doi.org/10.1002/3527600434.eap684.

Vetrone, F., Boyer, J.C., Capobianco, J.A. et al. (2004). Significance of Yb^{3+} concentration on the upconversion mechanisms in codoped Y_2O_3:Er^{3+},Yb^{3+} nanocrystals. *Journal of Applied Physics* 96 (1): 661–667. https://doi.org/10.1063/1.1739523.

Vetrone, F., Naccache, R., Mahalingam, V. et al. (2009). The active-core/active-shell approach: a strategy to enhance the upconversion luminescence in lanthanide-doped nanoparticles. *Advanced Functional Materials* 19 (18): 2924–2929. https://doi.org/10.1002/adfm.200900234.

Wang, F. and Liu, X. (2008). Upconversion multicolor fine-tuning: visible to near-infrared emission from lanthanide-doped $NaYF_4$ nanoparticles. *Journal of the American Chemical Society* 130 (17): 5642–5643. https://doi.org/10.1021/ja800868a.

Wang, F. and Liu, X. (2009). Recent advances in the chemistry of lanthanide-doped upconversion nanocrystals. *Chemical Society Reviews* 38 (4): 976–989. https://doi.org/10.1039/b809132n.

Wang, F. and Liu, X. (2014). Multicolor tuning of lanthanide-doped nanoparticles by single wavelength excitation. *Accounts of Chemical Research* 47 (4): 1378–1385. https://doi.org/10.1021/ar5000067.

Wang, F. and Liu, X. (2019). 1.18 rare-earth doped upconversion nanophosphors. *Comprehensive Nanoscience and Nanotechnology* 359: https://doi.org/10.1016/b978-0-12-812295-2.00146-x.

Wang, L., Yan, R., Huo, Z. et al. (2005). Fluorescence resonant energy transfer biosensor based on upconversion-luminescent nanoparticles. *Angewandte Chemie International Edition* 44 (37): 6054–6057. https://doi.org/10.1002/anie.200501907.

Wang, M., Mi, C.C., Wang, W.X. et al. (2009). Immunolabeling and NIR-excited fluorescent imaging of HeLa cells by using $NaYF_4$:Yb,Er upconversion nanoparticles. *ACS Nano* 3 (6): 1580–1586. https://doi.org/10.1021/nn900491j.

Wang, F., Banerjee, D., Liu, Y. et al. (2010a). Upconversion nanoparticles in biological labeling, imaging, and therapy. *Analyst* 135 (8): 1839–1854. https://doi.org/10.1039/c0an00144a.

Wang, G., Peng, Q., and Li, Y. (2010b). Luminescence tuning of upconversion nanocrystals. *Chemistry A European Journal* 16 (16): 4923–4931. https://doi.org/10.1002/chem.200903099.

Wang, F., Wang, J., and Liu, X. (2010c). Direct evidence of a surface quenching effect on size-dependent luminescence of upconversion nanoparticles. *Angewandte Chemie International Edition* 49 (41): 7456–7460. https://doi.org/10.1002/anie.201003959.

Wang, C., Cheng, L., and Liu, Z. (2011). Drug delivery with upconversion nanoparticles for multi-functional targeted cancer cell imaging and therapy. *Biomaterials* 32 (4): 1110–1120. https://doi.org/10.1016/j.biomaterials.2010.09.069.

Wang, C., Cheng, L., Xu, H., and Liu, Z. (2012). Towards whole-body imaging at the single cell level using ultra-sensitive stem cell labeling with oligo-arginine modified upconversion nanoparticles. *Biomaterials* 33 (19): 4872–4881. https://doi.org/10.1016/j.biomaterials.2012.03.047.

Wang, Y.F., Liu, G.Y., Sun, L.D. et al. (2013). Nd^{3+}-sensitized upconversion nanophosphors: efficient in vivo bioimaging probes with minimized heating effect. *ACS Nano* 7 (8): 7200–7206. https://doi.org/10.1021/nn402601d.

Weissleder, R. and Pittet, M.J. (2008). Imaging in the era of molecular oncology. *Nature* 452 (7187): 580–589. https://doi.org/10.1038/nature06917.

Willmann, J.K., Van Bruggen, N., Dinkelborg, L.M., and Gambhir, S.S. (2008). Molecular imaging in drug development. *Nature Reviews Drug Discovery* 7 (7): 591–607. https://doi.org/10.1038/nrd2290.

Wisser, M.D., Fischer, S., Siefe, C. et al. (2018). Improving quantum yield of upconverting nanoparticles in aqueous media via emission sensitization. *Nano Letters* 18 (4): 2689–2695. https://doi.org/10.1021/acs.nanolett.8b00634.

Wu, Y., Zhao, X., Zhang, Z. et al. (2021). Dual-mode dichromatic $SrBi_4Ti_4O_{15}$:Er^{3+} emitting phosphor for anti-counterfeiting application. *Ceramics International* 47 (11): 15067–15072. https://doi.org/10.1016/j.ceramint.2021.02.064.

Xiao, Q., Zheng, X., Bu, W. et al. (2013). A core/satellite multifunctional nanotheranostic for in vivo imaging and tumor eradication by radiation/photothermal synergistic therapy. *Journal of the American Chemical Society* 135 (35): 13041–13048. https://doi.org/10.1021/ja404985w.

Yadav, R.S., Verma, R.K., Bahadur, A., and Rai, S.B. (2015). Structural characterizations and intense green upconversion emission in Yb^{3+}, Pr^{3+} co-doped Y_2O_3 nano-phosphor. *Spectrochimica Acta Part A: Molecular and Biomolecular Spectroscopy* 137: 357–362. https://doi.org/10.1016/j.saa.2014.08.078.

Yang, L.W., Han, H.L., Zhang, Y.Y., and Zhong, J.X. (2009). White emission by frequency up-conversion in Yb^{3+}–Ho^{3+}–Tm^{3+} triply doped hexagonal $NaYF_4$ nanorods. *The Journal of Physical Chemistry C* 113 (44): 18995–18999. https://doi.org/10.1021/jp9021689.

Yang, D., Hou, Z., Cheng, Z. et al. (2015). Current advances in lanthanide ion (Ln^{3+})-based upconversion nanomaterials for drug delivery. *Chemical Society Reviews* 44 (6): 1416–1448. https://doi.org/10.1039/c4cs00155a.

Yang, M., Wang, H., Wang, Z. et al. (2019). A Nd^{3+} sensitized upconversion nanosystem with dual photosensitizers for improving photodynamic therapy efficacy. *Biomaterials Science* 7 (4): 1686–1695. https://doi.org/10.1039/c8bm01570h.

Yang, H., Li, X., Zhang, R. et al. (2021). Preparation and properties of Nd^{3+} doped Gd_2O_3 near-infrared phosphor. *Ceramics International* 47 (6): 8510–8517. https://doi.org/10.1016/j.ceramint.2020.11.218.

Yi, G., Lu, H., Zhao, S. et al. (2004). Synthesis, characterization, and biological application of size-controlled nanocrystalline $NaYF_4$:Yb,Er infrared-to-visible up-conversion phosphors. *Nano Letters* 4 (11): 2191–2196. https://doi.org/10.1021/nl048680h.

Yin, A., Zhang, Y., Sun, L., and Yan, C. (2010). Colloidal synthesis and blue based multicolor upconversion emissions of size and composition controlled monodisperse hexagonal $NaYF_4$:Yb,Tm nanocrystals. *Nanoscale* 2 (6): 953–959. https://doi.org/10.1039/b9nr00397e.

Yin, W., Zhao, L., Zhou, L. et al. (2012). Enhanced red emission from GdF3:Yb^{3+},Er^{3+} upconversion nanocrystals by Li+ doping and their application for bioimaging. *Chemistry A European Journal* 18 (30): 9239–9245. https://doi.org/10.1002/chem.201201053.

Yoon, B., Lee, J., Park, I.S. et al. (2013). Recent functional material based approaches to prevent and detect counterfeiting. *Journal of Materials Chemistry C* 1 (13): 2388–2403. https://doi.org/10.1039/c3tc00818e.

Zhan, Q., Zhang, X., Zhao, Y. et al. (2015). Tens of thousands-fold upconversion luminescence enhancement induced by a single gold nanorod. *Laser & Photonics Reviews* 9 (5): 479–487. https://doi.org/10.1002/lpor.201500013.

Zhang, F. (2015). *Photon Upconversion Nanomaterials*, vol. 416. Berlin: Springer https://doi.org/10.1007/978-3-662-45597-5.

Zhang, H., Li, Y., Ivanov, I.A. et al. (2010). Plasmonic modulation of the upconversion fluorescence in $NaYF_4$:Yb/Tm hexaplate nanocrystals using gold nanoparticles or nanoshells. *Angewandte Chemie International Edition* 49 (16): 2865–2868. https://doi.org/10.1002/anie.200905805.

Zhang, F., Shi, Q., Zhang, Y. et al. (2011). Fluorescence upconversion microbarcodes for multiplexed biological detection: nucleic acid encoding. *Advanced Materials* 23 (33): 3775–3779. https://doi.org/10.1002/adma.201101868.

Zhao, Q., Shao, B., Lü, W. et al. (2013). Doping alkaline-earth: a strategy of stabilizing hexagonal GdF_3 at room temperature. *Dalton Transactions* 42 (43): 15482–15488. https://doi.org/10.1039/c3dt51967h.

Zhou, J., Liu, Z., and Li, F. (2012). Upconversion nanophosphors for small-animal imaging. *Chemical Society Reviews* 41 (3): 1323–1349. https://doi.org/10.1039/c1cs15187h.

Zhu, X., Zhang, J., Liu, J., and Zhang, Y. (2019). Recent progress of rare-earth doped upconversion nanoparticles: synthesis, optimization, and applications. *Advanced Science* 6 (22): 1901358. https://doi.org/10.1002/advs.201901358.

Zou, W., Visser, C., Maduro, J.A. et al. (2012). Broadband dye-sensitized upconversion of near-infrared light. *Nature Photonics* 6 (8): 560. https://doi.org/10.1038/nphoton.2012.158.

8

Smart Upconverting Nanoparticles and New Types of Upconverting Nanoparticles

Akhilesh K. Singh

Department of Physical Sciences, Banasthali Vidyapith, Banasthali, Tonk, Rajasthan 304022, India

8.1 Introduction

Photon upconversion is a process to produce emission of high-energy visible/ultraviolet photon upon excitation with multiple low-energy near-infrared (NIR) photons (Haase and Schäfer 2011). This fascinating emission property finds applications in biosensing, light-triggered nanomedicine, optogenetics, solar energy harvesting, photocatalysis, luminescent-based temperature sensors, etc. (Wilhelm 2017). Photon upconversion can be achieved by using lanthanide-based upconverting nanoparticles (UCNPs) and triplet–triplet annihilation (TTA) in organic molecules (Zhou et al. 2015). UCNP systems used multiple long-lived, ladder-like electronic states of lanthanide ions doped in suitable hosts and can emit a photon with energy even two to four times higher than the incident photon's energy. However, because of the forbidden 4f–4f transitions in lanthanide ions, UCNPs have a low absorption cross section that limits their overall upconversion efficiencies (Wang and Liu 2009; Singh et al. 2013). Upconversion in TTA systems, on the other hand, is achieved through diffusion-mediated annihilation of two triplet excitons that result in one higher energy photon (Zhou et al. 2015). It possesses a high absorption cross section, with upconversion efficiency approaching unity. However, because of the two-photon upconversion process, the TTA-based upconverting materials exhibit a relatively low anti-Stokes shift (Zhou et al. 2015). Furthermore, the stability, particularly in oxygen and aqueous environment of TTA systems, is also a matter of concern. Therefore, the goal is to create a stable, smart, and a new type of UCNPs with high upconversion efficiencies to fulfill the necessities of various applications stated above.

To achieve this, researchers working in this area have identified several ways, for example, designing and synthesizing hybrid upconversion nanomaterials (Wen et al. 2019; Shahi et al. 2017), enhancing the upconversion emission through plasmonic effect (Park et al. 2015; Rai et al. 2015), upconversion-based metal–organic frameworks (MOFs), etc. (Pei et al. 2018). In this chapter, the endeavor will be highlighting these systems with their synthesis protocol, optical properties, and

Upconverting Nanoparticles: From Fundamentals to Applications, First Edition. Edited by Vineet K. Rai.

potentials applications in anticounterfeiting, bioimaging, luminescence-bases sensing, etc. This would be useful to attract minds of multidisciplinary research communities for the design of futuristic more efficient and stable UCNP systems.

8.2 Upconverting Core–Shell Nanostructures

UCNPs are typically synthesized at high temperatures (200–325 °C) using high boiling point organic solvents such as octadecene (Wang et al. 2011). Oleic acid is used to control the size of UCNPs by coordinating with nanoparticle surfaces. For biological applications, hydrophilic surface modification becomes inevitable that can be achieved by various surface modification processes (Liu et al. 2015). Because of low cytotoxicity and the ability for surface modification using various functional groups, UCNPs and silica core–shell (UCNP@SiO_2) nanostructure have attracted great interest in biological applications (Liu et al. 2015). Porous structure and high surface area of UCNPs@SiO_2 nanostructure ensure high loading of drugs. By judicious modification of its surface using suitable functional moieties, one can develop multifunctional nanomedical platforms for various biological applications.

Apart from the biological applications, core–shell UCNPs have also got the applications in anticounterfeiting, luminescent-based sensing, etc. (Zuo et al. 2018). Tan et al. have prepared UCNPs@carbon-dots@meso-SiO_2 multishell nanohybrids by encapsulating carbon dots (CDs) and UCNPs into mesoporous silica (mSiO_2) (Tan et al. 2019). These UCNPs@CDs@mSiO_2 nanohybrids possess a stable structure that improves their photostability and emission properties under an acidic environment. It was shown that by changing lanthanide ions and their composition, UCNPs can be prepared to emit green, red, and blue color light. This also enabled multimodal luminescence using UCNPs@CDs@mSiO_2 nanohybrids, i.e. red, green, and blue UC emission and blue downconversion (DC) luminescence (of CDs). These core–shell nanohybrids were demonstrated for developing various luminescent inks to produce 3D anticounterfeiting barcodes.

The schematic diagram given in Figure 8.1 shows that CDs and UCNPs were synthesized using a hydrothermal method and a thermal decomposition method, respectively. Thereafter, surface modification of UCNPs was carried out to convert its hydrophobic surface to the hydrophilic ones. After mixing of cetyltrimethylammonium bromide (CTAB)-captured UCNPs and CDs, the electrostatic interactions between them resulted in the absorption of CDs over CTAB-captured UCNPs. These CD-capped UCNPs can be encapsulated into mSiO_2 using the sol–gel method to obtain UCNPs@CDs@mSiO_2 nanohybrids. For the development of luminescent security ink, the UCNPs@CDs@mSiO_2 nanohybrids were dispersed in a mixture of deionized water, ethanol, and glycerol. This ink was further demonstrated for dual-mode fluorescent barcode designing using an inkjet printer.

Core–shell UCNPs have also been used for reversibly toggle photo-responsive organic compounds between their two states (see Figure 8.2) (Boyer et al. 2012). In this reversible "remote-control" photo-switching, organic chromophore does

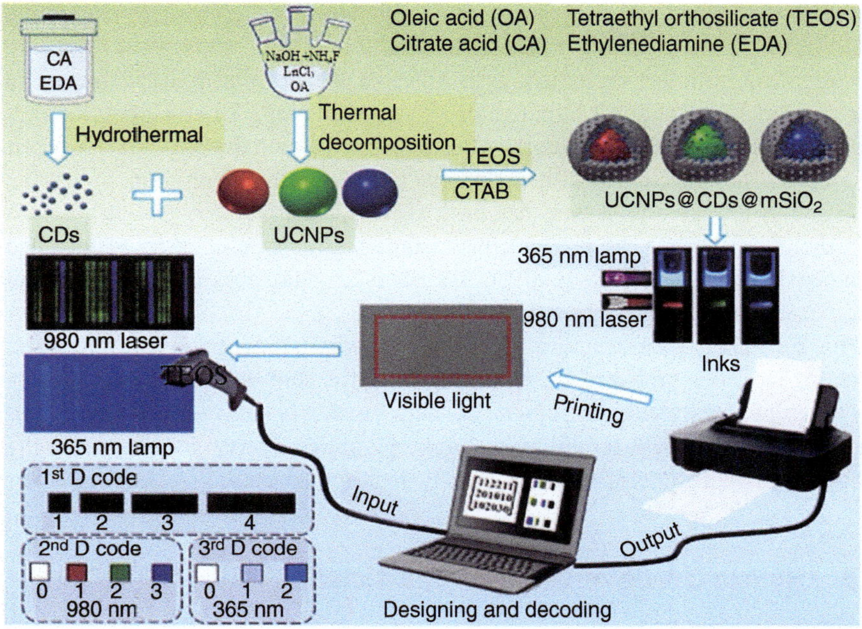

Figure 8.1 Schematic scheme showing the synthesis method and bimodal luminescence of UCNPs@CDs@mSiO$_2$ core–shell nanohybrids for anticounterfeiting barcodes. Source: Tan et al. (2019). Reprinted with permission of American Chemical Society.

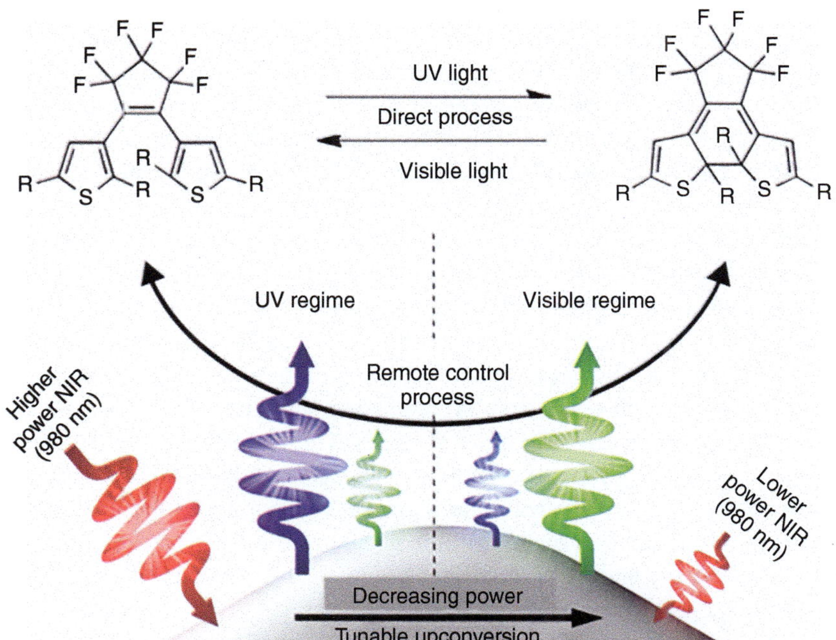

Figure 8.2 Demonstration of "direct" photoreactions of the DTE derivative triggered by core–shell UCNPs. The high-intensity 980 nm CW laser produces UV emission that triggers ring-closing and low-intensity beams produce visible emission that initiates the ring-opening reaction. The sizes of colored arrows qualitatively demonstrate the relative amount of excitation and emissions. Source: Boyer et al. (2012). Reprinted with permission of American Chemical Society.

not directly absorb the NIR excitation light. However, core–shell containing $NaYF_4$:Tm^{3+}/Yb^{3+} and $NaYF_4$:Er^{3+}/Yb^{3+} into separate layers emit lights of different colors under excitation with different powers that help in reversible photo-switching. Figure 8.2 demonstrates that at high excitation intensity, the leading emissions from UCNPs are in the ultraviolet (UV) region that is suitable to result in a ring-closing reaction of dithienylethene (DTE). On the other hand, the visible emission at low excitation densities from UCNPs causes a ring-opening reaction and therefore restores the original photoisomers. This type of remote-controlled photo-switching using core–shell UCNPs offers an adaptable method to control photochemical reactions with a single light source.

The core–shell nanostructure also helps in eliminating cross-relaxation and enabling judicious tuning of UC emission. Thus, the core–shell nanostructure is a new class of upconverting nanomaterials with tunable upconversion emission and has applications in high-resolution spectroscopy, anticounterfeiting, increasing the efficiency of photovoltaic cells, high-density data storage, etc.

8.3 Hybrid Upconverting Nanoparticles

UCNPs have various potential applications; however, extremely weak absorption of UCNPs in a rather narrow spectral range makes it quite inefficient. Further, as upconversion is a nonlinear optical phenomenon, sometimes, it requires a high-intensity light beam for the excitation process. $NaYF_4$ is considered as the best host for upconversion emission; however, the upconversion quantum yield in bulk $NaYF_4$:Er^{3+}/Yb^{3+} is found to be 5% and 3% only under 975 and 1522 nm NIR light irradiation, respectively (Zou et al. 2012). Moreover, nanoparticles of $NaYF_4$:Er^{3+}/Yb^{3+} show even lower upconversion quantum yields; for example, it is only 0.1% for 30 nm nanoparticles of $NaYF_4$:Er^{3+}/Yb^{3+}. Hybrid upconverting materials can overcome these limitations to a great extent by increasing the absorption bandwidth using organic NIR dyes as sensitizers. Here, dye molecules work as antennas; it absorbs NIR incident light and transfers excitation energy efficiently to the central activator ions present in the hybrid material (Safdar et al. 2020). In these hybrid materials, spin-triplet states of the dye molecules work as an intermediate level in the energy sensitization to the UCNPs. In this process, lanthanide ions, through the heavy atom effect, help in intersystem crossing (ISC) within the dye molecules (Singh et al. 2010; Shahi et al. 2015). These inorganic–organic hybrid materials possess strong NIR absorption in the broad spectral range, the ability to tune the absorption band by molecular bandgap engineering. In the literature, it has been demonstrated that dye-sensitized UCNPs result in 33 000 times increase in emission intensity and ~100 times increase in UC efficiency over bare UCNPs (Garfield et al. 2018).

In another type of hybrid materials, Marin et al. have synthesized a multiwavelength-responsive material that comprises the lanthanide (Ln^{3+})

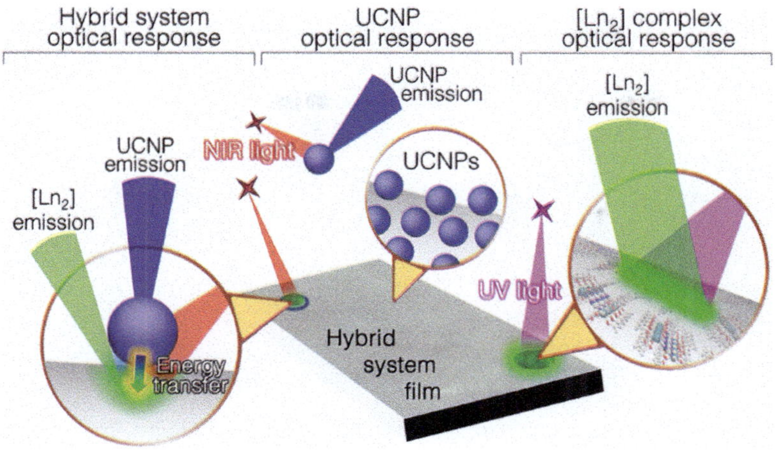

Figure 8.3 Schematic diagram of the hybrid material showing down-shifting and upconversion emission. Source: Marin et al. (2019). Reprinted with permission of American Chemical Society.

complex and UCNPs. In this hybrid material, the Ln^{3+} complex shows emission under UV as well as NIR light excitation (Marin et al. 2019). Under NIR light excitation, the emission of lanthanide complex was achieved through energy sensitization from UCNPs to lanthanide complex. In this hybrid material, UCNPs play a more valuable role, namely, the high surface-to-volume ratio of UCNPs favors surface interactions and showcase NIR operational capabilities. On the other hand, lanthanide complexes exhibit broadband absorption of organic ligands that through excitation-state energy transfer produces discrete characteristic emission of central Ln^{3+} ions, which is quite inert to the chemical environment. This invariability of the emission lines in these systems is of enormous interest for sensor and light conversion applications (Figure 8.3).

In another interesting work, Hemmer et al. have prepared dye-conjugated UCNPs and demonstrated temperature-induced energy transfer that can be used for nanothermometry (Hemmer et al. 2015). They modified the surface of UCNPs using thermo-responsive poly(*N*-isopropyl acrylamide) (pNIPAM) polymer and conjugated with FluoProbe532A dye to make a temperature-sensitive probe. In UCNPs-pNIPAM-dye hybrid materials, as the temperature exceeds 35 °C because of energy sensitization from UCNPs to the dye molecule, the emission intensity of the green peak of UCNPs gets suppressed. This characteristic temperature at which the energy sensitization process gets increased, 35 °C, is well matched with the transition temperature of thermo-responsive pNIPAM that is used as a coupler between the UCNPs and the dye molecule. As the temperature increases, the pNIPAM polymer chains get collapsed, resulting in nearby between the UCNP/FluoProbe pairs. This causes an increased energy transfer process that is reflected as a decrease in the green emission intensity of UCNPs. The schematic diagram showing this process is given in Figure 8.4.

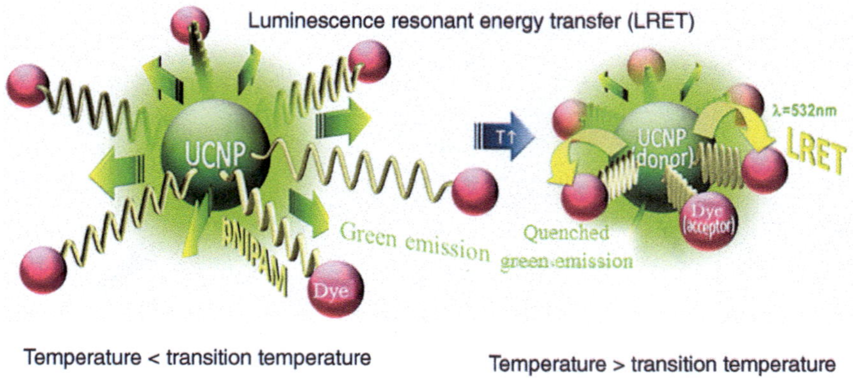

Figure 8.4 Scheme of the luminescence resonance energy transfer $NaGdF_4Er^{3+}/Yb^{3+}$ UCNPs and FluoProbe532A organic dye coupled with the pNIPAM thermo-responsive polymer. Source: Hemmer et al. (2015). Reprinted with permission of American Chemical Society.

8.4 Magnetic Upconverting Nanoparticles

The multifunctionality of magnetic UCNPs has attracted a great deal of interest, particularly for biomedical applications. These hybrid materials comprising UCNPs and magnetic nanoparticles (MNPs) exhibit upconversion emission along with magnetism. Upconversion emission can be altered with an external magnetic field that makes it attractive for various biological applications, namely, as a contrast agent in magnetic resonance imaging (MRI) (Dwivedi et al. 2014; Tian et al. 2012). The coexistence of upconversion luminescence and magnetism in a single material is very rarely observed. However, one can achieve this by making a composite of luminescent material and magnetic ones. One way of doing this could be by encapsulating UCNPs and iron oxide nanoparticles (IONPs) using amphiphilic polymers (UC-IO@Polymer). Xu et al. have shown that by introducing a water-soluble squaraine (SQ) dye while sample preparation, one can get a UC-IO@Polymer-SQ hybrid material that can be used for multimodal upconversion luminescence (UCL)/downconversion fluorescence (FL)/MRI (Xu et al. 2011).

Figure 8.5 shows a scheme for UC-IO@Polymer-SQ nanocomposite preparation with a size of ~200 nm. Because of its excellent magnetic and optical properties, the UC-IO@Polymer-SQ hybrid material was demonstrated in vivo cell tracking and imaging. The UC and FL-labeled cells were visible under different excitation wavelengths. The UCL/FL images of a mouse were again taken after the development of the tumor in 5 days post-injection of 4T1 cells that reveal largely retained UC and FL signals. UCL and FL emissions persist even after 10 days in the tumor position (Figure 8.6a). T_2-weighted MRI was also carried out 0 and 5 days post-injection, which reveals significant reduction in MR signals (Figure 8.6b,c). Compared to optical imaging, MR imaging was relatively insensitive; thus, the labeling contrast became weak with time.

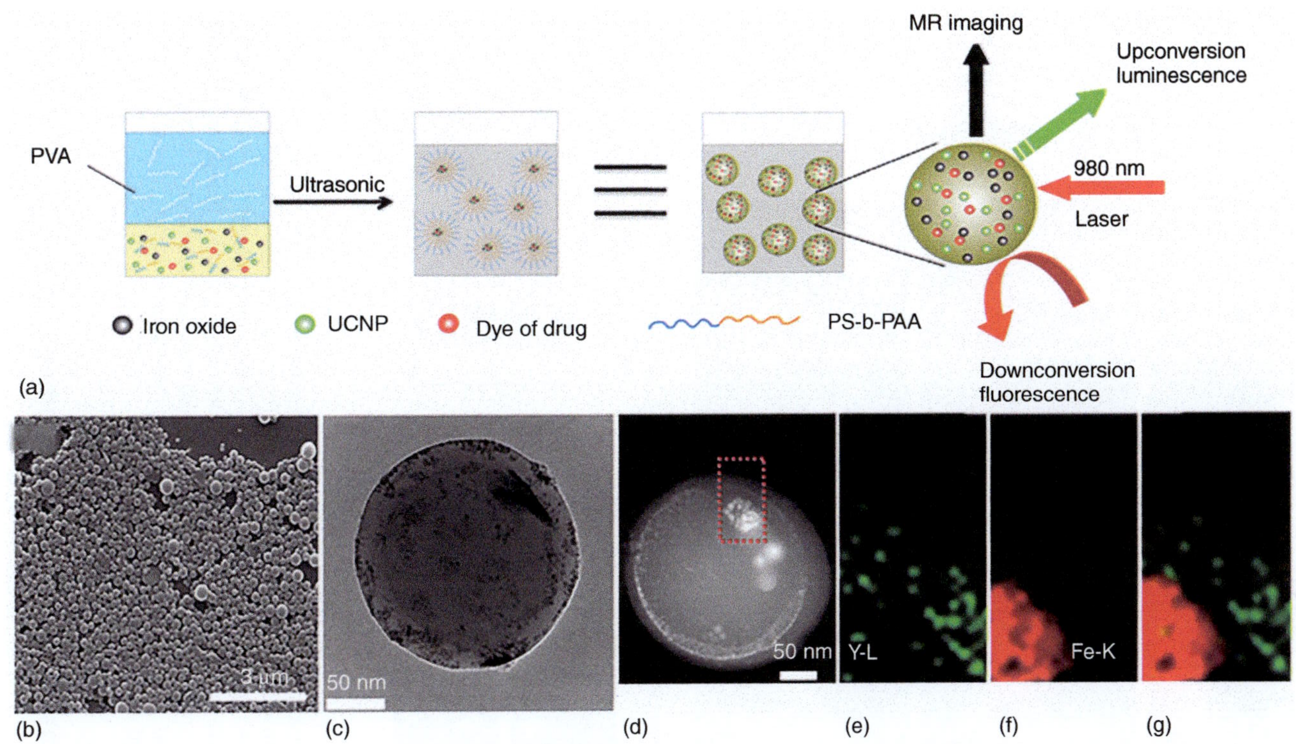

Figure 8.5 Synthesis and characterization of the UC-IO@Polymer nanohybrid. (a) Schematic diagrams showing synthesis and applications of the UC-IO@Polymer nanohybrid. (b) SEM of UC-IO@Polymer revealing an average particle size of ~200 nm. (c) TEM image of UC-IO@Polymer. (d–g) Energy dispersive X-ray spectroscopy (EDX/EDS) mapping of the UC-IO@Polymer nanohybrid. (d) Dark-field TEM of a single UC-IO@Polymer nanohybrid. The area represented under the red rectangle was further explored for EDX mapping. (e–g) EDS images of UC-IO@Polymer showing yttrium L edge, iron K edge, and the merged image, respectively. Source: Xu et al. (2011). Reprinted with permission of Elsevier.

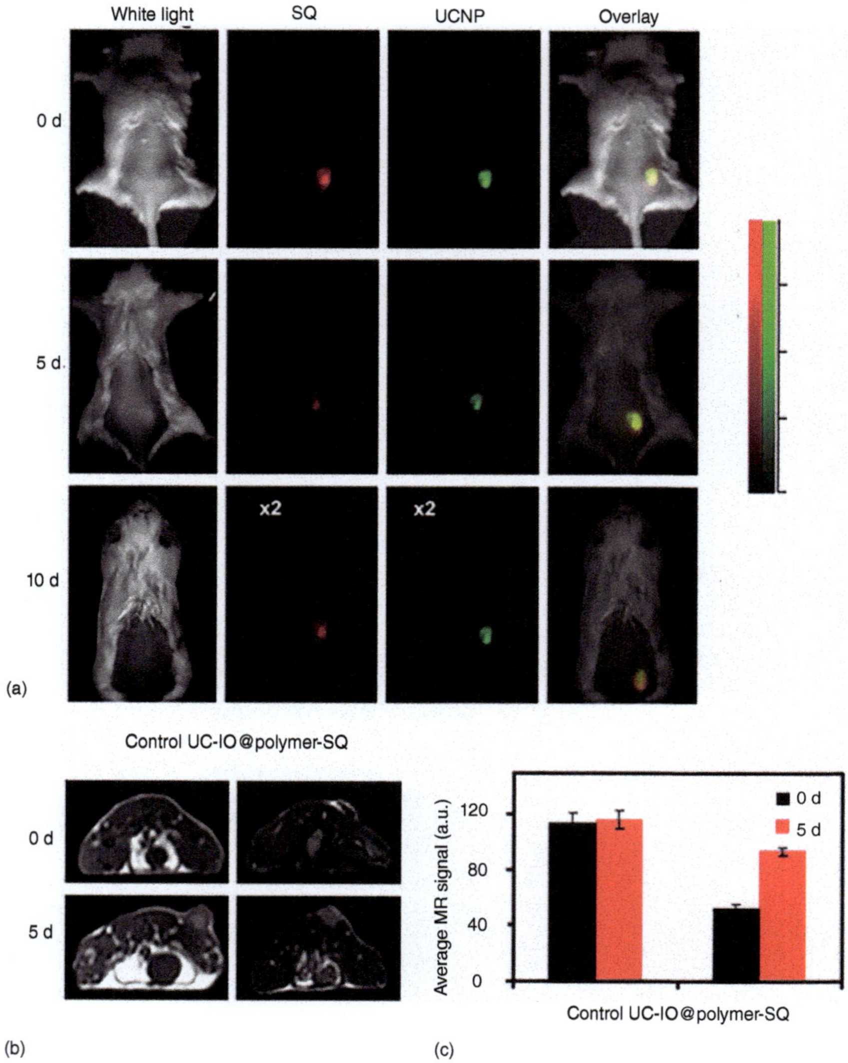

Figure 8.6 In vivo cell tracking. (a) Fluorescence/upconversion images of mice intravenously inoculated with UC-IO@Polymer-SQ-labeled 4T1 cells recorded 0, 5, and 10 days after injection. (b) T_2-weighted MR images of mice injected with UC-IO@Polymer-SQ-labeled 4T1 cells and control unlabeled cells after 0 and 5 days. (c) Plot of MR signals of tumors on mice immunized with UC-IO@Polymer-SQ-treated 4T1 cells and untreated cells after 0 and 5 days. Error bars were evaluated using three measurements. Source: Xu et al. (2011). Reprinted with permission of Elsevier.

8.5 UC-Based Metal–Organic Frameworks

We have discussed the upconverting core–shell nanoparticles, hybrid upconverting materials, and magnetic UCNPs; another appealing modification in UCNPs is the formation of MOFs based on UCNPs. Figure 8.7a shows a scheme for the growth of

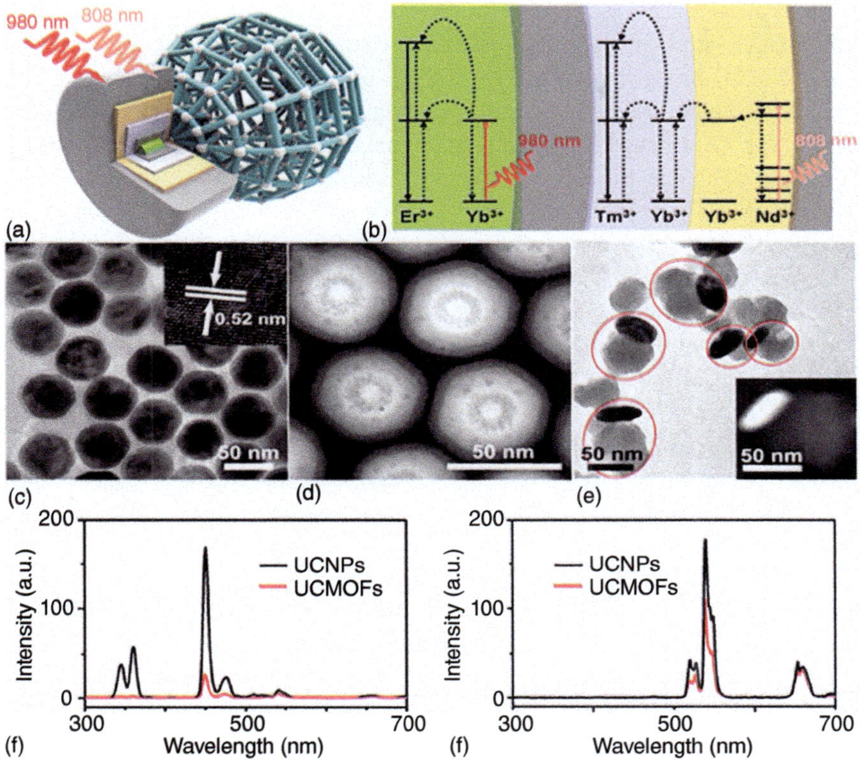

Figure 8.7 (a) Schematic diagram depicting the structure of UCMOFs obtained by growing nMOFs onto multishell UCNPs. (b) The mechanism for energy sensitization in the multishell UCNPs. (c) TEM, HRTEM, and (d) HAADF-STEM images of the multishell UCNPs. (e) TEM image of UCMOFs. Inset: HAADF-STEM image of a single UCMOF. UCL spectra of polyvinylpyrrolidon (PVP)-coated multishell UCNPs under (f) 808 nm and (g) 980 nm NIR laser excitations. Source: Li et al. (2017). Reprinted with permission of American Chemical Society.

Zr-based porphyrinic nanoscale metal–organic frameworks (nMOFs) onto UCNPs (upconverting metal organic framework [UCMOFs]). Through energy sensitization from UCNPs to nMOFs, UCMOFs can utilize photons outside the absorption range of porphyrinic MOFs. These heterogeneous UCMOFs can be used for NIR-induced tumor therapy. Li et al. have used $NaGdF_4$:Yb,Er@$NaYF_4$@$NaYF_4$:Yb, and Tm@$NaYbF_4$:Nd@$NaYF_4$ core–multishell with a large size to address the poor upconversion emission intensity in the ultrasmall UCNPs (<10 nm) for the growth of UCMOFs (Figure 8.7a) (Li et al. 2017). To realize orthogonal UC luminescence, activators (Er^{3+} and Tm^{3+}) and sensitizers (Nd^{3+} and Yb^{3+}) were doped in separate shells. These can be realized using continuous wave (CW) NIR laser light of wavelengths 808 and 980 nm (Figure 8.7b). As shown in Figure 8.7c, nanoplates have a uniform size of ~55 nm × 29 nm. An interplanar spacing of 0.52 nm in the high-resolution transmission electron microscopy (HRTEM) shown in the inset of Figure 8.7c was indexed as the (100) lattice fringe of the hexagonal phase of multishell UCNPs. The multishell structure of UCNPs was also confirmed by the high-angle annular dark-field scanning transmission electron microscopy (HAADF-STEM) image shown in Figure 8.7d. In this image, the darker regions represent the lighter element (Y) and the brighter regions to Gd, Yb, and Nd elements. The TEM image in Figure 8.7e reveals that UCMOFs were successfully realized with nMOF having a domain ranging from 30 to 80 nm.

As depicted in Figure 8.7f,g, multishell UCNPs exhibit orthogonal UC luminescence under 808 and 980 nm excitations, respectively. Under 980 nm excitation, UCNPs show characteristic emissions of the Er^{3+} ions in the green (522 and 541 nm) and red (654 nm) regions, whereas under excitation 808 nm, it shows the emission of Tm^{3+} in the blue (450 nm) and ultraviolet (360 nm) regions. The UC emission profile of the UCMOFs reveals a strong and moderate decrease in the emission intensity of Tm^{3+} and Er^{3+}, respectively. This indicates efficient energy sensitization from Tm^{3+} to Er^{3+}, which is resulted because of the larger distance between Er^{3+} and nMOFs in comparison to the distance between Tm^{3+} and nMOFs. These results indicate that UCMOFs could have applications in nanomedicine, increasing the efficiency of solar cells.

8.6 Smart UCNPs for Security Applications

Counterfeiting is spread in every walk of our life, hurting economies and quality of life across the globe (Singh et al. 2018; Qin et al. 2019; Hu et al. 2017). Therefore, it is imperative to develop advanced anticounterfeiting technologies to combat the menace of counterfeiting. Holography and quick response (QR) codes are currently being used as a security feature; however, high cost and complex fabrication process limit its applications. To further advance anticounterfeiting security features, optically active QR codes were also explored using quantum dots and plasmonic nanoparticles. However, poor photostability, self-absorption, and wide full-width at half maximum (FWHM) of their emission peak need further consideration. To overcome these, recently, optically active security labels and QR codes based on UCNPs were explored. Kaboli et al. have synthesized $NaYF_4$:Yb^{3+},Tm^{3+} and $NaYF_4$:Yb^{3+},Er^{3+} UCNPs using the thermal decomposition method (Kaboli et al. 2019). To fabricate different transparent patterns on nanocomposites, maskless photolithography was used. Figure 8.8 shows a demonstration of security codes based on UCNPs. Figure 8.8a shows a detection setup comprising a CW 980 nm laser to excite UCNP-based nanocomposites. A convex lens was used to focus the laser beam onto the sample and a set of perpendicular scanning galvo mirrors to scan the entire sample. A smartphone with a convex lens was used to capture the photoluminescence (PL) images.

Figure 8.8b shows the PL image of a portrait of Kharazmi (Al-Khwarizmi) under 1W@36 mm^2 laser light intensity. Here, the nanocomposite has only $NaYF_4$:Yb^{3+},Er^{3+} UCNPs. Figure 8.8c,d shows nanocomposites containing a mixture of two UCNPs. The QR code appear white (shown in Figure 8.8c) when galvo mirrors were off and it appear green when the sample was scanned using galvo

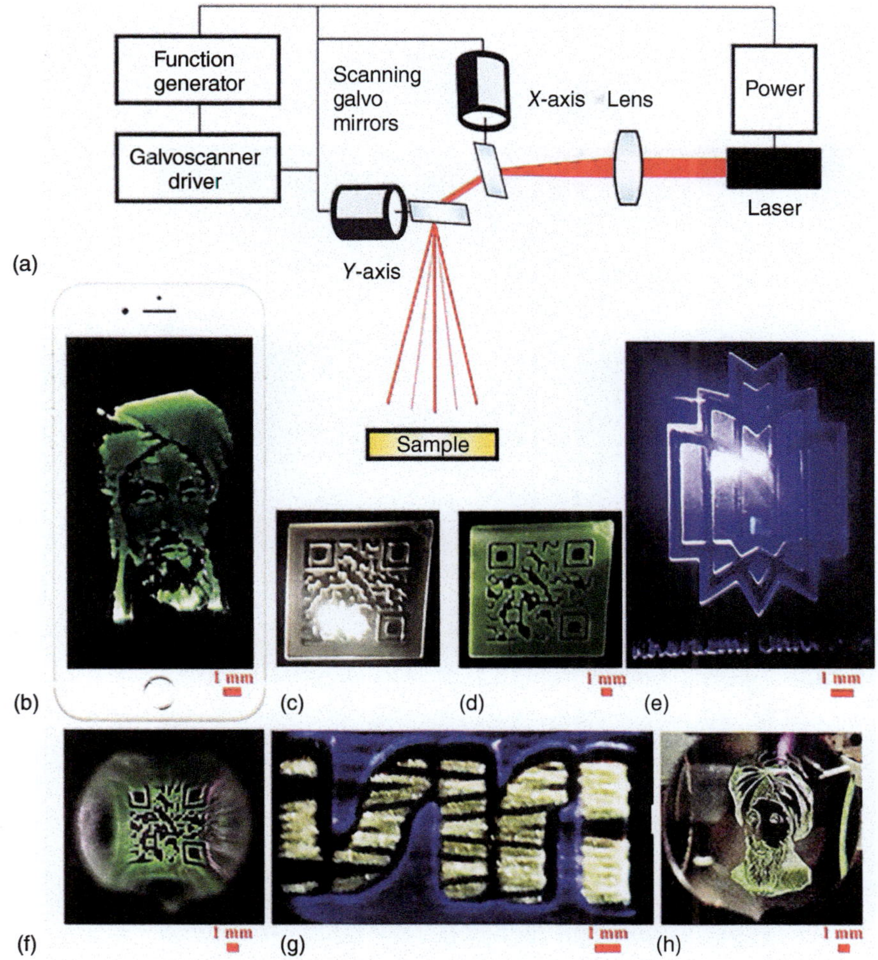

Figure 8.8 (a) Optical setup for excitation and imaging of patterns. (b) Portrait of Kharazmi (Al-Khwarizmi) printed with nanocomposites having $NaYF_4$:Yb^{3+},Er^{3+} UCNPs. (c) A QR code was prepared using two UCNPs and excited with static laser light. (d) A QR code having a mixture of two UCNPs and is excited with a dynamic laser beam. (e) Logo of the Kharazmi University having $NaYF_4$:Yb^{3+},Tm^{3+} UCNPs doped in the photoresist. (f) A QR code having $NaYF_4$:Yb^{3+},Er^{3+} UCNPs doped in the photoresist. (g) A bicolor pattern fabricated using maskless photolithography. (h) Portrait of Ibn al-Haytham comprising $NaYF_4$:Yb^{3+},Er^{3+} UCNPs doped in the photoresist. Source: Kaboli et al. (2019). Reprinted with permission of American Chemical Society.

mirrors. Figure 8.8e depicts a blue PL image of the logo of the Kharazmi University having $NaYF_4$:Yb^{3+},Tm^{3+} UCNPs. Figure 8.8f shows the QR code and (h) shows the portrait of Ibn al-Haytham, comprising $NaYF_4$:Yb^{3+},Er^{3+} UCNPs. Figure 8.8g shows a bicolor photograph that contains both UCNPs. Here, the yellow-colored pattern was first obtained using nanocomposites having both $NaYF_4$:Yb^{3+},Er^{3+} and $NaYF_4$:Yb^{3+},Er^{3+} UCNPs. Thus, the PL color can be changed with both dynamic and static excitation.

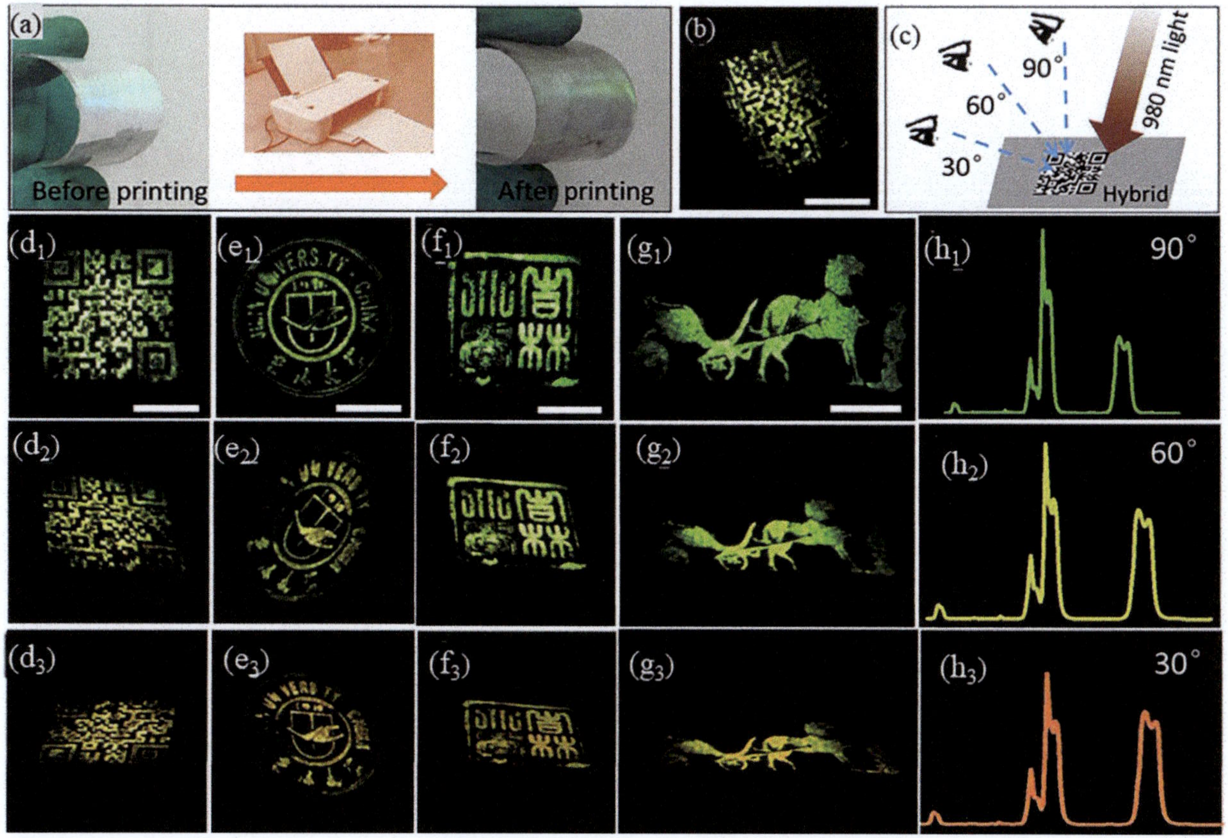

Figure 8.9 Scheme for producing anticounterfeiting hybrids. (a) Photograph of flexible OPC. (b) Two-dimensional code pattern developed on the bent flexible OPC. Scale bar: 0.5 cm. (c) Schematic diagram of angle-dependent anticounterfeiting. (d_1-d_3) Two-dimensional code pattern on the paper invoice recorded at 90°, 60°, and 30° observation angles, respectively, under 980 nm laser light irradiance. (e_1-e_3) Images of school badge on hybrids captured at 90°, 60°, and 30° observation angles, respectively, under 980 nm light irradiance. (f_1-f_3) Images of fluorescence patterns on hybrids captured at 90°, 60°, and 30° observation angles, respectively, under 980 nm light irradiance. (g_1-g_3) Images of fluorescence patterns on hybrids taken at 90°, 60°, and 30° observation angles, respectively, under 980 nm light irradiance. (h_1-h_3) Emission spectra of anticounterfeiting hybrids were recorded at 90°, 60°, and 30° observation angles, respectively. Source: Zhou et al. (2017). Reprinted with permission of American Chemical Society.

Despite its enormous potential applications, the necessity of a high-intensity light source to attain the emission in UCNPs poses challenges. Therefore, it is imperative to investigate novel mechanisms/materials for effective UC enhancement. Semiconductor plasmon NPs, such as CuS that show longitudinal surface plasmon resonance (LSPR) effect, can also possess nonlinear absorption under intense NIR light excitation, which could be used for designing the anticounterfeiting security feature (Ren et al. 2019). Bulk photonic crystals (PCs), with a periodically varying index, can control localized electric fields and thereby increase the emission intensity of UCNPs. Thus, the amalgamation of PC and LSPR could be effective to tune a localized electromagnetic field for varying UCL. Zhou et al. explored the nanoprinting method for producing CuS NPs and UCNPs on a flexible opal photonic crystal (OPC) film that results in good reproducibility and high recognition under 980 nm laser excitation (Zhou et al. 2017). To obtain this structure, poly (methyl methacrylate) (PMMA) OPCs were first manufactured onto positron emission tomography (PET), and then, $Cu_{2-x}S$ NPs and UCNPs were framed into inks and printed onto the OPCs.

Figure 8.9a shows the image of a flexible OPC film before printing, which possesses gradient structural color under normal illumination. The two-dimensional code pattern was stored in the film after the printing of CuS NPs and UCNPs. Figure 8.9b shows the observed pattern under 980 nm laser light illumination that is not affected by bending. The change in reflectivity and the photonic bandgap of OPCs with observation angle were demonstrated for anticounterfeiting applications (see Figure 8.9c). The OPC/$Cu_{2-x}S$/MoO_3/$NaYF_4$ hybrid with a 550 nm optical bandgap of OPC and $NaYF_4$:Yb^{3+}(40%), Er^{3+}(5%)@$NaYF_4$:Yb^{3+}(10%), and Nd^{3+}(10%) NPs were used for this purpose. Figure 8.9d shows a thin and flexible hybrid pasted over a paper invoice that shows a bright and clear two-dimensional code pattern visualized under infrared light. Figure 8.9e_1–e_3, f_1–f_3, and g_1–g_3 show fluorescence patterns of different images on hybrids captured at 90°, 60°, and 30° observation angles, respectively, under 980 nm light irradiance. The increased green emission compared with that of pure UCNPs is because of the OPC bandgap which is at 550 nm. When the angle of observation is at 60°, the bandgap of OPC changes to 515 nm, which results in a bright pattern with orange color. Further, at a 30° observation angle, the bandgap of OPC changes to 465 nm, which was far away from the green luminescence center wavelength. Thus, the changes of colors were recorded in the spectra of hybrids at 90°, 60°, and 30° observation angles (Figure 8.9h). This tactic is predominantly beneficial for application that necessitates high display quality, such as microdisplays, and security features.

8.7 Smart Upconverting Nanoparticles for Biological Applications

There is a great deal of interest in visualizing a target using different imaging methods, such as computed tomography (CT), MRI, optical imaging, and nuclear medicine imaging because of their different spatial resolutions, imaging depths, and areas of applications (Zhou et al. 2012). CT provides excellent anatomical information; however, it has low sensitivity. PET, on the other hand, provides high sensitivity (picomolar) but low (~mm) spatial resolution. Therefore, the grouping of PET/CT brings out high spatial resolution and sensitivity (Chen et al. 2016). Furthermore, because of low planar resolution, CT, PET, and MRI are inappropriate for visualizing living cells that can be resolved by fluorescent bioimaging. PL imaging has the highest spatial resolution (suitable for imaging living cells); however, its low penetration of light in the tissue makes it incapable to obtain anatomical and physiological details. Thus, for multilevel visualization, PL, radioactivity, and magnetic properties should be used together.

Liu et al. integrated UCL ($NaYF_4$:Yb^{3+},Er^{3+}), magnetism (Gd^{3+}), radioactivity (^{18}F), and targeted recognition (folic acid) to develop a multifunctional probe for bioimaging (Liu et al. 2011). To demonstrate multimodal imaging, in vivo UCL optical bioimaging was performed on Kunming mice through tail vein injection of these nanoparticles (200 μl, 0.5 mg/ml per animal). A representative UCL image is shown in Figure 8.10. The detailed analysis revealed that nanoparticle aggregates were distributed in the liver and spleen, which is the classical behavior of in vivo nanoparticles as the liver contains phagocytic cells, and the liver and spleen are the main organs for eliminating foreign particles. UCNPs improve penetration greatly; MRI provides an excellent spatial resolution and depth for in vivo imaging; and gadolinium (Gd^{3+}) is used as a positive contrast in MRI. MRI results shown in Figure 8.10 display the T_1-weighted relaxivity in the range of 0.016–0.125 mg/ml, which shows a signal that was enhanced by increasing (aminocaproic acid) the AA-Gd-UCNP concentrations. This demonstrates the utility of multimodal bioimaging to image cellular scale to whole-body evaluation.

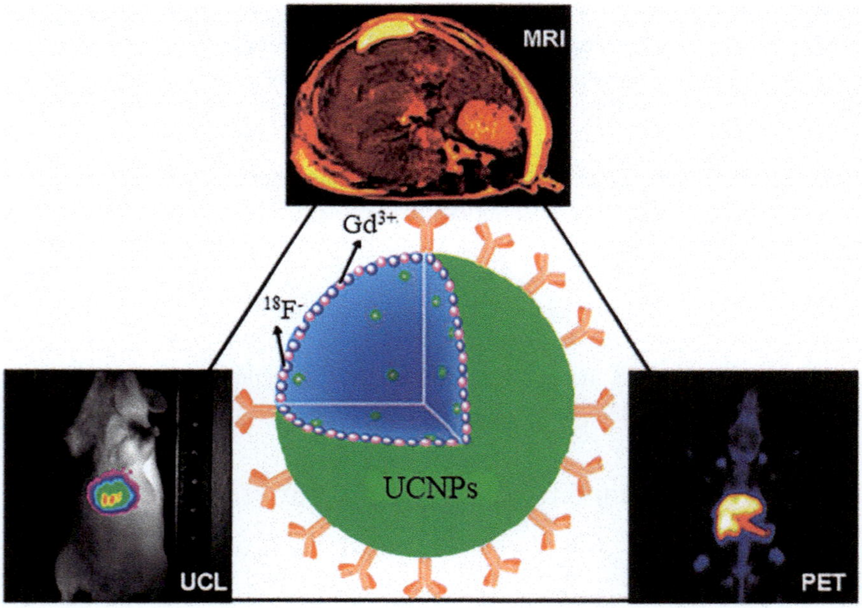

Figure 8.10 Scheme of multimodal imaging: a magnetic nanoparticle, a radionuclide, an optical tag, and a targeting moiety are integrated, which works as a multimodal molecular imaging platform. Source: Liu et al. (2011). Reprinted with permission of American Chemical Society.

8.8 Smart Upconverting Nanoparticles for Sensing

The UCNPs are widely used for temperature sensing, gas sensing, etc. (Achatz et al. 2011). Here, application of UCL materials for mechanical force sensing has been demonstrated. Mechanical performance is mostly characterized through external analytical methods that include atomic force microscopy (AFM), optical tweezing,

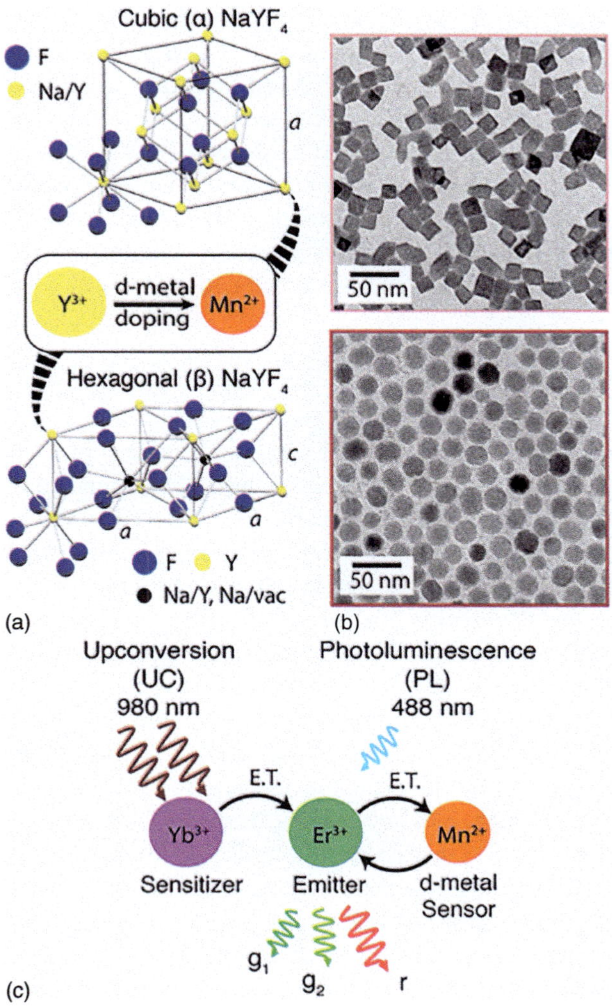

Figure 8.11 (a) Crystal structure of cubic α-$NaYF_4$ and hexagonal β-$NaYF_4$. Mn^{2+} occupies the Y^{3+} site in $NaYF_4$. (b) TEM images of Mn^{2+}-doped cubic (top) and hexagonal (bottom) nanoparticles. (c) Scheme depicting the interaction between lanthanide and d-metal ions: in UC process, Yb^{3+} ions absorb multiple NIR photons and transfer its excitation energy to Er^{3+}. Er^{3+} acts as an activator producing different emission peaks. Mn^{2+}-doping introduces additional crystal field sensitivity and facilitates forward as wells as backward energy transfer (ET) between Er^{3+} and Mn^{2+}. In PL, higher energy photons populate Er^{3+} and Mn^{2+} states to produce traditional Stokes emission. Source: Lay et al. (2017). Reprinted with permission of American Chemical Society.

and traction force microscopy (TFM). However, a low-dimensional platform is required to investigate forces inside electronic, material, and biological systems. As transition metal ions, such as Mn^{2+}, are sensitive to the external crystal field, if it is coupled with UCNPs, it can be used as a mechanical force sensor. The force sensor developed using this caters both color and intensity readouts in UC and down-shifting (DS) PL. To demonstrate force sensing, the $NaYF_4$ host lattice was used, as it is considered as one of the best upconversion hosts. Furthermore, the UC materials having a $NaYF_4$ host have two phases, cubic (α) and hexagonal (β), that possess different symmetries (see Figure 8.11a). Figure 8.11b shows TEM of cubic (upper image) and hexagonal (lower) phases of Mn^{2+} doped $NaYF_4$. For a bimodal luminescent material sensitizer, Yb^{3+} and emitter Er^{3+} are doped into the host lattice, where they occupy the Y^{3+} sites. To realize nN to μN sensitivity, Mn^{2+} ions were also substituted for Y^{3+} ions. As shown in Figure 8.11c, Mn^{2+} is integrated into the Er^{3+} green and red emission peaks. Thus, the external forces alter the crystal field of Mn^{2+} that also affects the energy sensitization processes, causing a change in the relative intensities of the emission spectra for both UC and PL.

8.9 Conclusion

In this chapter, various modifications in UCNPs, for example, making core–shell nanostructures, inorganic–organic hybrid UC material, UC-based MOFs, and magnetic UC nanoparticles, have been explored to increase the UC efficiency as well as to enable it for various strategic applications. The core–shell nanostructure helps in reducing the cross-relaxation and thereby increasing the UC efficiency. It also enables us in making multicolor UC or multimodal luminescent materials. The multishell UCNP nanosystems also open the way for multimodal imaging probes, which has a combination of UC luminescence, MRI, CT, and PET, and it is now possible within one single entity. The inorganic–organic hybrid material on the other hand increases the UC efficiency by complexing UCNPs with a suitable organic dye that absorbs NIR light efficiently and transfers its excitation energy to the activator ions and thereby increasing the UC efficiency by many folds. UC-based MOFs are also an example of inorganic–organic hybrid UC materials. In biomedical fields, the change in excitation wavelength to 800 nm in an inorganic–organic hybrid UC material is an important development in reducing the overheating that results from using relatively higher wavelength, i.e. 980 nm, which is strongly absorbed by water molecules. The integration of UC and magnetism brings multifunctionality in the material that is very appealing for biological imaging. In addition to this, the book chapter also highlights some specific applications of UC smart nanoparticles for anticounterfeiting, biological, and sensing.

The research in UC luminescence is getting advanced day-by-day, to which advancement in nanotechnology helps significantly. In the early and mid-2000s, it was possible to manufacture high-quality monodispersed UCNPs that opened up new promises for applications of UCNPs in biomedical applications. Now, the objective for the next few years could be a translation from academic research to real-world products. The synthesis and surface modification of UCNPs are at the core of this, and also, the standardized protocols and reference materials will help in enabling the comparison between the results of various groups worldwide.

References

Achatz, D.E., Meier, R.J., Fischer, L.H. et al. (2011). Luminescent sensing of oxygen using a quenchable probe and upconverting nanoparticles. *Angewandte Chemie International Edition* 50: 260–263. https://doi.org/10.1002/anie.201004902.

Boyer, J.C., Carling, C.J., Gates, B.D. et al. (2012). Two-way photoswitching using one type of near-infrared light, upconverting nanoparticles, and changing only the light intensity. *Journal of the American Chemical Society* 132: 15766–15772. https://doi.org/10.1021/ja107184z.

Chen, C., Li, C., and Shi, Z. (2016). Current advances in lanthanide-doped upconversion nanostructures for detection and bioapplication. *Advanced Science* 3: 1600029. https://doi.org/10.1002/advs.201600029.

Dwivedi, A., Singh, A.K., and Rai, S.B. (2014). Down-shifting and upconversion photoluminescence in Ho^{3+}/Yb^{3+} codoped $GdNbO_4$: effect of the Bi^{3+} ion and the magnetic field. *Dalton Transactions* 43: 15906–15914. https://doi.org/10.1039/c4dt01864h.

Garfield, D.J., Borys, N.J., Hamed, S.M. et al. (2018). Enrichment of molecular antenna triplets amplifies upconverting nanoparticle emission. *Nature Photonics* 12: 402–407. https://doi.org/10.1038/s41566-018-0156-x.

Haase, M. and Schäfer, H. (2011). Upconverting nanoparticles. *Angewandte Chemie International Edition* 50: 5808–5829. https://doi.org/10.1002/anie.201005159.

Hemmer, E., Quintanilla, M., Légaré, F. et al. (2015). Temperature-induced energy transfer in dye-conjugated upconverting nanoparticles: a new candidate for nanothermometry. *Chemistry of Materials* 27: 235–244. https://doi.org/10.1021/cm503799f.

Hu, L., Fan, Y., Liu, L. et al. (2017). Orthogonal multiplexed luminescence encoding with near-infrared rechargeable upconverting persistent luminescence composites. *Advanced Optical Materials* 5: 1700680. https://doi.org/10.1002/adom.201700680.

Kaboli, F., Ghazyani, N., Riahi, M. et al. (2019). Upconverting nanoengineered surfaces: maskless photolithography for security applications. *ACS Applied Nano Materials* 2: 3590–3596. https://doi.org/10.1021/acsanm.9b00549.

Lay, A., Wang, D.S., Wisser, M.D. et al. (2017). Upconverting nanoparticles as optical sensors of nano- to micronewton forces. *Nano Letters* 17: 4172–4177. https://doi.org/10.1021/acs.nanolett.7b00963.

Li, Y., Di, Z., Gao, J. et al. (2017). Heterodimers made of upconversion nanoparticles and metal organic frameworks. *Journal of the American Chemical Society* 139: 13804–13810. https://doi.org/10.1021/jacs.7b07302.

Liu, Q., Sun, Y., and Li, C. (2011). ^{18}F-labeled magnetic-upconversion nanophosphors via rare-earth cation-assisted ligand assembly. *ACS Nano* 5: 3146–3157. https://doi.org/10.1021/nn200298y.

Liu, J.N., Bu, W.B., and Shi, J.L. (2015). Silica coated upconversion nanoparticles: a versatile platform for the development of efficient theranostics. *Accounts of Chemical Research* 48: 1797–1805. https://doi.org/10.1021/acs.accounts.5b00078.

Marin, R., Halimi, I., Errulat, D. et al. (2019). Harnessing the synergy between upconverting nanoparticles and lanthanide complexes in a multiwavelength-responsive hybrid system. *ACS Photonics* 6: 436–445. https://doi.org/10.1021/acsphotonics.8b01381.

Park, W., Lu, D., and Ahn, S. (2015). Plasmon enhancement of luminescence upconversion. *Chemical Society Reviews* 44: 2940–2962. https://doi.org/10.1039/c5cs00050e.

Pei, W.B, Lau, R., Jing, Z. Y. et al. (2018). Nearly pure red color upconversion luminescence of Ln-doped Sc_2O_3 with unexpected RE-MOFs molecular alloys as precursor. *Inorganic Chemistry*, 57: 10511–10517. https://doi.org/10.1021/acs.inorgchem.7b02255

Qin, X., Xu, J., Wu, Y. et al. (2019). Energy-transfer editing in lanthanide-activated upconversion nanocrystals: a toolbox for emerging applications. *ACS Central Science* 5: 29–42. https://doi.org/10.1021/acscentsci.8b00827.

Rai, M., Mishra, K., Singh, S.K. et al. (2015). Enhanced red upconversion emission, magnetoluminescent behavior, and bioimaging application of $NaSc_{0.75}Er_{0.02}Yb_{0.18}Gd_{0.05}F_4$@AuNPs nanoparticles. *ACS Applied Materials & Interfaces* 7: 15339–15350. https://doi.org/10.1021/acsami.5b03218.

Ren, W., Lin, G., Clarke, C. et al. (2019). Optical nanomaterials and enabling technologies for high-security-level anticounterfeiting. *Advanced Materials* 32: 1901430. https://doi.org/10.1002/adma.201901430.

Safdar, M., Ghazy, A., and Lastusaari, M. (2020). Lanthanide-based inorganic–organic hybrid materials for photon-upconversion. *Journal of Materials Chemistry C* 8: 6946. https://doi.org/10.1039/d0tc01216e.

Shahi, P.K., Singh, A.K., Singh, S.K. et al. (2015). Revelation of the technological versatility of the $Eu(TTA)_3$Phen complex by demonstrating energy harvesting, ultraviolet light detection, temperature sensing, and laser applications. *ACS Applied Materials & Interfaces* 7: 18231–18239. https://doi.org/10.1021/acsami.5b06350.

Shahi, P.K., Singh, P., Singh, A.K. et al. (2017). A strategy to achieve efficient dual-mode luminescence in lanthanide-based magnetic hybrid nanostructure and its demonstration for the detection of latent fingerprints. *Journal of Colloid and Interface Science* 491: 199–206. http://dx.doi.org/10.1016/j.jcis.2016.12.034.

Singh, A.K., Singh, S.K., Mishra, H. et al. (2010). Structural, thermal, and fluorescence properties of $Eu(DBM)_3Phen_x$ complex doped in PMMA. *The Journal of Physical Chemistry B* 114: 13042. https://doi.org/10.1021/jp1050063.

Singh, A.K., Singh, S.K., Gupta, B.K. et al. (2013). Probing a highly efficient dual mode: down–upconversion luminescence and temperature sensing performance of rare-earth oxide phosphors. *Dalton Transactions* 42: 1065. https://doi.org/10.1039/c2dt32054a.

Singh, A.K., Singh, S., and Gupta, B.K. (2018). Highly efficient, chemically stable, and UV/blue-light-excitable biluminescent security ink to combat counterfeiting. *ACS Applied Materials & Interfaces* 10: 44570–44575. https://doi.org/10.1021/acsami.8b18997.

Tan, H., Xu, L., Gong, G. et al. (2019). Upconversion nanoparticles@carbondots@meso-SiO_2 sandwiched core–shell nanohybrids with tunable dual-mode luminescence for 3D anti-counterfeiting barcodes. *Langmuir* 35: 11503–11511. https://doi.org/10.1021/acs.langmuir.9b01919.

Tian, G., Gu, Z., Zhou, L. et al. (2012). Mn^{2+} dopant-controlled synthesis of $NaYF_4$:Yb/Er upconversion nanoparticles for in vivo imaging and drug delivery. *Advanced Materials* 24: 1226–1231. https://doi.org/10.1002/adma.201104741.

Wang, F. and Liu, X. (2009). Recent advances in the chemistry of lanthanide-doped upconversion nanocrystals. *Chemical Society Reviews* 38: 976–989. https://doi.org/10.1039/b809132n.

Wang, F., Deng, R., Wang, J. et al. (2011). Tuning upconversion through energy migration in core–shell nanoparticles. *Nature Materials* 10: 968–973. https://doi.org/10.1038/NMAT3149.

Wen, S., Zhou, J., Schuck, P.J. et al. (2019). Future and challenges for hybrid upconversion nanosystems. *Nature Photonics* 13: 828–838. https://doi.org/10.1038/s41566-019-0528-x.

Wilhelm, S. (2017). Perspectives for upconverting nanoparticles. *ACS Nano* 11: 10644–10653. https://doi.org/10.1021/acsnano.7b07120.

Xu, H., Cheng, L., Wang, C. et al. (2011). Polymer encapsulated upconversion nanoparticle/iron oxide nanocomposites for multimodal imaging and magnetic targeted drug delivery. *Biomaterials* 32: 9364–9373. https://doi.org/10.1016/j.biomaterials.2011.08.053.

Zhou, J., Liu, Z., and Li, F. (2012). Upconversion nanophosphors for small-animal imaging. *Chemical Society Reviews* 41: 1323–1349. https://doi.org/10.1039/c1cs15187h.

Zhou, J., Liu, Q., Feng, W. et al. (2015). Upconversion luminescent materials: advances and applications. *Chemical Reviews* 115: 395–465. https://doi.org/10.1021/cr400478f.

Zhou, D., Liu, D., Xu, W. et al. (2017). Synergistic upconversion enhancement induced by multiple physical effects and an angle-dependent anticounterfeit application. *Chemistry of Materials* 29: 6799–6809. https://doi.org/10.1021/acs.chemmater.7b01783.

Zou, W., Visser, C., and Maduro, J.A. (2012). Broadband dye-sensitized upconversion of near-infrared light. *Nature Photonics* 6: 560–564. https://doi.org/10.1038/NPHOTON.2012.158.

Zuo, J., Sun, D., Tu, L. et al. (2018). Precisely tailoring upconversion dynamics via energy migration in core–shell nanostructures. *Angewandte Chemie International Edition* 57: 3054–3058. https://doi.org/10.1002/anie.201711606.

9

Surface Modification and (Bio)Functionalization of Upconverting Nanoparticles

Yashashchandra Dwivedi

National Institute of Technology Kurukshetra, Department of Physics, Kurukshetra, Haryana 136119, India

9.1 Introduction

In the past decades, the surge of investigations surfaced in the development of light sources for the applications of lightening (fluorescence lamp, light emitting diode (LEDs) for household, offices, and street lightening), lasers, display devices, etc. The myriad of luminescent materials, resulting emission webbed over the entire visible to infrared regions, has been researched and explored by introducing lanthanide ions into oxide hosts. These materials are interesting for research because they exhibit the size depending on the properties having light thermal and chemical stability, which is a prerequisite for various applications such as optical coating, high-temperature materials, hydroresistance, etc. In recent years, various types of host materials are discovered and predicted to be suitable for hosting the lanthanide ions. Such host materials should possess superior mechanical and thermal properties with high coordination numbers, good ionic solubility, etc. Further, with the advent of newer synthesis techniques when such host materials are restricted in nanodimension, these materials are infused with the advantages of both, i.e. the host matrix and the electronic confinement effect. One of such materials is nanophosphors. The nanophosphors are a remarkable and surprising class of structures that have captivated the scientific community for decades (Dwivedi and Zilio 2014). Traversing over the disciplines of materials science and several engineering streams, the nanophosphors offer inquisitive challenges and astounding behaviors that bear significant investigations. Typically, the phosphors are in the form of crystalline powder with tiny size, having large absorption and emission cross section, and lifetime. The exceptional luminescence efficiency and the reduced radiative lifetime have rendered the nanophosphor compounds of significant intrigued. A phosphor requires essential ingredients, namely, a host material and an activator. In addition to the relative ionic sizes of the cations and anions, the sort of chemical bonding is one of the determinative variables in selecting the base material for phosphor preparation (Ebbesen et al. 1996). It is adequate to note that crystals with covalent, ionic, or blended ionic–covalent bindings

Upconverting Nanoparticles: From Fundamentals to Applications, First Edition. Edited by Vineet K. Rai.

are of utmost importance as base materials for phosphors. The dopant ions often produce luminescence emission; moreover, the impact of the host lattice cannot be underrated. It is conceivable to alter the optical properties of the luminescent ions by changing the host lattice. The main factors responsible for the distinctive optical properties of a luminescent ion in different host matrices are covalency and crystal field (Dwivedi and Rai 2017; Hänninen and Härmä 2011; de Bettencourt-Dias 2014). The covalency leads to alter the energy level position and the crystal field, which is associated with the electric field of the surroundings on the ion under consideration, leading to the splitting of energy levels. In general, the host materials ought to be transparent at the emission wavelengths of the luminescent ion, i.e. the bandgap of the host materials should be more than the emission energy. Further, to encourage the substitution of dopant ions and host cations, their sizes and the substituting cation's oxidation state ought to be preferentially the same as that of the dopant ion. The crystalline materials are comparatively favored over the host materials with the disordered structure for better luminescence intensity as the latter can cause an overall broadening in the emission peak (Dhoble et al. 2019). The phosphor characteristics are mainly common to all with little deviations from the ideal crystal structure because of tiny quantities of impurities or any structural defects. Thus, the key to the preparation of phosphors is the combination of a particular host matrix with a particular activator. Lanthanide (Ln) ions exhibit active f-shell electrons having fundamental importance as they impart unique optical and magnetic properties. Spectroscopic research on the lanthanide elements is a hot topic of investigation as it spawned vast commercial applications and helps to understand the basic mechanisms of several peculiar emissions dynamics. Ln elements are rich by several characteristic luminescent properties such as sharp emission lines, metastable states, and high quantum efficiency, which make them suitable for the fabrication of various optical components (Rai and Dwivedi 2012). Considering the emission range of the phosphors, the range is limited by the excitation energy as the emission energy is either the same or less than the excitation energy. However, there are several nonlinear processes through which we can produce the photons of high energy than the excitation energy; such process is referring as upconversion, and the materials possessing such emission is referred to upconverting materials. These upconverting materials are considered alternative fluorophores that have potential biomedical applications for in vitro and in vivo applications as the excitation energy may permeate in deeper depth, facilitating imaging with low-background noise. The following sections will briefly explain the key points related to the upconverting materials and processes involved.

9.2 Upconverting Nanomaterials

Photoluminescence (PL) is an optical emission process in which the active materials are excited by light itself and subdivided into fluorescence and phosphorescence. The optical excitation energy would be from the small energy sources such as a small lamp, arc, discharge, high energy sources such as lasers (continuous or pulsed lasers), etc. The optical excitation energy disturbs the electron distributions in the molecules. For this purpose, the energy of photons should have higher energies or the same value so that these can impart excess energy into the material in a process called photoexcitation. Further, the exciting energy may dissipate through the emission of photons or heat by transferring an electron from higher energy states to lower energy states. Overall, the intensity and spectral content (wavelength) of this photoluminescence directly measures the various material properties. Usually, the energy of an emitted photon is equal to or less than the excitation energy. However, as previously discussed, there are few processes through which one can produce high energy photons. These processes are nonlinear processes that involve the participation of the active dopant's virtual or real energy levels in the material. In general, the valance optical electron interacts with the lattice phonon or the transient virtual energy levels under the force field of high energy photons released by the suitable intense optical source such as lasers. In this process, the lifetimes of electrons in the excited states, charge carriers in the bands, and defect levels are among the few deciding parameters. Usually, the emission intensity from the fluorophore, after optical excitation, declines exponentially over time. The radiative lifetime characterizes the depopulation of the electronic level and the dynamics after ceasing the excitation. The spin multiplicities are the deterministic factor for the absorption and, subsequently, the emission process, either radiative or nonradiative. During the decay process, the excited electron may release energy radiatively, i.e. photon emission or nonradiatively, i.e. in the form of thermal energy. Both radiative and nonradiative lifetimes affect the observed luminescent intensity. In general, a photoluminescence decay rate is a sum of the radiative and nonradiative decay rates. Further, since nonradiative processes are thought to be more likely at the surface, the effective lifetime is often written as a combination of surface and bulk lifetimes (Gilliland 1997; Bunzli and Eliseeva 2010). Generally, the radiative lifetime of the optical electron is of the order of 10^{-9} (ns) seconds, and it further decreases for higher energy states. The lifetimes for various dynamical processes in the nanomaterials such as photon–electron, electron–electron, electron–phonon, and phonon–phonon interactions lie in the range of 10^{-15}–10^{-9} seconds. In contrast of continuous lasers, pulsed (nano-, pico-, or femtosecond) lasers generate nonequilibrium distribution of charge carriers and excitons because of high-intensity photons confined in nanosecond to femtosecond pulse. The use of intense laser pulses helps to excite the electrons to the targeted excited state without the intermediate states, which is difficult by low-intensity continuous wave (CW) lasers. The optical mixing of ps or fs pulses induces sum frequency generation (SFG) (see Figure 9.1) (Mukamel 2000; Niu and Marcus 2020; Shen 2020; Simon and Tittel 1997; Petrov and Noack 1995; Zhao et al. 2005). The spontaneous emission is mixed with another femtosecond pulse (gate) in a nonlinear crystal to generate the sum frequency of the gate and infrared (IR) pulse. The intensity of the generated signal is proportional to the intensity of the optical signal that is temporally overlapped with the gate pulse.

Several authors have investigated the frequency upconversion process by using the ultrashort lasers through intermediate energy states (Oliveira et al. 2021; Yao et al. 2016). Oliveira et al. reported qualitatively the similar emission spectrum by

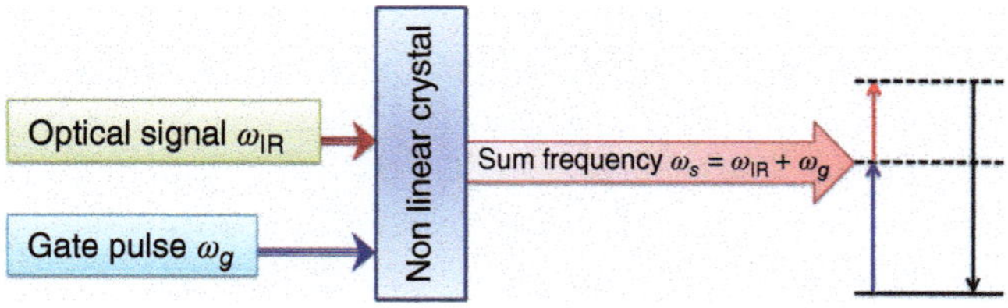

Figure 9.1 Pictorial representation of sum frequency generation emission process on ultrafast laser excitation.

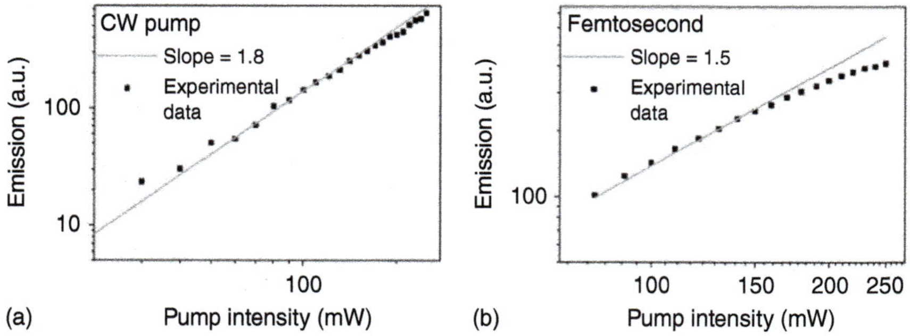

Figure 9.2 Upconversion emission intensities of upconversion nanoparticles (UCNPs) as a function of pump intensity: (a) CW and (b) femtosecond lasers (808 nm). Source: Modified from Yao et al. (2016). Reprinted with permission of American Chemical Society.

using CW or femtosecond excitation (808 nm), although their power dependence was found different (Oliveira et al. 2021). They observed that the femtosecond excitation power dependence slope was 1.5 in the log–log plot rather than 1.8 as estimated in the case of CW excitation, thereby indicating a different mechanism for filling the states at high-intensity pulse. Yao et al. investigated the nonresonant multiphoton absorption of femtosecond laser pulse, effect of laser phase, and polarization on the upconversion emission intensity (Yao et al. 2016) (see Figure 9.2).

However, in contrary to the discussed processes, there is another low-energy excitation upconversion process also in practice, which occurs mostly in lanthanide ions. Lanthanide (Ln) ions exhibit unique luminescent properties, including the ability to convert long-wavelength near-infrared (NIR) radiation into shorter wavelengths in the visible or ultraviolet (UV) region. Several exclusive optical upconversion phenomena were invented in Ln elements, e.g. APTE (Addition de Photons par Transfert d'Energie) (Auzel 1990), cooperative emission, multiphoton absorption, photon avalanche, etc. With the advent of nanotechnology, the efficiency of these processes, while confining these ions in nanodimensions, has been improved. Despite the myriad potential applications of Ln-doped materials, several perplexing problems are yet to be solved, which evoke much research interest among the researchers. The major technological challenges are to know the optical–magnetic interaction, optimizing black body emission (heat generation), single-phase material synthesis in bulk amount, attachment of biomolecules and the properties modifications, multifunctional luminescent materials compatible for biomedical purposes, etc. Further, the wide use of Ln elements and henceforth its ecological and biological impact, as true toxicity level of these ions, are still undetermined especially in nanomaterials.

9.3 Surface Modification

Several materials are enriched with superior thermal, mechanical, and optical transmission properties; however, they did not produce optical emissions. The problem related to optical emission and color modulation in such a matrix is one of the core issues of the research. By doping or dispersion of suitable active ions, materials may yield the emission, which varies with the doping ion. The emission may be tuned by altering the valence state of the dopants. However, to meet the industry demand and international perquisites for efficient and futuristic devices, a material emitting in multiple spectral regions is required, which can be realized by codoping of an adequate combination of the active ions in the matrix. Nevertheless, the issue pertaining to the high-end concentration limits the multicomponent codoping in the same lattice. Such optical problems are the real challenge to design the new materials and conceptualize the device accordingly. One of the possible remedies to improve the efficiency of the materials by avoiding the saturation limit and having the advantage of multispectral emission is to use the combination of two different synthesized materials or to use the individual material's combinations "commonly referred to as composite materials." The issues mentioned above may be resolved by using the multicomponent doping in the matrix or developing composite materials such as organic–inorganic and inorganic–inorganic (Rashmi and Dwivedi 2019a, b).

Advancement in the field of materials science has taken the novel lead since the advent of nanocomposites. The nanostructured materials, organic–inorganic composites, and inorganic–inorganic composite materials have become undoubtedly a dynamic research field in the area of materials science. However, the properties of each individual are constrained by constituent elements and crystal structure. Combining the properties of distinctive nanomaterial requires an understanding of the collective behavior of individuals. The composite materials possess the properties of all sub-units and, possibly, new properties arising from their interactions. Nanocomposites are ultrafine particles of two different materials (Pilato and Michno 1994). Nanocomposites originate from suitable combinations of two or more nanomaterials with some suitable technique. This results in materials with unique physical properties and the wide potential applications in diverse areas (Kumari and Dwivedi 2016). Novel properties of the nanocomposites can be determined from the effective combination of the characteristic properties of parent constituents into a resultant material. Nanocomposites, which are a viable combination of two or more inorganic materials, are anticipated to display unique properties emerging from the combination of each component (Kumari and Dwivedi 2018). Organic–inorganic composites have attracted special consideration, as they can combine the advantages of both components and may offer extraordinary properties through reinforcing or modifying each other (Dwivedi and Zilio 2013). Inorganic nanoparticles of distinctive nature and size can be combined with the organic dye, giving rise to a host of nanocomposites with interesting optical properties and imperative potential application. These nanocomposite materials are particularly vital because of their bridging role between the world of organic molecules and that of inorganic materials. Many challenges have been found that constrained the ability to extract the advantages of these materials with special and unique properties. Among them, the need for organic molecules coating on the surface of the nanoparticles, the small particle size, and relative high chemical sensitivity, makes deposition challenging. The term hybrid material is utilized for numerous distinctive frameworks spanning over a wide area of different materials (inorganic–organic) blended over the molecular scale. The hybrid constituents may interact with weak bonds such as hydrogen, van der Waals, or ionic bonds (class 1) or through strong chemical bonds such as covalent or ionocovalent bonds (class 1) (Romero and Sanchez 2006). This leads nearly no constrains to the combination of inorganic and organic components in the synthesis of hybrid materials. Therefore, the novel materials with unique compositions can be developed. The organic molecules may be uniformly coated or nonuniformly connected with the surface of the nanoparticles, depending on the chemical environment and the interaction between the species. The surface functionalization of nanoparticles helps to control precisely the overall physical and chemical properties, such as roughness (Patelli et al. 2018), surface charge (Singh et al. 2019), biocompatibility (Shim et al. 2002; Elsehly et al. 2020), and reactivity (Elsehly et al. 2020) of nanoparticles, which further help to be used for biomedical applications.

Several surface modification methods are reported for specific design of physicochemical, toxicological, and pharmacological properties to improve the biocompatibility of nanomaterials. In addition to the increase of the biocompatibility, the surface functionalization helps to modulate the immune response against the particles. The general mechanism of such response is the interaction of nanomaterials with the specific proteins, such as opsonin, complement proteins, immunoglobulin, fibronectin, and apolipoproteins, which trigger the immune response (Qie et al. 2016; Barbero et al. 2017; Boraschi et al. 2017). Primarily, the four major approaches are used for the surface functionalization, namely, (i) ligand engineering (Chen et al. 2014; Wang et al. 2010; Duong et al. 2018), (ii) layer-by-layer assembly (Zhang 2015), (iii) ligand attraction (Cheng et al. 2011), and (iv) surface polymerization (Wu et al. 2011; Alberti et al. 2015; Liu et al. 2013). The homo- or hetero-bifunctional cross-linkers are used to bind with solid surface by means of active terminal functional groups such as –OH, $–NH_2$, –COOH, –CH, –C=O, etc. Initially, the solid surface is coated with these organic linkers to provide the active binding site for biofunctionalization. In the case of silica nanoparticles, usually, aminosilane is introduced on the surface, while for metal nanoparticles, cross-linkers having thio-carboxylic acids or –SH or $–NH_2$ functional groups are used. In the case of metal oxides, the ligand exchange method is commonly used, in which the surface molecules are substituted with active functional groups (Korpany et al. 2016). One of the major problems to quantify the amount of exchange taking place has been addressed by Davis et al. (2016). Davis et al. have used a radiotracer technique to track the exchange of a radiolabeled ligand with hydrophilic ligands on the nanoparticle surface. Zhou et al. reported an oxide-assisted coating-etching process to remove the bioincompatible capping ligands from the surface of metal nanoparticles by means of a cuprous oxide layer growth on the nanoparticles, which competes with the coated ligands followed by the etching process (Zhou et al. 2020). They demonstrated that after coating and etching of Cu_2O on gold nanorods, the cetyltrimethylammonium bromide (CTAB) ligands attached with the surface can be readily replaced by a biocompatible ligand. Pristine carbon nanotubes possess hydrophobic and inert nature, resulting in low reactivity, poor dispersibility, and self-aggregation. Such properties limit their potential applications for biomedical purposes. However, the use of certain chemical radicals on the surface of nanoparticle provides enough hydrophilic nature for the biomedical applications (Elsehly et al. 2020). The surface functionalization (organo-trialkoxysilanes or organo-trichlorosilanes) helps to penetrate the hydrophobic drugs into *mesoporous silica* for targeted drug delivery and also delays the release of certain drugs from mesoporous channels (Ficai and Grumezescu 2017). The surface functionalization is a helpful drug delivery tool for disease such as cancer, in which a cancer receptor biomolecule or antibody is attached on the gold nanorods to target the growth factor receptors on cancer cells (Bose and Bandyopadhyay 2017; Black et al. 2012). Another emerging area of the surface biofunctionalization application is tissue engineering (for repair or replace damaged or injured tissues and organs) and regenerative medicine. The surface biofunctionalization promotes cell attachment, proliferation, and differentiation (Casanova et al. 2020). For regenerative purposes, the functional groups at the biomaterial surface act as recognition sites for cellular interaction and control to improve the cell growth and tissue repair and enable restoring the tissue function (Bishop et al. 2014).

9.4 Biofunctionalization of Upconverting Materials and Applications

Unique luminescent properties of the upconverting materials can be used for biodiagnosis by biolabeling, biosensing, bioimaging, and multiple imaging modality as well as therapeutic treatments including photothermal and photodynamic therapy (PDT), bioreductive chemotherapy, and drug delivery. Comprehensive and cogent reviews are available on the upconverting materials for such applications (Gee and Xu 2018; Sanità et al. 2020). The two crucial parameters, namely, toxicity and cellular uptake, are to be addressed before the use of nanomaterials for biomedical applications (Rees et al. 2019; Santos-Rasera et al. 2019; Zhang et al. 2019). Usually, the nanocarriers accumulate in cancer cells. When nanocarriers are in proximity of the tumor site (passive accumulation), they alter the permeability of tumor blood vessels (Maeda 2015). The other approach is the active targeting by means of the biofunctionalization having strong affinity and site selectivity on the tumor cells (Bertrand et al. 2014; Yoo et al. 2019). There are several biointerest molecules used for the functionalization purposes. Figure 9.3 represents a schematic of a few possible biofunctional species, which are in practice to manipulate the core/shell upconverting nanoparticle (UCNP) properties. These molecules would be peptide (a short chain of

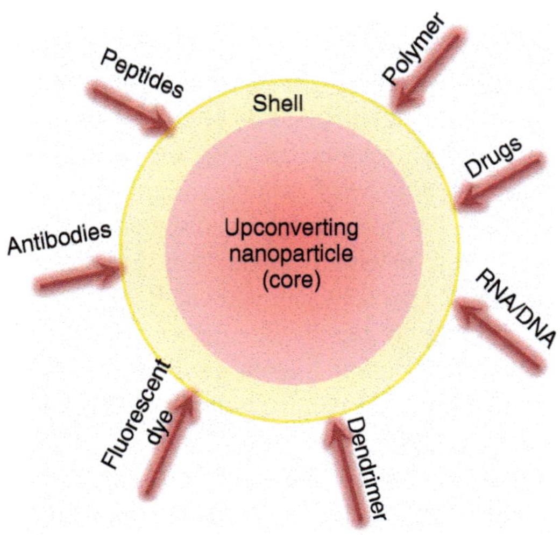

Figure 9.3 The schematic of possible biofunctional molecules and core@shell structure.

amino acids), fluorescent dyes, nucleic acids, enzymes, carbohydrates, dendrimers, biodegradable polymers, poly(ethylene glycol) (PEG), and many more (Conde et al. 2014). Among these, PEG is one of the very important surfactant as it possesses long half-life while propagating in blood stream or a particular organ (Kommareddy et al. 2005; Sperling and Parak 2010). This polymer not only stabilizes the nanoparticle but also prevents the other molecules to bind by steric effects and hence avoids binding to blood proteins and cells (Bertrand and Leroux 2012; Bridot et al. 2007).

The possibilities of tailoring UCNPs for biological labeling and targeting are the subject of main research (Bridot et al. 2007). One of the major issues with bioimaging is the absorption and scattering of UV/visible light by the biological tissues, which restrict the skin penetration depth for bioimaging (Figure 9.4). Usually, the light of wavelength 150–380 nm penetrates up to ~0.3–0.5 mm depth. Similarly, the light having wavelengths 545–600, 600–650, 650–950, and 950–1100 nm can penetrate up to ~0.5–1.0, ~1.0–2.0, 2–3, and 1 mm depth in skin. A comparison of the absorption spectra of hemoglobin (Hb) and water molecule reveals that the absorption of water at 800 nm is at least 20 times less than the 980 nm wavelength (Shen et al. 2013). Upconversion nanomaterials provide the partial solution of this problem, as it provides advantages such as background fluorescence interference, easy photolysis, weak penetration, less fluorescence time, and long-term toxicity over the conventional organic dyes and quantum dots (Gee and Xu 2018).

Keeping in mind that the quantum dots are photostable, yield bright emission, and possess broad absorption are useful for the multicolor detection. Issues such as photoblinking, nonradiant dark particles, and potential cytotoxicity limit the use of quantum dots in biomedical purposes (Nirmal et al. 1996; Yao et al. 2005; Derfus et al. 2004). However, surface passivation with organic ligands brings reduction in the emission intensity because of the interaction of electronic states of Ln ions with the high vibrational states of chemical bonds such as O–H, C–H, or N–H, leading to the nonradiative relaxation. Further, some of the chemicals used for the controlled growth of nanoparticles create hydrophobic surface, thereby reducing the

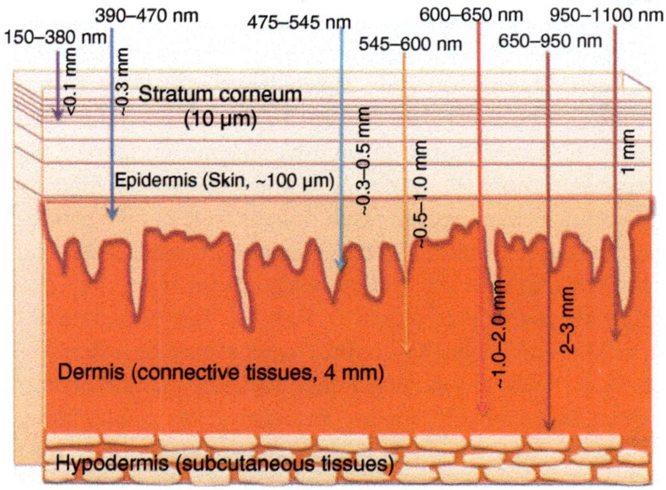

Figure 9.4 Penetration of various optical wavelength penetration in the skin (penetration depth not scaled).

bioanalytical application opportunity. Therefore, it becomes necessary to replace the hydrophobic ligands by the hydrophilic ligands. Although, the larger size of nanoparticles lowers the surface-to-volume ratio and the loss of emission quenching. Therefore, to meet the practical bioapplicability a balance in particle size and brightness is needed.

Polysiloxane-coated Gd_2O_3 nanoparticles conjugate to PEG molecules show at least two times higher relaxivities than that of Gd–DTPA (diethylene-triamine-penta acetic acid). The potential for these nanoparticles as imaging agents has been demonstrated in some studies (Fortin et al. 2007). Bridot et al. reported the multimodal Gd_2O_3 (~4.6 nm) nanoparticles encapsulated within a shell of fluorescent organic fluorophores (Bridot et al. 2007). Gd_2O_3 is considered as a very attractive choice as a host to dope with the different RE ions for achieving single-phase bifunctionality (i.e. paramagnetism and multicolor emission) (Das et al. 2012). The contrast agents derived from ultra-small nanotubes/nanorods have been reported to be highly promising for intracellular imaging (Lu et al. 2004; Kam et al. 2004). Nanotubes/nanorods of 20–100 nm size range are believed to be best suited for cellular uptake, biocompatibility, and ultimate elimination from the body (Sitharaman et al. 2005). Gd^{3+} ion-loaded single-walled carbon nanotubes (SWCNTs) for magnetic resonance imaging (MRI) have also been reported (Mackeyev et al. 2009). A variety of rare-earth hydroxide and oxide compound nanotubes, such as $Y(OH)_3$, $Dy(OH)_3$, Eu_2O_3, CeO_2, and $Tb(OH)_3$, have been synthesized through the hydrothermal process (Xu et al. 2003; Fang et al. 2003; Wang et al. 2003; Mao et al. 2008; Bai et al. 2005). Compared with other rare-earth compounds, there have been few reports on the synthesis of nanotubular gadolinium hydroxide, which can be used as catalysts, sorbents, MRI contrast agents, and precursors of Gd_2O_3 (Yang et al. 2007; Liu et al. 2007; Chang et al. 2006; Du and Tendeloo 2005; Lee et al. 2009).

It was well documented that the diagnostic and therapeutic effects of the nanoparticles were enhanced by using the appropriate combination of compounds having functional groups sensitive to binding with the selected enzymes or receptors, leading to biocatalytic action (Prinzen et al. 2007). Such customized multimodal nanoparticles are well suited for targeted drug delivery, bioimaging, and producing synergistic effects. Bioimaging techniques such as MRI and fluorescence imaging produce an excellent three-dimensional spatial resolution, as well as high screening sensitivity. Nanoparticle-based multimodality imaging provides additional information by producing simultaneous production of signals for more than one imaging technique using only a single injection. In light of this, nanoparticles combining fluorescent probes and magnetic contrast agents may be extremely useful for in vivo diagnosis of disease and surgical image guidance because of their ability to detect both optical and magnetic modes. Because of this combination, the issues related to the low sensitivity and anatomical background noise of MRI in fluorescence imaging can be synergistically compensated. Further, it leads to diagnostic accuracy and helps in the development of therapeutics. Considering the importance of multimodality nanoparticles, the integration of a fluorescent probe and an MRI contrast agent into a single hybrid imaging system is attracting the attention of researchers. Several new hybrid materials were developed, such as a Gd-fluorescent dye molecule coated on mesoporous silica nanoparticles (MSNs), Cy5.5-conjugated superparamagnetic iron oxide, quantum dots functionalized with Gd chelates using lipids, etc. (Liu et al. 2008; Tsai et al. 2008). The multifunctional nanoparticles containing MRI contrast agents along with conventional fluorescent downconverting materials such as dyes, quantum dots, semiconductor nanoparticles, etc., however, have some intrinsic limitations (Wang et al. 2005). These downconverting materials required excitation by UV/blue radiations, which restricts the deep skin penetration as the major proportion of these high-energy photons are absorbed in the skin. Additionally, the common problem with the use of organic dyes is the poor photostability, thermal stability, broad absorption and emission, autofluorescence, etc. Also, the toxicity of organic and inorganic quantum dots is not well explored yet. The one suggested solution of the problem is the use of NIR radiation on upconverting materials (Mai et al. 2006; Li and Zhang 2006). Most of the biomolecules possess least absorption in the near-infrared region; hence, the autofluorescence and infrared-induced photodamage is minimal. NIR light can penetrate into tissues more deeply because of these factors (Wang et al. 2005; Chatterjee et al. 2008; Jalil and Zhang 2008; Li et al. 2008). Furthermore, Ln-based upconverison nanomaterials are photostable, as these are well shielded because of the presence of an outer subshell. The contrast agents used in MRI, to distinguish normal and affected tissues, are composed of paramagnetic ion complexes (e.g. Gd^{3+} complexes) or superparamagnetic magnetite particles. The trivalent gadolinium ion is the preferred choice as it exhibits high magnetic moment and is most stable. The paramagnetic ion complex contrast agents shorten the T_1 (or longitudinal) and T_2 (or transverse) relaxation time. Although the T_1- and T_2-weighted contrast agents are in use, the MR signal produced by the T_2-weighted contrast agents is significantly affected by calcification or metal deposits (Na et al. 2007; Bulte and

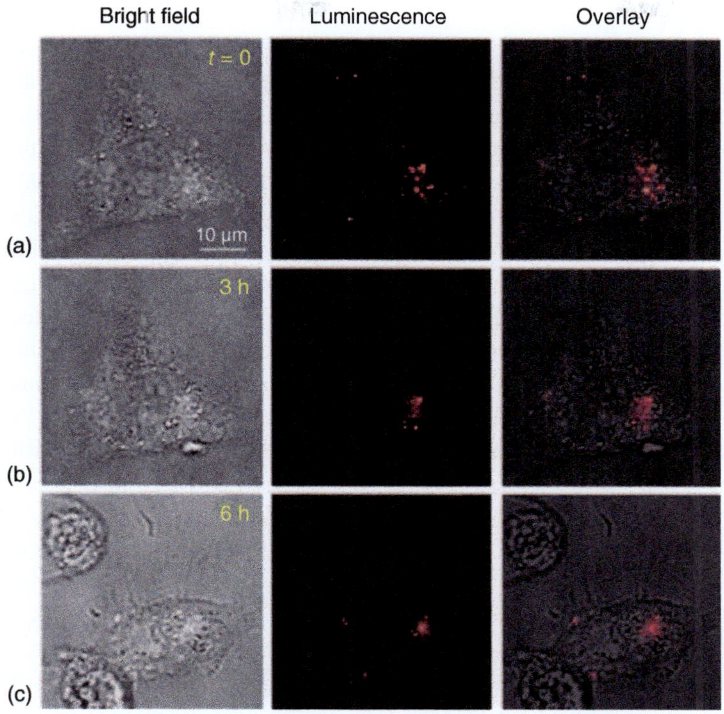

Figure 9.5 Live tracking images at (a) 0, (b) 3, and (c) 6 hours time intervals on 980 nm excitation. Source: Nam et al. (2011). Reprinted with permission of John Wiley and Sons.

Kraitchman 2004). The T_1-weighted MRI contrast agents were found to have higher efficiency with surface passivation when compared with the conventional contrast agents used (Rieter et al. 2007; Taylor et al. 2008 Abel et al. 2009). It is important to note that the molecular imaging modalities are useful to extract the complementary and rarely redundant information. Nam et al. attempted to track the real-time nanoparticles in HeLa cells (Nam et al. 2011). They used 30 nm sized Er:Yb:$NaYF_4$ UCNPs coated by PEG–phospholipids. Figure 9.5 shows some of the images taken from the real-time movie acquired continuously for six hours at the frame rate of 20 frames s^{-1}.

Lu et al. demonstrated the significant multiplexing capacity of optical time-domain coding by tuning the luminescence lifetimes of the $(Eu(TTFA)_3$ complex (europium tris(thenoyltrifluoroacetonate)) encapsulated in polystyrene microspheres (Lu et al. 2014). They developed a fast algorithm suitable for the microsecond region and compatible with the rapid scanning cytometry. This technique is applied for probing of pathogen DNA strands. The hexagonal (β) phase of $NaYF_4$ is reported to be one of the most efficient upconverting nanomaterials while doped with Er:Yb or Tm:Yb under infrared excitations. Wang et al. reported the detection of DNA by using the fluorescence probe $NaYF_4$:Yb:Er UCNPs (Wang and Li 2006; Wang et al. 2006). The magnetic-field-assisted biochemical separation and concentration technique is used for nucleic acid-bioconjugated nanoparticles to detect DNA up to 7.8 nM concentration. The magnetic nanoparticle was bioconjugated with the 59-propylthiol-terminated capture DNA, 39-AAT-TGA-GGA-GAAAGA-A10-(CH2)3-SH, while UCNPs were covalently linked to the 39-propylthiol-terminated DNA, 59-TCC-ATG-CAA-CTC-TAA-A10-(CH2)3-SH. Nyk et al. performed the PL imaging in vitro and in vivo using Tm:Yb:$NaYF_4$ UCNPs having crystalline size 20–30 nm (Nyk et al. 2008). They reported the animal imaging studies on mice injected intravenously with Tm:Yb:$NaYF_4$ UCNPs under 980 nm excitation and observed the high-contrast PL imaging in vivo. Further, Chatterjee et al. reported the animal imaging by using the polyethylenimine (PEI)-coated $NaYF_4$:Yb,Er UCNPs (see Figure 9.6) (Chatterjee et al. 2008). They found that the fluorescence intensity varied with tissues, the muscles with skin removed, showing a much stronger fluorescence from deep injections than intact skin at similar depths. Wang et al. synthesized the $NaYF_4$:Yb,Er UCNPs coated with thin layer of SiO_2 to form core@shell nanoparticles and modified with

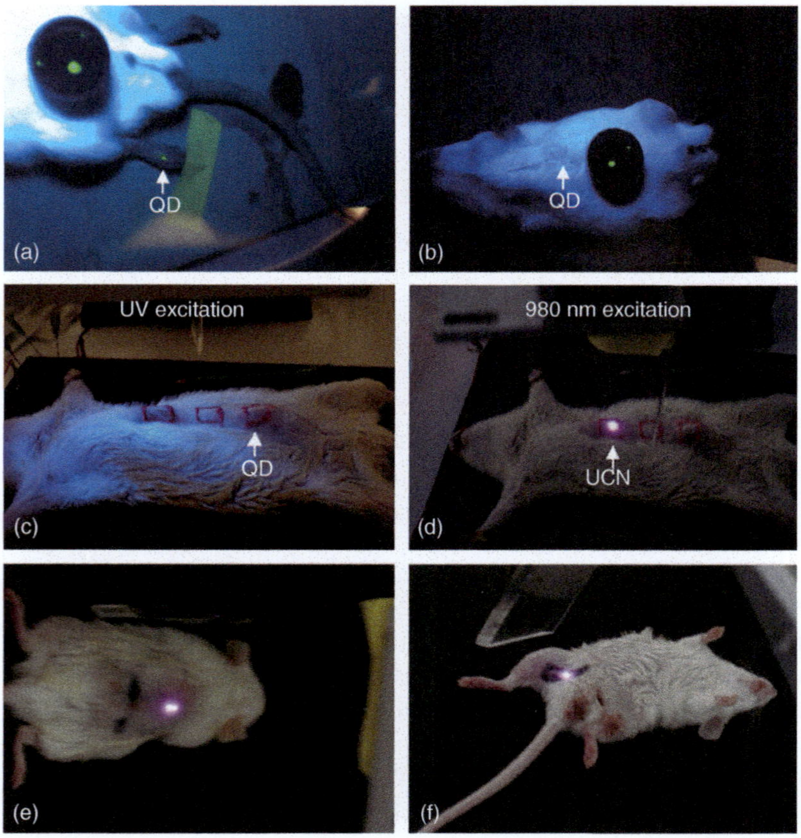

Figure 9.6 In vivo imaging of rat: quantum dots (QDs) injected into the translucent skin of foot (a) show fluorescence, but not through the thicker skin of back (b) or abdomen (c); PEI/$NaYF_4$:Yb,Er UCNPs injected below the abdominal skin (d), thigh muscles (e), or below the skin of back (f) show luminescence. QDs on a black disk in (a and b) are used as the control. Source: Chatterjee et al. (2008). Reprinted with permission of Elsevier.

amino groups (Wang et al. 2009). The addition of rabbit anti-CEA8 antibodies on the surface of nanoparticles is used as fluorescent biolabels for the detection of carcinoembryonic antigen. It was observed that the synthesized hybrid nanoparticles show affinity to the surface of the HeLa cells and yield bright green emission on 980 nm excitation. Zhang et al. developed Ag@SiO_2@Y_2O_3:Er nanostructures for potential applications in bioimaging (Zhang et al. 2010).

For the bioimgaing purposes, mainly three methods are in practice, namely: fluorescence imaging (high sensitivity and spatial resolution), nuclear medicine like positron emission tomography (PET) (use for whole-body imaging, picomolar sensitivity but low spatial resolution), and MRI (low sensitivity and high resolution), having different spatial resolutions, imaging depths, and areas of applications. Soukka et al. presented a comprehensive review on the importance of UCNPs for bioanalytical assays (Soukka et al. 2008). The UCNPs provide the advantages of narrow band emission, which is far from excitation to avoid any cross talk, and the organic ligand does not absorb the infrared excitation. Further, the efficiency of upconversion fluorescence resonance energy transfer is the function of distance and it can be manipulated as per the experimental requirement. Zhou et al. functionalized the Gd:Yb:Er:$NaYF_4$ nanoparticles with Fluorine-18 for multimodality PET, MRI, and laser scanning upconversion luminescence imaging (Zhou et al. 2011). The presence of high amount of Gd (60%) induces paramagnetic relaxivity for MRI and biolabeling, Fluorine-18 facilitates the use of three complementary techniques. The authors have performed in vivo PET/MR whole-body imaging of small animals and ex vivo upconversion imaging. Tu et al. reported 2-aminoethyl dihydrogen phosphate-functionalized Tb:Ce:$NaYF_4$ for time-resolved Förster resonance energy transfer (FRET) biosensor to detect avidin up to 5 nM (Tu et al. 2011). In this report, authors have selected the Tb lanthanide as it possess long lifetime (~few milliseconds), which eliminates the interference of short-lived background fluorescence while detecting by using the time-resolved emission technique. Authors have observed the significant reduction in avidin lifetime from 2.2 to 0.7 ms with increasing the concentration from 5 to 500 nM (see Figure 9.7).

Xia et al. have reported the application of tiny water-soluble $NaYF_4$:Yb:Tm@Fe_xO_y (20 nm core and 5 nm shell) nanocrystals for T_2-enhanced MR and NIR-to-NIR upconversion luminescent imaging (see Figure 9.8) in vivo of the lymphatic system and toxicity evaluation by the (3-(4,5-dimethylthiazol-2-yl)-2,5-diphenyltetrazolium bromide) tetrazolium (MTT) assay, half maximal (50%) inhibitory concentration (IC50) values, and histological analysis of lymphatic node sections (Xia et al. 2011). The value of IC50 was estimated to be about 295 and 190 mg/ml, after 24 and 48 hours, respectively.

Guller et al. evaluated the cytotoxicity of UCNPs coated with PEI by using six types of layered polymer coatings (Guller et al. 2018). It was observed that the coating of poly(acrylic acid-co-maleic acid) (PAMA), Na-salt dextran sulfate, dextran, sulfo-chitosan, and chitosan reduced the cytotoxicity to human skin keratinocytes under acute (24 hours) and subacute (120 hours) exposure. It was reported that without the coating of biodegradable polymers, nanoparticles significantly damage, i.e. ~50% and 85% after a 24 and 120 hour incubation of the transformed human keratinocyte

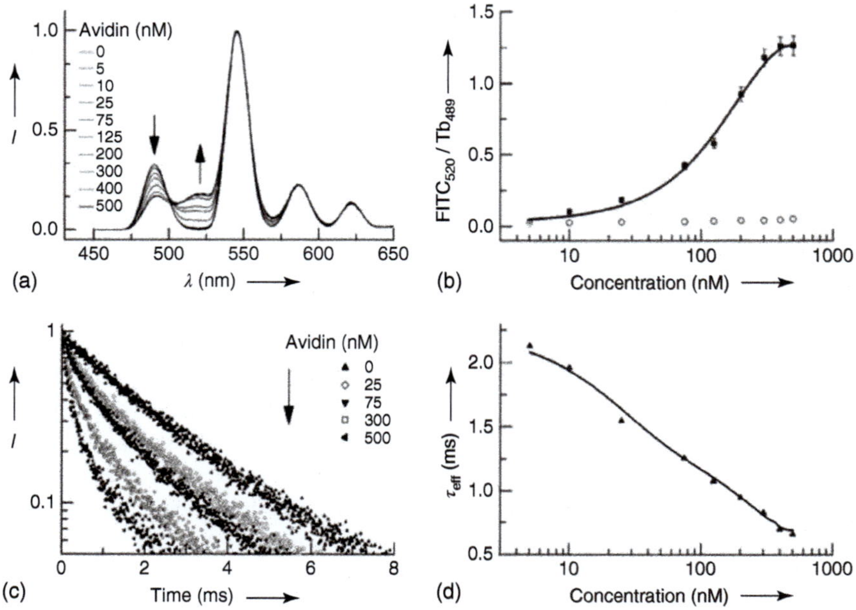

Figure 9.7 (a) Time-resolved fluorescence-Förster resonance energy transfer (TR-FRET) spectra of avidin at different concentrations; (b) calibration curve for the integrated PL intensity ratio $FITC_{520}/Tb_{489}$ versus avidin concentration. (c) Decay curves and (d) effective PL lifetime of $^5D_4$ of Tb^{3+} at different avidin concentrations. Source: Tu et al. (2011). Reprinted with permission of John Wiley and Sons.

cell (HaCaT) cells. However, the polymer coating leads to reduction in the emission intensity, and hence, a trade-off between cytotoxicity and brightness is observed.

The target-specific imaging was performed by Lee et al. (2013). Authors have used the radio-labeled arginine–glycine–aspartic acid functionalized Er:Yb:$NaGdF_4$ nanoparticles (~30 nm) on human-glioma (a type of tumor that occurs in the brain and spinal cord) U87MG cell, which is a commonly used cell line. These surface-functionalized nanoparticles were specifically localized on the surface of U87MG cells as confirmed by the T_1-weighted MR images, inductively coupled plasma mass spectrometry and transmission electron microscopy. Overall results indicate the high specificity for α(v)β(3) integrin expressing U87MG tumor cells and xenografted tumor models.

Similar to Er:Yb- and Tm:Yb-codoped systems, the Nd:Yb-codoped system is also predicted for its application in bioimaging, as trivalent Nd ions possess large absorption cross section at 808 nm ($1.2 \times 10^{19}\,cm^2$) as compared to that of the Yb^{3+} ($1.2 \times 10^{20}\,cm^2$ at 980 nm) ions. The excitation with 808 nm is beneficial as it is more penetrating than 980 nm as water molecules strongly absorb near this region, causing less penetration and NIR-induced heating in the tissues. Wang et al. addressed the issues of NIR-mediated heating effect as it damages the tissues (Wang et al. 2013). Authors developed the $NaGdF_4$:Nd,Yb shell of 4 nm thickness on the $NaGdF_4$:Er,Yb core (~13.5 nm) and evaluated the heating effect on human embryonic kidney (HEK) 293T cells. The cell vitality result indicates that under

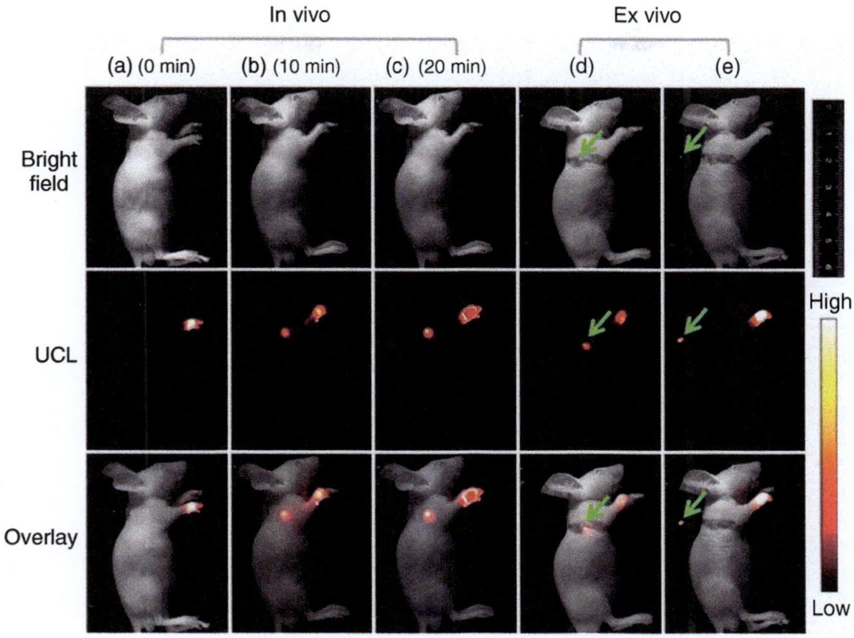

Figure 9.8 In vivo upconversion imaging after injection with $NaYF_4$:Yb:Tm@Fe_xO_y nanocrystals at various times: (a) 0 minutes, (b) 10 minutes, and (c) 20 minutes. (d and e) Upconversion luminescence (UCL) ex vivo images for 40 minutes after injection (green arrow indicate the lymphatic nodes). Source: Xia et al. (2011). Reprinted with permission of Elsevier.

identical laser irradiation and time interval, almost no cells were survived on 980 nm laser irradiation, which is in contrast to 808 nm excitation where most of the cells survived (Figure 9.9). Esipova et al. presented a comprehensive review on polyanionic dendrimers having multiple carboxylate sites at the periphery to be attached with nanoparticle surface (Esipova et al. 2012). Authors have used the upconversion emission to excite chromophore such as porphyrin dendrimers and polyglutamic dendritic via energy transfer and developed the ratiometric pH nanosensors. The in vivo experimental results indicate the imaging of cortical vasculature down to 400 μm under the tissue surface under low-power CW excitation sources.

Another interesting area, other than imaging, where upconverting nanomaterials show the potential to be applicable is PDT in which the photosensitizer is excited under IR exposure and resulting energy in visible region, which further cascades to the photosensitizer to produce cytotoxic reactive oxygen species (singlet oxygen ($^1O_2$)). The reactive oxygen causes photocytotoxicity-induced cell death either via apoptosis or necrosis. This technique is extremely useful for cancer-like diseases. For a detailed review on PDT, readers are referred to some of the cited references (Dolmans et al. 2003; Daniell et al. 1991; dos Santos et al. 2019; Benov 2015). Pham et al. carried out investigation on the dual photosensitizers (Rose bengal [RB] and Chlorin e6 [Ce6]) in silica-coated core–shell $LiYbF_4$:Er@$LiGdF_4$ nanoparticles

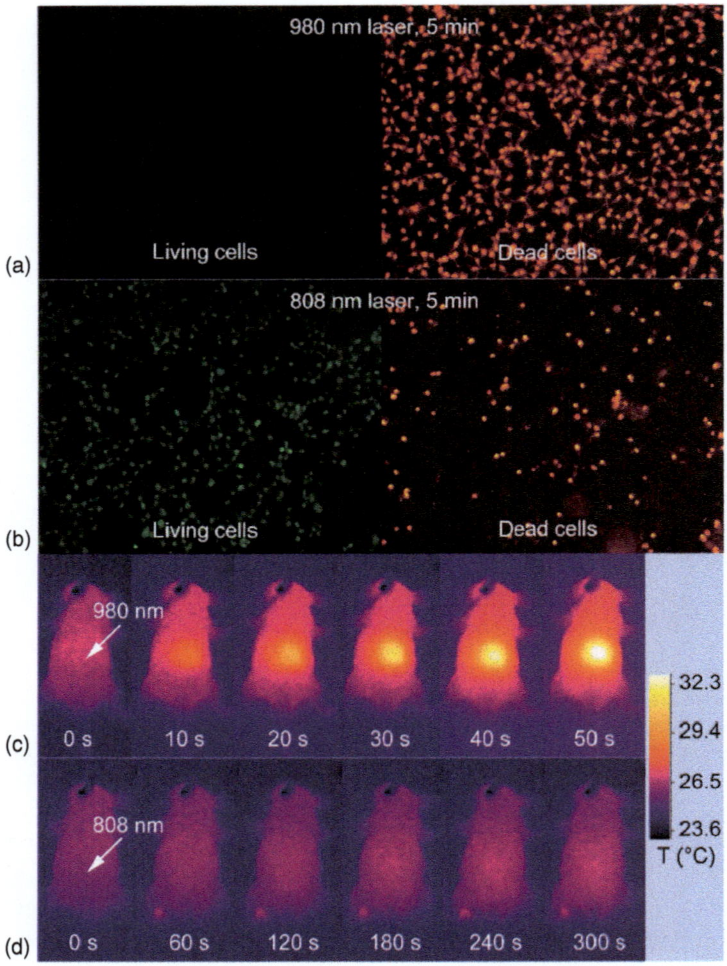

Figure 9.9 In vitro and in vivo images of HEK 293T cells after 5 minutes irradiation of 980 nm (a) and 808 nm laser (b) and infrared thermal image under (c) 980 nm laser irradiation for 50 seconds and (d) 808 nm laser irradiation for 300 seconds. Source: Wang et al. (2013). Reprinted with permission of ACS Publications.

(Pham et al. 2021). Authors have reported that the simultaneous triggering of dual-photosensitizer generates a good amount of $^1O_2$ species, resulting in boosted PDT efficacy under 1550 nm excitation via green and red upconversion emission by Er ion. Chien et al. addressed the issue to increase the concentration of the local effective therapeutic drug and to avoid the harmful consequences and side effects (Chien et al. 2013). Authors proposed a photocage having bioactive species protected by a photolabile group. Under adequate illumination, the active molecules release and serve the purpose. This technique is useful to improve the specific adsorption and cellular uptake for cell penetration peptides and tumor-homing agents. Authors considered the UCNPs ($NaYF_4$:Yb,Tm@SiO_2) as drug delivery vehicles and functionalized with the folic acid, which acts as a photoresponsive molecule. The assembly was loaded with anticancerous drug doxorubicin as well. The proposed hybrid assembly suggested for the improvement of selective targeting and possible reduction of adverse side effects from chemotherapy. Zhou et al. reported the method to analyze the multiple biomarkers in tumor cells and tissues by single-band upconversion nanoparticles (Zhou et al. 2015). In this method, authors have used a layer-by-layer strategy to synthesize the hybrid assembly. Nanocrystals of $NaGdF_4$:Yb and Er@$NaGdF_4$ were synthesized, and over which silica and amino-reactive organic dye doped silica layer was coated. Dye was used to filter the particular emissive color from the particles. Authors reported the determination of three biomarkers in breast cancer cells (estrogen receptors, progesterone receptors, and human epithelial growth factor receptor-2) with single-band upconversion nanoparticles by using the Western blotting and immune histochemical technologies. Chen et al. addressed the problem of short penetration depth of light and low extinction coefficient of the photosensitizers (Chen et al. 2016). Authors have developed a nanocomposite $NaYF_4$:Yb:Er conjugated with gold nanorods and methylene blue dye infused in a silica shell. It was reported that the visible emission from $NaYF_4$:Yb:Er excited the methylene blue (MB) to generate reactive oxygen species, and the plasmonic effect of gold nanorods enhances the upconversion efficiency. The NIR-triggered nanocomposites were further modified with folic acid to target oral epidermoid carcinoma, Meng-1 (OECM-1) oral cancer cells and evaluated by the cell apoptosis pathway.

References

Abel, K.A., Boyer, J., and van Veggel, F.C.J.M. (2009). Hard proof of the $NaYF_4$/$NaGdF_4$ nanocrystal core/shell structure. *Journal of the American Chemical Society* 131 (41): 14644–14645. https://doi.org/10.1021/ja906971y.

Alberti, S., Soler-Illia, G.J.A.A., and Azzaroni, O. (2015). Gated supramolecular chemistry in hybrid mesoporous silica nanoarchitectures: controlled delivery and molecular transport in response to chemical, physical and biological stimuli. *Chemical Communications* 51: 6050–6075. https://doi.org/10.1039/C4CC10414E.

Auzel, F. (1990). Upconversion processes in coupled ion systems. *Journal of Luminescence* 45: 341–345. https://doi.org/10.1016/0022-2313(90)90189-I.

Bai, X., Song, H.W., Yu, L.X. et al. (2005). Luminescent properties of pure cubic phase Y_2O_3/Eu^{3+} nanotubes/nanowires prepared by a hydrothermal method. *The Journal of Physical Chemistry B* 109: 15236. https://doi.org/10.1021/jp050652f.

Barbero, F., Russo, L., Vitali, M. et al. (2017). Formation of the protein corona: the interface between nanoparticles and the immune system. *Seminars in Immunology* 34: 52–60. https://doi.org/10.1016/j.smim.2017.10.001.

Benov, L. (2015). Photodynamic therapy: current status and future directions. *Medical Principles and Practice* 24 (suppl 1): 14–28. https://doi.org/10.1159/000362416.

Bertrand, N. and Leroux, J.C. (2012). The journey of a drug-carrier in the body: an anatomo-physiological perspective. *Journal of Controlled Release* 161: 152–163. https://doi.org/10.1016/j.jconrel.2011.09.098.

Bertrand, N., Wu, J., Xu, X. et al. (2014). Cancer nanotechnology: the impact of passive and active targeting in the era of modern cancer biology. *Advanced Drug Delivery Reviews* 66: 2–25. https://doi.org/10.1016/j.addr.2013.11.009.

de Bettencourt-Dias, A. (ed.) (2014). *Luminescence of Lanthanide Ions in Coordination Compounds and Nanomaterials.* Wiley ISBN: 978-1-119-95083-7.

Bishop, C.J., Kim, J., and Green, J.J. (2014). Biomolecule delivery to engineer the cellular microenvironment for regenerative medicine. *Annals of Biomedical Engineering* 42: 1557–1572. https://doi.org/10.1007/s10439-013-0932-1.

Black, K.C.L., Yi, J., Rivera, J.G. et al. (2012). Polydopamine-enabled surface functionalization of gold nanorods for cancer cell-targeted imaging and photothermal therapy. *Nanomedicine* 8: https://doi.org/10.2217/nnm.12.82.

Boraschi, D., Italiani, P., Palomba, R. et al. (2017). Nanoparticles and innate immunity: new perspectives on host defence. *Seminars in Immunology* 34: 33–51. https://doi.org/10.1016/j.smim.2017.08.013.

Bose, S. and Bandyopadhyay, A. (2017). *Materials and Devices for Bone Disorders.* Academic Press https://doi.org/10.1016/C2014-0-03361-8.

Bridot, J.L., Faure, A.C., Laurent, S. et al. (2007). Hybrid gadolinium oxide nanoparticles: multimodal contrast agents for in vivo imaging. *Journal of the American Chemical Society* 129: 5076–5084. https://doi.org/10.1021/ja068356j.

Bulte, J.W.M. and Kraitchman, D.L. (2004). Iron oxide MR contrast agents for molecular and cellular imaging. *NMR in Biomedicine* 17: 484–499. https://doi.org/10.1002/nbm.924.

Bunzli, J.-C.G. and Eliseeva, S.V. (2010). Basics of lanthanide photophysics. In: *Lanthanide Luminescence: Photophysical, Analytical and Biological Aspects*, Springer Series, Fluorescence (ed. P. Hanninen and H. Harma). Berlin/Heidelberg: Springer-Verlag https://doi.org/10.1007/978-3-642-21023-5.

Casanova, M.R., Reis, R.L., Martins, A., and Neves, N.M. (2020). Surface biofunctionalization to improve the efficacy of biomaterial substrates to be used in regenerative medicine. *Materials Horizons* 7: 2258–2275. https://doi.org/10.1039/D0MH00542H.

Chang, C.K., Zhang, Q.A., and Mao, D.L. (2006). The hydrothermal preparation, crystal structure and photoluminescent properties of GdOOH nanorods. *Nanotechnology* 17: 1981. https://doi.org/10.1088/0957-4484/17/8/031.

Chatterjee, D.K., Rufaihah, A.J., and Zhang, Y. (2008). Upconversion fluorescence imaging of cells and small animals using lanthanide doped nanocrystals. *Biomaterials* 29 (7): 937–943. https://doi.org/10.1016/j.biomaterials.2007.10.051.

Chen, G., Qiu, H., Prasad, P.N. et al. (2014). Upconversion nanoparticles: design, nanochemistry, and applications in theranostics. *Chemical Reviews* 114: 5161–5214. https://doi.org/10.1021/cr400425h.

Chen C. W., Chan Y. C., Hsiao M., Liu R. S. (2016). Plasmon-enhanced photodynamic cancer therapy by upconversion nanoparticles conjugated with Au nanorods. *ACS Applied Materials & Interfaces* 8(47), 32108–32119. https://doi.org/10.1021/acsami.6b07770.

Cheng, L., Yang, K., Shao, M. et al. (2011). Multicolor in vivo imaging of upconversion nanoparticles with emissions tuned by luminescence resonance energy transfer. *Journal of Physical Chemistry C* 115: 2686–2692. https://doi.org/10.1021/jp111006z.

Chien Y.H., Chou Y.L., Wang S. W., Hung S. T., et al. (2013). Near-infrared light photocontrolled targeting, bioimaging, and chemotherapy with caged upconversion nanoparticles in vitro and in vivo. *ACS Nano*, 7(10), 8516–8528. https://doi.org/10.1021/nn402399m.

Conde, J., Dias, J.T., Grazú, V. et al. (2014). Revisiting 30 years of biofunctionalization and surface chemistry of inorganic nanoparticles for nanomedicine. *Frontiers in Chemistry* 2: 2296–2646. https://doi.org/10.3389/fchem.2014.00048.

Daniell, M.D. and Hill, J.S. (1991). A history of photodynamic therapy. *Australian and New Zealand Journal of Surgery* 61 (5): 340–348. https://doi.org/10.1111/j.1445-2197.1991.tb00230.x.

Das G. K., Johnson N. J. J., Cramen J., Blasiak B., Latta P., Tomanek P., van Veggel F. C. J. M., (2012). $NaDyF_4$ nanoparticles as T_2 contrast agents for ultrahigh field magnetic resonance imaging. *Journal of Physical Chemistry Letters* 3(4), 524–529. https://doi.org/10.1021/jz201664h.

Davis K, Cole B., Ghelardini M. Powell B. A., Mefford O.T., (2016). Quantitative measurement of ligand exchange on iron oxides via radiolabeled oleic acid. *Langmuir* 32(51), 13716–13727. https://doi.org/10.1021/la502204g.

Derfus, A.M., Chan, W.C.W., and Bhatia, S.N. (2004). Probing the cytotoxicity of semiconductor quantum dots. *Nano Letters* 4: 11. https://doi.org/10.1021/nl0347334.

Dhoble, S., Pawade, V., Swart, H., and Chopra, V. (ed.) (2019). *Spectroscopy of Lanthanide Doped Oxide Materials*. Elsevier ISBN: 9780081029350.

Dolmans, D.E.J.G.J., Fukumura, D., and Jain, R.K. (2003). Photodynamic therapy for cancer. *Nature Reviews Cancer* 3 (5): 380–387. https://doi.org/10.1038/nrc1071.

dos Santos, A.F., de Almeida, D.R.Q., Terra, L.F. et al. (2019). Photodynamic therapy in cancer treatment - an update review. *Journal of Cancer Metastasis and Treatment* 5: 25. https://doi.org/10.20517/2394-4722.2018.83.

Du, G.H. and Tendeloo, G.V. (2005). Preparation and structure analysis of $Gd(OH)_3$ nanorods. *Nanotechnology* 16: 595. https://doi.org/10.1088/0957-4484/16/4/043.

Duong, H.T.T., Chen, Y., Tawfik, S.A. et al. (2018). Systematic investigation of functional ligands for colloidal stable upconversion nanoparticles. *RSC Advances* 8: 4842–4849. https://doi.org/10.1039/C7RA13765F.

Dwivedi, Y. and Rai, S.B. (ed.) (2017). *Advances in Applied Spectroscopy: Concepts and Techniques*. New York: Nova Science Publisher ISBN-13: 978-1536124392.

Dwivedi, Y. and Zilio, S.C. (2013). Metal enhanced upconversion emission in water dispersible $Y_8V_2O_{17}$:Er:Yb@Au nano-composites. *Science of Advanced Materials* 5 (10): 1514–1523. https://doi.org/10.1166/sam.2013.1613.

Dwivedi Y., Zilio S. C. (2014). Advances in rare earth spectroscopy and applications. *Journal of Nanoscience and Nanotechnology*, 14(2), 1578-96. https://doi.org/10.1166/jnn.2014.9104.

Ebbesen, T.W., Lezec, H.J., Hiura, H. et al. (1996). Electrical conductivity of individual carbon nanotubes. *Nature* 382: 54–56. https://doi.org/10.1038/382054a0.

Elsehly, E.M., Chechenin, N.G., Makunin, A.V. et al. (2020). Surface functionalization of multi-walled carbon nanotubes by ozone and the enhancement of their environmental applications. *Nano Express* 1: 020023. https://doi.org/10.1088/2632-959X/abaafd.

Esipova, T.V., Ye, X., Collins, J.E. et al. (2012). Dendritic upconverting nanoparticles enable in vivo multiphoton microscopy with low-power continuous wave sources. *Proceedings of the National Academy of Sciences of the United States of America* 109 (51): 20826–20831. https://doi.org/10.1073/pnas.1213291110.

Fang, Y.P., Xu, A.W., You, L.P. et al. (2003). Hydrothermal synthesis of rare earth (Tb, Y) hydroxide and oxide nanotubes. *Advanced Functional Materials* 13 (12): 955–960. https://doi.org/10.1002/adfm.200304470.

Ficai, A. and Grumezescu, A.M. (ed.) (2017). *Nanostructures for Cancer Therapy*. Elsevier https://doi.org/10.1016/C2015-0-04166-1.

Fortin, M.A., Petoral, R.M., Söderlind, F. et al. (2007). Polyethelene glycol-covered ultra-small Gd_2O_3 nanoparticles for positive contrast at 1.5T magnetic resonance clinical scanning. *Nanotechnology* 18: 1–9. https://doi.org/10.1088/0957-4484/18/39/395501.

Gee, A. and Xu, X. (2018). Surface functionalisation of upconversion nanoparticles with different moieties for biomedical applications. *Surfaces* 1 (1): 96–121. https://doi.org/10.3390/surfaces1010009.

Gilliland, G.D. (1997). Photoluminescence spectroscopy of crystalline semiconductors. *Materials Science and Engineering* 18: 99–399. https://doi.org/10.1016/S0927-796X(97)80003-4.

Guller, A.E., Nadort, A., Generalova, A.N. et al. (2018). Rational surface design of upconversion nanoparticles with polyethylenimine coating for biomedical applications: better safe than brighter? *ACS Biomaterials Science & Engineering* 4 (9): 3143–3153. https://doi.org/10.1021/acsbiomaterials.8b00633.

Hänninen, P. and Härmä, H. (ed.) (2011). *Lanthanide Luminescence: Photophysical, Analytical and Biological Aspects*. Springer https://doi.org/10.1007/978-3-642-21023-5.

Jalil, R.A. and Zhang, Y. (2008). Biocompatibility of silica coated NaYF(4) upconversion fluorescent nanocrystals. *Biomaterials* 29 (30): 4122–4128. https://doi.org/10.1016/j.biomaterials.2008.07.012.

Kam, N.W.S., Jessop, T.C., Wender, P.A., and Dai, H. (2004). Nanotube molecular transporters: internalization of carbon nanotube-protein conjugates into Mammalian cells. *Journal of the American Chemical Society* 126: 6850. https://doi.org/10.1021/ja0486059.

Kommareddy, S., Tiwari, S., and Amiji, M.M. (2005). Long-circulating nanovectors for tumor-specific gene delivery. *Technology in Cancer Research & Treatment* 4: 615–626. https://doi.org/10.1177/153303460500400605.

Korpany, K.V., Mottillo, C., Bachelder, J. et al. (2016). One-step ligand exchange and switching from hydrophobic to water-stable hydrophilic superparamagnetic iron oxide nanoparticles by mechanochemical milling. *Chemical Communications* 52: 3054–3057. https://doi.org/10.1039/C5CC07107K.

Kumari, P. and Dwivedi, Y. (2016). Structural and photophysical investigations of bright yellow emitting Dy:$ZnAl_2O_4$ nanophosphor. *Journal of Luminescence* 178: 407–413. https://doi.org/10.1016/j.jlumin.2016.06.027.

Kumari, P. and Dwivedi, Y. (2018). Bright emission via energy transfer from Dy to Tb in Bi_2SiO_5 nanophosphor. *Optical Materials* 75: 31–38. https://doi.org/10.1016/j.optmat.2017.10.017.

Lee, B.I., Lee, K.S., Lee, J.H. et al. (2009). Synthesis of colloidal aqueous suspensions of a layered gadolinium hydroxide: a potential MRI contrast agent. *Dalton Transactions* 14: 2490–2495. https://doi.org/10.1039/B823172A.

Lee J., Lee T. S., Ryu J., Hong S., Kang M., Im K., Kang J. H., Lim S. M., Park S., Song R. (2013). RGD peptide–conjugated multimodal $NaGdF_4$:Yb^{3+}/Er^{3+} nanophosphors for upconversion luminescence, MR, and PET imaging of tumor angiogenesis. *Journal of Nuclear Medicine*, 54 (1) 96-103. https://doi.org/10.2967/jnumed.112.108043.

Li, Z.Q. and Zhang, Y. (2006). Monodisperse silica-coated polyvinylpyrrolidone/$NaYF_4$ nanocrystals with multicolor upconversion fluorescence emission. *Angewandte Chemie International Edition* 45 (46): 7732–7735. https://doi.org/10.1002/anie.200602975.

Li, Z.Q., Zhang, Y., and Jiang, S. (2008). Multicolor core/shell-structured upconversion fluorescent nanoparticles. *Advanced Materials* 20: 4765–4769. https://doi.org/10.1002/adma.200801056.

Liu, L.Q., Ma, E., Li, R.F. et al. (2007). Effects of phonon confinement on the luminescence dynamics of Eu^{3+} in Gd_2O_3 nanotubes. *Nanotechnology* 18: 015403. https://doi.org/10.1088/0957-4484/18/1/015403.

Liu, Z.Y., Yi, G.S., Zhang, H.T. et al. (2008). Monodisperse silicananoparticles encapsulating upconversion fluorescent and superparamagnetic nanocrystals. *Chemical Communications* 694: https://doi.org/10.1039/B715402J.

Liu, F., Zhao, Q., You, H., and Wang, Z. (2013). Synthesis of stable carboxy-terminated $NaYF_4$:Yb^{3+},Er^{3+}@SiO_2 nanoparticles with ultrathin shell for biolabeling applications. *Nanoscale* 5: 1047–1053. https://doi.org/10.1039/c2nr33046f.

Lu, Q., Moore, J.M., Huang, G. et al. (2004). RNA polymer translocation with single-walled carbon nanotubes. *Nano Letters* 4 (12): 2473–2477. https://doi.org/10.1021/nl048326j.

Lu, Y., Lu, J., Zhao, J. et al. (2014). On-the-fly decoding luminescence lifetimes in the microsecond region for lanthanide-encoded suspension arrays. *Nature Communications* 5: 3741. https://doi.org/10.1038/ncomms4741.

Mackeyev, Y., Hartman, K.B., Ananta, J.S. et al. (2009). Catalytic synthesis of amino acid and peptide derivatized gadonanotubes. *Journal of the American Chemical Society* 131: 8342–8343. https://doi.org/10.1021/ja900918x.

Maeda, H. (2015). Toward a full understanding of the EPR effect in primary and metastatic tumors as well as issues related to its heterogeneity. *Advanced Drug Delivery Reviews* 91: 3–6. https://doi.org/10.1016/j.addr.2015.01.002. Epub 2015 Jan 9.

Mai, H.X., Zhang, Y.W., Si, R. et al. (2006). High-quality sodium rare-earth fluoride nanocrystals: controlled synthesis and optical properties. *Journal of the American Chemical Society* 128: 6426. https://doi.org/10.1021/ja060212h.

Mao, Y.B., Huang, J.Y., Ostroumov, R. et al. (2008). Synthesis and luminescence properties of erbium-doped Y_2O_3 nanotubes. *Journal of Physical Chemistry C* 112: 2278. https://doi.org/10.1021/jp0773738.

Marcin, N., Kumar, R., Ohulchanskyy, T.Y. et al. (2008). High contrast in vitro and in vivo photoluminescence bioimaging using near infrared to near infrared up-conversion in Tm^{3+} and Yb^{3+} doped fluoride nanophosphors. *Nano Letters* 8: 3834–3838. https://doi.org/10.1021/nl802223f.

Mukamel, S. (2000). *Principles of Nonlinear Optical Spectroscopy*. New York: Oxford University Press ISBN-13: 978-0195132915.

Na, H.B., Lee, J.H., An, K.J. et al. (2007). Development of a T_1 contrast agent for magnetic resonance imaging using MnO nanoparticles. *Angewandte Chemie International Edition* 46: 5397–5401. https://doi.org/10.1002/anie.200604775.

Nam, S.H., Bae, Y.M., Park, Y.I. et al. (2011). Long-term real-time tracking of lanthanide ion doped upconverting nanoparticles in living cells. *Angewandte Chemie International Edition* 50: 6093–6097. https://doi.org/10.1002/anie.201007979.

Nirmal, M., Dabbousi, B.O., Bawendi, M.G., and Macklin, J.J. (1996). Fluorescence intermittency in single cadmium selenide nanocrystals. *Nature* 383 (6603): 802–804. https://doi.org/10.1038/383802a0.

Niu, K. and Marcus, R.A. (2020). Sum frequency generation, calculation of absolute intensities, comparison with experiments, and two-field relaxation-based derivation. *Proceedings of the National Academy of Sciences of the United States of America* 117 (6): 2805–2814. https://doi.org/10.1073/pnas.1906243117.

Oliveira, G.H., Ferreira, F.S., Ferbonink, G.F. et al. (2021). Femtosecond laser induced luminescence in hierarchically structured NdIII, YbIII, ErIII co-doped upconversion nanoparticles: light-matter interaction mechanisms from experiments and simulations. *Journal of Luminescence* 234: 117953. https://doi.org/10.1016/j.jlumin.2021.117953.

Patelli, A., Mussano, F., Brun, P. et al. (2018). Nanoroughness, surface chemistry, and drug delivery control by atmospheric plasma jet on implantable devices. *ACS Applied Materials & Interfaces* 10 (46): 39512–39523. https://doi.org/10.1021/acsami.8b15886.

Petrov, V. and Noack, F. (1995). Frequency upconversion of tunable femtosecond pulses by parametric amplification and sum-frequency generation in a single nonlinear crystal. *Optics Letters* 20 (21): 2171–2173. https://doi.org/10.1364/OL.20.002171.

Pham, K.Y., Wang, L.C., Hsieh, C.C. et al. (2021). 1550 nm excitation-responsive upconversion nanoparticles to establish dual-photodynamic therapy against pancreatic tumors. *Journal of Materials Chemistry* B9: 694–709. https://doi.org/10.1039/D0TB02655G.

Pilato, L.A. and Michno, M.J. (1994). *Advanced Composite Materials*. Springer-Verlag https://doi.org/10.1007/978-3-662-35356-1.

Prinzen, L., Miserus, R., Dirksen, A. et al. (2007). Optical and magnetic resonance imaging of cell death and platelet activation using annexin a5-functionalized quantum dots. *Nano Letters* 7 (1): 93. https://doi.org/10.1021/nl062226r.

Qie, Y., Yuan, H., Von Roemeling, C.A. et al. (2016). Surface modification of nanoparticles enables selective evasion of phagocytic clearance by distinct macrophage phenotypes. *Scientific Reports* 6: 26269. https://doi.org/10.1038/srep26269.

Rai, S.B. and Dwivedi, Y. (ed.) (2012). *Synthesis, Characterization and Application of Multifunctional Materials*. New York: Nova Publishing Pvt. Ltd. ISBN: 978-1-61470-618-2.

Rashmi and Dwivedi, Y. (2019a). Optical interactions and white light emission in Eu:Y_2O_3/YAG:Ce nanophosphor. *Applied Physics A* 125 (8): https://doi.org/10.1007/s00339-019-2830-3.

Rashmi and Dwivedi, Y. (2019b). Tunable-color luminescence via energy-transfer in composite YAG:Ce/TTA for white light application. *Journal of Materials Science: Materials in Electronics* 30: 20506–20514. https://doi.org/10.1007/s10854-019-02397-4.

Rees, P., Wills, J.W., Brown, M.R. et al. (2019). The origin of heterogeneous nanoparticle uptake by cells. *Nature Communications* 10: 2341. https://doi.org/10.1038/s41467-019-10112-4.

Rieter, W.J., Kim, J.S., Taylor, K.M.L. et al. (2007). Hybrid silica nanoparticles for multimodal imaging. *Angewandte Chemie International Edition* 46: 3680–3682. https://doi.org/10.1002/anie.200604738.

Romero, P.G. and Sanchez, C. (ed.) (2006). *Functional Hybrid Materials*. Wiley ISBN: 978-3-527-30484-4.

Sanità, G., Carrese, B., and Lamberti, A. (2020). Nanoparticle surface functionalization: how to improve biocompatibility and cellular internalization. *Frontiers in Molecular Biosciences* 7: 381. https://doi.org/10.3389/fmolb.2020.587012.

Santos-Rasera, J.R., Sant'AnnaNeto, A., Rosim Monteiro, R.T. et al. (2019). Toxicity, bioaccumulation and biotransformation of Cu oxide nanoparticles in *Daphnia magna*. *Environmental Science. Nano* 6: 2897–2906. https://doi.org/10.1039/C9EN00280D.

Shen, Y.R. (2020). Revisiting the basic theory of sum-frequency generation. *The Journal of Chemical Physics* 153 (18): 180901. https://doi.org/10.1063/5.0030947.

Shen, J., Chen, G.Y., Vu, A.M. et al. (2013). Engineering the upconversion nanoparticle excitation wavelength: cascade sensitization of tri-doped upconversion colloidal nanoparticles at 800 nm. *Advanced Optical Materials* 1 (9): 644–650. https://doi.org/10.1002/adom.201300160.

Shim, M., Kam, N.W.S., Chen, R.J. et al. (2002). Functionalization of carbon nanotubes for biocompatibility and biomolecular recognition. *Nano Letters* 2 (4): 285–288. https://doi.org/10.1021/nl015692j.

Simon, U. and Tittel, F.K. (1997). *Experimental Methods in the Physical Sciences, Atomic, Molecular, and Optical Physics: Electromagnetic Radiation*. Elsevier ISBN: 978-0-12-475977-0.

Singh P., Srivastava S., Singh S. K., (2019). Nanosilica: recent progress in synthesis, functionalization, biocompatibility, and biomedical applications. *ACS Biomaterials Science & Engineering* 5(10), 4882–4898. https://doi.org/10.1021/acsbiomaterials.9b00464.

Sitharaman, B., Kissell, K.R., Hartman, K.B. et al. (2005). Superparamagnetic gadonanotubes are high-performance MRI contrast agents. *Chemical Communications* 31: 3915–3917. https://doi.org/10.1039/b504435a. Epub 2005 Jul 8.

Soukka, T., Rantanen, T., and Kuningas, K. (2008). Photon upconversion in homogeneous fluorescence-based bioanalytical assays. *Annals of the New York Academy of Sciences* 1130: 188–200. https://doi.org/10.1196/annals.1430.027.

Sperling, R.A. and Parak, W.J. (2010). Surface modification, functionalization and bioconjugation of colloidal inorganic nanoparticles. *Philosophical Transactions of the Royal Society A of London, Series A: Mathematical, Physical, and Engineering Sciences* 368: 1333–1383. https://doi.org/10.1098/rsta.2009.0273.

Taylor, K.M.L., Kim, J.S., Rieter, W.J. et al. (2008). Mesoporous silica nanospheres as highly efficient MRI contrast agents. *Journal of the American Chemical Society* 130 (7): 2154–2155. https://doi.org/10.1021/ja710193c.

Tsai, C.P., Hung, Y., Chou, Y.H. et al. (2008). High-contrast paramagnetic fluorescent mesoporous silica nanorods as a multifunctional cell-imaging probe. *Small* 4: 186. https://doi.org/10.1002/smll.200700457.

Tu, D., Liu, L., Ju, Q. et al. (2011). Time-resolved FRET biosensor based on amine-functionalized lanthanide-doped $NaYF_4$ nanocrystals. *Angewandte Chemie International Edition* 50: 6306–6310. https://doi.org/10.1002/anie.201100303.

Wang, L.Y. and Li, Y.D. (2006). Green upconversion nanocrystals for DNA detection. *Chemical Communications* 24: 2557–2559. https://doi.org/10.1039/B604871D.

Wang, X., Sun, X.M., Yu, D.P. et al. (2003). Rare earth compound nanotubes. *Advanced Materials* 15: 1442. https://doi.org/10.1002/adma.200305164.

Wang, L.L., Yan, R., Huo, Z. et al. (2005). Fluorescence resonant energy transfer biosensor based on upconversion-luminescent nanoparticles. *Angewandte Chemie International Edition* 44: 6054–6057. https://doi.org/10.1002/anie.200501907.

Wang, F., Tan, W.B., Zhang, Y. et al. (2006). Synthesis of polyethylenimine/$NaYF_4$ nanoparticles with upconversion fluorescence. *Nanotechnology* 17: R1. https://doi.org/10.1088/0957-4484/17/23/013.

Wang, M., Mi, C.C., Wang, W.X. et al. (2009). Immunolabeling and NIR-excited fluorescent imaging of HeLa cells by using $NaYF_4$:Yb,Er upconversion nanoparticles. *ACS Nano* 3 (6): 1580–1586. https://doi.org/10.1021/nn900491j.

Wang, F., Banerjee, D., Liu, Y. et al. (2010). Upconversion nanoparticles in biological labeling, imaging, and therapy. *Analyst* 135: 1839. https://doi.org/10.1039/C0AN00144A.

Wang, Y.F., Liu, G.Y., Sun, L.D. et al. (2013). Nd^{3+}-sensitized upconversion nanophosphors: efficient in vivo bioimaging probes with minimized heating effect. *ACS Nano* 7 (8): 7200–7206. https://doi.org/10.1021/nn402601d.

Wu, S.-H., Hung, Y., and Mou, C.-Y. (2011). Mesoporous silica nanoparticles as nanocarriers. *Chemical Communications* 47: 9972. https://doi.org/10.1021/ar3000986.

Xia, A.O., Gao, Y., Zhou, J. et al. (2011). Core–shell $NaYF_4$:Yb^{3+},Tm^{3+}@Fe_xO_y nanocrystals for dual-modality T_2-enhanced magnetic resonance and NIR-to-NIR upconversion luminescent imaging of small-animal lymphatic node. *Biomaterials* 32 (29): 7200–7208. https://doi.org/10.1016/j.biomaterials.2011.05.094.

Xu, A.W., Fang, Y.P., You, L.P., and Liu, H.Q. (2003). A simple method to synthesize $Dy(OH)_3$ and DY_2O_3 nanotubes. *Journal of the American Chemical Society* 125: 1494. https://doi.org/10.1021/ja029181q.

Yang, J., Li, C.X., Cheng, Z.Y. et al. (2007). Size-tailored synthesis and luminescent properties of one-dimensional Gd_2O_3:Eu^{3+} nanorods and microrods. *Journal of Physical Chemistry C* 111: 18148. https://doi.org/10.1021/jp0767112.

Yao, J., Larson, D.R., Vishwasrao, H.D. et al. (2005). Blinking and nonradiant dark fraction of water-soluble quantum dots in aqueous solution. *Proceedings of the National Academy of Sciences of the United States of America* 102: 14284. https://doi.org/10.1073pnas.0904792106.

Yao, Y., Xu, C., Zheng, Y. et al. (2016). Femtosecond laser-induced upconversion luminescence in rare-earth ions by nonresonant multiphoton absorption. *The Journal of Physical Chemistry. A* 12028: 5522–5526. https://doi.org/10.1021/acs.jpca.6b04444.

Yoo, J., Park, C., Yi, G. et al. (2019). Active targeting strategies using biological ligands for nanoparticle drug delivery systems. *Cancers* 11 (5): 640. https://doi.org/10.3390/cancers11050640.

Zhang, F. (2015). Photon upconversion nanomaterials. In: *Nanostructure Science and Technology* (ed. D.J. Lockwood). Berlin/Heidelberg: Springer https://doi.org/10.1007/978-3-662-45597-5.

Zhang, F., Braun, G.B., Shi, Y. et al. (2010). Fabrication of Ag@SiO_2@Y_2O_3:Er nanostructures for bioimaging: tuning of the upconversion fluorescence with silver nanoparticles. *Journal of the American Chemical Society* 132 (9): 2850–2851. https://doi.org/10.1021/ja909108x.

Zhang, C., Lu, J., Tian, F. et al. (2019). Regulation of the cellular uptake of nanoparticles by the orientation of helical polypeptides. *Nano Research* 12: 889–896. https://doi.org/10.1007/s12274-019-2319-6.

Zhao, L., Lustres, J.L.P., Farztdinov, V., and Ernsting, N.P. (2005). Femtosecond fluorescence spectroscopy by upconversion with tilted gate pulses. *Physical Chemistry Chemical Physics* 7: 1716–1725. https://doi.org/10.1039/B500108K.

Zhou, J., Yu, M., Sun, Y. et al. (2011). Fluorine-18-labeled Gd^{3+}/Yb^{3+}/Er^{3+} co-doped $NaYF_4$ nanophosphors for multimodality PET/MR/UCL imaging. *Biomaterials* 32: 1148–1156. https://doi.org/10.1016/j.biomaterials.2010.09.071.

Zhou, L., Wang, R., Yao, C. et al. (2015). Single-band upconversion nanoprobes for multiplexed simultaneous in situ molecular mapping of cancer biomarkers. *Nature Communications* 6: 6938. https://doi.org/10.1038/ncomms7938.

Zhou, C., Bai, Y., Yang, F. et al. (2020). Ligand exchange on noble metal nanocrystals assisted by coating and etching of cuprous oxide. *Materials Chemistry Frontiers* 4: 1614–1622. https://doi.org/10.1039/D0QM00092B.

10

Frequency Upconversion in Core@shell Nanoparticles

Raghumani S. Ningthoujam[1,2], *Rashmi Joshi*[1,2], *and Manas Srivastava*[1]

[1]*Bhabha Atomic Research Centre, Chemistry Division, Mumbai 400085, India*
[2]*Homi Bhabha National Institute, Mumbai 400094, India*

10.1 Introduction

Luminescence is a property of a material to produce light upon applying an external agency such as light, pressure, mechanical, chemical, radiation, heat, etc. If an external agency such as light is applied, term photoluminescence is used. Many terms such as mechanoluminescence (triboluminescence), chemiluminescence, and thermoluminescence are used if external agencies such as mechanical reaction, chemical reaction, and heating are applied, respectively (Ningthoujam 2012, p. 145). In the case of radiations such as when α, β, and γ are applied, term radioluminescence is used. There is a term used as Cerenkov luminescence when the phase velocity of light is less than that of a charged particle (say β-rays), which interacts with the tissue of living body or water (Yadav and Ningthoujam 2021, pp. 213–250; Yang et al. 2019, p. 2288).

Photoluminescence is of two types: downconversion and upconversion (UC) luminescence or emission (Figure 10.1).

10.1.1 Downconversion

In the case of downconversion luminescence, the excitation energy or frequency is more than that of emission ($E_{exc} > E_{em}$ or $\nu_{exc} > \nu_{em}$) (Ningthoujam 2012, p. 01). It indicates that the wavelength of excitation is less than that of emission ($\lambda_{exc} < \lambda_{em}$). When a semiconductor absorbs photon having energy exactly equal to the bandgap of the semiconductor or slightly more than its bandgap, electrons from the valence band of the semiconductor go to the conduction band. Excess electrons are stored in the conduction band, whereas excess holes are developed in the valence band. This electron–hole pair forms an exciton. Bi-exciton or tri-exciton can be created if the energy of excitation is much higher than that of the bandgap of the semiconductor. If the source of photon is removed, the excited electrons in the conduction band take two paths: (i) remain in the localized energy levels created by defects, which

Upconverting Nanoparticles: From Fundamentals to Applications, First Edition. Edited by Vineet K. Rai.

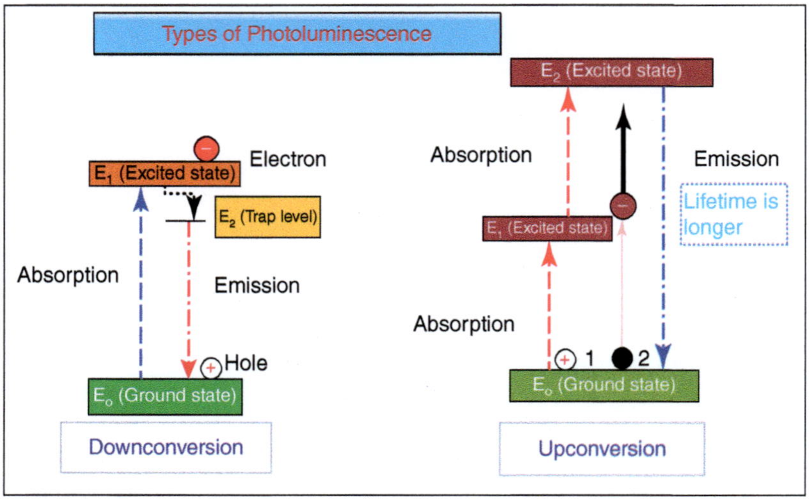

Figure 10.1 Schematic diagram showing frequency downconversion and upconversion after excitation at resonance frequency.

is just below the conduction band, a few meV, and then get decayed to the ground state (valence band) with a lifetime of nanoseconds to milliseconds to a few hours. (ii) Decay immediately and recombine with the holes in the valence band. The first one is considered as delayed or persistent luminescence if decay takes place with a few minutes to hours. Here, energy migration (EM) takes place in the localized energy levels. The second one is considered as fluorescence as decay takes place with a few nanoseconds to microseconds.

There are two terms in luminescence: donor and acceptor. The donor absorbs light, and this energy is transferred to the activator from where the emission occurs. The donor can be the host containing V-O charge transfer (CTA), Mo-O CTA, W-O CTA, or d^{n+} or f^{n+} ions. Activators are usually d^{n+} or f^{n+} ions. The d–d transitions are broad because the crystal field developed by the surrounding ligands to metal ions influences it (Ningthoujam 2012, p. 145; Yadav and Ningthoujam 2021, pp. 213–250). Depending on the crystal field, the transition can be shifted from lower to longer wavelength, and two or more transitions could be observed. The ground-state electronic configuration of 3d metal ions with 3+ charges has $1s^2$, $2s^22p^6$, and $3s^23p^63d^n$ ($n = 1$–10). In the case of 4f metal ions with 3+ charges, the ground-state electronic configuration has $1s^2$, $2s^22p^6$, $3s^23p^63d^{10}$, $4s^24p^64d^{10}4f^n$, and $5s^25p^6$ ($n = 1$–14). As $4f^n$ electrons are screened by outer shell $5s^25p^6$, the crystal field due to the ligand has little effect. Therefore, f–f transitions are generally sharp, and peak positions are almost unaffected by ligands (L) or crystal field or coordination number of metal ions (M) in polyhedron ML_n ($n = 1$–12). Lanthanides (Ln) can have common oxidation states of 2+, 3+, and 4+, e.g. Eu^{n+} ions can have 2+ and 3+ states in different chemical environments (Ningthoujam 2012, p. 145; Yadav and Ningthoujam 2021, pp. 213–250). Ce^{n+} ions can have 3+ and 4+ states. Tb^{n+} ions can have 3+ and 4+ states. In general, Ln has 3+ oxidation state. In some cases, electronic transitions such as 4f–5d transitions can take place in the UV region (absorption) and UV–visible region (emission). Also, Ln-O such as Eu-O

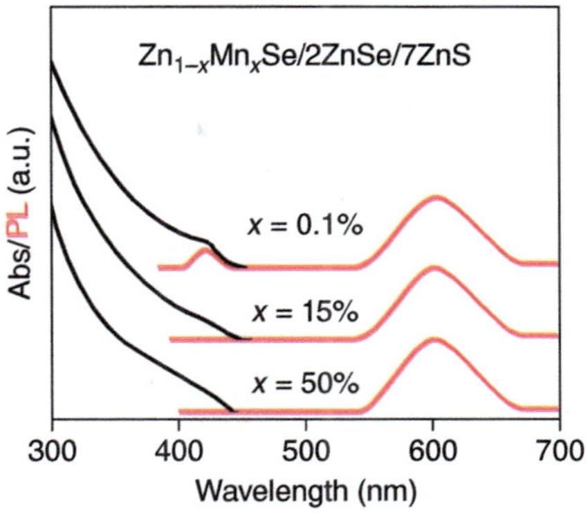

Figure 10.2 Absorption spectra (black lines) and photoluminescence (PL) spectra (red lines) of three typical samples: 0.1%, 15%, and 50% Mn substitution in $Zn_{1-x}Mn_xSe/2ZnSe/7ZnS$ (core@shell, 10 nm). Source: Yang et al. (2019), p. 2288. Reprinted with permission of American Chemical Society.

charge transfer (CTA) takes place in the UV region. In this way, by absorption of photons through the semiconductor or Ln-O CTA, the population of excited electrons can be increased, and these excited electrons can be transferred to the excited states of d or f metal ions, and then luminescence decay can take place. Generally, an enhanced luminescence from d or f metal ions can be observed.

Examples of semiconductors doped with d metal ions are $Zn_{1-x}Mn_xSe$, Cu-doped CdSe, Cu-doped ZnSe, Cr-doped TiO_2, V-doped SnO_2, etc. (Yang et al. 2019, p. 2288; Viswanatha 2015, p. 1055; Choudhury and Choudhury 2012, p. 1112; Ningthoujam et al. 2007a, p. 1293; Ningthoujam and Kulshreshtha 2009, p. 57). $Zn_{1-x}Mn_xSe/2ZnSe/7ZnS$ (core@shell, 10 nm) nanoparticles show absorption at 430 nm and emission at 600 nm. The 430 nm peak corresponds to the quantum dots (QDs) of the host and emission at 600 nm is due to the Mn^{2+} d–d transition ($^4T_1$–$^6A_1$), which is shown in Figure 10.2 (Yang et al. 2019, p. 2288). Cr^{3+}-doped $Bi_2Ga_{3.97}O_9$ shows broad absorption and emission peaks, and the peak intensity varies with temperature (Liu et al. 2015, p. 1876). This is due to the association of the surrounding ligand (or crystal field). In some cases, the defect-related emission is obtained.

If other metal ions (s, p, d, or f block elements) are doped in the semiconductor host, extra localized states are developed within the bandgap (Ningthoujam 2012, p. 145; Yadav and Ningthoujam 2021, pp. 213–250; Yang et al. 2019, p. 2288; Viswanatha 2015, p. 1055; Choudhury and Choudhury 2012, p. 1112; Ningthoujam et al. 2007a, p. 1293, 2009, p. 57; Liu et al. 2015, p. 1876; Curie 1963, p. 01; Blasse and Grabmaier 1994, p. 01). Thus, many emission peaks can be observed. Even, p- or n- typed semiconductors can be formed. In the case of d or f doping, d–d or f–f transitions can take place. However, these transitions are not allowed transitions (forbidden transitions).

Examples of oxide semiconductors and their doped ones with f block metal ions (Ln^{3+}) are SnO_2:Ln^{n+} (Ningthoujam et al. 2007b, p. 173113, 2008a, p. 1489, 2009, p. 2634; Ningthoujam 2010, p. 208), TiO_2:Ln^{n+} (Ningthoujam et al. 2009b, p. 864), ZnO:Ln^{n+} (Ningthoujam et al. 2008, p. 3059; Singh et al. 2009a, p. 466), ZrO_2:Ln^{n+} (Dhiren Meetei et al. 2012, p. 537), YVO_4:Ln^{n+} (Ningthoujam et al. 2009c, p. 782; Singh et al. 2009, p. 286; Ningombam et al. 2018, p. 6862; Singh and Ningthoujam 2010, p. 104304), $GdVO_4$:Ln^{n+} (Wangkhem et al. 2021, p. 150350; Singh et al. 2008b, p. 104307), $LaVO_4$:Ln^{n+} (Okram et al. 2014, p. 7204), YPO_4:Ln^{n+} (Prasad et al. 2013, p. 4885; Dutta et al. 2012, p. 042184; Parchur et al. 2010, p. 391;

Ningthoujam et al. 2015, p. 68234), $LaPO_4:Ln^{n+}$ (Phaomei et al. 2010, p. 616; 2011a, p. 11571; 2011b, p. 1164), $GdPO_4:Ln^{n+}$ (Yaiphaba et al. 2010a, p. 2682; 2010b, p. 034301; 2010c, p. 174; Sahu et al. 2014, p. 337), $CaMoO_4:Ln^{n+}$ (Wangkhem et al. 2018, p. 124303; Parchur et al. 2014, p. 53; Singh et al. 2014, p. 4770), $CaWO_4:Ln^{n+}$ (Ningombam et al. 2017, p. 249, 2020, p. 4217), $SrMoO_4:Ln^{n+}$ (Soni et al. 2018, p. 1), $Y_2O_3:Ln^{n+}$ (Prasad et al. 2018, p. 105310; Lahiri et al. 2010, p. 054316), $Gd_2O_3:Ln^{n+}$ (Ningthoujam et al. 2009d, p. 084304; Shukla et al. 2010, p. 843), $La_2O_3:Ln^{n+}$ (Loitongbam et al. 2013, p. 14), $Y_2Si_2O_7:Ln^{n+}$ (Prasad et al. 2018, p. 105310), $ZnGa_2O_4:Ln^{n+}$ (Manikyala Rao et al. 2008, p. 5776), $CaF_2:Ln^{n+}$ (Ghosh et al. 2012, p. 313), and $CdS:Ln^{n+}$ (Gajbhiye et al. 2008, p. 313). In order to modify the asymmetric nature of Ln^{n+} in LnM_n polyhedron, Li^+, Bi^{3+} are co-doped into hosts (Ningthoujam et al. 2007a, p. 1293, 2009, p. 084304, 2015, p. 68234; Curie 1963, p. 01; Gajbhiye et al. 2008, p. 313). The core–shell or composite nanoparticles such as SnO_2@SiO_2:Eu (Ningthoujam et al. 2008a, p. 1489, 2009, p. 2634), SnO_2:Eu@TiO_2 (Ningthoujam et al. 2007b, p. 173113), and SnO_2:Ln@Y_2O_3 (Lahiri et al. 2010, p. 054316) are prepared in order to improve luminescence as well as to modify the bandgap or dielectric medium of host. In some cases, the formation of quantum dots is difficult because of smaller exciton Bohr's radius. However, the similar size particles after dispersion in another dielectric medium or formation of core–shell become quantum dots (Ningthoujam et al. 2007b, p. 173113, 2008a, p. 1489, 2009, p. 2634). Examples of sulfide, tellurite, and selenide semiconductors are ZnS, CdS, PbSe, PbTe, CuS, Y_2O_2S, etc., and sometimes, Ln^{n+} ions are doped in order to get desired materials for a particular application (Ningthoujam 2012, p. 145; Curie 1963, p. 01; Blasse and Grabmaier 1994, p. 01; Ningthoujam et al. 2017, p. 2294; Wang and Wang 2006, p. 25). In nanosized particles, particle size, shape, and surface anisotropic effects can be observed. Figure 10.1a shows the downconversion process in the semiconductor. Figure 10.3a shows a schematic diagram illustrating effect of shell formation over core on luminescence and bandgap tuning in the semiconductor. Figure 10.3b shows the improvement of luminescence of Ln^{3+}

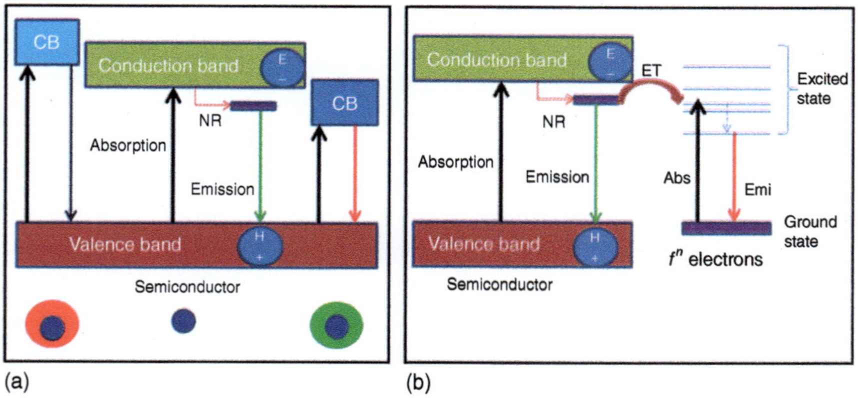

Figure 10.3 (a) Bandgap tuning after core@shell formation. Middle: core QDs, Left: core@shell (here, bandgap increases), right: core@shell (here, bandgap decreases). (b) Energy transfer from the semiconductor to Ln^{3+} ions to enhance the luminescence intensity of Ln^{3+} ions.

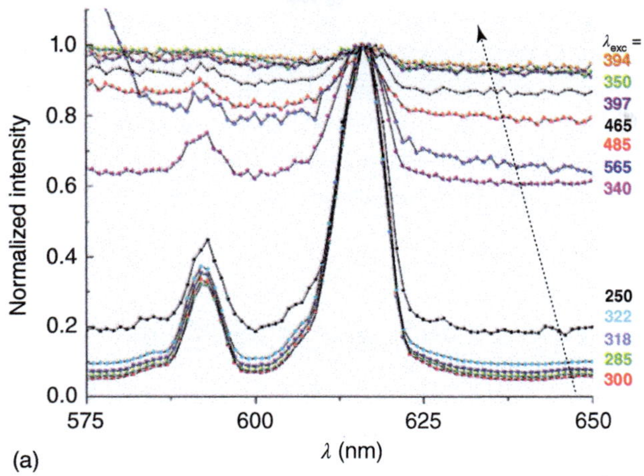

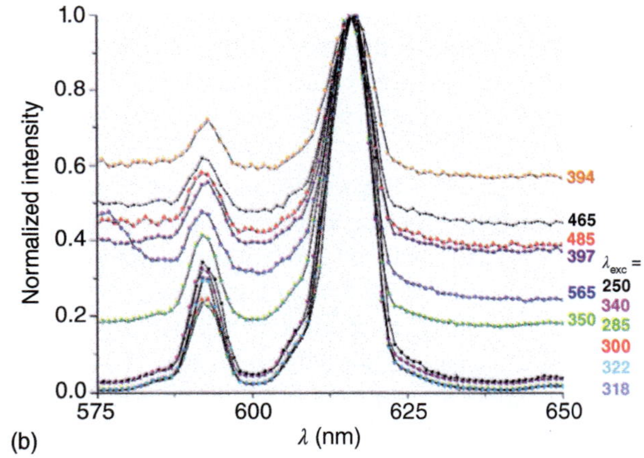

Figure 10.4 Photoluminescence spectra of (a) YVO_4:5Eu^{3+} and (b) YVO_4:5Eu^{3+}@YVO_4 at different excitation wavelengths. Source: Modified from Ningthoujam et al. (2009).

ions by energy transfer (ET) from the semiconductor host to the dopant Ln^{3+} ion. The luminescence intensity can be improved by the core@shell model. The luminescence intensity in YVO_4:Eu is improved by core@shell formation (with undoped YVO_4 or SiO_2) (Figure 10.4) (Ningthoujam et al. 2009c, p. 782).

10.1.2 Upconversion

In the case of upconversion (UC) luminescence, the excitation energy or frequency is less than that of emission ($E_{em} > E_{exc}$ or $\nu_{em} > \nu_{exc}$) (Ningthoujam 2012, p. 145). It indicates that the wavelength of excitation is more than that of emission ($\lambda_{em} < \lambda_{exc}$). Figure 10.1b shows the upconversion process. There are many ways to get upconversion luminescence (Auzel 2004, p. 139). One example of the best materials showing upconversion is $NaYF_4$:2 at.% Er^{3+}-10–20 at.% Yb^{3+}. However, its luminescence intensity decreases with the decrease of particle size down to nanosize because of the association of surface defects as well as the ligand or capping molecules, which act as a quencher in luminescence. One of the ways to improve the luminescence intensity is core@shell formation. With the increase of shell thickness, the luminescence intensity is increased because of the decrease in the non-radiative process from the quencher (Johnson et al. 2012, p. 11068). There are many ways to have core@shell nanostructures. In this chapter, we are focusing on the synthesis, characterization, optical properties, mechanisms, and applications of upconversion core@shell nanostructures.

10.2 Synthesis of Core@shell and Core@shell@shell UCNPs

Formation of a shell over core luminescent particles started to gain the following: (i) passivation of the core surface, (ii) decrease of energy migration from the core to the surface, (iii) enhanced radiative rate of active luminescent ions at the core, (iv) decrease in agglomeration of particles, (v) surface functionalization to shell so that it does not affect much the properties of core, (vi) easy control of luminescence intensity at a particular wavelength, (vii) tuning of absorption or emission wavelength, (viii) increase of quantum yield, etc.

In this section, the synthesis of core@shell nanoparticles will be discussed. This can provide the directions to the researchers for the preparation of other shapes of the nanoparticles. There are many ways to prepare core@shell nanoparticles using chemical and physical ways in the form of powder and thin films. Some of the chemical ways to prepare these are (i) thermolysis, (ii) hot injection, (iii) combined thermolysis and hot injection, (iv) cation exchange, (v) microemulsions, (vi) sol–gel, (vii) co-precipitation, (viii) polymerization, etc. (Johnson et al. 2012, p. 11068; Joshi et al. 2021, p. 3481; Choudhury and Choudhury 2012, p. 2373; Chatterjee et al. 2014, p. 8; Chiozzi and Rossi 2020, p. 5090; Zhou et al. 2014, p. 7666; Chen et al. 2015a, p. 1318; Soni et al. 2019, p. 7350; Abel et al. 2009, p. 14644; Dong et al. 2012, p. 1297).

10.2.1 Thermolysis Method

Thermolysis is a process of heating precursor(s) in a high boiling point solvent such as 1-octadecene, benzyl ether, etc. (Singh et al. 2021, p. 333). In this way, the metal precursor decomposes to give metal oxide, sulfide, fluorides, etc. During the heating, if a capping agent having long-chain fatty acids is added, monodispersed nanoparticles can be prepared.

In order to get monodispersed nanoparticles of core@shell nanostructures, the core was prepared at the initial stage. Then, the precursor for shell is added later. There are two ways of addition of shell (Johnson et al. 2012, p. 11068; Joshi et al. 2021, p. 3481). One way is the addition of precursors in the form of organic molecules or metal–organic molecules for the shell material. On heating, the precursors got decomposed and deposited on the surface of the core particles. This is generally known as the seed-mediated process. Another way is the preparation of small-sized particles for the precursor of shell, which will be added to the core particles (already prepared). In this, smaller sized particles get sacrificed, whereas the size of the core increases. The reaction temperature and duration are the important parameters to get control of the core@shell size. This is known as the Ostwald ripening process.

Let us take a few examples, where it undergoes the seed-mediated process for the formation of core@shell particles (Johnson 2012, p. 11068; Joshi et al. 2021, p. 3481).

(i) In the preparation of $NaYF_4$:2 at.% Er-20 at.% Yb@$NaYF_4$ (core-active@shell-inactive), firstly, core ($NaYF_4$:2 at.% Er-20 at.% Yb) is prepared by taking precursors (1 g of $Y(CH_3CO_2)_3{\cdot}H_2O$, 0.0331 g of $Er(CH_3CO_2)_3{\cdot}H_2O$, and 0.3374 g of $Yb(CH_3CO_2)_3{\cdot}H_2O$) in the three-neck round-bottom (RB) flask. Then, oleic acid (3–5 ml) and 1-octadecene (10–15 ml) are added into it and further heated at 120–150 °C for two hours in the Ar environment with vigorous stirring and cooled down to room temperature. In this way, the metal oleate (Y/Er/Yb-oleate) solution (clear solution) is formed. The NH_4F solution is prepared by the addition of NaOH (~0.1 g) and 10 mmol of NH_4F in methanol and added in the RB flask containing metal oleate. The reaction mixture is heated at 100 °C for one to three hours to remove methanol completely. The reaction is put forward at high temperature around 200–250 °C for one hour to get the cubic phase of $NaYF_4$ co-doped with Er^{3+}, Yb^{3+} ions (upconversion nanoparticles [UCNPs]) and further heated at 300–320 °C for two hours to get the hexagonal phase. UCNPs are recovered by the following steps: addition of ethanol, which enhances the coagulation and centrifugation at 5000–7000 RPM. Finally, it is stored in hexane medium for the further study.
In order to get the shell layer with different thicknesses, the metal precursors such as $Y(CH_3CO_2)_3{\cdot}H_2O$ and $Yb(CH_3CO_2)_3{\cdot}H_2O$ are dissolved in oleic acid and 1-octadecene and heated at 120 °C for one hour. This metal oleate solution is added to the solution of core at room temperature. However, core is kept without washing with ethanol and centrifugation. This is further heated up to 300 °C for one hour to get core@shell. In order to get core@shell-1@shell-2, the oleate of desired metal precursors of shell-2 will be added to the already prepared core@shell-1. Then, the entire solution will be heated at 300 °C for one hour. The core@shell-1@shell-2 will be collected by the addition of ethanol followed by centrifugation.
Similar to the above core@shell formation, many shells or multi-shells can be added (Figure 10.5). However, there will be an interface between the core and shell-1, shell-1, shell-2, and so on. If the duration of heat treatment increases, the interface phase will dominate over the core or shell composition because of the enhanced diffusion of metal ions between the interface layers. Sometimes, the presence of vacancies in lattice sites or impurities enhances the diffusion of metal ions. If the core and shells have the same lattice of crystal structure (say the hexagonal phase of $NaYF_4$), the growth process of layers is known as the epitaxial growth.

(ii) Core@shell formation (shell: inactive or active)
To prepare 1 mmol of C NPs (core $NaErF_4$:0.5 at.% Tm), 0.95 mmol of erbium acetate (III) and 0.05 mmol of thulium acetate (III) are mixed with 7 ml of

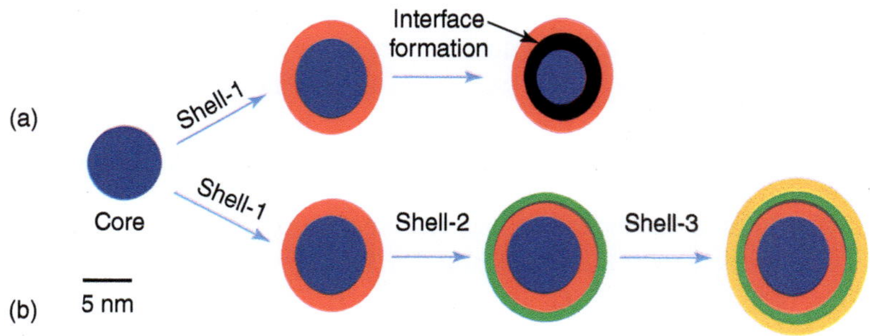

Figure 10.5 Schematic diagram showing the core@shell formation: (a) the formation of shell-1 over core and interface formation between core and shell-1 with the duration of heat treatment, (b) the formation of core@shell-1, core@shell-1@shell-2, core@shell-1@shell-2@shell-3, and so on. Interface formation is not shown here.

oleic acid and 14 ml of octadecene in a three-neck RB flask (Joshi et al. 2021, p. 3481). The content is heated under Ar atmosphere at 120–130 °C for one hour. In this process, metal oleates of Er and Tm are formed, and this solution is brought to room temperature. In another beaker, 2.5 mmol of NaOH is treated with 4 mmol of NH_4F in methanol, and this solution is added into the metal oleate solution. This is heated at 100 °C for one to three hours to remove methanol. Further, it is heated at 300 °C for one to two hours for the formation of the hexagonal phase (known as β-phase) of C NPs, and this is brought to room temperature. The particles are collected by mixing with ethanol and centrifugation. The particles obtained are dispersed in 10 ml of cyclohexane.

To prepare 1 mmol of core@inactive shell (C@Y NPs, $NaErF_4$:0.5 at.% Tm@$NaYF_4$), 1 mmol of yttrium acetate (III) is mixed with 7 ml of oleic acid and 14 ml of 1-octadecene in an RB flask, and this is heated at 120–130 °C under Ar atmosphere for one to two hours to prepare the Y^{3+} oleate. This solution is brought to room temperature. In another beaker, 2.5 mmol of NaOH is treated with 4 mmol of NH_4F in methanol, and this prepared solution is added to the RB flask containing Y^{3+} oleate. It is heated at 100 °C for two to three hours to remove methanol under an Ar atmosphere. This is added to core NPs (C NPs) in 10 ml of cyclohexane. In order to remove the cyclohexane, it is heated at 80–90 °C for one hour, and then the temperature is increased to 300 °C for two to three hours for the formation of core@inactive shell NPs. This gives the synthesis of core@inactive shell NPs, and the mole ratio of the core and shell is 1 : 1.

To prepare 1 mmol of the core@active shell (C@Yb NPs, $NaErF_4$:0.5 at.% Tm@$NaYF_4$:20 at.% Yb), a similar method of C@Y NPs is used. Ytterbium acetate (III) (20 at.% Yb^{3+}) and yttrium acetate (III) (80 at.% Y^{3+} ions) are used as metal precursors for the preparation of the reaction mixture.

(iii) NPs@SiO_2 (core@shell)

The nanoparticles (NPs) prepared through thermolysis are hydrophobic as every particle has oleic acid capping agents. In order to make hydrophilic surface, it needs to coat with organic ligands, polymers, or others. One way is SiO_2 coating, which has many OH groups (–Si–O–(Si–O)$_n$–Si–O–H) (Joshi et al. 2021, p. 3481). About 50–80 mg of NPs is dispersed in 100 ml of cyclohexane or hexane in a beaker. In another beaker, 500 μl of 2-{2-[4-(1,1,3,3-tetramethylbutyl) phenoxy] ethoxy} ethanol (IGEPAL CO-520) is dispersed in 10 ml of cyclohexane or hexane. Both are mixed and 1 ml of ammonia solution is added into it. Tetraethyl orthosilicate (TEOS, 50 μl) is added dropwise under stirring condition. For every 24 hours, 50 μl of TEOS is added. The amount of addition of TEOS will decide the thickness of the shell. In this way, NPs@SiO_2 will be obtained.

(iv) NPs@SiO_2-NH_2 (core@shell)
In order to functionalize NPs@SiO_2 with NH_2, the prepared NPs@SiO_2 can be functionalized with NH_2 so that it can help in further functionalization of other ligands such as antibody, polyethylene glycol (PEG), drug, luminescence molecules or particles, etc. For this, 1 ml of (3-aminopropyl)triethoxysilane (APTMS) is added to the solution of NPs@SiO_2 and stirred for 24 hours (Joshi et al. 2021, p. 3481). In this way, amine (NH_2)-functionalized silica-coated NPs is obtained (NP_s@SiO_2-NH_2). The particles are collected by mixing with ethanol, followed by centrifugation.

(v) NPs@SiO_2-NH_2-PEG (core@shell)
In order to do further surface functionalization with PEG diacid, NPs@SiO_2-NH_2 is mixed with 1 ml of PEG diacid in pH = 10–12 by the addition of ammonia solution. It is kept for 24 hours under stirring. Here, the amide bond (—NH—CO—) will be formed between NH_2 from nanoparticles and COOH from PEG diacid (Soni et al. 2019, p. 7350).

(vi) NPs@SiO_2-NH_2-folic acid (core@shell)
In order to prepare folic acid (FA) functionalized nanoparticles for targeted therapy to cancer cells (such as MCF 7), it requires a proper procedure. 1-Ethyl-3-(3-dimethylaminopropyl)carbodiimide (EDC, 0.1 mmol) and 0.1 mmol of *N*-hydroxysuccinimide (NHS) are dissolved in phosphate-buffered saline (PBS) at pH 5 under stirring for one hour (Joshi et al. 2021, p. 3481). In another beaker, 10–20 mg of folic acid is dissolved in PBS in pH = 5. Both solutions are mixed together, and folic acid is activated and mixed with 5 ml of 10 mg/ml NPs@SiO_2-NH_2 and is kept for 24 hours under stirring. In this way, folic acid coated NPs@SiO_2-NH_2 is prepared. The bonding between folic acid and NPs@SiO_2-NH_2 will be amide bond (—NH—CO—). These particles can be collected by centrifugation.

(vii) NPs@SiO_2-NH_2-drug (core@shell)
Many drug molecules such as anthracycline doxorubicin (DOX), cis-platin, can be loaded into NPs through chemical bonding or electrostatic interactions (Zhou et al. 2014, p. 7666). Targeted therapy and multi-model imaging can be done by using the upconversion nanoparticles (Chatterjee et al. 2014, p. 8; Chiozzi and Rossi 2020, p. 5090; Zhou et al. 2014, p. 7666; Chen et al. 2015a, p. 1318).

(viii) $NPs@SiO_2$-NH_2-Au (core@shell)
The proper procedure is needed to functionalize the UCNPs with Au nanoparticles for the purpose of radiosensitizers (such as X-rays or γ-rays) or photothermal heating (Zhou et al. 2014, p. 7666; Chen et al. 2015a, p. 1318; Soni et al. 2019, p. 7350). Initially, Au NPs are prepared. Typically, sodium citrate solution (2.2 mM per 250 ml of water) is prepared by dissolving sodium citrate (114 mg) in 250 ml of deionized water. This solution is heated in a three-neck RB flask at 60 °C for 30 minutes under stirring. A condenser is provided in order to prevent the evaporation of water. $HAuCl_4$ (20 mM, 4 ml) is injected to the reaction medium. The color of the solution changes from yellow to pink. This confirms the formation of Au NPs. The surface of Au NPs is negatively charged because of the presence of citrate ions. This solution of Au NPs is mixed with the solution of $NPs@SiO_2$-NH_2 to form $NPs@SiO_2$-NH_2-Au and chemical bonding will be amide linkage (–NH–CO–).

Let us take a few examples where it undergoes the Ostwald ripening for the formation of core@shell NPs. The process for the formation of core@shell is known as the Ostwald's repining method. The small size particles (about 2–3 nm) formed by metal oleate will be deposited over the core UCNPs (8–20 nm), and the small size particles get sacrificed and the core becomes core–shell. In order to prepare the uniform coating over the core, it is better to use small amount of metal oleate precursors.

(a) Hexagonal phase (core $NaYF_4$:Er-Yb, 10–16 nm) nanoparticles are prepared from metal precursors at 300 °C. Then, different amounts of the small-sized nanoparticles (5–6 nm) of cubic phase ($NaYF_4$) are added to the core in order to get different thicknesses of shells (Johnson et al. 2012, p. 11068).
(b) Similarly, the thickness can be increased by layer–layer epitaxial growth ($NaYF_4$:Er-Yb@$NaYF_4$@$NaYF_4$, $NaYF_4$:Er-Yb@$NaYF_4$@$NaYF_4$@$NaYF_4$). The size of core@shell increases from 20 to 24 nm (Johnson et al. 2012, p. 11068). The core@shell of different materials such as $NaYF_4$:Er-Yb@$NaGdF_4$ can be prepared (Johnson et al. 2012, p. 11068; Chen et al. 2015a, p. 1318; Abel et al. 2009, p. 14644).

10.2.2 Hot Injection

Hot injection is a process of injection of precursors at high temperature in a solvent or solution having high boiling point (Johnson et al. 2012, p. 11068; Chen et al. 2015a, p. 1318; Abel et al. 2009, p. 14644; Soni et al. 2021a, p. 483). The precursor gets decomposed and/or then interacted with materials/ions, which is already present in the solvent. In this way, the particles of oxides, sulfides, selenides, tellurides, metals, fluorides, etc., can be prepared. If the capping agent such as oleic acid or linoleic acid or oleylamine (fatty acids) is present, monodispersed particles can be prepared. In this way, nucleation and growth of nanoparticles can be separated.

In order to prepare the core@shell nanoparticles, core nanoparticles can be prepared by either thermolysis or hot injection method. Then, the precursors of the shell (in the form of metal oleate or small size particles) will be injected into the solution having core-monodispersed particles above room temperature (say 100–300 °C). The injection temperature can be varied depending on the type of compound formation required. Many UCNPs of core@shell such as $NaYF_4$:Er-Yb@$NaYF_4$ can be prepared.

10.2.3 Cation Exchange

The monodispersed core nanoparticles have been prepared. Here, the capping agent is fatty acid (oleic acid), and thus, particles are hydrophobic in nature. These capping agent molecule (fatty acids) nanoparticles are exchanged with water-soluble ligands and then dispersed in water (Dong et al. 2012, p. 1297). The dispersed particles interact with other metal ions in water medium. Then, they are heated at high temperature to form core@shell particles. An example is the formation of core@shell nanoparticles such as $NaYF_4$:Yb,Tm@$NaGdF_4$. Initially, oleate-stabilized $NaYF_4$:Yb(20%):Tm(2%) monodispersed particles are prepared by thermolysis. The particles are dispersed in dimethyl formamide (DMF) and dichloromethane, and then, polyvinylpyrrolidone-10 (PVP) is added. It is heated at 60 °C for 24 hours under stirring. In this way, a part of oleic acid is replaced by PVP. The PVP-stabilized particles are interacted with ethyl ether and collected by centrifugation. These particles are dispersed in water medium having different concentrations of $GdCl_3 \cdot 6H_2O$ and heated at different temperatures under an Ar atmosphere. During the heating, the Gd^{3+} ions occupy the surface of the $NaYF_4$:Yb,Tm core and exchange with Y^{3+}, Yb^{3+}, and Tm^{3+} ions to form $NaYF_4$:Yb,Tm@$NaGdF_4$.

10.2.4 Structural Characterizations

There are many possible ways to distinguish the core from core@shell nanostructures. Some methods are transmission electron microscopy (TEM), lattice fringes (high resolution transmission electron microscopy [HRTEM]), electron energy loss spectroscopy (EELS), scanning electron microscopy (SEM), X-ray photoelectron spectroscopy (XPS), high-angle annular dark-field (HAADF) images, secondary ion mass spectrometry (SIMS), and line scanning energy-dispersive X-ray spectroscopy (EDXS). There are limitations in the case of TEM and SEM images if the density of phase (core) is similar to that of the shell. If the thickness of shell is large (>50 nm), there is limitation for XPS because of the limitation of penetration depth. Sometimes, lattice fringes can predict the core@shell structure if the core and shell have different crystal structures. Otherwise, it is difficult to predict them. The line scanning EELS is a better choice among many techniques. Line scanning energy dispersive analysis of X-rays (EDAX) is one of the options to predict core@shell structure, and its resolution is less than that of EELS. SIMS is one of the ways to predict core@shell if particles have uniform size coating over core particle. Here, the upper layer in angstrom is removed by bombarding with heavy ions, and then m/z analysis is done. One of the indirect predictions for core@shell is the X-ray

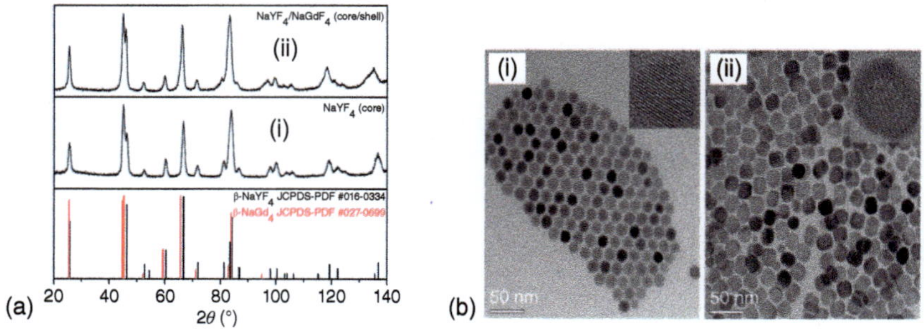

Figure 10.6 (a) XRD patterns of core: $NaYF_4$ (i) and core@shell ($NaYF_4$@$NaGdF_4$) (ii). (b) Respective TEM images. The inset shows the corresponding HRTEM images. Scale bar = 50 nm. Source: Abel et al. (2009).

absorption near-edge spectroscopy (XANES). The UC emission studies have been performed in different core@shell NPs by different researchers, a few of them are reported in the present chapter.

(i) $NaYF_4$@$NaGdF_4$: The X-ray diffraction (XRD) patterns of core $NaYF_4$ and core@shell $NaYF_4$@$NaGdF_4$ are shown in Figure 10.6a (Abel et al. 2009, p. 14644). Both core and shell show the hexagonal phase. It is difficult to distinguish whether it forms core@shell or alloy on the basis of lattice parameters. Here, alloy means solid solution having homogeneous distribution of metal/non-metal ions. The core@shell is a heterogeneous system. Figure 10.6b shows the TEM images of core $NaYF_4$ and core@shell $NaYF_4$@$NaGdF_4$. The core has uniform-sized spherical particles, and the average size is 16.2 ± 0.8 nm in diameter (Abel et al. 2009, p. 14644). Core–shell is 21 ± 1 nm in diameter, and its sizes are less uniform. Figure 10.7 shows the XPS spectra of the Y^{3+} (3d) and Gd^{3+} (4d) core levels at different kinetic energies or excitation energies. An increase in the kinetic energy enhances the penetration depth of the surface by X-rays (Abel et al. 2009, p. 14644). After proper correction, it is found that the Y^{3+}/Gd^{3+} ratio increases with excitation energy. It means that the Gd^{3+} ions are on the surface, and the Y^{3+} ions are inside. Even if alloys are formed, the Y^{3+}/Gd^{3+} ratio should be the same at different excitation energies. However, the interface between the core and shell could not be identified by XPS and HRTEM images as both have similar crystal structures.

The layer-by-layer epitaxial growth on upconverting nanocrystals has been prepared by scarifying small-sized shell particles over a larger core/seed particle (Johnson et al. 2012, p 11068). This is known as self-focusing through the Ostwald repining process. Firstly, small-sized 6.5 nm particles of α-$NaYF_4$ (cubic phase) are prepared. Different amounts of α-$NaYF_4$ dispersed in 1 ml of 1-octadecene are injected to the β-phase (hexagonal [β]) $NaYF_4$ upconverting core (2 mol% Er^{3+}, 15 mol% Yb^{3+} doped) at 300 °C. The size of the core particle is 18 nm. Within 10 minutes, α-$NaYF_4$ particles are found deposited over β-$NaYF_4$:Er-Yb and become complete β-phase formation (i.e. β-$NaYF_4$:Er-Yb (core)@$NaYF_4$ (shell)). With the increase in the amount of α-$NaYF_4$, the size of the particles increases because of

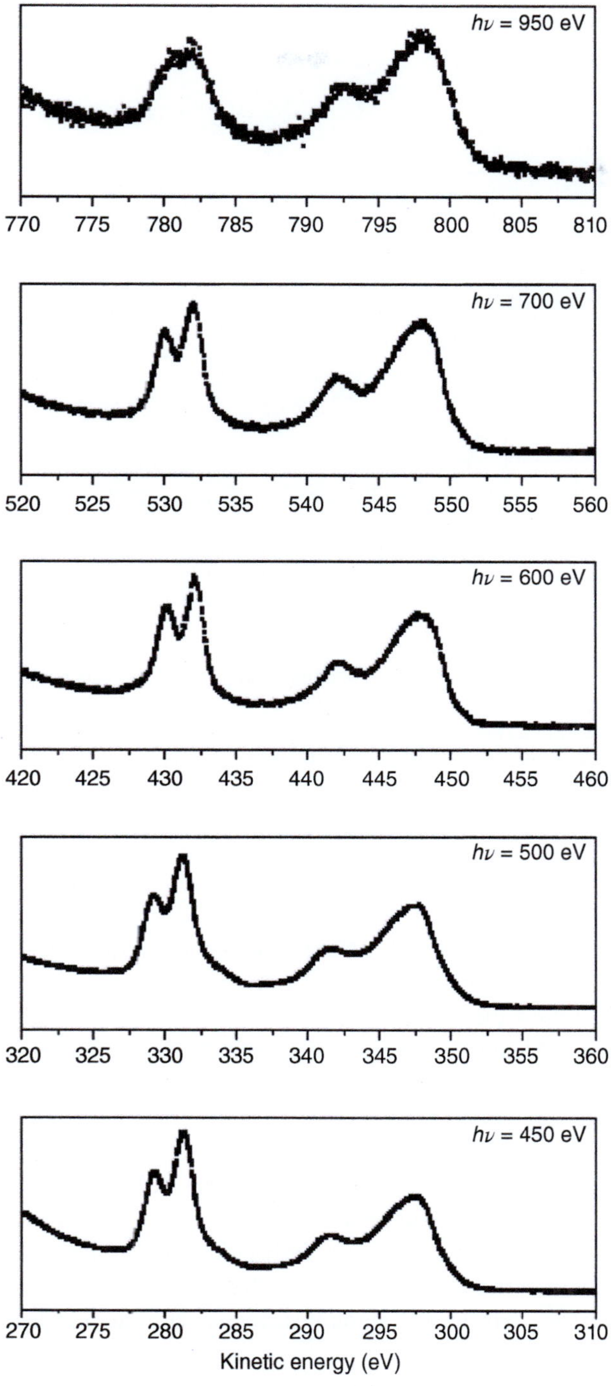

Figure 10.7 XPS spectra of core@shell ($NaYF_4$@$NaGdF_4$) at different excitation energies: 450, 500, 600, 700, and 950 eV. Source: Based on Abel et al. (2009), p. 14644.

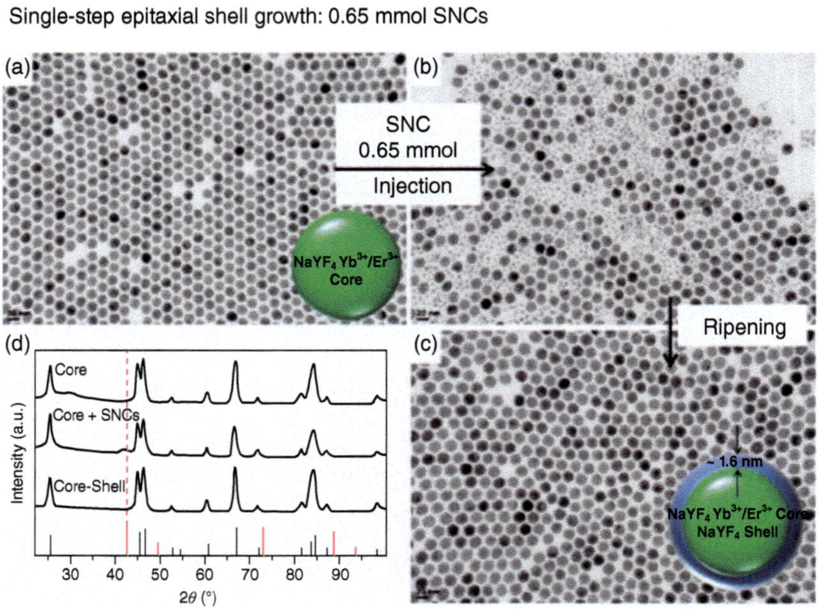

Figure 10.8 TEM images of (a) hexagonal β-$NaYF_4$:Yb^{3+}/Er^{3+} (15/2%) core NCs (at $t = 0$), (b) after injection of small size cubic α-$NaYF_4$ (at $t = 5$ seconds), and (c) after self-focusing β-$NaYF_4$:Yb^{3+}/Er^{3+} (15/2%) core@$NaYF_4$ shell (at $t = 10$ minutes). (d) XRD patterns of the above crystals along with the standard reference patterns of α-$NaYF_4$ (red color) and β-$NaYF_4$ (black color) (JCPDS – 06-0342 for α-$NaYF_4$ and 016-0334 for β-$NaYF_4$). The shell thickness increases to 1.6 nm after injection of 0.65 mmol of α-$NaYF_4$. Source: Modified from Johnson et al. (2012), p. 11068.

increase of extent of shell thickness. With the addition of α-$NaYF_4$ amount from 0.26 to 0.65 mmol, the thickness of the shell increases from 1 to 1.6 nm. At the initial stage of reaction between the core and α-$NaYF_4$, binomial distribution of particles is observed, and this is confirmed by the peak for cubic phase in the XRD pattern. Figures 10.8 and 10.9 show the increase of particle size in the TEM images after addition of cubic phase (0.65 and 0.25 mmol) to hexagonal phase (core) and their corresponding XRD patterns and particle size distribution of core, shell, and core@shell (Johnson et al. 2012, p. 11068).

By successive injection of 0.2 mmol of small-sized α-$NaYF_4$ at different times over core (β-$NaYF_4$:Er-Yb), the self-focusing monodispersed particles can be prepared, and the size of the particle increases with successive injection. This is referred to as layer-by-layer epitaxial growth (shown in Figure 10.10) (Johnson et al. 2012, p. 11068). ICP-MS study shows that the amounts of Er and Yb do not change in both the core and core@shell particles, but the Y^{3+} amount increases with the increase of shell thickness.

The core@shell formation is proved by taking core $NaYF_4$ and core@shell $NaYF_4$@$NaGdF_4$. The HAADF images of the core@shell $NaYF_4$@$NaGdF_4$ are shown in Figure 10.11 (Johnson et al. 2012, p. 11068). From that, EELS mapping confirms the presence of Gd in two different particles.

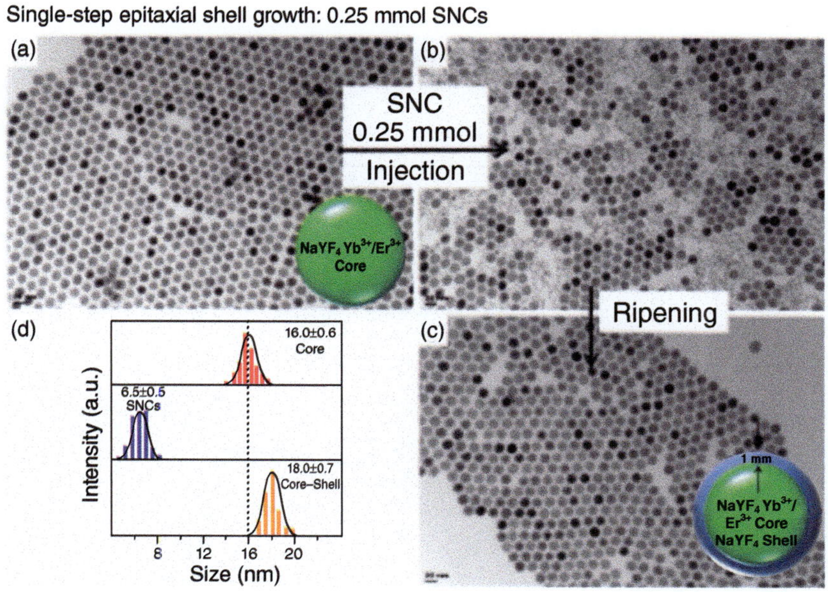

Figure 10.9 TEM images of (a) hexagonal β-$NaYF_4$:Yb^{3+}/Er^{3+} (15/2%) core NCs (at $t = 0$), (b) after injection of small size cubic α-$NaYF_4$ (at $t = 5$ seconds), and (c) after self-focusing β-$NaYF_4$:Yb^{3+}/Er^{3+} (15/2%) core@$NaYF_4$ shell (at $t = 10$ minutes). (d) Particle size distribution of core, shell, and core@shell. Shell thickness increases to 1.0 nm after injection of 0.25 mmol of α-$NaYF_4$. Source: Johnson et al. (2012), p. 11068. Reprinted with permission of American Chemical Society.

Core@shell formation is also proved by taking core $NaYF_4$ and core@shell $NaYF_4$@$NaGdF_4$. The HAADF images of core@shell and the corresponding EELS line scans are shown in Figure 10.12a–d, in which Gd is residing predominantly in the shell (Abel et al. 2011, p. 185). This is also supported by EDX line scanning (Figure 10.12e,f), in which the maximum peak corresponding to Y^{3+} is observed at the center of the particle, and maximum peaks corresponding to the Gd^{3+} ions are observed at both sides of the center of the particle (Abel et al. 2011, p. 185).

10.2.5 Optical Characterization

In the optical properties, inorganic monodispersed nanoparticles (NPs) will be discussed. Otherwise, there are many issues such as agglomeration, mixture of uncoated core, partially coated core, and core@shell or shell particles, etc. Depending on the nature of core and shell, frequency upconverting nanoparticles can be grouped into (a) core (crystalline) and shell (crystalline); (b) core (crystalline) and shell (non-crystalline); (c) core (non-crystalline) and shell (crystalline); (d) core (crystalline) and shell-1 (crystalline), shell-2 (crystalline), or multi-layer (epitaxial layer-by-layer); (e) core (crystalline) and shell-1 (crystalline), shell-2 (non-crystalline); etc. Many possible core@shell nanostructures are possible. Many examples of $3d^{n+}$ ion-doped unconverted core@shell nanoparticles are available.

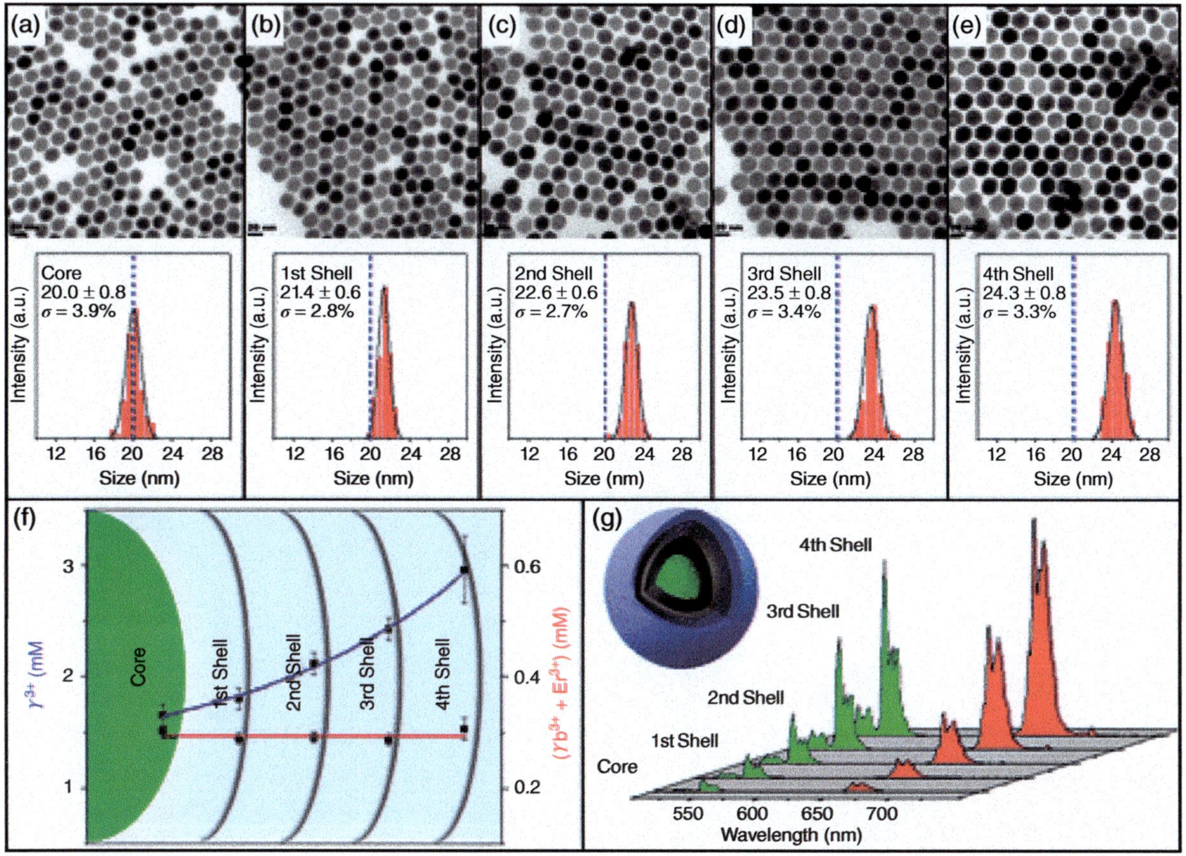

Figure 10.10 TEM images and particle size distribution of (a) hexagonal β-$NaYF_4$:Yb^{3+}/Er^{3+} (15/2%) core NCs (at $t = 0$). (b–e) The corresponding images and distribution of β-$NaYF_4$:Yb^{3+}/Er^{3+} (15/2%) core@$NaYF_4$ shell NCs after successive layer-by-layer epitaxial growth at $t = 5$, 10, 15, and 20 minutes, respectively. (f) ICP-MS elemental analysis of the core and core@shell NCs with the same number concentration of NCs. (g) Upconversion emission spectra of core and core@shell NCs dispersed in the hexane with the same number concentration of NCs under 980 nm laser excitation. Source: Johnson et al. (2012), p. 11068. Reprinted with permission of American Chemical Society.

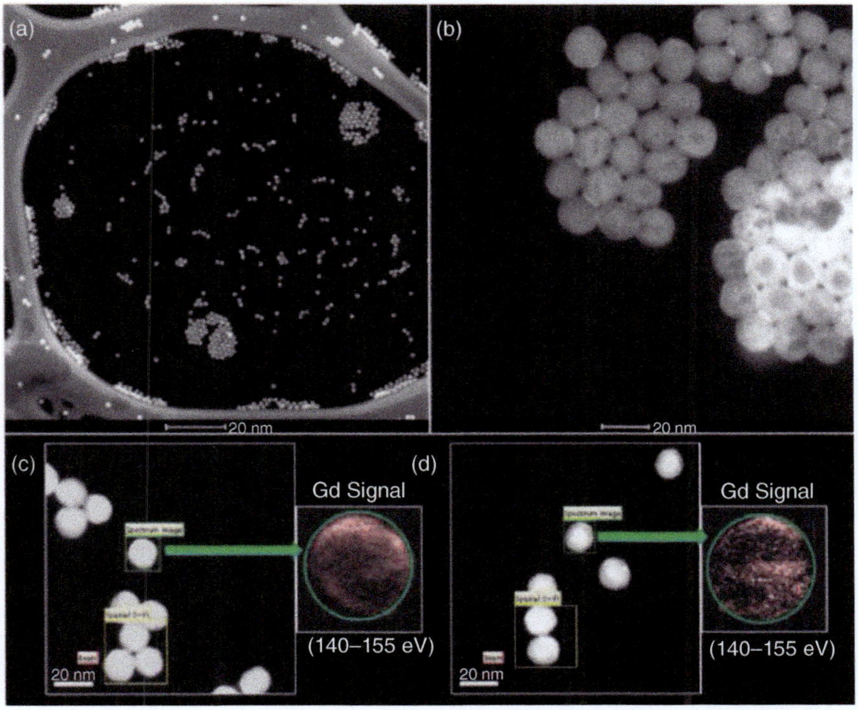

Figure 10.11 (a) HAADF TEM images of $NaYF_4$ core@$NaGdF_4$ shell NCs at low magnification, (b) HAADF TEM images of $NaYF_4$ core@$NaGdF_4$ shell NCs at high magnification, and (c, d) EELS mapping of Gd in two different randomly selected $NaYF_4$ core@$NaGdF_4$ shell NCs demonstrating the deposition of $NaGdF_4$ in shell. Source: Johnson et al. (2012), p. 11068. Reprinted with permission of American Chemical Society.

However, because of the limitation of space, lanthanide ion (Ln^{3+} = $4f^n$)-doped core@shell upconverted nanoparticles are discussed.

10.2.5.1 Normal Conversion Process in Ln-Doped Core@shell Nanoparticles

Let us discuss these groups in optical properties: (a) core (crystalline) and shell (crystalline): Core can be of $NaYF_4$:Er-Yb, which has two phases – cubic and hexagonal. It emits upconverted light at 546 nm (green) and 650 nm (red) by absorbing light at 980 nm. The Yb^{3+} ion absorbs light at 980 nm and transfers its excited energy to excited states/metastable levels of Er^{3+}. Er^{3+} emits light at 546 nm (green) and 650 nm (red) (Johnson et al. 2012, p. 11068; Soni et al. 2019, p. 7350). The luminescence intensity of Er^{3+} is higher in the case of hexagonal phase (β-phase) as compared to that of cubic phase (α-phase). This can be explained by the occupancy of low symmetry of Er^{3+} in the hexagonal phase and high symmetry of Er^{3+} in the cubic phase. It is to be noted that the cubic phase is kinetically stable, whereas the hexagonal phase is thermodynamically stable. When the undoped shell $NaYF_4$ is covered over core ($NaYF_4$:Er-Yb), the luminescence intensity is improved by the order depending on the thickness of the shell (Johnson et al. 2012, p. 11068). This is due to the decrease of non-radiative pathways. In the case of core, the surface or the

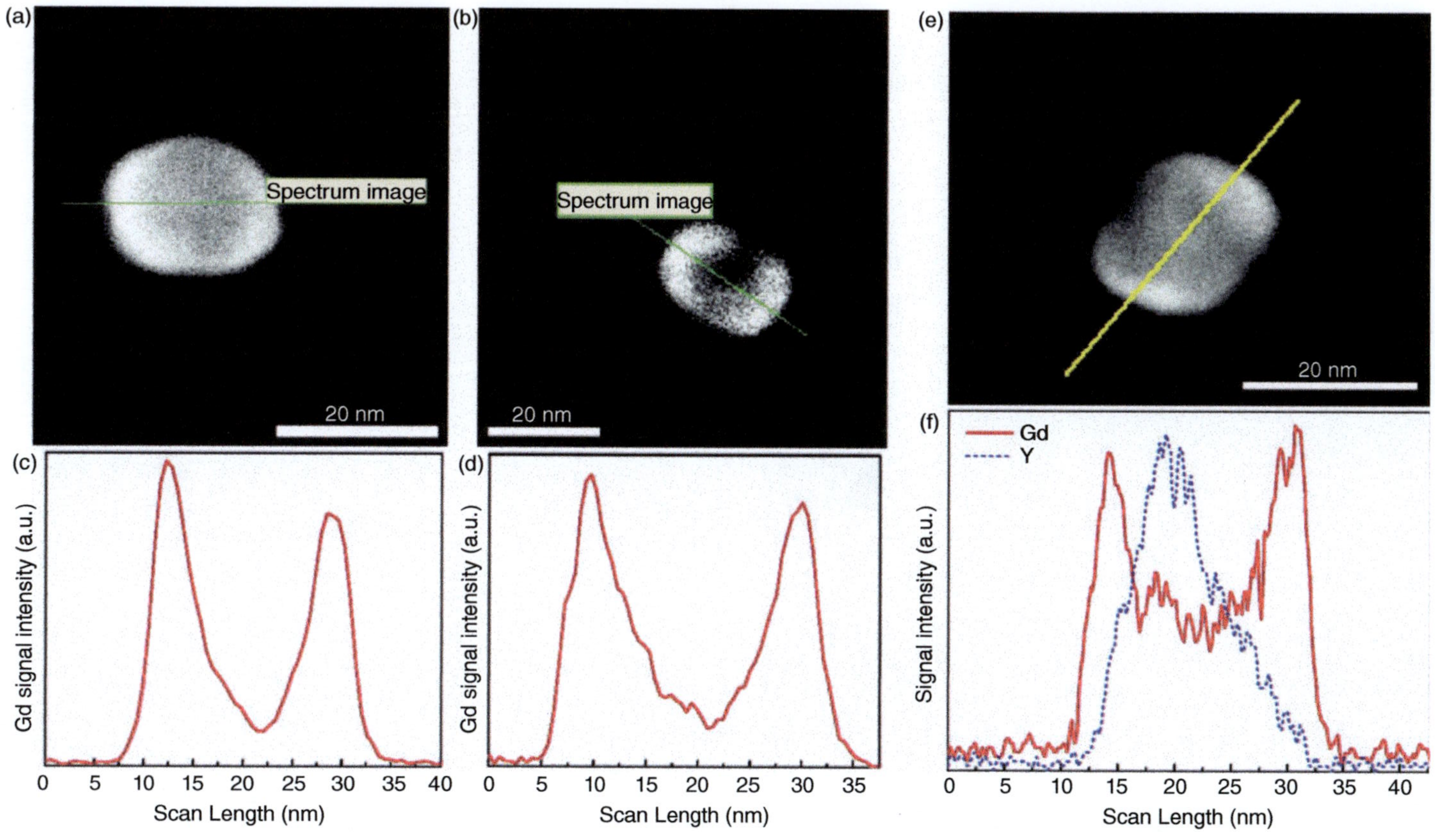

Figure 10.12 (a, b) HAADF images of single $NaYF_4$ core@$NaGdF_4$ shell NCs and corresponding (c, d) EELS line scans showing Gd residing predominantly in the shell. (e, f) EDS line scan across a single $NaYF_4$ core@$NaGdF_4$ shell NCs showing Y in the core of the particle and Gd located in the shell. Source: Abel et al. (2011), p. 185. Reprinted with permission of American Chemical Society.

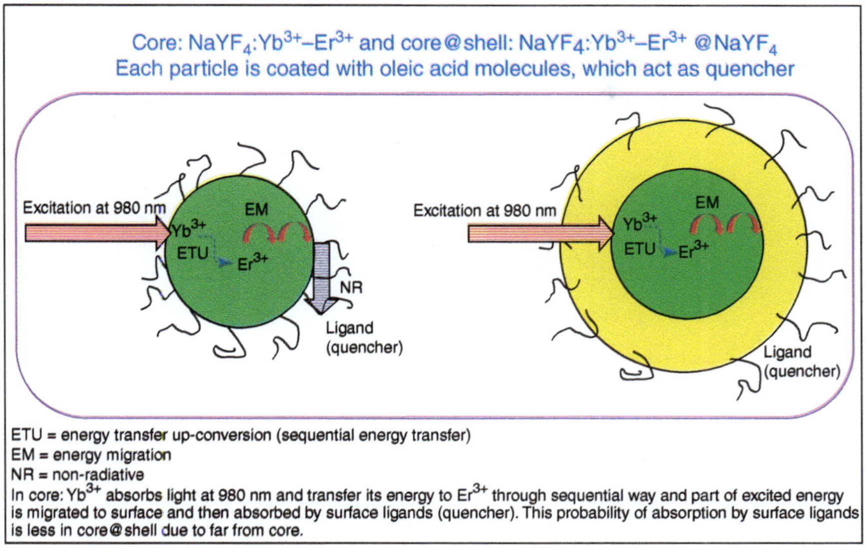

Figure 10.13 ETU and EM processes in core and core@shell particles. The core is represented by $NaYF_4$:Yb^{3+}-Er^{3+}. Core@shell is represented by $NaYF_4$:Yb^{3+}-Er^{3+}@$NaYF_4$.

surrounding solvent or ligands is the source of non-radiative pathways. When the shell is covered over the core, its surface dangling bonds are passivated (becomes perfect) and luminescent ions (active ions such as Er^{3+} and Yb^{3+}) are far from the solvent or ligands associated for the formation of particles or any design of experiment. Here, term-active ions mean that they can absorb light or emit light during the study, whereas term-inactive ions mean that they will not participate in absorption or emission but provide the platform during energy transfer or migration (Wang et al. 2011, p. 968; Wen et al. 2018, p. 1). In general, the amount of Er^{3+} is taken in less quantity (1–2 at.%), and the Yb^{3+} amount is taken 10–20 at.%. Figure 10.13 shows the schematic diagram showing energy migration (symbol = EM) from the activator core center to its surface and from the activator core center to the shell surface.

The luminescence intensity decreases when the concentration of the activator is more than that of the optimum value. However, it is reported that when high concentrations of dopants are used, the luminescence intensity can be improved by excitation at high power, which is more than the critical value (Joshi et al. 2021, p. 3481; Gargas et al. 2014, p. 300). By doping of Nd^{3+} in the host, excitation at 800 nm can improve the luminescence intensity. In the case of core@shell, the upconversion luminescence intensity is high even in highly doped materials.

One example of (a) core (crystalline) and shell (crystalline) is $NaErF_4$:0.5 at.% Tm@$NaYF_4$:20 at.% Yb. Here, 99.5 at.% Er^{3+} and 0.5 at.% Tm^{3+} occupy whole Y^{3+} sites in core $NaYF_4$ (Joshi et al. 2021 p. 3481). It means that the concentration of Er^{3+} is too high. The core can show luminescence at 480 nm (Tm^{3+}), 546 nm (Er^{3+}), and 650 nm (Er^{3+}) under 980 nm excitation, but their intensity is very weak. When shell $NaYF_4$ (undoped) is covered over core, enhanced luminescence intensity is observed. Further, if $NaYF_4$:20 at.% Yb (doped) is covered over core, 36 times

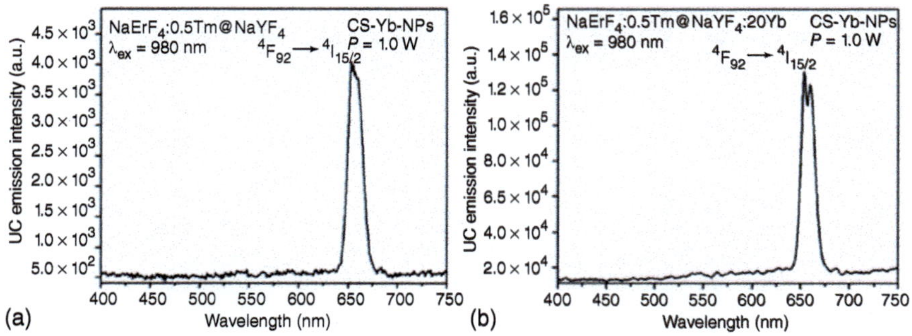

Figure 10.14 UC emission spectra of core@shell nanoparticles: (a) $NaErF_4$:0.5% Tm@$NaYF_4$ and (b) $NaErF_4$:0.5% Tm@$NaYF_4$:20% Yb. Source: Joshi et al. (2021), p. 3481. Reprinted with permission of American Chemical Society.

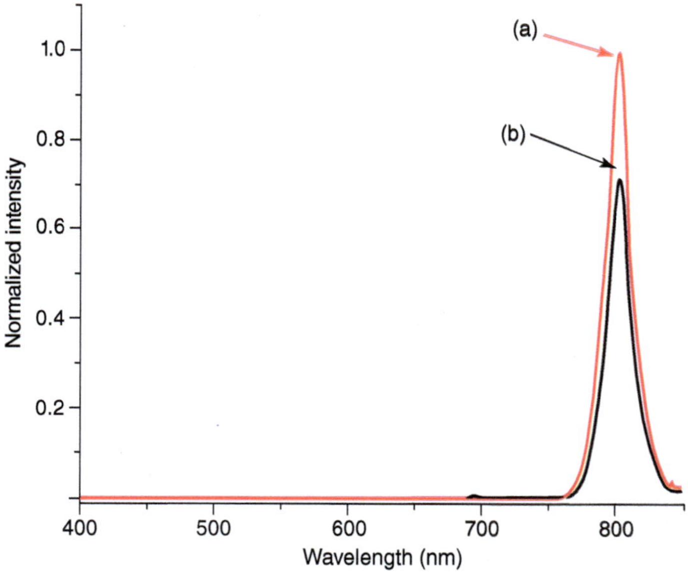

Figure 10.15 UC emission spectra of (a) core@shell $NaYF_4$:Yb-Tm@$NaGdF_4$ and (b) core $NaYF_4$:Yb-Tm. Source: Dong et al. (2012).

enhancement in the luminescence intensity is observed in $NaErF_4$:0.5 at.% Tm core@$NaYF_4$:20 at.% Yb shell as compared to that of $NaErF_4$:0.5 at.% Tm@$NaYF_4$ (Figure 10.14) (Joshi et al. 2021, p. 3481). Another example of improvement of luminescence intensity by core@sell is $NaYF_4$:Yb-Tm@NaGdF4. The presence of Tm^{3+} shows a luminescence peak at ~800 nm (shown in Figure 10.15) (Dong et al. 2012, p. 1297).

Other possible systems for (a) core (crystalline) and shell (crystalline) are $NaYF_4$:Er-Yb@NMs (NMs = noble metal nanoparticles Ag, Au, Pt, and Pd), $NaYF_4$:Er-Yb@QDs (QDs = ZnS, ZnTe, PbSe, and PbTe), and $NaYF_4$:Er-Yb@MOx (MOx = metal oxides such as MnO, Fe_3O_4, CaO, Ga_2O_3, and Bi_2O_3) (Choudhury and Choudhury 2012, p. 2373; Chatterjee et al. 2014, p. 8; Chiozzi and Rossi 2020, p. 5090; Zhou et al. 2014, p. 7666; Chen et al. 2015a, p. 1318; Kumar et al. 2013, Article: 672059; Ye et al. 2016, p. 1600302). In general, $NaYF_4$:Er-Yb@NMs show poor luminescence as compared to that of core $NaYF_4$:Er-Yb because of the decrease of penetration depth of the excitation light, which is protected by NMs, and sometimes, some light coming out from Er^{3+} is absorbed by surface plasmon resonance (SPR), which produces heating to surrounding, and this heat can reach to 42 °C, which is possible for killing of cancer cells, and this process is known as PTT (photothermal therapy) (Soni et al. 2019, p. 7350). Similarly, $NaYF_4$:Er-Yb@QDs and $NaYF_4$:Er-Yb@MOx can show the following: change of emission intensity ratio of green to red lights of Er^{3+} ions (because of the absorption of light coming out from Er^{3+} by QDs, Mn^{2+} ions) and addition of other emission peaks corresponding to QDs or MO_x (Ye et al. 2016, p. 1600302). Many examples of transition metal-involved photon upconversion including core@shell nanostructures are reported (Ye et al. 2016, p. 1600302).

One example of (b) core (crystalline) and shell (non-crystalline) is $NaYF_4$:Er-Yb@ SiO_2. In this case, the luminescence intensity of core–shell is less than that of core because the SiO_2 shell has many OH groups, which are the source of non-radiative pathways (Choudhury and Choudhury 2012, p. 2373; Chatterjee et al. 2014, p. 8; Chiozzi and Rossi 2020, p. 5090; Zhou et al. 2014, p. 7666; Chen et al. 2017, p. 01; Soni et al. 2019, p. 7350; Kumar et al. 2013, Article: 672059; Ye et al. 2016, p. 1600302).

One example of (c) core (non-crystalline) and shell (crystalline) is SiO_2@ $NaYF_4$:Er-Yb. In general, the luminescence intensity will be weak.

One example of (d) core (crystalline) and shell-1 (crystalline), shell-2 (crystalline), or multi-layer (epitaxial layer-by-layer) is $NaYF_4$:Er-Yb@$NaYF_4$@$NaYF_4$@$NaYF_4$. Here, the core is active, but shells are inactive. The luminescence intensity of Er^{3+} at 546 and 655 nm increases with the increase of shell layers under 980 nm excitation (Figure 10.10) (Johnson et al. 2012, p. 11068). Also, the luminescence intensity decreases if the shell thickness is more than the critical value because of the reduced penetration depth of excitation light.

Other examples of (d) core (crystalline) and shell-1 (crystalline), shell-2 (crystalline), or multi-layer (epitaxial layer-by-layer) are $NaYF_4$@$NaYF_4$:Er-Yb@$NaYF_4$, $NaYF_4$@$NaYF_4$:Er-Yb@$NaYF_4$ @$NaYF_4$@$NaYF_4$. Here, the core is inactive, the next shell is active, and the second or third shell layers are inactive. In this case, the luminescence intensity can be controlled by the layer thickness. Ag-Au@$NaYF_4$@$NaYF_4$:Er-Yb can enhance the luminescence intensity of Er^{3+} under 980 nm excitation. In general, the luminescence intensity decreases because of the absorption of light emitted by Er^{3+} by SPR from noble metals. In this, Ag–Au nanocages act as the core and surrounded by spacer $NaYF_4$ (inactive) and further surrounded by $NaYF_4$:Er-Yb. The SPR peak of Ag–Au nanocages show a broad peak centered at ~950 nm, which is matching with the excitation peak at ~980 nm. Thus, the absorption cross section is improved by the combination of SPR and excitation source. Also, by optimizing the thickness of a spacer, energy transfer from the excited state of Yb^{3+} to Er^{3+} is improved, and thus, the luminescence intensity is increased by 25-fold (Chen et al. 2017, p. 01).

The possible examples of (d) core (crystalline) and shell-1 (crystalline), shell-2 (crystalline), or multi-layer (epitaxial layer-by-layer) are $NaYF_4$:Er-Yb@$NaYF_4$@$NaYF_4$-RE, $NaYF_4$:Er-Yb@$NaYF_4$@$NaYF_4$-RE@$NaYF_4$, $NaYF_4$:Er-Yb@$NaYF_4$@$NaYF_4$-RE@$NaYF_4$@ $NaYF_4$-RE, etc. Here, RE stands for rare-earth ions such as Eu, Tb, Dy, Sm, Pr, Tm, Gd, and Nd. In this, many excitation routes through UV, visible, and NIR range can give emissions (downconversion and upconversion). The emission can be in UV or visible or NIR range or in combination. This can show multi-functional properties for multi-utilities.

The possible systems for (e) core (crystalline), shell-1 (crystalline), and shell-2 (non-crystalline) are $NaYF_4$:Er-Yb@NYF_4@SiO_2, $NaYF_4$:Er-Yb@NYF_4@SiO_2@antibody, $NaYF_4$:Er-Yb@NYF_4@SiO_2@dye, $NaYF_4$:Er-Yb@NYF_4@SiO_2@drug-molecule, $NaYF_4$:Er-Yb@NYF_4@SiO_2@-folic acid (FA), $NaYF_4$:Er-Yb@NYF_4@SiO_2@polymer, $NaYF_4$:Er-Yb@NYF_4@SiO_2@glucose, $NaYF_4$:Er-Yb@NYF_4@SiO_2@porphyrin, etc. The emitted light coming out from Er^{3+} can be absorbed by dye/porphyrin, and then, its e-h pair can be separated, and many photo-dissociation (presence of photocatalyst) such as water splitting, degradation, or production of singlet oxygen and reactive oxygen can occur (Ai et al. 2015, p. 10785). This extends the utilities of upconversion in different areas of H_2 generation, killing of germs or cancer cells through photodynamic therapy (PDT). The individual particles having antibody/glucose/FA can target cancer cells (Zhou et al. 2014, p. 7666). It is to be noted that the Nd^{3+} ions show low absorption cross section at 800 nm, and after conjugation with proper dye, which can absorb at 800 nm, it enhances the luminescence intensity (Joshi et al. 2021, 3481).

In the above examples, the $NaYF_4$ host is used as the core or shell in which activators or sensitizers or dopants are added. In addition, there are many hosts such as $LaYF_4$, $LiYF_4$, $NaGdF_4$, LnF_3, AF_2, metal oxides, metal sulfides, glasses, etc., which can be useful. Among these hosts, the hexagonal phase $NaYF_4$ host is one of the best hosts for the frequency upconvertors.

In general, the luminescence intensity of Ln^{3+} is poor because of the forbidden nature of f–f transitions, but peaks are very sharp because $4f^n$ electrons are screened from the surrounding ligands by outer electronic orbitals: $5s^25p^6$ (Ningthoujam 2012, p. 145). In addition, the luminescence intensity decreases when the amount of the Ln^{3+} ions is more than that of the optimum concentration. This is due to the cross-relaxation between the Ln^{3+} ions, which has many metastable levels in energy states. In this process, one ion gets to higher excited state, whereas another nearby ion is deactivated to lower energy state. This is known as the concentration quenching effect. When the particle size decreases to nanosize, the surface to volume becomes very high. The surface has dangling bonds, which is the source of quencher. In order to prepare the monodispersed particles, it requires long-chain molecules such as fatty acids, which can protect the surface of particles. Such a capping agent or a protector is the source of quencher. In order to achieve high luminescence intensity or high quantum yield, one way is the formation of core@shell nanostructure (Ningthoujam et al. 2009c, p. 782; Johnson et al. 2012, p. 11068). There is an energy migration from the activator to the surface in core, and this energy migration is stopped/suppressed by shell in core@shell structure (Wang et al. 2011, p. 968). In the luminescence, there is a donor or a sensitizer, which can absorb incoming light efficiently and then transfer its excitation energy to the activator through the exchange/electron energy transfer (Dexter type) or dipole–dipole energy transfer (Forster type) (Wen et al. 2018, p. 01). The luminescence intensity or the lifetime of the donor decreases with the increase of activator or acceptor ion concentration. In general, Dexter-type energy transfer is short range between donor and acceptor (less than 0.7 nm). The rate of energy transfer from donor (D) to activator (A) follows e^{-kR}, where k is a constant that depends on the inverse of the van der Waals radius of the atom and R is the D–A distance. Forster-type energy transfer can have up to 6 nm, and it follows $1/R^6$ (R = distance between the donor and activator). Thus, in the case of very high concentration of donor or activator, the energy transfer will be of Dexter type. There is another term "accumulator." In this, donor ions absorb the light efficiently, and then, the excited energy is accumulated to the nearby ions, from which there is an energy transfer to the activator/acceptor. This process can enhance the luminescence intensity at very high concentration of donor or accumulator or acceptor (activator). Let us take a few examples.

$NaGdF_4$:Yb/Tm (49/1%)@$NaGdF_4$:X (X = Tb, Eu, Dy, and Sm) shows the luminescence peaks for Gd^{3+} at 310 nm, Tm^{3+} at 280, 340, 450, and 475 nm, and the respective peaks for X (Wang et al. 2011, p. 968). The emission peaks at 475, 545 nm for X = Tb^{3+}, 590, 615 nm for X = Eu^{3+}, and 475, 575 nm for X = Dy^{3+} are observed. Here, the core ($NaGdF_4$:Yb/Tm) is doped with Yb^{3+} (49 at.%) and Tm^{3+} (1 at.%) with respect to Gd^{3+}, and shell ($NaGdF_4$:X) is doped with X. Here, the Yb^{3+} ions absorb 980 nm photons efficiently through the $^2F_{7/2} \rightarrow {}^2F_{5/2}$ transition, and the excited energy is transferred to the $^1I_6$ level of Tm^{3+} ions by the absorption of five number of photons. The excited energy is further accumulated to the excited energy level ($^6P_{7/2}$) of the Gd^{3+} ions. This energy is transferred to the excited energy levels of X, and finally, emission from X is observed. Some of the excited energy remains in the excited states of Tm^{3+} and Gd^{3+}, and then, their emission peaks are also observed along with the emission peaks of X. Here, the Yb^{3+} ions act as a sensitizer or donor. X ions act as an activator or a receptor. The Gd^{3+} ions act as an accumulator or a mediator. The process is known as energy-mediated upconversion (EMU), which is different from the ground-state absorption (GSA) or excited-state energy transfer or sequential energy transfer upconversion (Wang et al. 2011, p. 968).

10.2.5.2 Loop-Type and Avalanche-Type Upconversion Processes in Core@shell Nanoparticles

In the case of photon avalanche (PA), the upconverted luminescence intensity is improved by core@shell formation (Levy et al. 2016, p. 8423; Lee et al. 2021, p. 230). In this condition, the dopant (activators) amount is high, and non-resonance excitation takes place. When the excitation power is above the critical value, high intensity of emitted light occurs. The cross-relaxation among the dopant ions helps in producing a close loop of absorption of excitation light, and its high electron population at excitation levels from this absorption starts, e.g. $NaYF_4$:8–80% Tm@$NaGdF_4$.

Monodispersed core@shell nanoparticles of $NaY_{1-x}Tm_xF_4@NaY_{0.8}Gd_{0.02}F_4$ have been prepared by the thermolysis method. Here, shell acts as inert to the excitation wavelength of 1064 and 1450 nm and decreases the surface losses, whereas the core is active to these excitations. The amount of Tm^{3+} varies from 1, 4, 8, 20 to 100 at.%, and the lower concentrations (1, 4 at.% Tm^{3+}) can produce looping upconversion. To get PA, the concentration of Tm^{3+} increases to an amount of 8–100 at.%, and the critical power or threshold power to get PA is in the range of 4.9–32.8 kW/cm^2. The intensity (I_{em}) of emitted light is proportional to the power of excitation light (P^s, s = slope or number of photons involved in the upconversion process). Generally, $s = 2–3$ suggests for normal upconversion process, $s = 5–7$ for looping and $s > 10$ for PA. Here, emitted light at 800 nm is measured, and excitation light at 1064 or 1450 nm is used. Core ($NaY_{1-x}Tm_xF_4$) shows a mixture of spherical as well as prolate shapes in TEM images (Lee et al. 2021, p. 230). The thickness of the core ($NaY_xTm_yF_4$) varies from 10.2 to 17.5 nm, and the shell varies from 2.6 to 8.5 nm in core@shell ($NaY_{1-x}Tm_xF_4@NaY_{0.8}Gd_{0.02}F_4$), but the particles are of spherical shape (Lee et al. 2021, p. 230).

One example of core@shell NPs is $NaY_{1-x}Tm_xF_4@NaY_{0.8}Gd_{0.2}F_4$, which shows the PA mechanism. This shows avalanching only when the concentration of Tm^{3+} is more than 8 at.% (Levy et al. 2016, p. 8423). When light at 1064 nm is provided to core–shell NPs, there is no-resonance absorption. Some energy gets lost through phonon relaxation and excited electrons come down to the $^3H_5$ or $^3F_4$ level of Tm^{3+}. Thus, the GSA from $^3H_6$ is very weak. As continuous wave (CW) light is provided, excitation-state absorption (ESA) occurs, and this energy 1064 nm is matching well with the gap between the $^3F_4$ and $^3F_2$ levels through ESA. Then, the relaxation of excited energy occurs and comes back to $^3H_4$. Then, the emission at 800 nm occurs, and this transition is due to the $^3H_4$–$^3H_6$ transition. However, a part of populated excited electrons at $^3H_4$ gets relaxed to $^3F_4$. Then, energy is absorbed by the nearby Tm^{3+} ions, and electrons at the ground state go to the excited state and the population at $^3F_4$ increases. This energy transfer is known as cross-relaxation (CR or S_{31}) because the Tm^{3+} ion concentration is high. This process keeps on continuation if there are available Tm^{3+} ions as well as CW light. This makes a cycle or loop. Above the threshold excitation intensity (P_c), a very high unconverted emitted light at 800 nm occurs ($s > 10$). This is different from the standard or general energy transfer conversion (ETU) process, in which resonance-type GSA occurs. Generally, the Yb^{3+} ions absorb light at 980 nm and transfer their excited energy to metastable excited levels of Tm^{3+}, and then upconverted emitted light at 480 and 800 nm is observed in the Yb–Tm system.

The emitted intensity (I_{em}) at 800 nm against excitation intensity or power (P) shows the non-linear property in PA process under non-resonance excitation at 1064 or 1450 nm. Before threshold power (P_c), the slope of rise of emitted intensity with excitation power is below 4 ($s < 4$) and increases suddenly to $s = 15–31$ at P_c, and then it is saturated.

The intensity of emitted light versus excitation intensity or power for 1, 4, and 8 at.% Tm^{3+}-doped core–shell $NaY_{1-x}Tm_xF_4@NaY_{0.8}Gd_{0.2}F_4$ nanoparticles has been recorded (Lee et al. 2021, p. 230). The 1 and 4 at.% Tm^{3+}-doped samples can reach $s = 5.1–6.7$ nm at 1064 nm excitation, for 8 at.% Tm^{3+} $s = 22.2$ at 1064 nm excitation and $s = 14.9$ at 1450 nm excitation. At P_c, there is a rise time of emitted light, and it decreases with further increase of excitation power above P_c, and finally, the intensity of emitted light is saturated.

The energy looping process is also found in the hexagonal phase of $NaYF_4$:20 at.% Gd, 0.1–1.5 at.% Tm (Figure 10.16a) (Levy et al. 2016, p. 8423). Co-doping of Gd helps in getting the hexagonal phase instead of the cubic phase as the ionic size of Gd^{3+} is more than that of Y^{3+}. Here, excitation source is fixed at 1064 nm, and there is no GSA, but ESA occurs from the $^3F_4$ level to populate the $^3F_2$ level. Cross-relaxation among the Tm^{3+} ions populates the excited electrons at the $^3F_4$ level. The emission at 800 nm ($^3H_4$–$^3H_6$) is observed because of the looping process (shown in Figure 10.16b). Also, the TEM image of particles is provided as an inset of this figure (Levy et al. 2016, p. 8423). The slope value obtained from the emitted intensity versus excitation intensity is about 3 at 4–5 W/cm^2.

10.3 Frequency Upconversion and Its Various Mechanisms

Energy transfer (resonance or non-resonance type), energy migration from the center to the surface or within the active ion to another active ion, energy or photon trapping, back energy transfer, cross-relaxation due to high concentration of activators or donors, phonon-assisted energy transfer, probability of energy absorbed by the surrounding ligands, concentrations of donor and/or acceptor ions, metastable energy levels, and their lifetimes as well as the power of excitation energy are the main sources of down- or upconversion luminescence (Ningthoujam 2012, p. 145; Yadav and Ningthoujam 2021, pp. 213–250; Auzel 2004, p. 139; Johnson et al. 2012, p. 11068; Joshi et al. 2021, p. 3481; Soni et al. 2021a, p. 483; Wen et al. 2018, p. 1; Ye et al. 2016, p. 1600302; Levy et al. 2016, p. 8423; Lee et al. 2021, p. 230).

10.3.1 Inorganic-Based Upconversion

Here, Ln-based upconvertors from core@shell will be only considered because of the space limitation even though there are reports on transition metal ion-based upconvertors from core@shell. Ln ions are the source of activators and sensitizers or donors for frequency upconverting materials. $4f^n$ electrons of Ln are surrounded by $5s^2 5p^6$, and thereby during the 4f–4f electronic transition, there is no much effect from ligands and thus their transition is sharp. However, because of Stark splitting in the ground and excited states, some peaks can be seen in the limited wavelength range in highly crystalline materials or the presence of external electric field.

Let us take a few Ln^{3+}, which are mostly used in upconverting materials. The Yb^{3+} ion is used as a sensitizer or a donor. The ground-state electronic configuration of the Yb atom is [Xe] $6s^2$ $4f^{14}$ and its Yb^{3+} ion has [Xe] $4f^{13}$ (Figure 10.17a). Yb^{3+} ion has the metastable level because of spin-orbital coupling in 4f electrons say $^2F_{7/2}$ (ground state) and $^2F_{5/2}$ (excited state). In the case of Er^{3+} ($4f^{11}$), it has ground-state electronic configuration: [Xe] $4f^{11}$ (Figure 10.17b). Er^{3+} ion has the three metastable levels because of spin-orbital coupling in 4f electrons say $^4I_{15/2}$ (g.s.)

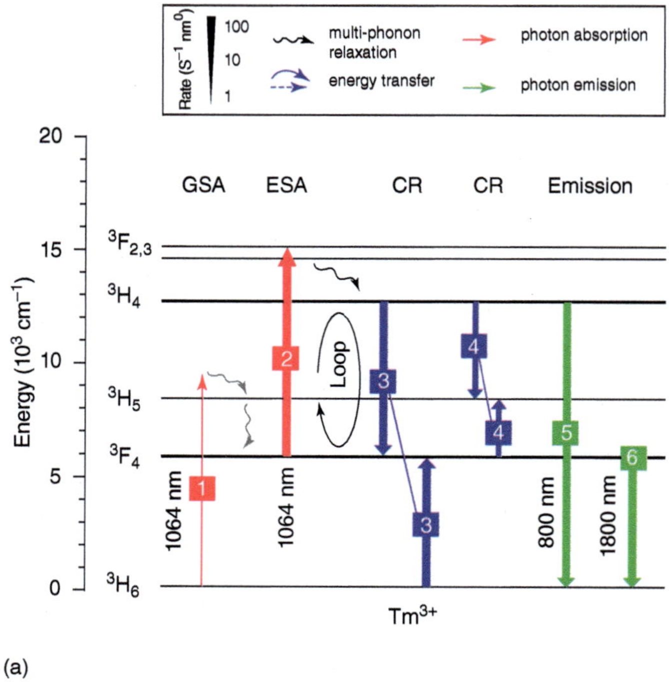

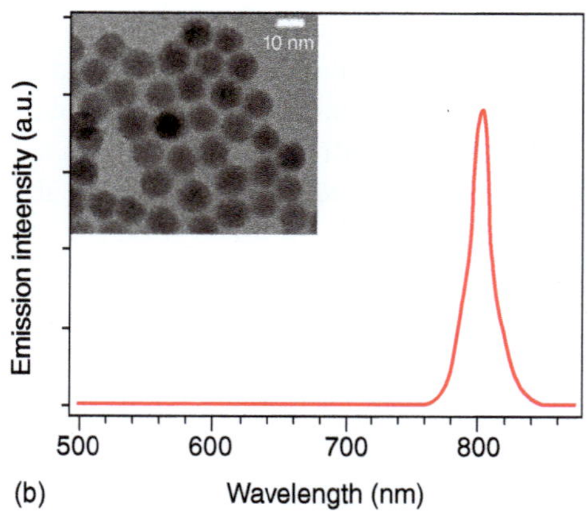

Figure 10.16 (a) Looping mechanism or process in $NaYF_4$:1 at.% Tm^{3+} at 1064 nm excitation (10^5 W/cm^2). Scheme showing energy levels of $4f^{12}$ manifolds of Tm^{3+}. The manifold-to-manifold transitions are provided by arrows. (b) Upconversion spectrum of $NaYF_4$:20at.% Gd-0.5 at.% Tm upon excitation at 1064 nm with a power of 10^5 W/cm^2 through the looping process. Inset: its TEM image. Source: Levy et al. (2016), p. 8423. Reprinted with permission of American Chemical Society.

and excited states $^4I_{13/2}$, $^4I_{11/2}$, and $^4I_{9/2}$. The splitting of peak in spin-orbital coupling energy level in a crystalline lattice doped with lanthanide ions such as Eu^{3+}, Er^{3+} is due to Stark splitting (Ningthoujam 2012, p. 145; Yadav and Ningthoujam 2021, pp. 213–250; Steinkemper et al. 2013, p. 053033). This is related to the electrostatic field of the host lattice or electric field gradient, and it can happen in the ground or excited levels. Term symbol with $j = 0$ does not have splitting.

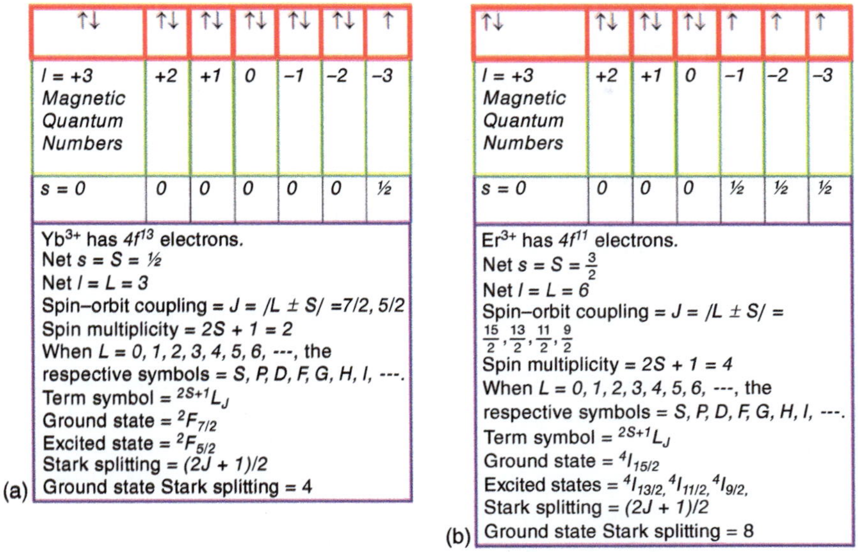

Figure 10.17 The ground-state electronic configurations of Yb^{3+} (a) and Er^{3+} (b). The spin–orbit coupling to give ground state (g.s.) and its excited state (e.s.) and possible Stark splitting.

The number of possible Stark levels (N_i) of each energy level depends on the quantum number of the total angular momentum j of the energy level in a particular term symbol ($^{2s+1}L_j$), which can be defined as

$$N_i = \begin{cases} 2j+1 & \text{for an even number of electrons,} \\ \dfrac{2j+1}{2} & \text{for an odd number of electrons} \end{cases}$$

For Yb^{3+} ions, there are four possible Stark splitting in the ground-state $^2F_{7/2}$, as it has an odd number of electrons. For Er^{3+} ions, there will be possible eight Stark levels in the ground-state $^4I_{15/2}$. It means that $^4I_{15/2}$ has eight energy levels. The excited electrons coming down to the ground state can have many peaks as there are many possible transitions. It is to be noted that transition from $j = 0$ (excited state) to $j = 0$ (ground state) is not allowed (Ningthoujam 2012, p. 145; Yadav and Ningthoujam 2021, pp. 213–250). The width of the peak and its height depend on site symmetry and defects associated with the lattice (Steinkemper et al. 2013, p. 053033; den Engelsen et al. 2018, p. 337). In addition, the peak shift to the longer or shorter wavelength and also the height of the peak depends on the temperature. The Y_2O_3 lattice has two site symmetries (C_2 and S_6 (C_{3i})) (Singh and Ningthoujam 2011, p. 120), and upon the Er^{3+} ion substitution in the Y^{3+} sites, the environment of O in Er^{3+} will be different in both sites.

The splitting of energy levels at ground state or excited state could not be observed properly because of very narrow gaps. This can be observed only when a high-resolution instrument is used. Otherwise, merge of peaks occurs. Some peaks are with hump. With the increase of resolution, the signal/noise ratio decreases, but a peak having two or more possible peaks can be separated. When the luminescence spectrum is recorded, the resolution can be done by decreasing the slit width of excitation or/and emission wavelength (monochromators). Generally, the upconversion luminescence spectrum is recorded using 980 nm laser (CW) with a spot size of 2–3 mm (in diameter) because the Yb^{3+} or Er^{3+} ions can absorb this energy easily. As the spot size is large, it is difficult to produce the resolution of spectrum. If the spot size is reduced to a few microns, the resolution of the spectrum can be observed. Here, energy separation among Stark levels is of a few cm^{-1}. Therefore, Stark splitting can be observed while using micron size laser spot. Similarly, many fluorescence or emission peaks due to Stark splitting can be observed in micro-Raman studies of Er^{3+}-doped Y_2O_3 (den Engelsen et al. 2018, p. 337). Stokes and anti-Stokes shifts can be observed above and below the excitation source (He-Ne laser = 633 nm).

In the Yb–Er system (Figure 10.18), the Yb^{3+} ion has one metastable excited state level ($^2F_{5/2}$), whereas the Er^{3+} ion has many metastable excited-state levels such as $^4I_{13/2}$, $^4I_{11/2}$, $^4I_{9/2}$, $^4F_{9/2}$, $^4S_{3/2}$, $^2H_{11/2}$, $^4F_{7/2}$, etc. The Yb^{3+} and Er^{3+} ions can absorb light at 980 nm (Auzel 2004, p. 139; Soni et al. 2019, p. 7350). However, the absorption cross section of the Yb^{3+} ions is more than that of the Er^{3+} ions. On CW laser excitation at 980 nm, the electron at the ground state ($^2F_{7/2}$) of Yb^{3+} ions goes to the next excited-state level ($^2F_{5/2}$). The lifetime of excited electron is about microsecond, whereas the time required to reach $^2F_{5/2}$ is about 10^{-12} seconds. As the CW laser is provided, another electron at $^2F_{7/2}$ goes to the excited state at $^2F_{5/2}$. The time required for absorption of electron from the ground state is much faster than that of the excited-state decay. In this way, excited electrons already occupied at $^2F_{5/2}$ transfer their excitation energy to Er^{3+}, and after absorbing this energy, another electron transits upward to the next higher state of Er^{3+}. The many relaxation phenomena occur at the metastable excited levels of Er^{3+}. Then, the main emitted peaks at 540 nm (green) and 650 nm (red) are observed, and these are assigned to the ($^2H_{11/2}$, $^4S_{3/2}$) → $^4I_{15/2}$ and $^4F_{9/2}$ → $^4I_{15/2}$ transitions, respectively. This is happened only in highly crystalline single or polycrystalline sample doped with Er^{3+} and Yb^{3+}. When the particle size goes to nanometer, the surface dangling bonds as well as the surface ligands or solvents are associated. These are the source of quenchers in luminescence. When such small size particles are coated with undoped one, the distance between the core and the surface increases, and hence, the luminescence intensity is enhanced. Sometimes, multi-phonon relaxation occurs in small-sized particles surrounded by water or capping ligands because excited energy is absorbed by the solvent or the ligand. That is why, core@shell formation helps in getting higher quantum yield or luminescence intensity. This is demonstrated in Figures 10.10 and 10.14. The examples of core and core@shell are $NaYF_4$:2 at.% Er^{3+}-10 at.% Yb^{3+} and $NaYF_4$:2 at.% Er^{3+}-10 at.% Yb^{3+}@$NaYF_4$, respectively. Here, the shell is an undoped host or an inactive host.

In the case of core@shell:$NaErF_4$:0.5% Tm@$NaYF_4$:20% Yb, it is supposed to observe luminescence peaks at 480, 546, and 655 nm. The peak at 480 nm is from Tm^{3+} ions, whereas peaks at 546 and 655 nm are from Er^{3+} ions. However, one luminescence peak at 650 nm (red) is observed under 980 nm excitation (Figure 10.14) (Joshi et al. 2021, p. 3481). Here, the core is with 99.5 at.% Er^{3+} and 0.5 at.% Tm^{3+}. At 980 nm excitation, the electrons from ground state go to the excited state of Yb^{3+}. There are two ways of energy transfer (ET) from Yb^{3+} to Er^{3+} at the excited

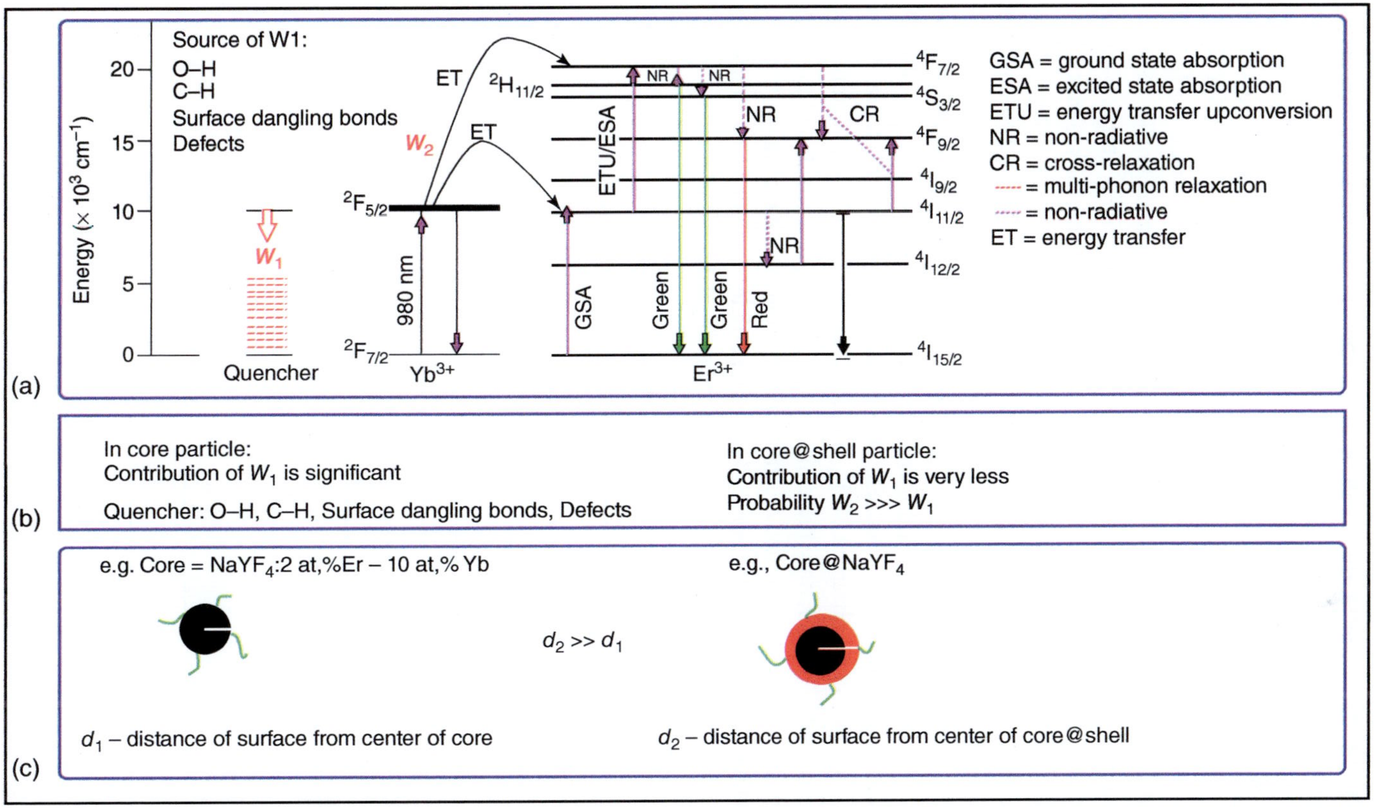

Figure 10.18 (a) Energy level diagram showing sequential energy transfer from Yb^{3+} to Er^{3+}. The green light at 546 nm and the red light at 650 nm are observed after excitation at 980 nm CW laser. In the core particle, its luminescence intensity is decided by probability terms W_1 (a quencher and the possible sources of the quencher are ligands or water defects) and W_2 (energy transfer from the excited state of Yb^{3+} to the excited states of Er^{3+}). (b) If W_2 is more than W_1, the luminescence intensity will be high. Otherwise, its luminescence intensity will be less. (c) In the case of core@shell, the distance between the center of core activator and surface (d_2) is longer than that between the center of core activator and surface (d_1). The luminescence intensity is improved by core@shell formation.

states: from $^2F_{5/2}$ (Yb^{3+}) to $^4F_{7/2}$ (Er^{3+}) and from $^2F_{5/2}$ (Yb^{3+}) to $^4I_{11/2}$ (Er^{3+}). If the probability of ET from $^2F_{5/2}$ (Yb^{3+}) to $^4F_{7/2}$ (Er^{3+}) is higher than that from $^2F_{5/2}$ (Yb^{3+}) to $^4I_{11/2}$ (Er^{3+}), there will be green and red emission after relaxation of excited-state energy. This is generally observed in very low concentration of Er^{3+} in $NaYF_4$. Otherwise, a chance of green emission is less. In another way, if $^2F_{5/2}$ (Yb^{3+}) to $^4I_{11/2}$ (Er^{3+}) has high probability, this energy can be transferred to $^4F_{9/2}$ through the following routes: trapped to $^3H_5$ level of Tm^{3+} as the energy gap between $^4I_{11/2}$ and $^3H_5$ is low and then this energy can be back to $^4I_{13/2}$ of Er^{3+} and finally transferred to $^4F_{9/2}$ because of continuous excitation at 980 nm. From this, the red emission is observed. In addition, if the core is surrounded by undoped shell ($NaYF_4$), the red emission is observed, but its intensity is much less than that of $NaErF_4$:0.5% Tm@$NaYF_4$:20% Yb. To explain this, Er^{3+} in $NaErF_4$:0.5% Tm@$NaYF_4$ is diluted after core@shell formation as well as the reduction from the surface effect. Er^{3+} ions undergo GSA followed by ESA. Because of ESA, this energy can be transferred to $^4F_{7/2}$ (Er^{3+}), which is denoted by W_2 and transferred to the $^3H_5$ level of Tm^{3+} (W_1), and then, this energy can be back to $^4I_{13/2}$ of Er^{3+} and finally transferred to $^4F_{9/2}$ because of continuous excitation at 980 nm. From this, the red emission is observed. If W_2 is more than W_1, green and red emissions can be observed. However, only red emission is observed. It is concluded that the red emission is due to trap excited electron at Tm^{3+} and then back energy transfer to Er^{3+} (Figure 10.19). In some core@shell nanoparticles, many activator ions are added to the core and different shell layers in order to get desired peaks.

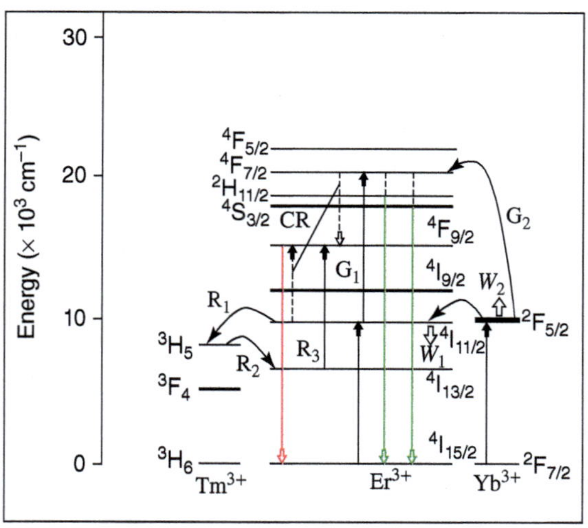

Figure 10.19 Possible energy transfer mechanism in highly doped core: $NaErF_4$:0.05 at.% Tm^{3+} and core@shell $NaErF_4$:0.05 at.% Tm^{3+}@$NaYF_4$ or $NaErF_4$:0.5 *at.% Tm*$^{3+}$@$NaYF_4$*:20* at.% Yb^{3+}. W_1 is the probability of energy transfer from Er^{3+}/Yb^{3+} to Tm^{3+}. W_2 is the probability of energy transfer from Yb^{3+} to Er^{3+} ($^4F_{7/2}$). R_1, R_2, and R_3 are possible energy transfer ways from Er^{3+} to Tm^{3+}, Tm^{3+} to Er^{3+}, and Er^{3+} to Er^{3+}, respectively, and this leads to the red emission. G_1 and G_2 are possible paths for green emission.

Besides the inorganic-based upconversion luminescence, agglomeration or donor–acceptor distance is an important parameter in organic-based upconversion luminescence. This will not be discussed here because so far, upconversion from the core–shell model for organic molecules is scarce.

10.4 Applications

10.4.1 Bioimaging Applications

The core@shell nanoparticles (NPs) are useful in biological applications. Nanoparticles dispersible in water, PBS, or saline are important parameter so that they can be consumed or injected in the human body for cure of disease. In addition, they need to get stable colloidal nanoparticles coated with biocompatible capping agents. This stability depends on the repulsive and attractive forces between particles in a medium (PBS). The main types of forces are steric repulsion (because of the sufficient thickness of the capping layer), electrostatic repulsion (high negative or positive surface charges, ± 30 mV or more), van der Walls attraction (weak force), electrostatic attraction, magnetic dipole attraction, induced-dipole attraction, etc. The capping layer is generally polymeric in nature (PEG, chitosan, glucose, etc.). If the repulsive force is more than that of the attractive force, nanoparticles are dispersible. However, it is suggested that NPs with high negative or positive surface charges, ± 30 mV or more in PBS medium (i.e. biological medium), are not good because it will affect other ions present in the blood upon administration. The stability in PBS can be brought by smaller size (less than 20 nm) as well as polymeric coating. For targeting to site in a living body, it needs to coat with the antibody (shown in Figure 10.20). Various formations of core@shell and core@shell-1@shell-2 can be possible. Even active ions such as Er^{3+}, Yb^{3+}, Nd^{3+}, Eu^{3+}, Dy^{3+}, Tb^{3+}, Sm^{3+}, and Gd^{3+} can occupy the core and shell or multi-shells as an individual or together. These nanoparticles can be converted to biomaterials by proper surface functionalization. Sometimes, PEG molecules are coated over nanoparticles to get more retention time in blood circulation.

10.4.1.1 Luminescence-Based Imaging

The core@shell particles of $NaErF_4$:0.5 at.% Tm@$NaYF_4$:20 at.% Yb and $NaErF_4$:0.5 at.% Tm@$NaYF_4$ show red emission in the visible region under the upconversion process and NIR emission under the downconversion process upon excitation at 980 nm (Figures 10.14 and 10.21a,b) (Joshi et al. 2021, p. 3481). The luminescence intensity is more in the case of $NaErF_4$:0.5 *at.% Tm@*$NaYF_4$*:20* at.% Yb as compared to that of $NaErF_4$:0.5 at.% Tm@$NaYF_4$. This is due to higher absorption cross section at 980 nm for Yb^{3+} as compared to that of Er^{3+}. When the particles are coated with SiO_2, it is found that the nanoparticles are dispersible in water. A part of particles is further coated with folic acid (FA). The particles are highly biocompatible with cancer cells (MCF 7). Nanoparticles (NPs) show broad emission with maximum in 520–550 nm (green) under UV-blue light excitation

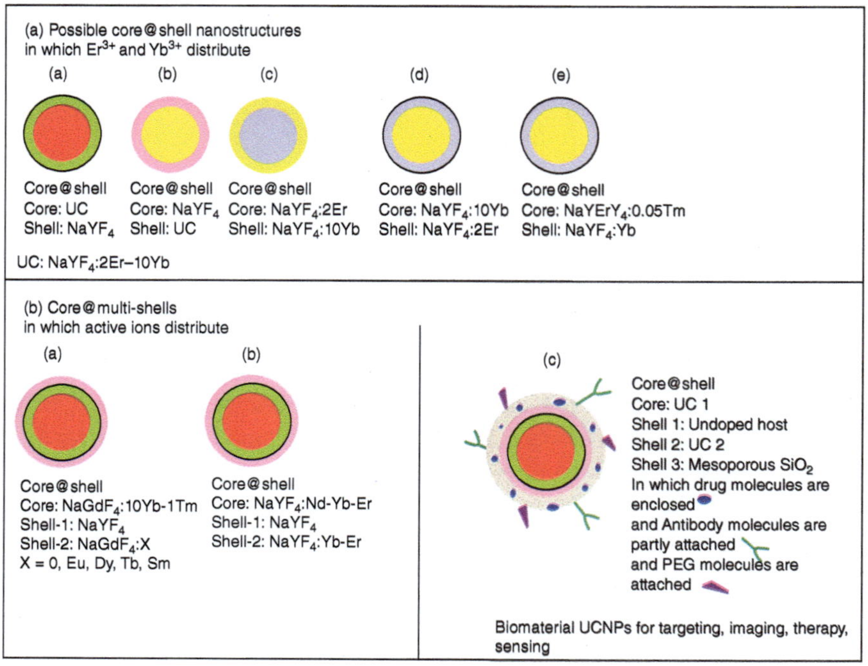

Figure 10.20 (A) Possible core@shell nanostructures in which Er^{3+} and Yb^{3+} ions are distributed. (B) Core@multi-shells and biomaterial UCNPs for targeting, imaging, therapy, and sensing.

(Figure 10.21c) (Joshi et al. 2021, p. 3481). Imaging of MCF 7 after interaction with 250 μg/ml of NPs shows the enhanced green emission from NPs coated with FA as compared to that without FA, indicating targeting to MCF 7 cancer cells, in which FA is overexpressed (Figure 10.22) (Joshi et al. 2021, p. 3481).

Nanoparticles of core–shell type ($NaYF_4$:20 at.% Gd^{3+}, 1.5 at.% Tm^{3+} core (11 nm)-$NaGdF_4$ (2 nm) shell) are functionalized to make water dispersible. The particles are treated with HeLa cells for 17 hours. It is found that the particles are uptaken by cells. The confocal microscopy images are shown in Figure 10.23 (Levy et al. 2016, p. 8423). There is no intensity at the spot outside the cell boundary (3), but spots (1) and (2) give luminescence intensity, indicating internalization of nanoparticles with cells.

The heterogeneously doped core@shell upconversion nanoparticles can overcome the concentration quenching (Wen et al. 2018, p. 1). Under 980 nm excitation, the Yb^{3+} ions can absorb light and transfer its energy to the Gd^{3+} ions and then further transfer its energy to the activator ions, which emit light. Under 800 nm excitation, heterogeneous doped core@shell $NaYF_4$:Yb/Er@$NaYF_4$:Yb/Nd shows enhanced luminescence as compared to that of homogeneous $NaYF_4$:Yb/Er/Nd. Under 800 nm excitation, the Nd^{3+} ions absorb, and their absorption coefficient is more than 980 nm excitation for the Yb^{3+} ions. Similarly, $NaYF_4$:Yb/Er@$NaYF_4$:Yb/Nd can be excited through 980 nm. Imaging though 800 and 980 nm can be done. With the increase of concentration of Nd^{3+} in the shell, the luminescence intensity

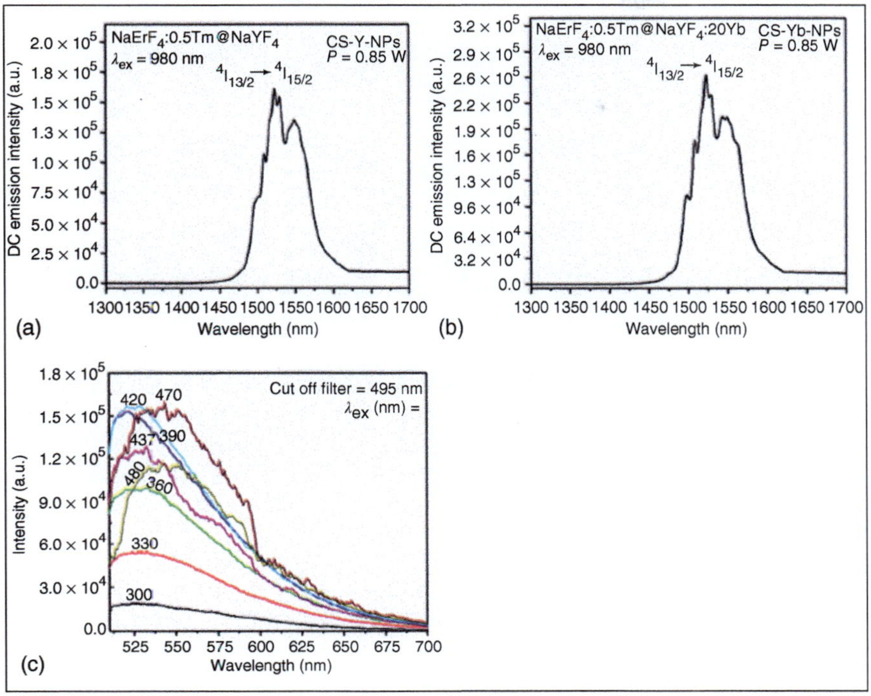

Figure 10.21 NIR III emission spectra of (a) $NaErF_4$:0.5 at.% Tm@$NaYF_4$ and (b) $NaErF_4$:0.5 *at.% Tm*@$NaYF_4$*:20* at.% Yb under 980 nm excitation and (c) visible emission spectra of $NaErF_4$:0.5 *at.% Tm*@$NaYF_4$*:20* at.% Yb under UV-blue light excitation. Source: Joshi et al. (2021), p. 3481. Reprinted with permission of American Chemical Society.

can be improved. Using a spacer (undoped = $NaYF_4$) between core and shell layers $NaYF_4$:Yb/Nd/Tm@$NaYF_4$@$NaYF_4$:Yb-Er, the luminescence intensity of peaks can be changed by excitation at 800 and 980 nm. The emitted lights at UV (365 nm) and visible (545 nm) regions can be employed for reversible isomerization of the spiropyran derivatives. The application of PA can also be extended to super-resolution imaging (Lee et al. 2021, p. 230).

10.4.1.2 Other Imaging Probes (MRI, CT, and SPECT)

Magnetic resonance imaging (MRI) is based on relaxation of ^1H of water in the presence of a contrast agent such as the Gd^{3+} complex $[Gd (DTPA)(H_2O)]^{2-}$, Gd_2O_3, and Fe_3O_4 NPs (Sun et al. 2013, p. 11290; Chen et al. 2015, p. 5249; Gautam and Komal 2018, p. 393). In the past 10 years, it was approached to prepare UCNPs, which are having high magnetic moment. The examples of UCNPs are $NaGdF_4$:Er-Yb, $NaYF_4$:Er-Yb@$NaGdF_4$, $NaYF_4$:Er-Yb@$NaHoF_4$, $NaYF_4$:Er-Yb@$NaDyF_4$, and $NaLuF_4$:Yb-Tm@$NaGdF_4$:^{153}Sm. Among Ln^{3+}, the highest magnetic moment belongs to Dy^{3+} (10.63 μ_B), followed by Ho^{3+}(10.60 μ_B). High magnetic moment is due to the contribution of spin and orbital angular momenta of f^n electrons. However, Gd^{3+} has a lower magnetic moment of 7.94 μ_B than that of Dy^{3+} or Ho^{3+} (Gautam and Komal 2018, p. 393). Even Gd^{3+} has seven unpaired electrons and

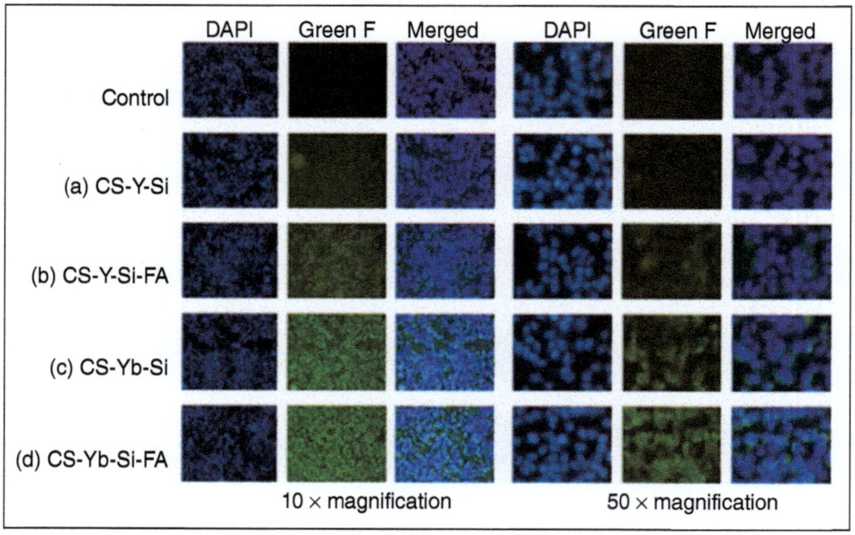

Figure 10.22 Imaging of MCF 7 after interaction with 250 μg/ml of NPs shows the enhanced green emission from NPs coated with FA as compared to that without FA. Control, (a) CS-Y-Si ($NaErF_4$:0.5% Tm@$NaYF_4$@SiO_2), (b) CS-Y-Si-FA (CS-Y-Si-NH-FA), (c) CS-Yb-Si ($NaErF_4$:0.5% Tm@$NaYF_4$:20Yb@SiO_2), and (d) CS-Yb-Si-FA (CS-Yb-Si-NH-FA). Source: Joshi et al. (2021), p. 3481. Reprinted with permission of American Chemical Society.

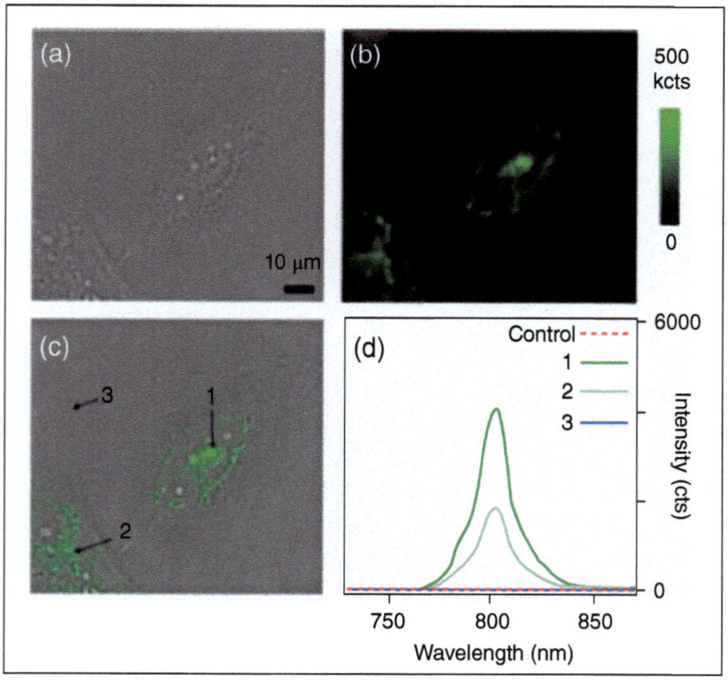

Figure 10.23 Internalization of NPs in HeLa cells. (a) Bright-field image, (b) integrated intensity in 740–870 nm, (c) overlap image between (a) and (b), and (d) luminescence spectra at three different spots located in (c). Source: Levy et al. (2016), p. 8423. Reprinted with permission of American Chemical Society.

does not have contribution in magnetic moment from orbital angular momentum. The ground-state term symbol of Gd^{3+} is $^4S_{7/2}$, which is contributed by spin moment only. Its Stark splitting will have 4, which has separate gaps among them. In the case of Dy^{3+} and Ho^{3+}, their respective ground-state term symbols are $^6H_{15/2}$ and $^5I_8$, and their respective Stark splitting will be 16 and 17. Thus, the energy gap between adjacent splitting will be very small. Thus, the magnetic effect of Gd^{3+} on ^1H of water is different from Dy^{3+} or Ho^{3+}. The Gd^{3+} ions have been used as a T_1 contrast agent, whereas Ho^{3+} and Dy^{3+} ions have been used as a T_2 contrast agent.

The magnetic moment (μ in Bohr magnetron) can be defined by

$$\mu = g_J\sqrt{J(J+1)}$$

where g_J is defined as

$$g_J = \frac{3}{2} + \frac{S(S+1) - L(L+1)}{2J(J+1)}$$

where S, L, and J are spin, orbital angular, and total angular momenta quantum numbers, respectively, for the electrons in an atom or ion.

X-ray computed tomography (CT) scan is used to get images of tumor, disease, or injured parts/spots in the body. In this, many contrast agents are used. UCNPs will be one of the contrast agents as it has high atomic density. The examples are $NaYF_4$:Er-Yb, $NaLuF_4$:Er-Yb, and $NaLuF_4$:Yb-Tm@$NaGdF_4$:^{153}Sm (Sun et al. 2013, p. 11290).

A single-photon emission computerized tomography (SPECT) scan has been used to get images of tumor, disease, and blood flow using the radioisotope contrast agent, which produces γ-rays. The radioisotope traces are technetium-99m, iodine-123, xenon-133, fluorine-18, and thallium-201. There are many isotopes of elements, which can produce the γ-rays with a half-life ($t_{1/2}$) of a few hours. ^{153}Sm emits γ-rays with $t_{1/2}$ = 46.3 hours, and the material containing this isotope is used in clinic for radiotherapy as well as the SPECT contrast agent. Nowadays, UCNPs doped with radioisotopes such as ^{153}Sm is used as a SPECT contrast agent. One example is $NaLuF_4$:Yb-Tm@$NaGdF_4$:^{153}Sm. The NPs of $NaLuF_4$:Yb-Tm@$NaGdF_4$:^{153}Sm are coated with citric acid, and such NPs have been used as contrast agents for luminescence-based imaging because of Yb–Tm pair under 980 nm laser excitation; the MRI contrast agent as the shell of the particle has Gd^{3+}; the CT contrast agent because of the presence of heavy and high atomic density elements such as Lu, Yb, Tm, and Gd; and the SPECT agent because of the presence of ^{153}Sm radioisotope (γ-rays). These core@shell UCNPs have multi-model probes for tumor angiogenesis imaging (Figure 10.24) (Sun et al. 2013, p. 11290). Here, the tumor-bearing nude mouse is used, and imaging is performed after one hour of intravenous injection of UCNPs in mice.

10.4.2 Photothermal Therapy (PTT)

Photothermal therapy (PTT) is a promising therapeutic approach in which laser light is used to excite molecules, nanoparticles, or photosensitive materials for generation

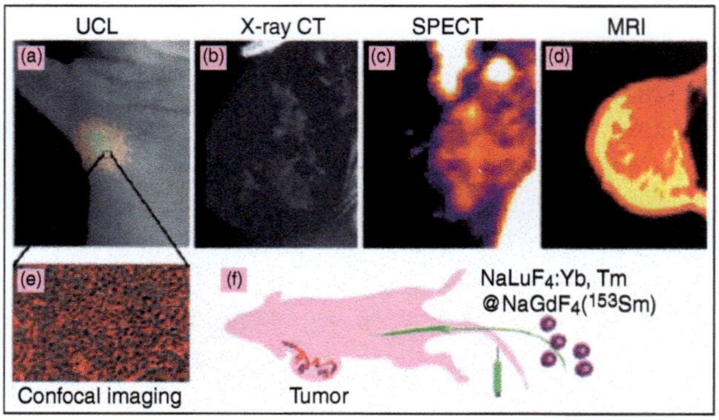

Figure 10.24 Four-modal imaging of the focused tumor from the tumor-bearing nude mouse after one hour IV (intravenous) injection of $NaLuF_4$:Yb-Tm@$NaGdF_4$:^{153}Sm coated with citric acid in mouse: (a) image of tumor through in vivo upconversion luminescence (UCL). (b) Image of tumor through X-ray CT scan. (c) Image through SPECT scan. (d) Image of tumor through MR. (e) The paraffin section of tumor tissue through UCL confocal. (f) Schematic diagram showing tumor angiogenesis imaging in mouse after IV injection of radioisotope NPs ($NaLuF_4$:Yb-Tm@$NaGdF_4$:^{153}Sm). Source: Sun et al. (2013), p. 11290. Reprinted with permission of American Chemical Society.

of heat into cells or tissues in order to cause thermal damage in the region of interest such as tumor (Soni et al. 2019, p. 7350). This approach has minimum side effects as compared to the conventional radiotherapy or chemotherapy. The wavelength of excitation is generally chosen in red – NIR windows I, II, and III regions in order to increase the penetration depth. The materials should have the following characteristics: (i) high absorption cross section, (ii) low toxicity, (iii) easy in functionalization with others such as antibody or drug, (iv) mixable with water, etc. In this way, cancer cells can be killed by the increase of temperature above the physiological temperature (say 42–43 °C). This is included in hyperthermia treatment.

Many organic molecules, noble metal nanoparticles (Au, Ag, and CuS), and 2D materials are the examples, which can absorb laser light in red – NIR windows efficiently. The noble metals can absorb the light through SPR. Upon laser excitation at NIR, UCNPs can produce light in the visible range. If noble metal nanoparticles (NMs) are associated with UCNPs, emitted light by UCNPs can be absorbed by NMs, and thereby, heat can be generated. After absorption, the excited electrons through SPR de-excited and come back to the ground state, leading to heat generation. UCNPs ($NaYF_4$:Er-Yb) upon excitation at 980 nm produce peaks at green and red regions. On the other hand, Au nanoparticles can produce the SPR peak in the green region. After the addition of Au nanoparticles into UCNPs, the intensity of green light decreases, and at the same time, heat is generated. This is due to fluorescence or Forster energy transfer (FRET) from UCNPs to Au nanoparticles (Figure 10.25) (Soni et al. 2019, p. 7350).

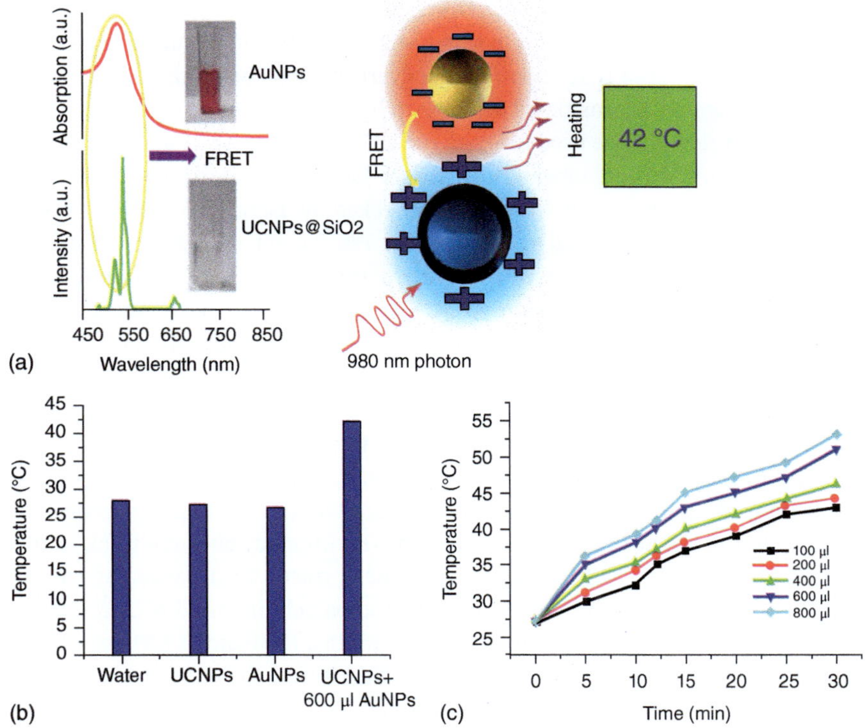

Figure 10.25 FRET mechanism for heat generation. (a) Spectral overlap of UC green band of UCNPs@SiO_2 with SPR absorption band of Au NPs. Schematic diagram illustrates the production of heat through FRET. (b) Rise of temperature from water, UCNPs, AuNPs, and UCNPs + AuNPs gain after irradiation of 980 nm laser for 15 minutes (power 600 mW). (c) Rise of temperature with time under irradiation of 980 nm CW laser at various concentrations of Au NPs in hybrid UCNPs@SiO_2@AuNPs. Source: Soni et al. (2019), p. 7350. Reprinted with permission of American Chemical Society.

10.4.3 Photodynamic Therapy (PDT)

UCNPs have been used in biological applications, particularly in PDT and imaging because they show unique properties such as high biocompatibility, large anti-Stokes shifts, tuning of emission from UV to visible to NIR region, deep penetration into biological tissues, reduction of quenching effects after core@shell formation, narrow emission bands, easy surface functionalization for targeted therapy, high spatial–temporal resolution, etc. (Ai et al. 2015, p. 10785).

In PDT, photosensitizer molecules absorb light in the visible range and produce the singlet oxygen ($^1O_2$) and other reactive oxygen species (ROS) that damage cancer cells, leading to cell death. As compared to the conventional chemotherapy or radiation therapy, PDT reduces the side effects to normal tissues, and localized killing to cancer cells is observed. Exposure at visible light reduces the penetration depth to tissues (say up to 1 mm). However, this penetration depth can be increased to 1–2 cm after exposure to NIR light. In this system, UCNPs can be one of materials, which absorb light at the NIR region and convert the absorbed light in the visible range, from where photosensitizers can absorb light efficiently. In general, 980 nm and 800 nm wavelengths are used to excite the UCNPs. However, in a 980 nm wavelength, the second overtone of water can absorb light and heat can be generated if the power of excitation is high. This can be checked by using both the light wavelengths. Can UCNPs be targeted to cancer cells? Yes, it can be possible to coat/load molecules, which are overexpressed in a particular cancer cell. In addition to therapy, how can we see internalization of UCNPs in cells? The above issues can be solved by core@shell formulation with proper surface functionalization.

The core@shell UCNPs such as $NaYbF_4$:Nd@$NaGdF_4$:Yb/Er@$NaGdF_4$ core@shell@shell coated with SiO_2-NH_2-Chlorin e6-Folic acid can perform simultaneous imaging and PDT (Ai et al. 2015, p. 10785). Here, $NaYbF_4$:Nd@$NaGdF_4$:Yb/Er @$NaGdF_4$ can produce visible light with peaks in 420 nm (violet), 550 nm (green), and 660 nm (red) under 800 nm and 980 nm laser excitations. Chlorin e6 molecule has absorption peaks at the above wavelengths (mentioned) but very weak at the green band. Folic acid (FA) molecules are overexpressed in a particular cancer cell such as KB (human mouth epidermal carcinoma cell line) and MCF 7 (breast cancer *cell line*). Under excitation at both wavelengths, the band at 650 nm will be used for the generation of singlet oxygen and ROS, whereas the band at 550 nm will be used for imaging. Two cancer cells are chosen: positive response cancer cells (FA receptor KB) and negative response cancer cells (A549, negative FA receptor).

At the initial stage, biocompatibility test of the above two cells (KB) and (A549) is performed at 800 and 980 nm at 1 W/cm^2 (Ai et al. 2015, p. 10785). With the increase of irradiation up to 30 minutes, it is found that the biocompatibility of cells decreases for 980 nm laser excitation, but it is 80% or more for 800 nm laser excitation. It means that 800 nm excitation shows high biocompatible nature. At 980 nm excitation, water molecules can absorb light, and then, heat is generated, and thereby, killing of cells takes place. The UCNPs show emission at 420, 550, and 650 nm under 800 and 980 nm excitations (Ai et al. 2015, p. 10785). Chlorin e6 molecule has tetrapyrrole pigments that are partially hydrogenated porphyrins and has absorption peaks at 400, 500, and 665 nm, which are almost matching with emitted bands from UCNPs, but its intensity at 550 nm is very weak (Ai et al. 2015, p. 10785). The UCNP nanoformulations (100 μg/ml) are interacted with KB and A549 cells. It is found that only KB (FR+) cells show green emission under 800 nm excitation (Ai et al. 2015, p. 10785). KB (FR−) and A549 (FR−) do not show the green emission. It means that there is an internalization of nanoparticle-coated FA with KB cells. Biocompatibility test for KB cells has been treated with 200 μg/ml of FA-coated nanoparticles under 800 nm laser irradiation (6 W/cm^2) for different durations of time (Ai et al. 2015, p. 10785). Upon laser excitation, cells have 80% or more biocompatibility up to 10 minutes. However, the combination of cells treated with FA-coated nanoparticles and laser shows a drastic decrease of biocompatibility. FA-coated nanoparticles (25–400 μg/ml) have high biocompatibility in the case of KB cells as compared to that of A549 cells (Ai et al. 2015, p. 10785). Here,

FA-coated nanoparticles means UCNPs coated with targeted folic acid molecule as well as loaded with the photosensitizer Chlorin e6 molecule. However, in the presence of laser at 800 nm irradiation for five minutes, the biocompatibility reduces significantly in the case of KB cells as compared to that of A549 cells. It is confirmed that FA-coated nanoparticles are more accepted by KB cells as compared to that of A549 cells. After 800 nm excitation, nanoparticles produce light at 650 nm, which is absorbed by Chlorin e6, and then singlet oxygen and ROS are produced. Such singlet oxygen and ROS destroy cells and thereby cells are killed.

In order to confirm the origin of killing of cells due to singlet oxygen, it is monitored with different amounts of FA-coated nanoparticles. After comparison of the standard molecule 1,3-diphenylisobenzofuran (DPBF), which can detect single oxygen, it is found that the amount of singlet oxygen increases with the addition of FA-coated nanoparticles. The penetration depth of 15 mm in pork muscle tissue using laser at 800 nm (6 W/cm^2 exposure density) can generate the singlet oxygen, but using 980 nm laser excitation, it cannot produce such singlet oxygen. Figure 10.26 shows the schematic diagram of killing of cancer cells using FA-coated nanoparticles under laser excitation and simultaneous imaging (Ai et al. 2015, p. 10785).

$NaYF_4$:Er-Yb UCNPs are coated with mesoporous SiO_2, which is further loaded with photosensitizers MC540 (merocyanine, which absorb light at green band) and ZnPc (zinc(II) phthalocyanine, which absorb light at red band): individual or both (co-loaded). UCNPs produce emission peaks at green and red bands, which will be absorbed by MC540 and ZnPc, respectively. In aqueous medium, it can produce

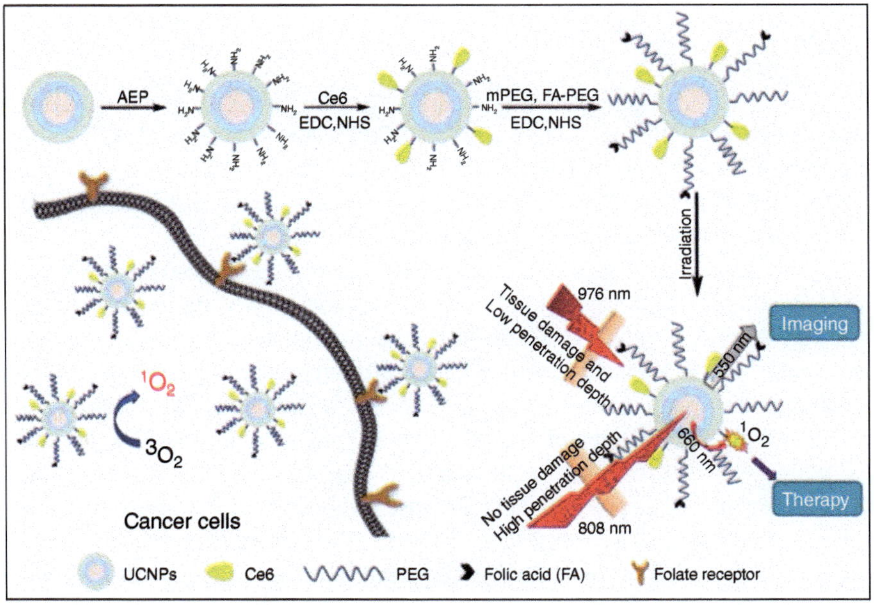

Figure 10.26 Schematic diagram of killing of cancer cells using FA-coated nanoparticles under laser excitation and simultaneous imaging. Source: Ai et al. (2015), p. 10785.

singlet oxygen as well ROS (Idris et al. 2012, p. 1580). Viability experiments of B16-F0 cells after interaction of materials such as no UCNPs (control), void UCNPs (absence of photosensitizer), ZnPc-UCNPs, MC540-UCNPs, and both ZnPc-MC540 co-loaded UCNPs are performed (Idris et al. 2012, p. 1580). It shows the highest killing of cells for both ZnPc-MC540 co-loaded UCNPs under 980 nm excitation. Also, ZnPc-UCNPs and MC540-UCNPs show the killing of cells. This killing is related to the generation of the singlet oxygen and ROS from photosensitizers (ZnPc and MC540). This is monitored using the standard molecule (ABDA = dye 9,10-anthracenediyl-bis(methylene)dimalonic acid (ABDA)), which can accept $^1O_2$ (singlet oxygen), and its fluorescence intensity at 407 nm decreases. The intensity of ABDA at 407 nm for ZnPc-UCNPs, MC540-UCNPs, or both ZnPc-MC540 coloaded UCNPs decreases with irradiation time at 980 nm (Idris et al. 2012, p. 1580). The lowest intensity is found for both ZnPc-MC540 co-loaded UCNPs.

Further, photosensitizer-loaded UCNPs is further coated with folic acid (FA) to target cancer cells (B16-F0 melanoma cells) in mice model. To increase the retentivity at blood circulation, nanoparticles are also coated with PEG. When such nanoformulation is injected to mice through tail, it is found that particles are more accumulated into tumor (B16-F0 melanoma cells). Under 980 nm laser excitation, the tumor size decreases significantly (Idris et al. 2012, p. 1580).

10.4.4 Temperature Sensor

In Boltzmann distribution law, the population of the quantized energy states (E_i and E_j) can be given by the following equation:

$$\frac{N_j}{N_i} = A\ \exp^{\left(\frac{E_i - E_j}{kT}\right)}$$

The symbols are defined in Figure 10.27, k is Boltzmann's constant, T is the temperature of measurement, and A is the pre-exponential factor/constant. With the increase of temperature, the population at higher energy level increases, whereas the population at lower energy level decreases.

Using the Boltzmann distribution law, the population of quantized energy states in Ln^{3+} can be predicted. In most Ln^{3+} ions, metastable excited states have energy gaps of UV to visible to NIR to IR. In Eu^{3+}, the ground-state term symbol is ^7F_j, where $j = 0, 1, 2, 3, 4, 5$, and 6 because of spin–orbit coupling, the lowest energy state $^7F_0$, which is very much close to $^7F_1$ ($\Delta_{1\text{-}0} \sim 291\ \text{cm}^{-1}$). The value of $\Delta_{1\text{-}0} \sim 291\ \text{cm}^{-1}$ is calculated from transitions of $^5D_0$ to $^7F_0$ = 581 nm and $^5D_0$ to $^7F_1$ = 591 nm ($17\,211.70\ \text{cm}^{-1} - 16\,920.47\ \text{cm}^{-1} = 291\ \text{cm}^{-1}$) (Ningthoujam 2012, p. 145; Yadav and Ningthoujam 2021, pp. 213–250). Because $\Delta_{1\text{-}0} \sim 291\ \text{cm}^{-1}$ is close to room temperature energy (0.025 eV or 201 cm^{-1}), there is an intermixing between $j = 0$ and $j = 1$. On the basis of $j = 0$, Eu^{3+} is supposed to be diamagnetic but becomes paramagnetic at room temperature. In the case of Stark splitting in the ground or excited state, energy gaps are about a few cm^{-1} (2–100 cm^{-1}), which can be dominated by the room temperature energy.

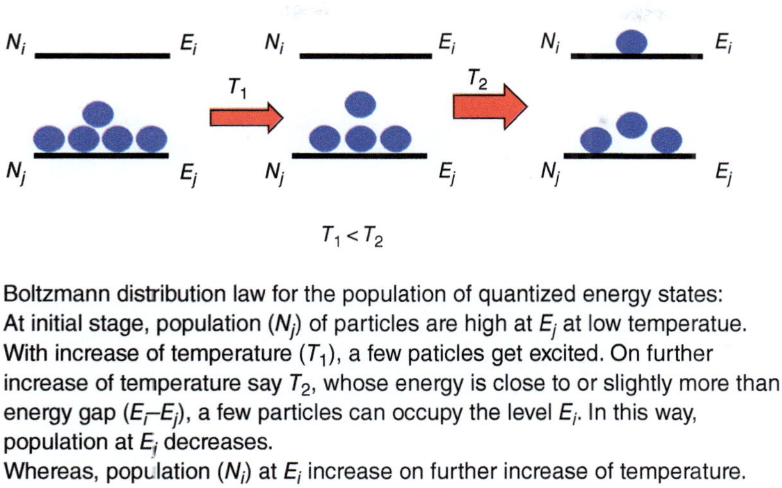

Figure 10.27 Schematic diagram showing Boltzmann distribution low for population of the quantized energy states.

In the Yb–Er system, the Yb^{3+} ions act as a sensitizer, and excited energy is transferred to Er^{3+} sequentially (Soni et al. 2019, p. 7350). Emission peaks at green and red regions occur. Please refer to Figure 10.25. The green emission has the main two peaks at 522 and 545 nm, which correspond to the transitions from excited energy level ${}^2H_{11/2}$ and ${}^4S_{3/2}$ to ${}^4I_{15/2}$, respectively, of Er^{3+}. The difference in energy between ${}^2H_{11/2}$ (522 nm = 19 157.09 cm^{-1}) and ${}^4S_{3/2}$ (545 nm = 18 348.62 cm^{-1}) is about 810 cm^{-1}. At room temperature, the emission peak at 545 nm is more than that at 522 nm. This population of excited-state ${}^2H_{11/2}$ can be increased by increasing the temperature. The ratio of peak at 522 nm to that at 545 nm is commonly known as fluorescence intensity ratio (FIR), which is given as

$$\mathrm{FIR} = \frac{N_j}{N_i} = A\exp^{\left(\frac{E_i - E_j}{kT}\right)}$$

where the populations of excited states at E_i = 522 nm and E_j = 545 nm are N_i and N_j, respectively. By taking logarithm, it can be written as

$$T = \frac{1}{\ln(A) - \ln(\mathrm{FIR})} \times \frac{\Delta E}{k}$$

By substituting the values of $\ln(A)$, k, ΔE ($E_{522\,\mathrm{nm}} - E_{545\,\mathrm{nm}}$), and FIR, the temperature of measurement can be calculated.

Figure 10.28a–c shows the UC emission spectra of YPO_4:Er-Yb under 980 nm CW laser excitation at different laser pump powers for one minute irradiation time each (Soni et al. 2021, p. 850). The temperature gets increased by increasing the FIR or laser power. The temperature is monitored by an IR thermal camera. The temperature in 28–375 °C range is calculated by the increase of the laser input power (0.081–3.07 W), which is almost matching with the thermal camera. The variations of emission peak intensity of 522 or 545 nm with different laser powers are shown in Figure 10.28d.

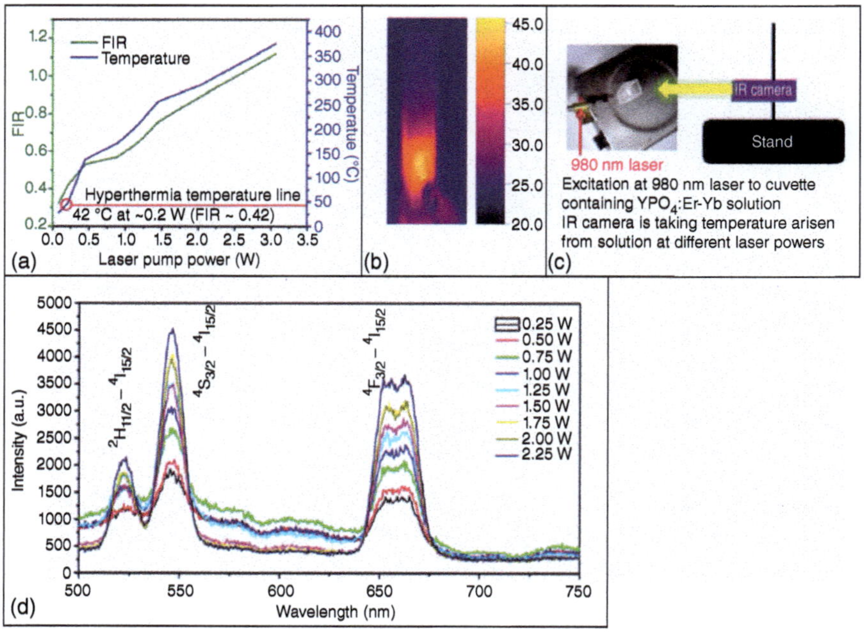

Figure 10.28 YPO4:Er-Yb UCNPs: (a) Rise of FIR and temperature with the increase of laser input power at 980 nm for one minute each of irradiation. (b) Digital photo showing the thermal image recorded from an IR thermal camera. (c) Experimental setup consisting of laser and cuvette with UCNPs and IR camera. (d) Luminescence spectra depicting the variation of peaks at 522 and 545 nm with different laser powers. Source: Soni et al. (2021), p. 850. Reprinted with permission of American Chemical Society.

In the case of 980 nm excitation, the heat from water by absorption through second overtone of water arises in higher excitation intensity. This effect can be reduced by excitation at 800 nm. The formation of core@shell UCNPs such as $NaGdF_4$:Yb-Er@$NaGdF_4$:Nd can measure the temperature rise. Here, excitation at wavelength 800 or 980 nm can produce heat. At 980 nm excitation, the Yb^{3+} ions absorb light and then transfer their excited energy to Er^{3+}, and emission at green is monitored. In the case of 800 nm, the Nd^{3+} ions absorb light and then transfer their excited energy to the Yb^{3+} ions, which is further transferred to Er^{3+} and then, emission at green band is monitored (Figure 10.29) (Shao et al. 2018, p. 188). With the increase of Nd^{3+} ion concentration, the luminescence intensity increases, and simultaneously, the heat rises. This heat rise is due to heat absorbed by the sample, which is dissipated to the surrounding solvent such as cyclohexane or water as well as many relaxation phenomena associated during the energy transfer process from Nd^{3+} to Yb^{3+} to Er^{3+} (Figure 10.30) (Shao et al. 2018, p. 188).

10.4.5 Security Ink

In anti-counterfeiting application, many phosphors such as organic dyes, inorganic nanoparticles, photonics, and quantum dots have been used. However, they used

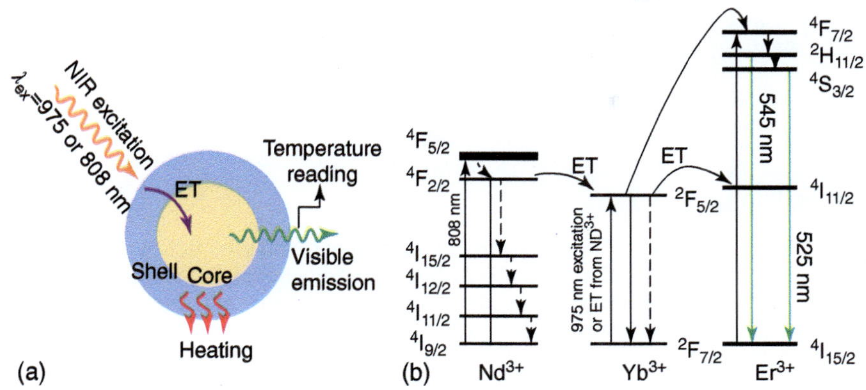

Figure 10.29 (a) Schematic diagram showing dual excitation at 800 or 980 nm and its heat rise, visible emission from core@shell UCNPs. (b) Energy-level diagram depicting the energy transfer process from Nd^{3+} to Yb^{3+} to Er^{3+} at 800 nm excitation and Yb^{3+} to Er^{3+} at 980 nm excitation. Source: Shao et al. (2018), p. 188.

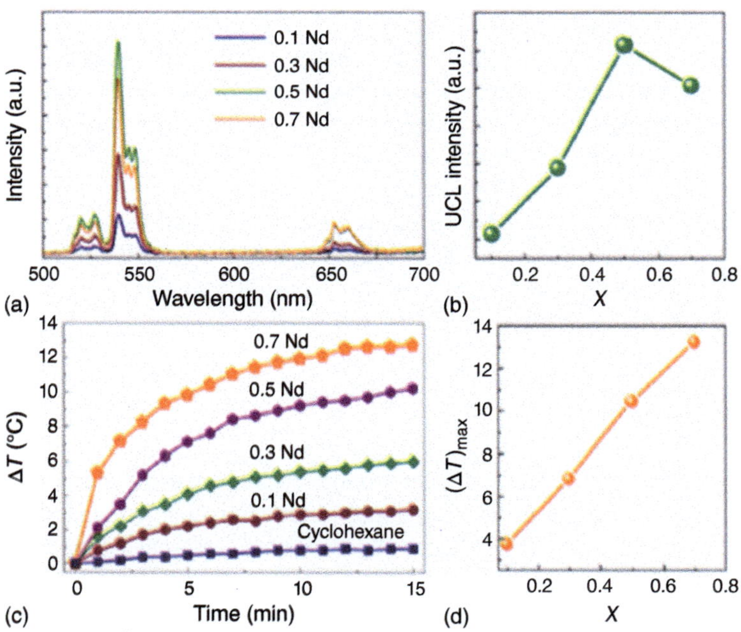

Figure 10.30 (a) Emission spectra of core@shell $NaGdF_4$:Yb-Er@$NaGdF_4$:xNd UCNPs under 800 nm excitation at different Nd^{3+} concentrations. (b) The corresponding variation of emission intensity with Nd^{3+} concentrations. (c) Rise in temperature with respect to room temperature with Nd^{3+} concentrations at different irradiation times. (d) Its maximum rise in temperature with respect to Nd^{3+} concentrations. Source: Shao et al. (2018), p. 188.

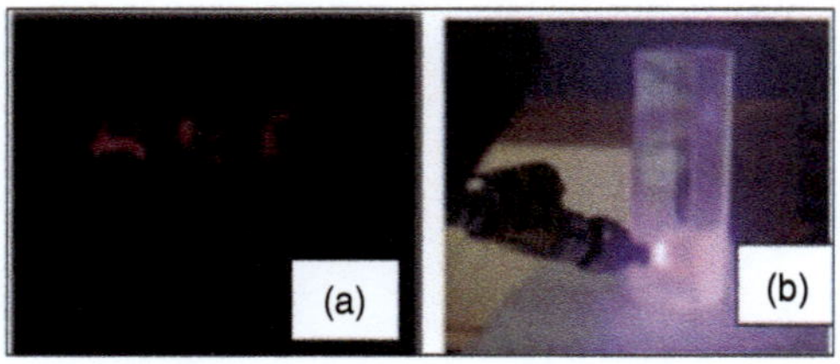

Figure 10.31 Digital photos illustrating the application of core@shell UCNPs $NaErF_4$:0.5 *at.% Tm*@$NaYF_4$:*20* at.% Yb (CS-Yb-NPs) in security ink: (a) Digital photo of letters such as ABC is recorded in dark mode under the CW 980 nm excitation with a band-pass filter. (b) Digital image of CS-Yb-NPs dispersed in cyclohexane is recorded under the CW 980 nm excitation in daytime without the band-pass filter. Source: Joshi et al. (2021), p. 3481. Reprinted with permission of American Chemical Society.

excitation sources such as X-rays or UV light. These sources are harmful to the surrounding environment. This may produce free radicals, which are highly dangerous to users. In order to replace such source, one option is to use the NIR light, which has larger penetration depth (up to 1–3 cm) than that of UV or visible light. Using NIR light, UCNPs can produce light in the visible or NIR region.

One example of core@shell UCNPs is $NaErF_4$:0.5 *at.% Tm*@$NaYF_4$:*20* at.% Yb (CS-Yb-NPs) (Joshi et al. 2021, p. 3481), which shows security ink application. Figure 10.31a shows digital photo of letters such as ABC, which is recorded in dark mode with proper band-pass filters. The letters are visible under 980 nm excitation and invisible without excitation at 980 nm. Figure 10.31b digital image of CS-Yb-NPs dispersed in cyclohexane is recorded in daytime under the CW 980 nm excitation without band-pass filters. Many UCNPs such as YPO_4:Er-Yb (Soni et al. 2021, p. 850), YVO_4:Ho^{3+}-Yb^{3+} (Perala et al. 2021a, p. 19471), and YVO_4:Ho^{3+}-Yb^{3+}-K^+ (Perala et al. 2021b, p. 19517) have been used as security ink as they show the various emitted colors in the visible range.

10.5 Conclusion

This chapter provides the basic concept of photoluminescence such as downconversion, upconversion, PA, energy transfer from donor to activator, energy migration, quencher, etc. The various synthesis methods of upconverting core@shell nanoparticles along with their photoluminescence behavior are provided and ways of their dispersion in water medium are demonstrated. The stability of such colloidal suspension in water depends on the repulsive and attractive forces between the nanoparticles. The main types of involved forces are steric repulsion (due to high thickness of capping), electrostatic repulsion (high negative or positive surface charges, $\pm$30 mV or more), van der Walls attraction (weak force), electrostatic attraction, magnetic dipole attraction, induced dipole attraction, etc. The optical properties of core@shell nanoparticles, energy transfer, and energy migration in the various core@shell models are mentioned. Last, applications of core@shell nanoparticles in various areas such as biological applications, imaging, photodynamic and photothermal therapy, security ink, etc., are discussed.

Acknowledgment

Dr. Manas Srivastava thanks CSIR, New Delhi (reference number CSIR-SRA 9146-A), for providing senior research associate fellowship.

References

Abel, K.A., Boyer, J.C., and Veggel, F.C. (2009). Hard proof of the $NaYF_4/NaGdF_4$ nanocrystal core/shell structure. *Journal of the American Chemical Society* 131 (41): 14644–14645.

Abel, K.A., Boyer, J.C., Andrei, C.M., and van Veggel, F.C.J.M. (2011). Analysis of the shell thickness distribution on $NaYF_4/NaGdF_4$ core/shell nanocrystals by EELS and EDS. *Journal of Physical Chemistry Letters* 2 (3): 185–189. https://doi.org/10.1021/jz101593g.

Ai, F., Ju, Q., Zhang, X. et al. (2015). A core–shell–shell nanoplatform upconverting near-infrared light at 808 nm for luminescence imaging and photodynamic therapy of cancer. *Scientific Reports* 5 (1): 10785. https://doi.org/10.1038/srep10785.

Auzel, F. (2004). Upconversion and anti-Stokes processes with f and d ions in solids. *Chemical Reviews* 104 (1): 139–173. https://doi.org/10.1021/cr020357g.

Blasse, G. and Grabmaier, B.C. (1994). *Luminescent Materials*. Springer.

Chatterjee, K., Sarkar, S., Jagajjanani Rao, K.J., and Paria, S. (2014). Core/shell nanoparticles in biomedical applications. *Advances in Colloid and Interface Science* 209: 8–39. https://doi.org/10.1016/j.cis.2013.12.008.

Chen, X., Peng, D., Ju, Q., and Wang, F. (2015a). Photon upconversion in core–shell nanoparticles. *Chemical Society Reviews* 44 (6): 1318–1330. https://doi.org/10.1039/c4cs00151f.

Chen, C., Kang, N., Xu, T. et al. (2015b). Core–shell hybrid upconversion nanoparticles carrying stable nitroxide radicals as potential multifunctional nanoprobes for upconversion luminescence and magnetic resonance dual-modality imaging. *Nanoscale* 7 (12): 5249–5261. https://doi.org/10.1039/c4nr07591a.

Chen, X., Zhou, D., Xu, W. et al. (2017). Fabrication of Au-Ag nanocage@ $NaYF_4$@ $NaYF_4$:Yb,Er core–shell hybrid and its tunable upconversion enhancement. *Scientific Reports* 7 (1): 41079. https://doi.org/10.1038/srep41079.

Chiozzi, V. and Rossi, F. (2020). Inorganic–organic core/shell nanoparticles: Progress and applications. *Nanoscale Advances* 2 (11): 5090–5105. https://doi.org/10.1039/D0NA00411A.

Choudhury, B. and Choudhury, A. (2012). Dopant induced changes in structural and optical properties of Cr^{3+} doped TiO_2 nanoparticles. *Materials Chemistry and Physics* 132 (2–3): 1112–1118. https://doi.org/10.1016/j.matchemphys.2011.12.083.

Curie, D. (1963). *Luminescence in Crystals*, 1e. Wiley.

Dhiren Meetei, S., Dorendrajit Singh, S., Shanta Singh, N. et al. (2012). Crystal structure and photoluminescence correlations in white emitting nanocrystalline $ZrO:Eu^{3+}$ phosphor: effect of doping and annealing. *Journal of Luminescence* 132 (2): 537–544. https://doi.org/10.1016/j.jlumin.2011.09.011.

Dong, C., Korinek, A., Blasiak, B. et al. (2012). Cation exchange: a facile method to make $NaYF_4$: Yb, Tm-$NaGdF_4$ core–shell nanoparticles with a thin, tunable, and uniform shell. *Chemistry of Materials* 24 (7): 1297.

Dutta, D.P., Ningthoujam, R.S., and Tyagi, A.K. (2012). Luminescence properties of Sm^{3+} doped YPO_4: effect of solvent, heat-treatment, Ca^{2+}/W^{6+}-co-doping and its hyperthermia application. *AIP Advances* 2 (4): 042184. https://doi.org/10.1063/1.4773443.

den Engelsen, D., Fern, G.R., Ireland, T.G., and Silver, J. (2018). Reassignment of electronic transitions in the laser-activated spectrum of nanocrystalline Y_2O_3:Er^{3+}. *Journal of Luminescence* 196 (196): 337–346. https://doi.org/10.1016/j.jlumin.2017.12.053.

Gajbhiye, N.S., Ningthoujam, R.S., Ahmed, A. et al. (2008). Re-dispersible Li_+ and Eu^{3+} co-doped CdS nanoparticles: luminescence studies. *Pramana* 70 (2): 313–321. https://doi.org/10.1007/s12043-008-0050-z.

Gargas, D.J., Chan, E.M., Ostrowski, A.D. et al. (2014). Engineering bright sub-10-nm upconverting nanocrystals for single-molecule imaging. *Nature Nanotechnology* 9 (4): 300–305. https://doi.org/10.1038/nnano.2014.29.

Gautam, A. and Komal, P. (2018). Probable ideal size of Ln^{3+}-based upconversion nanoparticles for single and multimodal imaging. *Coordination Chemistry Reviews* 376: 393–404. https://doi.org/10.1016/j.ccr.2018.08.008.

Ghosh, M., Ningthoujam, R.S., Patra, G.D. et al. (2012). Epitaxial growth of Eu doped CaF_2 thin film on CaF_2(111). *AIP Conference Proceedings* 1451 (1): 313–315.

Idris, N.M., Gnanasammandhan, M.K., Zhang, J. et al. (2012). In vivo photodynamic therapy using upconversion nanoparticles as remote-controlled nanotransducers. *Nature Medicine* 18 (10): 1580–1585. https://doi.org/10.1038/nm.2933.

Johnson, N.J., Korinek, A., Dong, C., and van Veggel, F.C. (2012). Self-focusing by Ostwald ripening: a strategy for layer-by-layer epitaxial growth on upconverting nanocrystals. *Journal of the American Chemical Society* 134 (27): 11068–11071. https://doi.org/10.1021/ja302717u.

Joshi, R., Perala, R.S., Shelar, S.B. et al. (2021). Super bright red upconversion in $NaErF_4$:0.5% Tm@$NaYF_4$:20% Yb nanoparticles for anti-counterfeit and bioimaging applications. *ACS Applied Materials and Interfaces* 13 (2): 3481–3490. https://doi.org/10.1021/acsami.0c21099.

Kumar, K.S., Kumar, V.B., and Paik, P. (2013). Recent advancement in functional core–shell nanoparticles of polymers: synthesis, physical properties, and applications in medical biotechnology. *Journal of Nanoparticles* 2013, Article ID: 672059. https://doi.org/10.1155/2013/672059.

Lahiri, D., Ningthoujam, R.S., Bhattacharyya, D., and Sharma, S.M. (2010). Structural understanding of the spectral characteristics of SnO_2:Eu:Y_2O_3, using extended X-ray absorption fine structure. *Journal of Applied Physics* 107 (5): 054316. https://doi.org/10.1063/1.3330705.

Lee, C., Xu, E.Z., Liu, Y. et al. (2021). Giant nonlinear optical responses from photon-avalanching nanoparticles. *Nature* 589 (7841): 230–235. https://doi.org/10.1038/s41586-020-03092-9.

Levy, E.S., Tajon, C.A., Bischof, T.S. et al. (2016). Energy-looping nanoparticles: harnessing excited-state absorption for deep-tissue imaging. *ACS Nano* 10 (9): 8423–8433. https://doi.org/10.1021/acsnano.6b03288.

Liu, C., Xia, Z., Chen, M. et al. (2015). Near-infrared luminescence and color tunable chromophores based on Cr^{3+}-doped mullite-type $Bi_2(Ga, Al)_4O_9$ solid solutions. *Inorganic Chemistry* 54 (4): 1876–1882. https://doi.org/10.1021/ic502779d.

Loitongbam, R.S., Singh, N.S., Singh, W.R., and Ningthoujam, R.S. (2013). Observation of exceptional strong emission transitions ^5D_j ($j = 1$–3) to ^7F_j ($j = 1$–3): multicolor from single Eu^{3+} ion doped La_2O_3 nanoparticles. *Journal of Luminescence* 134: 14–23.

Manikyala Rao, C., Sudarsan, V., Ningthoujam, R.S. et al. (2008). Luminescence studies on low temperature synthesized $ZnGa_2O_4$: Ln^{3+} (Ln = Tb and Eu) nanoparticles. *Journal of Nanoscience and Nanotechnology* 8 (11): 5776.

Ningombam, G.S., Singh, N.R., and Ningthoujam, R.S. (2017). Controlled synthesis of $CaWO_4$:Sm^{3+} microsphere particles by a reverse-micelle method and their energy transfer rate in luminescence. *Colloids and Surfaces A: Physicochemical and Engineering Aspects* 518: 249–262. https://doi.org/10.1016/j.colsurfa.2017.01.021.

Ningombam, G.S., Ningthoujam, R.S., Kalkura, S.N., and Singh, N.R. (2018). Induction heating efficiency of water-dispersible $Mn_{0.5}Fe_{2.5}O_4$@ YVO_4:Eu^{3+} magnetic-luminescent nanocomposites in an acceptable ac magnetic field: Hemocompatibility and cytotoxicity studies. *Journal of Physical Chemistry B* 122 (27): 6862–6871. https://doi.org/10.1021/acs.jpcb.8b02364.

Ningombam, G.S., Khundrakpam, N.S., Thiyam, D.S. et al. (2020). Salt-assisted size-controlled synthesis and luminescence studies of single phase $CaWO_4$:Dy^{3+}: an insight into its morphological evolution, energy transfer and colour evaluation. *New Journal of Chemistry* 44 (10): 4217–4228. https://doi.org/10.1039/C9NJ05677G.

Ningthoujam, R.S. (2010). Generation of exciton in two semiconductors interface: SnO_2:Eu–Y_2O_3. *Chemical Physics Letters* 497 (4–6): 208–212. https://doi.org/10.1016/j.cplett.2010.08.017.

Ningthoujam, R.S. (2012). Enhancement of luminescence by rare earth ions doping in semiconductor host. In: *Synthesis, Characterization and Applications of Multifunctional Materials* (ed. S.B. Rai and Y. Dwivedi), 145–184. Material Science and Technologies: Nova Science Publishers.

Ningthoujam, R.S. and Kulshreshtha, S.K. (2009). Nanocrystalline SnO_2 from thermal decomposition of tin citrate crystal: luminescence and Raman studies. *Materials Research Bulletin* 44 (1): 57–62. https://doi.org/10.1016/j.materresbull.2008.04.004.

Ningthoujam, R.S., Lahiri, D., Sudarsan, V. et al. (2007a). Nature of V^{n+} ions in SnO_2: EPR and photoluminescence studies. *Materials Research Bulletin* 42 (7): 1293–1300. https://doi.org/10.1016/j.materresbull.2006.10.006.

Ningthoujam, R.S., Sudarsan, V., Godbole, S.V. et al. (2007b). SnO_2:Eu^{3+} nanoparticles dispersed in TiO_2 matrix: improved energy transfer between semiconductor host and Eu^{3+} ions for the low temperature synthesized samples. *Applied Physics Letters* 90 (17): 173113. https://doi.org/10.1063/1.2731433.

Ningthoujam, R.S., Sudarsan, V., Vinu, A. et al. (2008a). Luminescence properties of SnO_2 nanoparticles dispersed in Eu^{3+} doped SiO_2 matrix. *Journal of Nanoscience and Nanotechnology* 8 (3): 1489–1493.

Ningthoujam, R.S., Gajbhiye, N.S., Ahmed, A. et al. (2008b). Re-dispersible Li^{+} and Eu^{3+} co-doped nanocrystalline ZnO: luminescence and EPR studies. *Journal of Nanoscience and Nanotechnology* 8 (6): 3059–3062. https://doi.org/10.1166/jnn.2008.152.

Ningthoujam, R.S., Vatsa, R.K., Vinu, A. et al. (2009a). Room temperature exciton formation in SnO_2 nanocrystals in SiO_2:Eu matrix: quantum dot system, heat-treatment effect. *Journal of Nanoscience and Nanotechnology* 9 (4): 2634–2638. https://doi.org/10.1166/jnn.2009.dk05.

Ningthoujam, R.S., Sudarsan, V., Vatsa, R.K. et al. (2009b). Photoluminescence studies on Eu doped TiO_2 nanoparticles. *Journal of Alloys and Compounds* 486 (1–2): 864–870. https://doi.org/10.1016/j.jallcom.2009.07.090.

Ningthoujam, R.S., Singh, L.R., Sudarsan, V., and Dorendrajit Singh, S.D. (2009c). Energy transfer process and optimum emission studies in luminescence of core–shell nanoparticles: YVO_4:Eu–YVO_4 and surface state analysis. *Journal of Alloys and Compounds* 484 (1–2): 782–789. https://doi.org/10.1016/j.jallcom.2009.05.044.

Ningthoujam, R.S., Shukla, R., Vatsa, R.K. et al. (2009d). Gd_2O_3:Eu^{3+} particles prepared by glycine-nitrate combustion: phase, concentration, annealing, and luminescence studies. *Journal of Applied Physics* 105 (8): 084304. https://doi.org/10.1063/1.3098253.

Ningthoujam, R.S., Sharma, A., Sharma, K.S. et al. (2015). Roles of solvent, annealing and Bi^{3+} co-doping on the crystal structure and luminescence properties of YPO_4:Eu^{3+} nanoparticles. *RSC Advances* 5 (83): 68234–68242. https://doi.org/10.1039/C5RA11587F.

Ningthoujam, R.S., Gautam, A., and Padma, N. (2017). Oleylamine as a reducing agent in syntheses of magic-size clusters and monodisperse quantum dots: optical and photoconductivity studies. *Physical Chemistry Chemical Physics* 19 (3): 2294–2303. https://doi.org/10.1039/c6cp07052c.

Okram, R., Yaiphaba, N., Ningthoujam, R.S., and Singh, N.R. (2014). Is higher ratio of monoclinic to tetragonal in $LaVO_4$ a better luminescence host? Redispersion and polymer film formation. *Inorganic Chemistry* 53 (14): 7204–7213. https://doi.org/10.1021/ic500828s.

Parchur, A.K., Okram, G.S., Singh, R.A. et al. (2010). Effect of EDTA on luminescence property of Eu^{3+} doped YPO_4 nanoparticles. *AIP Conference Proceedings* 1313 (1): 391–393.

Parchur, A.K., Ansari, A.A., Singh, B.P. et al. (2014). Enhanced luminescence of $CaMoO_4$:Eu by core@shell formation and its hyperthermia study after hybrid formation with Fe_3O_4: cytotoxicity assessment on human liver cancer cells and mesenchymal stem cells. *Integrative Biology: Quantitative Biosciences from Nano to Macro* 6 (1): 53–64. https://doi.org/10.1039/c3ib40148k.

Perala, R.S., Joshi, R., Singh, B.P. et al. (2021a). Brilliant nonlinear optical response of Ho^{3+} and Yb^{3+} activated YVO_4 nanophosphor and its conjugation with Fe_3O_4 for smart anticounterfeit and hyperthermia applications. *ACS Omega* 6 (30): 19471–19483. https://doi.org/10.1021/acsomega.1c01572.

Perala, R.S., Singh, B.P., Putta, V.N.K. et al. (2021b). Enrichment of crystal field modification via incorporation of alkali K+ ions in YVO_4:Ho^{3+}/Yb^{3+} nanophosphor and its hybrid with superparamagnetic iron oxide nanoparticles for optical, advanced anticounterfeiting, uranyl detection, and hyperthermia applications. *ACS Omega* 6 (30): 19517–19528. https://doi.org/10.1021/acsomega.1c01813.

Phaomei, G., Ningthoujam, R.S., Singh, W.R. et al. (2010). Low temperature synthesis and luminescence properties of re-dispersible Eu^{3+} doped $LaPO_4$ nanorods by ethylene glycol route. *Optical Materials* 32 (5): 616–622. https://doi.org/10.1016/j.optmat.2009.12.009.

Phaomei, G., Ningthoujam, R.S., Singh, W.R. et al. (2011a). Luminescence switching behavior through redox reaction in Ce^{3+} co-doped $LaPO_4$:Tb^{3+} nanorods: re-dispersible and polymer film. *Dalton Transactions* 40 (43): 11571–11580. https://doi.org/10.1039/c1dt11264c.

Phaomei, G., Rameshwor Singh, W.R., and Ningthoujam, R.S. (2011b). Solvent effect in monoclinic to hexagonal phase transformation in $LaPO_4$: RE (RE = Dy^{3+}, Sm^{3+}) nanoparticles: photoluminescence study. *Journal of Luminescence* 131 (6): 1164–1171. https://doi.org/10.1016/j.jlumin.2011.02.023.

Prasad, A.I., Parchur, A.K., Juluri, R.R. et al. (2013). Bi-functional properties of Fe_3O_4@YPO_4:Eu hybrid nanoparticles: hyperthermia application. *Dalton Transactions* 42 (14): 4885–4896. https://doi.org/10.1039/c2dt32508j.

Prasad, A.I., Singh, L.R., Joshi, R., and Ningthoujam, R.S. (2018). Luminescence study on crystalline phase of $Y_2Si_2O_7$ from mesoporous silica and Y_2O_3:Ln^{3+} at 900°C. *AIP Advances* 8 (10): 105310. https://doi.org/10.1063/1.5037184.

Sahu, N.K., Shanta Singh, N., Ningthoujam, R.S., and Bahadur, D. (2014). Ce^{3+}-sensitized $GdPO_4$:Tb^{3+} nanorods: an investigation on energy transfer, luminescence switching, and quantum yield. *ACS Photonics* 1 (4): 337–346. https://doi.org/10.1021/ph4001718.

Shao, Q., Yang, Z., Zhang, G. et al. (2018). Multifunctional lanthanide-doped core/shell nanoparticles: integration of upconversion luminescence, temperature sensing, and photothermal conversion properties. *ACS Omega* 3 (1): 188–197. https://doi.org/10.1021/acsomega.7b01581.

Shukla, R., Ningthoujam, R.S., Tyagi, A.K., and Vatsa, R.K. (2010). Luminescence properties of Dy^{3+} doped Gd_2O_3 nanoparticles prepared by glycine route: annealing effect. *International Journal of Nanotechnology* 7 (9–12): 843–851.

Singh, L.R. and Ningthoujam, R.S. (2010). Critical view on energy transfer, site symmetry, improvement in luminescence of Eu^{3+}, Dy^{3+} doped YVO_4 by core-shell formation. *Journal of Applied Physics* 107 (10): 104304. https://doi.org/10.1063/1.3373409.

Singh, L.R. and Ningthoujam, R.S. (2011). Critical view on luminescence properties of Y_2O_3:Eu^{3+} after dispersion in SiO_2. *Chemical Physics Letters* 510 (1–3): 120–125. https://doi.org/10.1016/j.cplett.2011.05.018.

Singh, L.R., Ningthoujam, R.S., Sudarsan, V. et al. (2008a). Probing of surface Eu^{3+} ions present in ZnO:Eu nanoparticles by covering ZnO:Eu core with Y_2O_3 shell: luminescence study. *Journal of Luminescence* 128 (9): 1544–1550. https://doi.org/10.1016/j.jlumin.2008.02.013.

Singh, N.S., Ningthoujam, R.S., Devi, L.R. et al. (2008b). Luminescence study of Eu^{3+} doped $GdVO_4$ nanoparticles: concentration, particle size, and core/shell effects. *Journal of Applied Physics* 104 (10): 104307. https://doi.org/10.1063/1.3026612.

Singh, L.R., Ningthoujam, R.S., and Singh, S.D. (2009a). Tuning of ultra-violet to green emission by choosing suitable excitation wavelength in ZnO: quantum dot, nanocrystals and bulk. *Journal of Alloys and Compounds* 487 (1–2): 466–471. https://doi.org/10.1016/j.jallcom.2009.07.166.

Singh, L.R., Ningthoujam, R.S., Singh, N.S., and Singh, S.D. (2009b). Probing Dy^{3+} ions on the surface of nanocrystalline YVO_4: luminescence study. *Optical Materials* 32 (2): 286–292. https://doi.org/10.1016/j.optmat.2009.08.005.

Singh, N.S., Ningthoujam, R.S., Yaiphaba, N. et al. (2009c). Lifetime and quantum yield studies of Dy^{3+} doped $GdVO_4$ nanoparticles: concentration and annealing effect. *Journal of Applied Physics* 105 (6): 064303. https://doi.org/10.1063/1.3082088.

Singh, B.P., Parchur, A.K., Ningthoujam, R.S. et al. (2014). Influence of Gd^{3+} co-doping on structural property of $CaMoO_4$: Eu nanoparticles. *Dalton Transactions* 43 (12): 4770–4778. https://doi.org/10.1039/c3dt52786g.

Singh, B.P., Perala, R.S., Srivastava, M., and Ningthoujam, R.S. (2021). Synthesis of nanostructured materials by thermolysis, Chapter 12. In: *Handbook on Synthesis Strategies for Advanced Materials* (ed. A.K. Tyagi and R.S. Ningthoujam), 333–382. Springer Nature.

Soni, A.K., Joshi, R., Jangid, K. et al. (2018). Low temperature synthesized $SrMoO_4$: Eu^{3+} nanophosphors functionalized with ethylene glycol: a comparative study of synthesize route, morphology, luminescence and annealing. *Materials Research Bulletin* 103: 1–12. https://doi.org/10.1016/j.materresbull.2018.03.002.

Soni, A.K., Joshi, R., Singh, B.P. et al. (2019). Near-infrared- and magnetic-field-responsive $NaYF_4$:Er^{3+}/Yb^{3+}@ SiO_2@AuNP@Fe_3O_4 nanocomposites for hyperthermia applications induced by fluorescence resonance energy transfer and surface plasmon absorption. *ACS Applied Nano Materials* 2 (11): 7350–7361. https://doi.org/10.1021/acsanm.9b01867.

Soni, A.K., Joshi, R., and Ningthoujam, R.S. (2021a). *Hot injection method for nanoparticle synthesis: basic concepts, examples and applications,* Chapter 13. In: *Handbook on Synthesis Strategies for Advanced Materials* (ed. A.K. Tyagi and R.S. Ningthoujam), 483–434. Springer Nature.

Soni, A.K., Yadav, K.K., Singh, B.P. et al. (2021b). Smart YPO_4:Er–Yb nanophosphor for optical heating, hyperthermia, security ink, cancer endoradiotherapy, and uranyl recovery. *ACS Applied Nano Materials* 4 (1): 850–860. https://doi.org/10.1021/acsanm.0c03198.

Steinkemper, H., Fischer, S., Hermle, M., and Goldschmidt, J.C. (2013). Stark level analysis of the spectral line shape of electronic transitions in rare earth ions embedded in host crystals. *New Journal of Physics* 15 (5): 053033. https://doi.org/10.1088/1367-2630/15/5/053033.

Sun, Y., Zhu, X., Peng, J., and Li, F. (2013). Core–shell lanthanide upconversion nanophosphors as four-modal probes for tumor angiogenesis imaging. *ACS Nano* 7 (12): 11290–11300. https://doi.org/10.1021/nn405082y.

Viswanatha, R. (2015). Effect of transition metal dopants on the optical and magnetic properties of semiconductor nanocrystals. *Pramana* 84 (6): 1055–1064. https://doi.org/10.1007/s12043-015-1001-0.

Wang, Y. and Wang, Z. (2006). Characterization of $Y_2O_2S:Eu^{3+},Mg^{2+},Ti^{4+}$ long-lasting phosphor synthesized by flux method. *Journal of Rare Earths* 24 (1): 25–28. https://doi.org/10.1016/S1002-0721(06)60059-8.

Wang, F., Deng, R., Wang, J. et al. (2011). Tuning upconversion through energy migration in core–shell nanoparticles. *Nature Materials* 10 (12): 968–973. https://doi.org/10.1038/nmat3149.

Wangkhem, R., Yaba, T., Shanta Singh, N., and Ningthoujam, R.S. (2018). Red emission enhancement from $CaMoO_4:Eu^{3+}$ by co-doping of Bi^{3+} for near UV/blue LED pumped white pcLEDs: energy transfer studies. *Journal of Applied Physics* 123 (12): 124303. https://doi.org/10.1063/1.5021285.

Wangkhem, R., Yaba, T., Singh, N.S., and Ningthoujam, R.S. (2021). Surface functionalized $GdVO_4:Eu^{3+}$ nanocrystals as turn-off luminescent probe for selective sensing of Cu^{2+} ions: role of pH. *Applied Surface Science* 563: 150350.

Wen, S., Zhou, J., Zheng, K. et al. (2018). Advances in highly doped upconversion nanoparticles. *Nature Communications* 9 (1): 2415. https://doi.org/10.1038/s41467-018-04813-5.

Yadav, R.S. and Ningthoujam, R.S. (2021). Synthesis and characterization of quantum cutting phosphor materials. In: *Handbook on Synthesis Strategies for Advanced Materials* (ed. A.K. Tyagi and R.S. Ningthoujam) 213–250. Springer nature. https://doi.org/10.1007/978-981-16-1892-5_7.

Yaiphaba, N., Ningthoujam, R.S., Singh, N.R., and Vatsa, R.K. (2010a). Luminescence properties of redispersible Tb^{3+}-doped $GdPO_4$ nanoparticles prepared by an ethylene glycol route. *European Journal of Inorganic Chemistry* 2010 (18): 2682–2687. https://doi.org/10.1002/ejic.200900968.

Yaiphaba, N., Ningthoujam, R.S., Singh, N.S. et al. (2010b). Luminescence, lifetime, and quantum yield studies of redispersible Eu^{3+}-doped $GdPO_4$ crystalline nanoneedles: core-shell and concentration effects. *Journal of Applied Physics* 107 (3): 034301. https://doi.org/10.1063/1.3294964.

Yaiphaba, N., Ningthoujam, R.S., Shanta Singh, N. et al. (2010c). Probing of inversion symmetry site in Eu^{3+}-doped $GdPO_4$ by luminescence study: concentration and annealing effect. *Journal of Luminescence* 130 (1): 174–180. https://doi.org/10.1016/j.jlumin.2009.08.008.

Yang, X., Pu, C., Qin, H. et al. (2019). Temperature-and Mn^{2+} concentration-dependent emission properties of Mn^{2+}-doped ZnSe nanocrystals. *Journal of the American Chemical Society* 141 (6): 2288–2298. https://doi.org/10.1021/jacs.8b08480.

Ye, S., Song, E.H., and Zhang, Q.Y. (2016). Transition metal-involved photon upconversion. *Advanced Science* 3 (12): 1600302. https://doi.org/10.1002/advs.201600302.

Zhou, L., Zheng, X., Gu, Z. et al. (2014). Mesoporous $NaYbF_4$@ $NaGdF_4$ core–shell up-conversion nanoparticles for targeted drug delivery and multimodal imaging. *Biomaterials* 35 (26): 7666–7678. https://doi.org/10.1016/j.biomaterials.2014.05.051.

11

UCNPs in Solar, Forensic, Security Ink, and Anti-counterfeiting Applications

Kaushal Kumar, Neeraj Kumar Mishra, and Kumar Shwetabh

Indian Institute of Technology (Indian School of Mines), Department of Physics, Optical Materials and Bio-imaging Research Laboratory, Dhanbad 826 004, India

11.1 Introduction

The basic underlying mechanisms and synthesis methods of upconversion nanoparticles (UCNPs) have already been discussed in previous chapters and hope that readers have got an idea about the special advantages of UCNPs over the conventional luminescent phosphors. Readers can be referred to works cited in references for further reading (Boyer et al. 2007; Pandey et al. 2013; Mahata et al. 2014; Kumar et al. 2015; Sinha et al. 2017). In this chapter, novel applications of UCNPs such as solar cell efficiency enhancement, fingerprint detection, invisible ink as an anti-counterfeiting agent, etc., are described. Few examples of high-contrast and multimodal bioimaging applications of UCNPs, which have revitalized the research on rare-earth upconversion emission for a long time, are also discussed (Sinha and Kumar 2018). The materials used in these applications were synthesized via chemical routes such as combustion, hydrothermal, co-precipitation, and thermal decomposition and were either in the form of powder or in the form of colloidal suspension (Pandey et al. 2013; Mondal et al. 2016; Kumar et al. 2016; Sinha et al., 2017a,b; Zhu et al. 2019). These synthesis routes are capable of producing nanometer size particles. Considering a particular application, colloidal stability is also a required parameter for the synthesized sample and can easily be achieved through chemical preparation routes. The UCNPs get excited with infrared radiation, and this is the beauty of these luminescent particles. The near-infrared (NIR) excitation is generally well suited for biological materials because of the relatively higher penetration inside the tissues. Moreover, used detection devices may be blind (insensitive) to the NIR radiation, and hence, chance of interference with detection wavelength minimizes. Most efficient UCNPs are excited either with 980 nm or with 800 nm radiation. However, the use of 980 nm excitation wavelength has got much interest as 980 nm radiation is strongly absorbed by the Yb^{3+} ions. The Yb^{3+} ions then efficiently transfer their energy to the sensitizer ions, and this strategy produces very high intense upconversion emission. The intense UCNP materials

Upconverting Nanoparticles: From Fundamentals to Applications, First Edition. Edited by Vineet K. Rai.

in the form of powder are applicable in persistence luminescence devices, lighting devices, temperature sensing devices, etc. However, colloidal dispersion UCNPs are suitable in the field of security ink and biological imaging. The UC materials in the form of thin films are found to be suitable in solar cell applications, display systems, sensors, etc. This chapter will discuss the applications of codoped UCNPs excited with NIR laser from a cost-effective diode laser.

11.2 UCNPs for Solar Cells

A solar cell is a photovoltaic device that converts solar energy into electrical energy. The as-much sun gives energy in one hour on the earth surface so that we can fulfill our energy requirement for almost an entire year. This is the kind of green energy that emits no greenhouse gases (Morton 2006). However, we are not able to utilize all this solar energy because of the limitation of our photovoltaic devices or solar cells. The major problem that limits the efficiency of a solar cell is due to the mismatch between the solar spectrum, which reaches to earth, and the energy bandgap of the material by which the solar cell is made of. The inability to harvest the entire solar spectrum reduces the efficiency of a solar cell. The solar spectrum that reaches to earth consists of a wavelength ranging from 300 to 2500 nm, and approximately 52% of this spectrum that earth receives has a wavelength in the NIR and IR regions. The as-said solar cells are a kind of photovoltaic devices that basically use some semiconductor materials. Various types of solar cells are available nowadays such as crystalline silicon (C-Si) solar cells, amorphous silicon solar cells, GaAs-based solar cells, dye-sensitized solar cells (DSSCs), and organic material-based solar cells, and among them, approximately 90% are C-Si-based solar cells (Shang et al. 2015). Crystalline silicon (C-Si) has a bandgap of 1.12 eV; hence, it cannot absorb light having a wavelength above 1100 nm (Shang et al. 2015). Similarly, other solar cell materials such as amorphous silicon and dyes have a bandgap of around 1.7 and 1.8 eV, respectively, and hence, all of these materials are limited to absorb a wavelength longer than 700 nm (de Wild et al. 2010). Because of this, a larger portion of the sun spectrum is unable to be utilized. If somehow we make some arrangement so that the excess portion, which is not absorbed by solar cell materials, could be absorbed and utilized by the cell materials, then there is a possibility to enhance the efficiency of solar cells. If we use such materials that absorb light in that forbidden region and transfer it to solar cell materials, then this can happen. There are two ways to do it: (i) If one material absorbs light in the forbidden region and emits the light having a wavelength that could be absorbed by the cell material (i.e. below 700 nm) or (ii) light absorption takes place by one material in the forbidden region and transfers it to the cell material without emitting any light. The former is a kind of radiative transfer, while the latter is a kind of nonradiative transfer called as Forster energy transfer or Forster resonance energy transfer (FRET). To enhance the efficiency of a solar cell, Gibart et al. have first proposed the use of upconversion materials with solar cells (Gibart et al. 1996). Upconversion materials are the materials that absorb light having lower energy, i.e. high wavelength, and emit higher energy, i.e. lesser wavelength. This upconversion process uses a chemically stable host material and a rare-earth material as the dopant. There are various methods such as thin film, coating, and powder method, etc., by which these upconversion materials are attached with different solar cell materials. By the use of upconversion materials along with solar cell materials, the absorption of the solar spectrum for solar cells increases in the region that was previously forbidden by photovoltaic materials alone. Gibart et al. have used GaAs solar cells along with Er^{3+} and Yb^{3+} ion-doped vitroceramic materials. The use of upconversion materials in various solar cells is discussed in the following sections.

11.2.1 C-Si Solar Cells

Lanthanide ion-doped upconversion materials are found to be promising candidates to enhance the C-Si solar cell efficiency. In 2005, Shalav et al. have made efforts to use Er^{3+} ion-doped $NaYF_4$ polycrystalline phosphor to enhance the solar cell efficiency (Shalav et al. 2004). A transparent coating of $NaYF_4$:20% Er^{3+} UC phosphor with an acrylic adhesive medium was applied on the rear side of a C-Si solar cell. The refractive index of the adhesive medium was kept the same as that of polycrystalline UC phosphor. A white reflecting layer was also painted on the rear of this bifacial solar cell system. By using 1523 nm excitation radiation, an increment of 2.5% in the external quantum efficiency of C-Si was observed. The 1523 nm radiation is absorbed by the Er^{3+} ions, and the excited Er^{3+} ions finally converted these radiations into green wavelength (520–560 nm). The C-Si cell is very efficient at green wavelength. This experiment has demonstrated that how unabsorbed infrared solar radiation could be converted into usable form to generate electricity; however, it was a laboratory demonstration. Similarly, Fischer et al. have also made investigations on the Er^{3+} ion-doped $NaYF_4$ upconverting phosphors using 1522 nm as the excitation wavelength. Silicon does not absorb this wavelength and hence shows transparency for this wavelength (Fischer et al. 2010). The UC efficiency was measured to be 5.1% while using 1522 nm excitation with a power density of 1880 W/m^2. This device had shown 0.34% increment in quantum efficiency at 1522 nm excitation wavelength and at a power density of 1090 W/m^2.

Apart from Er^{3+} ions, other rare-earth ions have also been explored to enhance the solar cell efficiency. Ho^{3+} ions have strong absorbance in the region 1150–1230 nm and possess emissions at 650 and 910 nm wavelengths. In 2008, Lahoz has used Ho^{3+}-doped glass ceramics as a substrate for C-Si solar cells (Lahoz 2008). The radiation of 1170 nm wavelength had shown considerable enhancement in the output current as compared to the bare glass precursor substrate. In addition, authors have also tried Yb^{3+} codoping along with Ho^{3+} ions in $NaYF_4$ to improve the efficiency of C-Si solar cells. Yb^{3+} ions have an absorbance around 980 nm wavelength, and hence, emission from Ho^{3+} ions can sensitize the Yb^{3+} ions; this absorption can further lead to an improvement in silicon solar cells. Some

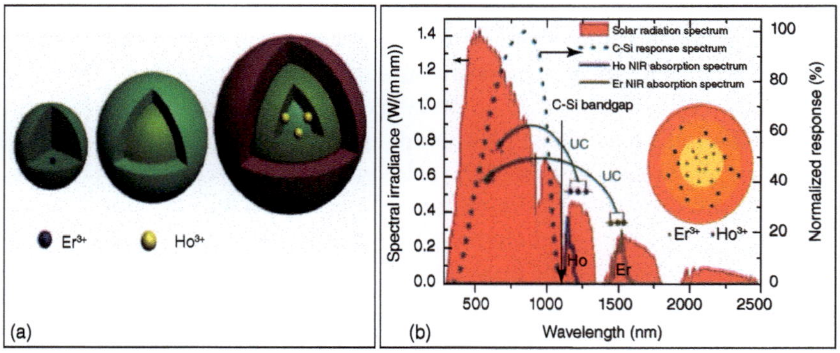

Figure 11.1 (a) A schematic illustration of core–shell–shell nanoparticle's structure: $NaGdF_4:Er^{3+}$ at the core, $NaGdF_4:Ho^{3+}$ as the first shell, and bare $NaGdF_4$ as the outer shell Source: Shang et al. (2015). Reprinted with permission of MDPI Journal. (b) A schematic illustration to show upconversion emission spectrum overlap of $NaGdF_4:Er^{3+}@NaGdF_4:Ho^{3+}@NaGdF_4$ nanocrystals with (i) bandgap of C-Si solar cells and (ii) sun irradiation spectrum. Source: Chen et al. (2018). Reprinted with permission of Elsevier.

geometrical modifications in upconversion phosphor-endorsed solar cell designs have also been proposed and included multi-layered or core/shell-type geometry. In 2012, Chen et al. have synthesized monodispersed $Er^{3+}:NaGdF_4$ $@Ho^{3+}:NaGdF_4$ @ $NaGdF_4$ core–shell–shell nanocrystals (Chen et al. 2012). During the synthesis process, authors had doped Er^{3+} ions in the core and Ho^{3+} ions in the first shell layer, as shown in Figure 11.1a. An illustration is also made in Figure 11.1b to show the spectrum overlap between the solar spectrum and the prepared core–shell–shell nanoparticles' emission spectrum.

In the above scheme, cross-relaxation between Er^{3+} and Ho^{3+} gets reduced because of the spatial separation between the two ions. Hence, a very intense upconversion emission is observed in their structure. However, here also, a limitation comes into picture, that is, emission intensity quenching due to interfacial relaxation in between every layer. To overcome this limitation, again a multi-layered core–shell structure is employed where an inert layer is added between the core and active shell layers. These inert layers prevent the emission intensity quenching by reducing the energy transfer between the layers. Shao et al. in 2015 had demonstrated a unique multi-layered structure of $NaYF_4:Er^{3+}@NaYF_4@NaYF_4:Ho^{3+}@NaYF_4@NaYF_4:Tm^{3+}@NaYF_4$ nanoparticles in which Er^{3+}, Ho^{3+}, and Tm^{3+} doping rates were 10%, 10%, and 1%, respectively (Shao et al. 2015). The undoped $NaYF_4$ layer acts as an inert layer to prohibit the cross-relaxation. Because of the tri-doping of lanthanide ions, this UC material had shown a broad absorbance range, i.e. from 1120 to 1190 nm (due to Ho^{3+}), 1190 to 1260 nm (due to Tm^{3+}), and 1450 to 1580 nm (due to Er^{3+}), and hence, it is very useful for C-Si and other solar cells. As an example, broad emission and absorption spectra obtained due to Ho^{3+}/Er^{3+} ions in $NaGdF_4$ core–shell structure are shown in Figure 11.2.

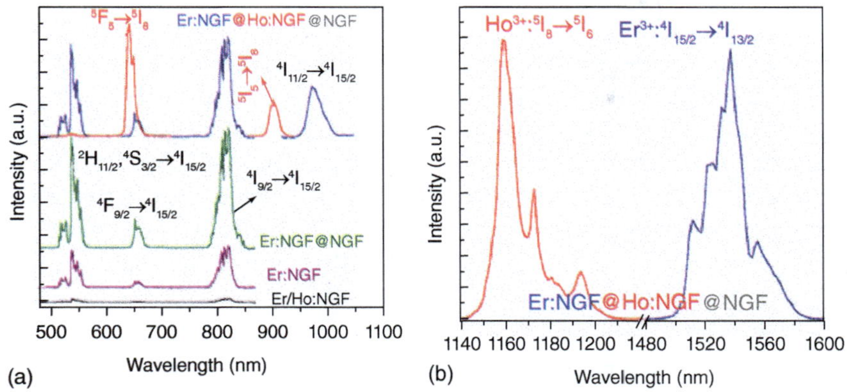

Figure 11.2 (a) Upconversion emission spectra for Er^{3+}/Ho^{3+}-codoped $NaGdF_4$ and single Er^{3+}-doped $NaGdF_4$ nanocrystal. (b) Upconversion excitation spectrum of the $NaGdF_4:Er^{3+}@NaGdF_4:Ho^{3+}@NaGdF_4$ core/shell/shell nanocrystals. Source: Huang (2015). Reprinted with permission of Elsevier.

11.2.2 Amorphous Silicon Solar Cells

Amorphous silicon solar cells can be a good alternative to C-Si solar cells because of their low cost, chemical stability, and easy preparation methods. Amorphous silicon has a bandgap of more than 1.8 eV, which prohibits it to absorb photons having wavelengths more than 700 nm. Hence, any material that can absorb photons above 700 nm and shows emission of photons below 700 nm can be much useful for such type of solar cells. To investigate this possibility, Zhang et al. have synthesized Er^{3+}/Yb^{3+}-doped $NaYF_4$ phosphors as an upconverter for amorphous silicon solar cells (Zhang et al. 2010). The doping amount of Er^{3+} and Yb^{3+} was 2% and 18%, respectively. This structure shows upconversion emissions at visible wavelengths of 655, 525, and 540 nm by absorbing 980 nm wavelength radiation. By the use of such scheme, an enhancement of 6.25% in short-circuit current density of solar cell was observed. In the same year, de Wild et al. had applied a 200–300 μm-thick coating of β-$NaYF_4:Er^{3+}/Yb^{3+}$ on the rear side of an amorphous silicon solar cell for improving the conversion efficiency of the solar cell (de Wild et al. 2010). For the coating of β-$NaYF_4:Er^{3+}/Yb^{3+}$ nanoparticles, the prepared upconversion sample was dispersed in polymethyl methacrylate. This arrangement had shown threefold enhancement in short-circuit current density, i.e. from 2.1 to 6.2 μA. This method of applying an upconverter layer as a back reflector coating has been further investigated by Yang et al. using lanthanide ion-doped oxyfluoride tellurite glass and by Sinha et al. by using $Gd_2(MoO_4)_3:Er^{3+}/Yb^{3+}$ upconversion phosphor (Yang et al. 2014; Sinha et al. 2016). Yang et al. had used TeO_2–PbF_2 glass doped with Er^{3+} and Yb^{3+} ions that had shown emissions at 527, 544, and 657 nm wavelengths under 980 nm wavelength excitation. Similarly, Sinha et al. had used Er^{3+}/Yb^{3+}-doped $Gd_2(MoO_4)_3$ upconversion materials to harvest a broad range of NIR solar spectra.

11.2.3 GaAs-Based Solar Cells

Gibart et al. had reported rare-earth-doped materials to enhance the efficiency of GaAs solar cells (Gibart et al. 1996). They had used a vitroceramic material doped with Er^{3+} and Yb^{3+} and attached it to a GaAs solar cell. The thickness of the solar cell was kept 100 μm. The rare-earth-doped vitroceramic material acts as an upconverting layer which on 891 nm excitation and 25.6 W/m^2 of irradiance had shown an increase of 2.5% in power conversion efficiency. Here, the used irradiance was much larger than the irradiance of sun (0.01–0.1 W/cm^2). Other such efforts had also been done by various researchers.

11.2.4 Dye-Sensitized Solar Cells (DSSCs)

DSSCs are the most promising alternative to C-Si solar cells because of their low cost, eco-friendliness, and flexibility. However, like other solar cell materials, they also have limitations to absorb high-wavelength radiation. Dye has bandgap of approximately 1.8 eV; hence, they cannot absorb wavelengths longer than 700 nm. Application of rare-earth-doped materials for harvesting the radiation beyond this limit can be much beneficial for DSSCs. In the year 2014, Yb^{3+} and Er^{3+} ions were doped into the $NaYF_4$ host to act as an upconversion layer in DSSCs (Dyck and Demopoulos 2014). Liu et al. had used Er^{3+}/Yb^{3+}-doped YAG ($Y_3Al_5O_{12}$) ceramic as a rear reflector layer to obtain reflection as well as NIR absorption as shown in Figure 11.3 (Liu et al. 2011). Miao et al. have synthesized Yb_2O_3 polycrystals having Er^{3+} ions in them and studied the upconversion luminescence properties by attaching the upconversion layer to a DSSC layer (Miao et al. 2013). By doing so, an enhancement of 10.9% and an overall efficiency of 9.21% were obtained. It is a remarkable achievement on overall solar cell efficiency. Also, a comparison between β-$NaYF_4$:Er^{3+}/Yb^{3+} nanoparticles and Er^{3+}-doped Yb_2O_3 polycrystals has been made by the authors. To add more, Li et al. had attached Er^{3+}/Yb^{3+}-doped YF_3 nanoparticles to the porous TiO_2 layer (Li et al. 2014a). Because of the very small size of YF_3:Er^{3+}/Yb^{3+} nanoparticles, these nanoparticles diffuse into a porous TiO_2 thin film and effectively interact with the dye. By using such arrangement, the power conversion efficiency of DSSCs was enhanced from 5.18% to 6.76%. A configuration of DSSCs consisting of different layers of DSSC, UCNP, and reflector is shown in Figure 11.3.

Some other arrangements including a double-shell layer have also been proposed. Liang et al. had used a silica layer in between the UCNP layer and TiO_2 layer (Liang

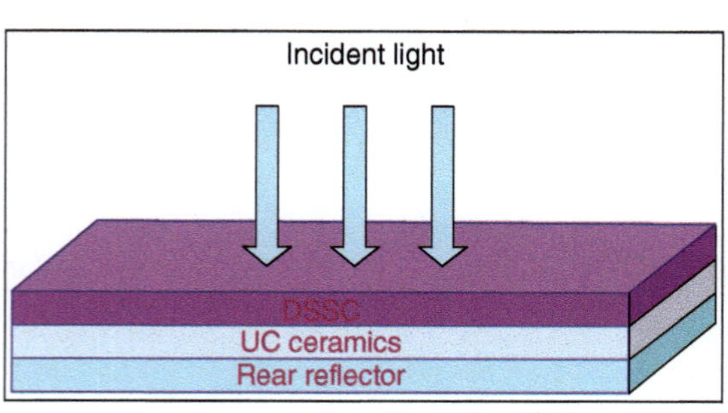

Figure 11.3 A schematic of DSSCs attached with the upconverter layer. Source: Liu et al. (2011). Reprinted with permission of Elsevier.

et al. 2013). By doing so, they have claimed to achieve 29.4% increment in efficiency. Thus, lots of research has been performed to increase the efficiency of solar cells using UCNPs, but further possibilities are yet to be investigated. Above are few examples of the use of upconversion phenomenon in enhancing the quantum efficiency of various types of solar cells. The research is under process, and in future, further improvement in the efficiency of solar cells is expected with the help of advanced upconversion phosphors.

11.3 Forensic, Security Printing, and Anti-counterfeiting Applications

Lanthanide ion (Ln^{3+})-doped UCNPs have sparked great attention for their novel applications as invisible ink for anti-counterfeiting and security codes, latent fingermark detection in forensic science, and similar other areas. It can be seen from Figure 11.4 that there is an exponential increase in research interest on upconversion luminescent materials.

Here, we can sort two-valued examples where UCNPs can do much. (i) High demand of consumer needs has also pushed the penetration of various counterfeiting stuffs, product tempering into the market, and hence, to resist it, various anti-counterfeiting models were introduced including the use of fluorescent materials. Encoding of specific product information into an independent unit has

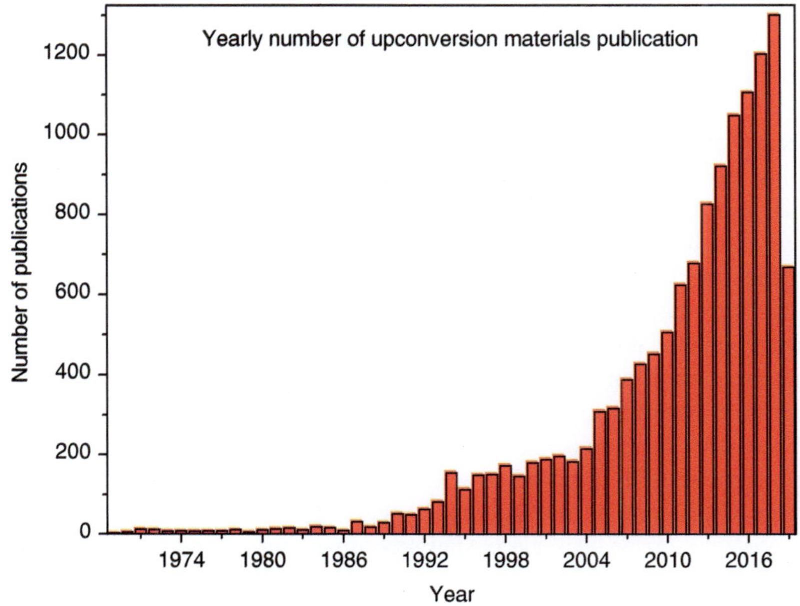

Figure 11.4 Number of upconversion material-based paper publications per year (search done using Scopus up to year 2019).

proved its significance to improve the information security level, and emission color of fluorescent materials has especially been found to be suitable for encoding information and optical multiplexing (Kim et al. 2011). We have been witnessed with colorful code printed as labels with traditional downconversion (DC) materials that convert short-wavelength radiation into longer wavelength radiation. (ii) Forensic science experts are dependent on certain tools that help to record data at the crime scene to identify the suspect, and fingerprint recording is one of the tools used by forensic experts. At many occasions, fingerprints are either poorly visible or completely invisible to the detectives (Choi et al. 2007) and hence needs to be visible for the recognition of criminals. Traditional methods for fingerprint detection are ninhydrin spraying (Oden and Von Hofsten 1954), cyanoacrylate fuming (Tahtouh et al. 2011), and powder dusting (Sodhi and Kaur 2001; Choi et al. 2007). Traditional DC phosphors have proved their job satisfactorily, but at many crucial circumstances, DC phosphors give poor results. In a latent fingerprinting case, DC phosphors fail completely on colored and porous surfaces. Hence, scientists were forced to look for alternative techniques, and fortunately, UCNPs, because of their unique intrinsic properties, have come forward to tackle the limitations that are faced by DC phosphors and other traditional methods (Wang et al. 2015; Liu et al. 2017; Baride et al. 2019). UCNPs can be prepared to produce various colors such as red, yellow, green, blue, purple, and white to produce many codes, and materials can be delivered in the form of powder, plastic masterbatch, yarn, label, colloidal suspension, etc. (Sametband et al. 2007; Jin et al. 2008; Theaker et al. 2008; Gao et al. 2011). Recently, a company named microtrace based in Minnesota, USA, came into existence that has started selling of upconversion-based materials for consumer packaging (Wang et al. 2010). However, much research has to be done to deliver best upconversion phosphors for such innovative applications. In the following, few examples are discussed to show the potential of UCNPs in the above application areas.

For the successful application of upconversion phosphors in latent fingermarks detection, one needs to have a phosphor with good upconversion efficiency, uniform particle morphology and size, dusty nature, and stability against the humid environment. Singh et al. have reported $Gd_2O_3:Er^{3+}/Yb^{3+}$ phosphor material to detect the fingermark by the powder dusting technique (Singh et al. 2009). With this phosphor material, they have used glass surfaces to detect the fingermark. The powder dusting method is basically used in forensic science to develop fingermarks. The upconversion emission intensity depends on the particle size, and it is known that the emission intensity in micron size particles is higher than the nanosized particles. Therefore, the micron size (1–2 μm) particles are best suited for detecting the latent fingermarks. In a typical powder dusting method, a small amount of dusty powder is sprinkled in the fingermark region. A soft brush is then used to spread the powder all over the region. After spreading, excess powder is removed from the surface using a light air flow. Some phosphor particles got stick to the ridges of the fingermarks. A diode laser emitting at 980 nm is then scanned over the area, and a general purpose camera is used to record the pattern. A scheme of the recoding pattern is shown in Figure 11.5. Kumar et al. (2015) have studied in

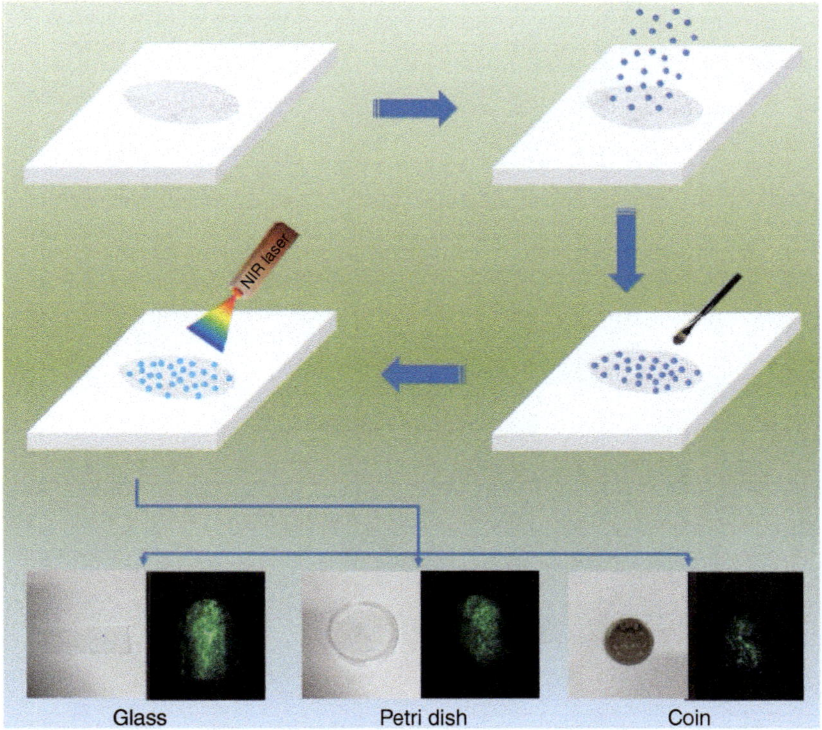

Figure 11.5 Schematic representation of steps of latent fingermark recording. Source: Du et al. (2017). Reprinted with permission of Elsevier.

detail La_2O_3:Er^{3+}/Yb^{3+}, Gd_2O_3:Ho^{3+}/Yb^{3+}, and other UC phosphors and demonstrated the latent fingermark detection with micron size particles. Authors have reported the successful recording of finger patterns on colored 100 rupee Indian banknotes. Authors also have shown that the synthesis method of upconversion phosphors plays an important role in achieving good results. They have compared La_2O_3:Er^{3+}/Yb^{3+} phosphor prepared via combustion and chemical precipitation methods and found that fingermark clarity for combustion-synthesized sample is better than the sample synthesized via a chemical precipitation route.

Wang et al. have used $NaYF_4$ doped with Er^{3+} and Yb^{3+} UCNPs to detect latent fingerprints (Wang et al. 2015). They have demonstrated the successful detection of fingerprint on various surfaces, viz, glass, marble, ceramic tiles, etc., under the 980 nm wavelength excitation source. Ma et al. have used commercially available UPNPs such as $NaYF_4$:Yb,Er (Ma et al. 2011) and YVO_4:Yb, Er (Ma et al. 2012) for the latent fingerprint detection on different semiporous and nonporous surfaces. Among various fluoride-based UPNPs, $NaYF_4$ having the hexagonal phase is the most efficient host for the rare-earth-doped ions because of its low vibrational energy and high radiative emission rate (Gao et al. 2013; Ding et al. 2015). Zhang et al. have used the rare-earth-doped hexagonal phase of $NaYF_4$ for the latent finger detection (Zhang et al. 2019). They have taken the latent fingerprint of the same volunteer on different

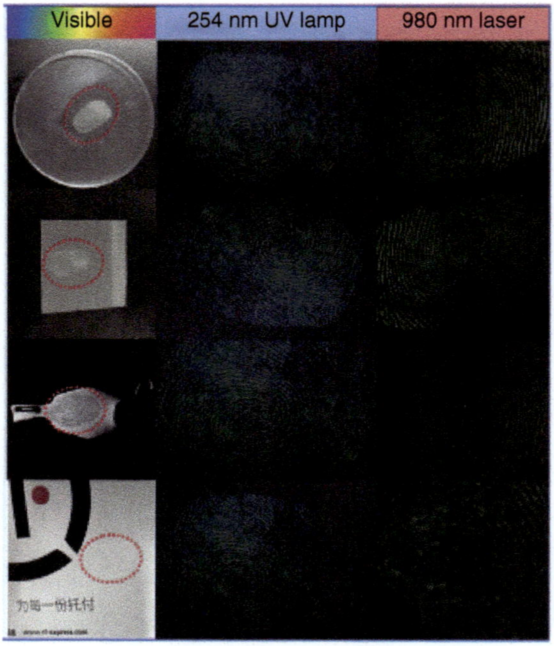

Figure 11.6 Identification of latent fingerprints using hexagonal $NaYF_4$:Ce^{3+},Tb^{3+} and $NaYF_4$ Yb^{3+},Er^{3+} powders. Downshifting and upconverting fluorescent images of fingerprints marked on glass slide, plastic culture dish, metal key, and document envelope are shown (top to bottom). Source: Zhang et al. (2019). Reprinted with permission of Elsevier.

surfaces such as glass slide, plastic culture dish, metal key, and document envelope, and then, they uniformly sprinkled synthesized powder (sample) over the latent fingerprint with the help of a feather brush. They have synthesized two types of hexagonal $NaYF_4$:Yb^{3+}/Er^{3+} and $NaYF_4$:Ce^{3+}/Tb^{3+} up- and downconverting phosphors, respectively, and then applied these powders for the identification of latent fingerprints using 980 and 254 nm excitation sources, respectively. In Figure 11.6, the developed fingermarks by the authors are shown. Kumar et al. in 2019 have demonstrated the use of YF_3:Er^{3+}/Yb^{3+} UCNPs for fingerprint detection, security printing, as well as anti-counterfeiting, which is shown in Figure 11.7 (Kumar et al. 2019). To demonstrate the security printing and anti-counterfeiting applications, a colloidal solution of YF_3:Er^{3+}/Yb^{3+} UCNPs was prepared and then filled in a cylindrical refill. Writing was made with this refill and then text was illuminated with NIR light. The emission from the particles was recorded, and written words are as shown in Figure 11.7e,f. To increase the contrast, the emission intensity of different geometries of UCNPs is employed as a core–shell structure. In 2019, Baride et al. had used the core–shell structure of β-$NaYF_4$ and β-$NaYF_4$:Tm^{3+}/Yb^{3+} (Baride et al. 2019).

At the crime scene, blood fingerprints are also the important evidence against the killer. Liu et al. have used $NaYF_4$:Yb^{3+}/Er^{3+}/Gd^{3+} conjugated with anti-hHb (UCNRs/anti-hHb) UCNPs for fingerprint detection, which is formed by the blood of human, chicken, and porcine, and they have found that $NaYF_4$:Yb^{3+}/Er^{3+}/Gd^{3+} conjugated with anti-hHb produces better contrast of blood fingerprint of human than porcine and chicken (Liu et al. 2018). Figure 11.8 shows the detailed mechanism for the recognition of blood fingerprint using UCNPs.

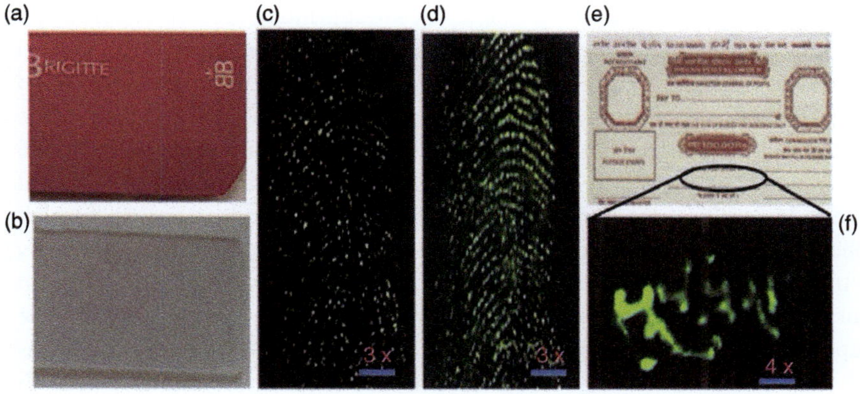

Figure 11.7 Fingerprint demonstration on (a) paper surface and (b) transparent glass surface and fingerprint deposited on both surface (c) and (d) security printing demonstration donor name with or without illumination of NIR light (e) and (f), respectively. Source: Kumar et al. (2019). Reprinted with permission of Elsevier.

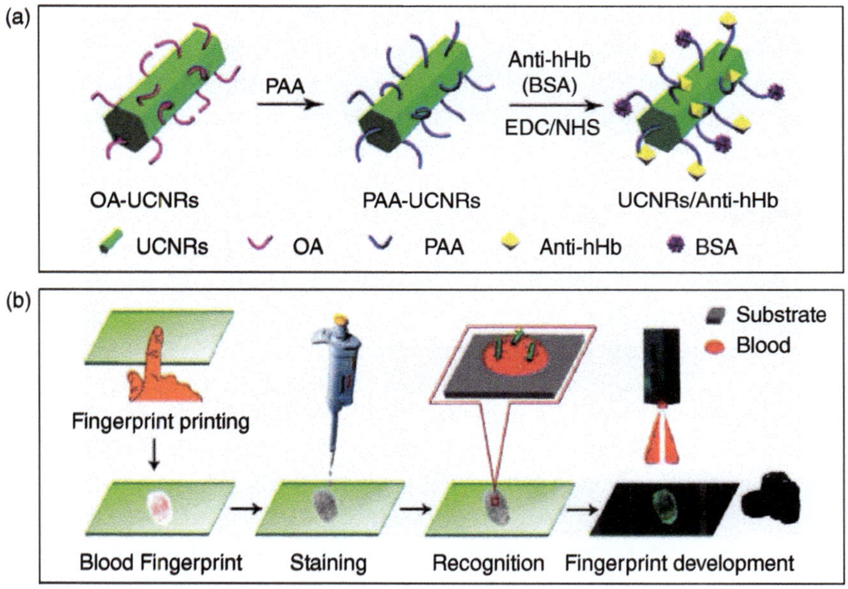

Figure 11.8 Schematic representation of (a) methods of formation of UCNPs conjugated with anti-hHb and (b) specific identification of blood fingerprint. Source: Liu et al. (2018). Reprinted with permission of Elsevier.

Counterfeit documents such as checks, currency, certificates, identity cards, etc., are global epidemic causing corruption and fraud around the world. Therefore, development of an anti-counterfeiting technology has high demand globally. Traditional anti-counterfeiting technologies such as quick response (QR) codes, holography, radiofrequency identification (RFID), etc., have disadvantages of high cost, complex manufacturing process, and easy infringement. Several printing technologies such as mask photolithography, inject printing (You et al. 2016), and aerosol jet printing have been developed using UPNPs (Kim et al. 2009). Kaboli et al. have used $NaYF_4$:Yb^{3+},Er^{3+} and $NaYF_4$:Yb^{3+},Tm^{3+} UPNPs for the execution of maskless photolithography. They have also shown that the detectability of the anti-counterfeiting approach is performed just by using a smartphone (Kaboli et al. 2019). Zhou et al. have reported the synergistic upconversion enhancement induced by multiple physical effects and an angle-dependent anti-counterfeit application. They have described that $Cu_{2-x}S$ nanoparticles exhibit not only the tunable localized surface plasmon resonance (LSPR) but also a two-photon absorption effect under infrared light pumping, which depends strongly on the composition, size, and bandgap of the NPs. The interaction of these nanoparticles with the $NaYF_4$:Yb^{3+},Er^{3+}@$NaYF_4$:Yb^{3+},Nd^{3+} UCNPs under multiwavelength excitations at 808, 980, and 1540 nm has been studied. On the basis of the developed hybrids, Zhou et al. have successfully performed the novel angle-dependent infrared anti-counterfeiting (Zhou et al. 2017). The detailed processes adopted for the angle-dependent anti-counterfeiting are shown in Figure 11.9.

Hu et al. have prepared double-perovskite tungstate $LiYMgWO_6$:Er^{3+} UC phosphors through a high-temperature solid-state reaction method and observed green and red double-color emissions under 1550 nm excitation. Authors have adjusted the emission color of $LiYMgWO_6$:Er^{3+} from yellow-green to enhance red by codoping of Ho^{3+} ions (Hu et al. 2020). To demonstrate anti-counterfeiting applications, authors had dispersed a known amount of prepared powder phosphor in a mixture of ethanol (40 wt%), deionized water (40 wt%), and glycerol (20 wt%). The viscous ink was formed and then used to write text on a plane paper. The hidden words were then illuminated with 1550 nm to read the words. In Figure 11.10, few hidden Chinese words are shown in yellow and red colors when illuminated with 1550 nm radiation. Two multicolor flowers such as patterns were also designed by the authors via a combination of yellow and red fluorescent inks as shown in Figure 11.10e,f. Results show that the developed UC fluorescent ink possesses excellent hiddenness and security and has real potential for anti-counterfeiting applications. The work of Hu et al. is promising in the sense that they have used 1550 nm excitation, which is safer to the human eye than the mostly used 980 and 808 nm wavelengths (Hu et al. 2020).

In explosive detection also, UCNPs are used, e.g. detection of TNT (trinitrotoluene). TNT is a nitroaromatic explosive that is very stable and toxic and often used by military and terrorists. These are electron-deficient molecules that act as an electron acceptor, and hence, by reacting to any electron donor, they form a complex called Meisenheimer complexes that reduce the UCNP emission. Ma and Wang had synthesized $NaYF_4$:Er^{3+}/Yb^{3+} nanocrystals and observed the upconversion luminescence quenching caused by TNT and TNP (trinitrophenol), which made possible to detect the TNT in aqueous solution (Ma and Wang 2014). Liu et al. have used the core–shell structure of $NaYF_4$ nanoparticles for detection of nitroaromatic compounds in aqueous solution (Liu et al. 2019). They had prepared a sandwiched core–shell structure of $NaYF_4$:Er^{3+}/Yb^{3+}@$NaYF_4$ UCNPs and used the surface modification to attach the NH_2 molecule on particle's surface to act as an electron donor and detected TNT and TNP in aqueous media. A diagrammatic illustration is

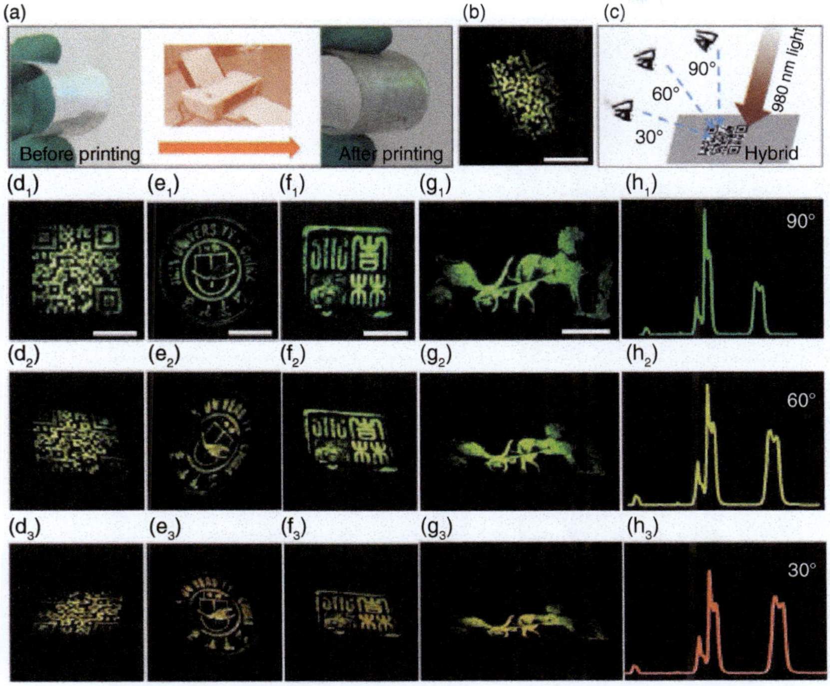

Figure 11.9 (a) Schematic diagram of constructing anti-counterfeiting hybrids (flexible OPC and hybrid). (b) Image of the two-dimensional code pattern on the flexible OPC. Scale bar: 0.5 cm. (c) Sketch of the angle-dependent anti-counterfeiting. (d) (d_1-d_3) Image of the fluorescence of a two-dimensional code pattern on the paper recorded at an angle of 90°, 60°, and 30° under the excitation of 980 nm, respectively. (e) (e_1-e_3) Image of the school logo pattern on the hybrid at the angles of 90°, 60°, and 30° under the excitation of 980 nm. (f) (f_1-f_3) Image of fluorescence pattern of a specific logo on hybrid recorded at angles of 90°, 60°, and 30° under the excitation of 980 nm, respectively. (g) (g_1-g_3) Image of the fluorescence pattern of a specific logo on the hybrid recorded at angles of 90°, 60°, and 30° under the excitation of 980 nm, respectively. (h) (h_1-h_2) Upconversion emission spectra of anti-counterfeiting hybrids recorded at 90°, 60°, and 30° under the excitation of 980 nm, respectively. Source: Zhou et al. (2017). Reprinted with permission of American Chemical Society.

made in Figure 11.11 to understand the explosive detection using surface-modified UCNPs based on FRET and luminescent resonance energy transfer (LRET) principle.

11.4 Biomedicals

Lanthanide-doped upconverting materials have also added a new dimension in the area of biomedicals. The UCNPs have redefined in vitro and in vivo bioimaging with multimodal capability. These UC materials have strong property of deep penetration depth, absence of autofluorescence, and high resistance to photobleaching (Singh-Rachford et al. 2008; Liu et al. 2013; Zhao et al. 2013). Such

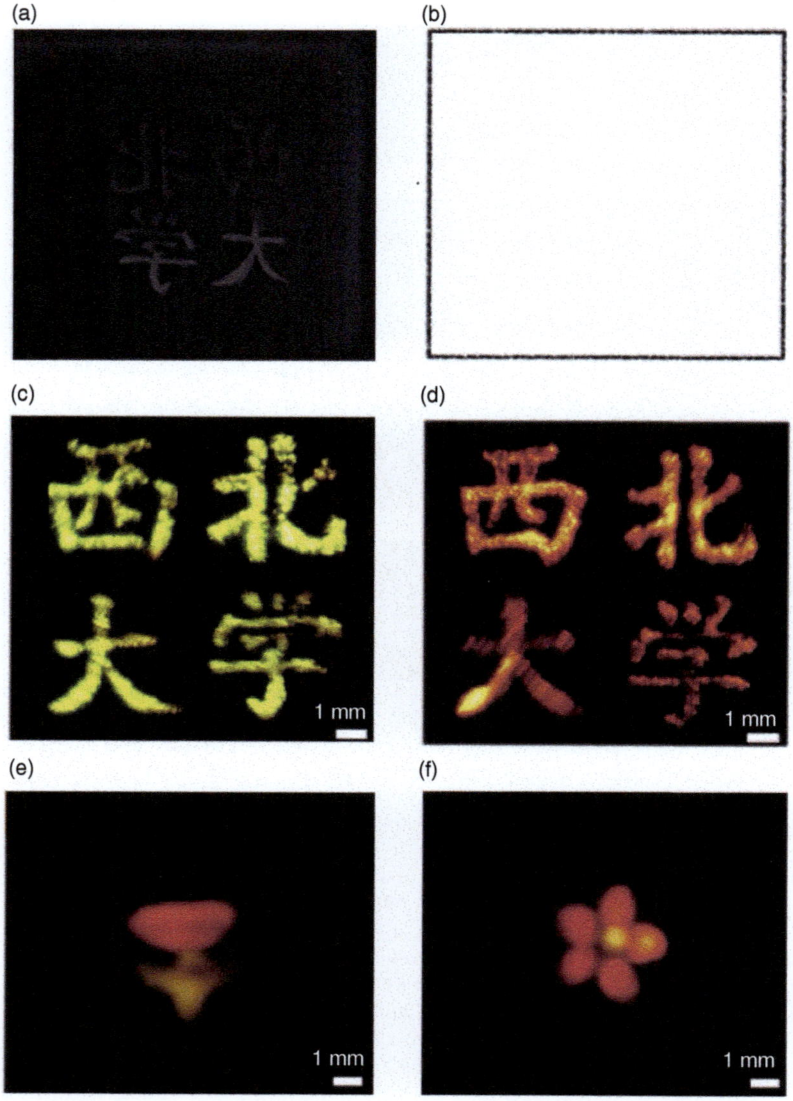

Figure 11.10 (a) Stamp of Chinese words used to write on plain paper; (b) hidden words written using $LiYMgWO_6:Er^{3+}$ phosphor ink on plain paper; (c, d) yellow and red color images of printed words when illuminated with 1550 nm radiation. (e, f) Multicolor flower patterns formed by combining green phosphor ($LiYMgWO_6:Er^{3+}$) and red phosphor ($LiYMgWO_6:Er^{3+}/Ho^{3+}$) ink. Source: Hu et al. (2014). Reprinted with permission of Elsevier.

kind of characteristics of UC materials are highly desired for biological imaging, biosensing, photodynamic therapy (PDT), etc. However, the immense applications offered by UC materials are limited by low upconversion efficiency of UC particles whose size is less than 8 nm because particles having size less than 8 nm can pass from the blood into the interstitial tissue without any difficulty. Along with this,

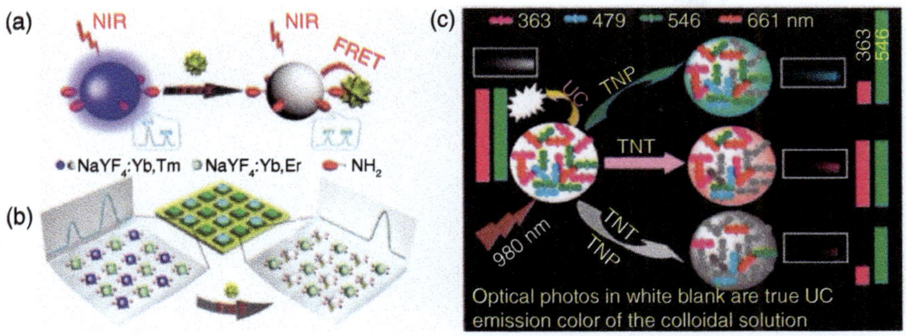

Figure 11.11 A schematic illustration of the (a, b) working principle and detection of TNT based on FRET. Source: Hu et al. (2014). (c) Colloidal $NaYF_4$ UCNPs and the upconversion-based probe for TNT or TNP detection. Source: Ma and Wang (2014).

one of the most important properties of UC materials should be concerned, which is nontoxic in nature for biological applications.

11.4.1 Bioimaging

Because of the extraordinary temporal and spatial resolution property of fluorescence imaging, it is one of the important techniques for the studies of biological tissues (Rao et al. 2007). Earlier fluorescence bioimaging was based on the down-conversion phenomenon, in which the photon of low energy is emitted on excitation with the high-energy source of light. There is limitation for the use of high-energy light as the excitation source as it may cause substantial autofluorescence of biological tissues, DNA/cell damage on longtime irradiation, and less penetration depth in biotissues. However, fluorescence imaging is based on the upconversion phenomenon, in which higher energy photon (visible emission) is observed on the absorption two or more photons of low energy generally (NIR) radiations (Schenke-Layland et al. 2006). Also, Ln^{3+}-doped UC materials are less toxic in nature than the semiconductor-based quantum dots (QDs). UC materials, in which the infrared excitation sources are used for bioimaging applications, have other advantages, viz, (i) minimizes the autofluorescence, (ii) less harmful to cell/DNA, and (iii) penetrates the biological tissues to a greater level. Researchers have been using UC materials for the detection of ovarian cancer cells (Boyer et al. 2010), breast carcinoma cells (Lin et al. 2012), colon cancer cells (Chatterjee et al. 2008), HeLa cells (Cheng et al. 2011), etc. Chatterjee et al. have used polyethylene-imine (PEI)-coated Er^{3+}/Yb^{3+}-doped $NaYF_4$ UCNPs for in vivo imaging of tissues and in vitro imaging of cancer cells (Chatterjee et al. 2008). To understand the applicability of UCNPs in biological imaging of tissues, they injected PEI/$NaYF_4$:Yb,Er nanoparticles (4.4 mg/ml) at the groin and upper leg portion of Wistar rats. Green color emitting quantum dots (QDs) were injected into the transparent parts of the foot, which shows fluorescence on excitation with the UV source. However, this green color emission is not seen through the thicker skin of the back as shown in Figure 11.12. Moreover, the emission of light can be seen through the

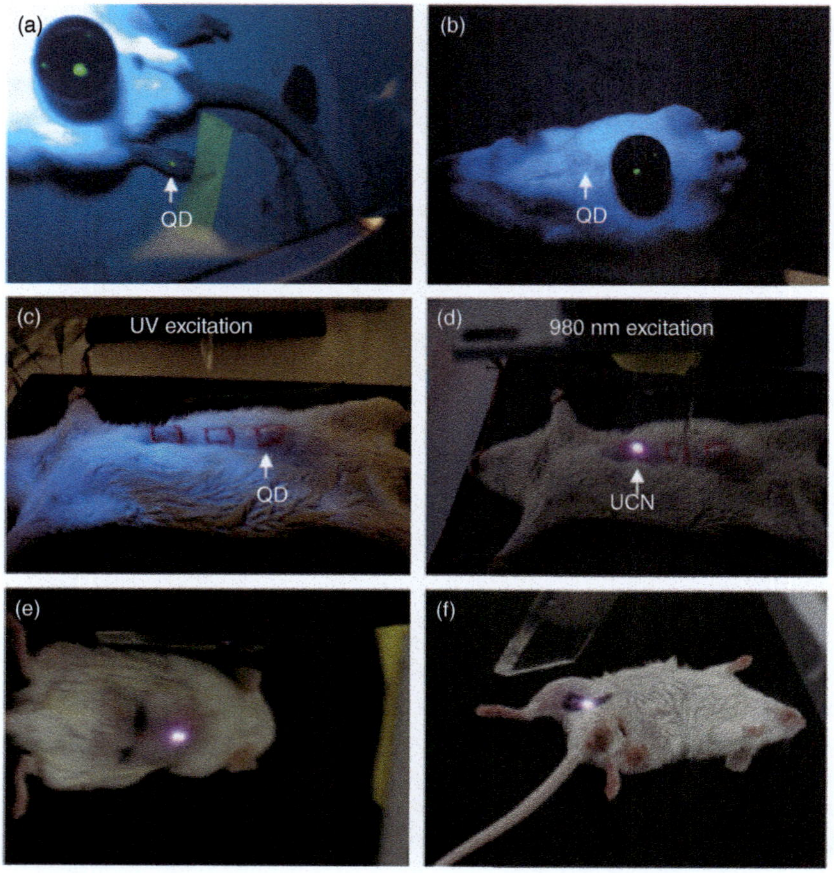

Figure 11.12 In vivo and in vitro imaging of rats: quantum dots and PEI-coated $NaYF_4$:Yb,Er nanoparticles injected into the rats' body, respectively: (a) green color emission can be seen, not through the thicker portion of back (b) or abdomen. (c) PEI-coated $NaYF_4$:Yb,Er upconversion nanoparticles injected into the abdomen skin of the rats, (d) thigh portion, (e) or lower portion of back. (f) Luminescence can be seen through QDs. Source: Chatterjee et al. (2008). Reprinted with permission of Elsevier.

thicker skin because of injected UCNPs on excitation with the 980 nm, as shown in Figure 11.12d. Thus, UCNPs could be applicable for the detection of tumors and exploration of pathologies without any sacrifice of animals.

The beauty of UCNPs is that such particles could be tailored as per the requirement and hence can be utilized in many applications. The UCNPs can be used as a multimodal/multifunctional way, for example, $NaGdF_4:Er^{3+}/Yb^{3+}$ phosphors can show two properties: (i) frequency upconversion optical emission and (ii) magnetic property. The first property belongs to the optical functionality, and the second property belongs to the magnetic functionality, and both can be explored at the same time. Gadolinium (Gd^{3+}) host-based compounds are commonly used as contrast agents for magnetic resonance imaging (MRI). Gd^{3+} is among those four elements that has magnetism at room temperature (gadolinium, nickel, iron, and cobalt). It offers alternatives to high-energy ionizing radiation methods of tomographic imaging such as computed tomography and positron emission tomography. Interestingly, UCNPs can be used to enhance the depth and contrast of images recorded by optical coherence tomography (OCT). OCT has evolved as an important imaging technology for biomedical imaging of the skin, blood vessels (angiography), malignant tumors, and other tissues. The OCT uses NIR radiation to provide good spatial resolution without the sacrifice of penetration depth. Swept source (varying wavelength)-based optical coherence tomography (SSOCT) works on the principle of Fourier transform interferometry and has increased the sensitivity over the conventional OCT. SSOCT at many occasions also produces lower contrast, and hence, researchers are trying to enhance the image contrast with various additive methods. The use of contrast agents is one method to improve the contrast of image. The idea is that when UCNPs come closer to the cell surface, then there is a mismatch in the refractive index, and it results phase contrast between two different regions. The refractive index mismatch can induce light scattering and may enhance the contrast. The gold (with visible wavelength plasmon band) and other scattering nanoparticles were used for this purpose, and the idea has worked but low penetration of visible light inside the tissues has limited the use. If one tries to select the excitation within the biological transparency window (650–1350 nm), then more depth resolution can be achieved. This NIR window could be used efficiently with Er^{3+}/Yb^{3+}- or Ho^{3+}/Yb^{3+}-doped UCNPs excited with 980 nm wavelength. Ryu et al. have investigated the $NaGdF_4:Er^{3+}/Yb^{3+}$ nanoparticles on HeLa cells for in vitro imaging application (Ryu et al. 2010). Authors have observed that the colloidal stability of UCNPs is an important factor, and the presence of polar functional groups, e.g. -OH, -COOH, and $-NH_2$, enhance the colloidal stability. Maurya et al. have prepared $NaGdF_4:Er^{3+}/Yb^{3+}$ through the solvothermal method using ethylene glycol (EG) and oleic acid (OA) as the reaction medium (Maurya et al. 2020). Two samples, viz, polyvinylpyrrolidone (PVP)-coated $NaGdF_4:Er^{3+}/Yb^{3+}$ (sample named as NaGdF-I) and oleic acid coated $NaGdF_4:Er^{3+}/Yb^{3+}$ (sample named as NaGdF-II), were prepared for this purpose. Two thin slices of chicken breast tissues were prepared for in vitro tissue imaging. NaGdF-I and -II solutions were prepared separately using pure ethanol. A tissue sample without UCNPs was scanned first using the SSOCT system (called control experiment). Following that, few drops of NaGdF-I and -II colloidal solutions were placed on the tissue surface and then allowed to penetrate inside the tissue layers. Imaging of tissue was done using 1000 nm wavelength, but this wavelength does not excite the UCNPs and hence 976 nm wavelength radiation was used to excite the UCNPs during the imaging. OCT images were recorded at the intervals of 0, 5, 10, 20, and 30 minutes after placing drops of UCNPs at the surface. Images were recorded with and without the use of 976 nm excitation, which means that one image of tissue was taken without UCNP excitation and another image was taken with 976 nm excitation. The SSOCT B-scan pictures of the control sample are shown in Figure 11.13a,d. Figure 11.13b,e indicates the SSOCT image taken 30 minute after NaGdF-I treatment and with no excitation. Figure 11.13c,f shows a high-contrast SSOCT image taken while NaGdF-I was excited at 976 nm. SSOCT

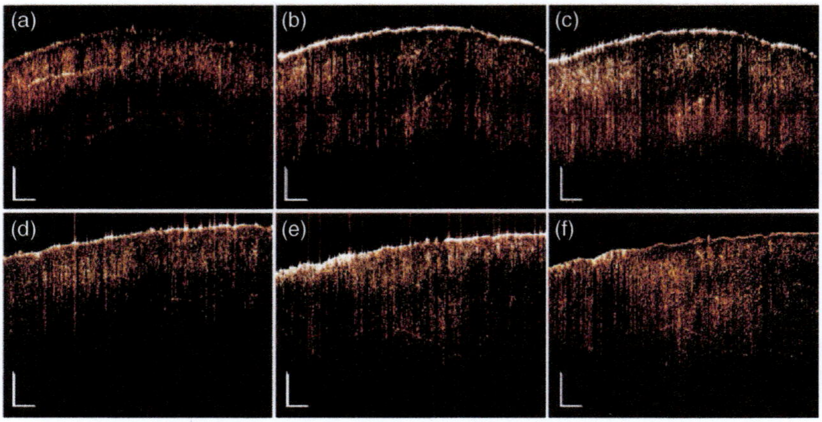

Figure 11.13 SSOCT B-scan images of chicken breast tissue; (a) control sample for the first sample; (b) without excitation of NaGdF-I UCNPs after 30 minutes of application; (c) excitation of NaGdF-I UCNPs after 30 minutes of application; (d) control sample for the second experiment; (e) without excitation of NaGdF-II UCNPs after 30 minutes of application; and (f) with excitation of NaGdF-II UCNPs after 30 minutes of application (Scale: 200 μm). Source: Maurya et al. (2020). Reprinted with permission of American Chemical Society.

images establish increased depth of penetration and improved image contrast, signifying the diffusion of UCNPs deeper into tissues after a 30-minute period.

The B-scan pictures are hard to read, and hence, A-scan pictures were created from the B-scans. Figure 11.14 shows the A-scan graph extracted from Figure 11.13. From the figure, it can be observed that as the time increases, the second peak increases in height and shifts toward a higher depth. It indicates the incremental penetration of UCNPs with time. Moreover, the intensity of the second peak is found higher when the sample is excited with 976 nm (Figure 11.14b). The increase in intensity is due to the increased scattering of imaging wavelength (1000 nm) from excited UCNPs. Comparison of Figure 11.14b,c also indicates that NaGdF-I UCNPs are better over NaGdF-II UCNPs, as the intensity is found to be higher for earlier case. The studies made here are just the initial efforts to extend the application of UCNPs in OCT imaging, and the results are inspiring. It is expected that further investigations may lead to real-world applications.

11.4.2 Biosensing

Biological sensing can also be done with UCNPs. There are several sensing techniques out of which fluorescent-based biological sensing attracted the attention of researchers because of its high sensitivity, simplicity, and versatility in finding the different information of biological systems. UC material-based biosensors have important position in fluorescence-based optical biosensing owing to their photostability and efficient emission intensity on excitation with the NIR radiation, which is less harmful to the biological system (Liang et al. 2017). Chen et al. have used streptavidin-coupled based UCNPs for DNA sensor (Chen et al. 2008). The design of the mechanism of a biosensor is shown in the Figure 11.15. The

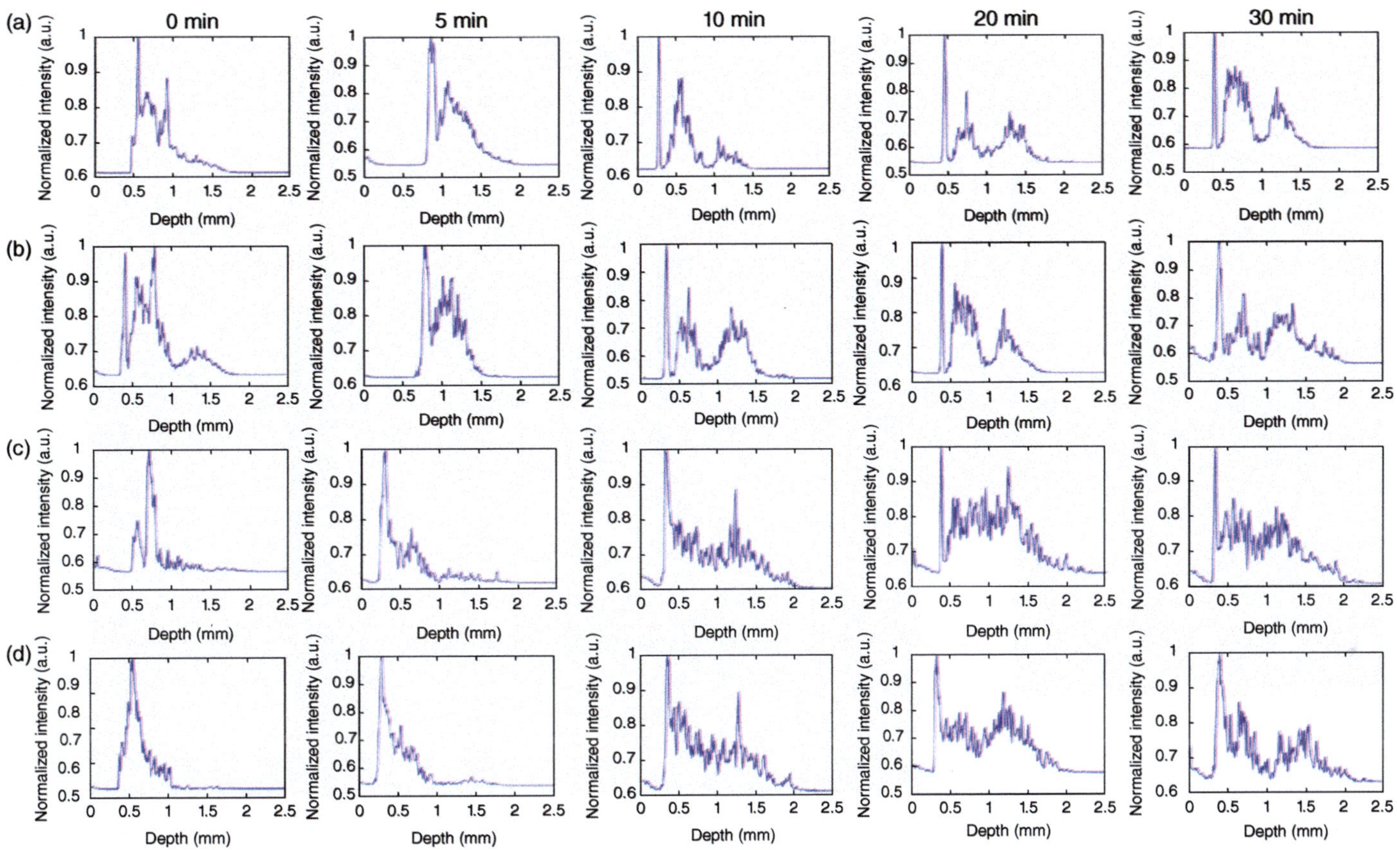

Figure 11.14 Averaged A-scan profile plot of the SSOCT image over the region of interest at different intervals of time 0, 5, 10, 20, and 30 minutes. A-scan for sample with (a) NaGdF-I without excitation; (b) NaGdF-I with excitation; (c) NaGdF-II without excitation; and (d) NaGdF-II with excitation. The *X*-axis in all spectra represents depth in millimeters, whereas the *Y*-axis represents the normalized intensity in arbitrary units (a.u.). Source: Maurya et al. (2020). Reprinted with permission of American Chemical Society.

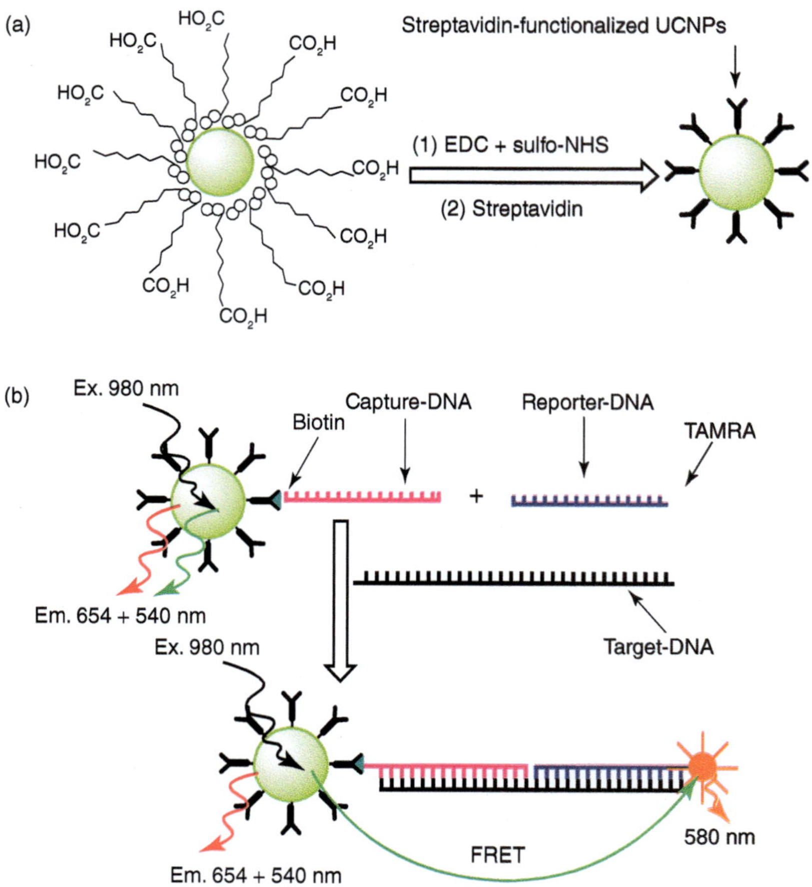

Figure 11.15 (a) Synthesis of streptavidin-functionalized $NaYF_4$:Yb^{3+},Er^{3+} nanoparticles and (b) diagram representing the $NaYF_4$:Yb^{3+},Er^{3+} nanoparticle-based DNA sensor. Source: Chen et al. (2008). Reprinted with permission of American Chemical Society.

shorter oligonucleotide is labeled capture DNA (DNA1-biotin) and another one is labeled reporter DNA, which is considered as TAMRA (DNA2-TAMRA). The excitation band of TAMRA lies within the emission band of $NaYF_4$:Yb^{3+},Er^{3+} in the green region. Figure 11.16 shows that a composition of DNA1-biotin and streptavidin-functionalized $NaYF_4$:Yb^{3+},Er^{3+} emits luminescent bands under the excitation of 980 nm laser source, and this predicts the absence of DNA2-TAMRA. However, in the presence of target DNA2-TAMRA, an additional band around 580 nm appears gradually corresponding to the TAMRA emission. Moreover, on increasing the concentration of target DNA2-TAMRA, the intensity of bands around 580 nm starts increasing and the intensities of bands 514 and 560 nm that are attributed to the Er^{3+} start decreasing. However, the intensity of red emission bands 635 and 680 nm of Er^{3+} remains unchanged up to a small increment of target DNA. Therefore, there is energy transfer from green bands of $NaYF_4$:Yb^{3+},Er^{3+} particles to the TAMRA. Along with this, the intensity ratio of 580 and 540 nm and

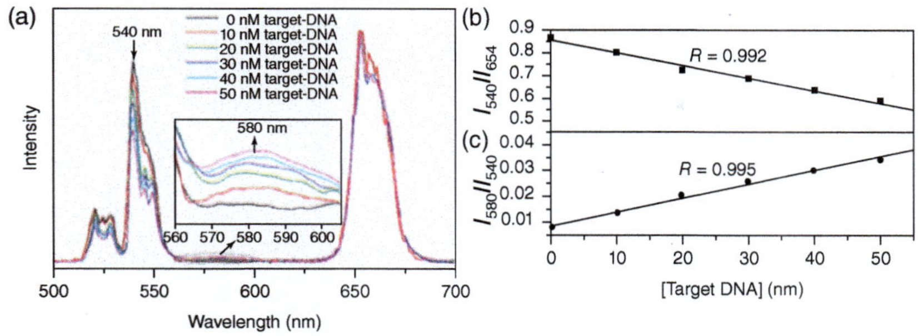

Figure 11.16 (a) Upconversion luminesce spectra of composition of streptavidin-functionalized $NaYF_4$:Yb^{3+},Er^{3+} nanoparticles, capture, and reporter DNA in the presence of various concentrations of target DNA under the excitation wavelength of 980 nm, (b) linear relationship between the intensity ratio 540 and 654 nm, and (c) 580 and 540 nm with target DNA concentration. Source: Chen et al. (2008). Reprinted with permission of American Chemical Society.

540 and 654 nm changes linearly on increasing the concentration of target DNA. Thus, we can quantitatively sense/detect the amount of target DNA.

11.5 Display and Lighting Purposes

Material scientists have tough challenges to develop a material that is capable of emitting color in a broad range of spectra on using the external stimuli pressure, magnetic, electric field, and light. The development of materials that can emit multicolor may bring the revolutionary applications in the research field of displays, graphic imaging, information storage, biological sensing, etc. (Matsubara et al. 2008; Lee et al. 2014). Conventional methods to get multicolor emission are used to synthesize special organic dyes or semiconducting dots, and the emitting color can be modulated by varying the excitation wavelengths or with changing the power of excitations (Sun et al. 2006). However, there are some drawbacks of these methods, such as the need of multiple sources of excitations, and it is hard to provide multicolor emissions with steady brightness. From the past decades, the development in upconversion materials opens the new opportunity for multicolor emitting materials using a single low-energy excitation source (Heer et al. 2004; Rapaport et al. 2006; Li et al. 2014). Downing et al. have used the upconverting materials for the design of three colors emission, solid state, and 3D display (Downing et al. 1996). Deng et al. have prepared multicolor tunable upconversion nanocrystals; for this, they have synthesized the hexagonal phase of $NaYF_4$ with the doping of an appropriate amount of five lanthanides Yb^{3+}, Nd^{3+}, Tm^{3+}, Ho^{3+}, and Ce^{3+} ions in the layer-by-layer form (Deng et al. 2015), wherein the Yb^{3+} and Nd^{3+} ions were used as sensitizers for excitations at 980 and 808 nm, respectively. Near the inner shell of the nanocrystals, Tm^{3+} was confined to give possible blue emission, and then, the outer shell containing Ho^{3+} was confined to give green and red color. Ce^{3+} ions were also codoped with the Ho^{3+} ions in the same layer to enhance the red color emission of Ho^{3+} through energy transfer

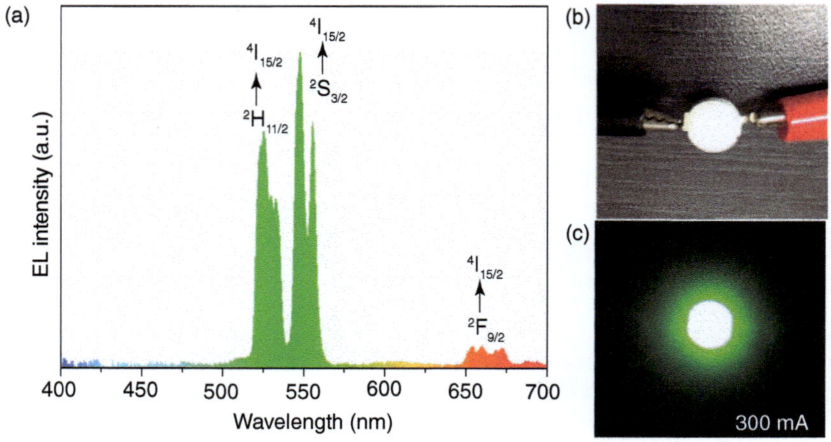

Figure 11.17 (a) Electroluminescence emission spectra of Yb^{3+}/Er^{3+}-codoped $Y_2Mo_4O_{15}$-based LED device recorded at 300 mA in forward biased. (b) Fabricated LED device. (c) Green color emission from the fabricated LED device on applying 300 mA in forward biased. Source: Du and Yu (2018). Reprinted with permission of American Chemical Society.

from the Ce^{3+} to Ho^{3+} ions. Du et al. have used Yb^{3+}/Er^{3+}-codoped $Y_2Mo_4O_{15}$ upconverting materials for the fabrication of green color emitting diode device (Du and Yu 2018). Figure 11.17 shows the fabricated green color emitting LED device.

References

Baride, A., Sigdel, G., Cross, W.M. et al. (2019). Near infrared-to-near infrared upconversion nanocrystals for latent fingerprint development. *ACS Applied Nano Materials* 2 (7): 4518–4527. https://doi.org/10.1021/acsanm.9b00890.

Boyer, J.C., Cuccia, L.A., and Capobianco, J.A. (2007). Synthesis of colloidal upconverting $NaYF_4$:Er^{3+}/Yb^{3+} and Tm^{3+}/Yb^{3+} monodisperse nanocrystals. *Nano Letters* 7 (3): 847–852. https://doi.org/10.1021/nl070235.

Boyer, J.C., Manseau, M.P., Murray, J.I., and Van Veggel, F.C. (2010). Surface modification of upconverting $NaYF_4$ nanoparticles with PEG-phosphate ligands for NIR (800 nm) biolabeling within the biological window. *Langmuir* 26 (2): 1157–1164. https://doi.org/10.1021/la902260j.

Chatterjee, D.K., Rufaihah, A.J., and Zhang, Y. (2008). Upconversion fluorescence imaging of cells and small animals using lanthanide doped nanocrystals. *Biomaterials* 29 (7): 937–943. https://doi.org/10.1016/j.biomaterials.2007.10.051.

Chen, Z., Chen, H., Hu, H. et al. (2008). Versatile synthesis strategy for carboxylic acid – functionalized upconverting nanophosphors as biological labels. *Journal of the American Chemical Society* 130 (10): 3023–3029. https://doi.org/10.1021/ja076151k.

Chen, D., Lei, L., Yang, A. et al. (2012). Ultra-broadband near-infrared excitable upconversion core/shell nanocrystals. *Chemical Communications* 48 (47): 5898–5900. https://doi.org/10.1039/C2CC32102E.

Chen, T., Shang, Y., Hao, S. et al. (2018). Enhancement of dye sensitized solar cell efficiency through introducing concurrent upconversion/downconversion core/shell nanoparticles as spectral converters. *Electrochimica Acta* 282: 743–749. https://doi.org/10.1016/j.electacta.2018.06.111.

Cheng, L., Yang, K., Li, Y. et al. (2011). Facile preparation of multifunctional upconversion nanoprobes for multimodal imaging and dual-targeted photothermal therapy. *Angewandte Chemie* 123 (32): 7523–7528. https://doi.org/10.1002/anie.201101447.

Choi, M.J., Smoother, T., Martin, A.A. et al. (2007). Fluorescent TiO_2 powders prepared using a new perylene diimide dye: applications in latent fingermark detection. *Forensic Science International* 173 (2–3): 154–160. https://doi.org/10.1016/j.forsciint.2006.09.014.

Deng, R., Qin, F., Chen, R. et al. (2015). Temporal full-colour tuning through non-steady-state upconversion. *Nature Nanotechnology* 10 (3): 237–242. https://doi.org/10.1038/nnano.2014.317.

Ding, M., Chen, D., Yin, S. et al. (2015). Simultaneous morphology manipulation and upconversion luminescence enhancement of β-$NaYF_4$:Yb^{3+}/Er^{3+} microcrystals by simply tuning the KF dosage. *Scientific Reports* 5 (1): 1–14. https://doi.org/10.1038/srep12745.

Downing, E., Hesselink, L., Ralston, J., and Macfarlane, R. (1996). A three-color, solid-state, three-dimensional display. *Science* 273 (5279): 1185–1189. https://doi.org/10.1126/science.273.5279.1185.

Du, P. and Yu, J.S. (2018). Near-infrared light-triggered visible upconversion emissions in Er^{3+}/Yb^{3+}-codoped $Y_2Mo_4O_{15}$ microparticles for simultaneous noncontact optical thermometry and solid-state lighting. *Industrial & Engineering Chemistry Research* 57 (39): 13077–13086. https://doi.org/10.1021/acs.iecr.8b02938.

Du, P., Zhang, P., Kang, S.H., and Yu, J.S. (2017). Hydrothermal synthesis and application of Ho^{3+}-activated $NaYbF_4$ bifunctional upconverting nanoparticles for in vitro cell imaging and latent fingerprint detection. *Sensors and Actuators B: Chemical* 252: 584–591. https://doi.org/10.1016/j.snb.2017.06.032.

Dyck, N.C. and Demopoulos, G.P. (2014). Integration of upconverting β-$NaYF_4$:Yb^{3+},Er^{3+}@TiO_2 composites as light harvesting layers in dye-sensitized solar cells. *RSC Advances* 4 (95): 52694–52701. https://doi.org/10.1039/C4RA08775E.

Fischer, S., Goldschmidt, J.C., Löper, P. et al. (2010). Enhancement of silicon solar cell efficiency by upconversion: optical and electrical characterization. *Journal of Applied Physics* 108 (4): 044912. https://doi.org/10.1063/1.3478742.

Gao, F., Han, J., Zhang, J. et al. (2011). The synthesis of newly modified CdTe quantum dots and their application for improvement of latent fingerprint detection. *Nanotechnology* 22 (7): 075705. https://doi.org/10.1088/0957-4484/22/7/075705.

Gao, D., Zhang, X., and Gao, W. (2013). Formation of bundle-shaped β-$NaYF_4$ upconversion microtubes via Ostwald ripening. *ACS Applied Materials & Interfaces* 5 (19): 9732–9739. https://doi.org/10.1021/am402843h.

Gibart, P., Auzel, F., Guillaume, J.-C.G.J.-C., and Zahraman, K.Z.K. (1996). Below band-gap IR response of substrate-free GaAs solar cells using two-photon up-conversion. *Japanese Journal of Applied Physics* 35 (8R): 4401. https://doi.org/10.1143/JJAP.35.4401.

Heer, S., Kömpe, K., Güdel, H.U., and Haase, M. (2004). Highly efficient multicolour upconversion emission in transparent colloids of lanthanide-doped $NaYF_4$ nanocrystals. *Advanced Materials* 16 (23–24): 2102–2105. https://doi.org/10.1002/adma.200400772.

Hu, X., Wei, T. et al. (2014). Near-infrared-light mediated ratiometric luminescent sensor for multimode visualized assays of explosives. *Analytical Chemistry* 86 (20): 10484–10491. https://doi.org/10.1021/ac5032308.

Hu, W., Li, T., Liu, X. et al. (2020). 1550 nm pumped upconversion chromaticity modulation in Er^{3+} doped double perovskite LiYMgWO6 for anti-counterfeiting. *Journal of Alloys and Compounds* 818: 152933. https://doi.org/10.1016/j.jallcom.2019.152933.

Huang, X. (2015). Dual-model upconversion luminescence from $NaGdF_4$:Nd/Yb/Tm@$NaGdF_4$:Eu/Tb core–shell nanoparticles. *Journal of Alloys and Compounds* 628: 240–244. https://doi.org/10.1016/j.jallcom.2014.12.178.

Jin, Y.J., Luo, Y.J., Li, G.P. et al. (2008). Application of photoluminescent CdS/PAMAM nanocomposites in fingerprint detection. *Forensic Science International* 179 (1): 34–38. https://doi.org/10.1016/j.forsciint.2008.04.010.

Kaboli, F., Ghazyani, N., Riahi, M. et al. (2019). Upconverting nanoengineered surfaces: maskless photolithography for security applications. *ACS Applied Nano Materials* 2 (6): 3590–3596. https://doi.org/10.1021/acsanm.9b00549.

Kim, W.J., Nyk, M., and Prasad, P.N. (2009). Color-coded multilayer photopatterned microstructures using lanthanide(III) ion co-doped $NaYF_4$ nanoparticles with upconversion luminescence for possible applications in security. *Nanotechnology* 20 (18): 185301. https://doi.org/10.1088/0957-4484/20/18/185301.

Kim, T.H., Cho, K.S., Lee, E.K. et al. (2011). Full-colour quantum dot displays fabricated by transfer printing. *Nature Photonics* 5 (3): 176–182. https://doi.org/10.1038/nphoton.2011.12.

Kumar, A., Tiwari, S.P., and Kumar, K. (2015). Synthesis and upconversion of Er^{3+}/Yb^{3+} doped $NaGdF_4$ phosphor for security applications. *Advanced Science Letters* 21 (8): 2632–2634. https://doi.org/10.1166/asl.2015.6427.

Kumar, A., Tiwari, S.P., Kumar, K., and Rai, V.K. (2016). Structural and optical properties of thermal decomposition assisted Gd_2O_3:Ho^{3+}/Yb^{3+} upconversion phosphor annealed at different temperatures. *Spectrochimica Acta Part A: Molecular and Biomolecular Spectroscopy* 167: 134–141. https://doi.org/10.1016/j.saa.2016.05.010.

Kumar, A., Tiwari, S. P. et al. (2019). Infrared interceded YF_3: Er^{3+}/Yb^{3+} upconversion phosphor for crime scene and anti-counterfeiting applications. *Optical Materials* 92: 347–351. https://doi.org/10.1016/J.OPTMAT.2019.04.050

Lahoz, F. (2008). Ho^{3+}-doped nanophase glass ceramics for efficiency enhancement in silicon solar cells. *Optics Letters* 33 (24): 2982–2984. https://doi.org/10.1364/OL.33.002982.

Lee, J., Bisso, P.W., Srinivas, R.L. et al. (2014). Universal process-inert encoding architecture for polymer microparticles. *Nature Materials* 13 (5): 524–529. https://doi.org/10.1038/nmat3938.

Li, L., Yang, Y., Fan, R. et al. (2014a). A simple modification of near-infrared photon-to-electron response with fluorescence resonance energy transfer for dye-sensitized solar cells. *Journal of Power Sources* 264: 254–261. https://doi.org/10.1016/j.jpowsour.2014.04.100.

Li, X., Wang, R., Zhang, F., and Zhao, D. (2014b). Engineering homogeneous doping in single nanoparticle to enhance upconversion efficiency. *Nano Letters* 14 (6): 3634–3639. https://doi.org/10.1021/nl501366x.

Liang, L., Liu, Y., and Zhao, X.Z. (2013). Double-shell β-$NaYF_4$:Yb^{3+}, Er^{3+}/SiO_2/TiO_2 submicroplates as a scattering and upconverting layer for efficient dye-sensitized solar cells. *Chemical Communications* 49 (38): 3958–3960. https://doi.org/10.1039/C3CC41252K.

Liang, L., Zvyagin, A.V., and Zhang, R. (2017). *Development and Applications of Upconversion Nanoparticle in Biosensing*, 49–79. International Frequency Sensor Association (IFSA). ISBN: 9788769734681/9788469734674.

Lin, M., Zhao, Y., Wang, S. et al. (2012). Recent advances in synthesis and surface modification of lanthanide-doped upconversion nanoparticles for biomedical applications. *Biotechnology Advances* 30 (6): 1551–1561. https://doi.org/10.1016/j.biotechadv.2012.04.009.

Liu, M., Lu, Y., Xie, Z.B., and Chow, G.M. (2011). Enhancing near-infrared solar cell response using upconverting transparent ceramics. *Solar Energy Materials and Solar Cells* 95 (2): 800–803. https://doi.org/10.1016/j.solmat.2010.09.018.

Liu, Q., Feng, W., Yang, T. et al. (2013). Upconversion luminescence imaging of cells and small animals. *Nature Protocols* 8 (10): 2033–2044. https://doi.org/10.1038/nprot.2013.114.

Liu, X., Wang, Y., Li, X. et al. (2017). Binary temporal upconversion codes of Mn^{2+}-activated nanoparticles for multilevel anti-counterfeiting. *Nature Communications* 8 (1): 1–7. https://doi.org/10.1038/s41467-017-00916-7.

Liu, C.M., Zhang, L.Y., Li, L. et al. (2018). Specific detection of latent human blood fingerprints using antibody modified $NaYF_4$:Yb,Er,Gd fluorescent upconversion nanorods. *Dyes and Pigments* 149: 822–829. https://doi.org/10.1016/j.dyepig.2017.11.050.

Liu, L., Hua, R., Chen, B. et al. (2019). Detection of nitroaromatics in aqueous media based on luminescence resonance energy transfer using upconversion nanoparticles as energy donors. *Nanotechnology* 30 (37): 375703. https://doi.org/10.1088/1361-6528/ab26dd.

Ma, Y. and Wang, L. (2014). Upconversion luminescence nanosensor for TNT selective and label-free quantification in the mixture of nitroaromatic explosives. *Talanta* 120: 100–105. https://doi.org/10.1016/j.talanta.2013.12.009.

Ma, R., Bullock, E., Maynard, P. et al. (2011). Fingermark detection on non-porous and semi-porous surfaces using $NaYF_4$:Er,Yb up-converter particles. *Forensic Science International* 207 (1–3): 145–149. https://doi.org/10.1016/j.forsciint.2010.09.020.

Ma, R., Shimmon, R., McDonagh, A. et al. (2012). Fingermark detection on non-porous and semi-porous surfaces using YVO_4:Er,Yb luminescent upconverting particles. *Forensic Science International* 217 (1–3): e23–e26. https://doi.org/10.1016/j.forsciint.2011.10.033.

Mahata, M.K., Kumar, K., and Rai, V.K. (2014). Structural and optical properties of Er^{3+}/Yb^{3+} doped barium titanate phosphor prepared by co-precipitation method. *Spectrochimica Acta Part A: Molecular and Biomolecular Spectroscopy* 124: 285–291. https://doi.org/10.1016/j.saa.2014.01.014.

Matsubara, H., Yoshimoto, S., Saito, H. et al. (2008). GaN photonic-crystal surface-emitting laser at blue-violet wavelengths. *Science* 319 (5862): 445–447. https://doi.org/10.1126/science.1150413.

Maurya, S.K., Mohan, M., Poddar, R. et al. (2020). Synthesis of $NaGdF_4:Er^{3+}/Yb^{3+}$ upconversion particles as exogenous contrast agent for swept-source optical coherence tomography: in vitro animal tissue imaging. *The Journal of Physical Chemistry C* 124 (33): 18366–18378. https://doi.org/10.1021/acs.jpcc.0c05786.

Miao, C., Liu, T., Zhu, Y. et al. (2013). Super-intense white upconversion emission of Yb_2O_3 polycrystals and its application on luminescence converter of dye-sensitized solar cells. *Optics Letters* 38 (17): 3340–3343. https://doi.org/10.1364/OL.38.003340.

Mondal, M., Kumar Rai, V., Srivastava, C. et al. (2016). Enhanced frequency upconversion in $Ho^{3+}/Yb^{3+}/Li^{+}:YMoO_4$ nanophosphors for photonic and security ink applications. *Journal of Applied Physics* 120 (23): 233101. https://doi.org/10.1063/1.4971966.

Morton, O. (2006). Solar energy: a new day dawning? Silicon Valley sunrise. *Nature* 443 (7107): 19–22. https://doi.org/10.1038/443019a.

Oden, S. and Von Hofsten, B. (1954). Detection of fingerprints by the ninhydrin reaction. *Nature* 173 (4401): 449–450. https://doi.org/10.1038/173449a0.

Pandey, A., Rai, V.K., Dey, R., and Kumar, K. (2013). Enriched green upconversion emission in combustion synthesized $Y_2O_3:Ho^{3+}–Yb^{3+}$ phosphor. *Materials Chemistry and Physics* 139 (2–3): 483–488. https://doi.org/10.1016/j.matchemphys.2013.01.043.

Rao, J., Dragulescu-Andrasi, A., and Yao, H. (2007). Fluorescence imaging in vivo: recent advances. *Current Opinion in Biotechnology* 18 (1): 17–25. https://doi.org/10.1016/j.copbio.2007.01.003.

Rapaport, A., Milliez, J., Bass, M. et al. (2006). Review of the properties of up-conversion phosphors for new emissive displays. *Journal of Display Technology* 2 (1): 68. https://doi.org/10.1109/JDT.2005.863781.

Ryu, J., Park, H.Y., Kim, K. et al. (2010). Facile synthesis of ultrasmall and hexagonal $NaGdF_4:Yb^{3+},Er^{3+}$ nanoparticles with magnetic and upconversion imaging properties. *The Journal of Physical Chemistry C* 114 (49): 21077–21082. https://doi.org/10.1021/jp107725r.

Sametband, M., Shweky, I., Banin, U. et al. (2007). Application of nanoparticles for the enhancement of latent fingerprints. *Chemical Communications* 11: 1142–1144. https://doi.org/10.1039/B618966K.

Schenke-Layland, K., Riemann, I., Damour, O. et al. (2006). Two-photon microscopes and in vivo multiphoton tomographs—powerful diagnostic tools for tissue engineering and drug delivery. *Advanced Drug Delivery Reviews* 58 (7): 878–896. https://doi.org/10.1016/j.addr.2006.07.004.

Shalav, A., Richards, B.S., Trupke, T. et al. (2004). Application of $NaYF_4:Er^{3+}$ up-converting phosphors for enhanced near-infrared silicon solar cell response. *Applied Physics Letters* 86 (1): 013505. https://doi.org/10.1063/1.1844592.

Shang, Y., Hao, S., Yang, C., and Chen, G. (2015). Enhancing solar cell efficiency using photon upconversion materials. *Nanomaterials (Basel, Switzerland)* 5 (4): 1782–1809. https://doi.org/10.3390/nano5041782.

Shao, W., Chen, G., Ohulchanskyy, T.Y. et al. (2015). Lanthanide-doped fluoride core/multishell nanoparticles for broadband upconversion of infrared light. *Advanced Optical Materials* 3 (4): 575–582. https://doi.org/10.1002/adom.201400404.

Singh, S.K., Kumar, K., and Rai, S.B. (2009). Multifunctional Er^{3+}–Yb^{3+} codoped Gd_2O_3 nanocrystalline phosphor synthesized through optimized combustion route. *Applied Physics B* 94 (1): 165–173. https://doi.org/10.1007/s00340-008-3261-6.

Singh-Rachford, T.N., Haefele, A., Ziessel, R., and Castellano, F.N. (2008). Boron dipyrromethene chromophores: next generation triplet acceptors/annihilators for low power upconversion schemes. *Journal of the American Chemical Society* 130 (48): 16164–16165. https://doi.org/10.1021/ja807056a.

Sinha, S. and Kumar, K. (2018). Studies on up/down-conversion emission of Yb^{3+} sensitized Er^{3+} doped $MLa_2(MoO_4)_4$ (M = Ba, Sr and Ca) phosphors for thermometry and optical heating. *Optical Materials* 75: 770–780. https://doi.org/10.1016/j.optmat.2017.10.036.

Sinha, S., Mahata, M.K., and Kumar, K. (2016). Up/down-converted green luminescence of Er^{3+}–Yb^{3+} doped paramagnetic gadolinium molybdate: a highly sensitive thermographic phosphor for multifunctional applications. *RSC Advances* 6 (92): 89642–89654. https://doi.org/10.1039/C6RA20332A.

Sinha, S., Mahata, M.K., Swart, H.C. et al. (2017a). Enhancement of upconversion, temperature sensing and cathodoluminescence in the K^+/Na^+ compensated $CaMoO_4:Er^{3+}/Yb^{3+}$ nanophosphor. *New Journal of Chemistry* 41 (13): 5362–5372. https://doi.org/10.1039/C7NJ00086C.

Sinha, S., Mahata, M. K. et al. (2017b). Dualistic temperature sensing in Er^{3+}/Yb^{3+} doped $CaMoO_4$ upconversion phosphor. *Spectrochimica Acta Part A: Molecular and Biomolecular Spectroscopy* 173: 369–375. https://doi.org/10.1016/j.saa.2016.09.039

Sodhi, G.S. and Kaur, J. (2001). Powder method for detecting latent fingerprints: a review. *Forensic Science International* 120 (3): 172–176. https://doi.org/10.1016/s0379-0738(00)00465-5.

Sun, Y.P., Zhou, B., Lin, Y. et al. (2006). Quantum-sized carbon dots for bright and colorful photoluminescence. *Journal of the American Chemical Society* 128 (24): 7756–7757. https://doi.org/10.1021/ja062677d.

Tahtouh, M., Kalman, J.R., and Reedy, B.J. (2011). Synthesis and characterization of four alkyl 2-cyanoacrylate monomers and their precursors for use in latent fingerprint detection. *Journal of Polymer Science Part A: Polymer Chemistry* 49 (1): 257–277. https://doi.org/10.1002/pola.24449.

Theaker, B.J., Hudson, K.E., and Rowell, F.J. (2008). Doped hydrophobic silica nano-and micro-particles as novel agents for developing latent fingerprints. *Forensic Science International* 174 (1): 26–34. https://doi.org/10.1016/j.forsciint.2007.02.030.

Wang, F., Han, Y., Lim, C.S. et al. (2010). Simultaneous phase and size control of upconversion nanocrystals through lanthanide doping. *Nature* 463 (7284): 1061–1065. https://doi.org/10.1038/nature08777.

Wang, M., Zhu, Y., and Mao, C. (2015). Synthesis of NIR-responsive $NaYF_4$:Yb,Er upconversion fluorescent nanoparticles using an optimized solvothermal method and their applications in enhanced development of latent fingerprints on various smooth substrates. *Langmuir* 31 (25): 7084–7090. https://doi.org/10.1021/acs.langmuir.5b01151.

de Wild, J., Meijerink, A., Rath, J.K. et al. (2010). Towards upconversion for amorphous silicon solar cells. *Solar Energy Materials and Solar Cells* 94 (11): 1919–1922. https://doi.org/10.1016/j.solmat.2010.06.006.

Yang, F., Liu, C., Wei, D. et al. (2014). Er^{3+}–Yb^{3+} co-doped TeO_2–PbF_2 oxyhalide tellurite glasses for amorphous silicon solar cells. *Optical Materials* 36 (6): 1040–1043. https://doi.org/10.1016/j.optmat.2014.01.020.

You, M., Lin, M., Wang, S. et al. (2016). Three-dimensional quick response code based on inkjet printing of upconversion fluorescent nanoparticles for drug anti-counterfeiting. *Nanoscale* 8 (19): 10096–10104. https://doi.org/10.1039/C6NR01353H.

Zhang, X.D., Jin, X., Wang, D.F. et al. (2010). Synthesis of $NaYF_4$:Yb,Er nanocrystals and its application in silicon thin film solar cells. *Physica Status Solidi C* 7 (3–4): 1128–1131. https://doi.org/10.1002/pssc.200982762.

Zhang, D., Ding, M. et al. (2019). Hexagonal sodium yttrium fluoride mesocrystals: one-pot hydrothermal synthesis, formation mechanism and multicolor up-/down-converted luminescence for anti-counterfeiting and fingerprint detection. *Ceramics International* 45 (16): 20307–20315. https://doi.org/10.1016/J.CERAMINT.2019.07.001.

Zhao, J., Wu, W., Sun, J., and Guo, S. (2013). Triplet photosensitizers: from molecular design to applications. *Chemical Society Reviews* 42 (12): 5323–5351. https://doi.org/10.1039/C3CS35531D.

Zhou, D., Liu, D., Xu, W. et al. (2017). Synergistic upconversion enhancement induced by multiple physical effects and an angle-dependent anticounterfeit application. *Chemistry of Materials* 29 (16): 6799–6809. https://doi.org/10.1021/acs.chemmater.7b01783.

Zhu, X., Zhang, J., Liu, J., and Zhang, Y. (2019). Recent progress of rare-earth doped upconversion nanoparticles: synthesis, optimization, and applications. *Advanced Science* 6 (22): 1901358. https://doi.org/10.1002/advs.201901358.

12

Application of Upconversion in Photocatalysis and Photodetectors

Priyam Singh, Sachin Singh, Prabhakar Singh, and Sunil Kumar Singh

Indian Institute of Technology (BHU), Department of Physics, Varanasi 221005, India

12.1 Introduction

The lanthanide-doped upconversion nanoparticles (UCNPs) have attracted great deal of interest because of their unique optical conversion properties. They efficiently convert the near-infrared light into visible and ultraviolet (UV) radiation through sequential multiphoton processes, which make them the backbone of optoelectronic devices, particularly for the conversion of the infrared part of solar or artificial light (Idris et al. 2015; Wen et al. 2018; Chen et al. 2014a, 2016a; Xie and Liu 2012). The converted light in visible or UV radiation could be utilized by wide bandgap semiconductors such as crystalline silicon (c-Si), TiO_2, GaAs, etc., for different applications. In recent years, optoelectronic devices with enhanced performance had been realized by integrating UCNPs and semiconductors into one device/system (Yang et al. 2018; Lian et al. 2021; He et al. 2016; Singh et al. 2017a; Gupta et al. 2021). The efficient spectral converting capability of UCNPs allows facile photon management, and therefore, it opens the possibility of many applications that cannot be realized using conventional materials. For example, UCNPs capture the NIR portion of solar radiation and convert it into the visible region and thereby minimizes the non-absorption energy losses by the photovoltaic devices (Gupta et al. 2021; Niu et al. 2014; Sun et al. 2007; Luo et al. 2016; Goldschmidt and Fischer 2015; Atabaev and Molkenova 2019). A similar and important application of UCNPs has been found in the field of photocatalysis. UCNPs can be used as spectral converters in photocatalysis devices for enhancing their performance (Atabaev and Molkenova 2019; Dong et al. 2015a; Cui et al. 2017; Ge et al. 2018; Ke et al. 2017; Reddy et al. 2019).

Lanthanides are a group of 15 elements starting from atomic number 57 (lanthanum) to 71 (lutetium) and are characterized by filling up the 4f orbital, from $4f^0$ to $4f^{14}$. The two additional elements scandium and yttrium are chemically similar and are collectively called as rare-earth metals. The luminescence spectra of trivalent lanthanide ions arise from intra f–f transitions and are observed in UV, visible, and NIR spectral regions (Rink et al. 1991). The f-orbitals of lanthanide ions

Upconverting Nanoparticles: From Fundamentals to Applications, First Edition. Edited by Vineet K. Rai.

are well shielded with outer 5s and 5p orbitals that perturb the electron–phonon coupling strengths, and therefore, 4f electrons remain highly localized (Liu 2015). Therefore, the emission spectra of lanthanides are generally long lived, sharp, and narrow. In addition, the ladder-like energy states of lanthanide ions provide flexibility of photon absorption, multistep energy transfer, and cross-relaxation processes (Singh et al. 2016, 2019, 2020a,b). Because of these exceptional optical and chemical properties, lanthanide-based UCNPs play an important role in the field of both photocatalysis and photodetector (PD) applications (Singh et al. 2016; Nie et al. 2019; Chen et al. 2014b; Ding et al. 2020; Ke et al. 2017; Kwon et al. 2018).

Upconversion is a nonlinear anti-Stokes process that converts two or more than two low-frequency photons into one high-frequency photon. Usually, upconverting materials consist of a low-phonon host doped with activator ions, mostly trivalent lanthanide ions. The UC emission is obtained by optical transition within the partially filled 4f orbital of lanthanides (Ronda 2007, pp. 1–20). The long-lived excited levels of lanthanide ions act as energy storage for excited ions. To date, many UC mechanisms have been identified, such as ground-state absorption (GSA), excited-state absorption (ESA), cooperative energy transfer (CET), cross-relaxation (CR), and photon avalanche (PA). The detailed mechanism of these UC processes can be easily found in the literature (Singh et al. 2021, p. 389; Dong et al. 2015b). The trivalent lanthanides Er^{3+}, Tm^{3+}, and Ho^{3+} ions are mostly taken as activator ions either singly or even in combination of more than one. These ions show ladder-like energy levels, which facilitate the multistep photon absorption as well as the energy transfer processes.

The efficiency of UC process can be enhanced by using sensitizer ions. Ytterbium (Yb^{3+}) is an excellent sensitizer, which is used because of its large absorption cross section for 980 nm radiation, corresponding to the ${}^2F_{5/2} \rightarrow {}^2F_{7/2}$ transition. The ${}^2F_{5/2} \rightarrow {}^2F_{7/2}$ transition of Yb^{3+} easily gets a resonance level in all the three important activators (Er^{3+}, Tm^{3+}, and Ho^{3+}), which facilitates the energy transfer and diminishes the low NIR photon absorption issue by other rare-earth ions. Thus, Er^{3+}–Yb^{3+}, Tm^{3+}–Yb^{3+}, and Ho^{3+}–Yb^{3+} are the most commonly used co-dopants in UC systems (Shi et al. 2018). Figure 12.1 shows the energy transfer mechanism

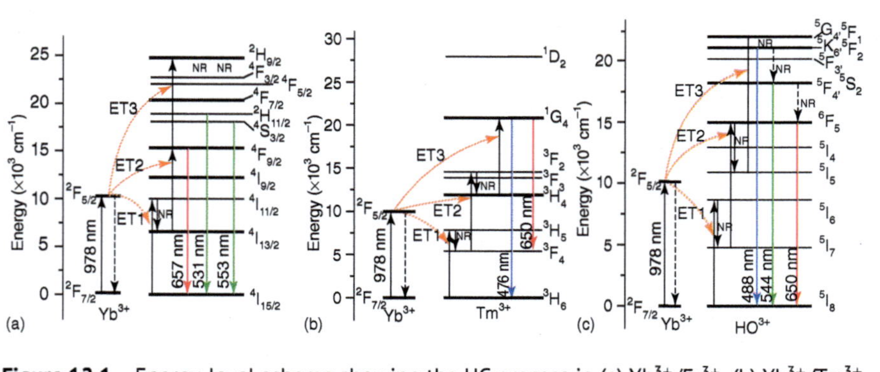

Figure 12.1 Energy-level scheme showing the UC process in (a) Yb^{3+}/Er^{3+}, (b) Yb^{3+}/Tm^{3+}, and (c) Yb^{3+}/Ho^{3+}. Source: Shi et al. (2018). Reprinted with permission of American Chemical Society.

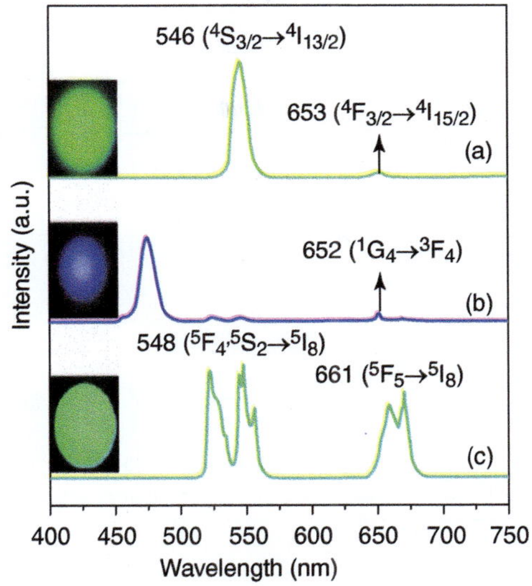

Figure 12.2 UC emission spectra of (a) La_2O_3:Yb^{3+}/Er^{3+}, (b) La_2O_3:Yb^{3+}/Tm^{3+}, and (c) La_2O_3:Yb^{3+}/Ho^{3+} microrods. Source: Zhang et al. (2012a). Reprinted with permission of American Chemical Society.

in these systems. Yb^{3+} ions absorb the 980 nm photon energy and transfer their excitation energy to the nearby Er^{3+}/Tm^{3+}/Ho^{3+} ions to excite them into their higher energy levels. Radiative transitions from these excited levels of activator ions produce visible emission (see Figure 12.2) (Zhang et al. 2012a). As the emission spectra of these lanthanides lie in UV and visible regions, it could be utilized by wide bandgap semiconductor materials. Herein, we summarize some recent investigations carried out in the field of these rare-earth ion-activated UCNPs for exploring the performance of semiconductor-based photodetector and photocatalytic devices.

12.2 Photocatalysis

Environmental pollution, which is an important alarming issue of global concern, has received great attention for research and development in the area of environmental and energy-related fields. A wide range of biological and chemical treatments have been proposed/used to remove the toxic compounds in wastewater, air, and soil. Among all, photocatalysis is a convenient, economical, and environment-friendly approach to remove different pollutants from water, air, and soil. Photocatalytic processes using semiconductors such as TiO_2, ZnO, and SnO_2 have been widely applied for the photodegradation of various hazardous organic compounds (Atabaev and Molkenova 2019; Dong et al. 2015a; Cui et al. 2017; Ge et al. 2018; Ke et al. 2017; Reddy et al. 2019).

In a photocatalytic process, initially, a semiconductor absorbs the incident sunlight, wavelengths that match with the bandgap, and converts solar energy into chemical energy. The photocatalytic material generates electron–hole pairs to produce hydroxyl radicals and superoxide ions. These radicals are highly reactive in nature and oxidize the pollutants to CO_2 and H_2O. For example, titania (TiO_2) is the most widely used common photocatalyst because of its low cost, high chemical

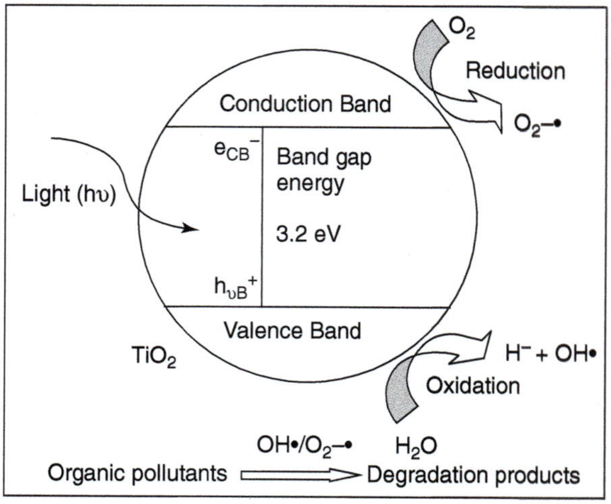

Figure 12.3 Mechanism of photocatalytic process based on a TiO_2 semiconductor. Source: Sujatha et al. (2020). Open access license 2020.

and photostability, and biocompatibility (Woan et al. 2009; Anderson and Bard 1995; Borges et al. 2016; Lee and Park 2013; Sujatha et al. 2020). Figure 12.3 demonstrates the basic principle of photocatalytic activity in a TiO_2 semiconductor to generate free radicals (Sujatha et al. 2020). In the first step, TiO_2 absorbs the incident UV radiation, whose energy matches to the bandgap, and electron–hole pairs are created through excitation of valence electrons to the conduction band. These electron–hole pairs are diffused at the surface of TiO_2. Next, the electron from the pair participates in the reduction process. It reacts with oxygen molecules to form superoxide radical anions (O_2^-). Meanwhile, the holes react with H_2O molecules, adsorbed on the surface of TiO_2, to create hydroxyl radicals (OH). These hydroxyl radicals and superoxide ions, being highly reactive, bind together with harmful organic compounds and break them into CO_2 and H_2O molecules.

Inappropriately, these semiconductor materials have large bandgaps and may absorb only UV radiation (4–5% of solar irradiation). Many efforts have been devoted to enhance the photocatalytic activity of semiconductors. These methods include doping of the TiO_2 host lattice with metals and non-metals, coupling with organic dyes, quantum dots, upconversion phosphors, graphene, and narrow bandgap semiconductors to extend the absorption range in the visible as well NIR region (Zhang et al. 2011, 2012b; Du et al. 2020; Liu et al. 2019; Akhavan et al. 2010). Among these, integration of upconversion phosphors with semiconductors appears to be a highly efficient method to extend the absorption range of semiconductors and hence photocatalytic property (Tian et al. 2019). Herein, UCNPs serve as a frequency converter that harvests the incident NIR photons and converts them into UV/visible photons. These converted photons are further absorbed by the semiconductors, and thereby, they generate the electron–hole pairs.

In 2005, Wang et al. firstly developed the photocatalyst utilizing visible light, which was based on a combination of upconversion materials, i.e. Er_2O_3 and TiO_2 semiconductors (Wang et al. 2005). Later on, in 2010, Qin et al. developed a NIR light-based photocatalyst composed of TiO_2-coated YF_3:Yb/Tm nanoparticles (Qin et al. 2010). In this combination, the $YF_3:Yb^{3+},Tm^{3+}$ core converts NIR photons into UV and visible photons, which can be absorbed by the TiO_2 layer to extend its absorptivity in the NIR region. In 2013, the same group synthesized a composite of $NaYF_4$:Yb,Tm and TiO_2 for NIR light-driven photocatalysis degradation of methylene blue (MB) compound, see Figure 12.4a (Tang et al. 2013). They have demonstrated that the energy transfer between $NaYF_4$:Yb,Tm NPs and TiO_2 plays an important role to enhance the photocatalytic activity. The degradation rate of MB increases with the irradiation time up to 65% in 14 hours in the presence of

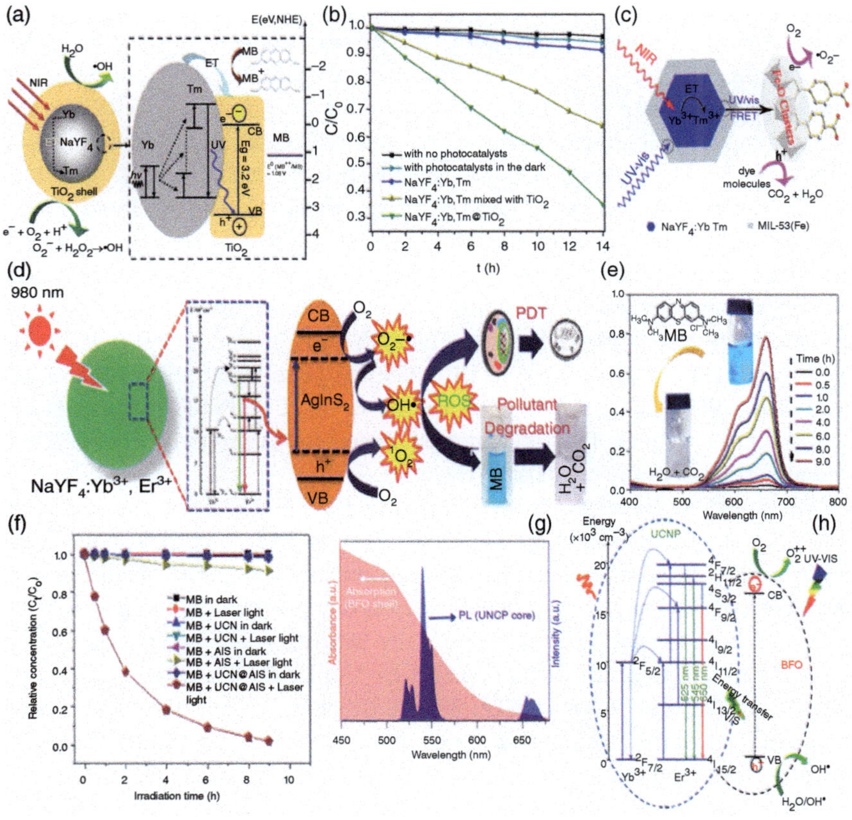

Figure 12.4 (a) The proposed energy transfer mechanism of $NaYF_4$:Yb,Tm@TiO_2 under NIR exposure and (b) the degradation rate of MB with respect to time for different as-prepared samples. Source: Tang et al. (2013). Reprinted with permission of American Chemical Society. (c) Schematic illustration of photocatalysis in UCNPs@MIL-53 (Fe). Source: Li et al. (2017a). Reprinted with permission of American Chemical Society. (d) Schematic illustration of the NIR active MB degradation using UCN@AIS for wastewater treatment, (e) intensity variation of absorption spectra and corresponding digital photograph for MB dye photodegradation under 980 nm NIR light irradiation, and (f) variation of relative concentration (C_t/C_0) (665 nm) with respect to time for the MB degradation for different as-prepared samples under 980 nm light irradiation. Source: Maji and Kim (2018). Reprinted with permission of American Chemical Society. (g) Spectral overlapping of BFO shell and UCNP core and (h) energy-level diagram illustrating the UC mechanism, energy transfer from UCNP to BFO, and the photocatalytic activity in BFO. Source: Zhang et al. (2017a). Reprinted with permission of American Chemical Society).

$NaYF_4$:Yb,Tm@TiO_2 particles, which is much higher than the physical mixture of $NaYF_4$:Yb,Tm/TiO_2 (about 30%), as shown in Figure 12.4b. They proposed that the compact interface between the core–shell structure of $NaYF_4$:Yb,Tm and TiO_2 facilitates the fluorescence resonance energy transfer (FRET) process between them, which is not possible in the case of physical mixture of these two species.

Many other upconversion material–semiconductor-based composites have been developed for photocatalytic applications. The photocatalytic activity of core–shell-type UC materials depends on the core size and shell thickness. In 2014, Su et al. synthesized $NaYF_4$:Yb,Tm@TiO_2 nanoparticles, which show excellent photocatalytic activities (for degradation of RhB dyes) under both UV and NIR excitations (Su et al. 2014). They demonstrated that the photocatalytic activity of the material depends on the core size and is maximum for a 50 nm-thick layer of UCNPs. Further, Zhang et al. demonstrated the effect of shell thickness on the photocatalytic activity of the $NaYF_4$:Yb,Tm@TiO_2 core–shell composite (Zhang and Hong 2013). They reported that the optimized shell thickness probes excellent photocatalytic activity. When the thickness of the TiO_2 shell is small, the absorption of UV radiation by the TiO_2 shell becomes poor. However, a thick TiO_2 shell discourages the upconversion emission. Along with TiO_2, other semiconductor–upconversion composite structures, for example, UCNP-BiOX (X = Cl, Br, and I) (Wu et al. 2017), UCNP–$AgInS_2$ (Maji and Kim 2018), UCNP–MOFs (Li et al. 2017a), UCNP–ZnO (Zhang et al. 2019a), UCNP–CdS (Balaji et al. 2017), UCNP–BFO (Zhang et al. 2017a), etc., have been reported for enhanced photocatalytic property of the pristine semiconductor used in the composite system.

Li et al. integrated UCNPs with MIL-53 (Fe) to develop a NIR-responsive photocatalyst with controllable shell thickness, see Figure 12.4c (Li et al. 2017a). They further extended the absorption of the metal–organic framework (MOF) shell in the visible region by adding $-NH_2$ groups. Maji et al. demonstrated the photocatalytic degradation of an organic dye (MB) under the 980 nm NIR laser light exposure. They used a hybrid structure of $NaYF_4/Yb^{3+},Er^{3+}$ (UCNPs) coated with a ternary semiconductor ($AgInS_2$, AIS), see Figure 12.4d (Maji and Kim 2018). The absorption spectra peak of the MB solution almost disappears after nine hours of illumination of 980 nm light, see Figure 12.4e. The relative concentration (C_t/C_0) versus time (t) plot reveals the superiority of UCN@AIS NPs for photocatalytic degradation of MB, see Figure 12.4f. Zhang et al. synthesized an assembly of bismuth ferrite (BFO) coated on UCNPs and used it for the photocatalytic degradation of methylene orange and 4-chlorophenol both under visible and NIR irradiation (Zhang et al. 2017a). By recording the absorption spectra of BFO, it was found that there is a good overlapping between the absorption spectra of BFO and the upconversion spectra of $NaGdF_4:Er^{3+},Yb^{3+}$, see Figure 12.4g. The spectral overlapping and an effective coupling between the UCNPs and BFO facilitate the efficient energy transfer (via FRET) from UCNPs to BFO, see Figure 12.4h.

However, the core–shell structure of UCNPs and the photocatalytic semiconductor material extended the absorption edge up to the NIR region, and the major drawback of these core–shell structures is high lattice mismatching between the crystalline core and the amorphous shell. To tackle out this limitation, an amorphous interlayer (such as SiO_2, carbon, PVP, etc.) can be introduced to eliminate the high crystal mismatching between the semiconductor and UCNPs. In 2014, Wang et al. synthesized a double-shell structure, i.e. β-$NaYF_4$:Yb^{3+},Tm^{3+}/Er^{3+}@SiO_2@TiO_2, and used it for the photocatalytic degradation of RhB dye under NIR light (Wang et al. 2014).

The outer TiO_2 nanocrystals absorb the UV and visible light. However, the UCNPs absorb the NIR photons and convert them into UV and visible photons, which can be easily absorbed by TiO_2. Therefore, the composite structure absorbed the full spectrum of light for photocatalysis, see Figure 12.5a. The double-shell structure of β-$NaYF_4$:Yb^{3+},Tm^{3+}/Er^{3+}@SiO_2@TiO_2 can be clearly observed in the TEM image, as shown in Figure 12.5b,c. An effective energy transfer from UCNPs to TiO_2 is observed. They further reported that the photocatalytic activity depends on the NIR power density, see Figure 12.5d (Wang et al. 2014). In addition to this, the photocatalytic activity of β-$NaYF_4$:Yb^{3+},Tm^{3+}@SiO_2@TiO_2 is superior than

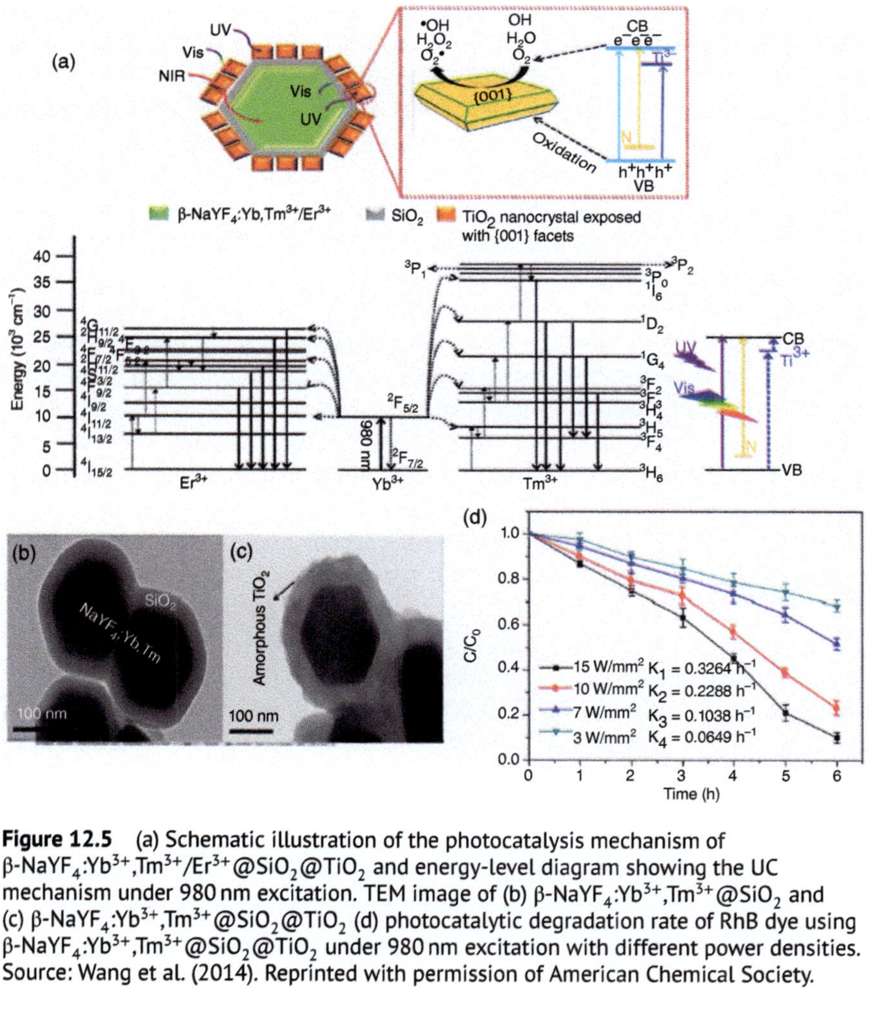

Figure 12.5 (a) Schematic illustration of the photocatalysis mechanism of β-$NaYF_4$:Yb^{3+},Tm^{3+}/Er^{3+}@SiO_2@TiO_2 and energy-level diagram showing the UC mechanism under 980 nm excitation. TEM image of (b) β-$NaYF_4$:Yb^{3+},Tm^{3+}@SiO_2 and (c) β-$NaYF_4$:Yb^{3+},Tm^{3+}@SiO_2@TiO_2 (d) photocatalytic degradation rate of RhB dye using β-$NaYF_4$:Yb^{3+},Tm^{3+}@SiO_2@TiO_2 under 980 nm excitation with different power densities. Source: Wang et al. (2014). Reprinted with permission of American Chemical Society.

β-$NaYF_4$:Yb^{3+},Er^{3+}@SiO_2@TiO_2. This clearly indicates that the emission range of upconversion materials is an important parameter for improved photocatalytic activity.

In 2016, Tou et al. introduced a carbon intermediate layer between $NaYF_4$:Yb,Tm and CdS (Tou et al. 2016). They observed an enhanced photocatalytic activity under the exposure of visible–NIR light because of strong adsorption capacity of the carbon shell. To get enhanced photocatalytic activity, along with a core–shell structure, direct doping of lanthanides in the semiconductor structure is also reported in the literature (Wang et al. 2010; Zhanga et al. 2009; Li et al. 2011; Zheng et al. 2015). Zheng et al. prepared the $CaIn_2O_4$:Yb^{3+},Tb^{3+} phosphor, which shows significant photocatalytic efficiency (about 76.2% decolorization efficiency) for MO dye under NIR excitation (Zheng et al. 2015). Li et al. synthesized a 3D flower-like hierarchical architecture (HAs) of BiOBr co-doped with the Yb^{3+}/Er^{3+} ions (Li et al. 2018a). They found that the pH value significantly affects the morphology of the prepared samples. The 3D flower-like microarchitectures are observed for pH 6, see Figure 12.6a. The UC emission intensity of the sample increases with increasing pH value, see Figure 12.6b. The author predicted that with decreasing thickness of the nanosheet (increasing the pH value), a relatively strong and oriented built-in IEF (internal electric field) is generated, which causes an enhanced UC intensity, as shown in Figure 12.6c. For the practical applications of the nanostructure, the photocatalytic degradation of RhB as well as stability investigation has been carried out under simulated solar light exposure, as shown in Figure 12.6d,e. The author concluded that 3D flower-like microarchitectures show very high (85%) and stable photocatalytic activity.

In spite of the above-mentioned superior features, there are some limitations of UCNPs such as low quantum yield and narrow absorption spectra because of the dipole-forbidden transitions. A well-established way to boost the UC efficiency is coupling of novel metal nanoparticles (such as AgNPs and AuNPs) with the UCNPs (Singh et al. 2017b, 2020a; 2020b; Xu et al. 2015). In the vicinity of plasmonic nanoparticles, the electric field distribution is manipulated, which can enhance the absorption as well as the emission efficiency of UCNPs. Integration of metal nanoparticles in semiconductor–UCNP combination can create a NIR light-triggered photocatalysis with efficient performance (Xu et al. 2015; Li et al. 2018b; Chilkalwar and Rayalu 2018; Zhang et al. 2017c). Zhang et al. designed a plasmon (AuNPs) and upconversion ($NaYF_4$:Yb^{3+}, Er^{3+}, and Tm^{3+})-enhanced broadband graphitic C3N4 (g-C3N4) nanosheet (Au-NYF/g-C3N4) for photocatalytic application (Zhang et al. 2017b). Under UV light exposure, electrons and holes are created in the CB and VB of g-C3N4 respectively, see Figure 12.7a. When AuNPs are in close contact with g-C3N4, it acts as an electron reservoir and reacts with O_2 to produce O_2^- radicals. At the same time, the holes in the VB of g-C3N4 can react with OH^-/H_2O to create OH radicals. When the samples are exposed with visible light, the surface plasmon resonance (SPR) band of AuNPs becomes the major source for the charge carriers, as shown in Figure 12.7b. However, for NIR excitation source, UCNPs become a major source for charge carriers. Therefore, a combination

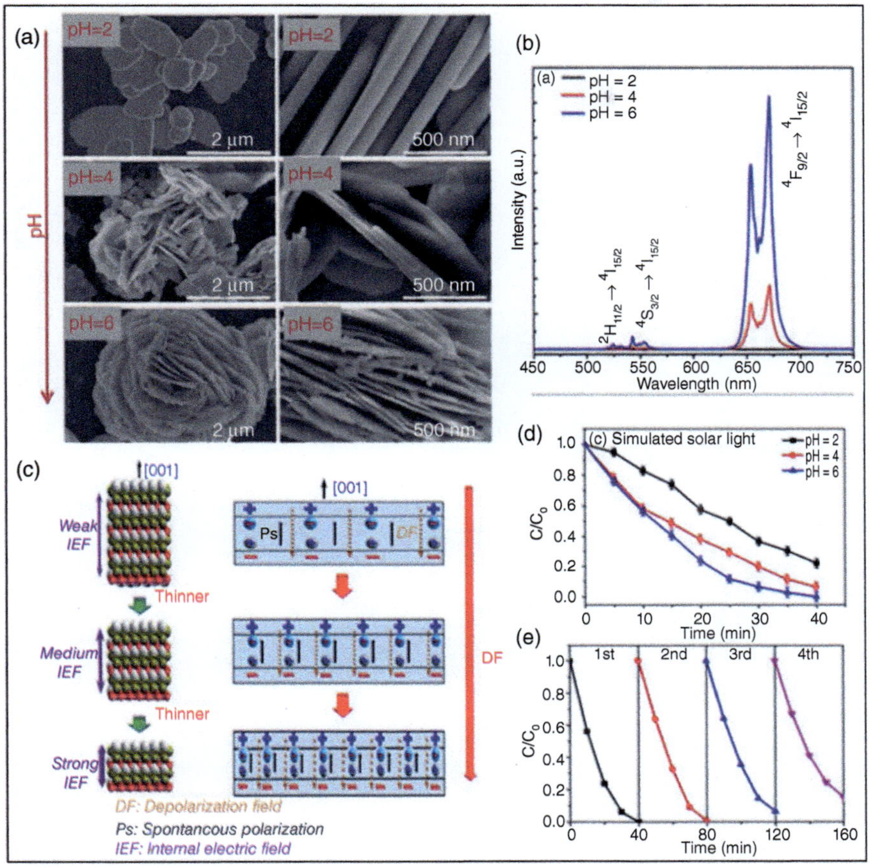

Figure 12.6 (a) FE-SEM image of BiOBr:Yb^{3+}/Er^{3+} nano/microcrystals for different pH values, (b) UC spectra of BiOBr:Yb^{3+}/Er^{3+} for different pH values, (c) schematic representation of the IEF (internal electric field) and DF (depolarization field) with decreasing thickness of the sample, (d) degradation rate of RhB, and (e) recycling performance of 3D flower-like BiOBr:Yb^{3+}/Er^{3+} HAs under simulated solar light exposure. Source: Li et al. (2018a). Reprinted with permission of American Chemical Society.

of plasmonic AuNPs, UCNPs, and g-C3N4 nanosheets into a single structure can harvest UV, visible, and NIR light simultaneously for efficient photocatalysis.

Xu et al. demonstrated an efficient NIR-active photocatalytic structure that consists of a UCNP shell decorated on a core of Au/dark TiO_2 and enables TiO_2-based photocatalysis for biomedical applications, as shown in Figure 12.7c (Xu et al. 2020). The D-TiO_2/Au core shows strong and broadband absorption in the 350–800 nm region, which can strongly overlap with the green emission band of $NaYF_4$:Er/Yb. Therefore, the combination of D-TiO_2 and AuNPs in core structure serves as an outstanding absorber for green emission to generate free radicals under NIR excitation. The author demonstrated the application of this composite structure as the photocatalytic drug release platform, for example, to release loaded AMP (a bactericide)

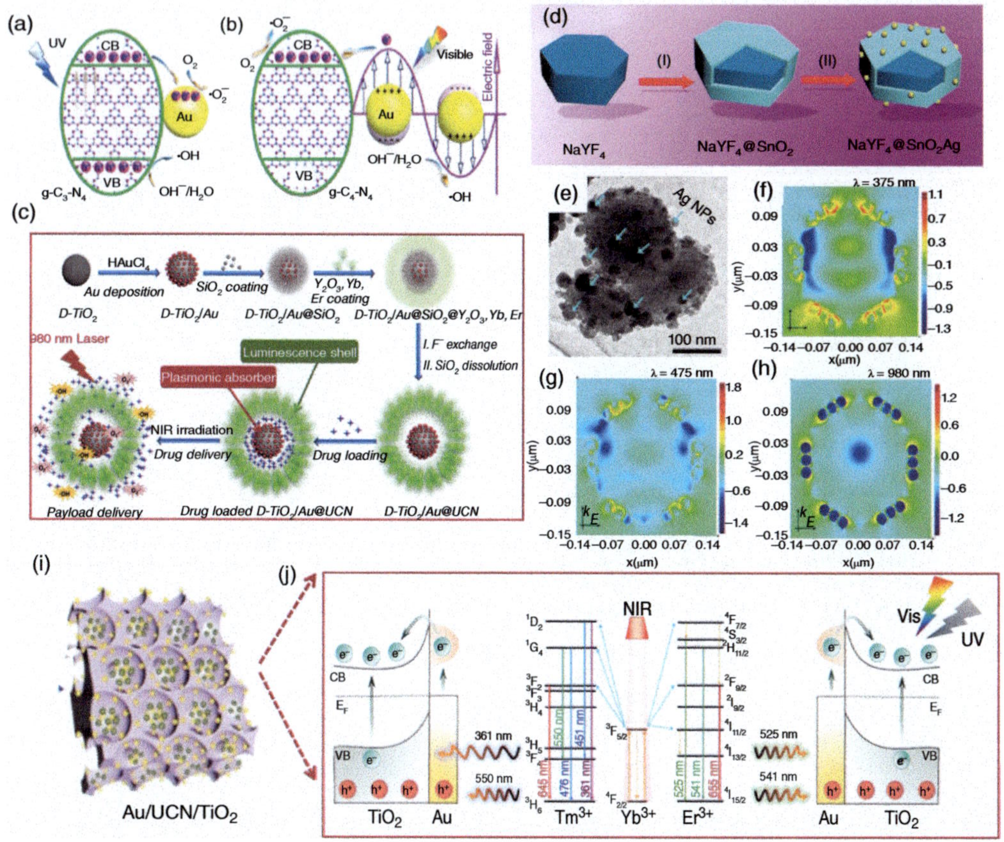

Figure 12.7 Photocatalytic mechanism of Au-UCNPs/g-C3N4 under (a) UV and (b) visible light excitation. Source: Modified from Zhang et al. (2017b). Reprinted with permission of American Chemical Society. (c) Schematic illustration of NIR light-controlled drug release D-TiO_2/Au@UCN nanostructure. Source: Xu et al. (2020). Reprinted with permission of American Chemical Society. (d) Schematic representation for the synthesis of $NaYF_4$@SnO_2@AgNPs, (e) TEM image of $NaYF_4$@SnO_2@AgNPs, and (f–h) FDTD-calculated local electric field enhancement for $NaYF_4$@SnO_2@Ag nanoparticles under different incident wavelengths (375, 475, and 980 nm). Source: Tian et al. (2017). Reprinted with permission of American Chemical Society. (i) Schematic illustration of the rationally designed TiO_2 inverse opal nanostructures loaded with AuNPs and UCNPs. (j) Photocatalytic mechanism with Er-UCN- and Tm-UCN-based ternary photocatalysts (Au/UCN/TiO_2) under UV–vis–NIR light exposure. Source: Boppella et al. (2019). Reprinted with permission of American Chemical Society.

in a fully functional state and its antibacterial effect. NIR excitation provide deep penetration depth and able to inactivate bacteria even through 2 mm of skin tissue.

Tian et al. designed a composite by integrating β-$NaYF_4$:18% Yb^{3+}, 2% Tm^{3+}, SnO_2, and AgNPs (denoted as $NaYF_4$@SnO_2@AgNPs) to get a photocatalyst that gives response in a wide spectral range (from the UV–vis–NIR region) based on photon upconversion and plasmonic effects, as shown in Figure 12.7d (Tian et al. 2017). The TEM image in Figure 12.7e shows the effective attachment of AgNPs (marked with blue) on the surface of $NaYF_4$@SnO_2. Further, they simulated the EM field intensity of AgNPs under UV, visible, and NIR light exposure by using finite difference time domain (FDTD) analysis, Figure 12.7f–h. The SPR effect of AgNPs generates a large number of photocarrier and are responsible for the enhancement of photocatalytic performance. Further, Boppella et al. designed a composite photoanode by integrating an upconversion nanocrystal ($NaYF_4$:Yb^{3+} UCNP co-doped with Er^{3+}/Tm^{3+} ions) and AuNPs into a 3D-TiO_2 inverse opal nanostructure, Figure 12.7i (Boppella et al. 2019). The rational design of this architecture provides an indirect utilization of NIR light by both Au and TiO_2 to enhance the photocatalytic activity, see Figure 12.7j. In summary, generation of visible and UV photons using infrared light helps in extending the absorption range of semiconductors, which enhances the photocatalyst properties by increasing the generation of electron–hole pairs. Further, coupling of novel metal nanoparticles (such as AgNPs and AuNPs) with UCNPs can enhance the absorption as well as emission efficiency of UCNPs. Thus, integration of metal nanoparticles in semiconductor–UCNP combination creates a NIR light-triggered photocatalysis with efficient performance.

12.3 Photodetector

Infrared photodetectors (PDs) that convert the NIR portion of electromagnetic energy into electrical energy have been widely used in the field of security, night vision, communications, medical imaging, environmental monitoring, etc. (Downs and Vandervelde 2013; Krishna 2005; Koppens et al. 2014; Bao et al. 2018; Kelley et al. 2014). To date, commercial PDs are based on semiconductor materials including group IV semiconductors (Ge, Si, and carbon nanotubes), III–V compound semiconductors (such as GaAs and InP), and some ternary semiconductors (such as CdSeTe and HgCdTe). The main challenges with such PDs are their high manufacturing cost, complicated operation, low-temperature operation, mechanical inflexibility, and environmental issues because of the use of highly toxic heavy metal elements in them. In contrast, on the other hand, solution-processed colloidal nanocrystals such as quantum dots and metal halide perovskites are excellent material candidates for photodetector applications (Yan et al. 2010; Hong et al. 2015; Chen et al. 2016b; Saidaminov et al. 2016; Konstantatos and Sargent 2010; Zhang et al. 2017c). They show advantages because of their remarkable properties such as low cost, tunable absorption, high flexibility, easy operation, large-scale production, etc.

In 2010, Yan et al. used a seeded growth method to construct a nanoheterostructure consisting of lanthanide-doped $NaYF_4$ UCNPs decorated with CdSe

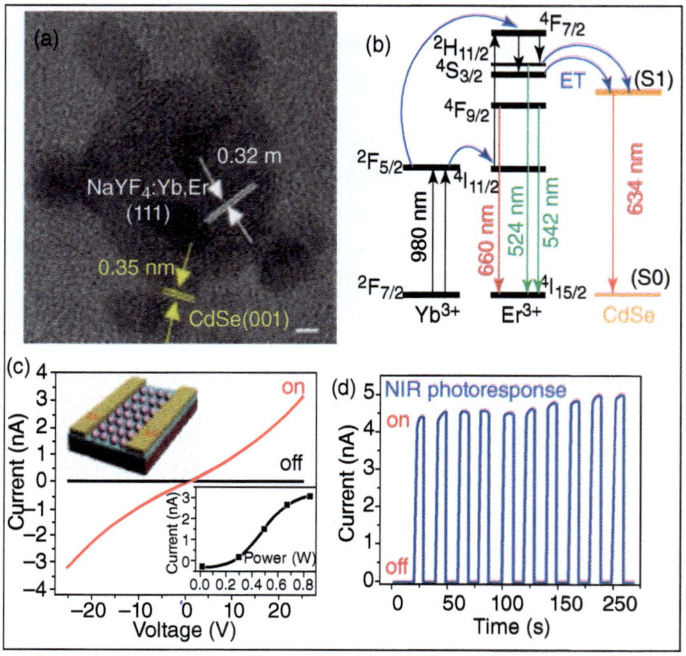

Figure 12.8 (a) TEM images of UCNPs-CdSe nanostructure, (b) schematic representation of energy transfer mechanism from UCNPs to QDs, (c) I–V characteristics in the dark (black line) and under 980 nm excitation (red line), inset shows the laser power dependence of the photocurrent, and (d) ON–OFF switching characteristics of the UCNP-CdSe nanostructure. Source: Yan et al. (2010). Reprinted with permission of American Chemical Society.

quantum dots, see Figure 12.8a (Yan et al. 2010). This architecture combines the optical properties of UCNPs with semiconductor QDs. This materializes the sub-bandgap photoconductivity, which would be useful for reversible and stable NIR photoconductivity switch (Yan et al. 2010). Herein, the $NaYF_4$:Yb,Er core converts the incident 980 nm photon into the green region, which is transferred to CdSe QDs through the energy transfer process, see Figure 12.8b. In effect, charge carriers are created in an external biased field that generates the photocurrent. Therefore, the photoresponse of the synthesized material is a combined effect of upconversion and ET from UCNPs to QDs. The authors fabricated a device that contains Si/SiO_2 substrates pre-patterned with Au electrodes and its surface is coated (through spin coating) with a solution of nanoparticles in toluene. In the OFF state, they observed very low current of the order of 14 pA (Figure 12.8c). In the ON state (under 980 nm excitation), they measured a much higher current of the order of 5 nA. The switching between the ON and OFF states was reversible over a large number of cycles (Figure 12.8d).

In 2015, Hong et al. again prepared a photodetector based on the core–shell structure of UCNPs and QDs (Hong et al. 2015). They used intense green-emitting β-$NaYF_4$:Yb; Er/β-$NaYF_4$ core/shell UCNPs and Cd-free $CuInS_2$/ZnS QDs instead of CdSe QDs. This structure shows a broad absorption band in the visible green spectral region along with a high quantum efficiency of up to 60%. Under 980 nm excitation, $CuInS_2/ZnS$ QDs mixed with UCNPs emit in the broad orange spectral region because of the efficient energy transfer from UCNPs to QDs. They fabricated NIR-responsive photodetectors by separate deposition of the QD layer and the UCNPs layer. Under an applied bias of 25 V, under the switching between OFF and ON states, the value of dark current was around 7.1 μA, while the photocurrent was 13.3 μA, respectively. The authors claimed for highly stable and reversible ON–OFF switching states. In addition to this, the photocurrent value of a QD–UCNP bilayer-based device was superior than any of the previously reported Cd-based QD–UCNP nanostructure device (Yan et al. 2010; Hong et al. 2015).

Along with these, perovskite-based photodetectors have also been widely explored (Chen et al. 2016b; Saidaminov et al. 2016; Konstantatos and Sargent 2010). Particularly, methyl ammonium lead halide perovskite ($MAPbX_3$, X = Cl, I, and Br)-based photodetectors show excellent photodetection ability because of high absorption coefficient, flexibility, large-scale production ability, high mobility, long carrier diffusion length, etc. (Chen et al. 2016b; Saidaminov et al. 2016; Konstantatos and Sargent 2010). However, all these perovskite-based photodetectors absorb only in the visible region because of their wide bandgap greater than 1.5 eV. To develop perovskite-based IR photodetectors, integration of lanthanide-based UCNPs with perovskite materials is an excellent approach. In 2017, Zhang et al. firstly demonstrated the novel perovskite-EYS (single-crystalline erbium ytterbium silicate) nanosheet hybrid waveguide photodetector architecture. They reported an excellent photoresponse at the near-infrared telecommunication wavelengths (≈1.54 μm), which is much better than the pristine perovskite (Zhang et al. 2017c). The photoresponsivity of the device is high and is in close match with the reported microscale silicon-based photodetectors. The authors claimed for superior photoswitching characteristics having a response time of ≈900 μs, which is much faster than the silicon-based hot electron photodetectors as well as non-avalanche semiconductor photodetectors at the telecommunication wavelengths. The excellent photodetection property is largely because of an efficient energy transfer between the EYS waveguide cavity and the perovskite film. In the same year, Li et al. also reported a flexible planar photodetector that includes integration of $MAPbI_3$ microarrays with UCNPs ($NaYF_4$:Yb/Er) (Li et al. 2017b). Figure 12.9a schematically represents the fabricated device structure that contains a flexible poly(ethylene terephthalate) (PET) substrate layered with top gold electrodes.

The SEM image in Figure 12.9b clearly demonstrates the formation of a monolayer of $NaYF_4$:Yb/Er UCNPs on the MAPbI3 film. Figure 12.9c demonstrates a detailed energy transition mechanism from the $NaYF_4$:Yb/Er UCNPs to the MAPbI3 film where incident NIR photons are absorbed by the UCNPs and upconverted in the visible region, which is further absorbed by the perovskites to generate photocurrent. The author reported the responsivity (R) of the device in the range of 0.55–0.86 A/W, and the maximum value of detectivity (D*) is 5.9×10^{12} Jones in the visible spectrum, which is much higher than the perovskite film-based planar photodetectors (Figure 12.9d). The external quantum efficiency (EQE) is found to be in the range of 120–180%. The inset of Figure 12.9d shows the high flexibility of the fabricated device.

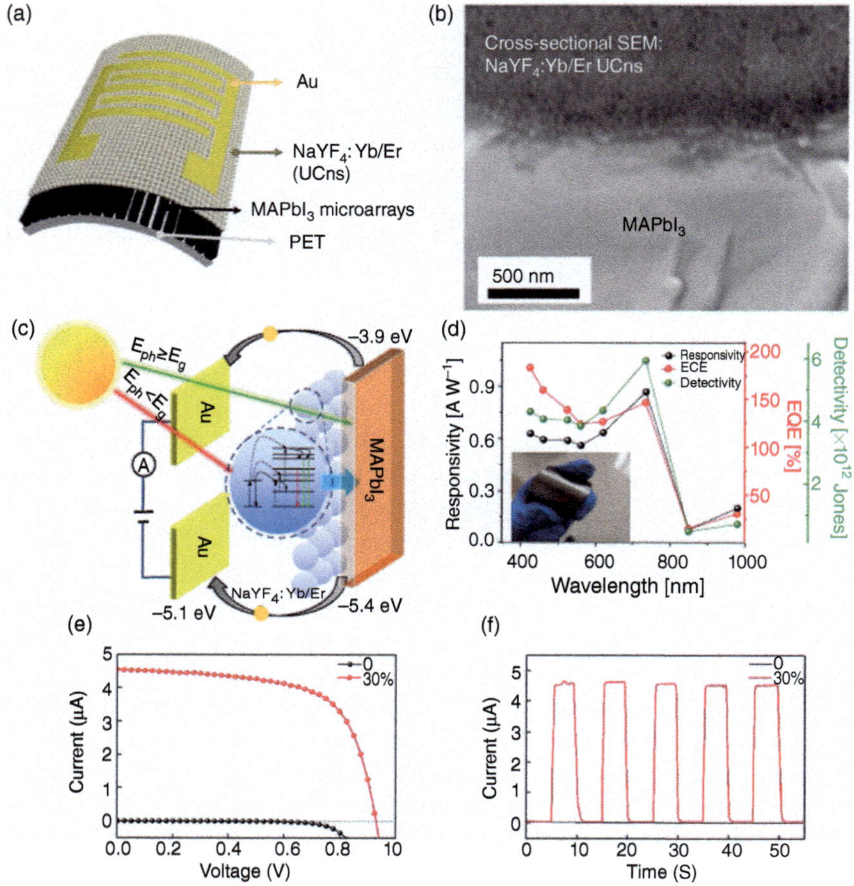

Figure 12.9 (a) Schematic illustration, (b) SEM image of the $MAPbI_3$/UCNPs bilayer photodetector, (c) detailed energy transfer mechanism from UCNPs to MAPbI3, and (d) photoresponse parameters of the bilayer photodetector as a function of wavelength. Inset shows the photo of the flexible integrated photodetector. Source: Li et al. (2017b). Reprinted with permission of American Chemical Society. (e) I–V curves of C-PSCs and C-PSCs with 30% of NYFS under the excitation of 980 nm laser and (f) photocurrent responses of devices with and without NYFS NPs under a 980 nm excitation. Source: Zhang et al. (2019b). Reprinted with permission of American Chemical Society.

In 2019, Zhang et al. reported the photodetection property of hexagonal $NaYF_4$:Yb^{3+},Tm^{3+}@SiO_2 (NYFS) upconversion nanoprism introduced into hole-transport-layer free carbon-based perovskite solar cells (C-PSC) (Zhang et al. 2019b). They measured the short-circuit current of C-PSCs with 30% of NYFS, which is about 4.54 μA (Figure 12.9e). The C-PSC with 30% NYFS shows detecting ON/OFF (interval of 10 seconds) property under the exposure of 980 nm laser light (Figure 12.9f). In 2019, Ji et al. employed the plasmon semiconductor Cs_xWO_3 to enhance the upconversion luminescence and further integrated it to $MAPbI_3$ to get NIR responsive photodetector (Ji et al. 2019). They observed ~144-fold enhanced UC intensity after employment of the plasmonic semiconductor. They constructed

$MAPbI3/Cs_xWO_3/NaYF_4/NaYF_4:Yb^{3+},Er^{3+}@NaYF_4:Yb^{3+},Tm^{3+}$ hybrid photodetector with high responsivity of ~0.33 A/W, detectivity of $\sim 4.5 \times 10^{10}$ Jones, and short response time of ~180 ms.

Along with QDs and halide perovskites, organic semiconductors loaded with lanthanide-based UCNPs are another important class of materials for NIR photodetectors (Sun et al. 2007; Zhou et al. 2014; Tong et al. 2015). Organic semiconductors are organic materials with semiconducting properties that possess weak van der Waals forces between the molecules (Sun et al. 2007). The presence of weak van der Waals forces provides a very soft and flexible device structure. In 2014, Zhou et al. developed a photonic flash memory device composed of Er^{3+}, Yb^{3+} co-doped $NaYF_4$ UCNPs blended with the poly(3-hexylthiophene) (P3HT) semiconducting polymer (SCS) (Zhou et al. 2014). By doping various concentrations of UCNPs in P3HT, the photonic memory characteristics can be tuned. Along with this, the authors also demonstrated the bending test to confirm the high flexibility of memory. With the assistance of IR, multilevel memory cells were obtained, which could store more than one single bit data. Later on in 2015, Tong et al. used the same combination of semiconductor polymer and UCNPs and claim for a significant enhancement of photocurrent, which is ascribed to the high UC efficiency of UCNPs (Zhou et al. 2014; Tong et al. 2015).

In 2019, Zhao et al. designed a planar-type photodetector for multiple wavelengths (808, 975, and 1532 nm) composed of SCS and Er^{3+}-doped $NaYF_4$ core–shell NPs (Zhao et al. 2019). The outer undoped $NaYF_4$ shell eliminates the quenching centers and provides a shielding effect for emitting centers within the core to boost the upconversion efficiency (Figure 12.10a). The Er^{3+} ion absorbed the incident 1532 nm photon and upconverted it into higher energy photons (e.g. 654 and 808 nm). These upconverted photons are further absorbed by the diketopyrrolopyrrole-quaterthiophene terpolymer (PDPP4T) to generate photocurrent (Figure 12.10a). A schematic of the RE-SCP-based planar photodetector architecture and the digital photograph of the RE-SCP photodetector showing its flexibility after bending is shown in Figure 12.10b,c. The planar-type RE-SCP composite-based photodetector device could detect shortwave infrared (SWIR) source (eye safe, at 1532 nm) with a responsivity of 0.02 A/W and an EQE ~2%.

To measure the stability and robustness of the device current density, responsivity and EQE were further recorded under different bending conditions (Figure 12.10d,e). No significant difference has been observed in detection parameters for different curvatures, showing the potential of device for a high-performance flexible SWIR photodetector. In the same year, Xiang et al. also demonstrated a photodiode device architecture for a SWIR photodetector. This conjugates a polymer/small-molecule bulk heterojunction (BHJs, host) with Er^{3+}-doped UCNPs (Figure 12.10f) (Xiang et al. 2019). In order to utilize the different upconversion peaks of Er^{3+} ions, a BHJ blend of polythieno[3,2-b]thiophene-diketopyrrolopyrrole-co-thiophene (DPPTT-T) and PC70BM has been employed. The used organic blend absorbs the upconversion fluorescence of UCNPs and creates electron–hole pairs, which later on transported toward their respective electrodes (Figure 12.10g). Under $\lambda = 1.5\ \mu m$ photon illumination, the fabricated device offers a photoresponsivity of

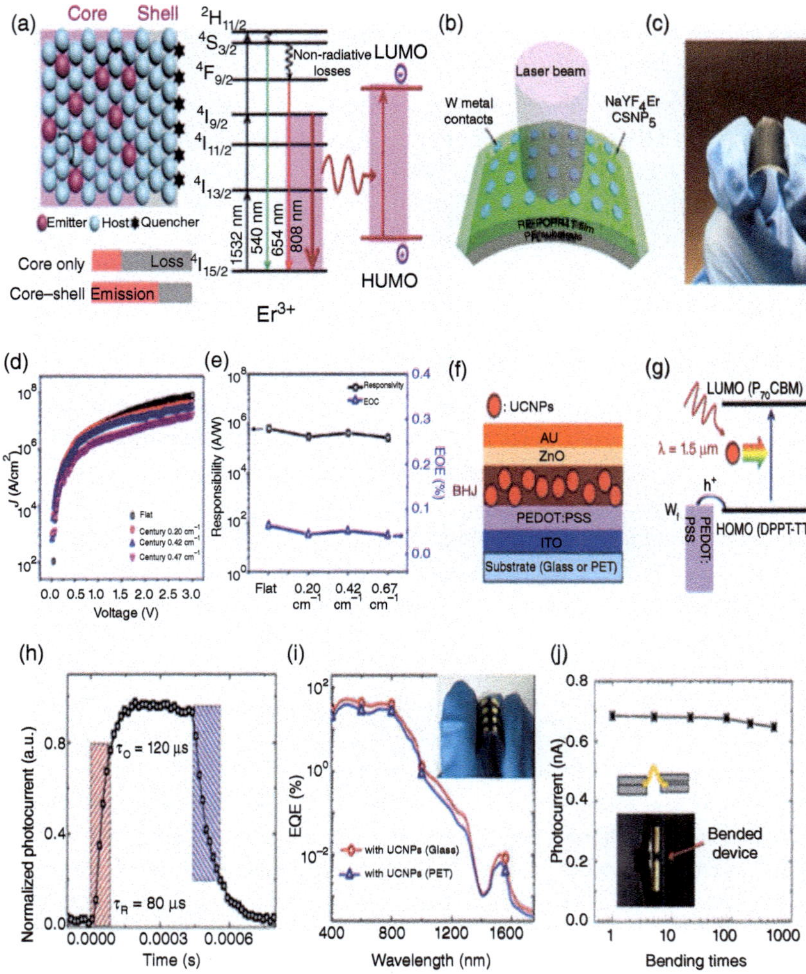

Figure 12.10 (a) Core–shell structure efficiently converted the incident SWIR light to higher photon energy emissions (i.e. 654 and 808 nm), which are efficiently absorbed by PDPP4T to generate electrical signals. (b) Schematic of the RE-SCP-based planar photodetector architecture, (c) digital photograph of the RE-SCP photodetector showing flexibility after bending, (d) measured J–V characteristics, and (e) calculated R and EQE of RE-SCP photodetector at different bending configurations. Source: Zhao et al. (2019). Reprinted with permission of American Chemical Society, (f) Schematic representation of the device architecture, (g) schematic of the operating principle for $\lambda = 1.5\,\mu m$ photon illumination of hybrid BHJ(DPPTT-T/PC70BM)/UCNP photodetector device, (h) response time study on a single representative light/dark cycle (τ_R = rise time and τ_D = decay time), (i) EQE of the hybrid photodetector, inset shows the image of a flexible hybrid photodetector, (j) bending test of a flexible hybrid BHJ(DPPTT-T/PC70BM)/UCNP photodetector built on a ITO/PET substrate. Source: Xiang et al. (2019). Reprinted with permission of American Chemical Society.

0.73 mA/W for the devices built on rigid ITO/glass substrates, while it is 0.44 mA/W for flexible ITO/PET substrates. Time-dependent photoresponse of the hybrid device in Figure 12.10h demonstrated the short photocurrent rise time down to 80 μs. Figure 12.10i reveals the high external quantum efficiencies up to 10^{-2}% at λ = 1.5 μm. Mechanical bending tests revealed the robustness of the fabricated device. The time-dependent photoresponse under λ = 1.5 μm illumination of the flexible device does not show significant change after 200 bending times, suggesting a highly reproducible photodetector performance (Figure 12.10j).

Light–matter interactions in 2D materials such as hexagonal boron nitride (h-BN), molybdenum disulfide (MoS_2), black phosphorus (BP), and graphene open up a wide range of applications of these nanomaterials in electronics, optoelectronics, and photonic devices (Niu et al. 2014; Kataria et al. 2018a,b; Zhai et al. 2018; Zhou et al. 2017). Especially, high room temperature carrier mobility, flexibility, optical transparency, light weight, and environment-friendly nature of graphene make it an ideal 2D material for a high-speed photodetector. However, the performance of graphene as a photodetector is limited by its low light absorption capability (≈2% of normal incident light) (Niu et al. 2014; Kataria et al. 2018a,b). Therefore, integration of graphene with UCNPs is a smart choice to develop an infrared photodetector (Niu et al. 2014; Kataria et al. 2018a,b).

Niu et al. demonstrated a reduced graphene oxide (rGO)-based near-infrared photodetector coupled with upconversion and gold plasmon (Niu et al. 2014). The authors showed that rGO absorbed the upconverted shorter wavelength light through the FRET process, and therefore, photoresponsivity to 980 nm light is improved more than three times as compared to control the device with rGO alone. Further coupling of rGO-UCNPs with a thin plasmonic layer enhances the upconversion emission intensity. The author observed an overall improvement in the photoresponsivity by a factor of ≈10, which is a combined effect of both UCNPs and gold plasmon. Kataria et al. fabricated a filter less, light-weighted, visible blind, and wearable NIR photodetector composed of UCNPs, graphene, and micropyramidical poly(dimethylsiloxane) (PDMS) film (Kataria et al. 2018b).

Figure 12.11a demonstrates a photodetector made up of UCNPs and graphene hybrid micropyramidical structure. Regular arrangement of square pyramids can be seen in the SEM image, as shown in Figure 12.11b(i). The graphene layer is transferred successfully over the micropyramidical PDMS structure, as shown in Figure 12.11b (ii) and (iii). Further deposition of UCNPs can be seen in Figure 12.11b(iv). UCNPs absorbed the incident 980 nm photons and produced photogenerated charge carriers, Figure 12.11c.

The metastable energy levels of Tm^{3+} ions prevent the recombination of excited photons and trap the bulk number of charge carriers and hence support the charge transfer from UCNPs to graphene layer. Photogenerated charge carriers get an outstanding conduction path because of the graphene layer, and hence, responsivity is increased significantly. The micropyramidical structure efficiently trapped the light on the surface of the photodetector and therefore catalyzes the process of charge carrier generation, Figure 12.11d. The prepared NIR-responsive photodetector provides very high photoresponsivity of ~800 A/W. Along with

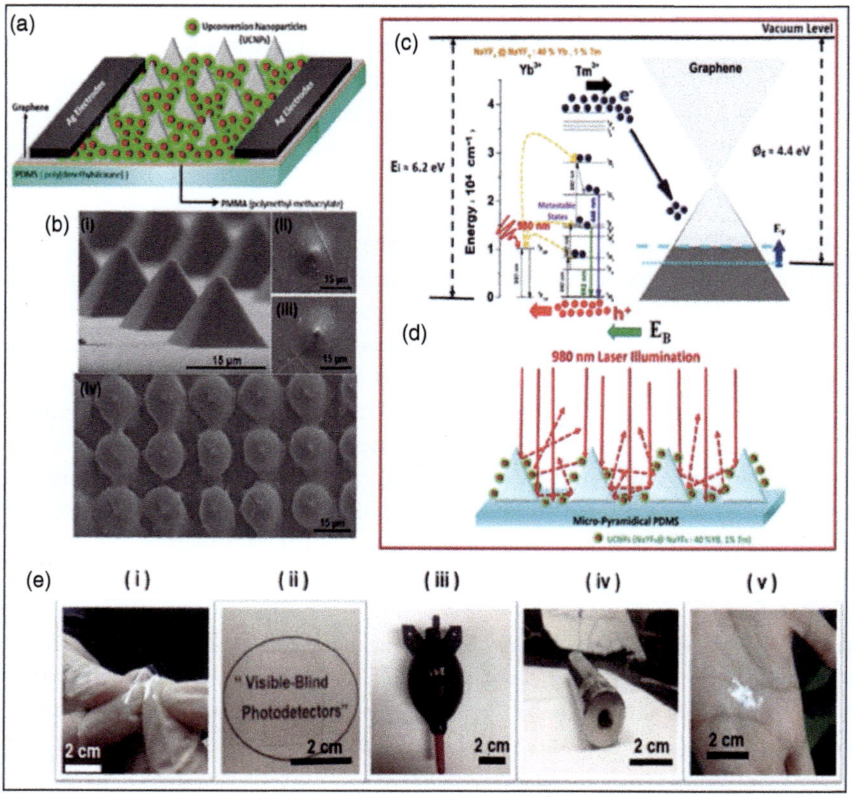

Figure 12.11 (a) Schematic diagram of the UCNPs/graphene hybrid micropyramidical photodetector device and (b) (i) SEM image of the micropyramidical structure PDMS. (ii and iii) SEM image of the micropyramidical structure PDMS with graphene. (iv) SEM image of the micropyramidical structure PDMS with UCNPs coating. (c) Schematic energy band diagram of the UCNPs and graphene showing the efficient flow of photogenerated charge carriers. (d) Schematic diagram of the trapping process of incident 980 nm light in micropyramidical structure. (e) Photograph showing (i) flexible; (ii) transparency; and (iii), (iv), and (v) wearable property of the UCNP/graphene hybrid micropyramidical photodetector. Source: Kataria et al. (2018b). Reprinted with permission of American Chemical Society.

this, the photodetector shows excellent flexibility, transparency, and exceptional wearability, Figure 12.11e. Similarly, MoS_2 is another excellent 2D material for photodetector applications (Zhai et al. 2018; Zhou et al. 2017). Zhai et al. reported a NIR photonic resistive switching memory based on MoS_2–$NaYF_4$:Yb^{3+},Er^{3+} UCNP nanocomposites (Zhai et al. 2018). The device shows excellent controllability over cycle-to-cycle as well as device-to-device consistency and high compatibility. Zhou et al. also designed a narrowband wavelength-selective near-infrared photodetector that contains lanthanide-doped UCNPs and monolayer MoS_2 (Zhou et al. 2017).

12.4 Conclusion

In conclusion, this chapter describes the use of upconversion luminescence materials for photocatalysis and photodetector applications. Generation of visible and UV photons using infrared light helps in extending the absorption range of semiconductors used for photocatalytic applications. This enhances the photocatalyst properties by increasing the generation of electron–hole pairs. In this context, integration of UCNPs with TiO_2, composites of UCNs and different semiconductors, e.g. UCNPs–BiOX (X = Cl, Br, and I), UCNPs–$AgInS_2$, UCNPs–MOFs, UCNPs–ZnO, UCNPs–CdS, UCNPs–BFO, and composites of UCNPs, plasmonic nanoparticles, and 2D material etc., shows excellent increase in the photocatalytic activity. However, there are few limitations, e.g. (i) the core–shell structure of UCNPs and the photocatalytic semiconductor material possess a major drawback of high lattice mismatching between the crystalline core and the amorphous shell. To tackle out this limitation, the concept of an amorphous interlayer (such as SiO_2, carbon, PVP, etc.) has been suggested. (ii) UCNPs possess low quantum yield and narrow absorption spectra because of the dipole-forbidden transitions. In order to meet this, a well-established way to boost the UC efficiency has been adopted, i.e. the coupling of novel metal nanoparticles (such as AgNPs and AuNPs) with the UCNPs. In the vicinity of plasmonic nanoparticles, the electric field distribution is manipulated, which can enhance the absorption as well as emission efficiency of UCNPs. Integration of metal nanoparticles in the semiconductor–UCNPs combination creates a NIR light-triggered photocatalysis with efficient performance. However, above all, it would be interesting to see the upconversion nanoparticles that could convert the infrared part of sunlight directly usable for photocatalytic applications.

On the other hand, the study shows that the integration of UCNPs with solution-processed QDs (such as CdSe, ZnS, $CuInS_2$, etc.), halide perovskite QDs (such as $MAPbX_3$, X = Cl, I, Br), organic semiconductors (e.g. P3HT), semiconductor polymers (e.g. PDPP4T), and 2D materials (such as h-BN, molybdenum disulfide (MoS_2), BP, graphene, etc.) can supersede the commercial photodetectors based on semiconductor materials including group IV semiconductors (Ge, Si, and carbon nanotubes), III–V semiconductors (such as GaAs and InP), and some ternary semiconductors (such as CdSeTe and HgCdTe). These material-based PDs show remarkable properties such as low cost, tunable and wide absorption range, high flexibility, easy operation, large-scale production, etc.

References

Akhavan, O., Abdolahad, M., Esfandiar, A., and Mohatashamifar, M. (2010). Photodegradation of graphene oxide sheets by TiO_2 nanoparticles after a photocatalytic reduction. *Journal of Physical Chemistry C* 114 (30): 12955–12959. https://doi.org/10.1021/jp103472c.

Anderson, C. and Bard, A.J. (1995). An improved photocatalyst of TiO_2/SiO_2 prepared by a sol–gel synthesis. *Journal of Physical Chemistry* 99 (24): 9882–9885. https://doi.org/10.1021/j100024a033.

Atabaev, T.S. and Molkenova, A. (2019). Upconversion optical nanomaterials applied for photocatalysis and photovoltaics: recent advances and perspectives. *Frontiers of Materials Science* 13: 335–341. https://doi.org/10.1007/s11706-019-0482-z.

Balaji, R., Kumar, S., Reddy, K.L. et al. (2017). Near-infrared driven photocatalytic performance of lanthanide-doped $NaYF_4$@CdS core–shell nanostructures with enhanced upconversion properties. *Journal of Alloys and Compounds* 724 (15): 481–491. https://doi.org/10.1016/j.jallcom.2017.07.050.

Bao, C.X., Yang, J., Bai, S. et al. (2018). High performance and stable all-inorganic metal halide perovskite-based photodetectors for optical communication applications. *Advanced Materials* 30: 1803422. (8p). https://doi.org/10.1002/adma.201803422.

Boppella, R., Mota, F.M., Lim, J.W. et al. (2019). Plasmon and upconversion mediated broadband spectral response in TiO_2 inverse opal photocatalysts for enhanced photoelectrochemical water splitting. *ACS Applied Energy Materials* 2: 3780–3790. https://doi.org/10.1021/acsaem.9b00469.

Borges, M.E., Sierra, M., Cuevas, E., and García, R.D. (2016). Photocatalysis with solar energy: sunlight-responsive photocatalyst based on TiO_2 loaded on a natural material for wastewater treatment. *Solar Energy* 135: 527–535. https://doi.org/10.1016/j.solener.2016.06.022.

Chen, G., Qiu, H., Prasad, P.N., and Chen, X. (2014a). Upconversion nanoparticles: design, nanochemistry, and applications in theranostics. *Chemical Reviews* 114 (10): 5161–5214. https://doi.org/10.1021/cr400425h.

Chen, Y., Mishra, S., Ledoux, G. et al. (2014b). Direct synthesis of hexagonal NaGdF4 nanocrystals from a single-source precursor: upconverting NaGdF4:Yb^{3+},Tm^{3+} and its composites with TiO_2 for near-IR-driven photocatalysis. *An Asian Journal of Chemistry* 9: 2415–2421. https://doi.org/10.1002/asia.201402347.

Chen, X., Jin, L., Kong, W. et al. (2016a). Confining energy migration in upconversion nanoparticles towards deep ultraviolet lasing. *Nature Communications* 7: 10304. https://doi.org/10.1038/ncomms10304.

Chen, S., Teng, C.J., Zhang, M. et al. (2016b). A flexible UV–vis–NIR photodetector based on a perovskite/conjugated-polymer composite. *Advanced Materials* 28: 5969. (1–6). https://doi.org/10.1364/OL.40.004959.

Chilkalwar, A.A. and Rayalu, S.S. (2018). Synergistic plasmonic and upconversion effect of the (Yb,Er)NYF-TiO_2/Au composite for photocatalytic hydrogen generation. *Journal of Physical Chemistry C* 122 (46): 26307–26314. https://doi.org/10.1021/acs.jpcc.8b05480.

Cui, C., Tou, M., Li, M. et al. (2017). Heterogeneous semiconductor shells sequentially coated on upconversion nanoplates for NIR-light enhanced photocatalysis. *Inorganic Chemistry* 56: 2328–2336. https://doi.org/10.1021/acs.inorgchem.6b03079.

Ding, N., Xu, W., Zhou, D. et al. (2020). Upconversion ladder enabled super-sensitive narrowband near-infrared photodetectors based on rare earth doped florine perovskite nanocrystals. *Nano Energy* 76: 105103. https://doi.org/10.1016/j.nanoen.2020.105103.

Dong, Y., Choi, J., Jeong, H.K., and Son, D.H. (2015a). Hot electrons generated from doped quantum dots via upconversion of excitons to hot charge carriers for enhanced photocatalysis. *Journal of the American Chemical Society* 137 (16): 5549–5554. https://doi.org/10.1021/jacs.5b02026.

Dong, H., Sun, L.D., and Yan, C.H. (2015b). Energy transfer in lanthanide upconversion studies for extended optical applications. *Chemical Society Reviews* 44: 1608–1634. https://doi.org/10.1039/C4CS00188E.

Downs, C. and Vandervelde, T.E. (2013). Progress in infrared photodetectors since 2000. *Sensors* 13: 5054–5098. https://doi.org/10.3390/s130405054.

Du, Y., Liu, H., Tian, Y. et al. (2020). Recyclable SERS-based immunoassay guided by photocatalytic performance of Fe_3O_4@TiO_2@Au nanocomposites. *Biosensors* 10 (3): 25. (15p). https://doi.org/10.3390/bios10030025.

Ge, W., Li, Z., Chen, T. et al. (2018). Extended near-infrared photoactivity of $Bi_6Fe_{1.9}Co_{0.1}Ti_3O_{18}$ by upconversion nanoparticles. *Nanomaterials* 8 (7): 534. (12p). https://doi.org/10.3390/nano8070534.

Goldschmidt, J.C. and Fischer, S. (2015). Upconversion for photovoltaics-a review of materials, devices and concepts for performance enhancement. *Advanced Optical Materials* 3 (4): 510–535. https://doi.org/10.1002/adom.201500024.

Gupta, A., Thakur, M.K., Effendi, T.A. et al. (2021). Metallo-graphene enhanced upconversion luminescence for broadband photodetection under polychromatic illumination. *Chemical Engineering Journal* 420 (2): 127608. https://doi.org/10.1016/j.cej.2020.127608.

He, M., Pang, X., Liu, X. et al. (2016). Monodisperse dual-functional upconversion nanoparticles enabled near-infrared organolead halide perovskite solar cells. *Angewandte Chemie International Edition* 128 (13): 4352–4356. https://doi.org/10.1002/ange.201600702.

Hong, A.R., Kim, J., Kim, S.Y. et al. (2015). Core/shell-structured upconversion nanophosphor and cadmium-free quantum-dot bilayer-based near-infrared photodetectors. *Optics Letters* 40 (21): 4959–4962. https://doi.org/10.1364/OL.40.004959.

Idris, N.M., Jayakumar, M.K.G., Bansal, A., and Zhang, Y. (2015). Upconversion nanoparticles as versatile light nano-transducers for photoactivation applications. *Chemical Society Reviews* 44 (6): 1449–1478. https://doi.org/10.1039/C4CS00158C.

Ji, Y., Xu, W., Li, D. et al. (2019). Semiconductor plasmon enhanced monolayer upconversion nanoparticles for high performance narrowband near-infrared photodetection. *Nano Energy* 61: 211–220. https://doi.org/10.1016/j.nanoen.2019.04.064.

Kataria, M., Yadav, K., Haider, G. et al. (2018a). Transparent, wearable, broadband, and highly sensitive upconversion nanoparticles and graphene-based hybrid photodetectors. *ACS Photonics* 5: 2336–2347. https://doi.org/10.1021/acsphotonics.8b00141.

Kataria, M., Yadav, K., Cai, S.Y. et al. (2018b). Highly sensitive, visible blind, wearable, and omnidirectional near-infrared photodetectors. *ACS Nano* 12 (9): 9596–9607. https://doi.org/10.1021/acsnano.8b05582.

Ke, J., Li, X., Zhao, Q. et al. (2017). Upconversion carbon quantum dots as visible light responsive component for efficient enhancement of photocatalytic performance. *Journal of Colloid and Interface Science* 496: 425–433. https://doi.org/10.1016/j.jcis.2017.01.121.

Kelley, S.O., Mirkin, C.A., Walt, D.R. et al. (2014). Advancing the speed, sensitivity and accuracy of biomolecular detection using multi-length-scale engineering. *Nature Nanotechnology* 9: 969–980. https://doi.org/10.1038/nnano.2014.261.

Konstantatos, G. and Sargent, E.H. (2010). Nanostructured materials for photon detection. *Nature Nanotechnology* 5: 391–400. https://doi.org/10.1038/nnano.2010.78.

Koppens, F.H.L., Mueller, T., Avouris, P. et al. (2014). Photodetectors based on graphene, other two-dimensional materials and hybrid systems. *Nature Nanotechnology* 9: 780–793. https://doi.org/10.1038/nnano.2014.215.

Krishna, S. (2005). Quantum dots-in-a-well infrared photodetectors. *Journal of Physics D: Applied Physics* 38: 2142–2150. https://doi.org/10.1088/0022-727/38/13/010.

Kwon, H., Mota, F.M., Chung, K. et al. (2018). Enhancing solar light-driven photocatalytic activity of mesoporous carbon–TiO_2 hybrid films via upconversion coupling. *ACS Sustainable Chemistry & Engineering* 6 (1): 1310–1317. https://doi.org/10.1021/acssuschemeng.7b03658.

Lee, S.Y. and Park, S.J. (2013). TiO_2 photocatalyst for water treatment applications. *Journal of Industrial and Engineering Chemistry* 19 (6): 1761–1759. https://doi.org/10.1016/j.jiec.2013.07.012.

Li, Z.X., Shi, F.B., Zhang, T. et al. (2011). Ytterbium stabilized ordered mesoporous titania for near-infrared photocatalysis. *Chemical Communications* 47: 8109–8111. https://doi.org/10.1039/C1CC12539G.

Li, M., Zheng, Z., Zheng, Y. et al. (2017a). Controlled growth of metal-organic framework on upconversion nanocrystals for NIR-enhanced photocatalysis. *ACS Applied Materials & Interfaces* 9: 2899–2905. https://doi.org/10.1021/acsami.6b15792.

Li, J., Shen, Y., Liu, Y. et al. (2017b). Stable high-performance flexible photodetector based on upconversion nanoparticles/perovskite microarrays composite. *ACS Applied Materials & Interfaces* 9 (22): 19176–19183. https://doi.org/10.1021/acsami.7b03229.

Li, Y., Xu, D., Yao, L. et al. (2018a). Enhanced upconversion luminescence in controllable self-assembled BiOBr:Yb^{3+}/Er^{3+} 3D hierarchical architectures and their application in NIR photocatalysis. *Industrial and Engineering Chemistry Research* 57 (50): 17161–17169. https://doi.org/10.1021/acs.iecr.8b04608.

Li, D., Yu, S.H., and Jiang, H.L. (2018b). From UV to near-infrared light-responsive metal–organic framework composites: plasmon and upconversion enhanced photocatalysis. *Advanced Materials* 30: 1707377. (7P). https://doi.org/10.1002/adma.201707377.

Lian, H., Liao, Q., Yang, B. et al. (2021). Optoelectronic synaptic transistors based on upconverting nanoparticles. *Journal of Materials Chemistry C* 9: 640–648. https://doi.org/10.1039/D0TC04115G.

Liu, G. (2015). Advances in the theoretical understanding of photon upconversion in rare-earth activated nanophosphors. *Chemical Society Reviews* 44: 1635–1652. https://doi.org/10.1039/C4CS00187G.

Liu, Q., Tan, X., Wang, S. et al. (2019). MXene as a non-metal charge mediator in 2D layered CdS@Ti_3C_2@TiO_2 composites with superior Z-scheme visible light-driven photocatalytic activity. *Environmental Science Nano* 6: 3158–3169. https://doi.org/10.1039/C9EN00567F.

Luo, L.B., Zou, Y.F., Ge, C.W. et al. (2016). A surface plasmon enhanced near-infrared nanophotodetector. *Advanced Optical Materials* 4: 763–771. https://doi.org/10.1002/adom.201500701.

Maji, S.K. and Kim, D.H. (2018). $AgInS_2$-coated upconversion nanoparticle as a photocatalyst for near-infrared light-activated photodynamic therapy of cancer cells. *ACS Applied Bio Materials* 1 (5): 1628–1638. https://doi.org/10.1021/acsabm.8b00467.

Nie, Z., Ke, X., Li, D. et al. (2019). $NaYF_4$:Yb,Er,Nd@$NaYF_4$:Nd upconversion nanocrystals capped with Mn:TiO_2 for 808 nm NIR-triggered photocatalytic applications. *Journal of Physical Chemistry C* 123: 22959–22970. https://doi.org/10.1021/acs.jpcc.9b05234.

Niu, W., Chen, H., Chen, R. et al. (2014). Synergetically enhanced near-infrared photoresponse of reduced graphene oxide by upconversion and gold plasmon. *Small* 10 (18): 3637–3643. https://doi.org/10.1002/smll.201400400.

Qin, W., Zhang, D., Zhao, D. et al. (2010). Near-infrared photocatalysis based on YF3:Yb^{3+},Tm^{3+}/TiO_2 core/shell nanoparticles. *Chemical Communications* 46: 2304–2306. https://doi.org/10.1039/B924052G.

Reddy, K.L., Kumar, S., Kumar, A., and Krishnan, V. (2019). Wide spectrum photocatalytic activity in lanthanide-doped upconversion nanophosphors coated with porous TiO_2 and Ag–Cu bimetallic nanoparticles. *Journal of Hazardous Materials* 367: 694–705. https://doi.org/10.1016/j.jhazmat.2019.01.004.

Rink, S.S., Varma, C.M., and Levi, A.F.J. (1991). Excitation mechanisms and optical properties of rare-earth ions in semiconductors. *Physical Review Letters* 66: 2782. https://doi.org/10.1103/PhysRevLett.66.2782.

Ronda, C.R. (2007). *Luminescence: From Theory to Applications*, vol. 1. Weinheim: WILEY-VCH Verlag GmbH & Co. KGaA. ISBN: 978-3-527-31402-7.

Saidaminov, M.I., Haque, M.A., Savoie, M. et al. (2016). Perovskite photodetectors operating in both narrowband and broadband regimes. *Advanced Materials* 28: 8144. https://doi.org/10.1002/adma.201601235.

Shi, X., Molokeev, M.S., Wang, X. et al. (2018). Crystal structure of $NaLuW_2O_8 \cdot 2H_2O$ and down/upconversion luminescence of the derived $NaLu(WO_4)_2$:Yb/Ln phosphors (Ln = Ho, Er, Tm). *Inorganic Chemistry* 57 (17): 10791–10801. https://doi.org/10.1021/acs.inorgchem.8b01427.

Singh, P., Shahi, P.K., Singh, S.K., and Rai, S.B. (2016). Photoluminescence, upconversion and quantum-cutting emission in Tm/Tb/Pr and Yb co-doped oxide phosphor: a comparative study. *Journal of Alloys and Compounds* 681: 477–485. https://doi.org/10.1016/j.jallcom.2016.04.250.

Singh, P., Shahi, P.K., Singh, S.K. et al. (2017a). Lanthanide doped ultrafine hybrid nanostructures: multicolour luminescence, upconversion based energy transfer and luminescent solar collector applications. *Nanoscale* 9: 696–705. https://doi.org/10.1039/C6NR07250J.

Singh, P., Shahi, P.K., Prakash, R., and Rai, S.B. (2017b). An assembly and interaction of upconversion and plasmonic nanoparticles on organometallic nanofibers: enhanced multicolor upconversion, downshifting emission and the plasmonic effect. *Nanotechnology* 28: 415701–415712. https://doi.org/10.1088/1361-6528/aa8066.

Singh, P., Singh, P., Prakash, R., and Rai, S.B. (2019). Photo-physical studies of ultrasmall upconversion nanoparticles embedded organometallic complexes: probing a dual mode optical sensor for hydrogen peroxide. *Optical Materials* 98: 109459. (7p). https://doi.org/10.1016/j.optmat.2019.109459.

Singh, P., Singh, P., Prakash, R. et al. (2020a). Colour tunability in a bimodal fluorescent hybrid nanostructure UCNPs@AuNPs@QDs. *Current Applied Physics* 20: 1150–1155. https://doi.org/10.1016/j.cap.2020.08.001.

Singh, P., Singh, S.K., Singh, P. et al. (2020b). Generation of red-NIR bi-modal fluorescence in hybrid nanostructure. *Materials Research Bulletin* 122: 110663. https://doi.org/10.1016/j.materresbull.2019.110663.

Singh, D.K., Pradhan, M., and Materny, A. (2021). *Modern Techniques of Spectroscopy: Basics, Instrumentation, and Applications*, vol. 1. Springer Nature https://doi.org/10.1007/978-981-33-6084-6.

Su, W., Zheng, M., Li, L. et al. (2014). Directly coat TiO_2 on hydrophobic $NaYF_4$:Yb,Tm nanoplates and regulate their photocatalytic activities with the core size. *Journal of Materials Chemistry A* 2: 13486–13491. https://doi.org/10.1039/C4TA02756F.

Sujatha, G., Shanthakumar, S., and Chiampo, F. (2020). UV light-irradiated photocatalytic degradation of coffee processing wastewater using TiO_2 as a catalyst. *Environments* 7: 47. https://doi.org/10.3390/environments7060047.

Sun, C.J., Xu, Z., Hu, B. et al. (2007). Application of $NaYF_4$:Yb, $NaYF_4$:Yb, Er upconversion fluorescence nanocrystals for solution-processed near infrared photodetectors. *Applied Physics Letters* 91: 191113. (3p). https://doi.org/10.1063/1.2807841.

Tang, Y., Di, W., Zhai, X. et al. (2013). NIR-responsive photocatalytic activity and mechanism of $NaYF_4$:Yb,Tm@TiO_2 core–shell nanoparticles. *ACS Catalysis* 3: 405–412. https://doi.org/10.1021/cs300808r.

Tian, Q., Yao, W., Wu, W. et al. (2017). Efficient UV-vis-NIR responsive upconversion and plasmonic-enhanced photocatalyst based on lanthanide-doped $NaYF_4$/SnO_2/Ag. *ACS Sustainable Chemistry & Engineering* 5: 10889–10899. https://doi.org/10.1021/acssuschemeng.7b02806.

Tian, Q., Yao, W., Wu, W., and Jiang, C. (2019). NIR light-activated upconversion semiconductor photocatalysts. *Nanoscale Horizons* 4: 10–25. https://doi.org/10.1039/C8NH00154E.

Tong, Y., Zhao, X., Tan, M.C., and Zhao, R. (2015). Cost-effective and highly photoresponsive nanophosphor-P3HT photoconductive nanocomposite for near-infrared detection. *Scientific Reports* 5: 16761. (9p). https://doi.org/10.1038/srep16761.

Tou, M.J., Mei, Y.Y., Bai, S. et al. (2016). Depositing CdS nanoclusters on carbon-modified $NaYF_4$:Yb,Tm upconversion nanocrystals for NIR-light enhanced photocatalysis. *Nanoscale* 8: 553–562. https://doi.org/10.1039/C5NR06806A.

Wang, J., Wen, F., Zhang, Z. et al. (2005). Degradation of dyestuff wastewater using visible light in the presence of a novel nano TiO_2 catalyst doped with upconversion luminescence agent. *Journal of Environmental Sciences* 17: 727–730. PMID: 16312992.

Wang, W., Shang, Q., Zheng, W. et al. (2010). A novel near-infrared antibacterial material depending on the upconverting property of Er^{3+}–Yb^{3+}–Fe^{3+} tridoped TiO_2 nanopowder. *Journal of Physical Chemistry C* 114: 3663–13669. https://doi.org/10.1021/jp102320x.

Wang, W., Huang, W., Ni, Y. et al. (2014). Different upconversion properties of β-$NaYF_4$:Yb^{3+},Tm^{3+}/Er^{3+} in affecting the near-infrared-driven photocatalytic activity of high-reactive TiO_2. *ACS Applied Materials & Interfaces* 6: 340–348. https://doi.org/10.1021/am404389g.

Wen, S., Zhou, J., Zheng, K. et al. (2018). Advances in highly doped upconversion nanoparticles. *Nature Communications* 9: 2415. (12p). https://doi.org/10.1038/s41467-018-04813-5.

Woan, K., Pyrgiotakis, G., and Sigmund, W. (2009). Photocatalytic carbon-nanotube-TiO_2 composites. *Advanced Materials* 21: 2233–2239. https://doi.org/10.1002/adma.200802738.

Wu, X., Zhang, K., Zhang, G., and Yin, S. (2017). Facile preparation of BiOX (X=Cl, Br, I) nanoparticles and up-conversion phosphors/BiOBr composites for efficient degradation of NO gas: oxygen vacancy effect and near infrared light responsive mechanism. *Chemical Engineering Journal* 325: 59–70. https://doi.org/10.1016/j.cej.2017.05.044.

Xiang, H., Hu, Z., Billot, L. et al. (2019). Heavy-metal-free flexible hybrid polymer-nanocrystal photodetectors sensitive to 1.5 μm wavelength. *ACS Applied Materials and Interfaces* 11: 42571–42579. https://doi.org/10.1021/acsami.9b14034.

Xie, X. and Liu, X. (2012). Upconversion goes broadband. *Nature Materials* 11: 842–843. https://doi.org/10.1038/nmat3426.

Xu, Z., Quintanilla, M., Vetrone, F. et al. (2015). Harvesting lost photons: plasmon and upconversion enhanced broadband photocatalytic activity in core@shell microspheres based on lanthanide-doped $NaYF_4$, TiO_2, and Au. *Advanced Functional Materials* 25: 2950–2960. https://doi.org/10.1002/adfm.201500810.

Xu, J., Liu, N., Wu, D. et al. (2020). Upconversion nanoparticle-assisted payload delivery from TiO_2 under near-infrared light irradiation for bacterial inactivation. *ACS Nano* 14: 337–346. https://doi.org/10.1021/acsnano.9b05386.

Yan, C., Dadvand, A., Rosei, F., and Perepichka, D.F. (2010). Near-IR photoresponse in new up-converting CdSe/$NaYF_4$:Yb,Er nanoheterostructures. *Journal of the American Chemical Society* 132 (26): 8868–8869. https://doi.org/10.1021/ja103743t.

Yang, B., Wang, Y., Wei, T. et al. (2018). Solution-processable near-infrared-responsive composite of perovskite nanowires and photon-upconversion nanoparticles. *Advanced Functional Materials* 28 (31): 1801782. (7p). https://doi.org/10.1002/adfm.201801782.

Zhai, Y., Yang, X., Wang, F. et al. (2018). Infrared-sensitive memory based on direct-grown MoS_2–upconversion-nanoparticle heterostructure. *Advanced Materials* 30 (49): 1803563. (9p). https://doi.org/10.1002/adma.201803563.

Zhang, Y. and Hong, Z. (2013). Synthesis of lanthanide-doped $NaYF_4$@TiO_2 core–shell composites with highly crystalline and tunable TiO_2 shells under mild conditions and their upconversion-based photocatalysis. *Nanoscale* 5: 8930–8933. https://doi.org/10.1039/C3NR03051B.

Zhang, Q., Lima, D.Q., Lee, I. et al. (2011). A highly active titanium dioxide based visible-light photocatalyst with nonmetal doping and plasmonic metal decoration. *Angewandte Chemie International Edition* 50: 7088–7092. https://doi.org/10.1002/anie.201101969.

Zhang, X., Yang, P., Wang, D. et al. (2012a). $La(OH)_3$:Ln^{3+} and La_2O_3:Ln^{3+} (Ln = Yb/Er,Yb/Tm,Yb/Ho) microrods: synthesis and up-conversion luminescence properties. *Crystal Growth and Design* 12: 306–312. https://doi.org/10.1021/cg201091u.

Zhang, X., Veikko, U., Mao, J. et al. (2012b). Visible-light-induced photocatalytic hydrogen production over binuclear RuII–Bipyridyl dye-sensitized TiO_2 without noble metal loading. *Chemistry A European Journal* 18: 12103–12111. https://doi.org/10.1002/chem.201200725.

Zhang, J., Huang, Y., Jin, L. et al. (2017a). Efficient upconverting multiferroic core@shell photocatalysts: visible-to-near-infrared photon harvesting. *ACS Applied Materials & Interfaces* 9: 8142–8150. https://doi.org/10.1021/acsami.7b00158.

Zhang, Q., Deng, J., Xu, Z. et al. (2017b). High-efficiency broadband C3N4 photocatalysts: synergistic effects from upconversion and plasmons. *ACS Catalysis* 7: 6225–6234. https://doi.org/10.1021/acscatal.7b02013.

Zhang, X., Yang, S., Zhou, H. et al. (2017c). Perovskite–erbium silicate nanosheet hybrid waveguide photodetectors at the near-infrared telecommunication band. *Advanced Materials* 29: 1604431. (8p). https://doi.org/10.1002/adma.201604431.

Zhang, J., Zhao, S., Xu, Z. et al. (2019a). Near-infrared light-driven photocatalytic $NaYF_4$:Yb,Tm@ZnO core/shell nanomaterials and their performance. *RSC Advances* 9: 3688–3692. https://doi.org/10.1039/C8RA07861K.

Zhang, H., Xiao, Y., Qi, F. et al. (2019b). Near-infrared light-sensitive hole-transport-layer free perovskite solar cells and photodetectors with hexagonal $NaYF_4$:Yb^{3+},Tm^{3+}@SiO_2 upconversion nanoprism-modified TiO_2 scaffold. *ACS Sustainable Chemistry & Engineering* 7 (9): 8236–8244. https://doi.org/10.1021/acssuschemeng.8b06606.

Zhanga, Z., Zhang, J., Zhou, W. et al. (2009). Structure and up-conversion luminescence properties of Er^{3+}–Yb^{3+} co-doped TiO_2 nanocrystals prepared by sol–gel method. *Advanced Materials Research* 66: 167–170. https://doi.org/10.4028/www.scientific.net/AMR.66.167.

Zhao, X., Song, L., Zhao, R., and Tan, M.C. (2019). High-performance and flexible shortwave infrared photodetectors using composites of rare earth-doped nanoparticles. *ACS Applied Materials and Interfaces* 11: 2344–2351. https://doi.org/10.1021/acsami.8b16978.

Zheng, B., Guo, Q., Wang, D. et al. (2015). Energy-transfer modulation for enhanced photocatalytic activity of near-infrared upconversion photocatalyst. *Journal of the American Ceramic Society* 98 (1): 136–140. https://doi.org/10.1111/jace.13232.

Zhou, Y., Han, S.T., Chen, X. et al. (2014). An upconverted photonic nonvolatile memory. *Nature Communications* 5: 4720. (8p). https://doi.org/10.1038/ncomms5720.

Zhou, N., Xu, B., Gan, L. et al. (2017). Narrowband spectrally selective near-infrared photodetector based on up-conversion nanoparticles used in a 2D hybrid device. *Journal of Materials Chemistry C* 5: 1591–1595. https://doi.org/10.1039/C6TC05113H.

13

UCNPs in Lighting and Displays

Riya Dey

106 Wedgewood Drive, Apartment 11, Morgantown, WV 26505, USA

13.1 Introduction

Extensive research has been going on for the past couple of decades on rare-earth-doped upconversion nanoparticle (UCNP) synthesis and the study of their optical behavior under near-infrared (NIR) excitations. These UCNPs are a special class of nanomaterials in which the rare-earth elements are the key player for their optical behavior. In rare-earth-doped upconverting luminescent materials, two or more low-energy photons are being used to generate one higher energy photon. For the generation of the higher energy photon from these materials, the 4f electronic band structure associated with the rare-earth elements serve as the fountainhead. The rare-earth elements when doped into the host matrices exist in the most stable triply ionized state. The triply ionized rare-earth elements are characterized by their partially filled intra 4f shell shielded by the completely/partially filled 5s–5p orbitals and the transitions within, which results in sharp emission bands. These unique optical properties of these materials enable for their wide variety of applications in diverse field spanning from bioimaging, background-free biological sensing, solar cell harvesting, solid-state lighting (SSL), photocatalysis, displays, sensing, pollutant degradation, and hydrogen generation for security purposes (Liang et al. 2020; Feng et al. 2013; Song et al. 2012; Corstjens et al. 2003; Seng et al. 2015; Naccache et al. 2013; Chen et al. 2020; Yang et al. 2014; Atabaev and Molkenova 2019; Shin et al. 2018; Dey and Rai 2014; Dey et al. 2015; Zhou et al. 2020; Chilkalwar and Rayalu 2018; Dey et al. 2014).

13.2 Major Factors that Affect the UC Emission Efficiency

Although the outstanding upconversion properties of such materials increases their importance in real-world applications, one of the major challenges related to these materials is to enhance the emission efficiencies (Wang et al. 2010a). In order to

Upconverting Nanoparticles: From Fundamentals to Applications, First Edition. Edited by Vineet K. Rai.

get enhanced upconversion efficiencies, the codoping of sensitizer ions alongside the activator ion plays an important role. In order to choose proper sensitizer ions, the intermediate energy level for both sensitizer and activator ions should be well matched. There are several factors that are considered to be the deciding factors in order to improve the upconversion emission. The quantum efficiency of UCNPs is a very important factor in order to increase its importance in real-world applications. Different research groups have chosen different approaches in order to increase the quantum efficiency. The concentration of the doped rare earth is one of the crucial factors. Concentration quenching is a major obstacle that hinders the UC intensity. In 1954, Dexter and Schulman identified that when the concentration of the activator ions reaches 10^{-3}–10^{-2}, then the luminescence intensity in bulk materials reduces significantly (Dexter and Schulman 1954). The optimum concentration of the dopants results the optimum emission intensities where other factors are invariable. In the case of UCNPs, the high doping concentration results reduced the UC emission intensity. This reduction in emission intensity arises because of two factors: (i) migration of the excited-state electrons to the surface where a large number of quenchers are present (Wang et al. 2010) and (ii) due to the cross-relaxation process in between the dopants (Tu et al. 2015). Therefore, the proper management of the doping concentration results enhanced UC emission. For efficient UC emission with high emission efficiency, the doping level in UCNPs should be kept below 10^{-2} M (Haase and Schäfer 2011).

The host matrix with low-phonon energies is considered to be a smart choice for synthesizing efficient UC materials. Apart from the low-phonon energy, the unit cell size of the host nanocrystal is also an important parameter that can monitor the UC efficiency. The nanocrystalline host with large unit cell size is ideal to reduce the concentration quenching and thus enhance the emission intensities (Wang et al. 2014). Among different host matrices, $NaYF_4$ is considered as an efficient host by different research groups for synthesizing UCNPs (Boyer and van Veggel 2010; Zhan et al. 2011; Tian et al. 2012; Guo et al. 2019).

Another effective approach to enhance the upconversion efficiency is the core–shell structure of the synthesized material. Using the rare-earth-doped UCNPs as the core and introducing the doped/undoped metal fluorides as the shell, the quantum efficiency of the resultant core@shell upconverting materials can be enhanced significantly. In the nanosized UCNPs, a major proportion of the doped ions exists on the surface as a less symmetric crystal field (Chen et al. 2015a). Therefore, the energy associated with the emission spectrum may easily get attenuated because of nonradiative relaxation and high energy oscillations caused by impurities, surface defects, ligands, and solvents (Li et al. 2014; Zhang et al. 2014). Energy loss can be circumvented and the surface quenching can be reduced with the introduction of a uniform shell that subsequently strengthens the quantum yield (Chen et al. 2015; Zeng et al. 2013; Qian and Zhang 2008). Huang coated the $NaYF_4$:Nd^{3+}/Yb^{3+} shell around $NaYF_4$:Nd^{3+}/Yb^{3+}/Ho^{3+}nanoparticles, and an enhancement of 990 factors has been observed under 808 nm excitation (Huang 2015). Yi's group showed that the coating of the undoped $NaYF_4$ shell around Er-, Tm-, and Yb-doped $NaYF_4$ nanoparticles results an enhancement of 7.4 times

Table 13.1 Upconversion quantum yield in different upconverting nanoparticles.

SL. No.	Material composition	Phase	Size of nanoparticles	Excitation wavelength	Excitation power density (W/cm^2)	Quantum yield (%)	References
1	Dye-sensitized $NaYbF_4$:Tm^{3+} 0.5% @$NaYF_4$:Nd^{3+}	Hexagonal	54 nm	800 nm	10	4.8	Chen et al. (2015)
2	$\beta NaY_{0.8-2x}Gd_xLu_xF_4$:$Yb_{0.18}Er_{0.02}$	Hexagonal	25 nm	980 nm	63	0.074	Wisser et al. (2016)
3	$NaYbF_4$/Tm@CaF_2 (0.5 for Tm)	Cubic	27	980 nm	0.3	0.60 ± 0.1	Chen et al. (2012)
4	$NaYF_4$/Yb/Er@$NaYF_4$	Hexagonal	9	980 nm	10^3	0.18 ± 0.01	Ostrowski et al. (2012)
5	$LiLuF_4$/Yb/Tm@$LiLuF_4$	Tetragonal	50	980 nm	127	7.6	Huang et al. (2014)
6	$LiLuF_4$/Yb/Er@$LiLuF_4$	Tetragonal	50	980 nm	127	5.0	Huang et al. (2014)
7	$NaYF_4$/Yb/Er@$NaLuF_4$	Hexagonal	~50	980 nm	~63	4 ± 0.5	Fischer et al. (2016)
8	$NaYF_4$/Yb/Er@$NaYF_4$	Hexagonal	30	980 nm	150	0.30 ± 0.10	Boyer and van Veggel (2010)
9	$NaGdF_4$/Yb/Er@$NaYF_4$	Hexagonal	17	980 nm	100	0.51 ± 0.08	Li et al. (2013)
10	$NaYF_4$/Yb/Tm@$NaYF_4$	Hexagonal	43	975	78	3.5	Xu et al. (2012)
11	α-$NaYF_4$:Yb(80%), Er(2%)@CaF_2	Cubic	—	980	10	3.2 ± 0.1	Punjabi et al. (2014)
12	β-$NaYF_4$:Er^{3+}@β-$NaLuF_4$	—	38.8 ± 2.1	1523	0.43 ± 0.03	0.71 ± 0.08	Fischer et al. (2015)

compared to Er, Yb $NaYF_4@NaYF_4$ and 29.6 times from Tm, Yb $NaYF_4@NaYF_4$ (Yi and Chow 2007). Wang et al. have explored an enhancement of 450-fold of anti-Stokes emission under 980 nm excitation in $NaGdF_4$:Yb,Tm@$NaGdF_4$ nanoparticles (Wang et al. 2010b). Several research groups have explored that core@shell-structured upconverting nanoparticles show good quantum yield, which have been tabulated in Table 13.1.

13.3 UC Mechanisms with Rate Equations

To date, different possible UC mechanisms have been recognized in rare-earth-doped luminescent materials: (i) excited-state absorption (ESA), (ii) energy-transfer upconversion (ETU), (iii) cross-relaxation (CR), and (iv) cooperative sensitization upconversion (CSU), which have been discussed by several research groups in detail (Soni et al. 2015; Kumar and Bahadur 2021; Tiwari et al. 2018; Dey et al. 2012). The most important UC mechanisms for UCNPs are ESA and ETU process. The ESA and ETU processes have been discussed with the help of an energy-level diagram (Figure 13.1).

The three energy levels are defined by their respective energies E_1, E_2, and E_3 as shown in Figure 13.1. The numbers of ions present in these three levels are given by N_1, N_2, and N_3, respectively. Before going into a detailed discussion, three basic and essential assumptions are as follows:

i. The ground-state population N_1 is kept fixed.
ii. The pumping process is continuous, i.e. the intermediate level E_2 of this system is continuously pumped.
iii. There is no pump depletion.

The rate equations representing the population of the three levels (Figure 13.1) can be written as follows:

$$\frac{dN_2}{dt} = \sigma_1 \rho N_1 - \sigma_2 \rho N_2 - 2W_2 N_2^2 - A_2 N_2 \tag{13.1}$$

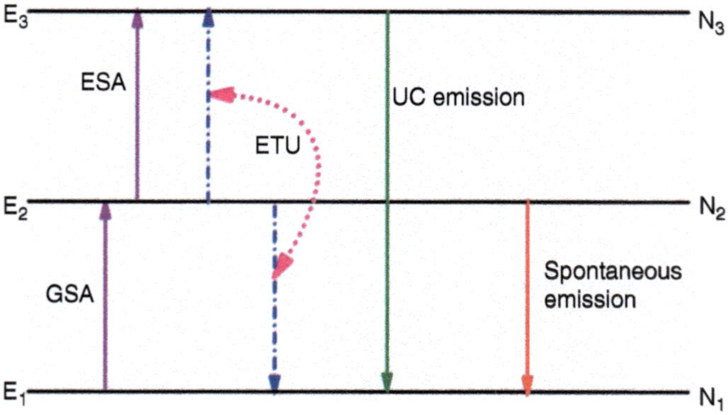

Figure 13.1 The energy-level scheme of a three-level system and possible upconversion mechanisms. Source: Based on Dey 2015.

$$\frac{dN_3}{dt} = \sigma_2 \rho N_2 + W_2 N_2^2 - A_3 N_3 \tag{13.2}$$

$$[A_3 = A_{32} + A_{31}]$$

where "σ" is the absorption cross section corresponding to the 980 nm photon and W_i is the relaxation rate of ith level caused by energy transfer. A_i is the emission rate of the ith level because of spontaneous emission and ρ indicates the pump rate. In Eq. (13.1), the first term in the right hand side signifies the ground state absorption (GSA) process for increasing the population of level 2. The second term stands for the ESA process. The third and fourth terms represent the ETU and spontaneous emissions, respectively. The negative signs in the equations indicate the decrease in the population of that particular level caused by different mechanisms. Now, in Eq. (13.2), the first and second terms in the right hand side are positive, which indicate the increase in population because of the pump absorption by ESA and the energy transfer processes from level 1, respectively. The probability of reduction in the population of level 3 is due to the spontaneous emission from level 3 to level 2 and level 1, and these are represented by the last term in Eq. (13.2). The relation between the pump rate and the pump power can be expressed by Eq. (13.3):

$$\rho = \frac{\lambda P}{hc\pi\omega_P^2} \tag{13.3}$$

where "P" is the pump power, "λ" is the wavelength of the pump photon, "h" is Planck's constant, "c" is the velocity of light in vacuum, and "ω_P" is the pump beam waist.

13.3.1 Pump Power Dependence in the Case of Dominant ETU-Assisted Upconversion over ESA

In a case where the energy transfer process is dominant over the ESA, we can modify Eqs. (13.1) and (13.2) by considering the steady-state condition and using Eq. (13.3) as

$$\frac{dN_2}{dt} = \sigma_1 \rho N_1 - 2W_2 N_2^2 - A_2 N_2 = 0 \tag{13.4}$$

and

$$\frac{dN_3}{dt} = W_2 N_2^2 - A_3 N_3 = 0 \tag{13.5}$$

Equations (13.4) and (13.5) can be written as

$$\sigma_1 \rho N_1 = 2W_2 N_2^2 + A_2 N_2 \tag{13.6}$$

$$W_2 N_2^2 = A_3 N_3 \tag{13.7}$$

Equation (13.8) implies that

$$N_3 \propto N_2^2 \tag{13.8}$$

Now, if the spontaneous relaxation dominates over the upconversion, then $2W_2N_2^2 \ll A_2N_2$, which gives $N_2 \propto \rho \propto P$. Finally, from Eq. (13.8), we can have $N_3 \propto N_2^2 \propto P^2$.

The other possibility is that if the level E_2 is metastable, i.e. it has longer lifetime, then the upconversion process is dominant over the linear decay. In that case, $2W_2N_2^2 \gg A_2N_2$. Therefore, $N_2^2 \propto \rho \propto P$, which finally gives the dependency of the UC emission on the pump power as$N_3 \propto N_2^2 \propto P$.

13.3.2 Pump Power Dependence in the Case of Dominant ESA-Assisted Upconversion over ETU

When the ESA process is dominant, the terms containing $W_2N_2^2$ in Eq. (13.1) and Eq. (13.2) can be neglected, and by considering the steady-state condition, Eqs. (13.1) and (13.2) can be modified as

$$\frac{dN_2}{dt} = \sigma_1\rho N_1 - \sigma_2\rho N_2 - A_2N_2 = 0 \tag{13.9}$$

and

$$\frac{dN_3}{dt} = \sigma_2\rho N_2 - A_3N_3 = 0 \tag{13.10}$$

Now, if spontaneous decay is dominant over upconversion (i.e. $A_2N_2 \gg \sigma_2\rho N_2$), then Eqs. (13.9) and (13.10) can be modified as

$$\sigma_1\rho N_1 = A_2N_2 \tag{13.11}$$

$$\sigma_2\rho N_2 = A_3N_3 \tag{13.12}$$

Equation (13.12) implies that

$$N_3 \propto \rho N_2 \tag{13.13}$$

From Eq. (13.11), it is observed that $N_2 \propto \rho \propto P$, which finally implies that $N_3 \propto \rho^2 \propto P^2$.

Now, when the upconversion is dominant over linear decay (i.e. $A_2N_2 \ll \sigma_2\rho N_2$), then from Eq. (13.9), it is clear that N_2 is proportional to N_1 but independent of ρ while Eq. (13.10) implies that $N_3 \propto \rho \propto P$.

For generalization of the above-discussed processes, we can choose the $(n+1)$th level system, and then the upconversion dependency on the pump power varies $N_{n+1} \propto N_n^2 \propto P^n$ to $N_{n+1} \propto N_n^2 \propto P$.

13.4 UCNPs in Solid-State Laser

Rare-earth-doped upconverting nanoparticles can also be considered as efficient optical gain media for lasing applications. Different research groups have devoted many efforts in this specific area and reported interesting results associated with upconversion lasing. UCNP-based lasing is very attractive as it does not need rigid control of optical alignment, temperature, and humidity (Jin et al. 2017; Zhu et al. 2018). Zhu and his group demonstrated the first upconversion multiband lasing where they have used $NaYF_4$:Yb/Er@$NaYF_4$ nanoparticles as the gain media (Zhu et al. 2013). For the excitation purpose, they used a 980 nm three-pulse pumping system with pulse width and time delay between the adjacent pulses 6 and 10 ns, respectively, and the repetition rate was 10 Hz. The Fabry–Perot cavity containing UCNPs dispersed in cyclohexane solution was built. The UC lasing emission spectra have been shown in Figure 13.2a. Figure 13.2b represents the output intensity versus the pump power plot.

Figure 13.2c represents the measured and fitted values of the threshold power versus the diameter of the cavity for the three lasing bands around 407, 540, and 655 nm.

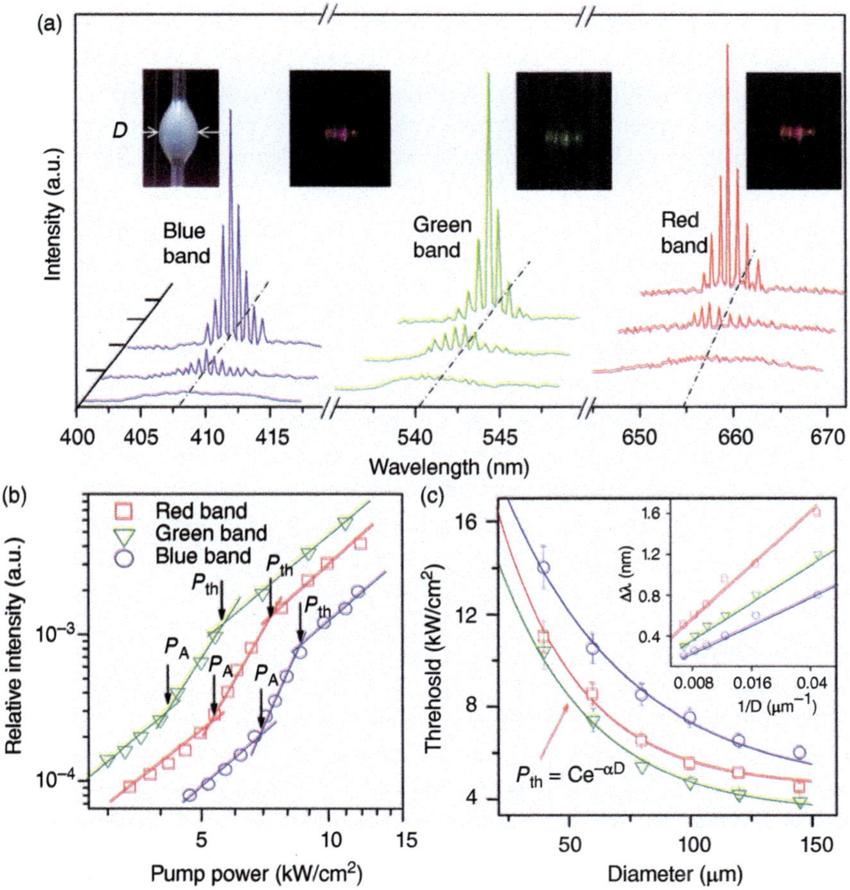

Figure 13.2 (a) Lasing spectra of the microcavity with a bottle-like geometry of diameter equal to 80 μm under a three-pulse excitation scheme at room temperature. The insets show the images of the microcavity under different excitation power. (b) Log–log plot of output power versus pump power for the three-color bands of the microcavity. (c) Measured and fitted (solid lines) values of P_{th} and $\Delta\lambda$ (inset figure) versus D for the three-color bands of the microcavity. Source: Zhu et al. (2013). Reprinted with permission of American Chemical Society.

Amplified spontaneous emission associated with Er^{3+} transitions was gained under sequential high pump power pulse excitation. More excitingly, they fabricated a bottle-like microcavity by coating silica resin dispersed with the UCNPs onto an optical fiber. Finally, they concluded that the lasing mechanism of the microcavity was due to the whispering gallery mode (WGM). Chen with his colleagues reported the deep ultraviolet upconversion lasing emission from core@shell@shell-structured $NaYF_4$@$NaYbF_4$:Tm/Gd@$NaYF_4$ UCNPs (Chen et al. 2016). To realize the lasing emission, the five-pulse excitation scheme was used. In this case also, they had fabricated the laser cavity by coating the silica resin containing the UCNPs onto an optical fiber. In this study also, the bottle-like cavity structure was fabricated for upconversion lasing. With increasing pump power of the excitation source, the output intensity was plotted against the input pump, and the plotted graph represented an S-like curve. This nonlinear trend with two threshold pump power values of the curve was divided into three regions. The S-like trend further indicates the transition from spontaneous emission to an amplified spontaneous emission to finally lasing emission. The mode spacing was calculated and found to lie within 0.25 to 0.27 nm, which is in good agreement with the theoretical value, i.e. 0.26 nm. Furthermore, single mode lasing was obtained by using a microresonator of diameter D_m = 20 μm. The calculated Q factor (~2800) also indicates a high quality of ultraviolet upconversion lasing from the material. Apart from bottle-like microcavities, there are many other structures that have proven effective for generating upconversion lasing. Peng and coworkers had developed a planer microcavity containing a $NaYF_4$:Yb, Tm nanocrystal film of 300 μm thickness, sandwiched in between a quartz plate and an Al-coated glass substrate (Peng et al. 2018). For the excitation process, 980 nm CW laser was used, and the lasing output was obtained with a peak wavelength of ~345 and ~474 nm. The presence of the mirrors enabled that confinement of the longitudinal modes of the lighting, which finally resulted in the optical feedback along the laser microcavities. The laser emission was detected from the side of the quartz mirror. Shang et al. have reported the ultra-low threshold lasing in β-$NaYF_4$:20% Yb, x% Tm^{3+} (Shang et al. 2020). The upconverted lasing threshold of 150 W/cm^2 was obtained from this material under the excitation with CW laser. Multimode lasing emission under 100 kW/cm^2 has been represented and compared with the simulated optical modes in Figure 13.3a as shown by Shang et al. (2020). Using the Lorentz fitting function, they have calculated the full width half maxima (FWHM) of the sharp lasing peak, which was found to be 0.54 nm and the calculated Q-factor was found to be ~1500. They observed that both TE and TM mode lasing emissions exhibit linear polarization with orthogonal periodicity as shown in Figure 13.3b. In Figure 13.3c, excitation power-dependent emission spectra have been shown. The presence of the narrow peak at a threshold value is the indication of increased coherence. Figure 13.3d represents the nonlinear change characteristic at an ultra-low pumping threshold of 150 W/cm^2, and Figure 13.3e shows the Tm^{3+} concentration-dependent lasing threshold.

Fernandez-Bravo and his group have reported the continuous wave upconverting lasing from β-$NaYF_4$:Er^{3+}–Yb^{3+}@$NaYF_4$ core–shell-structured UCNPs

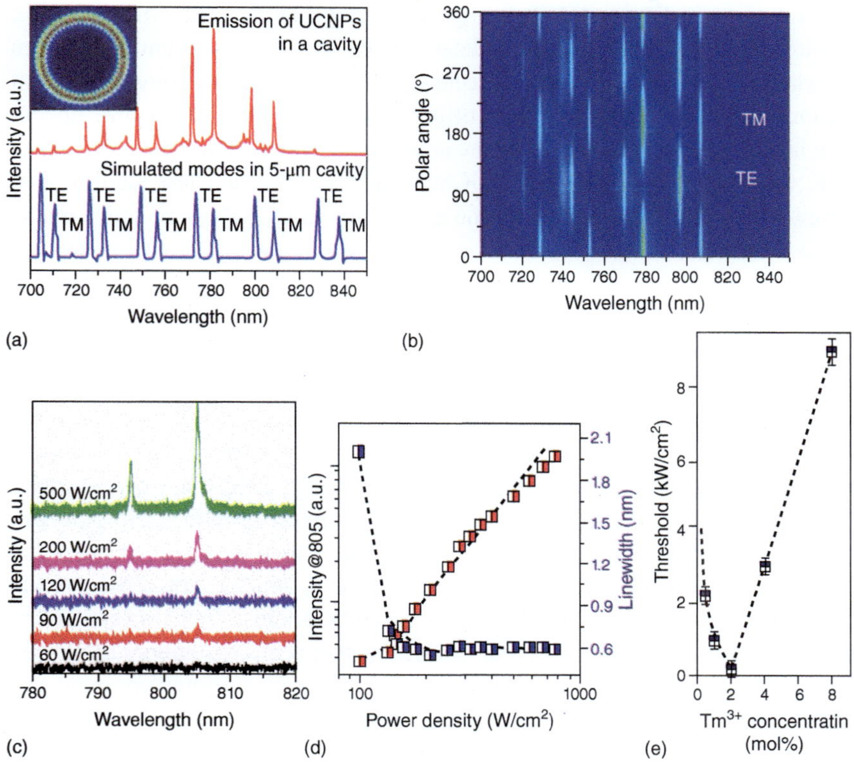

Figure 13.3 (a) Numerical simulations of resonance spectrum and experimental emission spectrum. The inset shows a numerical simulation of the electrical field distributions at 800 nm within a major plane. (b) Emission polarization angle-dependent intensities of the lasing peaks. (c) Power-dependent upconversion emission spectra showing the gradual appearance of lasing peaks. (d) The pumping power-dependent plots of emission intensities and spectral linewidth narrowing, showing the onset of upconversion lasing emissions. (e) Tm^{3+}concentration-dependent thresholds for the onsets of upconversion lasing emissions. Source: Shang et al. (2020). Reprinted with permission of Springer Nature.

(Fernandez-Bravo et al. 2019). They obtained selective, single-mode lasing from the core–shell-structured UCNPs coated on Ag nanopillar arrays (80 nm diameter, 50 nm height). Excitingly, they achieved CW lasing under both CW and pulse laser excitation. More interestingly, they observed an exceptionally stable lasing, which lasts more than six hours with a low lasing threshold of 29 W/cm^2. The same group has also reported the CW mode lasing where core@shell-structured $NaYF_4$:Gd/Tm@$NaGdF_4$ UCNPs are used as the gain media (Fernandez-Bravo et al. 2018). They have successfully obtained the upconverted lasing under the excitation with 1064 nm CW laser and with low power density 14 kW/cm^2. They also reported that the coupling of these energy-looping nanoparticles to the WGMs of polystyrene microsphere results stable lasing in blue and NIR region simultaneously for as long as five hour. Apart from using bottle-like microcavities, there are several other structures that have been employed successfully for lasing purposes (Wang et al. 2017, 2018).

In addition to visible and NIR lasing, UV lasing has its great importance in photonics, physics, and biological applications. Besides the lasing emissions from the lanthanide elements that occurs because of the 4f electronic transition, 5d–4f transition within Ce^{3+} ions has been utilized by Wang's group to realize upconverting lasing (Chen et al. 2017). For the gain media, core@shell@shell-structured $NaYbF_4$:Gd/Tm@$NaGdF_4$@CaF_2:Ce UCNPs have been used. These studies open up a new and promising way toward the application of UCNPs as an active media in lasing purposes.

13.5 UCNPs in Solid-State Lighting and Displays

Two major issues that are of main concern of the mankind are the pollution and energy shortage of the conventional energy resources. The current society is mostly dependent on the fossil fuel-based conventional energy resources. However, with the current demand of the society, the conventional energy resources are approaching toward its end. In addition to this, combustion of this fossil fuels releases the hazardous gas such as SO_x and CO_x in the environment, which eventually cause the air pollution (Zhang et al. 2009). Therefore, researchers across the world are paying great attention to find some alternate reliable nonconventional energy resources, which are free from the above-mentioned shortcomings. In a fluorescent lamp and incandescent lights, a huge amount of energy losses occur because of the high temperature and the association of Stokes shift (Xie and Hirosaki 2007). SSL is considered to be a next-generation light source. As SSL is the direct conversion of electricity into visible light unlike fluorescent lighting or incandescent, therefore, the efficiency of such lighting sources is much higher. The SSL resources are used in a variety of applications as they possess several outstanding features over conventional lighting sources.

- The life span of these lighting sources is much higher compared to incandescent light bulbs and they are very durable.
- They are eco-friendly as they do not cause environmental pollution by emitting hazardous gases such as SO_2 and CO_2.
- The light they produce is of better quality.
- They are safe to use.
- They are much compact and flexible.

13.5.1 Requirements for LED Applications

The basic requirements to fabricate light emitting diode (LED) devices are as follows:

1. Low-cost material usage
2. Strong absorption in blue and near-UV region
3. High thermal and chemical stability
4. High emission intensity and quantum efficiency

5. Short decay time associated with the emission
6. Environment-friendly composition
7. High luminescence stability.
8. High quantum efficiency.

SSL relies on the conversion of electricity to visible light using solid materials. When talking about the adaptability of UCNPs for display purposes, the energy flux density requirement is much higher for LEDs compared to the fluorescent lamps (Lv et al. 2014). Thermal stability is another factor that determines whether the material is suitable for use in LED applications or not. UCNPs with lower thermal quenching are considered to be ideal for the above-said purposes. Nanomaterials that can be excited in NIR I and NIR II regions are gaining much interest day by day (Tiwari and Yakhmi 2015). The two reasons for thermal quenching are lattice vibrations and activation energy. Mondal et al. have investigated the thermal stability of core@shell-structured Er^{3+}–Yb^{3+}:$YMoO_4$@SiO_2 UCNPs and explained its acceptability for LED packaging (Mondal et al. 2017). They have discussed the thermal quenching effect on both core-only nanoparticles and core@shell nanoparticles as shown in Figure 13.4. In their study, they have shown that the core@shell-structured nanoparticles is more promising compared to their core-only nanoparticle, which is clearly observed in Figure 13.4.

They have calculated the activation energy for the core@shell nanoparticles using the Arrhenius equation (Chen et al. 2015) and found that their calculated value is ideal for LED applications. In this regard, Du et al. have also investigated the thermal stability of Ho^{3+}/Yb^{3+}/Zn^{2+}:Gd_2O_3 nanoparticles and calculated the activation energy of the material. The theoretically calculated value for the doped Gd_2O_3 upconverting nanoparticle was found to be 0.18 eV (Du et al. 2018). In order to prove the potential of their synthesized material in SSL, they had further fabricated an LED device by coating the doped nanoparticle on the surface of an NIR chip. Then, they

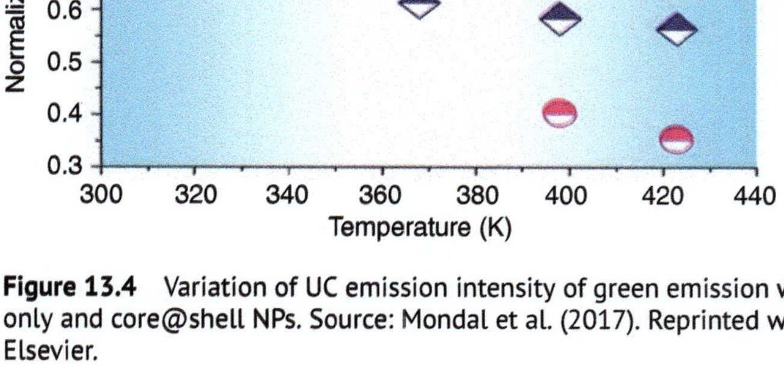

Figure 13.4 Variation of UC emission intensity of green emission with temperature of core only and core@shell NPs. Source: Mondal et al. (2017). Reprinted with permission of Elsevier.

recorded the electroluminescence spectrum under a forward bias current of 500 mA, and their results conclude that the $Ho^{3+}/Yb^{3+}/Zn^{2+}$:Gd_2O_3 nanoparticles could be used in NIR chip-based LED bulb fabrication.

Mahata and his group have reported the applicability of Er^{3+}/Yb^{3+}:$BaTiO_3$ and $Er^{3+}/Yb^{3+}/Zn^{2+}$:$BaTiO_3$ nanophosphors in field emission display (FEDs) applications (Mahata et al. 2015). They studied the cathodoluminescence spectra of both the nanophosphors as shown in Figure 13.5. The cathodoluminescence spectra were recorded at low-energy electron beam excitation (5 keV accelerating voltage and 1.08 μA filament current). The presence of Zn^{2+} ions improves the cathodoluminescence intensity and therefore makes $Er^{3+}/Yb^{3+}/Zn^{2+}$:$BaTiO_3$ nanophosphor more promising in FED application. Wang and his group have reported a new strategy to fabricate upconverted white light-emitting diodes (WLEDs) under NIR excitation (Wang et al. 2021). For this purpose, they used $NaYF_4$:Tm/Yb@$NaYF_4$ core–shell-structured UCNPs and encapsulated into the ZIF-8 thin film along with 4-(dicyanomethylene)-2-methyl-6-(*p*-dimethyla-minostyryl)-4*H*-pyran (DCM) and coumarin derivatives (C_6). They have calculated the Commission Internationale de l'Elcairage (CIE) coordinates, color-correlated temperature (CCT), color rendering index (CRI), and the quantum yield of the optimized film, and the values were found to be (0.33, 0.33), 5567 K, 93, and 1.25×10^{-2}%, respectively. The fabricated WLED device emits bright white light with a luminous efficiency of 126 lm/W.

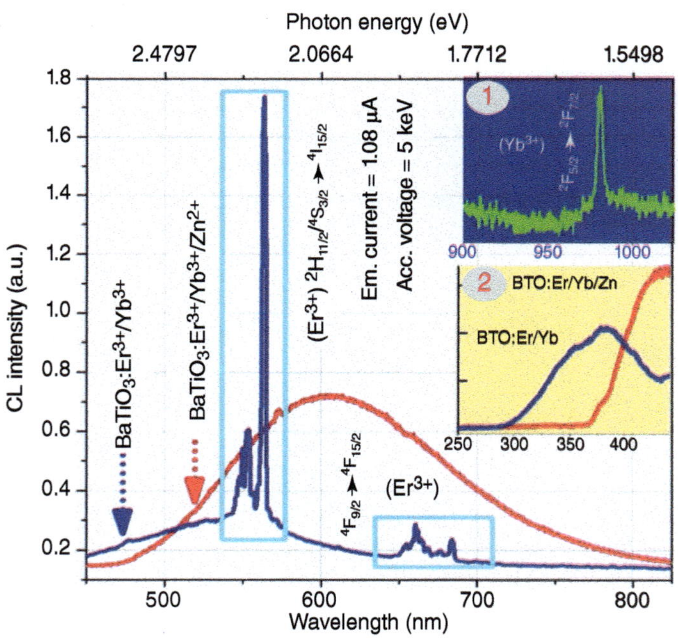

Figure 13.5 Cathodoluminescence spectra of $BaTiO_3$:Er^{3+}/Yb^{3+} and $BaTiO_3$:$Er^{3+}/Yb^{3+}/Zn^{2+}$ phosphors; inset (1) is the emission of Yb^{3+} ions of $BaTiO_3$:Er^{3+}/Yb^{3+} phosphor and inset (2) is the comparative spectra in the lower wavelength region. CL, cathodoluminescence; BTO, $BaTiO_3$; A_{cc}, voltage (accelerating voltage). Source: Mahata et al. (2015). Reprinted with permission of Royal Society of Chemistry.

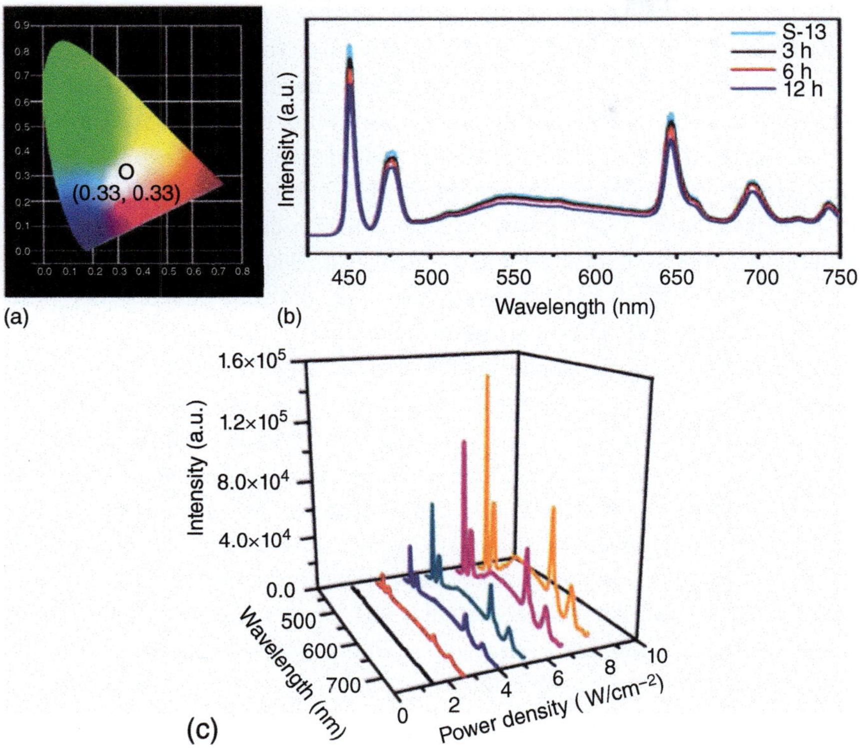

Figure 13.6 (a) The CIE chromaticity diagram associated with the emission color of the UCNP/DCM/C-6@ZIF-8 thin film under 980 nm excitation. (b) The PL stability of the UCNP/DCM/C6@ZIF-8 thin film treated at 80 °C under 980 nm laser excitation. (c) PL spectra of the UCNP/DCM/C6@ZIF-8 thin film excited at 980 nm with different power densities. Source: Wang et al. (2021). Reprinted with permission of Elsevier.

The PL stability along with the chromaticity diagram and the pump power-dependent PL spectra of the designed material under 980 nm excitation has been shown in Figure 13.6. Meng et al. have reported the white light emission from core@shell@shell-structured $NaYbF_4:Er^{3+}$@$NaYbF_4:Tm^{3+}$@$NaYF_4$ upconverting nanoparticles (Meng et al. 2020). The simultaneous emission of red, green, and blue light under NIR excitation made it possible to realize resultant white light emission. They reported the warm white light emission with excitation 20.9 W/cm^2. They prepared the white emission LED by encapsulating the UCNPs on a NIR chip.

Zhang et al. have designed the $NaGdF_4$:Yb/Tm/Er@$NaGdF_4$:Eu@$NaYF_4$ nanostructure in order to generate white light by upconversion under 980 nm laser excitation (Zhang et al. 2015). The set of selected lanthanides enables red–green–blue (RGB)-based white light emission. With the variation of laser power density, the tuning in emitted color was observed. With the variation in excitation power density from 3 to 30 W/cm^2, the emitted color changed from green to cyan to white. The nanocrystal was found to emit very bright white light with an excitation power density of 15 W/cm^2. They have presented a prototype concept of a flat-panel display in which the synthesized nanocrystal coating was used as a substrate. A display was fabricated, which consisted an array of colored dots (nanocrystals were loaded into etched microwells on a piece of silicon wafer). They further used the colloidal nanocrystal as colorless ink, and different color motifs were appeared on paper under different excitation densities. Chen and his colleagues investigated the applicability of Na(Yb/Y)F_4 nanocrystals in SSL purposes (Chen et al. 2019). They have introduced a prototype consisted of Er/Tm-doped monolithic glasses containing Na(Yb/Y)F_4 nanocrystals and coupled with 980 nm laser. They reported bright upconverted white light emission from their designed system. They have calculated the luminous efficiency and the energy efficiency associated with the bright white light emission, which were found to be 1 lm/W and 0.5%, respectively. Mahalingam and his group reported bright white light upconversion emission from $Tm^{3+}/Yb^{3+}/Er^{3+}$-doped $Lu_3Ga_5O_{12}$ nanocrystals under 980 nm excitation (Mahalingam et al. 2008). The white light emission from their synthesized material was the resultant of the RGB emission. The material is able to emit bright white light with the excitation power less than 3.4 W/cm^2. The calculated CIE coordinates found to fall within the standard white region. Furthermore, the emitted color found to be almost invariant with excitation power. These two characteristics along with the thermal stability associated with lutetium garnet make the material a potential candidate for making WLEDs.

References

Atabaev, T. S., Molkenova, A. (2019). Upconversion optical nanomaterials applied for photocatalysis and photovoltaics: recent advances and perspectives. *Frontiers of Materials Science* 13: 335–341. https://doi.org/10.1007/s11706-019-0482-z.

Boyer, J.C. and van Veggel, F.C.J.M. (2010). Absolute quantum yield measurements of colloidal $NaYF_4$:Er^{3+},Yb^{3+} upconverting nanoparticles. *Nanoscale* 2: 1417–1419. https://doi.org/10.1039/C0NR00253D.

Chen, G., Shen, J., Ohulchanskyy, T.Y. et al. (2012). (α-$NaYbF_4$:Tm^{3+})/CaF_2 core/shell nanoparticles with efficient near-infrared to near-infrared upconversion for high-contrast deep tissue bioimaging. *ACS Nano* 6 (9): 8280–8287. https://doi.org/10.1021/nn302972r.

Chen, G., Ågren, H., Ohulchanskyy, T. Y. et al. (2015a). Light upconverting core–shell nanostructures: nanophotonic control for emerging applications. *Chemical Society Reviews* 44: 1680–1713. https://doi.org/10.1039/C4CS00170B.

Chen, G., Damasco, J., Qiu, H. et al. (2015b). Energy-cascaded upconversion in an organic dye-sensitized core/shell fluoride nanocrystal. *Nano Letters* 15 (11): 7400–7407. https://doi.org/10.1021/acs.nanolett.5b02830.

Chen, J., Liu, Y., Mei, L. et al. (2015c). Crystal structure and temperature-dependent luminescence characteristics of $KMg_4(PO_4)_3$:Eu^{2+} phosphor for white light-emitting diodes. *Scientific Reports* 5: 9673. https://doi.org/10.1038/srep09673.

Chen, X., Jin, L., Kong, W. et al. (2016). Confining energy migration in upconversion nanoparticles towards deep ultraviolet lasing. *Nature Communications* 7 (10304): 1. https://doi.org/10.1038/ncomms10304.

Chen, X., Jin, L.M., Sun, T.Y. et al. (2017). Energy migration upconversion in Ce(III)-doped heterogeneous core–shell–shell nanoparticles. *Small* 13 (43): 1701479. https://doi.org/10.1002/smll.201701479.

Chen, J., Wang, S., Li, S. et al. (2019). Yttrium-dopants-induced phase-controllable and luminescence-tunable lanthanide-doped α/β-$NaYbF_4$ nanocrystals in glass for laser-driven upconverted lighting. *Journal of the European Ceramic Society* 39 (16): 5364–5372. https://doi.org/10.1016/j.jeurceramsoc.2019.08.035.

Chen, D., Zhang, L., Liang, Y. et al. (2020). Yolk–shell structured Bi_2SiO_5:Yb^{3+},Ln^{3+} (Ln = Er, Ho, Tm) upconversion nanophosphors for optical thermometry and solid-state lighting. *CrystEngComm* 22: 4438–4448. https://doi.org/10.1039/D0CE00610F.

Chilkalwar, A.A. and Rayalu, S.S. (2018). Synergetic plasmonic and upconversion effect of the (Yb, Er) NYF–TiO_2/Au composite for photocatalytic hydrogen generation. *The Journal of Physical Chemistry C* 122 (46): 26307–26314. https://doi.org/10.1021/acs.jpcc.8b05480.

Corstjens, P.L.A.M., Zuiderwijk, M., Nilsson, M. et al. (2003). Lateral-flow and up-converting phosphor reporters to detect single-stranded nucleic acids in a sandwich-hybridization assay. *Analytical Biochemistry* 312 (2): 191–200. https://doi.org/10.1016/s0003-2697(02)00505-5.

Dexter, D.L. and Schulman, J.H. (1954). Theory of concentration quenching in inorganic phosphors. *The Journal of Chemical Physics* 22: 1063. https://doi.org/10.1063/1.1740265.

Dey, R. (2015). Some rare earth doped upconverting phosphors and their multiutility. PhD thesis. Indian School of Mines Dhanbad, India.

Dey, R. and Rai, V.K. (2014). Yb^{3+} sensitized Er^{3+} doped La_2O_3 phosphor in temperature sensors and display devices. *Dalton Transactions* 43: 111–118. https://doi.org/10.1039/C3DT51773J.

Dey, R., Rai, V.K., and Pandey, A. (2012). Green upconversion emission in Nd^{3+}–Yb^{3+}–Zn^{2+}:Y_2O_3 phosphor. *Spectrochimica Acta Part A: Molecular and Biomolecular Spectroscopy* 99: 288–291. https://doi.org/10.1016/j.saa.2012.09.001.

Dey, R., Pandey, A., and Rai, V.K. (2014). The Er^{3+}–Yb^{3+} codoped La_2O_3 phosphor in finger print detection and optical heating. *Spectrochimica Acta Part A: Molecular and Biomolecular Spectroscopy* 128: 508–513. https://doi.org/10.1016/j.saa.2014.02.175.

Dey, R., Kumari, A., Soni, A.K. et al. (2015). $CaMoO_4$:Ho^{3+}–Yb^{3+}–Mg^{2+} upconverting phosphor for application in lighting devices and optical temperature sensing. *Sensors and Actuators B: Chemical* 210: 581–588. https://doi.org/10.1016/j.snb.2015.01.007.

Du, P., Kim, E., and Yu, J.S. (2018). Local symmetry distortion-induced enhancement of upconversion luminescence in Gd_2O_3:Ho^{3+}/Yb^{3+}/Zn^{2+} nanoparticles for solid-state lighting and bioimaging. *Current Applied Physics* 18 (3): 310–316. https://doi.org/10.1016/j.cap.2018.01.004.

Feng, W., Zhu, X., and Li, F. (2013). Recent advances in the optimization and functionalization of upconversion nanomaterials for in vivo bioapplications. *NPG Asia Materials* 5: e75. https://doi.org/10.1038/am.2013.63.

Fernandez-Bravo, A., Yao, K., Barnard, E.S. et al. (2018). Continuous-wave upconverting nanoparticle microlasers. *Nature Nanotechnology* 13: 572–577. https://doi.org/10.1038/s41565-018-0161-8.

Fernandez-Bravo, A., Wang, D., Barnard, E.S. et al. (2019). Ultra-threshold, continuous-wave upconverting lasing from subwavelength plasmons. *Nature Materials* 18: 1172–1176. https://doi.org/10.1038/s41563-019-0482-5.

Fischer, S., Johnson, N.J.J., Pichaandi, J. et al. (2015). Upconverting core-shell nanocrystals with high quantum yield under low irradiance: on the role of isotropic and thick shells. *Journal of Applied Physics* 118: 193105. https://doi.org/10.1063/1.4936119.

Fischer, S., Bronstein, N.D., Swabeck, J.K. et al. (2016). Precise tuning of surface quenching for luminescence enhancement in core–shell lanthanide-doped nanocrystals. *Nano Letters* 16 (11): 7241–7247. https://doi.org/10.1021/acs.nanolett.6b03683.

Guo, Q., Wu, J., Yang, Y. et al. (2019). High performance perovskite solar cells based on β-$NaYF_4$:Yb^{3+}/Er^{3+}/Sc^{3+}@$NaYF_4$ core-shell upconversion nanoparticles. *Journal of Power Sources* 426: 178–187. https://doi.org/10.1016/j.jpowsour.2019.04.039.

Haase, M. and Schäfer, H. (2011). Upconverting nanoparticles. *Angewandte Chemie International Edition* 50 (26): 5808–5829. https://doi.org/10.1002/anie.201005159.

Huang, X. (2015). Giant enhancement of upconversion emission in ($NaYF_4$:Nd^{3+}/Yb^{3+}/Ho^{3+})/($NaYF_4$:Nd^{3+}/Yb^{3+}) core/shell nanoparticles excited at 808 nm. *Optics Letters* 40 (15): 3599–3602. https://doi.org/10.1364/OL.40.003599.

Huang, P., Zheng, W., Zhou, S. et al. (2014). Lanthanide-doped $LiLuF_4$ upconversion nanoprobes for the detection of disease biomarkers. *Angewandte Chemie International Edition* 53 (5): 1252–1257. https://doi.org/10.1002/anie.201309503.

Jin, L.M., Chen, X., Siu, C.K. et al. (2017). Enhancing multiphoton upconversion from $NaYF_4$:Yb/Tm@$NaYF_4$ core–shell nanoparticles via the use of laser cavity. *ACS Nano* 11: 843–849. https://doi.org/10.1021/acsnano.6b07322.

Kumar, A. and Bahadur, A. (2021). Intense green upconversion emission by photon avalanche process from Er^{3+}/Yb^{3+} co-doped $NaBi(WO_4)_2$ phosphor. *Journal of Alloys and Compounds* 857: 158196. https://doi.org/10.1016/j.jallcom.2020.158196.

Li, X., Shen, D., Yang, J. et al. (2013). Successive layer-by-layer strategy for multi-shell epitaxial growth: shell thickness and doping position dependence in upconverting optical properties. *Chemistry of Materials* 25 (1): 106–112. https://doi.org/10.1021/cm3033498.

Li, X., Zhang, F., and Zhao, D. (2014). Lab on upconversion nanoparticles: optical properties and applications engineering via designed nanostructure. *Chemical Society Reviews* 44: 1346–1378. https://doi.org/10.1039/C4CS00163J.

Liang, G., Wang, H., Shi, H. et al. (2020). Recent progress in the development of upconversion nanomaterials in bioimaging and disease treatment. *Journal of Nanobiotechnology* 18: 154. https://doi.org/10.1186/s12951-020-00713-3.

Lv, W., Jia, Y., Zhao, Q. et al. (2014). Synthesis, structure and luminescence properties of $K_2Ba_7Si_{16}O_{40}$:Eu^{2+} for white light-emitting diodes. *The Journal of Physical Chemistry C* 118: 4649–4655. https://doi.org/10.1021/jp500662a.

Mahalingam, V., Mangiarini, F., Vetrone, F. et al. (2008). Bright white upconversion emission from Tm^{3+}/Yb^{3+}/Er^{3+}-doped $Lu_3Ga_5O_{12}$ nanocrystals. *The Journal of Physical Chemistry C* 112 (46): 17745–17749. https://doi.org/10.1021/jp8076479.

Mahata, M.K., Koppe, T., Mondal, T. et al. (2015). Incorporation of Zn^{2+} ions into $BaTiO_3$:Er^{3+}/Yb^{3+} nanophosphor: an effective way to enhance upconversion, defect luminescence and temperature sensing. *Physical Chemistry Chemical Physics* 17: 20741–20753. https://doi.org/10.1039/C5CP01874A.

Meng, Z., Zhang, S., Wu, S. (2020). Power density dependent upconversion properties of $NaYbF_4$:Er^{3+}@$NaYbF_4$:Tm^{3+}@$NaYF_4$ nanoparticles and their application in white-light emission LED. *Journal of Luminescence* 227: 117566. https://doi.org/10.1016/j.jlumin.2020.117566.

Mondal, M., Rai, V.K., and Srivastava, C. (2017). Influence of silica surface coating on optical properties of Er^{3+}–Yb^{3+}:$YMoO_4$ upconverting nanoparticles. *Chemical Engineering Journal* 327: 838–848. https://doi.org/10.1016/j.cej.2017.06.166.

Naccache, R., Vetrone, F., and Capobianco, J.A. (2013). Lanthanide-doped upconverting nanoparticles: harvesting light for solar cells. *ChemSusChem* 6 (8): 1308–1311. https://doi.org/10.1002/cssc.201300362.

Ostrowski, A.D., Chan, E.M., Gargas, D.J. et al. (2012). Controlled synthesis and single-particle imaging of bright, sub-10 nm lanthanide-doped upconverting nanocrystals. *ACS Nano* 6 (3): 2686–2692. https://doi.org/10.1021/nn3000737.

Peng, Y., Lu, Y., Ren, P. et al. (2018). Integration of nanoscale light emitters: an efficient ultraviolet and blue random lasing from $NaYF_4$:Yb/Tm hexagonal nanocrystals. *Photonics Research* 6: 943–947. https://doi.org/10.1364/PRJ.6.000943.

Punjabi, A., Wu, X., Tokatli-Apollon, A. et al. (2014). Amplifying the red-emission of upconverting nanoparticles for biocompatible clinically used prodrug-induced photodynamic therapy. *ACS Nano* 8 (10): 10621–10630. https://doi.org/10.1021/nn505051d.

Qian, H.S. and Zhang, Y. (2008). Synthesis of hexagonal-phase core–shell $NaYF_4$ nanocrystals with tunable upconversion fluorescence. *Langmuir* 24 (21): 12123–12125. https://doi.org/10.1021/la802343f.

Seng, Y., Hao, S., Yang, C. et al. (2015). Enhancing solar cell efficiency using photon upconversion materials. *Nanomaterials* 5 (4): 1782–1809. https://doi.org/10.3390/nano5041782.

Shang, Y., Zhou, J., Cai, Y. et al. (2020). Low threshold lasing emissions from a single upconversion nanocrystal. *Nature Communications* 11: 6156. https://doi.org/10.1038/s41467-020-19797-4.

Shin, J., Kyhm, J., Hong, A. et al. (2018). Multicolor tunable upconversion luminescence from sensitized seed-mediated grown $LiGdF_4$:Yb,Tm-based core/triple–shell nanophosphors for transparent displays. *Chemistry of Materials* 30(23): 8457–8464. https://doi.org/10.1021/acs.chemmater.8b02497.

Song, Z., Anissimov, Y.G., Zhao, J. et al. (2012). Background free imaging of upconversion nanoparticle distribution in human skin. *Journal of Biomedical Optics* 18 (6): 061215. https://doi.org/10.1117/1.JBO.18.6.061215.

Soni, A., Dey, R., and Rai, V.K. (2015). Stark sublevels in Tm^{3+}–Yb^{3+} codoped $Na_2Y_2B_2O_7$ nanophosphor for multifunctional applications. *RSC Advances* 5: 34999–35009. https://doi.org/10.1039/C4RA15891A.

Tian, G., Gu, Z., Zhou, L. et al. (2012). Mn^{2+} dopant-controlled synthesis of $NaYF_4$:Yb/Er upconversion nanoparticles for in vivo imaging and drug delivery. *Advanced Materials* 24 (9): 1226–1231. https://doi.org/10.1002/adma.201104741.

Tiwari, S. and Yakhmi, J.V. (2015). Recent advances in luminescent nanomaterials for solid state lighting applications. *Defect and Diffusion Forum* 361: 15–68. https://doi .org/10.4028/www.scientific.net/DDF.361.15.

Tiwari, S.P., Maurya, S.K., Yadav, R.S. et al. (2018). Future prospects of fluoride based upconversion nanoparticles for emerging applications in biomedical and energy harvesting. *Journal of Vacuum Science & Technology B* 36: 060801. https://doi.org/10 .1116/1.5044596.

Tu, L., Liu, X., Wu, F. et al. (2015). Excitation energy migration dynamics in upconversion nanomaterials. *Chemical Society Reviews* 44: 1331–1345. https://doi .org/10.1039/C4CS00168K.

Wang, F., Han, Y., Lim, C.S. et al. (2010a). Simultaneous phase and size control of upconversion nanocrystals through lanthanide doping. *Nature* 463: 1061–1065. https://doi.org/10.1038/nature08777.

Wang, F., Wang, J., and Liu, X. (2010b). Direct evidence of a surface quenching effect on size-dependent luminescence of upconversion nanoparticles. *Angewandte Chemie International Edition* 49 (41): 7456–7460. https://doi.org/10.1002/anie.201003959.

Wang, J., Deng, R., MacDonald, M.A. et al. (2014). Enhancing multiphoton upconversion through energy clustering at sublattice level. *Nature Materials* 13: 157–162. https://doi.org/10.1038/nmat3804.

Wang, T., Yu, H., Siu, C.K. et al. (2017). White-light whispering-gallery-mode lasing from lanthanide-doped upconversion $NaYF_4$ hexagonal microrods. *ACS Photonics* 4 (6): 1539–1543. https://doi.org/10.1021/acsphotonics.7b00301.

Wang, T., Siu, C.K., Yu, H. et al. (2018). Influence of plasmonic effect on the upconversion emission characteristics of $NaYF_4$ hexagonal microrods. *Inorganic Chemistry* 57 (14): 8200–8204. https://doi.org/10.1021/acs.inorgchem.8b00690.

Wang, X., Li, C., Li, Z. et al. (2021). Near-infrared-light emitting diode driven white light emission: upconversion nanoparticles decorated metal–organic frame-works thin film. *Chemical Engineering Journal* 409: 128220. https://doi.org/10.1016/j.cej.2020 .128220.

Wisser, M.D., Fischer, S., Maurer, P.C. et al. (2016). Enhancing quantum yield via local symmetry distortion in lanthanide-based upconverting nanoparticles. *ACS Photonics* 3 (8): 1523–1530. https://doi.org/10.1021/acsphotonics.6b00166.

Xie, R.J. and Hirosaki, N. (2007). Silicon based oxynitride and nitride phosphors for white LEDs – a review. *Science and Technology of Advanced Materials* 8 (7, 8): 588–600. https://doi.org/10.1016/j.stam.2007.08.005.

Xu, C.T., Svenmarker, P., Liu, H. et al. (2012). High-resolution fluorescence diffuse optical tomography developed with nonlinear upconverting nanoparticles. *ACS Nano* 6 (6): 4788–4795. https://doi.org/10.1021/nn3015807.

Yang, W., Li, X., Chi, D. et al. (2014). Lanthanide-doped upconversion materials: emerging applications for photovoltaics and photocatalysis. *Nanotechnology* 25 (48): 482001. https://doi:10.1088/0957-4484/25/48/482001.

Yi, G.S. and Chow, G.M. (2007). Water-soluble $NaYF_4$:Yb,Er(Tm)/$NaYF_4$/polymer core/shell/shell nanoparticles with significant enhancement of upconversion fluorescence. *Chemistry of Materials* 19 (3): 341–343. https://doi.org/10.1021/cm062447y.

Zeng, Q., Xue, B., Zhang, Y. et al. (2013). Facile synthesis of $NaYF_4$:Yb, Ln/$NaYF_4$:Yb core/shell upconversion nanoparticles via successive ion layer adsorption and one-pot reaction technique. *CrystEngComm* 15:4765–4772. https://doi.org/10.1039/C3CE40216A.

Zhan, Q., Qian, J., Liang, H. et al. (2011). Using 915 nm laser excited Tm^{3+}/Er^{3+}/Ho^{3+}-doped $NaYbF_4$ upconversion nanoparticles for in vitro and deeper in vivo bioimaging without overheating irradiation. *ACS Nano* 5 (5): 3744–3757. https://doi.org/10.1021/nn200110j.

Zhang, Q.F., Dandeneau, C.S., Zhou, X.Y. et al. (2009). ZnO nanostructure for dye-sensitized solar cells. *Advanced Materials* 21 (41): 4087–4108. https://doi.org/10.1002/adma.200803827.

Zhang, Y., Wei, W., Das, G.K. et al. (2014). Engineering lanthanide-based materials for nanomedicine. *Journal of Photochemistry and Photobiology C: Photochemistry Reviews* 20: 71–96. https://doi.org/10.1016/j.jphotochemrev.2014.06.001.

Zhang, C., Yang, L., Zhao, J. et al. (2015). White-light emission from an integrated upconversion nanostructure: toward multicolor displays modulated by laser power. *Angewandte Chemie International Edition* 54: 11531–11535. https://doi.org/10.1002/anie.201504518.

Zhou, Y., Wu, S., Wang, F. et al. (2020). Assessing the toxicity in vitro of degradation products from deoxynivalenol photocatalytic degradation by using upconversion nanoparticles@TiO_2 composite. *Chemosphere* 238: 124648. https://doi.org/10.1016/j.chemosphere.2019.124648.

Zhu, H., Chen, X., Jin, L.M. et al. (2013). Amplified spontaneous emission and lasing from lanthanide-doped up-conversion nanocrystals. *ACS Nano* 7: 11420–11426. https://doi.org/10.1021/nn405387t.

Zhu, H., Chen, A.Q., Wu, Y.Y. et al. (2018). Seven-photon-excited upconversion lasing at room temperature. *Advanced Optical Materials* 6 (16): 1800518. https://doi.org/10.1002/adom.201800518.

14

Upconversion Nanoparticles in pH Sensing Applications

Manoj Kumar Mahata[1,3], Ranjit De[2,3], and Kang Taek Lee[3]*

[1]*Georg-August-Universität Göttingen, III.Physical Institute- Biophysics, Friedrich-Hund-Platz 1, 37077 Göttingen, Germany*
[2]*Pohang University of Science and Technology (POSTECH), Department of Life Sciences, 77 Cheongam-ro, Nam-gu, Pohang 37673, South Korea*
[3]*Gwangju Institute of Science and Technology (GIST), Department of Chemistry, 123 Cheomdan-gwagiro, Gwangju 61005, South Korea*

14.1 Introduction

Lanthanide (Ln)-doped upconversion nanoparticles (UCNPs) can convert low-energy infrared (IR) photons into visible or ultraviolet (UV) photons through successive absorption. These UCNPs exhibit many advantages and may overcome the issues of quantum dots and fluorescent dyes (Wang et al. 2010; Mahata et al. 2020). The UCNPs have an extensive field of applications, including solar cell concentrators, 3-D volumetric displays, bioimaging, gas sensing, temperature sensing, pH sensing, biomolecule sensing, etc. (Dong et al. 2015; Mahata and Lee 2019; Mahata et al. 2016a,b, 2017, 2021). Photon upconversion emission is usually obtained from the rare-earth (RE)-doped crystal host materials. Recently, stochastic photon emission in micro- and nanosecond scales in UCNPs has also been observed (Lee et al. 2017). The upconversion emission is obtained with low-power, readily available lasers at low cost and benefits over a second harmonic generation or simultaneous two-photon absorption attained from expensive high-intense laser sources.

On the other hand, optical sensing and imaging are valuable tools for the characterization of biological samples. Recently, many optical detections and sensing systems have been developed, which are applicable in a biological environment. For instance, fluorescent probes can be employed to know the localization of proteins and to monitor biological processes (Ong et al. 2014; Drees et al. 2016). The quantum dots could also be used as probes for in vivo biological imaging and sensing (Michalet et al. 2005; Schäferling 2016), but they cannot meet the challenges in terms of penetration depth in the target samples. The other drawbacks of common fluorescent probes are high autofluorescence, low intensity, high photobleaching, light scattering, etc. Monitoring pH and temperature in real time within living cells are

* Manoj Kumar Mahata is the main corresponding author.

Upconverting Nanoparticles: From Fundamentals to Applications, First Edition. Edited by Vineet K. Rai.

demanded to understand cellular dynamics. However, measuring concentrations of carbon dioxide, oxygen, and ammonia generated through biochemical reactions are also important in biomedical research and diagnostics (Mader and Wolfbeis 2010). Thus, developing nanomaterials with multifunctional features for clinical research is urgently required.

Intracellular pH quantification with high precision is certainly needed in biomedical research. In recent years, drug delivery systems based on pH-sensitive nanomaterials with surface ligands attached to the nanoparticles are being developed (Liu et al. 2014). However, we still do not know what happens after the cellular internalization of the nanoparticles. Few studies consider that the pH-based drug delivery system elapses in the acidic compartment, although the complete process is not confirmed yet. It is worth mentioning that the surface modification of the nanoparticles has a pivotal role in cellular internalization via the plasma membrane (Sedlmeier and Gorris 2015; Gu et al. 2016). For imaging and sensing pH, several chemical and optical probes have been reported in the past two decades. Knowing the pH value of the cytoplasm is necessary for successful therapy, and pH monitoring is useful for understanding phagocytosis and endocytosis processes. On the other hand, the pH of blood should be known to understand respiratory acidosis/alkalosis, where excess carbon dioxide decreases the blood pH below 7.35. Apart from those, cell cycle progression and cell death are related to the pH change (Marches et al. 2001). Thus, the determination of pH is essential for better understanding the cellular processes.

So far, some optical systems have been reported for pH sensing (Guo et al. 2014; Kim et al. 2013; Lee et al. 2015; Yin et al. 2015). In the ratiometric method, the pH sensors work based on two fluorescence emissions, and their main benefits are minimum dependence on the experimental environment and conditions. In general, various pH-sensitive dyes are attached with the nanoparticles, such as silica (core–cell), quantum dots (CdSe/ZnSe/ZnS), latex polymer dots, etc. (Burns et al. 2006; Liu et al. 2007; Hornig et al. 2008). However, they are not suitable because of their excitation with UV light, and molecular dyes have poor photostability. These issues can be solved by utilizing UCNPs, which are favorable to be applied in biological environments (Wang et al. 2010, 2011). The UCNPs' excitation wavelength is 700–1000 nm, where the absorption rate of hemoglobin, water, and other biomatters is low (Mader et al. 2010). For instance, a pH nanoprobe using $NaYF_4$:Tm^{3+}/Yb^{3+} and fluorescein (fluorescein isothiocyanate [FITC]) was developed. For ratiometric pH determination, the upconversion emissions, upon near-infrared (NIR) (980 nm) laser light excitation, in the blue (475 nm) and red (645 nm) bands were used as response and reference signals, respectively (Li et al. 2016). A pH probe bound to the UCNPs' surface was used to obtain energy transfer (resonance) from UCNPs to the dye pHrodo™ Red, which can absorb the green (550 nm) upconversion emission of $NaYF_4$:Er^{3+}/Yb^{3+} as an excitation source (Näreoja et al. 2017). These studies mostly show surface-modified UCNPs with a pH indicator. A rare report designs the sensors without pH-sensitive dyes and uses the quenching and recovery of luminescence from UCNPs prepared with some quenchers (Yan et al. 2014; Xu et al. 2017). Additionally, the thin films consisting of pH nanoprobes are required for designing the practical sensors for flowing liquids. In a recent work, to detect pH in implanted medical devices, a pH sensor was developed on the basis of interaction of UCNPs and Bromocresol green-doped silica (Wang et al. 2014a). However, the response time, reversibility, and testing of the molecular pH sensors are important issues.

Recently, developments in optical pH sensors with various luminescent nanomaterials except UCNPs have been summarized by Shi et al. (2014), Schäferling (2016), and Wencel et al. (2014). This chapter presents a comprehensive overview of the fundamental aspects of upconversion phenomena and its different mechanisms followed by the various approaches employed to manipulate the upconversion emissions. This chapter will also help the readers to perform modeling and understanding the optical properties of upconversion nanomaterials and their recent advances in pH sensing capability.

14.2 Basic Properties of UCNPs

The concept of photon upconversion was initially adapted by Bloembergen while in 1959 (Bloembergen 1959). In the next year, F. Auzel noticed IR to visible conversion, and the mechanism was illustrated for Yb^{3+}-doped Tm^{3+} and Er^{3+} systems. In 2004, a detailed review of the upconversion mechanism was put forward (Auzel 2004). However, at that time, the future applications of UCNPs were not predicted because of the lack of control over the particle's morphology. With progress in nanotechnology in recent years, upconversion is being reinvestigated. The RE elements with similar configuration ([Xe] $4f^{n-1}5d^{0-1}6s^2$) allows many electronic transitions. The RE^{3+} ions in which 4f electrons are shielded by the filled $5s^25p^6$ subshells make them less active in the host lattice. Moreover, they show longer lifetimes, sharp emission bands, and large anti-Stokes shifts.

UCNPs consist of inorganic materials doped with low concentration of RE ions. Here, careful choosing of the REs is important to achieve energy transfer. The upconversion efficiency varies with host compounds mainly because of their coordination, dopant distribution, and energy transfer (Schäfer et al. 2007, 2009). Low-phonon energy materials are suitable in this respect because of less non-radiative rates. Fluoride materials ($NaXF_4$; X = Y, Gd, and La) have low phonon energies, which make them suitable hosts for RE doping. However, the chemical and thermal stabilities of these UCNPs are also important (Haase and Schäfer 2011; Mahata et al. 2016b). The hexagonal $NaXF_4$ crystals are known to show efficient upconversion than their cubic ones. The Laporte forbidden intra-4f transitions are increased in a less-symmetrical crystal field around REs in hexagonal form, and finally, 4f electrons' coupling increases. Some other examples of other effective host matrices are vanadates ($GdVO_4$, $LaVO_4$, and YVO_4), phosphates (YPO_4, $GdPO_4$, and $LuPO_4$), oxides (ZrO_2 and Y_2O_3), etc. (Mahata et al. 2016b).

The RE environment inside the host matrix has a reasonable impact for monitoring the luminescence properties. For the construction of efficient UCNPs, lesser non-radiative decay, even distribution of REs, long-lived excited states, and population of the excited states are prerequisites. It is to be mentioned that upconversion emissions may be decreased by quenching processes that reduce the number of excited states. Non-radiative transitions are involved with the impurities and neighboring RE ions in the locality. The ground and excited states are bridged by phonons when the mediation of the energy gap is possible by five to six phonons (Vennerberg and Lin 2011; Gulzar et al. 2017). In the case of a higher number of the required phonons, the radiative luminescence mechanism becomes dominant. The UCNPs with Y^{3+}, Na^{+}, and Ca^{2+} as core elements are usually chosen because of their closer ionic radii to the RE; thus, it creates lesser lattice defects and strains without compromising for the efficient upconversion (UC) emission (Chen et al. 2007; Wang et al. 2009; Zhang et al. 2010).

To obtain high upconversion efficiency, the activator and sensitizer ions have very important roles as luminescent centers in the UCNPs. The ladder-like energy levels of RE^{3+} ions are useful to cover a wide spectrum of luminescence. The Er^{3+}, Ho^{3+}, and Tm^{3+} ions are mostly used as dopants to achieve good emission intensity desired for light emission in the electromagnetic spectrum. Various UCNPs (Heer et al. 2004; Liu and Chen 2007; Wang et al. 2007; Naccache et al. 2009; Mahata et al. 2014a, Mahata et al. 2013, 2014b, 2015a,b,c; Sinha et al. 2016, 2017a,b; Soni et al. 2017) with their upconversion emission bands upon NIR light excitation are presented in Table 14.1. The energy levels of Er^{3+} ($^4I_{15/2}$, $^4I_{13/2}$, $^4I_{11/2}$, $^4F_{9/2}$, and $^2H_{11/2}/^4S_{3/2}$), Ho^{3+} ($^5I_8$, $^5I_7$, $^5I_6$, $^5F_5$, $^5F_4$, and $^5S_2$), and Tm^{3+} ($^3H_6$, $^3F_4$, $^3H_4$, $^1G_4$, and $^1D_2$ and $^5I_8$, $^5I_7$, $^5I_6$, $^5F_5$, $^5F_4$, and $^5S_2$) are suitable for excitation with 980 nm or 800 nm laser light. The Dy^{3+}, Pr^{3+}, Tb^{3+}, and Eu^{3+} ions are sometimes used as activators in designing UCNPs (Liu 2015). The concentrations of activator ions are also important, and the optimized amount of these ions is chosen for which the upconversion emission is the highest. Beyond that, the upconversion emission decreases because of the looping of energy transfer within the activators themselves. In this situation, the

Table 14.1 Emission bands of various upconversion materials.

Host material	Dopant ions	Major emission bands			References
		Blue	Green	Red	
$CaMoO_4$	Yb^{3+}/Er^{3+}	475	530, 552	656	Sinha et al. (2017a)
$NaYF_4$	Yb^{3+}/Tm^{3+}	460, 477		650	Heer et al. (2004)
$NaGdF_4$	Ho^{3+}/Yb^{3+}	487	541	647, 751	Naccache et al. (2009)
Y_2O_3	Yb^{3+}/Er^{3+}	487	522, 554	664	Mahata et al. (2013)
LaF_3	Yb^{3+}/Tm^{3+}	475		800	Liu and Chen (2007)
LaF_3	Yb^{3+}/Er^{3+}		521, 545	660	Wang et al. (2007)
$Na_2Y_2B_2O_7$	Yb^{3+}/Er^{3+}	476, 488	525, 550	660	Soni et al. (2017)
YVO_4	Yb^{3+}/Er^{3+}		525, 554	661	Mahata et al. (2014b, 2015b)
YVO_4	Yb^{3+}/Ho^{3+}		538, 548	655	Mahata et al. (2015b)
$BaTiO_3$	Yb^{3+}/Er^{3+}	476	563	650	Mahata et al. (2014b)
$BaTiO_3$	Yb^{3+}/Ho^{3+}		538, 548	655	Mahata et al. (2020)
$Gd_2Mo_3O_9$	Yb^{3+}/Er^{3+}		527, 547	660	Sinha et al. (2016)

sensitizer ions are introduced for transferring energy to the activator ions to enhance the upconversion under NIR laser light excitation. However, most activators possess low absorption at the NIR excitation wavelength, giving low upconversion efficiency. With the doping of sensitizer ions, the efficiency of the upconversion systems can be increased by several folds. In this regard, the Yb^{3+} ion is a well-known sensitizer for energy transfer under 980 nm for the Er^{3+}, Eu^{3+}, Ho^{3+}, Tm^{3+}, etc., activators (Chen et al. 2015). Non-RE ions also sometimes tailor the luminescence properties; for example, Zn^{2+}, K^{+}, and Na^{+} have shown enhanced upconversion in Y_2O_3, $BaTiO_3$, $CaMoO_4$, etc., hosts (Pandey and Rai 2013; Mahata et al. 2015c; Sinha et al. 2017b). Surface modification of the UCNPs with polymeric macromolecules is also important to biomedical applications of UCNPs (Näreoja et al. 2017). Therefore, several parameters such as choosing activators, sensitizers, their concentrations, and surface modification should be considered for boosting the upconversion processes.

The five basic upconversion processes (Sun et al. 2015; Zheng et al. 2015; Mahata et al. 2016b; Nadort et al. 2016), viz, excited-state absorption (ESA), energy transfer upconversion (ETU), photon avalanche (PA), cooperative upconversion (CU), and energy migration-mediated upconversion (EMU), are depicted in Figure 14.1.

ESA occurs because of the absorption of one or more photons, sequentially from the ground state to the excited state (Figure 14.1a). The ETU process (Figure 14.1b) occurs as an energy transfer between two (similar or dissimilar) types of luminescent centers (sensitizer and activator), and it is more efficient than ESA. After the excitation of sensitizer ions, the energy is transferred to the nearby activators, and upconversion emission is obtained during the relaxation of the activator ions to lower excited states or ground states. To achieve a large amount of energy transfer between the activator and sensitizer, they should have resonant energy levels, and thus, the slower excitation time enhances the chance of upconversion events (Sun et al. 2015; Nadort et al. 2016). The PA (Figure 14.1c) takes place after a certain pump power density, and it is a complex process. At very high pump density, the intermediate energy levels of some REs get populated, and the population of the upconversion emitting level occurs through ETU or ESA from other excited ions. Following that, cross-relaxation occurs within the ions at the ground and excited states. Thus, the population of the intermediate reservoir and upconversion

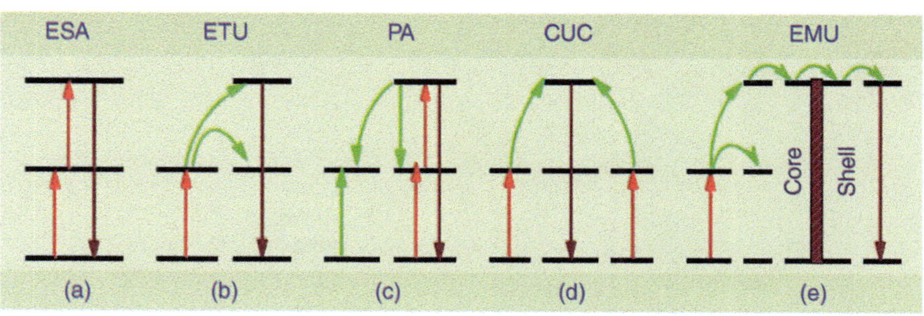

Figure 14.1 (a) excited state absorption, (b) energy transfer upconversion, (c) photon avalanche, (d) cooperative upconversion and (e) energy migration-mediated upconversion. Source: Mahata et al. (2016b).

emitting states increases, resulting in an "avalanche" of generating many excited ions via feedback looping (Zheng et al. 2015). The CU (Figure 14.1d) is primarily observed in the Yb^{3+}/RE^{3+} pairs. Through cooperative energy transfer by two nearby sensitizer ions, the upconversion emitting level of activators is populated. The EMU (Figure 14.1e) involves a set of four types of ions: sensitizer, accumulator, migrator, and activator. The activator and sensitizer/accumulator remain at separate layers and are connected by migrators. The sensitizer raises the accumulator to higher levels through energy transfer. The excited accumulator then migrates the energy to the migrators through the core–shell interface of UCNPs. Finally, the migrated energy is trapped by the activator in the shell, and upconversion emission occurs.

14.3 Working Principle of UCNP-Based pH Sensor

UCNPs have a broad range of emission wavelengths, and they are considered as good candidates to facilitate self-referenced and reversible pH sensing. Among the developed pH sensors, a few were similar to the pH sensing membranes, and others are nanohybrids or nanoparticles. The pH-sensitive molecular dyes absorb one of the upconversion emission bands, and pH is determined by considering the variation of absolute luminescence or ratiometric intensity change of emission in the bands. A few numbers of pH sensors have been developed using UCNPs and quenchers without including pH-sensitive dyes. In this case, pH is measured based on the variation of UC intensity with pH. The schematic presentation of the working principle of upconversion-based pH nanoprobes is shown in Figure 14.2.

The pH of a solution is defined as the negative logarithm of hydronium ions (H_3O^+).

$$\mathrm{pH} = -\log{[\mathrm{H_3O}]^+} f_{h^+}$$

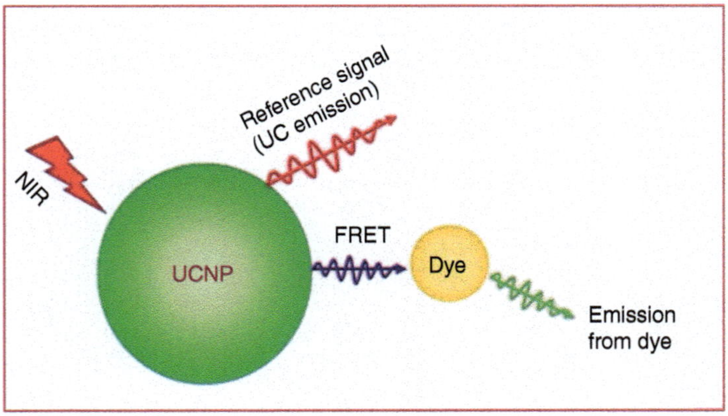

Figure 14.2 Schematic diagram of an upconversion-sensitized pH sensor. The pH-sensitive dye emission is compared with the upconversion emission. The FRET refers to the Förster resonance energy transfer from the UCNP to the pH-sensitive dye.

where $f_h{}^+$ refers to the activity coefficient, and it is close to unity for an ideal solution. pH is measured according to the Henderson–Hasselbalch equation

$$\mathrm{pH} = \mathrm{p}K_\mathrm{a} + \log \frac{[\mathrm{A}^{C-1}]}{[\mathrm{HA}^{C}]}$$

where $\mathrm{p}K_\mathrm{a}$ is the negative logarithmic of the acidic dissociation constant K_a. $[\mathrm{HA}^C]$ and $[\mathrm{A}^{C-1}]$ are the concentrations of acidic and basic forms of the dye, respectively.

The above equation also provides the range of pH measuring of a dye. The maximum pH sensitivity is considered close to $\mathrm{p}K_\mathrm{a}$. The value of $\mathrm{p}K_\mathrm{a}$ is dependent on the interaction of the pH indicator with its environment and temperature (Wencel et al. 2014). Thus, a standard pH value is required for the pH-responsive dyes to use them in an intercellular medium. To design a highly sensitive sensor, the signal response should be prominent with a minute change in pH. The extracellular pH range is 7.35 to 7.45, and endosome, lysosome, and cytosol have a pH range within pH 4.4–6.0, 4.5–5.0, and 6.8–7.4, respectively. Therefore, the design of pH probes with high sensitivity instead of a wide range with low responsivity is more demanded (Schäferling 2016). Additionally, a standard method for intercellular pH calibration should be undertaken to calibrate the pH inside the living cells. The dual-wavelength-based optical fiber pH sensors show high precision of pH, but a similar precision is unlikely to be attained inside the cells. Moreover, the response time is a matter of concern.

14.4 Photon Upconversion-Based pH Sensing Systems

The pH sensors should have performance above ~500 nm wavelength; otherwise, some problems such as autofluorescence and absorption by biomatters may encounter (Sun et al. 2009). The available reports on UCNP-based pH sensing are tabulated in Table 14.2.

Sun et al. in 2009 introduced for the first time the UCNPs for pH sensing. Since then, rapid progress has been noticed in UCNP-based pH sensing research (Sun et al. 2009). The $NaYF_4$:Er^{3+}/Yb^{3+} nanorods with bromothymol blue, a pH-sensitive dye, were used to design the first pH sensor. Even though the UCNPs were pH insensitive, they still exhibited pH-dependent upconversion because of their energy transfer to the dye.

14.4.1 Upconversion Nanoparticles as pH Sensors

Mahata and Lee (2019) have recently designed an upconversion system (Figure 14.3) that combines $NaYF_4$:Yb^{3+}/Tm^{3+}@$NaYF_4$:Yb^{3+}@$NaYF_4$:Yb^{3+}/Nd^{3+} with fluorescein-5-isothiocyanate (FITC) and its pH sensing upon 980 and 808 nm laser light excitations (Figure 14.3). The Tm^{3+} upconversion luminescence bands in the blue region were found to vary with pH because of the absorption of blue-upconverted light by the conjugated dye molecules, while the other luminescence bands at the red or NIR regions remained unchanged. Thus, a self-referenced ratiometric sensor was designed using 474, 643, and 802 nm upconversion emissions upon

Table 14.2 Upconversion-based pH sensors with their response time, range, and method.

pH probe					
UCNPs	Indicator	Response time	pH range	Method	References
$NaYF_4:Yb^{3+}/Tm^{3+}$@$NaYF_4:Yb^{3+}$@$NaYF_4:Yb^{3+}/Nd^{3+}$	Fluorescein-5-isothiocyanate	N/A	3–8	Ratio of 474, 643, and 802 nm	Mahata and Lee (2019)
$NaYF_4:Er^{3+}/Yb^{3+}$	Bromothymol blue	N/A	2–11	Absolute intensity of 548 and 658 nm bands	Sun et al. (2009)
$Y_2O_2S:Er^{3+}/Yb^{3+}$ (film)	Bromocresol green	~ 10 min	5–10	Ratio of 671 to 661 nm	Wang et al. (2014b)
$NaYF_4:Er^{3+}/Yb^{3+}$ with graphene oxide (film)	(No dye)	1 min	5–8	Changes in emission intensity of 540 nm	Yan et al. (2014)
$NaYF_4:Er^{3+}/Yb^{3+}$ aminosilane coated	pHrodo Red	N/A	2.5–7.2	Ratio of UC emission at 540 nm to the dye emission at 590 nm	Arppe et al. (2014)
$NaYF_4:Er^{3+}/Yb^{3+}$ – PEI	pHrodo Red	N/A	5–7	Ratio of UC emission at 540 nm to the dye emission at 590 nm	Näreoja et al. (2017)
$NaYF_4:Tm^{3+}/Yb^{3+}$	Fluorescein isothiocyanate	N/A	3–7	Ratio of UC emission at 475 to 645 nm	Li et al. (2016)
$Y_2O_3:Tm^{3+}$@SiO_2	Xylenol orange	N/A	4–8	Ratio of UC emission at 450 and 646 nm	Ma et al. (2015)
$NaGd(WO_4)_2:Eu^{3+}$	Fluorescein isothiocyanate allyl amine hydrochloride	N/A	4–10	Ratio of fluorescein emission at 512 nm to RE emission at 611 nm	Laguna et al. (2017)
$NaYF_4:Er^{3+}/Yb^{3+}$	Bromothymol blue	N/A	6–8	Absolute UC emission at 650 nm	Song Wang et al. (2014b)
Commercial phosphor:Er^{3+} (film)	Neutral red	23 s	5–8.5	Ratio of green to red UC emission	Meier et al. (2014)
$NaYF_4:Er^{3+}/Yb^{3+}$@$NaYF_4$@Ni	(No dye)	N/A	5–7.4	Quenching and recovery of green and red UC bands	Xu et al. (2017)
$NaGdF_4$@$NaYF_4:Yb^{3+}/Tm^{3+}$@$NaYF_4$-PAA	Hemicyanine		6.8–9	Ratio of 650 to 513 nm	Li et al. (2017)
$NaYF_4:Er^{3+}/Yb^{3+}$(film)	ETH 5418	N/A	6–11	Ratio of UC emission bands at 656 to 542 nm	Xie et al. (2012)
$NaGdF_4:Tm^{3+}/Yb^{3+}$@$NaGdF_4:Nd^{3+}/Yb^{3+}$@$NaYF_4$-PEI	Fluorescein	N/A	4.7–7.4	Ratio of dye emission at 515 nm to UC emission at 645 nm	Du et al. (2017)

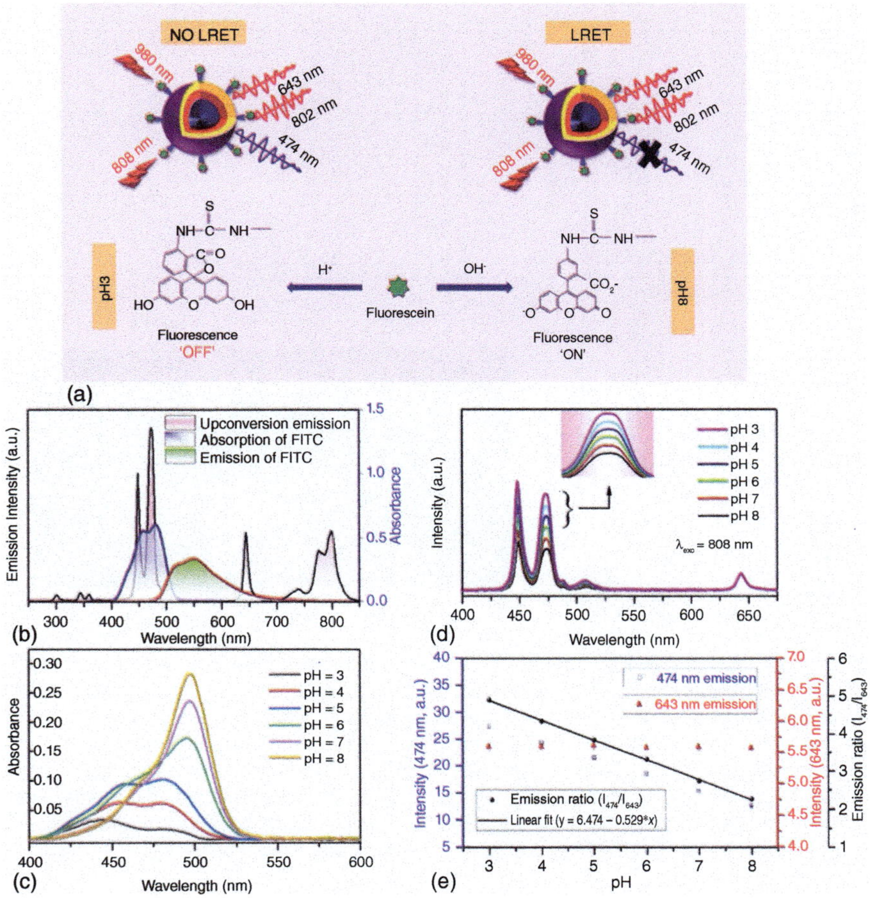

Figure 14.3 (a) Schematic description of the developed nanoprobe and its working principle. (b) Black line, upconversion emission upon 980 nm; blue line, absorption spectrum of FITC; and red line, FITC emission upon 488 nm laser excitation. (c) Absorption spectra of FITC in various pH values. (d) Variation of 808 nm excited upconversion emission spectra of FITC-conjugated core–shell–shell nanoprobes with pH; inset shows the 474 nm band. (e) Variation of 474 and 643 nm and their ratio (474–643 nm) with pH values ranging from 3 to 8. LRET = luminescence resonance energy transfer. Source: Mahata and Lee (2019). Reprinted with permission of Royal Society of Chemistry.

long-wavelength light excitations. This approach will open pH sensing in the biomedical fields (Mahata and Lee 2019).

Aminosilane-coated $NaYF_4$; Er^{3+}/Yb^{3+} coupled with a pHrodo Red indicator was used for UCNPs to dye energy transfer (Arppe et al. 2014). The upconversion emission at 550 nm upon 980 nm light excitation was transferred to excite the pHrodo Red, emitting luminescence at 590 nm. In this case, luminescence from the dye was compared with a reference upconversion signal to calibrate the pH sensor. The pH-dependent (pH 3.23 to pH 7.62) luminescence spectra show that the indicator's luminescence increases in acidic conditions because of protonation of the rhodamine chromophore, while the upconversion at 550 nm remained

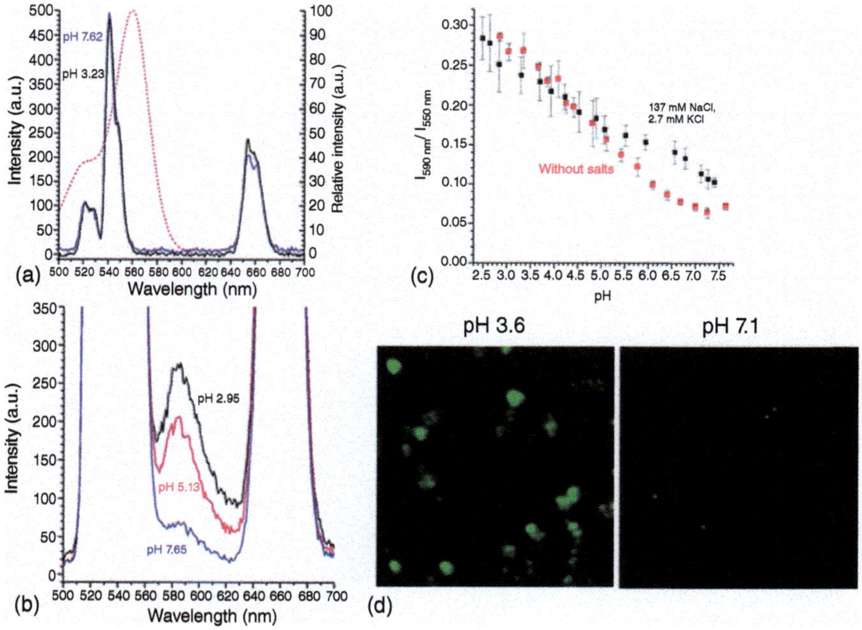

Figure 14.4 (a) Upconversion luminescence spectrum of dye-conjugated UCNPs in pH 7.62 and 3.23. The excitation spectrum of pHrodo Red (red dots, right y-axis) is overlapped with the upconversion emission spectrum. (b) Upconversion-sensitized luminescence emission spectrum of pHrodo Red in pH 7.65, 5.13, and 2.95. (c) Ratio of the intensities of the sensitized luminescence emission of pHrodo Red at 590 nm and the upconversion emission at 550 nm with (black) and without (red) a physiological salt concentration. (d) Fluorescence microscopy images of the green luminescence of pHrodo-conjugated UCNPs at different pHs. Source: Arppe et al. (2014). Reprinted with permission of Royal Society of Chemistry.

unchanged as (Figure 14.4a,b). Only a few upconversion emitters are supposed to transfer energy to the dye, and a large number of nanoparticles are unaffected. Thus, the intensity of the 550 nm emission band does not decrease. At a higher dye concentration, it is likely that attenuation of UCNP emission at 550 nm will be noticeable. The calibration curve within pH 3 to 6.7 exhibits linear response with the change in the ratio of dye emission (590 nm) to upconversion emission (550 nm) and a non-linear behavior in the region pH 2.5 to 7.2. The confocal microscopy images at various pHs of 550 nm emission of dye-conjugated UCNPs with a concentration of 0.02 mg/ml in saline help to infer that at low pH (pH 3.6), aggregation of particles increases, and high pH (pH 7.1) aggregation decreases (Figure 14.4d).

However, to achieve effective energy transfer from the UCNPs to dye, the work was further expanded, and with a major advancement, the surface of the nanoparticles was improved through surface modification by polyethyleneimine (PEI) (Näreoja et al. 2017). The cytotoxicity of PEI was diminished after their binding to the surface of UCNPs (Godbey et al. 1999). Modification through PEI is advantageous because cellular uptake increases by PEI modification via the formation of positive charges on the UCNP surface, and thus, the colloidal stability of UCNPs increases. Apart from these, resonance energy transfer to the dye increases because of the presence of strong covalent bonds by amino groups of PEI.

It has been noticed that a poor acceptor–donor pair is one of the main reasons for obtaining low pH sensitivity because of the lack of the overlap of the absorption and emission bands of the indicator and UCNPs. The long distance between the acceptor and donor is not suitable for energy transfer. The Förster resonance energy transfer (FRET) was increased sufficiently in a PEI-modified $NaYF_4:Tm^{3+}/Yb^{3+}$-fluorescein isothiocyanate pH sensor (Li et al. 2016), which shows a ratio change by 3.56 unit per unit change in pH within the 3–7 pH range in intracellular pH calibration. The sensor uses the change in Tm^{3+} upconversion luminescence at 475 nm and a reference signal at 650 nm.

Another pH sensor based on PAA-modified $NaGdF_4:Tm^{3+}/Yb^{3+}@NaYF_4$ conjugated with hemicyanine was successfully developed (Li et al. 2017). The dye-conjugated UCNP exhibits a decrease of upconversion emission at 450 and 475 nm in an acidic medium because of the absorption of the dye at these wavelengths, while in the neutral and basic medium, the emissions at 475 and 513 nm decrease. Meanwhile, the emission band at 655 nm remains unaffected. This probe is reversible, and imaging of a live cell was performed with these UCNPs to monitor the pH within pH 6.8–9.

An alternative hybrid upconverting nanostructure $NaYF_4:Er^{3+}/Yb^{3+}@NaYF_4@Ni$ was developed based on quenching and recovery of upconversion emissions in basic and acidic solutions (Xu et al. 2017). The hybrid nanoparticles (molar ratio, UCNPs/Ni = 1 : 0.28) in phosphate buffer solutions of pH 5.0, pH 6.0, pH 6.5, and pH 7.4 exhibit that the upconversion emission is quenched at pH 7.4 (basic), and the luminescence gets recovered slowly with changing the buffer toward pH 5.0 (acidic). The upconversion luminescence decreases because of the result of decomposition of Ni nanoparticles. Figure 14.5 shows the spectra at different pHs with recovery times for upconversion emissions in four buffer solutions.

14.4.2 Upconversion-Based pH Sensing Membranes

The pH-sensitive membranes are useful for measuring the pH in liquids (e.g. blood and serum). The pH sensor films are required for the detection of bacterial infections in medically implanted devices. A biofilm generally forms on the implanted devices, which damages the devices, and hence, early detection and treatment of infections with these devices are difficult. Therefore, the development of a non-invasive pH sensor film for bacterial growth detection in the implanted medical devices is demanding in medical treatment. A study (Ma et al. 2015) with xylenol orange (dye)-conjugated, silica-coated UCNP-doped Tm^{3+} presents quenching and recovery of 450 nm emission (Tm^{3+}: $^1D_2 \rightarrow {}^3F_4$) at various pHs. The ratio of 450 nm emission to a reference emission (646 nm) at low pH decreases, while the opposite was observed at high pH because of the low absorption of the dye at the blue region, with a linear variation of ratio within pH 4–8. A ~12 μm-thick sensor membrane was designed by using the UCNPs. The membrane shows pH sensitivity, and a major change in upconversion emissions is observed at pH 10 compared to pH 6.

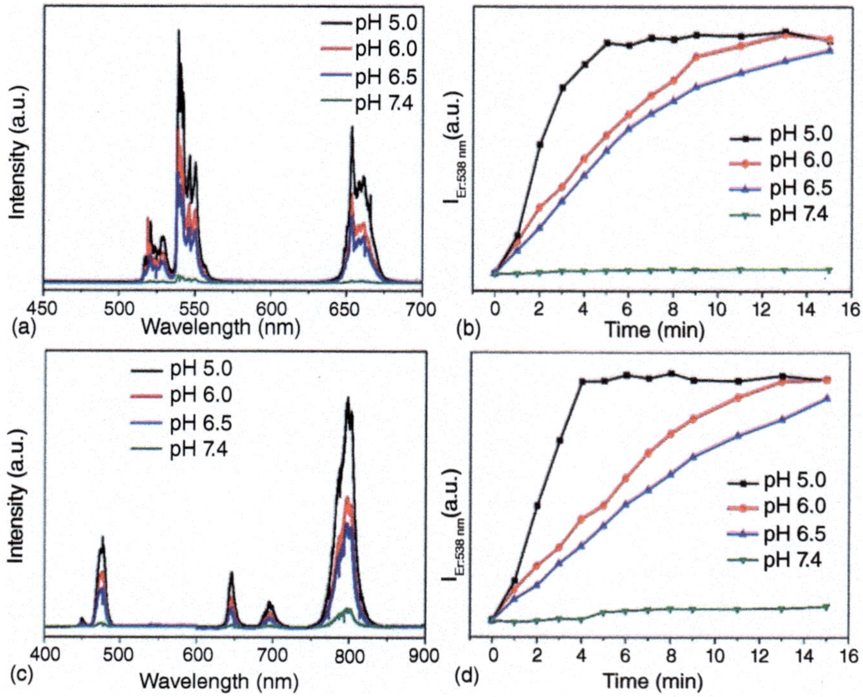

Figure 14.5 Upconversion response of (a) $NaYF_4:Er^{3+}/Yb^{3+}$@$NaYF_4$@Ni and (c) $NaYF_4:Tm^{3+}/Yb^{3+}$@$NaYF_4$@Ni nanoparticles in different pH solutions. The plot of upconversion emission intensity at (b) 538 nm of Er^{3+} ions and (d) 800 nm of Tm^{3+} ions doped with Ni-modified UCNPs in different pH buffers versus reaction time. Source: Xu et al. (2017). Reprinted with permission of Royal Society of Chemistry.

A proof of concept of sensor films, which consisted of an upconversion luminescence layer ($Y_2O_2S:Er^{3+}/Yb^{3+}$) and a pH indicator (Bromocresol green) layer, was introduced by the group of Anker et al. (Wang et al. 2014b). The developed film is schematically shown in the inset (right side) of Figure 14.6b with the optical absorption and upconversion spectra of the pH indicator and $Y_2O_2S:Er^{3+}/Yb^{3+}$ nanoparticles in Figure 14.6a. The luminescence emission changes with pH, but it is not noticeable because of the low emission intensity of the dye. The intensity variation due to protonation and deprotonation of the dye at various pHs is observed in absorption spectra. The film response time for pH determination was tested for 10 minutes, which is less than the time for bacterial acidification. The test was carried out in porcine tissues with the developed pH sensor film, which exhibited pH-dependent variation of the intensity ratio of 671 nm over 661 nm, with sufficient response in pH 5–10. The calibration curves in the presence or absence of biological tissues do not show any significant difference (Wang et al. 2014b).

The sensor was then tested in real time to monitor the pH variation caused by bacteria on a tryptic soy agar (TSA) plate (Figure 14.6c). The Staphylococcus epidermidis 35 984 bacteria were found to make colonies in the film and change pH from 7.4 to 5.7 after 40 hours. The color of the film changes from green (neutral) to

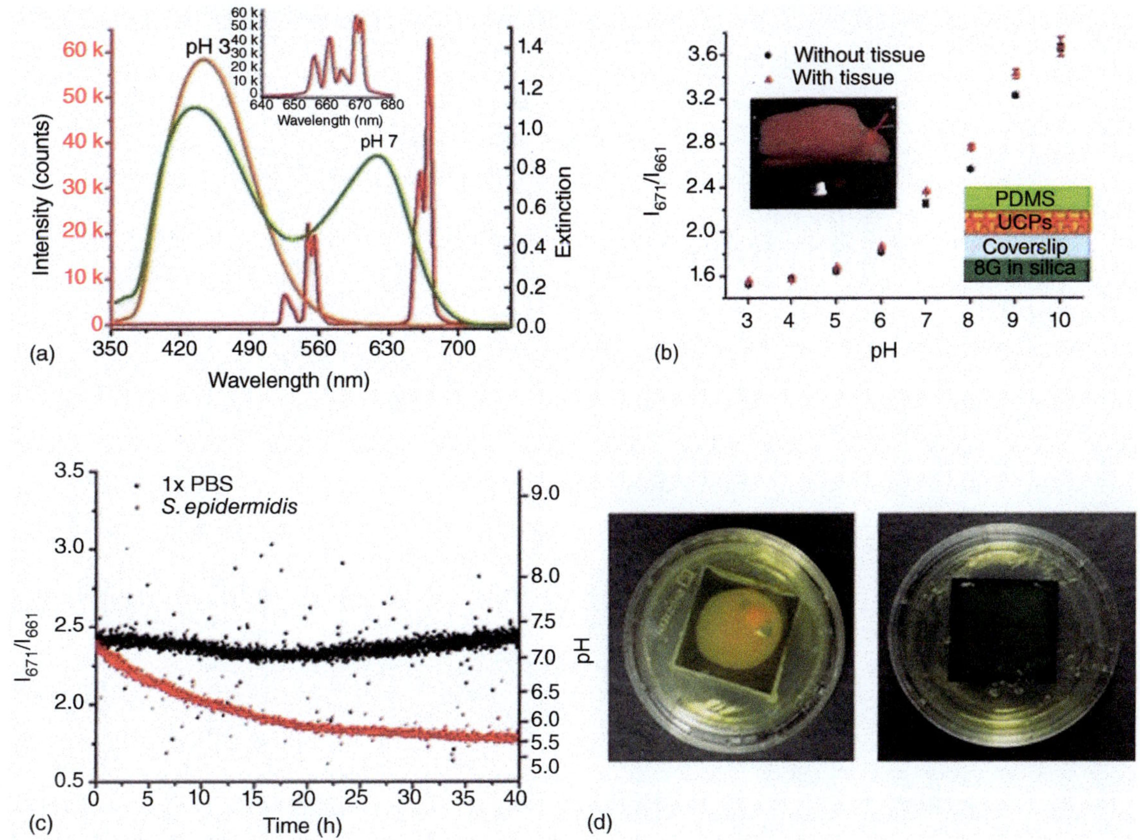

Figure 14.6 (a) The upconversion spectrum of $Y_2O_2S:Er^{3+}/Yb^{3+}$ UCNPs (red line, left *y*-axis) and the extinction spectra (right *y*-axis) of bromocresol green-doped silica films in pH 3.0 (orange) and buffer 7.0 (green). Inset shows the red luminescence with higher resolution. (b) The calibration curves of bromocresol green-conjugated $Y_2O_2S:Er^{3+}/Yb^{3+}$ films, without (black dots) and with (red points) passing through porcine tissue. Left inset shows the photograph of the sensor film sandwiched between two porcine tissues for pH calibration. The inset in the right side shows the schematic representation of the pH sensor. (c) pH determination in real time with the sensor film passing through porcine tissues. The red dots exhibit the variation of pH because of the bacterial development at the interface of the TSA plate and the pH sensor; the black dots show the variation of pH of the controlled sample with the same volume of phosphate buffer solution. The corresponding pH values, calculated according to the calibration curve, are shown in the right *y*-axis. (d) The photograph at the left side shows the pH sensor film by the end of the real-time experiment with *Staphylococcus epidermidis* (i.e. *S. epidermidis*) and at the right side shows the film of the controlled sample by the end of acquisition. PDMS = polydimethylsiloxane. Source: Wang et al. (2014b). Reprinted with permission of Royal Society of Chemistry.

yellow (acidic) (Figure 14.6d). Thus, it shows that upconversion-based ratiometric pH sensors could be a potential tool to detect bacterial infection on implanted medical devices in real time (Wang et al. 2014b).

Another work by the group Yan et al. has used nanohybrids for pH measurement (Yan et al. 2014). As graphene oxide is a quencher (Morales-Narváez and Merkoçi 2012; Yan et al. 2014) for quantum dots and dyes, the sensor film was composed of graphene oxide (as an acceptor) and an upconversion material (as a donor) (Figure 14.7) (Yan et al. 2014).

In a separate study, an inexpensive and straight design of a pH sensor film consisting of upconversion phosphors and a pH-sensitive dye hosted on a hydrogel matrix was developed by the researchers (Meier et al. 2014). The sensor film's color changes with pH. As it uses green and red emissions, estimation of pH can be done by the naked eye. Apart from that, it has the ability to sense and image pH through ratiometric and referenced red green blue (RGB) technique by a digital color camera. Figure 14.8a exhibits the calibration of RGB-pH imaging and pH determination in human serum using the sensor film.

Another multifunctional sensor for metal ions and pH determination in biological specimens was designed (Xie et al. 2012). The film was embedded in a

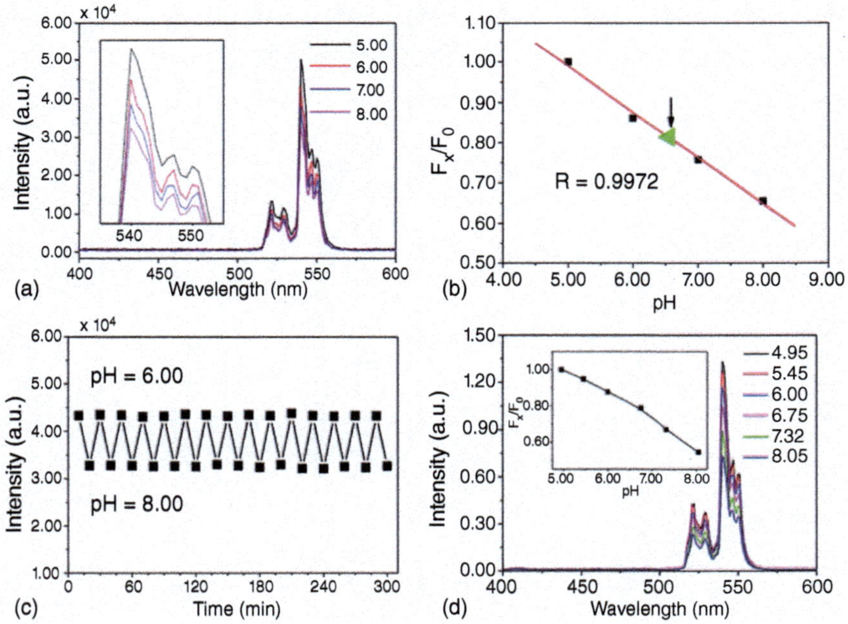

Figure 14.7 (a) The upconversion spectra of the graphene oxide–PEI-$NaYF_4$;Er^{3+}/Yb^{3+} film in buffer solutions upon excitation at 980 nm. (b) Ratio of green emission intensity at pH 5 to green emission intensity at other pH values. (c) Repetitive cycles of the pH sensor between pH 6.00 and pH 8.00, showing response time. (d) Upconversion spectra of graphene oxide–PEI-$NaYF_4$:Er^{3+}/Yb^{3+} in the urine of mice. Inset shows the ratio of green emission intensity at pH 5 to green emission intensity at other pH values plotted against the pH values in diluted urine. Source: Yan et al. (2014). Reprinted with permission of Royal Society of Chemistry.

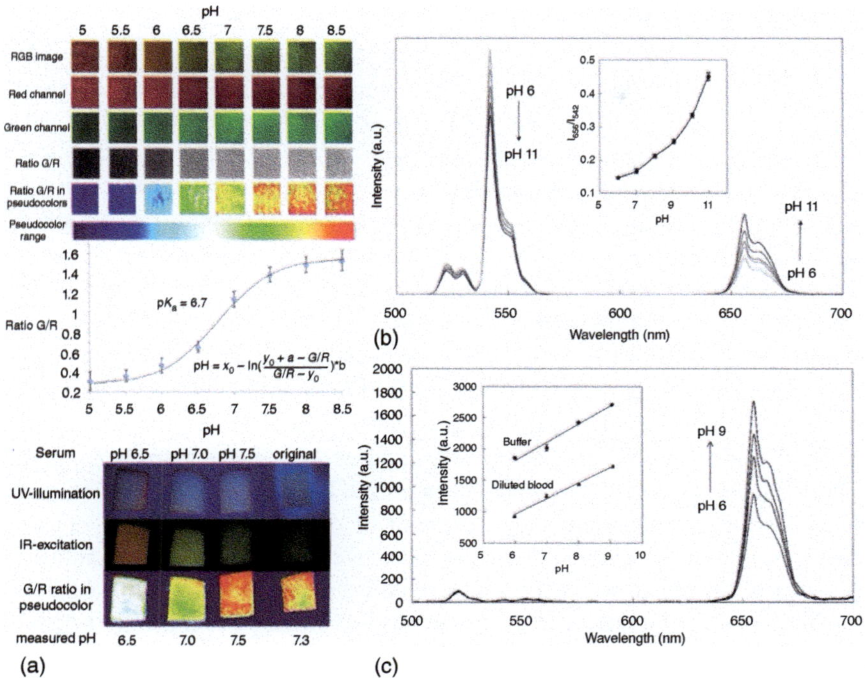

Figure 14.8 (a) Imaging calibration row (RGB) of the membrane. The graph at the middle of (a) shows the ratio G/R values versus the pH for the calibration point. At the bottom, four sensor spots were impregnated with serum specimens. The first three samples have adjusted pH values at pH 6.5, pH 7.0, pH 7.0, and pH 7.5 and the fourth one is original. Autofluorescence from the background is seen under UV illumination (first row). IR excitation and RGB imaging of the real serum sample remains unaffected from the background (second row). The G/R ratio is shown in pseudocolor (bottom row) and converted into the corresponding pH values. Source: Meier et al. (2014). Reprinted with permission of American Chemical Society. (b) The 980 nm excited upconversion spectra of the pH sensor film containing UCNPs ($NaYF_4$:Er^{3+}/Yb^{3+}) and chromoionophore (ETH 5418), within pH values 6–11. The inset shows an intensity ratio of 656 nm to 542 nm emission band as a function of pH. (c) Upconversion spectra of the pH sensor film containing UCNPs and ETH 5418 in diluted blood. The inset shows the intensity variation of 656 nm emission band of a sensor film in buffer and diluted blood as a function of pH. Source: Xie et al. (2012). Reprinted with permission of American Chemical Society.

plasticized polyvinyl chloride matrix and consisted of UCNPs ($NaYF_4$:Er^{3+}/Yb^{3+}) and chromoionophore (ETH 5418). The pH indicator shows a clear isosbestic point at 577 nm, and the absorption maxima changes from green (525 nm) to red (680 nm) because of deprotonation and protonation. Upon 980 nm excitation, the green emission ($^2H_{11/2}$/$^4S_{3/2}$ → $^4I_{15/2}$) decreases, and the red emission ($^4F_{9/2}$ → $^4I_{15/2}$) increases with increasing pH (Figure 14.8b). The sensor film's evaluation in blood specimen shows that the green emission decreases highly, and the red emission was weakened by 50%, and the pH was measured through intensity variation. As shown in Figure 14.8c, the red emission linearly increases with the pH of blood, thus confirming that it can measure pH in the complex environment of blood (Xie et al. 2012).

14.5 Conclusion

In conclusion, we present the basic properties of upconversion in the RE ion-doped UC materials with their pH sensing applications. The UCNPs are useful in different fields spanning from bioimaging, biosensing, drug delivery technologies to photovoltaics. To design a better pH sensor for living cells, several criteria should be fulfilled, e.g. high cellular uptake, good colloidal stability, photostability, strong brightness, and non-cytotoxicity. The common problem that appeared to the researchers while internalizing the nanoprobes inside a cell is that because of internalization through the endocytosis process, the nanoparticles are not homogeneously distributed inside; rather, they remain in specific parts of the cells (e.g. in acidic vesicles), which are unwanted sites for the experimental goal and could disable to determine the endosomal compartments correctly. Control and monitor of exact pH in different cellular components are necessary for the living being. Cell proliferation and movements can also be organized by monitoring the pH values. Cellular dysfunction generally occurs because of uncontrolled variation of pH. Therefore, the internalization of the nanoprobes in cells should be further studied and developed to fulfill a successive conclusion for the application of nanoparticles.

In the biological approach, UCNPs are favorable compared to the conventional quantum dots or organic dyes because of their many advantages, including the absence of background autofluorescence, high resistance to photobleaching, non-photoblinking, distinct emissions from the NIR excitation wavelength, etc. Despite that, there are still some limitations at the basic level related to the use of upconversion materials in practical applications. For instance, the internal quantum yield of upconversion is very low under NIR excitation. Apart from these remarkable features, many aspects associated with UCNPs remain a great concern, such as surface modification, efficiency enhancement, emission color tuning, and manipulation of energy transfer pathways to achieve desired upconversion emissions for their specific applications. Tremendous expansion in the active research in UCNPs has been started in the past decade, and a vast number of papers are witnessing this activity in the nanoscience community. Very recently, the UCNPs have been studied to optically stimulate the brain of a mammalian system by activating the light-gated ion channels. However, one of the major challenges is the design of the potent nanoparticles so that they can have (i) bright, (ii) photostable, (iii) long-term functionality that does not fade away over time and (iv) non-toxic properties. Another challenge for UCNPs is that its efficiency decreases with size. Additionally, in biological applications, smaller nanoparticles are preferred for making low-cytotoxic materials. Manipulating the local environment around the RE ions and treatment with plasmons, quantum dots, molecular dyes, and polymers might be effective ways to sort out these problems. Finally, more significant and thriving research is yet to come in the following years to understand the practical aspects of photon upconverting nanomaterials in pH sensing.

References

Arppe, R., Näreoja, T., Nylund, S. et al. (2014). Photon upconversion sensitized nanoprobes for sensing and imaging of pH. *Nanoscale* 6 (12): 6837–6843. https://doi.org/10.1039/C4NR00461B.

Auzel, F. (2004). Upconversion and anti-stokes processes with f and d ions in solids. *Chemical Reviews* 104 (1): 139–174. https://doi.org/10.1021/CR020357G.

Bloembergen, N. (1959). Solid state infrared quantum counters. *Physical Review Letters* 2 (3): 84. https://doi.org/10.1103/PhysRevLett.2.84.

Burns, A., Sengupta, P., Zedayko, T. et al. (2006). Core/shell fluorescent silica nanoparticles for chemical sensing: towards single-particle laboratories. *Small* 2 (6): 723–726. https://doi.org/10.1002/SMLL.200600017.

Chen, D., Wang, Y., Yu, Y., and Huang, P. (2007). Intense ultraviolet upconversion luminescence from Tm^{3+}/Yb^{3+}:β-YF_3 nanocrystals embedded glass ceramic. *Applied Physics Letters* 91 (5): 051920. https://doi.org/10.1063/1.2767988.

Chen, X., Peng, D., Ju, Q., and Wang, F. (2015). Photon upconversion in core–shell nanoparticles. *Chemical Society Reviews* 44 (6): 1318–1330. https://doi.org/10.1039/C4CS00151F.

Dong, H., Sun, L.D., and Yan, C.H. (2015). Energy transfer in lanthanide upconversion studies for extended optical applications. *Chemical Society Reviews* 44 (6): 1608–1634. https://doi.org/10.1039/C4CS00188E.

Drees, C., Raj, A.N., Kurre, R. et al. (2016). Engineered upconversion nanoparticles for resolving protein interactions inside living cells. *Angewandte Chemie International Edition* 55 (38): 11668–11672. https://doi.org/10.1002/ANIE.201603028.

Du, S., Hernández-Gil, J., Dong, H. et al. (2017). Design and validation of a new ratiometric intracellular pH imaging probe using lanthanide-doped upconverting nanoparticles. *Dalton Transactions* 46 (40): 13957–13965. https://doi.org/10.1039/C7DT02418E.

Godbey, W.T., Wu, K.K., and Mikos, A.G. (1999). Poly (ethylenimine) and its role in gene delivery. *Journal of Controlled Release* 60 (2, 3): 149–160. https://doi.org/10.1016/S0168-3659(99)00090-5.

Gu, Y., Qiao, X., Zhang, J. et al. (2016). Effects of surface modification of upconversion nanoparticles on cellular uptake and cytotoxicity. *Chemical Research in Chinese Universities* 32 (3): 474–479. https://doi.org/10.1007/s40242-016-6026-5.

Gulzar, A., Xu, J., Yang, P. et al. (2017). Upconversion processes: versatile biological applications and biosafety. *Nanoscale* 9 (34): 12248–12282. https://doi.org/10.1039/C7NR01836C.

Guo, Z., Park, S., Yoon, J., and Shin, I. (2014). Recent progress in the development of near-infrared fluorescent probes for bioimaging applications. *Chemical Society Reviews* 43 (1): 16–29. https://doi.org/10.1039/C3CS60271K.

Haase, M. and Schäfer, H. (2011). Upconverting nanoparticles. *Angewandte Chemie International Edition* 50 (26): 5808–5829. https://doi.org/10.1002/ANIE.201005159.

Heer, S., Kömpe, K., Güdel, H.U., and Haase, M. (2004). Highly efficient multicolour upconversion emission in transparent colloids of lanthanide-doped $NaYF_4$ nanocrystals. *Advanced Materials* 16 (23, 24): 2102–2105. https://doi.org/10.1002/adma.200400772.

Hornig, S., Biskup, C., Gräfe, A. et al. (2008). Biocompatible fluorescent nanoparticles for pH-sensoring. *Soft Matter* 4 (6): 1169–1172. https://doi.org/10.1039/B800276B.

Kim, H.J., Heo, C.H., and Kim, H.M. (2013). Benzimidazole-based ratiometric two-photon fluorescent probes for acidic pH in live cells and tissues. *Journal of the American Chemical Society* 135 (47): 17969–17977. https://doi.org/10.1021/JA409971K.

Laguna, M., Escudero, A., Núñez, N.O. et al. (2017). Europium-doped $NaGd(WO_4)_2$ nanophosphors: synthesis, luminescence and their coating with fluorescein for pH sensing. *Dalton Transactions* 46 (35): 11575–11583. https://doi.org/10.1039/C7DT01986F.

Lee, M.H., Kim, J.S., and Sessler, J.L. (2015). Small molecule-based ratiometric fluorescence probes for cations, anions, and biomolecules. *Chemical Society Reviews* 44 (13): 4185. https://doi.org/10.1039/c4cs00280f.

Lee, E., Jung, M., Han, Y. et al. (2017). Stochastic photon emission from nonblinking upconversion nanoparticles. *The Journal of Physical Chemistry C* 121 (38): 21073–21079. https://doi.org/10.1021/ACS.JPCC.7B08509.

Li, C., Zuo, J., Zhang, L. et al. (2016). Accurate quantitative sensing of intracellular pH based on self-ratiometric upconversion luminescent nanoprobe. *Scientific Reports* 6 (1): 1–9. https://doi.org/10.1038/srep38617.

Li, H., Dong, H., Yu, M. et al. (2017). NIR ratiometric luminescence detection of pH fluctuation in living cells with hemicyanine derivative-assembled upconversion nanophosphors. *Analytical Chemistry* 89 (17): 8863–8869. https://doi.org/10.1021/ACS.ANALCHEM.7B01324.

Liu, G. (2015). Advances in the theoretical understanding of photon upconversion in rare-earth activated nanophosphors. *Chemical Society Reviews* 44 (6): 1635–1652. https://doi.org/10.1039/C4CS00187G.

Liu, C. and Chen, D. (2007). Controlled synthesis of hexagon shaped lanthanide-doped LaF_3 nanoplates with multicolor upconversion fluorescence. *Journal of Materials Chemistry* 17 (37): 3875–3880. https://doi.org/10.1039/B707927C.

Liu, Y.S., Sun, Y., Vernier, P.T. et al. (2007). pH-sensitive photoluminescence of CdSe/ZnSe/ZnS quantum dots in human ovarian cancer cells. *The Journal of Physical Chemistry C* 111 (7): 2872–2878. https://doi.org/10.1021/JP0654718.

Liu, J., Huang, Y., Kumar, A. et al. (2014). pH-sensitive nano-systems for drug delivery in cancer therapy. *Biotechnology Advances* 32 (4): 693–710. https://doi.org/10.1016/J.BIOTECHADV.2013.11.009.

Ma, T., Ma, Y., Liu, S. et al. (2015). Dye-conjugated upconversion nanoparticles for ratiometric imaging of intracellular pH values. *Journal of Materials Chemistry C* 3 (26): 6616–6620. https://doi.org/10.1039/C5TC00849B.

Mader, H.S. and Wolfbeis, O.S. (2010). Optical ammonia sensor based on upconverting luminescent nanoparticles. *Analytical Chemistry* 82 (12): 5002–5004. https://doi.org/10.1021/AC1007283.

Mader, H.S., Kele, P., Saleh, S.M., and Wolfbeis, O.S. (2010). Upconverting luminescent nanoparticles for use in bioconjugation and bioimaging. *Current Opinion in Chemical Biology* 14 (5): 582–596. https://doi.org/10.1016/J.CBPA.2010.08.014.

Mahata, M.K. and Lee, K.T. (2019). Development of near-infrared sensitized core–shell–shell upconverting nanoparticles as pH-responsive probes. *Nanoscale Advances* 1 (6): 2372–2381. https://doi.org/10.1039/C9NA00088G.

Mahata, M. K., Kumari, A., Rai, V. K., & Kumar, K. (2013). Er^{3+}, Yb^{3+} doped yttrium oxide phosphor as a temperature sensor. In AIP Conference Proceedings (Vol. 1536, no. 1, pp. 1270–1271). American Institute of Physics. https://doi.org/10.1063/1.4810704.

Mahata, M.K., Kumar, K., and Rai, V.K. (2014a). Structural and optical properties of Er^{3+}/Yb^{3+} doped barium titanate phosphor prepared by co-precipitation method. *Spectrochimica Acta Part A: Molecular and Biomolecular Spectroscopy* 124: 285–291. https://doi.org/10.1016/J.SAA.2014.01.014.

Mahata, M.K., Tiwari, S.P., Mukherjee, S. et al. (2014b). $YVO_4:Er^{3+}/Yb^{3+}$ phosphor for multifunctional applications. *Journal of the Optical Society of America B* 31 (8): 1814–1821. https://doi.org/10.1364/JOSAB.31.001814.

Mahata, M.K., Kumar, K., and Rai, V.K. (2015a). Er^{3+}–Yb^{3+} doped vanadate nanocrystals: a highly sensitive thermographic phosphor and its optical nanoheater behavior. *Sensors and Actuators B: Chemical* 209: 775–780. https://doi.org/10.1016/J.SNB.2014.12.039.

Mahata, M.K., Koppe, T., Hofsäss, H. et al. (2015b). Host sensitized luminescence and time-resolved spectroscopy of $YVO_4:Ho^{3+}$ nanocrystals. *Physics Procedia* 76: 125–131. https://doi.org/10.1016/j.phpro.2015.10.023.

Mahata, M.K., Koppe, T., Mondal, T. et al. (2015c). Incorporation of Zn^{2+} ions into $BaTiO_3:Er^{3+}/Yb^{3+}$ nanophosphor: an effective way to enhance upconversion, defect luminescence and temperature sensing. *Physical Chemistry Chemical Physics* 17 (32): 20741–20753. https://doi.org/10.1039/C5CP01874A.

Mahata, M.K., Koppe, T., Kumar, K. et al. (2016a). Demonstration of temperature dependent energy migration in dual-mode $YVO_4:Ho^{3+}/Yb^{3+}$ nanocrystals for low temperature thermometry. *Scientific Reports* 6 (1): 1–11. https://doi.org/10.1038/srep36342.

Mahata, M.K., Hofsäss, H.C., and Vetter, U. (2016b). Photon-upconverting materials: advances and prospects for various emerging applications. In: *Luminescence – An Outlook on the Phenomena and Their Applications* (ed. J. Thirumalai), 109–131. Rijeka: InTech. https://doi.org/10.5772/65118.

Mahata, M.K., Bae, H., and Lee, K.T. (2017). Upconversion luminescence sensitized pH-nanoprobes. *Molecules* 22 (12): 2064. https://doi.org/10.3390/MOLECULES22122064.

Mahata, M.K., Koppe, T., Kumar, K. et al. (2020). Upconversion photoluminescence of Ho^{3+}–Yb^{3+} doped barium titanate nanocrystallites: optical tools for structural phase detection and temperature probing. *Scientific Reports* 10 (1): 1–12. https://doi.org/10.1038/s41598-020-65149-z.

Mahata, M.K., De, R., and Lee, K.T. (2021). Near-infrared-triggered upconverting nanoparticles for biomedicine applications. *Biomedicine* 9 (7): 756. https://doi.org/10.3390/BIOMEDICINES9070756.

Marches, R., Vitetta, E.S., and Uhr, J.W. (2001). A role for intracellular pH in membrane IgM-mediated cell death of human B lymphomas. *Proceedings of the National Academy of Sciences of the United States of America* 98 (6): 3434–3439. https://doi.org/10.1073/PNAS.061028998.

Meier, R.J., Simbürger, J.M., Soukka, T., and Schaferling, M. (2014). Background-free referenced luminescence sensing and imaging of pH using upconverting phosphors and color camera read-out. *Analytical Chemistry* 86 (11): 5535–5540. https://doi.org/10.1021/AC5009207.

Michalet, X., Pinaud, F.F., Bentolila, L.A. et al. (2005). Quantum dots for live cells, in vivo imaging, and diagnostics. *Science* 307 (5709): 538–544. https://doi.org/10.1126/SCIENCE.1104274.

Morales-Narváez, E. and Merkoçi, A. (2012). Graphene oxide as an optical biosensing platform. *Advanced Materials* 24 (25): 3298–3308. https://doi.org/10.1002/ADMA.201200373.

Naccache, R., Vetrone, F., Mahalingam, V. et al. (2009). Controlled synthesis and water dispersibility of hexagonal phase $NaGdF_4$:Ho^{3+}/Yb^{3+} nanoparticles. *Chemistry of Materials* 21 (4): 717–723. https://doi.org/10.1021/CM803151Y.

Nadort, A., Zhao, J., and Goldys, E.M. (2016). Lanthanide upconversion luminescence at the nanoscale: fundamentals and optical properties. *Nanoscale* 8 (27): 13099–13130. https://doi.org/10.1039/C5NR08477F.

Näreoja, T., Deguchi, T., Christ, S. et al. (2017). Ratiometric sensing and imaging of intracellular pH using polyethylenimine-coated photon upconversion nanoprobes. *Analytical Chemistry* 89 (3): 1501–1508. https://doi.org/10.1021/ACS.ANALCHEM.6B03223.

Ong, L.C., Ang, L.Y., Alonso, S., and Zhang, Y. (2014). Bacterial imaging with photostable upconversion fluorescent nanoparticles. *Biomaterials* 35 (9): 2987–2998. https://doi.org/10.1016/J.BIOMATERIALS.2013.12.060.

Pandey, A. and Rai, V.K. (2013). Improved luminescence and temperature sensing performance of Ho^{3+}–Yb^{3+}–Zn^{2+}:Y_2O_3 phosphor. *Dalton Transactions* 42 (30): 11005–11011. https://doi.org/10.1039/C3DT50592H.

Schäfer, H., Ptacek, P., Kömpe, K., and Haase, M. (2007). Lanthanide-doped $NaYF_4$ nanocrystals in aqueous solution displaying strong up-conversion emission. *Chemistry of Materials* 19 (6): 1396–1400. https://doi.org/10.1021/CM062385B.

Schäfer, H., Ptacek, P., Eickmeier, H., and Haase, M. (2009). Synthesis of hexagonal Yb^{3+}, Er^{3+}-doped $NaYF_4$ nanocrystals at low temperature. *Advanced Functional Materials* 19 (19): 3091–3097. https://doi.org/10.1002/adfm.200900642.

Schäferling, M. (2016). Nanoparticle-based luminescent probes for intracellular sensing and imaging of pH. *Wiley Interdisciplinary Reviews: Nanomedicine and Nanobiotechnology* 8 (3): 378–413. https://doi.org/10.1002/WNAN.1366.

Sedlmeier, A. and Gorris, H.H. (2015). Surface modification and characterization of photon-upconverting nanoparticles for bioanalytical applications. *Chemical Society Reviews* 44 (6): 1526–1560. https://doi.org/10.1039/C4CS00186A.

Shi, W., Li, X., and Ma, H. (2014). Fluorescent probes and nanoparticles for intracellular sensing of pH values. *Methods and Applications in Fluorescence* 2 (4): 042001. https://doi.org/10.1088/2050-6120/2/4/042001.

Sinha, S., Mahata, M.K., and Kumar, K. (2016). Up/down-converted green luminescence of Er^{3+}–Yb^{3+} doped paramagnetic gadolinium molybdate: a highly sensitive thermographic phosphor for multifunctional applications. *RSC Advances* 6 (92): 89642–89654. https://doi.org/10.1039/C6RA20332A.

Sinha, S., Mahata, M.K., Kumar, K. et al. (2017a). Dualistic temperature sensing in Er^{3+}/Yb^{3+} doped $CaMoO_4$ upconversion phosphor. *Spectrochimica Acta Part A: Molecular and Biomolecular Spectroscopy* 173: 369–375. https://doi.org/10.1016/J.SAA.2016.09.039.

Sinha, S., Mahata, M.K., Swart, H.C. et al. (2017b). Enhancement of upconversion, temperature sensing and cathodoluminescence in the K^+/Na^+ compensated $CaMoO_4$:Er^{3+}/Yb^{3+} nanophosphor. *New Journal of Chemistry* 41 (13): 5362–5372. https://doi.org/10.1039/C7NJ00086C.

Soni, A.K., Rai, V.K., and Mahata, M.K. (2017). Yb^{3+} sensitized $Na_2Y_2B_2O_7$:Er^{3+} phosphors in enhanced frequency upconversion, temperature sensing and field emission display. *Materials Research Bulletin* 89: 116–124. https://doi.org/10.1016/J.MATERRESBULL.2017.01.009.

Sun, L.N., Peng, H., Stich, M.I. et al. (2009). pH sensor based on upconverting luminescent lanthanide nanorods. *Chemical Communications* 33: 5000–5002. https://doi.org/10.1039/B907822C.

Sun, L.D., Dong, H., Zhang, P.Z., and Yan, C.H. (2015). Upconversion of rare earth nanomaterials. *Annual Review of Physical Chemistry* 66: 619–642. https://doi.org/10.1146/ANNUREV-PHYSCHEM-040214-121344.

Vennerberg, D. and Lin, Z. (2011). Upconversion nanocrystals: synthesis, properties, assembly and applications. *Science of Advanced Materials* 3 (1): 26–40. https://doi.org/10.1166/sam.2011.1137.

Wang, J., Hu, J., Tang, D. et al. (2007). Oleic acid (OA)-modified LaF_3:Er,Yb nanocrystals and their polymer hybrid materials for potential optical-amplification applications. *Journal of Materials Chemistry* 17 (16): 1597–1601. https://doi.org/10.1039/B617754A.

Wang, G., Peng, Q., and Li, Y. (2009). Upconversion luminescence of monodisperse CaF_2:Yb^{3+}/Er^{3+} nanocrystals. *Journal of the American Chemical Society* 131 (40): 14200–14201. https://doi.org/10.1021/JA906732Y.

Wang, F., Banerjee, D., Liu, Y. et al. (2010). Upconversion nanoparticles in biological labeling, imaging, and therapy. *Analyst* 135 (8): 1839–1854. https://doi.org/10.1039/C0AN00144A.

Wang, M., Abbineni, G., Clevenger, A. et al. (2011). Upconversion nanoparticles: synthesis, surface modification and biological applications. *Nanomedicine: Nanotechnology, Biology and Medicine* 7 (6): 710–729. https://doi.org/10.1016/J.NANO.2011.02.013.

Wang, F., Raval, Y., Chen, H. et al. (2014a). Development of luminescent pH sensor films for monitoring bacterial growth through tissue. *Advanced Healthcare Materials* 3 (2): 197–204. https://doi.org/10.1002/ADHM.201300101.

Wang, S., Feng, J., Song, S., and Zhang, H. (2014b). A long-wave optical pH sensor based on red upconversion luminescence of $NaGdF_4$ nanotubes. *RSC Advances* 4 (99): 55897–55899. https://doi.org/10.1039/C4RA09686J.

Wencel, D., Abel, T., and McDonagh, C. (2014). Optical chemical pH sensors. *Analytical Chemistry* 86 (1): 15–29. https://doi.org/10.1021/AC4035168.

Xie, L., Qin, Y., and Chen, H.Y. (2012). Polymeric optodes based on upconverting nanorods for fluorescent measurements of pH and metal ions in blood samples. *Analytical Chemistry* 84 (4): 1969–1974. https://doi.org/10.1021/AC203003W.

Xu, X., Long, Y., Lei, P. et al. (2017). A pH-responsive assembly based on upconversion nanocrystals and ultrasmall nickel nanoparticles. *Journal of Materials Chemistry C* 5 (37): 9666–9672. https://doi.org/10.1039/C7TC02665J.

Yan, L., Chang, Y.N., Yin, W. et al. (2014). Biocompatible and flexible graphene oxide/upconversion nanoparticle hybrid film for optical pH sensing. *Physical Chemistry Chemical Physics* 16 (4): 1576–1582. https://doi.org/10.1039/C3CP54317J.

Yin, J., Hu, Y., and Yoon, J. (2015). Fluorescent probes and bioimaging: alkali metals, alkaline earth metals and pH. *Chemical Society Reviews* 44 (14): 4619–4644. https://doi.org/10.1039/C4CS00275J.

Zhang, H., Li, Y., Ivanov, I.A. et al. (2010). Plasmonic modulation of the upconversion fluorescence in $NaYF_4$:Yb/Tm hexaplate nanocrystals using gold nanoparticles or nanoshells. *Angewandte Chemie International Edition* 49 (16): 2865–2868. https://doi.org/10.1002/ANIE.200905805.

Zheng, W., Huang, P., Tu, D. et al. (2015). Lanthanide-doped upconversion nano-bioprobes: electronic structures, optical properties, and biodetection. *Chemical Society Reviews* 44 (6): 1379–1415. https://doi.org/10.1039/C4CS00178H.

15

Upconversion Nanoparticles in Temperature Sensing and Optical Heating Applications

Praveen K. Shahi[1] *and Shyam B. Rai*[2]

[1] *SPM Government Degree College (University of Allahabad), Department of Physics, Phaphamau, Allahabad, 211013, India*
[2] *Banaras Hindu University, Institute of Science, Department of Physics, Lanka, Varanasi, 221005, India*

15.1 Introduction

Temperature (T) is a fundamental physical quantity, and the statistical physics defines it as the "temperature of a body is the inverse of the change of its entropy S, with respect to the internal energy U, $T^{-1} = \partial S/\partial U$"(Reif and Scott 1998; Hartmann 2006). The entropy of the body quantitatively measures the atomic disorderness, whereas the temperature measures the intensity of submicroscopic random movements of the body's constituents. This physical parameter plays a vital role in almost every area, such as automotive, aerospace and defense, metrology, climate, marine research, biomedicine and nanomedicine, electronics, heating and cooling devices, production plants, food storage, etc. (Allison and Gillies 1997; Childs et al. 2000; Khalid and Kontis 2008; Wolfbeis 2008; McCabe and Hernandez 2010; Jaque et al. 2014a,b; Bünzli 2016, p. 141; Uchiyama et al. 2017).

The precise and accurate measurement of temperature is a key factor. Generally, the temperature reading is achieved from an invasive probe material in direct physical contact with the body for which temperature is unknown. The world market is dominated by thermometers such as thermocouples and thermistors. However, these thermometers do not work properly for small-scale (particle size below 10 μm) samples (Uchiyama et al. 2006; Brites et al. 2010, 2012; Jaque and Vetrone 2012; Wang et al. 2013; Dramićanin 2016; Laha et al. 2017; Quintanilla and Liz-Marzán 2018). Furthermore, these conventional contact thermometers depend on an electrical link, thereby hampering their applications where the electromagnetic noise is significant and sparks are dangerous (Allison and Gillies 1997). Hence, such restrictions of contact thermometers, for systems of small dimensions, have been stimulating the development of new noncontact precise thermometers with spatial resolution at the micrometric and nanometric scales, which is an exciting research topic under continuous expansion in the past decade (Jaque and Vetrone 2012; Brites et al. 2012; Wang et al. 2013; Dramićanin 2016, Quintanilla and Liz-Marzán 2018; Yue and Wang 2012). High-resolution noncontact thermometers operating at these

Upconverting Nanoparticles: From Fundamentals to Applications, First Edition. Edited by Vineet K. Rai.

scales have been grouped using distinct criteria, as whether they make use of optical or electrical signals or they are based on near-field or far-field techniques. However, each method possesses several advantages as well as drawbacks and exhibits different spatial, temporal, and temperature resolution (Bünzli 2016, p. 141)

Among the upconversion nanoparticles (UCNPs), the lanthanide-based nanomaterials or phosphor-based thermometry is a noninvasive spectroscopic technique for temperature measurement based on phosphor emission temperature dependence (e.g. peak energy and intensity, band shape, and excited state lifetimes). This technique combines high relative thermal sensitivity (S_r) (>1% K^{-1}) and spatial resolution (<10 μm) in short acquisition times (<1 ms) and, as it operates remotely, works even in biological fluids, fast-moving objects, and strong electromagnetic fields (Jaque and Vetrone 2012; Brites et al. 2012; Wang et al. 2013; Dramićanin 2016). The most used method to calculate the absolute temperature through thermographic phosphor thermometry is by measuring the luminescence intensities of two electronic transitions (occasionally involving Stark components of an excited electronic state) in thermal equilibrium. Moreover, different phosphors have been examined for providing a contactless thermal reading through their light emission properties, e.g. polymers (Uchiyama et al. 2003, 2004; Okabe et al. 2012); DNA or protein-conjugated systems (Donner et al. 2012); organic dyes (Löw et al. 2008; Yang et al. 2016; Paviolo et al. 2013); quantum dots (QDs) (Lee et al. 2005; Kalytchuk et al. 2016); Cr^{3+}-based materials (Back et al. 2016); and trivalent lanthanide (Ln^{3+}) ions (Brites et al. 2010).

Organic dyes are the most available and used thermal probes; for an exhaustive review of the subject, see the works of Hoogenboom and coworkers (Pietsch et al. 2011; De La Rosa et al. 2016). However, recently, QDs and Ln^{3+}-based materials are gaining importance because of their higher photostability. For instance, QDs were employed in submicrometer thermometry because of its temperature-dependent luminescence features (intensity changes or emission peak shifts) (Li et al. 2007; Haro-González et al. 2013). One of the most appealing application areas for QDs is nanomedicine as its bioconjugation can make them target selective. Nevertheless, QD usage in future use in clinical trials may be difficult, as they generally include highly toxic elements (e.g. Cd) (Cho et al. 2007; Gnach et al. 2015).

The Ln^{3+}-based materials are stable and narrow band emitters covering the entire electromagnetic spectrum with high emission quantum yields (>50% in the visible) (Shahi et al. 2015a,b, 2017a,b; Bünzli 2010, 2015; Carlos et al. 2009; Binnemans 2009; Feng and Zhang 2013; Chen et al. 2013a,b). In the past decade, many Ln^{3+}-based thermometers have been reported covering a wide temperature range, from cryogenic ($T < 100$ K) to physiological (298–323 K) values, and including lanthanide complexes (Uchiyama et al. 2006; Brites et al. 2010; Shahi et al. 2015a,b), lanthanide organic frameworks (LOFs) (Wang et al. 2015a,b; Cui et al. 2015; Cadiau et al. 2013; Rocha et al. 2016), polymers (Shahi et al. 2019; Cerõn et al. 2015), organic–inorganic hybrids (Rocha et al. 2016; Shahi et al. 2019; Cerõn et al. 2015; Singh et al. 2017a,b), upconverting (Savchuk et al. 2014; Singh et al. 2017a,b; Lojpur et al. 2014; Vetrone et al. 2010) and downshifting (DS) (Gavrilović et al. 2014; Shahi et al. 2017a,b) nanoparticles (NPs), and multifunctional heater–thermometer nanoplatforms (Jalani et al. 2016; Debasu et al. 2016; Zhu et al. 2016). The implementation of these Ln^{3+}-based materials as ratiometric thermometers in diverse applications was extensively revised in the past decade (Singh et al. 2009, 2015, 2016; Yu et al. 2010; Fischer et al. 2007; Jaque et al. 2014a,b; Millan et al. 2016, p. 237; Quintanilla et al. 2016, p. 124; Hemmer et al. 2017; Liu et al. 2017; Shahi et al. 2018).

Schena et al. have used fluorescence-based thermal sensors for tumor removal (Schena et al. 2016), and Musolino et al. have used fluorescence-based thermal sensors for in vivo measurements of brain temperature (Musolino et al. 2016). Wang et al. have used luminescent NPs for thermometry for the first time (Wang et al. 2002). The $ZnS:Mn^{2+},Eu^{3+}$ NPs were the first ratiometric luminescent thermometers, in which the temperature-dependent ratio of the emission intensities of the two dopants were used, and this emission intensity ratio was named as the fluorescence intensity ratio (FIR). Furthermore, this concept became a driving force for the development of lanthanide NP-based thermometers (e.g. $BaTiO_3:Er^{3+}$ NPs) (Alencar et al. 2004). From 2010 and onward, lanthanide-based thermometers have drawn major attention among researchers because of the enormous potential of their applications in nanotechnology and nanomedicine. A few Ln^{3+}-doped materials report from the literature are as follows:

- Phonon-induced heat generation and simultaneous noncontact temperature sensing using Yb^{3+}/Er^{3+}-based UCNPs (Tikhomirov et al. 2009; Rai et al. 2015).
- Intracellular thermometers using Yb^{3+}/Er^{3+}-doped UCNPs (Vetrone et al. 2010; Zhan et al. 2011).
- Eu^{3+}-doped organic–inorganic hybrid NPs for temperature probing in the physiological range (Shahi et al. 2015a,b).
- Temperature sensing using $Tm^{3+}/Er^{3+}/Ho^{3+}$-doped (Dong et al. 2012) and poly-acrylic acid-coated Mo-sensitized Yb^{3+}/Er^{3+}-doped NPs (Kusama et al. 1976) and temperature-stimuli polymers (Wei et al. 2019).

The interaction of radiation with phosphor materials arises two possibilities: (i) some part of radiation is absorbed, and consequently, radiative emission is observed, whereas (ii) interaction of other part of radiation generates heat into the system, and this heating is known as optical heating (Wen et al. 2018; Dey and Rai 2017; Du et al. 2019). The optical heating is relatively a newly observed phenomenon in the field of lanthanide-based phosphors. The optical heating might be due to the interaction of phosphor materials with radiation source. If the source is a laser, this effect is termed as laser-induced optical heating. It is a function of pump power of the excitation source via nonradiative (NR) relaxation channels in the material (Choudhary et al. 2018). As the pump power of the excitation source is increased, the phenomenon of laser-induced optical heating becomes more prominent. The phonon frequency of the host material also plays a very crucial role in the NR relaxation, which causes heating in the material (Dwivedi et al. 2020). The other parameters that can affect the optical emissive properties are wavelength, photon density, pump power, etc. (Pandey et al. 2015).

15.2 Classification of Temperature Sensors: Primary and Secondary Thermometers

Thermometers may be classified in many ways. Our approach to classify thermometers will be based on the intensity to temperature calibration relation and physical contact with the probe. Thus, thermometers are divided into two parts: (i) contact (evasive, e.g. thermocouples and thermistors) and (ii) noncontact (minimally evasive, e.g. infrared pyrometers). Contact thermometers are well suited for routine measurements of static systems, whereas noncontact thermometers are appropriate for temperature measurements or mappings on dynamic/moving objects or on objects in hazardous locations (e.g. thermal inspection of devices). The performance of a temperature sensor is related to some thermal parameters, such as operating range, temperature uncertainty, thermal sensitivity, and spatiotemporal resolution. Furthermore, it is also affected by the materials characteristics, such as physical state, simple and easy processable synthesis, facility to be implemented, and mechanical and thermal stability.

Considering the calibration relation, the temperature probes can be classified into primary and secondary thermometers. The temperature determination in a primary thermometer is carried out directly by using the thermodynamic laws and quantities. However, temperature determination in secondary thermometers cannot be done by using the thermodynamic equation of state and hence requires calibration by using a reference thermal probe (Balabhadra et al. 2017; Quinn 2013), thus known as secondary thermometers. In general, the primary thermometers are complex, and temperature determination in primary thermometers requires five thermodynamic measurable quantities: (i) the speed of sound in a monatomic gas, (ii) the pressure of a gas in a constant volume, (iii) the black-body radiation, (iv) the dielectric constant of a gas, and (v) the power spectral density of Johnson-noise in an electrical resistor (Fischer et al. 2011; Collins et al. 1998).

Majority of the secondary thermometers are electrical probes, such as silicon diodes, thermistors, capacitance thermometers, thermocouples, and platinum resistance thermometers (Lojpur et al. 2014). Although the complexity of secondary thermometers is less as compared with the primary one, it demands continuous calibration for functioning in a different medium. This is tedious and time taking for cell level biological applications.

15.3 Performance of Temperature Sensors

The quantitative comparison of the performance of any temperature probe is critical to evaluate its use and for the comparison between distinct techniques. The performance of a temperature sensor can be evaluated based on its relative thermal sensitivity, temperature uncertainty, repeatability, reproducibility, and spatiotemporal resolution. Bünzli has given detailed analysis about the all performing parameters for luminescent thermometers (Bünzli 2016, p. 141).

15.3.1 Thermal Sensitivity

The thermal sensitivity is defined as the rate of change of the thermometric parameter (represented by θ), with respect to the change in temperature. The absolute thermal sensitivity (S_a) is expressed as (Alicki and Leitner 2015)

$$S_a = \frac{\partial\theta}{\partial T} \tag{15.1}$$

where S_a solely depends on the degree of the thermally induced variations in θ. It may be different for different types of thermometers. Thus, the quantitative comparison of the thermal sensitivity of different types of thermometers is meaningless as their operating nature is different in physical principles, such as optical, electrical, or mechanical thermometers. In order to compare the performance of different types of thermometers (irrespective of their nature or the materials employed), the relative thermal sensitivity (S_r) can be a more suitable parameter:

$$S_r = \frac{1}{\theta}\frac{|\partial\theta|}{|\partial T|} \tag{15.2}$$

Collins et al. had introduced this parameter (Collins et al. 1998), and further, it has been widely accepted and used for the comparison of the thermometer's performance. S_r is usually expressed in units of % change per degree of temperature change (% K^{-1}), and S_m is used for the maximum value of S_r for an operating temperature T_m (Bünzli 2016, p. 141).

15.3.2 Thermal Uncertainty *(δT)*

The thermal uncertainty (or temperature resolution) δT is defined as the smallest measured temperature by the thermometer, which can be resolved properly. This depends on the material used as well as the experimental setup. The thermal uncertainty in the temperature arises from several factors associated with the experimentation. These factors are (i) the experimental detection setup, (ii) the environmental conditions, and (iii) the signal-to-noise ratio. Usually, δT depends on the time-dependent output fluctuations of the thermometer and is estimated by calculating the effect of the temporal fluctuations on the thermometric parameter. A calibration curve is used to calculate the temperature that corresponds to each θ; this data is used for constructing a histogram of the temperature readouts during a certain time interval. The experimental δT of the thermometer is the standard deviation of the resulting temperature histogram. When this strategy is not possible to be implemented, an estimate of the temperature uncertainty is given by (Bünzli 2016, p. 141)

$$\delta T_r = \frac{1}{S_r}\frac{\partial\theta}{\theta} \tag{15.3}$$

where $\frac{\partial\theta}{\theta}$ is the relative uncertainty in the determination of the thermometric parameter, and it depends on the acquisition setup and calculated from the errors in θ. This value is controlled by adjusting the signal-to-noise ratio of the emission spectrum used to calculate each θ value, e.g. larger integration times and/or averaging consecutive measurements. Thus, drop in δT increases the acquisition time, i.e. a minimum uncertainty in temperature measurement desires the longer acquisition time.

Alicki and Leitner have given a new strategy to minimize the temperature uncertainty of any thermometer by applying the spin-boson model along with the size- and system-dependent properties (Alicki and Leitner 2015). If the number of atoms in the sample (N_A) and its Debye temperature (T_D) is known for the solid-state nanothermometers, then the relative fluctuation in temperature can be calculated by using the following Eq. (15.4):

$$\delta T = \left(\frac{4T}{3\sqrt{3T_D}} e^{\frac{3T_D}{8T}} \right) \frac{1}{\sqrt{N_A}} T \tag{15.4}$$

For T_D in the range 100–2000 K, the term in parenthesis changes between 0.9 and 1.3, indicating that the order of magnitude of the temperature uncertainty is determined by (Alicki and Leitner 2015)

$$\delta T = \frac{T}{\sqrt{N_A}} \tag{15.5}$$

Thus, the size of the thermal probe plays an important role for achieving minimum δT. Thus, large number of atoms in the thermal probe minimizes the temperature fluctuations and vice versa. For nonentangled particles, the precision ($\delta\eta$) of a general quantity η in quantum metrology is inversely proportional to the number of particles (N_P) (Tóth and Apellaniz 2014):

$$\delta\eta = \frac{1}{N_P} \tag{15.6}$$

Equation (15.6) is known as shot noise scaling. It is obvious that N_P is proportional to N_A; hence, Eq. (15.6) supports the result of the model derived by Alicki and Leitner (Eq. (15.5)).

There are very few reports in the literature that exhibit that the thermal resolution and the size of the luminescent thermometers are correlated. Alaulamie has carried out temperature sensing experiment on Er^{3+}-doped UCNP clusters of different sizes (ranging from 1 to 9 μm) and found that temperature uncertainty/thermal resolution is a function of particle size (Alaulamie 2017). As the cluster size increases, the signal-to-noise ratio also increases, resulting in small fluctuation/uncertainty in temperature (low standard deviation value). The experiment also verifies the theoretical formulation of Eqs. (15.5) and (15.6). Equations (15.3) and (15.5) are only compatible if the decrease in $1/S_r$ is less than the increment of $\delta\theta/\theta$, resulting in an overall increase of δT with the decrease of N_A.

15.3.3 Reproducibility and Repeatability

The thermal reproducibility can be defined as the change of the same measurement carried out under different conditions, such as the use of various equipment, different measuring techniques, different experimentalists, etc.

The repeatability, R, is the ability of the thermometer to reproduce repeatedly the same result under the same circumstances and is computed by

$$R = 1 - \frac{\max |\theta_c - \theta_i|}{\theta_c} \tag{15.7}$$

where θ_c represents the thermometric parameter's mean value and θ_i represents the thermometric parameter measured at different temperatures (T_i).

15.4 Temperature Sensing with Luminescence

The temperature is a physical quantity that affects the luminescence internally as well as externally. This internal and external effect depends on the domination of radiative and NR processes over each other. Upon exposing luminescent materials with pumping source, if NR relaxation dominates the radiative relaxation, the luminescent material generates heat very fast and becomes warm, whereas the luminescence intensity observed is negligible; this process is known as self-induced heating or optical heating. On the other hand, if the external temperature is increased and still the radiative process dominates over the NR process, a sufficient amount of luminescence intensity is observed, and it decreases with the rise of temperature. This method is used for the determination of temperature and termed as temperature sensing technique. This temperature measurement technique uses the following features of luminescence: (i) the integrated emission intensity of a single transition or of a pair of transitions; (ii) the spectral shift, band shape, or bandwidth of a given transition; and (iii) lifetime measurements, using the time–decay intensity profiles of emitting excited states. It has been shown in Figure 15.1.

Luminescence thermometry is carried out by exploiting the temperature-induced changes in the emission intensity or band shape/size, known as "time-integrated scheme." Further, for the calculation of the thermometric parameter (θ) based on

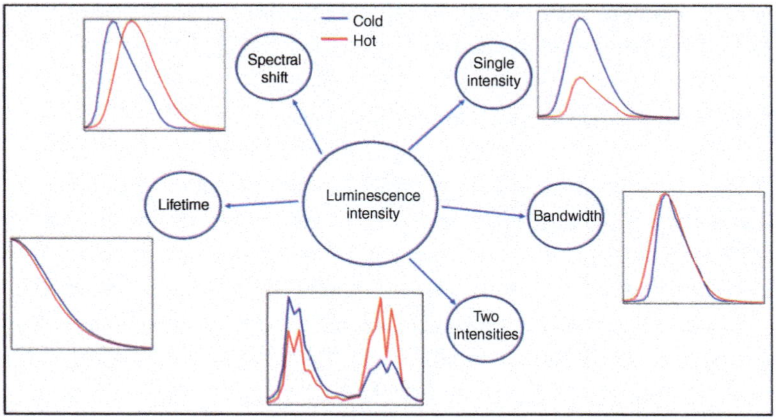

Figure 15.1 Schematic diagram of the effect of temperature on the emission spectrum of the phosphor materials.

integrated intensities, spectral shifts, or bandwidths, the emission spectra must be represented as a function of energy to circumvent misleading and incorrect conclusions. For this correction, the photon flux per constant wavelength interval function, $\varphi(\lambda)$, must be converted to the photon flux per energy interval, $\varphi(E)$, according to (Blasse and Grabmaier 1994; Mooney and Kambhampati 2013) the following equation:

$$\varphi(E) = \varphi(\lambda)\frac{\mathrm{d}\lambda}{\mathrm{d}E} = \varphi(\lambda)\frac{\mathrm{d}}{\mathrm{d}E}\left(\frac{hc}{E}\right) = -\varphi(\lambda)\frac{hc}{E^2} \tag{15.8}$$

where h is the Planck constant, c is the speed of light, and the minus sign indicates the distinct directions of integration in λ and E. This correction (known as Jacobian transformation) was seldom performed in the field of luminescence thermometry (in fact, more generally in luminescence spectroscopy) but can be critical, especially when the spectrum displays multiple peaks over a wide energy range and with different bandwidths (Mooney and Kambhampati 2013).

15.4.1 Time-Integrated Schemes

The radiative transitions among the energy states of trivalent lanthanides (Ln^{3+}) are sensitive to the temperature. Thus, temperature also affects the intensity of emission in Ln^{3+} because of the following processes:

- Population redistribution over electronic levels by following the Boltzmann statistics;
- Temperature-activated quenching mechanisms (e.g. cross-relaxation between electronic levels);
- Thermal (nonradiative) deactivations (the electrons collide to each other, causing heat generation during transitions);
- Phonon-assisted Auger conversion processes;
- Luminescent thermometers based on the intensity of a single transition are highly dependent on phonon-assisted oscillations, signal-to-noise ratio, absorption cross sections, scattering, and local fluctuations on the phosphor concentration.

15.4.1.1 Fluorescence Intensity Ratio (FIR) or Band Shape

FIR is defined as the relative change in the population (thus, intensity ratio) of two independent and closely separated energy levels because of radiative transitions. In the band shape-based nanothermometry, the emission lines can be generated from a single luminescent center, termed as single-center thermometers, or they can result from two distinct emitting centers, termed as dual-center thermometers (Jaque and Vetrone 2012; Brites et al. 2012). The use of single-center thermometers was firstly introduced by Kusama et al. (1976). Further, Wade et al. have given detailed review of the technique (Wade et al. 2003).

In single-center ratiometric thermometers, θ (or FIR) is defined by using the emission intensities of the $|2\rangle \rightarrow |0\rangle$ ($I_{02} \equiv I_2$) and $|1\rangle \rightarrow |0\rangle$ ($I_{01} \equiv I_1$) transitions, where $|0\rangle$ denotes the ground level and $|1\rangle$ and $|2\rangle$ denote the two thermally coupled excited levels (level $|2\rangle$ is more energetic than level $|1\rangle$) (Shinn et al. 1983; León-Luis et al. 2012; Khalid and Kontis 2009):

$$\theta = \frac{I_2}{I_1} = \frac{A_{02}h\nu_{02}N_2}{A_{01}h\nu_{01}N_1} \tag{15.9}$$

where N_1 and N_2 are the populations of the $|1\rangle$ and $|2\rangle$ energy levels, ν_{01} and ν_{02} are the frequencies of the $|1\rangle \rightarrow |0\rangle$ and $|2\rangle \rightarrow |0\rangle$ transitions, and A_{01} and A_{02} are the total spontaneous emission rates from energy levels $|1\rangle$ and $|2\rangle$ to level $|0\rangle$, respectively. Further, if the depopulation from the $|1\rangle$ and $|2\rangle$ energy levels do not occur to the energy level $|0\rangle$ and it takes place through some other intermediate level, then Eq. (15.9) needs to be corrected by the β_2/β_1 ratio, where β_i ($i = 1, 2$) is the branching ratio of the $|i\rangle$ level (i.e. the percentage of the total emission from the thermalized levels ($|1\rangle$ or $|2\rangle$) to the final $|0\rangle$ state). If the two high energetic levels are in thermal equilibrium (called as "thermally coupled levels," with energetic separations of the order of thermal energy $k_\beta T$), N_1 and N_2 are related by

$$N_2 = \frac{g_2}{g_1} N_1 \exp\left(-\frac{\delta E}{k_\beta T}\right) \tag{15.10}$$

where g_1 and g_2 are the degeneracies of the two energy levels and δE is the energy gap between the barycenters of the $|1\rangle \rightarrow |0\rangle$ and $|2\rangle \rightarrow |0\rangle$ emission bands. Thus, Eq. (15.9) can be written as

$$\theta = \frac{g_2A_{02}h\nu_{02}}{g_1A_{01}h\nu_{01}} \exp\left(-\frac{\delta E}{k_\beta T}\right) = \mathrm{B} \exp\left(-\frac{\delta E}{k_\beta T}\right) \tag{15.11}$$

where $B = \frac{g_2A_{02}h\nu_{02}}{g_1A_{01}h\nu_{01}}$. In order to observe thermalization, the two energy levels must not be widely separated; otherwise, this effect cannot be detected. Typically, these closely spaced two levels are known as "thermally coupled levels" (e.g. in a thermodynamically quasi equilibrium state). The energy gap (δE) between these two levels should lie in the range of 200–2000 cm^{-1} (Carlson et al. 2011).

Until now, most of the luminescent nanothermometers are single-center thermometers that require the FIR of thermally coupled energy levels for the determination of temperature, and the parameters such as δE and B are calculated from a fitting, which demands for an external thermal calibration from an independent thermal source (i.e. a thermocouple or an infrared thermal mapping camera) in order to estimate the thermometric parameters for the intensity to temperature conversion. The alternate medium and different environments (pressure, pH, ionic strength or atmospheric compositions, etc.) disturb the thermometric parameters and thus the calibration. Hence, this technique is medium and environment dependent and requires a new calibration curve for a new condition, making this technique tedious.

However, recently, Balabhadra et al. have calculated δE and B without using any experimental calibration curve (Balabhadra et al. 2017). δE is calculated by using an envelope fitting to the I_1 and I_2 transitions, which requires no experimental calibration, and the parameter B is empirically inferred from the plot of θ versus laser excitation power, provided that the laser-induced heating is negligible, which is possible near null excitation power (Bünzli 2016, p. 141; Wei et al. 2019). Then, the temperature (T_0) corresponds to the room temperature (Carlson et al. 2011), and the thermometric parameter θ_0 is expressed by

$$\theta_0 = B \exp\left(-\frac{\delta E}{k_\beta \mathrm{T}_0}\right) \tag{15.12}$$

The magnitude of absolute temperature is calculated by taking the ratio between Eqs. (15.11) and (15.12), giving the ratio θ/θ_0, which is represented in the form of Eq. (15.13)

$$\frac{1}{T} = \frac{1}{T_0} - \frac{k_\beta}{\delta E} \ln \frac{\theta}{\theta_0} \tag{15.13}$$

Balabhadra et al. have compared the temperature calculated through Eq. (15.13) with the readings of conventional thermometers kept in contact with the particles (for powders) or immersed in aqueous suspensions and found a quite remarkable agreement between theoretical (calculated using Eq. (15.13)) and experimentally observed temperature, irrespective of the NPs' size and of the dispersion media (Balabhadra et al. 2017). It is notable that this approach is general and is extensible to any other thermometer based on two "thermally coupled" levels. The parameter B can also be calculated by using the Judd–Ofelt theory (Judd 1962; Ofelt 1962). For that, let us consider the integrated coefficient of spontaneous emission of a transition between J and J' manifolds as (Binnemans 2009; de Sá et al. 2000; Malta and Carlos 2003; Carnall et al. 1977)

$$A_{JJ'} = \frac{64\pi^4 e^2 v_i^3}{3hc^3}\left[\frac{n\left(n^2+2\right)^2}{9} S_{\mathrm{ed}} + n^3 S_{\mathrm{md}}\right] \tag{15.14}$$

where e is the charge of electron and n is the refractive index of the medium. The electric (S_{ed}) and magnetic (S_{md}) dipole strengths are given (in units of e^2) by

$$S_{\mathrm{ed}} = \frac{1}{(2J+1)} \sum_{\lambda=2,4,6} \Omega_\lambda |J'\|U^\lambda\|J|^2 \tag{15.15}$$

$$S_{\mathrm{md}} = \frac{h^2}{16\pi^2 mc^2} |0\|U^\lambda\|i|^2 \tag{15.16}$$

where the quantities Ω_λ ($\lambda = 2, 4,$ and 6) are called Judd–Ofelt intensity parameters (Judd 1962; Ofelt 1962) and m is the mass of electron. $\langle\|U(\lambda)\|\rangle$ and $\langle\|L+2S\|\rangle$ are reduced matrix elements (tabulated by Carnall et al. 1977), and the angular operators L and S are in units of h. The $\Omega_{2,4,6}$ parameters in Eq. (15.15) are phenomenologically obtained from the absorption (Gorller-Walrand and Binnemans 1998, p. 101).

15.4.1.2 Bandwidth

Generally, the full width at half maxima (FWHM) of spectral emitting line is temperature dependent, and the bandwidth is broadened as the temperature increases. This thermal broadening also depends on the materials characteristics; a homogeneous broadening is due to the phonon frequencies (as known as lattice vibrations) of the material and its high temperature dependence. On the other hand, inhomogeneous broadening is due to the presence of different optical centers and materials defects. The temperature dependence of the bandwidth (W) of emission and absorption bands is described by Henderson and Imbusch (1989) as

$$W(T) = W_0 \sqrt{\coth \frac{h\Omega}{2k_\beta T}} \tag{15.17}$$

where W_0 is the FWHM of the band at 0 K and $h\Omega$ is the phonon energy (lattice vibration) that interacts with the electronic transitions. There are few reports using intra-4f line's emission bandwidth to measure the temperature, despite that all of them present the functional form predicted by Eq. (15.17). Peng et al. have applied Eq. (15.17) for the estimation of temperature with broadening by using the $^5D_0 \rightarrow {}^7F_2$ transition in the Y_2O_3:Eu^{3+} phosphors in the temperature range 10–670 K. Interestingly, the transition bandwidth remains unchanged below 70 K, and linear increment in broadening was observed over 70 K at a rate of 0.078 cm^{-1} K^{-1} with a maximum relative sensitivity (S_r) of ~ 0.78% K^{-1} (at 70 K) (Peng et al. 2003).

15.4.2 Lifetime Technique

Unlike the luminescence intensity methods, the lifetime of a luminescent material is unaffected by the size, geometry, and the concentration of the luminescent probe, and it is also independent of physical parameters such as reflection, refraction, scattering, or intensity fluctuations of the excitation source. These parameters suggest that lifetime is a better technique for measuring thermal luminescent behavior. However, requirement of long lifetime determination needs a pulsed excitation source with long illumination and acquisition time, which makes it an expensive and time-consuming technique. It is also difficult to carry out measurement for high temperature as temperature gradients at time intervals shorter than or equal to the lifetime of the luminescence make the technique impractical. If we make a comparison between the FIR and lifetime schemes, we can conclude that the lifetime technique requires more complex detection systems and the outcome of thermal sensitivity is the same as that of the FIR technique. Therefore, the wisest approach to measure temperature variation is the ratiometric intensity method.

15.5 Upconversion (UC) and UC-Based Thermal Sensor of Ln^{3+} Ions

15.5.1 Upconversion (UC) and Upconverting Nanoparticles (UCNPs)

Upconversion (UC) emission is a nonlinear and radiative process, in which one gets blue-shifted (anti-Stoke's like) emission via real energy levels (Bloembergen 1959; Auzel 1973). A detailed review article by Prodi et al. explains all mechanisms involved in UC emission and its application as thermal sensor (Prodi et al. 2015). The UC emission can be classified on the basis of UC emission signals. A single-center emission is generated by using a single Ln^{3+} ion, and a multicenter emission is generated by using a combination of different Ln^{3+} ions.

15.5.2 Single-Center UC Nanothermometers and Multicenter UC Nanothermometers

The most commonly used Ln^{3+} ions for UC emissions are Er^{3+}, Ho^{3+}, Tm^{3+}, Tb^{3+}, and Eu^{3+} along with the Yb^{3+} ions. Here, Yb^{3+} ions play the role of a sensitizer because of higher absorption cross section and well-matched energy levels with 980 nm pumping source, while other Ln^{3+} ions play the role of activators, which give UC emission. Furthermore, the excited-state energy of Er^{3+}, Ho^{3+}, Tm^{3+}, Tb^{3+}, and Eu^{3+} matches well with the excited states of Yb^{3+}, thus allowing an efficient resonant energy transfer. As an activator, Er^{3+} ion is one of the most used upconverting lanthanides, and it is also one of the most used lanthanide for temperature sensing in green region arising from the ${}^2H_{11/2} \rightarrow {}^4I_{15/2}$ (515–525 nm) and ${}^4S_{3/2} \rightarrow {}^4I_{15/2}$ (535–545 nm) transitions (Mooney and Kambhampati 2013; Shinn et al. 1983). Based on that emission and strong thermal dependence, Rodrigues et al. measured one of the highest relative sensitivity (S_r) values for UCNPs through thermal sensing by using β-$NaGd_{0.94}Pr_{0.02}Er_{0.02}Yb_{0.02}F_4$@$3NaY_{0.8}Yb_{0.2}F_4$ core@shell NPs. The visible emission of Er^{3+} shows a relative thermal sensitivity in the 83–323 K range under the exposure of 980 nm pumping, and the value of S_m is 9.52% K^{-1} at 83 K (Rodrigues et al. 2018). Nigoghossian et al. have developed a complex system (CS) by merging the $(tta)_3$ complex with silica-coated $NaGdF_4$:Yb^{3+}/Er^{3+} UCNPs. The Eu^{3+} complex is excited by UV source (352 nm) to give intense emission because of the ${}^5D_0 \rightarrow {}^7F_2$ (603–643 nm) line and later is excited with 980 nm laser to give emission in the green region. The temperature sensing is measured in the physiological temperature range (293–323 K), which reveals thermal parameter (S_m) at 303 K, $S_m = 2.48\%\ K^{-1}$, $\delta T = 0.47$ K for 980 nm excitation and $S_m = 2.67\%\ K^{-1}$, $\underline{\delta} T = 0.06$ K for 352 nm (UV) excitation (Nigoghossian et al. 2017).

The major limitations of nanothermometers are their functioning at higher temperatures (>400 K). There are very few reports of nanothermometers with proper functioning above 400 K, in which Er^{3+}, Nd^{3+}, Tm^{3+}, and Ho^{3+} are used as activator ions (Geitenbeek et al. 2017; Kolesnikov et al. 2017; Sedlmeier et al. 2012; Lojpur et al. 2013; Dong et al. 2011; Marciniak et al. 2017a,b). Geitenbeek et al. have recorded temperature sensing in the temperature range of 300–600 K in bare $NaYF_4$:Yb^{3+}/Er^{3+} NPs (as shown in Figure 15.2), and beyond 600 K, this bare material starts deteriorating. In order to further extend the temperature sensing, $NaYF_4$:Yb^{3+}/Er^{3+} NPs was coated with silica core and thus reported temperature sensing up to 900 K with S_r values decreasing from 1.02% K^{-1} at 300 K to 0.13% K^{-1} at 900 K (Geitenbeek et al. 2017). Kolesnikov et al. have also designed a nanothermometer in YVO_4:Nd^{3+} NPs up to about 900 K (Kolesnikov et al. 2017).

The relative sensitivity of thermally coupled levels is the intensity ratio taken from two transitions arising from coupled levels; the thermometric parameter (S_r) can be calculated by using Eqs. (15.11) and (15.2), resulting

$$S_r = \frac{1}{B\exp\left(-\frac{\delta E}{k_\beta T}\right)} \times \left| \frac{\partial B}{\partial T}\exp\left(-\frac{\delta E}{k_\beta T}\right) - B\frac{\delta E}{k_\beta T^2}\exp\left(-\frac{\delta E}{k_\beta T}\right) \right|$$
$$= \left| \frac{1}{B}\frac{\partial B}{\partial T} - \frac{\delta E}{k_\beta T^2} \right| \qquad (15.18)$$

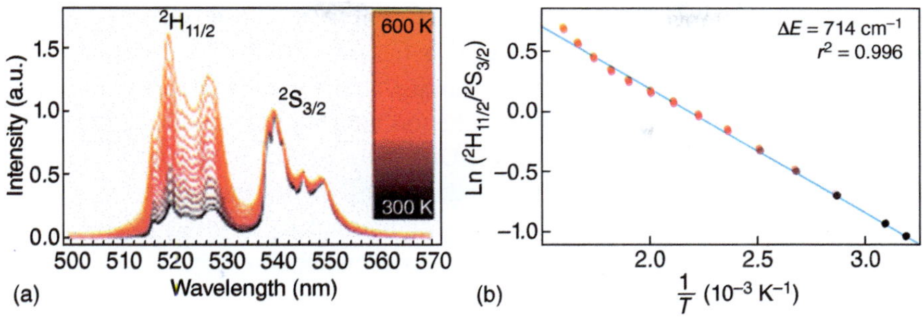

Figure 15.2 Emission spectra of $NaYF_4:Yb^{3+}/Er^{3+}$ UCNPs upon excitation at 980 nm for temperatures ranging from 300 to 600 K recorded with steps of 25 K (a). The linear relation between the natural logarithm of the integrated intensity ratio of the $^2H_{11/2}$ and $^4S_{3/2}$ peak versus $1/T$ (b). The r-squared value for the fit, 0.996, is close to unity, and the steepness of the fit corresponds to a ΔE of 714 cm^{-1}. Five spectra were measured at every temperature interval to estimate the error in the measurements; this error is within the dot size of the data points. Source: Geitenbeek et al. (2017). Reprinted with permission of American Chemical Society.

Therefore, Eq. (15.18) shows that B and δE are two possible temperature-dependent parameters, which create two possibilities: (i) If B is temperature independent, then the S_r value is solely dependent on δE and S_r decreases with the increase in temperature. (ii) If B is also a function of temperature, S_r is also affected by the temperature dependence of the branching ratios of the two thermally coupled levels. The variation in S_r value is also affected by the sample parameter, such as phonon frequency, doping concentration of Ln^{3+}, size, and morphology. Alencar et al. and Marciniak et al. have discussed about the impact of the size of nanoparticles on S_r value in their works (Alencar et al. 2004; Marciniak et al. 2017a,b). Alencar et al. have worked on nanothermometry of $BaTiO_3:Er^{3+}$ nanocrystals, in which the intensity ratio between two thermally coupled levels, $^4S_{3/2} \rightarrow {}^4I1_{5/2}$ and $^2H_{11/2} \rightarrow {}^4I_{15/2}$ of Er^{3+} transitions (the energy gap [δE] of these transitions remained constant), were monitored by varying the crystal size in the range of 60–26 nm as a function of temperature, and they found firstly the dependence of size of the nanoparticles on S_r value (Alencar et al. 2004).

A similar tendency is observed by Marciniak et al. in two host matrices $NaYF_4:Yb^{3+}/Er^{3+}$ and $LiLaP_4O_{12}:Yb^{3+}/Er^{3+}$. On varying the particle size in the range of 200 to 8 nm in the $NaYF_4:Yb^{3+}/Er^{3+}$ nanophosphors, the S_m values raise from 1.1% to 2.1% K^{-1} at 200 K (Marciniak et al. 2017a,b). Similarly, in the case of $LiLaP_4O_{12}:Yb^{3+}/Er^{3+}$, the variation in S_m value was from 1.1% to 1.8% K^{-1} at 200 K on changing particle size in the range of 240 to 20 nm. The increase in S_m value on decreasing particle size is due to the larger surface area of smaller particle size and enhancement in the number of optically active ions on the surface. This contributes to an increase in optical activity, which is affected by the electron–phonon interactions and phonon density that increases NR radiations on increasing the temperature (Alencar et al. 2004; Marciniak et al. 2017a,b). However, Brandão-Silva et al. have not observed any significant change on S_r values on decreasing the particle size of the $Y_2O_3:Er^{3+}$ nanophosphors from 86 to 21 nm. Thus, the effect of particle size of nanoparticles on relative thermal sensitivity (S_r) values for nanothermometers cannot be clearly concluded. It is probably dependent on the host matrix, phonon energy, or electron phonon interactions (Brandão-Silva et al. 2018). It is still an area of deep research. One more limitation of single-centered UC nanothermometer is the strict restriction on δE values that lies from 200 to 2000 cm^{-1}. If δE is greater than 2000 cm^{-1}, the thermalization effect of thermally coupled levels starts disappearing, resulting poor emission intensity.

Thus, in order to increase the S_r value, a new idea of multicenter UC thermometry based on "moderately coupled levels" or "fully decoupled levels" has been considered. The moderately coupled levels are the levels for which the FIR of emission intensity of two thermally decoupled levels of same Ln^{3+} ions are considered for observing the thermalization effect. This is also known as indirect thermal coupling, as energy transfer is carried out by host matrices or ligands. The fully decoupled levels are the levels for which the FIR of emission intensity of two levels of distinct Ln^{3+} ions is chosen for observing the thermalization effect. However, there are very few reports available on multicenter UC thermometry (Marin et al. 2018; Yang et al. 2018; Zhou et al. 2014; Zheng et al. 2014; Xu et al. 2016; Pandey and Rai 2013;Savchuk et al. 2018).

The core@shell structure technique is an efficient technique to transfer energy by incorporating different dopants (Ln^{3+}), and this technique has been reported for noncontact thermal sensing (Skripka et al. 2017; Marciniak et al. 2016; Labrador-Páez et al. 2018). Xu et al. have developed core@shell nanophosphors ($Yb/Ho/Ce:NaGdF_4@Yb/Tm:NaYF_4$) and recorded the thermalization effect with and without Ce^{3+}. In order to measure S_m values, the ratio of thermally decoupled energy levels of red (Ho^{3+}:$^5F_5 \rightarrow {}^5I_8$ and Tm^{3+}:$^1G_4 \rightarrow {}^3F_4$ transitions) and green (Ho^{3+}: $^5S_2$, $^5F_4 \rightarrow {}^5I_8$ transition) bands has been taken. The value of S_m was found to be ~2.4% K^{-1} (at 298 K) in the absence of Ce^{3+}. The presence of Ce^{3+} in the core tunes S_m between 0.7% and 4.4% K^{-1}, and this is due to the cross-relaxation processes between the Ce^{3+} and Ho^{3+} ions (Xu et al. 2016). A report of a nanothermometer on $KLuWO_4:Tm^{3+}/Ho^{3+}$ NPs under 808 nm excitation has been published by Savchuk et al. They have taken the intensity ratio of the Tm^{3+} $^3F_{2,3} \rightarrow {}^3H_6$ line (at 696 nm) and that of Ho^{3+} $^5S_2$ and $^5F_4 \rightarrow {}^5I_7$ (at 755 nm) and reported $S_m = 2.8\%\ K^{-1}$ and $\delta T = 0.2$ K (at 300 K) (Savchuk et al. 2018).

15.5.3 Complex Systems

This section will deal with the temperature sensing application of complex systems such as core@shell nanostructures (Skripka et al. 2017; Marciniak et al. 2016; Labrador-Páez et al. 2018), polymer- and hybrid-based materials (Huang et al. 2015, 2016; Chen et al. 2013a,b; Xiao et al. 2014; Wang et al. 2015a,b), and heater–thermometer nanoplatforms (Rohani et al. 2015). The multiple core@shell systems have shown massive potential to be used for multiple center nanothermometry as it combines multiple Ln^{3+}, i.e. the multiple activators in a single luminescent system. Marciniak et al. have developed the active-core/active-shell (ACAS)

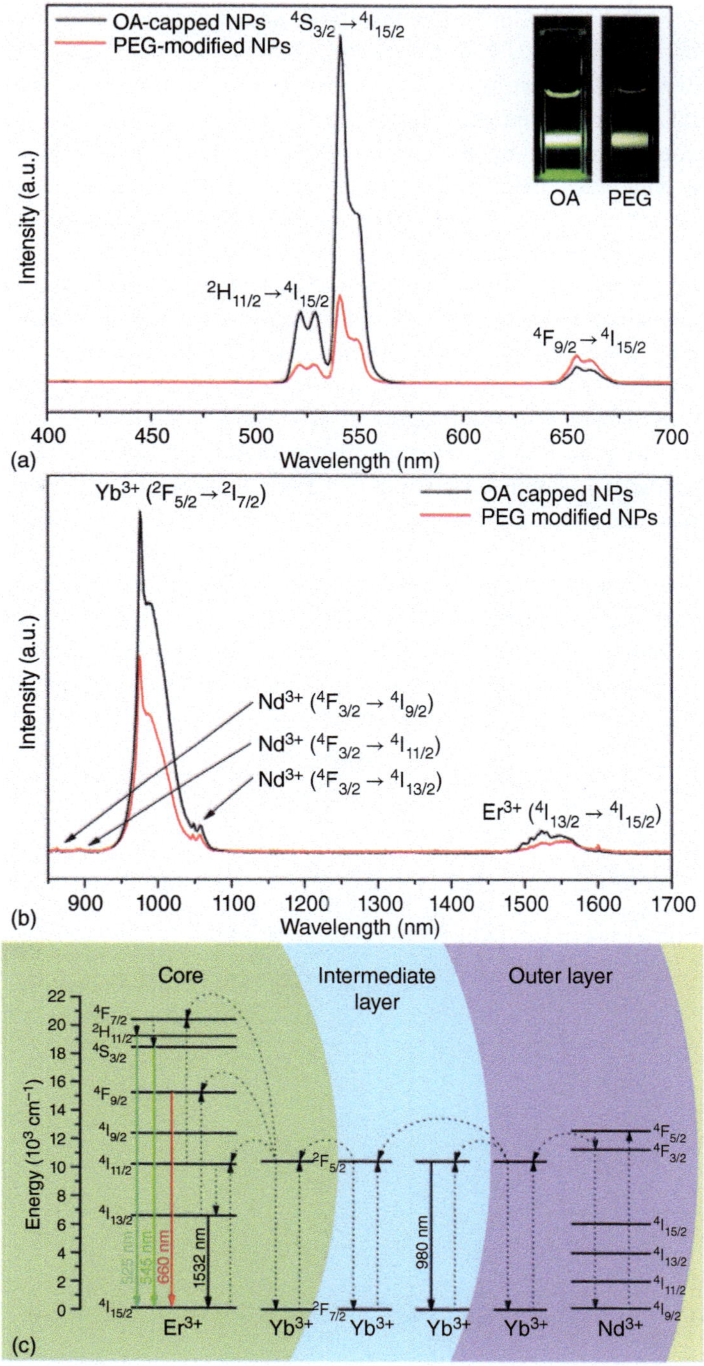

Figure 15.3 (a) UC luminescence spectra of OA-capped C@S1@S2@S3 NPs in cyclohexane and PEG-modified C@S1@S2@S3 NPs in water. Luminescence photographs of the corresponding samples are shown in the insets. (b) DC luminescence spectra of OA-capped C@S1@S2@S3 NPs in cyclohexane and PEG-modified C@S1@S2@S3 NPs in water. (c) Proposed UC and DC mechanisms in the core–shell system. Source: Wang et al. (2015a,b). Reprinted with permission of Royal Society of Chemistry.

(β-$NaYF_4$:Yb^{3+},Er^{3+}/β-$NaYF_4$:Yb^{3+},Nd^{3+}) nanoparticles and recorded temperature sensing for the Er^{3+} UC and Yb^{3+}/Nd^{3+} DS (downshifting) emissions (Marciniak et al. 2016). The values of relative maximum sensitivity (S_m) were measured to be 2.0% K^{-1} (at 200 K, Er^{3+}-based core thermometer) and 2.5% K^{-1} (at 365 K, Yb^{3+}/Nd^{3+}-based shell thermometer) (Marciniak et al. 2016). Another complex was developed with the core of $NaLuF_4$:Gd^{3+}/Yb^{3+}/Er^{3+} and coating of three layers as $NaLuF_4$:Gd^{3+}/Yb^{3+}/Er^{3+}@$NaLuF_4$:Yb^{3+}@$NaLuF_4$:Nd^{3+}/Yb^{3+}@$NaLuF_4$core@shell system for both UC and DC emissions under 808 nm excitation. The emission spectra and energy level diagram are shown in Figure 15.3. The sensitivity increases gradually with the temperature rising from 223 to 423 K, and the maximum sensitivity was calculated as 0.002 52 K^{-1} at 423 K (as shown in Figure 15.4). Further, the intensity ratio of the Yb^{3+} 980 nm ($^2F_{5/2} \rightarrow {}^2F_{7/2}$) and the Er^{3+} 1532 nm transitions ($^4I_{13/2} \rightarrow {}^4I_{15/2}$) was used for in vivo NIR imaging studies (Wang et al. 2015a,b).

A new luminescent-based fully noncontact nanothermometer has been developed by combining UCNPs with plasmonic nanostructures. These types of nanostructures combine with the heat realization and temperature monitoring in a controlled single system. Rohani et al. have combined gold (Au) nanorods with $NaGdF_4$:Yb^{3+}/Er^{3+} UCNPs and tuned the longitudinal surface plasmon resonance of the nanorods at 980 nm, which is obviously resonant with the Yb^{3+} absorption wavelength. Thus, luminescence thermometry and heating can be simultaneously observed as well as

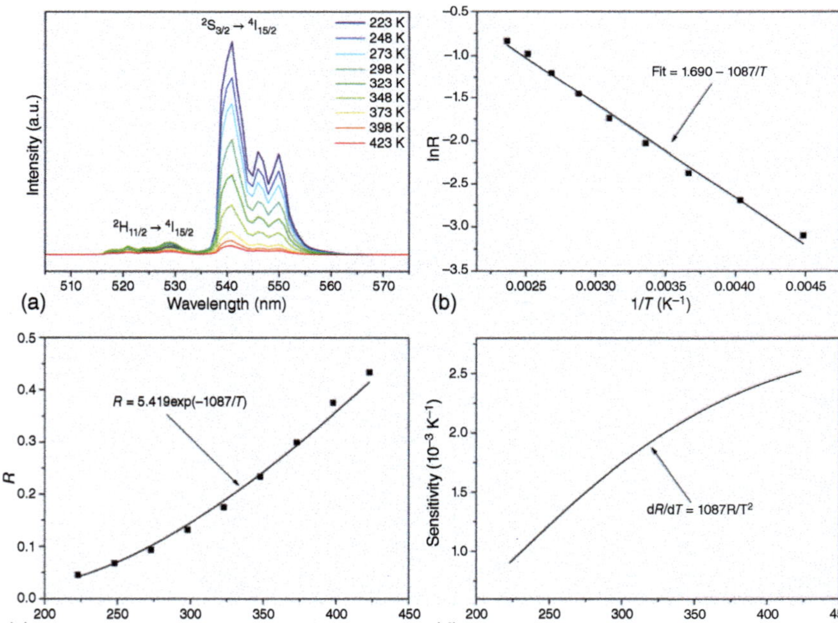

Figure 15.4 (a) UC luminescence spectra of Er^{3+} from $^2H_{11/2}$ and $^4S_{3/2}$ levels to $^4I_{15/2}$ level at different temperatures. (b) The monolog plot of *R* as a function of inverse absolute temperature. (c) *R* relative to the absolute temperature. (d) The sensitivity as a function of the absolute temperature. Source: Wang et al. (2015a,b). Reprinted with permission of Royal Society of Chemistry.

measured, which enhances the emission of the Er^{3+} ions ninefolds. In this process, temperature is controlled by the excitation power of 980 nm laser, and the measurement of temperature is carried out by the thermal-sensitive emission of Er^{3+} ions (Rohani et al. 2015).

15.6 Optical Heating

Beyond optical thermometry, the photothermal (PT) effect is also simultaneously generated in UCNPs via phonon-assisted NR channels. This effect is an inherent property of UCNPs or phosphor materials containing lanthanide ions and varies from material to material. The primary reason is the presence of abundant intermediate energy levels with small energy gaps in trivalent lanthanides, which result in fast relaxation of electrons/atoms, causing prominent heating inside the material. Thus, the UCNPs may play the important role of optical heaters, which have great potential applications in the local thermal treatment or drilling nanoholes in organics and soft materials (Wen et al. 2018; Dey and Rai 2017; Du et al. 2019). Therefore, characterization and evaluation of the PT properties of these UCNPs are important for their future applications as optical heaters.

Du et al. have studied optical heating in the $Er^{3+}/Yb^{3+}/Fe^{3+}$-tridoped $NaBiF_4$ nanoparticles and reported a maximum internal temperature of ~ 399 K at 260 mW of 980 nm diode laser power (Du et al. 2019). Choudhary et al. (2018) have studied and compared the laser-induced optical heating in $CaAl_{(0.793)2}Er_{0.007}Yb_{0.05}Zn_xO_4$ phosphors (where x = 0, 0.5, 0.10, 0.15, and 0.20) by varying the concentration of Zn^{2+} ions and reported maximum optical heating at 15% concentration Zn^{2+} ions (as shown in Table 15.1). They have reported heating over 400 K above 1 W

Table 15.1 Comparison between the calculated value and measured value of temperature at different input pump power of laser for CAO:Er,Yb and CAO:Er,Yb, Zn phosphors.

	Temperature (K) for CAO:Er,Yb		Temperature (K) for CAO:Er,Yb, Zn	
Laser power (W)	Calculated value	Measured value	Calculated value	Measured value
0.4	298.528 99	309	312.057 05	309
0.5	312.112 14	311	329.698 31	315
0.6	327.451 34	316	347.430 11	321
0.7	342.389 11	319	368.003 45	325
0.8	355.488 35	321	387.805 61	329
0.9	366.841 23	326	402.608 73	335
1.0	375.044 58	329	426.081 76	339
1.1	386.552 5	332	447.805 48	345
1.2	396.139 38	337	473.247 24	351
1.3	405.400 28	341	502.773 88	354

Source: Choudhary et al. (2018). Reprinted with permission of Institute of Physics Publishing.

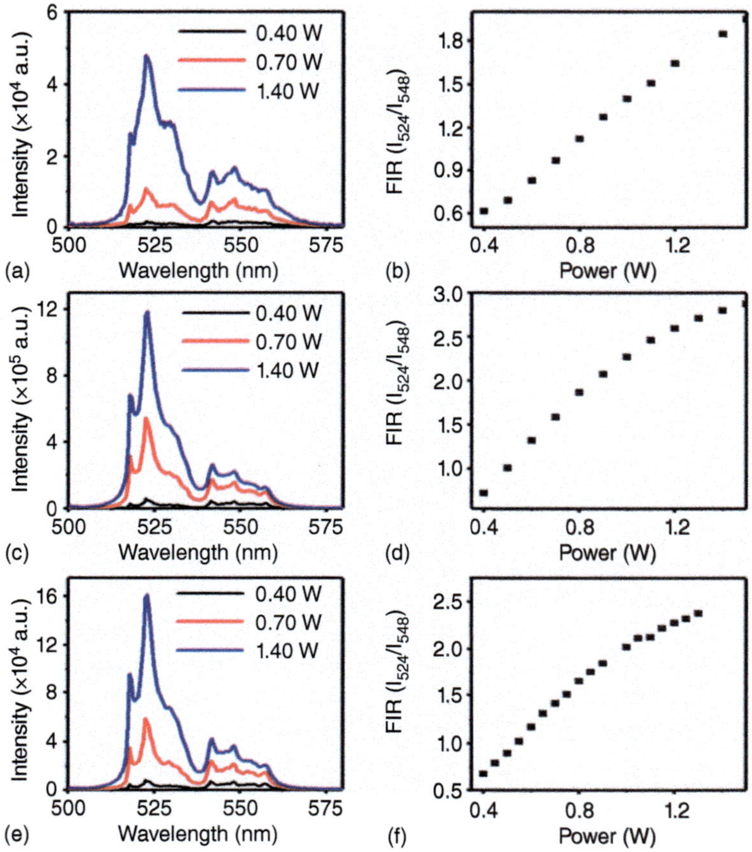

Figure 15.5 A variation in the UC emission intensity of $^2H_{11/2} \rightarrow {}^4I_{15/2}$ (524 nm) and $^4S_{3/2} \rightarrow {}^4I_{15/2}$ (548 nm) transitions and their intensity ratio (FIR) with a variation in input pump power of diode laser, respectively, for (a, b) CAO:Er,Yb at 1473 K, (c, d) CAO:Er,Yb at 1623 K, and (e, f) CAO:Er,Yb, Zn at 1473 K. Source: Choudhary et al. (2018). Reprinted with permission of Institute of Physics Publishing.

pump power (shown in Figure 15.5) (Choudhary et al. 2018). Dwivedi et al. have studied the effect of phonon frequency on optical heating phenomenon in the Tm^{3+}/Yb^{3+}-doped XVO_4, X_2O_3 (where X = Y or Gd) and $CaAl_2O_3$. The variation in the emission intensities of Tm^{3+} in different host matrices are monitored as a function of time, and it has been shown in Figures 15.6–15.8 (Dwivedi et al. 2020). The highest optical heating was observed (as shown in Table 15.2) in self-activated host matrix $GdVO_4$ (having highest phonon frequency). Pandey et al. have carried out power-dependent temperature sensing and optical heating in the Er^{3+}/Yb^{3+}-doped $SrWO_4$ phosphor. They have concluded the optical heating using optical thermometry and reported heat generation in the phosphor up to 417 K (Pandey et al. 2015). Yadav et al. have also studied the optical heating through the three-photon UC process in the Tm^{3+}-, Yb^{3+}-, and Zn^{2+}-doped $ZnWO_4$ (Yadav et al. 2018).

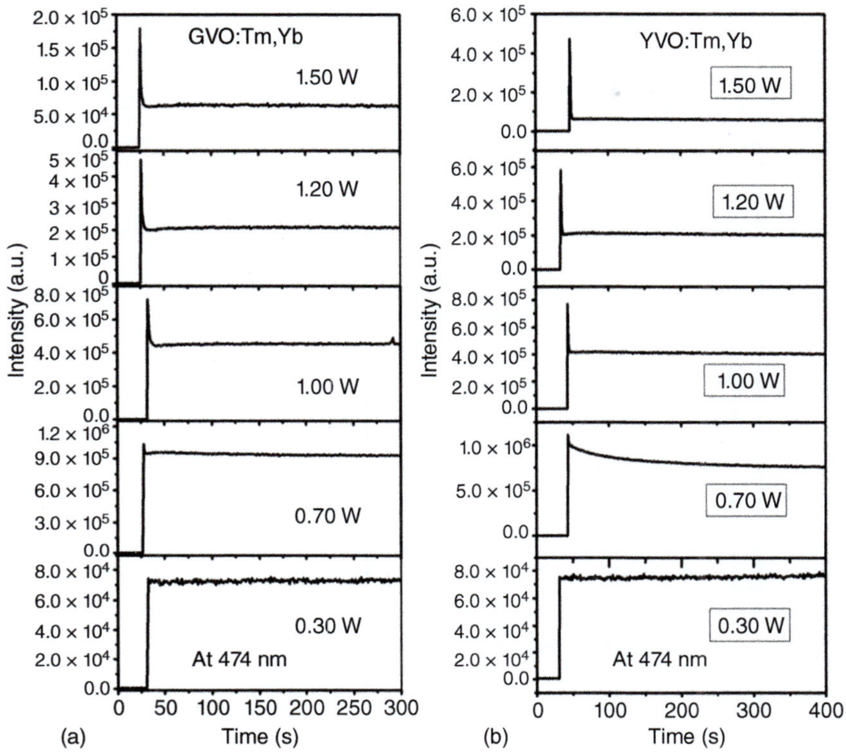

Figure 15.6 Variation of blue UC emission (at 474 nm) intensity with time (in Sec.) for (a) GVO:Tm,Yb and (b) YVO:Tm,Yb phosphors. Source: Dwivedi et al. (2020). Reprinted with permission of Elsevier.

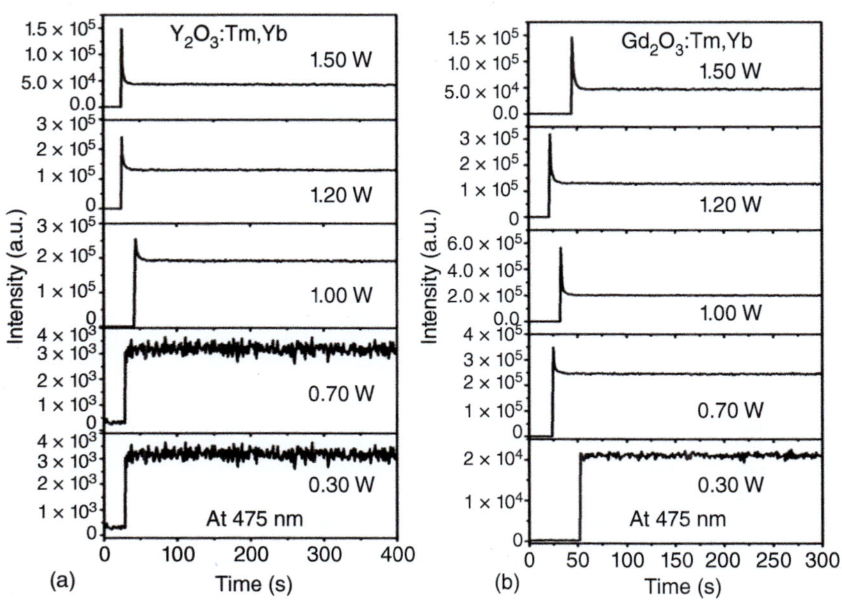

Figure 15.7 Variation of blue UC emission (at 475 nm) intensity with time (in Seconds) for (a) Y_2O_3:Tm,Yb and (b) Gd_2O_3:Tm,Yb phosphors. Source: Dwivedi et al. (2020). Reprinted with permission of Elsevier.

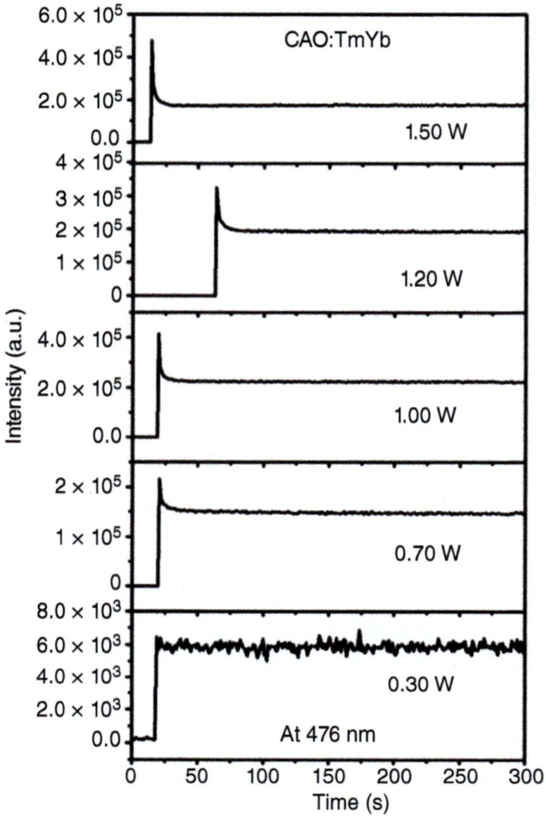

Figure 15.8 Variation of blue UC emission (at 476 nm) intensity with time (in Sec.) for CAO:Tm,Yb phosphor. Source: Dwivedi et al. (2020). Reprinted with permission of Elsevier.

Table 15.2 The temperature generated in different phosphor samples prepared exactly in the same conditions at different pump power.

S. no.	Pump power (in W)	Temperature (in K)				
		GVO:Tm,Yb	YVO:Tm,Yb	GdO:Tm,Yb	YO:Tm,Yb	CAO:Tm,Yb
1	0.00	303	303	303	303	303
2	0.50	308	306	305	304	304
3	0.70	311	309	306	305	305
4	1.00	315	314	310	307	308
5	1.20	319	316	312	309	309
6	1.50	325	319	315	311	311
7	1.70	329	322	318	314	312
8	1.90	331	325	320	316	314

Source: Dwivedi et al. (2020). Reprinted with permission of Elsevier.

However, few groups are working in this area, and most of them are working on microscale particles, only very few are working on nanoscale particles. Thus, a lot of effort is still required for full-fledged explanation and application of this field due to immense possibilities in the field of bioimaging and biomedical applications (Lin et al. 2012; Zhou et al. 2015).

References

Alaulamie, A.A. (2017). Nanoscale temperature measurement of phase transition in water using novel optical thermometry techniques. Ph.D. Thesis. http://rave.ohiolink.edu/etdc/view?acc_num=ohiou1493904666617419.

Alencar, M.A.R.C., Maciel, G.S., De Araújo, C.B., and Patra, A. (2004). Er^{3+}-doped $BaTiO_3$ nanocrystals for thermometry: influence of nanoenvironment on the sensitivity of a fluorescence-based temperature sensor. *Applied Physics Letters* 84 (23): 4753–4755. https://doi.org/10.1063/1.1760882.

Alicki, R. and Leitner, D.M. (2015). Size-dependent accuracy of nanoscale thermometers. *The Journal of Physical Chemistry. B* 119 (29): 9000–9005. https://doi.org/10.1021/jp508047q.

Allison, S.W. and Gillies, G.T. (1997). Remote thermometry with thermographic phosphors: instrumentation and applications. *The Review of Scientific Instruments* 68: 2615–2650. https://doi.org/10.1063/1.1148174.

Auzel, F.E. (1973). Materials and devices using double-pumped phosphors with energy transfer. *Proceedings of the IEEE* 61: 758–786. https://doi.org/10.1109/PROC.1973.9155.

Back, M., Trave, E., Ueda, J., and Tanabe, S. (2016). Ratiometric optical thermometer based on dual near-infrared emission in Cr^{3+}-doped bismuth-based gallate host. *Chemistry of Materials* 28 (22): 8347–8356. https://doi.org/10.1021/acs.chemmater.6b03625.

Balabhadra, S., Debasu, M.L., Brites, C.D.S. et al. (2017). Upconverting nanoparticles working as primary thermometers in different media. *Journal of Physical Chemistry C* 121 (25): 13962–13968. https://doi.org/10.1021/acs.jpcc.7b04827.

Binnemans, K. (2009). Lanthanide-based luminescent hybrid materials. *Chemical Reviews* 109 (9): 4283–4374. https://doi.org/10.1021/cr8003983.

Blasse, G. and Grabmaier, B.C. (1994). A general introduction to luminescent materials. In: *Luminescent Materials* (ed. G. Blasse and B.C. Grabmaier), 1–9. Springer-Verlag. https://doi.org/10.1007/978-3-642-79017-1.

Bloembergen, N. (1959). Solid state infrared quantum counters. *Physical Review Letters* 2 (3): 84–85. https://doi.org/10.1103/PhysRevLett.2.84.

Brandão-Silva, A.C., Gomes, M.A., Novais, S.M.V. et al. (2018). Size influence on temperature sensing of erbium-doped yttrium oxide nanocrystals exploiting thermally coupled and uncoupled levels' pairs. *Journal of Alloys and Compounds* 731: 478–488. https://doi.org/10.1016/j.jallcom.2017.09.156.

Brites, C.D.S., Lima, P.P., Silva, N.J.O. et al. (2010). A luminescent molecular thermometer for long-term absolute temperature measurements at the nanoscale. *Advanced Materials* 22: 4499–4504. https://doi.org/10.1002/adma.201001780.13.

Brites, C.D.S., Lima, P.P., Silva, N.J.O. et al. (2012). Thermometry at the nanoscale. *Nanoscale* 4: 4799–4829. https://doi.org/10.1039/c2nr30663h.

Bünzli, J.C.G. (2010). Lanthanide luminescence for biomedical analyses and imaging. *Chemical Reviews* 110 (5): 2729–2755. https://doi.org/10.1021/cr900362e.

Bünzli, J.C.G. (2015). On the design of highly luminescent lanthanide complexes. *Coordination Chemistry Reviews* 293–294: 19–47. https://doi.org/10.1016/j.ccr.2014.10.013.

Bünzli, J.C.G. (2016). Lanthanide luminescence: from a mystery to rationalization, understanding, and applications. In: *Handbook on the Physics and Chemistry of Rare Earths*, vol. 50, 141–176. Elsevier. https://doi.org/10.1016/bs.hpcre.2016.08.003.

Cadiau, A., Brites, C.D.S., Costa, P.M.F.J. et al. (2013). Ratiometric nanothermometer based on an emissive Ln^{3+}-organic framework. *ACS Nano* 7 (8): 7213–7218. https://doi.org/10.1021/nn402608w.

Carlos, L.D., Ferreira, R.A.S., De Zea Bermudez, V., and Ribeiro, S.J.L. (2009). Lanthanide-containing light-emitting organic-inorganic hybrids: a bet on the future. *Advanced Materials* 21 (5): 509–534. https://doi.org/10.1002/adma.200801635.

Carlson, M.T., Khan, A., and Richardson, H.H. (2011). Local temperature determination of optically excited nanoparticles and nanodots. *Nano Letters* 11 (3): 1061–1069. https://doi.org/10.1021/nl103938u.

Carnall, W.T., Crosswhite, H., and Crosswhite, H.M. (1977). Energy level structure and transition probabilities in the spectra of the trivalent Lanthanides in LaF3. 78-XX–95. https://doi.org/10.2172/6417825.

Cerõn, E.N., Ortgies, D.H., Del Rosal, B. et al. (2015). Hybrid nanostructures for high-sensitivity luminescence nanothermometry in the second biological window. *Advanced Materials* 27 (32): 4781–4787. https://doi.org/10.1002/adma.201501014.

Chen, G., Yang, C., and Prasad, P.N. (2013a). Nanophotonics and nanochemistry: controlling the excitation dynamics for frequency up- and down-conversion in lanthanide-doped nanoparticles. *Accounts of Chemical Research* 46 (7): 1474–1486. https://doi.org/10.1021/ar300270y.

Chen, R., Ta, V.D., Xiao, F. et al. (2013b). Multicolor hybrid upconversion nanoparticles and their improved performance as luminescence temperature sensors due to energy transfer. *Small* 9 (7): 1052–1057. https://doi.org/10.1002/smll.201202287.

Childs, P.R.N., Greenwood, J.R., and Long, C.A. (2000). Review of temperature measurement. *The Review of Scientific Instruments* 71: 2959–2978. https://doi.org/10.1063/1.1305516.

Cho, S.J., Maysinger, D., Jain, M. et al. (2007). Long-term exposure to CdTe quantum dots causes functional impairments in live cells. *Langmuir* 23 (4): 1974–1980. https://doi.org/10.1021/la060093j.

Choudhary, A.K., Singh, S.K., Dwivedi, A. et al. (2018). Enhanced upconversion emission of Er^{3+}/Yb^{3+} and $Er^{3+}/Yb^{3+}/Zn^{2+}$ doped calcium aluminate for use in optical thermometry and laser induced optical heating. *Methods and Applications in Fluorescence* 6: 035014. https://doi.org/10.1038/s41467-018-04813-5.

Collins, S.F., Baxter, G.W., Wade, S.A. et al. (1998). Comparison of fluorescence-based temperature sensor schemes: theoretical analysis and experimental validation. *Journal of Applied Physics* 84 (9): 4649–4654. https://doi.org/10.1016/C2013-0-11338-4.

Cui, Y., Zhu, F., Chen, B., and Qian, G. (2015). Metal-organic frameworks for luminescence thermometry. *Chemical Communications* 51 (35): 7420–7431. https://doi.org/10.1039/C5CC00718F.

De La Rosa, V.R., Woisel, P., and Hoogenboom, R. (2016). Supramolecular control over thermoresponsive polymers. *Materials Today* 19 (1): 44–55. https://doi.org/10.1016/j.mattod.2015.06.013.

Debasu, M.L., Brites, C.D.S., Balabhadra, S. et al. (2016). Nanoplatforms for plasmon-induced heating and thermometry. *ChemNanoMat* 2: 520–527. https://doi.org/10.1002/cnma.201600061.

Dey, R. and Rai, V.K. (2017). Er^{3+}-Tm^{3+}-Yb^{3+}:$CaMoO_4$ phosphor as an outstanding upconversionbased optical temperature sensor and optical heater. *Methods and Applications in Fluorescence* 5 (1): 15006. https://doi.org/10.1088/2050-6120/aa5e31.

Dong, N.N., Pedroni, M., Piccinelli, F. et al. (2011). NIR-to-NIR two-photon excited CaF_2:Tm^{3+},Yb^{3+} nanoparticles: multifunctional nanoprobes for highly penetrating fluorescence bio-imaging. *ACS Nano* 5 (11): 8665–8671. https://doi.org/10.1021/nn202490m.

Dong, B., Cao, B., He, Y. et al. (2012). Temperature sensing and in vivo imaging by molybdenum sensitized visible upconversion luminescence of rare-earth oxides. *Advanced Materials* 24 (15): 1987–1993. https://doi.org/10.1002/adma.201200431.

Donner, J.S., Thompson, S.A., Kreuzer, M.P. et al. (2012). Mapping intracellular temperature using green fluorescent protein. *Nano Letters* 12 (4): 2107–2111. https://doi.org/10.1021/nl300389y.

Dramićanin, M.D. (2016). Sensing temperature via downshifting emissions of lanthanide-doped metal oxides and salts. A review. *Methods and Applications in Fluorescence* 4: 042001. https://doi.org/10.1039/c3cs60102a.

Du, P., Zhang, Q., Wang, X. et al. (2019). Upconversion luminescence, temperature sensing and internal heating behaviors of $Er^{3+}/Yb^{3+}/Fe^{3+}$-tridoped $NaBiF_4$ nanoparticles. *Journal of Alloys and Compounds* 805: 171–179. https://doi.org/10.1016/j.jallcom.2019.07.054.

Dwivedi, A., Kumar, D., Rai, S.B., and Rai, A.K. (2020). Effect of host on the radiative (upconversion emission) as well as non-radiative relaxation (laser induced optical heating) in Tm^{3+}/Yb^{3+} co-doped phosphors. *Journal of Luminescence* 226: 117421. https://doi.org/10.1016/j.jlumin.2020.117421.

Feng, J. and Zhang, H. (2013). Hybrid materials based on lanthanide organic complexes: a review. *Chemical Society Reviews* 42 (1): 387–410. https://doi.org/10.1039/c2cs35069f.

Fischer, J., Gerasimov, S., Hill, K.D. et al. (2007). Preparative steps towards the new definition of the kelvin in terms of the Boltzmann constant. *International Journal of Thermophysics* 28 (6): 1753–1765. https://doi.org/10.1007/s10765-007-0253-4.

Fischer, L.H., Harms, G.S., and Wolfbeis, O.S. (2011). Upconverting nanoparticles for nanoscale thermometry. *Angewandte Chemie, International Edition* 50 (20): 4546–4551. https://doi.org/10.1002/anie.201006835.

Gavrilović, T.V., Jovanović, D.J., Lojpur, V., and Dramićanin, M.D. (2014). Multifunctional Eu^{3+} and Er^{3+}/Yb^{3+}-doped $GdVO_4$ nanoparticles synthesized by reverse micelle method. *Scientific Reports* 4: 1–9. https://doi.org/10.1038/srep04209.

Geitenbeek, R.G., Prins, P.T., Albrecht, W. et al. (2017). $NaYF_4:Er^{3+},Yb^{3+}/SiO_2$ core/shell upconverting nanocrystals for luminescence thermometry up to 900 K. *Journal of Physical Chemistry C* 121 (6): 3503–3510. https://doi.org/10.1021/acs.jpcc.6b10279.

Gnach, A., Lipinski, T., Bednarkiewicz, A. et al. (2015). Upconverting nanoparticles: assessing the toxicity. *Chemical Society Reviews* 44 (6): 1561–1584. https://doi.org/10.1039/c4cs00177j.

Gorller-Walrand, C. and Binnemans, K. (1998). *Handbook on the Physics and Chemistry of Rare Earths*, vol. 25, 101. Amsterdam: North-Holland Publishers. ISBN: 0-444-82871-0.

Haro-González, P., Ramsay, W.T., Maestro, L.M. et al. (2013). Quantum dot-based thermal spectroscopy and imaging of optically trapped microspheres and single cells. *Small* 9 (12): 2162–2170. https://doi.org/10.1002/smll.201201740.

Hartmann, M. (2006). Minimal length scales for the existence of local temperature. *Contemporary Physics* 47: 89–102. https://doi.org/10.1080/00107510600581136.

Hemmer, E., Acosta-Mora, P., Méndez-Ramos, J., and Fischer, S. (2017). Optical nanoprobes for biomedical applications: shining a light on upconverting and near-infrared emitting nanoparticles for imaging, thermal sensing, and photodynamic therapy. *Journal of Materials Chemistry B* 5 (23): 4365–4392. https://doi.org/10.1039/C7TB00403F.

Henderson, B. and Imbusch, G.F. (1989). *Optical Spectroscopy of Inorganic Solids*. Oxford Sci. Publ. Univ. Press. ISBN: 9780199298624.

Huang, J., He, B., Cheng, Z., and Zhou, L. (2015). Upconverting PAAm/PNIPAM/$NaYF_4$:Yb:Er hydrogel with enhanced luminescence temperature sensitivity. *Journal of Luminescence* 160: 254–257. https://doi.org/10.1016/j.jlumin.2014.12.039.

Huang, Y., Hemmer, E., Rosei, F., and Vetrone, F. (2016). Multifunctional liposome nanocarriers combining upconverting nanoparticles and anticancer drugs. *The Journal of Physical Chemistry. B* 120 (22): 4992–5001. https://doi.org/10.1021/acs.jpcb.6b02013.

Jalani, G., Naccache, R., Rosenzweig, D.H. et al. (2016). Photocleavable hydrogel-coated upconverting nanoparticles: a multifunctional theranostic platform for NIR imaging and on-demand macromolecular delivery. *Journal of the American Chemical Society* 138 (3): 1078–1083. https://doi.org/10.1021/jacs.5b12357.

Jaque, D. and Vetrone, F. (2012). Luminescence nanothermometry. *Nanoscale* 4: 4301–4326. https://doi.org/10.1039/c2nr30764b.

Jaque, D., Martínez Maestro, L., Del Rosal, B. et al. (2014a). Nanoparticles for photothermal therapies. *Nanoscale* 6: 9494–9530. https://doi.org/10.1039/c4nr00708e.

Jaque, D., Martín, E., Martínez, L., and Haro-, P. (2014b). Fluorescent nanothermometers for intracellular thermal sensing. *Nanomedicine* 9: 1047–1062. https://doi.org/10.2217/NNM.14.59.

Judd, B.R. (1962). Optical absorption intensities of rare-earth ions. *Physics Review* 127 (3): 750–761. https://doi.org/10.1103/PhysRev.127.750.

Kalytchuk, S., Zhovtiuk, O., Kershaw, S.V. et al. (2016). Temperature-dependent exciton and trap-related photoluminescence of CdTe quantum dots embedded in a NaCl matrix: implication in thermometry. *Small* 12 (4): 466–476. https://doi.org/10.1002/smll.201501984.

Khalid, A.H. and Kontis, K. (2008). Thermographic phosphors for high temperature measurements: principles, current state of the art and recent applications. *Sensors* 8: 5673–5744. https://doi.org/10.3390/s8095673.

Khalid, A.H. and Kontis, K. (2009). 2D surface thermal imaging using rise-time analysis from laser-induced luminescence phosphor thermometry. *Measurement Science and Technology* 20: 025305. https://doi.org/10.1088/0957-0233/20/2/025305.

Kolesnikov, I.E., Kalinichev, A.A., Kurochkin, M.A. et al. (2017). YVO4:Nd^{3+} nanophosphors as NIR-to-NIR thermal sensors in wide temperature range. *Scientific Reports* 7 (1): 1–8. https://doi.org/10.1038/s41598-017-18295-w.

Kusama, H., Sovers, O.J., and Yoshioka, T. (1976). Line shift method for phosphor temperature measurements. *Japanese Journal of Applied Physics* 15 (12): 2349–2358. https://iopscience.iop.org/article/10.1143/JJAP.15.2349.

Labrador-Páez, L., Ximendes, E.C., Rodríguez-Sevilla, P. et al. (2018). Core-shell rare-earth-doped nanostructures in biomedicine. *Nanoscale* 10: 12935–12956. https://doi.org/10.1039/C8NR02307G.

Laha, S.S., Naik, A.R., Kuhn, E.R. et al. (2017). Nanothermometry measure of muscle efficiency. *Nano Letters* 17 (2): 1262–1268. https://doi.org/10.1021/acs.nanolett.6b05092.

Lee, J., Govorov, A.O., and Kotov, N.A. (2005). Nanoparticle assemblies with molecular springs: a nanoscale thermometer. *Angewandte Chemie, International Edition* 44 (45): 7439–7442. https://doi.org/10.1002/anie.200501264.

León-Luis, S.F., Rodríguez-Mendoza, U.R., Haro-González, P. et al. (2012). Role of the host matrix on the thermal sensitivity of Er^{3+} luminescence in optical temperature sensors. *Sensors and Actuators B: Chemical* 174: 176–186. https://doi.org/10.1016/j.snb.2012.08.019.

Li, S., Zhang, K., Yang, J.M. et al. (2007). Single quantum dots as local temperature markers. *Nano Letters* 7 (10): 3102–3105. https://doi.org/10.1021/nl071606p.

Lin, M., Zhao, Y., Wang, S.Q. et al. (2012). Recent advances in synthesis and surface modification of lanthanide-doped upconversion nanoparticles for biomedical applications. *Biotechnology Advances* 30 (6): 1551–1561. https://doi.org/10.1016/j.biotechadv.2012.04.009.

Liu, B., Li, C., Yang, P. et al. (2017). 808-nm-light-excited lanthanide-doped nanoparticles: rational design, luminescence control and theranostic applications. *Advanced Materials* 29: 1605434. https://doi.org/10.1002/adma.201605434.

Lojpur, V., Nikolic, M., Mancic, L. et al. (2013). Y_2O_3:Yb,Tm and Y_2O_3:Yb,Ho powders for low-temperature thermometry based on up-conversion fluorescence. *Ceramics International* 39 (2): 1129–1134. https://doi.org/10.1016/j.ceramint.2012.07.036.

Lojpur, V., Nikolić, G., and Dramićanin, M.D. (2014). Luminescence thermometry below room temperature via up-conversion emission of Y_2O_3:Yb^{3+},Er^{3+} nanophosphors. *Journal of Applied Physics* 115: 203106. https://doi.org/10.1063/1.4880158.

Löw, P., Kim, B., Takama, N., and Bergaud, C. (2008). High-spatial-resolution surface-temperature mapping using fluorescent thermometry. *Small* 4 (7): 908–914. https://doi.org/10.1002/smll.200700581.

Malta, O.L. and Carlos, L.D. (2003). Intensities of 4f–4f transitions in glass materials. *Quimica Nova* 26 (6): 889–895. https://doi.org/10.1590/S0100-40422003000600018.

Marciniak, L., Prorok, K., Francés-Soriano, L. et al. (2016). A broadening temperature sensitivity range with a core-shell YbEr@YbNd double ratiometric optical nanothermometer. *Nanoscale* 8 (9): 5037–5042. https://doi.org/10.1039/C5NR08223D.

Marciniak, L., Prorok, K., and Bednarkiewicz, A. (2017a). Size dependent sensitivity of Yb^{3+},Er^{3+} up-converting luminescent nano-thermometers. *Journal of Materials Chemistry C* 5 (31): 7890–7897. https://doi.org/10.1039/C7TC02322G.

Marciniak, L., Bednarkiewicz, A., and Strek, W. (2017b). The impact of nanocrystals size on luminescent properties and thermometry capabilities of Cr, Nd doped nanophosphors. *Sensors and Actuators B: Chemical* 238: 381–386. https://doi.org/10.1016/j.snb.2016.07.080.

Marin, R., Labrador-Paéz, L., Skripka, A. et al. (2018). Upconverting nanoparticle to quantum dot Förster resonance energy transfer: increasing the efficiency through donor design. *ACS Photonics* 5 (6): 2261–2270. https://doi.org/10.1021/acsphotonics.8b00112.

McCabe, K.M. and Hernandez, M. (2010). Molecular thermometry. *Pediatric Research* 67: 469–475. https://doi.org/10.1203/PDR.0b013e3181d68cef.

Millan, A., Carlos, L.D., Brites, C.D.S. et al. (2016). *Thermometry at the Nanoscale: Techniques and Selected Applications*, vol. 38 (ed. L.D. Carlos and F. Palacio), 237. Oxfordshire: The Royal Society of Chemistry. https://doi.org/10.1039/9781782622031.

Mooney, J. and Kambhampati, P. (2013). Get the basics right: Jacobian conversion of wavelength and energy scales for quantitative analysis of emission spectra. *Journal of Physical Chemistry Letters* 4 (19): 3316–3318. https://doi.org/10.1021/jz401508t.

Musolino, S., Schartner, E.P., Tsiminis, G. et al. (2016). Portable optical fiber probe for in vivo brain temperature measurements. *Biomedical Optics Express* 7 (8): 3069. https://doi.org/10.1364/BOE.7.003069.

Nigoghossian, K., Messaddeq, Y., Boudreau, D., and Ribeiro, S.J.L. (2017). UV and temperature-sensing based on NaGdF4:Yb^{3+}:Er^{3+}@SiO_2-Eu(tta)3. *ACS Omega* 2 (5): 2065–2071. https://doi.org/10.1021/acsomega.7b00056.

Ofelt, G.S. (1962). Intensities of crystal spectra of rare-earth ions. *The Journal of Chemical Physics* 37 (3): 511–520. https://doi.org/10.1063/1.1701366.

Okabe, K., Inada, N., Gota, C. et al. (2012). Intracellular temperature mapping with a fluorescent polymeric thermometer and fluorescence lifetime imaging microscopy. *Nature Communications* 3: 1–9. https://doi.org/10.1038/ncomms1714.

Pandey, A. and Rai, V.K. (2013). Optical thermometry using FIR of two close lying levels of different ions in Y_2O_3:Ho^{3+}-Tm^{3+}-Yb^{3+} phosphor. *Applied Physics B: Lasers and Optics* 113 (2): 221–225. https://doi.org/10.1007/s00340-013-5460-z.

Pandey, A., Rai, V.K., Kumar, V. et al. (2015). Upconversion based temperature sensing ability of Er^{3+}-Yb^{3+}codoped $SrWO_4$: an optical heating phosphor. *Sensors and Actuators B: Chemical* 209: 352–358. https://doi.org/10.1016/j.snb.2014.11.126.

Paviolo, C., Clayton, A.H.A., Mcarthur, S.L., and Stoddart, P.R. (2013). Temperature measurement in the microscopic regime: a comparison between fluorescence lifetime- and intensity-based methods. *Journal of Microscopy* 250 (3): 179–188. https://doi.org/10.1111/jmi.12033.

Peng, H., Song, H., Chen, B. et al. (2003). Temperature dependence of luminescent spectra and dynamics in nanocrystalline Y_2O_3:Eu^{3+}. *The Journal of Chemical Physics* 118 (7): 3277–3282. https://doi.org/10.1063/1.1538181.

Pietsch, C., Schubert, U.S., and Hoogenboom, R. (2011). Aqueous polymeric sensors based on temperature-induced polymer phase transitions and solvatochromic dyes. *Chemical Communications* 47: 8750–8765. https://doi.org/10.1039/c1cc11940k.

Prodi, L., Rampazzo, E., Rastrelli, F. et al. (2015). Imaging agents based on lanthanide doped nanoparticles. *Chemical Society Reviews* 44 (14): 4922–4952. https://doi.org/10.1039/c4cs00394b.

Quinn, T.J. (2013). *Temperature*. Academic Press. https://doi.org/10.1016/C2013-0-11338-4.

Quintanilla, M. and Liz-Marzán, L.M. (2018). Guiding rules for selecting a nanothermometer. *Nano Today* 19: 126–145. https://doi.org/10.1016/j.nantod.2018.02.012.

Quintanilla, M., Benayas, A., Naccache, R., and Vetrone, F. (2016). *Thermometry at the Nanoscale: Techniques and Selected Applications*, vol. 38, 124. Oxfordshire: The Royal Society of Chemistry. https://doi.org/10.1039/9781782622031.

Rai, M., Singh, S.K., Singh, A.K. et al. (2015). Enhanced red upconversion emission, magnetoluminescent behavior, and bioimaging application of NaSc0.75Er0.02Yb0.18 Gd0.05F4@AuNPs nanoparticles. *ACS Applied Materials & Interfaces* 7 (28): 15339–15350. https://doi.org/10.1021/acsami.5b03218.

Reif, F. and Scott, H.L. (1998). Fundamentals of statistical and thermal physics. *American Journal of Physics* 66: 164–167. https://doi.org/10.1119/1.19073.

Rocha, J., Brites, C.D.S., and Carlos, L.D. (2016). Lanthanide organic framework luminescent thermometers. *Chemistry - A European Journal* 22: 14782–14795. https://doi.org/10.1002/chem.201600860.

Rodrigues, E.M., Gálico, D.A., Lemes, M.A. et al. (2018). One pot synthesis and systematic study of the photophysical and magnetic properties and thermal sensing of α and β-phase $NaLnF_4$ and β-phase core@shell nanoparticles. *New Journal of Chemistry* 42 (16): 13393–13405. https://doi.org/10.1039/C8NJ02471E.

Rohani, S., Quintanilla, M., Tuccio, S. et al. (2015). Enhanced luminescence, collective heating, and nanothermometry in an ensemble system composed of lanthanide-doped upconverting nanoparticles and gold nanorods. *Advanced Optical Materials* 3 (11): 1606–1613. https://doi.org/10.1002/adom.201500380.

de Sá, G.F., Malta, O.L., de Mello Donega, C. et al. (2000). Spectroscopic properties and design of highly luminescent lanthanide coordination complexes. *Coordination Chemistry Reviews* 196: 165–195. https://doi.org/10.1016/S0010-8545(99).00054-5.

Savchuk, O.A., Haro-González, P., Carvajal, J.J. et al. (2014). Er:Yb:NaY_2F_5O up-converting nanoparticles for sub-tissue fluorescence lifetime thermal sensing. *Nanoscale* 6 (16): 9727–9733. https://doi.org/10.1039/C4NR02305F.

Savchuk, O.A., Carvajal, J.J., Brites, C.D.S. et al. (2018). Upconversion thermometry: a new tool to measure the thermal resistance of nanoparticles. *Nanoscale* 10 (14): 6602–6610. https://doi.org/10.1039/C7NR08758F.

Schena, E., Tosi, D., Saccomandi, P. et al. (2016). Fiber optic sensors for temperature monitoring during thermal treatments: an overview. *Sensors (Switzerland)* 16 (7): 1–20. https://doi.org/10.3390/s16071144.

Sedlmeier, A., Achatz, D.E., Fischer, L.H. et al. (2012). Photon upconverting nanoparticles for luminescent sensing of temperature. *Nanoscale* 4 (22): 7090–7096. https://doi.org/10.1039/c2nr32314a.

Shahi, P.K., Singh, A.K., Singh, S.K. et al. (2015a). Revelation of the technological versatility of the Eu(TTA).3Phen complex by demonstrating energy harvesting, ultraviolet light detection, temperature sensing, and laser applications. *ACS Applied Materials & Interfaces* 7 (33): 18231–18239. https://doi.org/10.1021/acsami.5b06350.

Shahi, P.K., Singh, A.K., Rai, S.B., and Ullrich, B. (2015b). Lanthanide complexes for temperature sensing, UV light detection, and laser applications. *Sensors and Actuators A: Physical* 222: 255–261. https://doi.org/10.1016/j.sna.2014.12.021.

Shahi, P.K., Singh, P., Singh, A.K. et al. (2017a). A strategy to achieve efficient dual-mode luminescence in lanthanide-based magnetic hybrid nanostructure and its demonstration for the detection of latent fingerprints. *Journal of Colloid and Interface Science* 491: 199–206. https://doi.org/10.1016/j.jcis.2016.12.034.

Shahi, P.K., Singh, P., and Rai, S.B. (2017b). Sunlight activated lanthanide complex for luminescent solar collector applications: effect of waveguide matrix. *Journal of Physics D: Applied Physics* 50: 075501. https://iopscience.iop.org/article/10.1088/1361-6463/aa4ecd/meta.

Shahi, P.K., Prakash, R., and Rai, S.B. (2018). Silver nanoparticles embedded hybrid organometallic complexes: structural interactions, photo-induced energy transfer, plasmonic effect and optical thermometry. *AIP Advances* 8: 065117. https://doi.org/10.1063/1.5020812.

Shahi, P.K., Kumar, B., Prakash, R., and Rai, S.B. (2019). Investigation of optical properties and energy transfer in Eu(III) and Tb(III) based composite compound dispersed in polar, non-polar solvents and polymer matrix. *Materials Research Express* 6: 046204. https://doi.org/10.1088/2053-591/aafae8.

Shinn, M.D., Sibley, W.A., Drexhage, M.G., and Brown, R.N. (1983). Optical transitions of Er^{3+} ions in fluorozirconate glass. *Physical Review B* 27 (11): 6635–6648. https://doi.org/10.1103/PhysRevB.27.6635.

Singh, S.K., Kumar, K., and Rai, S.B. (2009). Er^{3+}/Yb^{3+} codoped Gd_2O_3 nano-phosphor for optical thermometry. *Sensors and Actuators A: Physical* 149 (1): 16–20. https://doi.org/10.1016/j.sna.2008.09.019.

Singh, A.K., Shahi, P.K., Rai, S.B., and Ullrich, B. (2015). Host matrix impact on Er^{3+} upconversion emission and its temperature dependence. *RSC Advances* 5: 16067. https://doi.org/10.1039/C4RA12637H.

Singh, P., Shahi, P.K., Rai, A. et al. (2016). Effect of Li^+ ion sensitization and optical temperature sensing in Gd_2O_3: Ho^{3+}/Yb^{3+}. *Optical Materials* 58: 432–438. https://doi.org/10.1016/j.optmat.2016.06.010.

Singh, P., Shahi, P.K., Prakash, R., and Bahadur Rai, S. (2017a). An assembly and interaction of upconversion and plasmonic nanoparticles on organometallic nanofibers: enhanced multicolor upconversion, downshifting emission and the plasmonic effect. *Nanotechnology* 28: 415701. https://doi.org/10.1088/1361-6528/aa8066/meta.

Singh, P., Shahi, P.K., Singh, S.K. et al. (2017b). Lanthanide doped ultrafine hybrid nanostructures: multicolour luminescence, upconversion based energy transfer and luminescent solar collector applications. *Nanoscale* 9: 696. https://doi.org/10.1039/C6NR07250J.

Skripka, A., Benayas, A., Marin, R. et al. (2017). Double rare-earth nanothermometer in aqueous media: opening the third optical transparency window to temperature sensing. *Nanoscale* 9 (9): 3079–3085. https://doi.org/10.1039/C6NR08472A.

Tikhomirov, V.K., Driesen, K., Rodriguez, V.D. et al. (2009). Optical nanoheater based on the Yb^{3+}-Er^{3+} co-doped nanoparticles. *Optics Express* 17 (14): 11794. https://doi.org/10.1364/oe.17.011794.

Tóth, G. and Apellaniz, I. (2014). Quantum metrology from a quantum information science perspective. *Journal of Physics A: Mathematical and Theoretical* 47 (42): https://doi.org/10.1088/1751-8113/47/42/424006.

Uchiyama, S., Matsumura, Y., De Silva, A.P., and Iwai, K. (2003). Fluorescent molecular thermometers based on polymers showing temperature-induced phase transitions and labeled with polarity-responsive benzofurazans. *Analytical Chemistry* 75 (21): 5926–5935. https://doi.org/10.1021/ac0346914.

Uchiyama, S., Kawai, N., De Silva, A.P., and Iwai, K. (2004). Fluorescent polymeric AND logic gate with temperature and pH as inputs. *Journal of the American Chemical Society* 126 (10): 3032–3033. https://doi.org/10.1021/ja039697p.

Uchiyama, S., de Silva, A.P., and Iwai, K. (2006). Luminescent molecular thermometers. *Journal of Chemical Education* 83: 720–727. https://doi.org/10.1021/ed083p720.

Uchiyama, S., Gota, C., Tsuji, T., and Inada, N. (2017). Intracellular temperature measurements with fluorescent polymeric thermometers. *Chemical Communications* 53: 10976–10992. https://doi.org/10.1039/c7cc06203f.

Vetrone, F., Naccache, R., Zamarrón, A. et al. (2010). Temperature sensing using fluorescent nanothermometers. *ACS Nano* 4 (6): 3254–3258. https://doi.org/10.1021/nn100244a.

Wade, S.A., Collins, S.F., and Baxter, G.W. (2003). Fluorescence intensity ratio technique for optical fiber point temperature sensing. *Journal of Applied Physics* 94 (8): 4743–4756. https://doi.org/10.1063/1.1606526.

Wang, S., Westcott, S., and Chen, W. (2002). Nanoparticle luminescence thermometry. *The Journal of Physical Chemistry. B* 106 (43): 11203–11209. https://doi.org/10.1021/jp026445m.

Wang, X.D., Wolfbeis, O.S., and Meier, R.J. (2013). Luminescent probes and sensors for temperature. *Chemical Society Reviews* 42 (19): 7834–7869. https://doi.org/10.1039/c3cs60102a.

Wang, Z., Ananias, D., Carné-Sánchez, A. et al. (2015a). Lanthanide-organic framework nanothermometers prepared by spray-drying. *Advanced Functional Materials* 25 (19): 2824–2830. https://doi.org/10.1002/adfm.201500518.

Wang, Z., Zhang, P., Yuan, Q. et al. (2015b). Nd^{3+}-sensitized $NaLuF_4$ luminescent nanoparticles for multimodal imaging and temperature sensing under 808 nm excitation. *Nanoscale* 7 (42): 17861–17870. https://doi.org/10.1039/C5NR04889C.

Wei, T., Yang, F., Jia, B. et al. (2019). High performance temperature sensing and optical heating of Tm^{3+}- and Yb^{3+}- codoped $SrBi_4Ti_4O_{15}$ up-conversion luminescence nanoparticles. *Ceramics International* 45 (14): 18084–18090. https://doi.org/10.1016/j.ceramint.2019.06.030.

Wen, S., Zhou, J., Zheng, K. et al. (2018). Advances in highly doped upconversion nanoparticles. *Nature Communications* 9 (1): 1. https://doi.org/10.1038/s41467-018-04813-5.

Wolfbeis, O.S. (2008). Sensor paints. *Advanced Materials* 20: 3759–3763. https://doi.org/10.1002/adma.200702276.

Xiao, Q., Li, Y., Li, F. et al. (2014). Rational design of a thermalresponsive-polymer-switchable FRET system for enhancing the temperature sensitivity of upconversion nanophosphors. *Nanoscale* 6 (17): 10179–10186. https://doi.org/10.1002/smll.201202287.

Xu, M., Chen, D., Huang, P. et al. (2016). A dual-functional upconversion core@shell nanostructure for white-light-emission and temperature sensing. *Journal of Materials Chemistry C* 4 (27): 6516–6524. https://doi.org/10.1039/C6TC02218A.

Yadav, R.S., Dhoble, S.J., and Rai, S.B. (2018). Enhanced photoluminescence in Tm^{3+}, Yb^{3+}, Mg^{2+} tri-doped $ZnWO_4$ phosphor: three photon upconversion, laser induced optical heating and temperature sensing. *Sensors and Actuators B: Chemical* 273 (March): 1425–1434. https://doi.org/10.1016/j.snb.2018.07.049.

Yang, G., Liu, X., Feng, J. et al. (2016). *Thermometry at the Nanoscale: Techniques and Selected Applications*, Nanoscience and Nanotechnology Series, vol. 38, 167. The Royal Society of Chemistry. https://doi.org/10.1039/9781782622031.

Yang, F., Skripka, A., Benayas, A. et al. (2018). An integrated multifunctional nanoplatform for deep-tissue dual-mode imaging. *Advanced Functional Materials* 28 (11): 1–9. https://doi.org/10.1002/adfm.201706235.

Yu, J., Sun, L., Peng, H., and Stich, M.I.J. (2010). Luminescent terbium and europium probes for lifetime based sensing of temperature between 0 and 70 °C. *Journal of Materials Chemistry* 20 (33): 6975–6981. https://doi.org/10.1039/c0jm01069c.

Yue, Y. and Wang, X. (2012). Nanoscale thermal probing. *Nano Reviews* 3 (1): 11586. https://doi.org/10.3402/nano.v3i0.11586.

Zhan, Q., Qian, J., Liang, H. et al. (2011). Using 915 nm laser excited $Tm^{3+}/Er^{3+}/Ho^{3+}$-doped $NaYbF_4$ upconversion nanoparticles for in vitro and deeper in vivo bioimaging without overheating irradiation. *ACS Nano* 5 (5): 3744–3757. https://doi.org/10.1021/nn200110j.

Zheng, S., Chen, W., Tan, D. et al. (2014). Lanthanide-doped $NaGdF_4$ core-shell nanoparticles for non-contact self-referencing temperature sensors. *Nanoscale* 6 (11): 5675–5679. https://doi.org/10.1039/c4nr00432a.

Zhou, S., Jiang, G., Li, X. et al. (2014). Strategy for thermometry via Tm^{3+}-doped $NaYF_4$ core-shell nanoparticles. *Optics Letters* 39 (23): 6687. https://doi.org/10.1364/OL.39.006687.

Zhou, B., Shi, B., Jin, D., and Liu, X. (2015). Controlling upconversion nanocrystals for emerging applications. *Nature Nanotechnology* 10 (11): 924–936. https://doi.org/10.1038/NNANO.2015.251.

Zhu, X., Feng, W., Chang, J. et al. (2016). Temperature-feedback upconversion nanocomposite for accurate photothermal therapy at facile temperature. *Nature Communications* 7: 10437. https://doi.org/10.1038/ncomms10437.

16

Upconverting Nanoparticles in Pollutant Degradation and Hydrogen Generation

Wanni Wang, Zhaoyou Chu, Benjin Chen, and Haisheng Qian

Anhui Medical University, School of Biomedical Engineering, Research and Engineering Center of Biomedical Materials, 81 Meishan Road, Shushan District, Hefei 230032, P.R. China

16.1 Introduction

The phenomenon of upconversion (UC) was proposed by Auzel, Ovsyankin, and Feofilov in the mid-1960s (Auzel 2004). Upconversion luminescence is an anti-Stokes luminescence (Zhu et al. 2006). As the name suggests, it is contrary to the traditional Stokes law. It can emit high-energy light (short wavelength) when excited by low-energy light (long wavelength). High-energy light (short wavelength) can be radiated when excited by low-energy light (long wavelength). Upconversion luminescent materials generally occur in compounds containing rare-earth ions, which can be prepared by thermal decomposition (Cheng et al. 2012; Na et al. 2013; Chen et al. 2016), water (solvent) thermal method (Zhou et al. 2013; Mi et al. 2011), sol–gel method (Chen et al. 2013), and co-precipitation method (Lee and Kim 2011). Rare-earth ion-doped upconversion luminescent materials mainly consist of substrates, activators, and sensitizers. The substrate itself has no luminescence, but it provides the crystal field for the activator, which determines the ion spacing, relative spatial position, coordination number, and the type of surrounding anions. It has a great influence on the upconversion fluorescence performance. Currently, fluorides with stable structure and low phonon energy are mostly used as matrix materials, such as BaF_2, LaF_2, KYF_4, CaF_2, $NaYF_4$, $NaLuF_4$, $NaGaF_4$, YF_3, etc. (Liu et al. 2011; Renero-Lecuna et al. 2011; Schafer et al. 2008; Dong et al. 2011; Wang et al. 2014a; Chen et al. 2014; Li et al. 2013). These matrix materials can effectively reduce the polyphonon relaxation in the luminescence and weaken the passive emission transition, so as to obtain higher fluorescence efficiency. The upconversion luminescence efficiency is low with only one ion doped. In order to improve the efficiency, the sensitizer ions and the activator ions are usually co-doped, and the sensitizer (such as Yb^{3+}, Nd^{3+}, and other rare-earth ions with a large absorption cross section to the excited light) is responsible for absorbing the excited light. Then, the absorbed energy is transferred to the activator (such as Er^{3+}, Tm^{3+}, Ho^{3+}, Tb^{3+}, Pr^{3+}, and other

Upconverting Nanoparticles: From Fundamentals to Applications, First Edition. Edited by Vineet K. Rai.

rare-earth ions with upconversion luminescence properties) through resonance transfer. This is the process of sensitization luminescence. Upconversion fluorescent materials have many advantages such as stable chemical properties, low toxicity, high and stable emission intensity, narrow emission band, adjustable emission wavelength, and so on, which have attracted extensive attention from researchers in biological imaging (Yuan et al. 2012; Liu et al. 2012; Rao et al. 2016; DaCosta et al. 2014; Xu et al. 2019a), cancer treatment (He et al. 2020; Shao et al. 2020; Wang et al. 2019; Dong et al. 2019; Jin et al. 2019; Liu et al. 2019a; Xu et al. 2017, 2018; Yao et al. 2018), temperature sensor (Runowski et al. 2019, 2020; Du et al. 2016; Tian et al. 2016; Qiu et al. 2020; Zhou et al. 2013), and photocatalytic technology (Yang et al. 2014, 2018; Wang et al. 2013a; Zhang et al. 2019a, 2021). Upconversion fluorescent materials can be combined with other catalysts to convert captured long-wavelength visible and near-infrared (NIR) incident light into high-energy short-wavelength visible or ultraviolet (UV) light by constructing a fluorescence resonance energy transfer (FRET) system (Tian et al. 2019). It indirectly broadens the absorption range of the catalyst, improves the utilization rate of sunlight, and has a broad application prospect in the traditional broad-spectrum photoactivation photocatalytic materials. Based on the recent work and literature reports, this chapter reviews the research progress of photocatalytic materials with upconversion effect in photodecomposition of aquatic hydrogen and degradation of pollutants. It may stimulate interest in the rational design and preparation of upconversion materials in highly efficient full-spectrum photocatalytic systems and their extensive applications.

16.2 Degradation of Organic Pollutants

Photocatalytic technology has achieved good results in treating sewage, purifying air, and producing clean energy and has a broad development prospect (Bushuyev et al. 2018; Ren et al. 2017; Spasiano et al. 2015; Li et al. 2016a; Anwer et al. 2019). In spite of this, photocatalytic technology is still in the stage of simulation test and has hardly been applied in the industry. One of the main reasons is that the two characteristics of semiconductor photocatalysts itself affect the activity of the photocatalyst, leading to the reduction of photocatalytic reaction efficiency. Firstly, semiconductor photocatalysts have different bandgap widths, which can only absorb light of a part of the wavelength and cannot use all wavelengths of light in the whole solar spectrum range. For example, wide-bandgap semiconductor photocatalysts such as TiO_2 and ZnO can only absorb UV light, which accounts for less than 5.0% of the solar spectrum. This restricts photocatalysts from making full use of sunlight. Second, the photogenerated carriers are easy to recombine, resulting in a low quantum yield, which limits the photocatalytic reaction rate. Therefore, it should be made from these two aspects to design more efficient semiconductor photocatalysts. It has become the core problem in the development of photocatalytic technology.

Upconversion luminescent materials can convert long-wavelength photon energy into short-wavelength emission light; that is, under the excitation of sunlight, they can absorb NIR light and emit visible light or UV light through energy transfer and transition, thus realizing the photocatalytic degradation under NIR light (Qin et al. 2010). The construction of upconversion photocatalytic materials, which generally consist of rare earths, rare-earth bearing matrix materials, and semiconductor photocatalytic materials, is a key link to realize the utilization of the full spectrum of sunlight. Generally, rare-earth elements such as Er, Tm, and Yb are selected because their 4f shells are not filled and have rich energy level structure and 4f electron transition characteristics (Zhou et al. 2015). The semiconductor photocatalytic material itself can be used as the matrix material for carrying rare-earth ions, but this will lead to low upconversion luminescence efficiency. It is necessary to choose a more suitable matrix material for carrying rare-earth ions. According to the theory of non-radiative relaxation and energy transfer of rare-earth ions, the lower the phonon energy of the luminescent matrix material, the lower the probability of non-radiative relaxation and the higher the upconversion luminescence efficiency (Wang and Liu 2009). Combined with factors such as upconversion luminescence efficiency and material structure stability, researchers chose fluorides with low phonon energy (such as BaF_2, LaF_2, $NaYF_4$, $NaLuF_4$, $NaCaF_4$, YF_3 etc.) to carry rare-earth ions. The upconversion photocatalytic material of rare-earth-doped fluoride and semiconductor composite has higher absorption efficiency and upconversion efficiency for NIR light, thus giving it better photocatalytic performance. The photocatalyst with upconversion effect has set off a boom in the field of organic pollutant degradation. Semiconductor-coupled upconversion materials have been widely used in the field of organic pollution treatment, especially in dye treatment.

16.2.1 Degradation of RhB

With the rapid development of modern printing and dyeing industry, the types and quantities of compounds entering water through various ways have increased sharply, and organic wastewater contains many toxic pollutants that are difficult to degrade. Rhodamine B (RhB) poses a serious threat to the ecosystem and human health because of its high biodegradability and easy accumulation (Rochat et al. 1978). Therefore, to reduce environmental pollution and protect human living environment, the treatment of organic wastewater is a long-term and important problem to be solved.

ZnS is one of the most promising photocatalytic materials, but its wide bandgap requires excitation of UV light, which limits its application in visible and NIR wavelengths. Li et al. (2019) added YF_3:Yb^{3+},Er^{3+} (YYE) nanoparticles with different mass prepared by a simple hydrothermal method in the process of hydrothermal formation of ZnS and then synthesized different proportions of YYE/ZnS heterojunction photocatalysts. Figure 16.1a,b shows the energy transfer process between YYE and ZnS interfaces. Figure 16.1c shows the full analysis of the photocatalytic mechanism of YYE/ZnS. Yb^{3+} ions can quickly transition from the ground state ($^2F_{7/2}$) to the excited state ($^2F_{5/2}$) and then induce the Er^{3+} ions to higher energy levels ($^2H_{11/2}$, $^4S_{3/2}$, and $^4F_{9/2}$) by energy transfer. YYE converted the NIR light of sunlight into shorter wavelength light and then sensitized the ZnS via energy transfer. Superoxide radical ($^{\bullet}O_2^-$) and hydroxyl radical ($^{\bullet}OH$) were the main active substances. After

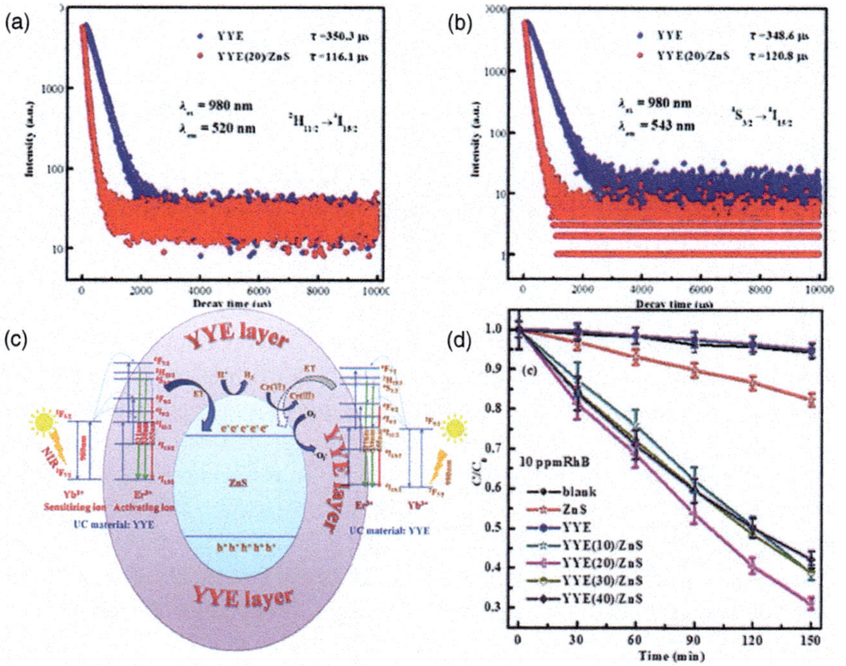

Figure 16.1 The luminescence decays of the excited-state levels of Er^{3+} for the YYE and YYE(20)/ZnS samples at (a) 520 and (b) 543 nm, respectively; (c) proposed photocatalytic mechanism for the YYE/ZnS heterojunction photocatalyst; (d) photocatalytic performance of the YYE/ZnS composites for the degradation of RhB. Source: Li et al. (2019). Reprinted with permission of Royal Society of Chemistry.

irradiated by a metal halide lamp for 150 minutes, UC materials played an irreplaceable role in the photocatalytic reaction; the YYE/ZnS catalyst could degrade RhB to 70%, while ZnS could only degrade to 18% (Figure 16.1d).

Electrospinning technology is a simple and efficient method to achieve large-scale synthesis of one-dimensional nanocomposites by blending various functional components with the polymer matrix under the action of electrostatic field (Chen et al. 2018a; Wen et al. 2017; Xue et al. 2019). Electrospinning technology is used to integrate upconversion nanoparticles (UCNPs) and semiconductors, which can effectively avoid the lattice mismatch and greatly reduce the difficulty of synthesis. Compared with other methods, electrospinning technology has the advantages of simple equipment, high efficiency, low cost, strong universality, and simple post-processing. Zhang et al. (2016a) successfully realized the large-scale synthesis of $NaYF_4:Yb^{3+},Tm^{3+}@NaYF_4/TiO_2$ ($UCNPs/TiO_2$) nanofibers via electrospinning technology by using tetrabutyl titanate (TBT) as the titanium source and polyvinylpyrrolidone (PVP) as the matrix. Under the irradiation of NIR light, the UV light generated by the $^1I_6 \rightarrow {}^3F_4$, $^1D_2 \rightarrow {}^3H_6$, and $^1D_2 \rightarrow {}^3F_4$ transitions from Tm^{3+} can activate the TiO_2 matrix by FRET or irradiative energy transfer (IET). Activated TiO_2 can produce singlet oxygen ($^1O_2$), $^{\cdot}O_2^-$, $^{\cdot}OH$, and other reactive oxygen species (ROS) that can degrade the organic pollutant RhB. Under the irradiation of 980 nm laser, 38% of the RhB molecules were degraded within six hours in the presence of UCNPs/TiO_2 nanofibers. However, in the presence of only TiO_2 or no catalyst, RhB molecules were almost not degraded.

It has been shown that the fluorescence generated by the radiation transition of Tm^{3+} in UCNPs/TiO_2 nanofibers at $^1I_6 \rightarrow {}^3F_4$, $^1D_2 \rightarrow {}^3H_6$, and $^1D_2 \rightarrow {}^3F_4$ was greatly quenched. However, in UCNPs/TiO_2 nanofibers, the radiation transition of Tm^{3+} at $^1G_4 \rightarrow {}^3H_6$ can still produce strong blue light radiation. The direct bandgap width of TiO_2 was 3.0–3.2 eV, and the energy of $^1G_4$ level was lower than the conduction band energy of TiO_2, so the radiation transition at this point cannot be absorbed. Zhang et al. (2016) introduced CdS nanoparticles that can absorb visible light and successfully constructed $NaYF_4$:Yb^{3+},Tm^{3+}@$NaYF_4$/CdS/TiO_2 (UCNPs/CdS/TiO_2) composite nanofibers by using the electrostatic spinning technology (Figure 16.2). Compared with UCNPs/TiO_2 composite nanofibers, the production performance of ROS such as $^{\bullet}OH$ was obviously improved under light. The fluorescence peaks generated by UCNPs were obviously quenched by the radiation transitions of Tm^{3+} at $^1I_6 \rightarrow {}^3F_4$, $^1D_2 \rightarrow {}^3H_6$, $^1D_2 \rightarrow {}^3F_4$, and $^1G_4 \rightarrow {}^3H_6$, and the energy was transferred to CdS nanoparticles and TiO_2 matrix by FRET and IET. The design of the ternary composite full-spectrum photocatalyst significantly improved the degradation efficiency of RhB under light, especially NIR light.

The lattice mismatch between the main lattice planes between $NaYF_4$ and ZnO is as high as 20%, so the lattice matching is the main obstacle to the direct bonding between the upconversion materials and ZnO. Wang et al. (2017a) prepared ZnO/β-$NaYF_4$:Yb^{3+},Tm^{3+}@$NaYF_4$ (ZnO/UCNPs) composite nanofibers by using electrospinning as a large-scale equipment technology with zinc acetate as the zinc source and hexadecyl trimethyl ammonium bromide (CTAB) as the source of preparing porous fibers. In the composite nanofibers prepared after electrospinning, the Tm^{3+} excitation radiation levels of $^1I_6 \rightarrow {}^3F_4$ and $^1D_2 \rightarrow {}^3H_6$ completely disappeared, which was due to the energy transfer of UCNPs to ZnO via FRET or IET process. After the irradiation of NIR light, ZnO/UCNP composite fibers with different contents of UCNPs showed better degradation effect of RhB than simple ZnO nanofibers.

Zhang et al. (2017a) adopted a simple solution-phase process and then calcined to obtain $NaYF_4$:Yb^{3+},Tm^{3+}@$NaYF_4$@ZnO (UCNPs/ZnO) core–shell nanoparticles. Under the irradiation of NIR light, the sensitized ion Yb^{3+} in UCNPs transmitted the absorbed energy to the activated ion Tm^{3+} through the process of energy resonance transfer to generate fluorescence radiation in the UV–vis region, in which the UV light partially activated the shell ZnO, making the electrons transition from its VB to CB. These electrons in the CB can reduce O_2 in the solution to form $^{\bullet}O_2^-$, and the holes formed by electron migration in the VB oxidized OH^- ions to form $^{\bullet}OH$. Under the irradiation of NIR light, nearly 100% of RhB in the solution was completely decomposed by UCNPs@ZnO core–shell nanoparticles within 100 minutes, and the mixture of core and shell showed a stronger photodegradation effect than ZnO.

Wang et al. developed a simple template-assisted hydrothermal method to prepare the yolk–shell structure combining $NaYF_4$:Yb^{3+},Tm^{3+}@$NaYF_4$:Yb^{3+},Er^{3+} nanoparticles (UCNPs) and alloy $Zn_xCd_{1-x}S$ (Wang et al. 2018a). By activating UCNPs

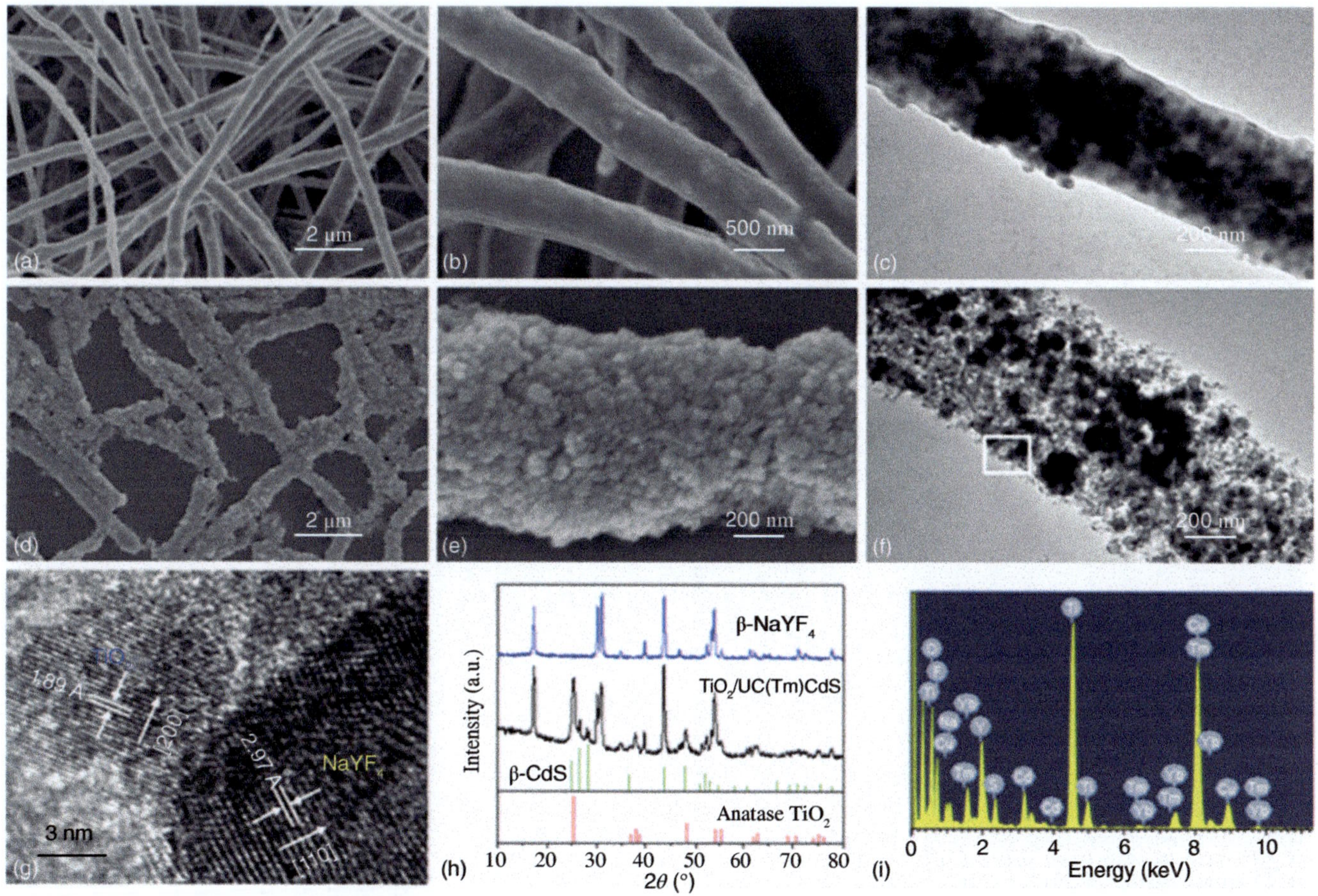

Figure 16.2 (a and b) SEM images of the UCNPs/CdS/PVP/TBT nanofibers developed by using the electrospinning technique; (c) TEM image of a single nanofiber of PVP/TBT/UCNPs/CdS; (d and e) SEM images of UCNPs/CdS/TiO_2 nanofibers obtained after hydrothermal treatment of the PVP/TBT/UCNPs/CdS nanofibers; (f) TEM image of a single UCNPs/CdS/TiO_2 nanofiber; (g) high resolution transmission electron (HRTEM) image for the selected region shown in Figure 16.2f; (h) X-ray diffraction (XRD) of the final product after hydrothermal treatment of the PVP/TBT/UCNPs/CdS nanofiber mat at 180 °C for 12 hours; and (i) energy dispersive X-ray (EDX) analysis of the as-prepared nanofibers of UCNPs/CdS/TiO_2. Source: Zhang (2016a). Reprinted with permission of John Wiley & Sons, Inc.

under NIR light excitation, UCNPs can excite $Zn_xCd_{1-x}S$ via FRET or IET process to generate photogenerated electrons and holes, resulting in the decomposition of the organic dye RhB by large amounts of $^{\bullet}OH$ and $^1O_2$. The development of these UC photocatalysts provides new opportunities for the application of NIR photocatalytic technology in the degradation of RhB, which is expected to be applied to a large-scale advanced wastewater treatment (Table 16.1) (Li et al. 2010, 2016, 2019; Zhang et al. 2016a,b, 2017, 2018; Wang et al. 2014; Wang et al. 2017a,b; Wang et al. 2018, 2020; Cheng et al. 2017; Liu et al. 2019; Zhao et al. 2018a; Chatti et al. 2016; Huang et al. 2019; Xu et al. 2019; Chen et al. 2015; Su et al. 2014; Tou et al. 2017; Guo et al. 2013).

16.2.2 Degradation of MB

Some common UC photocatalysts that are capable of degrading methylene blue (MB) are listed in Table 16.2 (Zhang et al. 2018b, 2019; Tang et al. 2013; Guo et al. 2014; Wu et al. 2015; Huang et al. 2015; Zhou et al. 2018; Chen et al. 2018b). Zhang et al. (2018) prepared the $NaYF_4:Yb^{3+},Tm^{3+}@Cu_2O$ ($UCNPs@Cu_2O$) core–shell photocatalyst for the first time and used it to degrade MB. The FRET process was effective in the $UCNPs@Cu_2O$ core–shell composites, so that more energy can be transferred from the excited state of the Tm^{3+} ion to Cu_2O. Compared with UCNPs (9%), Cu_2O (~0%), and the mixture of UCNPs and Cu_2O (39%), $UCNPs@Cu_2O$ core–shell composites showed significantly improved catalytic activity for MB degradation at 980 nm (76%).

In 2013, Tang et al. developed a method to directly coat UCNPs with TiO_2. $NaYF_4:Yb^{3+},Tm^{3+}$ nanoparticles (UCNPs) were prepared by solvothermal method (Tang et al. 2013). Then, the nanoparticles were dispersed in the ethanol phase and added into the solution with $Ti(OBU)_4$. The amorphous TiO_2 shell was obtained after hydrolysis. Finally, the crystalline TiO_2 was obtained by hydrothermal treatment. The results showed that the energy transfer pathways of $UCNPs@TiO_2$ core–shell composites and the mixture of UCNPs and TiO_2 were different. For the two different photocatalysts, FRET and radiation reabsorption mechanisms have been shown to be dominant in NIR photocatalysis, respectively. In addition, the availability of FRET allowed additional energy to be transferred from the excited state of Tm^{3+} ions to the TiO_2 shell. Therefore, $UCNPs@TiO_2$ core–shell composites had higher photocatalytic degradation of MB than the mixture of UCNPs and TiO_2.

The bandgap of TiO_2 is 3.0–3.2 eV, and the absorption is mainly in the UV region. If the absorption range can be further expanded, the FRET efficiency will be improved again, and the catalytic performance will also become stronger. Based on this, in 2014, Guo et al. introduced CdS nanoparticles with a bandgap of about 2.42 eV having absorption in the visible region and developed a method to combine UCNPs with TiO_2 and CdS (Guo et al. 2014). Firstly, $NaYF_4:Yb^{3+},Tm^{3+}$ nanoparticles, CdS nanoparticles, and TiO_2 were prepared. Then, CdS and TiO_2 were coupled with UCNPs in the form of heterojunction by hydrothermal method

Table 16.1 Degradation of RhB by typical UC photocatalysts.

Photocatalysts	Energy-level transitions	Light source	Photocatalyst amount	Aqueous solution	Reaction time	Removal percentage (%)	References
YF_3:Yb^{3+},Er^{3+}/ZnS	$^2H_{11/2} \rightarrow {}^4I_{15/2}$, $^4S_{3/2} \rightarrow {}^4I_{15/2}$, $^4F_{9/2} \rightarrow {}^4I_{15/2}$	<365 nm 420–760 nm 980 nm	30 mg	50 ml of aqueous solution containing RhB (10 ppm)	150 min	70%	Li et al. (2019)
$NaYF_4$:Yb^{3+}, Tm^{3+}@$NaYF_4$/TiO_2	$^1I_6 \rightarrow {}^3F_4$, $^1D_2 \rightarrow {}^3H_6$, $^1D_2 \rightarrow {}^3F_4$	980 nm, 8 W	3 mg	5 ml of RhB solution with a concentration of 3.0×10^{-6} M (1.43 mg/l)	6 h	38%	Zhang et al. (2016)
β-$NaYF_4$:Yb^{3+}, Tm^{3+}@$NaYF_4$/CdS/TiO_2	$^1I_6 \rightarrow {}^3H_6$, $^1I_6 \rightarrow {}^3F_4$, $^1D_2 \rightarrow {}^3H_6$, $^1D_2 \rightarrow {}^3F_4$, $^1G_4 \rightarrow {}^3H_6$	IR irradiation	5 mg	50 ml of the RhB solution with an initial concentration (C_0) of 2.5 mg/l	25 min	53%	Zhang et al. (2016)
ZnO/β-$NaYF_4$:Yb^{3+}, Tm^{3+}@$NaYF_4$	$^1I_6 \rightarrow {}^3H_6$, $^1I_6 \rightarrow {}^3F_4$, $^1D_2 \rightarrow {}^3H_6$, $^1D_2 \rightarrow {}^3F_4$	780–2500 nm	30 mg	RhB (50 ml, 2.5 mg/l)	150 min	Nearly 100%	Wang et al. (2017)
$NaYF_4$:Yb^{3+}, Tm^{3+}@$NaYF_4$@ZnO	$^1I_6 \rightarrow {}^3H_6$, $^1I_6 \rightarrow {}^3F_4$, $^1D_2 \rightarrow {}^3H_6$, $^1D_2 \rightarrow {}^3F_4$	780–2500 nm	20 mg	RhB (50 ml, 5 mg/l)	100 min	Nearly 100%	Zhang et al. (2017)
$NaYF_4$:Yb^{3+}, Tm^{3+}@$NaYF_4$:Yb^{3+}, Er^{3+}@$Zn_xCd_{1-x}S$	$^1I_6 \rightarrow {}^3F_4$, $^1D_2 \rightarrow {}^3H_6$, $^1D_2 \rightarrow {}^3F_4$, $^1G_4 \rightarrow {}^3H_6$, $^2H_{11/2} \rightarrow {}^4I_{15/2}$, $^4S_{3/2} \rightarrow {}^4I_{15/2}$	1500 mW/cm^2 Xe lamp with n UV–vis filter	15 mg	50 ml of RhB solution (5.0 mg/l)	30 min	65%	Wang et al. (2018)
$NaYF_4$,Yb^{3+}, Tm^{3+}@TiO_2@g-C_3N_4	$^1I_6 \rightarrow {}^3F_4$, $^1D_2 \rightarrow {}^3H_6$, $^1D_2 \rightarrow {}^3F_4$, $^1G_4 \rightarrow {}^3H_6$	Vis–NIR light	10 mg	RhB (1×10^{-5} M, 50 ml)	70 min	~100%	Cheng et al. (2017)
$NaYF_4$:Yb^{3+},Tm^{3+}/CdS	$^1D_2 \rightarrow {}^3F_4$, $^1G_4 \rightarrow {}^3H_6$, $^3P_0 \rightarrow {}^3F_4$, $^1D_2 \rightarrow {}^3H_6$, $^1G_4 \rightarrow {}^3F_4$	976 nm, 2 W	20 mg	10 ml aqueous solution of RhB (10 ppm)	3 h	Show high photoactivity	Li et al. (2010)
β-$NaYF_4$:Tm^{3+}, Yb^{3+}@$NaLuF_4$@mSiO_2-Ag_3PO_4	$^1G_4 \rightarrow {}^3H_6$, $^1D_2 \rightarrow {}^3F_4$	Simulated sunlight; NIR light	8 mg	15 ml RhB aqueous solution (10 mg/l)	40 min 120 min	91% 85%	Liu et al. (2019)

$NaLuF_4$:Gd^{3+},Yb^{3+}, Tm^{3+}@$NaLuF_4$:Gd^{3+}, Yb^{3+}@SiO_2@Ag/g-C_3N_4	$^1G_4 \rightarrow {}^3H_6$, $^1D_2 \rightarrow {}^3F_4$, $^1D_2 \rightarrow {}^3H_6$, $^1I_6 \rightarrow {}^3F_4$, $^6P_{7/2} \rightarrow {}^8S_{7/2}$	Simulated solar light	20 mg	50 ml of RhB aqueous solution (20 mg/l)	150 min	~100%	Zhao et al. (2018)
$NaYF_4$:Yb^{3+},Er^{3+}@TiO_2	$^4G_{11/2} \rightarrow {}^4I_{15/2}$, $^2H_{9/2} \rightarrow {}^4I_{15/2}$, $^2H_{11/2} \rightarrow {}^4I_{15/2}$, $^4S_{3/2} \rightarrow {}^4I_{15/2}$, $^4F_{9/2} \rightarrow {}^4I_{15/2}$	NIR light irradiation ($\lambda > 800$ nm)	80 mg	RhB aqueous solution (50 ml, 5 mg/l)	10 h	79.7%	Zhang et al. (2018)
$NaGdF_4$:Yb^{3+}, Tm^{3+}@$NaGdF_4$@TiO_2	$^6P_{7/2} \rightarrow {}^8S_{7/2}$, $^1I_6 \rightarrow {}^3F_4$, $^1D_2 \rightarrow {}^3H_6$, $^1D_2 \rightarrow {}^3F_4$, $^1G_4 \rightarrow {}^3H_6$	980 nm, 2.5 W/cm^2	1 mg	1 ml of RhB aqueous solution (20 μg/ml)	5 h	Revealed a k value of 11.7×10^{-3} min^{-1}	Wang et al. (2020)
MoS_2-$NaYF_4$:Yb^{3+},Er^{3+}	$^4G_{11/2} \rightarrow {}^4I_{15/2}$, $^2H_{9/2} \rightarrow {}^4I_{15/2}$, $^2H_{11/2} \rightarrow {}^4I_{15/2}$, $^4S_{3/2} \rightarrow {}^4I_{15/2}$, $^4F_{9/2} \rightarrow {}^4I_{15/2}$	980 nm	20 mg	10 ml RhB solution (25 mg/l)	12 h	61%	Chatti et al. (2016)
$NaYF_4$:Yb^{3+}, Tm^{3+}@$NaYF_4$:Yb^{3+}, Nd^{3+}@TiO_2	$^3H_6 \rightarrow {}^3H_5$, $^3F_4 \rightarrow {}^3F_2$, $^3H_4 \rightarrow {}^1G_4$, $^1G_4 \rightarrow {}^1D_2$, $^1D_2 \rightarrow {}^1I_6$	980 nm (3 W) 808 nm (3 W)	10 mg 50 mg	3 ml of RhB aqueous solution (40 mg/l) 100 μl C_2H_4	150 min	85% 94%	Huang et al. (2019)
Ag@SiO_2@YF_3:Ho^{3+}@TiO_2		A 1000-W Xe lamp ($\lambda \geq 400$ nm)	0.20 g	200 ml of RhB solution (5 mg/l)	10 h	76%	Xu et al. (2019)
g-C_3N_4/$NaYF_4$:Yb^{3+},Tm^{3+}	$^1I_6 \rightarrow {}^3F_4$, $^1D_2 \rightarrow {}^3H_6$, $^1D_2 \rightarrow {}^3F_4$, $^1G_4 \rightarrow {}^3H_6$	Visible: 400~780 nm NIR: 780~2500 nm	10 mg	100 ml of RhB solution (10 mg/l)	105 min	Decreases dramatically	Chen et al. (2015)

(continued)

Table 16.1 (Continued)

Photocatalysts	Energy-level transitions	Light source	Photocatalyst amount	Aqueous solution	Reaction time	Removal percentage (%)	References
$NaYF_4$:Yb^{3+},Tm^{3+}@TiO_2	$^1I_6 \rightarrow {}^3F_4$, $^1D_2 \rightarrow {}^3H_6$, $^1D_2 \rightarrow {}^3F_4$, $^1G_4 \rightarrow {}^3H_6$	300–400 nm, 400–780 nm, 780–2500 nm	7 mg	50 ml of RhB solution (5×10^{-5} M)	30 min	A fast degradation	Su et al. (2014)
$NaYF_4$:Yb^{3+},Tm^{3+}@SiO_2-ZnO	$^1I_6 \rightarrow {}^3F_4$, $^1D_2 \rightarrow {}^3H_6$, $^1D_2 \rightarrow {}^3F_4$, $^1G_4 \rightarrow {}^3H_6$	300–400 nm, 400–780 nm, 780–2500 nm	50 mg	50 ml of RhB (2×10^{-5} M)	7 h	Showed good performance	Tou et al. (2017)
(1) β-$NaYF_4$:Yb^{3+}, Tm^{3+}@SiO_2@TiO_2 (2) β-$NaYF_4$:Yb^{3+}, Er^{3+}@SiO_2@TiO_2	(1) $^1I_6 \rightarrow {}^3H_6$, $^1D_2 \rightarrow {}^3H_6$, $^1D_2 \rightarrow {}^3F_4$, $^1G_4 \rightarrow {}^3H_6$, $^1G_4 \rightarrow {}^3F_4$, $^3F_3 \rightarrow {}^3H_6$ (2) $^4G_{11/2} \rightarrow {}^4I_{15/2}$, $^2H_{9/2} \rightarrow {}^4I_{15/2}$, $^2H_{11/2} \rightarrow {}^4I_{15/2}$, $^4S_{3/2} \rightarrow {}^4I_{15/2}$, $^4F_{9/2} \rightarrow {}^4I_{15/2}$	980 nm, 15 W/mm^2	10 mg	10 ml of RhB aqueous solution (10^{-5} M)	6 h	k(Er) = 0.2912 h^{-1} k(Tm) = 0.3624 h^{-1}	Wang et al. (2014b)
β-$NaYF_4$:Yb^{3+},Tm^{3+}@ZnO	$^1I_6 \rightarrow {}^3H_6$, $^1I_6 \rightarrow {}^3F_4$, $^1D_2 \rightarrow {}^3H_6$, $^1D_2 \rightarrow {}^3F_4$, $^1G_4 \rightarrow {}^3H_6$, $^1G_4 \rightarrow {}^3F_4$	980 nm, 2.0 W/cm^2	0.5 mg	0.5 ml 20 mg/l of RhB aqueous solution	30 h	65%	Guo et al. (2013)
$NaYF_4$:Yb^{3+}, Tm^{3+}@$NaYF_4$@mSiO_2/CdS	$^1I_6 \rightarrow {}^3F_4$, $^1D_2 \rightarrow {}^3H_6$, $^1D_2 \rightarrow {}^3F_4$, $^1G_4 \rightarrow {}^3H_6$	1500 mW/cm^2 xenon lamp equipped with an IR filter	30 mg	30 ml of RhB solution (1 mg/l)	300 min	90%	Li et al. (2016)
$NaYF_4$:Yb^{3+}, Tm^{3+}@$NaYF_4$:Yb^{3+}, Er^{3+}/SiO_2/CdS/TiO_2	$^1I_6 \rightarrow {}^3F_4$, $^1D_2 \rightarrow {}^3H_6$, $^1D_2 \rightarrow {}^3F_4$, $^1G_4 \rightarrow {}^3H_6$, $^2H_{11/2} \rightarrow {}^4I_{15/2}$, $^4S_{3/2} \rightarrow {}^4I_{15/2}$	1500 mW/cm^2 Xe lamp with a UV–vis filter	20 mg	50 ml of RhB solution with an initial concentration of 4.7 mg/l	90 min	Enhanced photocat-alytic performance	Wang et al. (2017)

Source: Based on Li et al. 2019; Zhang et al. 2016; Zhang et al. 2016; Wang et al. 2017; Zhang et al. 2017; Wang et al. 2018; Cheng et al. 2017; Li et al. 2010; Liu et al. 2019; Zhao et al. 2018; Zhang et al. 2018; Wang et al. 2020; Chatti et al. 2016; Huang et al. 2019; Xu et al. 2019; Chen et al. 2015; Su et al. 2014; Tou et al. 2017; Wang et al. 2014; Guo et al. 2013; Li et al. 2016 and Wang et al. 2017.

Table 16.2 Degradation of MB by typical UC photocatalysts.

Photocatalysts	Energy-level transitions	Light source	Photocatalyst amount	Aqueous solution	Reaction time (h)	Removal percentage (%)	References
$NaYF_4$:Yb^{3+},Tm^{3+}@Cu_2O	$^1I_6 \rightarrow {}^3H_6$, $^1I_6 \rightarrow {}^3F_4$, $^1D_2 \rightarrow {}^3H_6$, $^1D_2 \rightarrow {}^3F_4$, $^1G_4 \rightarrow {}^3H_6$, $^1G_4 \rightarrow {}^3F_4$	980 nm, 10 W/cm^2	0.5 mg	A cuvette containing MB aqueous solution (0.5 ml 15 mg/l)	10	76%	Zhang et al. (2018)
$NaYF_4$:Yb^{3+},Tm^{3+}@TiO_2	$^1I_6 \rightarrow {}^3H_6$, $^1I_6 \rightarrow {}^3F_4$, $^1D_2 \rightarrow {}^3H_6$, $^1D_2 \rightarrow {}^3F_4$, $^1G_4 \rightarrow {}^3H_6$	980 nm, 10 W/cm^2	0.5 mg	0.5 ml of MB aqueous solution (15 mg/l)	14	65%	Tang et al. (2013)
$NaYF_4$:Yb^{3+},Tm^{3+}/CdS/TiO_2	$^1I_6 \rightarrow {}^3H_6$, $^1I_6 \rightarrow {}^3F_4$, $^1D_2 \rightarrow {}^3F_4$, $^1G_4 \rightarrow {}^3H_6$, $^1G_4 \rightarrow {}^3H_4$	980 nm, 2.0 W/cm^2	0.5 mg	0.5 ml 15 mg/l MB aqueous solution	50	$K = 9.94 \times 10^{-4}$ min^{-1}	Guo et al. (2014)
Y_2SiO_5:Pr^{3+},Li^+/TiO_2	$^3H_4 \rightarrow {}^3P_0 \rightarrow {}^4f_{5d}$	A 500 W xenon lamp with a cutoff filter ($l > 400$ nm)	1.15 g	80 ml 10 mg/l MB aqueous solution	80	80%	Wu et al. (2015)
β-$NaYF_4$:Yb^{3+},Tm^{3+}/g-C_3N_4	$^1I_6 \rightarrow {}^3F_4$, $^1D_2 \rightarrow {}^3H_6$, $^1D_2 \rightarrow {}^3F_4$, $^1G_4 \rightarrow {}^3H_6$, $^1G_4 \rightarrow {}^3F_4$, $^3F_3 \rightarrow {}^3H_6$	NIR laser ($k = 980 \pm 10$ nm, power = 1 W, fiber diameter = 100 μm)	1 mg	0.5 ml of MB aqueous solution (15 ppm)	6	83%	Huang et al. (2015)
$NaYF_4$:Yb^{3+},Er^{3+}@BiOCl	$^2H_{11/2} \rightarrow {}^4I_{15/2}$, $^4S_{3/2} \rightarrow {}^4I_{15/2}$, $^4F_{9/2} \rightarrow {}^4I_{15/2}$	A 500 W long-arc xenon lamp with an optical filter ($\lambda > 700$ nm)			10	78%	Zhou et al. (2018)
β-$NaYF_4$:Yb^{3+}, Tm^{3+}/TiO_2/Fe_3O_4@SiO_2	$^1I_6 \rightarrow {}^3F_4$, $^1D_2 \rightarrow {}^3H_6$, $^1D_2 \rightarrow {}^3F_4$, $^1G_4 \rightarrow {}^3H_6$, $^1D_2 \rightarrow {}^3H_5$, $^1D_2 \rightarrow {}^3H_4$, $^1G_4 \rightarrow {}^3F_4$, $^3F_3 \rightarrow {}^3H_6$	980 nm (1.0 W/cm^2)	2.0 mg	0.5 ml of MB solution (10 ppm)	24	86.69%	Chen et al. (2018b)
$NaYF_4$:Yb^{3+},Tm^{3+}@ZnO	$^1I_6 \rightarrow {}^3H_6$, $^1I_6 \rightarrow {}^3F_4$, $^1D_2 \rightarrow {}^3H_6$, $^1D_2 \rightarrow {}^3F_4$, $^1G_4 \rightarrow {}^3H_6$, $^1G_4 \rightarrow {}^3F_4$,	980 nm, 10 W/cm^2	0.5 mg	MB aqueous solution (0.5 ml 15 mg/l)	25	68.7%	Zhang et al. (2019)

Source: Based on Zhang et al. 2018; Tang et al. 2013; Guo et al. 2014; Wu et al. 2015; Huang et al. 2015; Zhou et al. 2018; Chen et al. 2018 and Zhang et al. 2019.

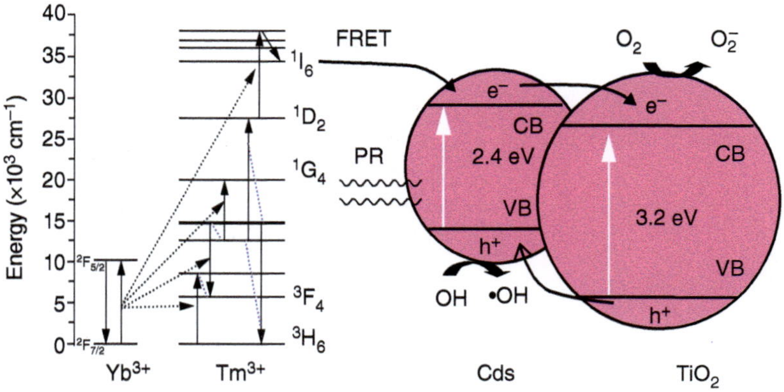

Figure 16.3 Diagrams of the energy levels of Yb^{3+}–Tm^{3+} and the upconversion luminescence processes in a Yb^{3+}–Tm^{3+}-co-doped system upon 980 nm excitation. Source: Guo et al. (2014). Reprinted with permission of Royal Society of Chemistry.

to prepare UCNPs/CdS/TiO_2 composites. As shown in Figure 16.3, the CB position of CdS was higher than that of TiO_2, so the photogenerated electrons on CdS can be directly injected into the CB band of TiO_2. At the same time, the VB position of TiO_2 was lower than that of CdS, so the photogenerated holes in TiO_2 can be easily transferred to CdS through the developed interface. It made charge separation more efficient and reduced the possibility of recombination, which lead to the enhancement in the photocatalytic activity. The results showed that compared with UCNPs@TiO_2 core–shell nanoparticles, UCNPs/CdS/TiO_2 composites had more obvious fluorescence quenching and higher FRET efficiency, which lead to the improvement in the photocatalytic performance of MB.

16.2.3 Degradation of MO

In addition to RhB and MB, UC photocatalysts are also used to degrade methyl orange (MO), a common organic pollutant (Table 16.3) (Huang et al. 2014; Xu et al. 2015; Zhang et al. 2017; Cui et al. 2016). Huang et al. (2014) synthesized a conversion photocatalyst with dendritic structure composed of $BiVO_4/CaF_2:Er^{3+},Tm^{3+},Yb^{3+}$ (BVO/CF) by a two-step hydrothermal method (Figure 16.4a). The carbon fiber particles were uniformly deposited on the surface of BVO/CF composites, which provided the conditions for the efficient upconversion emission. In BVO/CF composites, Bi^{3+} ions can improve their emission intensity by changing the crystal symmetry of the Er^{3+} and Tm^{3+} ions. The absorption edge of the BVO/CF composites was 506 nm, so the BVO will collect the upconversion emission of UV (361 and 379 nm), violet (408 nm), and blue (485 nm) light. Therefore, a hole (h^+) in VB and an electron (e^-) in CB can be generated to decompose methyl orange (MO) molecules. Strong oxidizing holes (h^+) can react with H_2O to produce hydroxyl radical (·OH), which was a major contribution to the degradation of organic pollutants (Figure 16.4b). As shown in Figure 16.4c, under the irradiation of 980 nm

Table 16.3 Degradation of MO by typical UC photocatalysts.

Photocatalysts	Energy-level transitions	Light source	Photocatalyst amount (mg)	Aqueous solution	Reaction time	Removal percentage (%)	References
$BiVO_4/CaF_2:Er^{3+},Tm^{3+},Yb^{3+}$	$^1G_4 \rightarrow {}^3H_6$, $^2H_{11/2} \rightarrow {}^4I_{15/2}$, $^4S_{3/2} \rightarrow {}^4I_{15/2}$, $^4F_{9/2} \rightarrow {}^4I_{15/2}$, $^1G_4 \rightarrow {}^3F_4$, $^2H_{9/2} \rightarrow {}^4I_{15/2}$, $^4G_{11/2} \rightarrow {}^4I_{15/2}$, $^1D_2 \rightarrow {}^3H_6$	980 nm (power = 2.0 A); 1000 W high-pressure mercury lamp	20	MO aqueous solutions (10 mg/l, 10 ml)	6 h; 3 h	Average removal rate constants of 0.113×10^{-1} h^{-1}; the corresponding average removal rate is 3.60×10^{-2} min^{-1}	Huang et al. (2014)
$NaYF_4:Yb^{3+},Er^{3+},Tm^{3+}@TiO_2$-Au	$^1I_6 \rightarrow {}^3F_4$, $^1D_2 \rightarrow {}^3H_6$, $^1D_2 \rightarrow {}^3F_4$, $^1G_4 \rightarrow {}^3H_6$, $^2H_{11/2} \rightarrow {}^4I_{15/2}$, $^4S_{3/2} \rightarrow {}^4I_{15/2}$	UV, visible, NIR, simulated sunlight irradiation	10	10 ml of MO solution (20 mg/l)	70 min; 8 h; 20 h; 120 min	Higher than that of the benchmark P25	Xu et al. (2015)
Au-$NaYF_4:Yb^{3+},Er^{3+},Tm^{3+}$/g-$C_3N_4$	$^1I_6 \rightarrow {}^3F_4$, $^1D_2 \rightarrow {}^3H_6$, $^1D_2 \rightarrow {}^3F_4$, $^1G_4 \rightarrow {}^3H_6$, $^2H_{11/2} \rightarrow {}^4I_{15/2}$, $^4S_{3/2} \rightarrow {}^4I_{15/2}$	980 nm, 2 W	10	25 ml of MO solution (10 mg/l)	200 min	The highest normalized reaction rate constant (0.0035/h/mg)	Zhang et al. (2017)
β-$NaYF_4:Yb^{3+},Tm^{3+}@CeO_2/ZnO$	$^1I_6 \rightarrow {}^3F_4$, $^1D_2 \rightarrow {}^3H_6$, $^1D_2 \rightarrow {}^3F_4$, $^1G_4 \rightarrow {}^3H_6$	Simulated sunlight supplied by a Xe lamp (100 mW/cm^2)	~50	50 ml of MO solution (10^{-5} M)	80 min	Show a very fast decoloration rate at the first 40 min	Cui et al. (2016)

Source: Based on Huang et al. 2014; Xu et al. 2015; Zhang et al. 2017 and Cui et al. 2016.

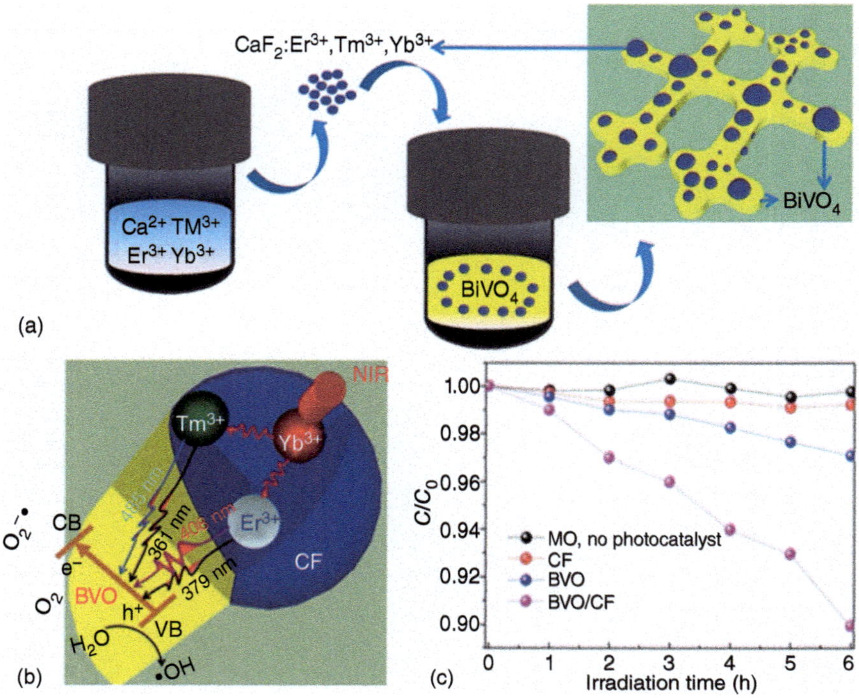

Figure 16.4 (a) Schematic illustrations for the preparation of the BVO/CF composites; (b) schematic illustration of the photocatalytic processes involved in the BVO/CF composites; and (c) photocatalytic degradation rates of MO for the samples under 980 nm NIR irradiation (power = 2.0 A). Source: Huang et al. (2014). Reprinted with permission of Royal Society of Chemistry.

laser, the upconversion luminescence intensity of the BVO/CF composite material was eight times higher than that of pure CF.

Xu et al. reported the novel $NaYF_4$:Yb^{3+},Er^{3+},Tm^{3+}-core@porous-TiO_2-shell (denoted as NYF@TiO_2-Au) microspheres modified by plasmonic Au nanoparticles (Xu et al. 2015). The synthesized NYF@TiO_2-Au microspheres can capture photons in a wide range of wavelengths from UV, visible to NIR and exhibit excellent MO degradation activity, which was significantly better than the P25 TiO_2 reference photocatalyst. This enhanced activity was attributed to the synergistic effect of the nanocomponents arraying in the nanostructure, which facilitated the effective charge/energy transfer between the nanocomponents and greatly reduced the charge recombination. This research provided a promising approach to improve the solar energy efficiency and MO degradation rate by combining nanocomponents with different properties, such as plasmonic, upconversion, and photocatalytic properties, to design their structures properly.

16.2.4 Degradation of Various Organic Pollutants

In addition to the degradation of single pollutants, it has also been pointed out that UC photocatalytic materials can also be used for the degradation of a variety of pollutants (Table 16.4) (Zhang et al. 2017, 2020; Li et al. 2020; Duo et al. 2016;

Table 16.4 Degradation of various organic pollutants by typical UC photocatalysts.

Photocatalysts	Energy-level transitions	Light source	Photocatalyst amount (mg)	Aqueous solution	Reaction time	Removal percentage (%)	References
$NaYF_4$:Yb^{3+},Er^{3+}/Ag_2CrO_4	$^2H_{11/2} \rightarrow {}^4I_{15/2}$, $^4S_{3/2} \rightarrow {}^4I_{15/2}$, $^4F_{9/2} \rightarrow {}^4I_{15/2}$	400 W metal halide lamp, consisting of vis and NIR light source with a UV filter	30	50 ml of dyes (MO, MB, RhB, and tetracyclines (TCs), 10 mg/l)	150 min	62% 87% 55% 57%	Zhang et al. (2020)
SrF_2-Bi_2O_3-B_2O_3/Yb^{3+},Tb^{3+}/BiOBr		NIR light ($\lambda \geq 780$ nm, a 1000 W mercury lamp)	20	NOR (20 ml, 20 mg/l) and MO (20 ml, 20 mg/l)	180 min 30 min	56% 88%	Li et al. (2020)
$NaGdF_4$:Er^{3+},Yb^{3+}@$BiFeO_3$	$^2H_{11/2} \rightarrow {}^4I_{15/2}$, $^4S_{3/2} \rightarrow {}^4I_{15/2}$, $^4F_{9/2} \rightarrow {}^4I_{15/2}$	$\lambda > 420$ nm, ~85 mW/cm^2; 980 nm, ~0.08 W/mm^2	20	20 ml of a 10 mg/l MO and 4-CP aqueous solution	3.5 h 15 h	~80% ~55%	Zhang et al. (2017)
β-$NaYF_4$:Yb^{3+},Tm^{3+}@TiO_2@Au	$^1D_2 \rightarrow {}^3H_6$, $^1D_2 \rightarrow {}^3F_4$, $^1G_4 \rightarrow {}^3H_6$, $^1D_2 \rightarrow {}^3H_5$, $^1G_4 \rightarrow {}^3F_4$, $^3F_3 \rightarrow {}^3H_6$	1 KW Xe lamp with the simulated sunlight	20	50 ml of MB solution (15 mg/l) or 50 ml of RhB solution (10 mg/l)	4.5 h	73% 80%	Duo et al. (2016)
$NaYF_4$:Yb^{3+},Tm^{3+}@C@CdS	$^1I_6 \rightarrow {}^3F_4$, $^1D_2 \rightarrow {}^3H_6$, $^1D_2 \rightarrow {}^3F_4$, $^1G_4 \rightarrow {}^3H_6$, $^1G_4 \rightarrow {}^3F_4$	vis bandpass (400–780 nm); vis–NIR bandpass (400–2500 nm)	50	50 ml of RhB solution or MB solution (5×10^{-5} M)	12 h	Showing the good catalytic activity	Tou et al. (2016)
$NaYF_4$:Yb^{3+}/Er^{3+}@CdS	$^2H_{11/2} \rightarrow {}^4I_{15/2}$, $^4S_{3/2} \rightarrow {}^4I_{15/2}$, $^4F_{9/2} \rightarrow {}^4I_{15/2}$	NIR light	10	25 ml of a 5×10^{-5} M aqueous solution of the pollutant (MO/CZ)	6 h	52% 64%	Balaji et al. (2017)

(*continued*)

Table 16.4 (Continued)

Photocatalysts	Energy-level transitions	Light source	Photocatalyst amount (mg)	Aqueous solution	Reaction time	Removal percentage (%)	References
α-$NaYF_4$:Yb^{3+},Tm^{3+}@TiO_2/rGO	$^1D_2 \rightarrow {}^3H_6$, $^1D_2 \rightarrow {}^3F_4$, $^1G_4 \rightarrow {}^3H_6$	2 W 980 nm diode laser	5	5 ml of MB (10 ppm), MO (15 ppm) or phenol (20 ppm) aqueous solution	12 h	81.1% exhibits much enhanced photocatalytic activity	Wang et al. (2015)
$NaYF_4$:Yb^{3+},Tm^{3+}-BiOCl	$^1I_6 \rightarrow {}^3F_4$, $^1D_2 \rightarrow {}^3H_6$, $^1D_2 \rightarrow {}^3F_4$, $^1G_4 \rightarrow {}^3H_6$	NIR light ($\lambda > 780$ nm), 18 mW/cm^2	20	50 ml of 2×10^{-5} mol/l RhB and MB aqueous solution	160 min	7.63×10^{-3} gRhB/gcat/h	Bai et al. (2016)
$NaYF_4$:Yb^{3+},Tm^{3+}@TiO_2	$^1D_2 \rightarrow {}^3H_6$, $^1G_4 \rightarrow {}^3H_6$, $^1D_2 \rightarrow {}^3H_5$, $^1G_4 \rightarrow {}^3F_4$, $^3F_3 \rightarrow {}^3H_6$	980 nm laser irradiation	20	10 ml RhB or phenol aqueous solution (10^{-5} M)	60 min	22% 17%	Wang et al. (2013)
β-$NaYF_4$:Yb^{3+},Tm^{3+}@TiO_2	$^1I_6 \rightarrow {}^3H_6$, $^1I_6 \rightarrow {}^3F_4$, $^1D_2 \rightarrow {}^3H_6$, $^1D_2 \rightarrow {}^3F_4$, $^1G_4 \rightarrow {}^3H_6$, $^1D_2 \rightarrow {}^3H_5$, $^1D_2 \rightarrow {}^3H_4$, $^1G_4 \rightarrow {}^3F_4$, $^3F_3 \rightarrow {}^3H_6$	980 nm, 1 W	5	0.5 ml of MB, MO, and RhB aqueous solution (20 ppm)	25 h	68.37% 69.34% 75.75%	Xu et al. (2013)

Source: Based on Zhang et al. 2020; Li et al. 2020; Zhang et al. 2017; Duo et al. 2016; Tou et al. 2016; Balaji et al. 2017; Wang et al. 2015; Bai et al. 2016; Wang et al. 2013; Xu et al. 2013.

Tou et al. 2016; Balaji et al. 2017; Wang et al. 2013, 2015; Bai et al. 2016; Xu et al. 2013). Zhang et al. successfully prepared the novel narrow bandgap photocatalysts $NaYF_4:Yb^{3+},Er^{3+}/Ag_2CrO_4$ (NYF/AC) by hydrothermal and precipitation methods (Zhang et al. 2020) (Figure 16.5a). The whole photocatalytic process can be divided into two schemes: (i) the energy transfer process between Er^{3+} and Ag_2CrO_4

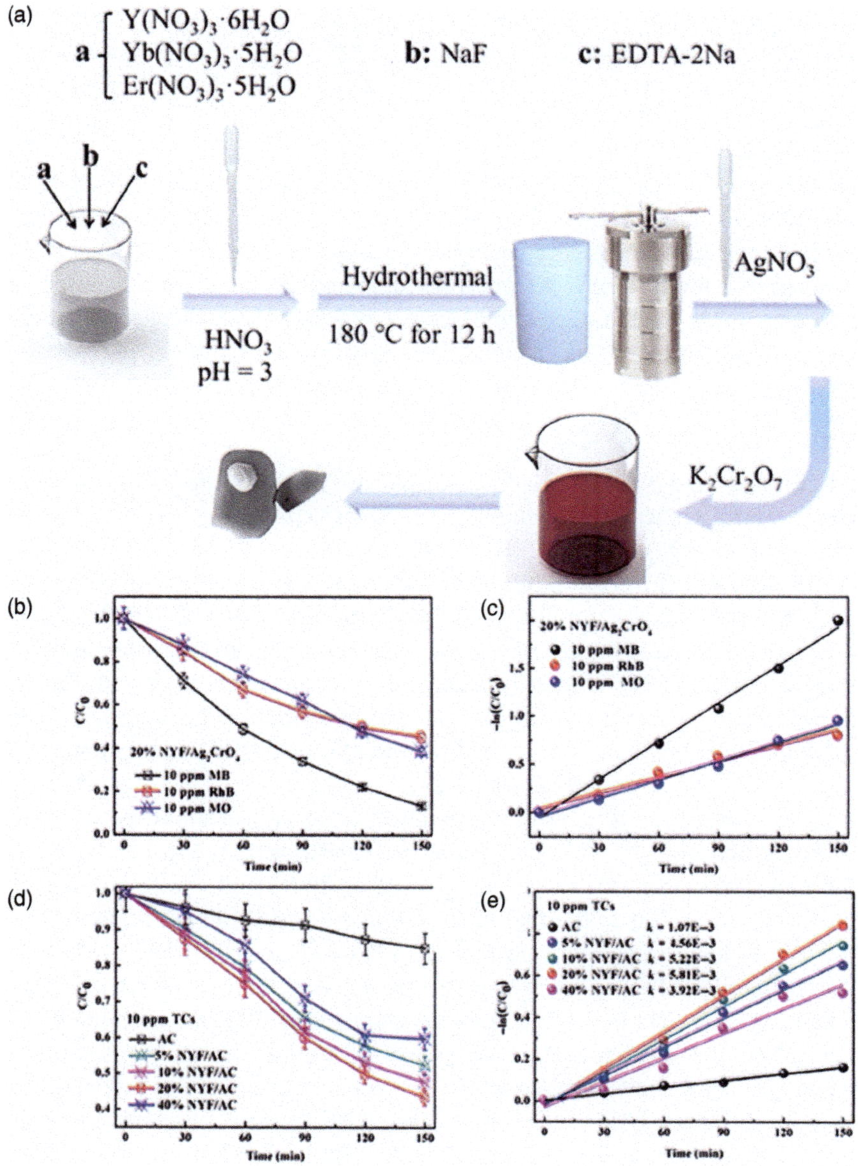

Figure 16.5 (a) Schematic illustration of the NYF/AC composites synthesized by the hydrothermal and precipitation methods; photodegradation performance for the pollutants over AC and NYF/AC samples under visible–NIR light irradiation. (b) and (c) for MO, RhB, and MB and (d) and (e) for TCs, respectively. Source: Zhang et al. (2020). Reprinted with permission of Springer Nature.

semiconductor and (ii) the photocatalytic process between reactants and Ag_2CrO_4 semiconductor. Yb^{3+} ions, as sensitizers in the material, can be pumped from the ground state ($^2F_{7/2}$ level) to the excited state ($^2F_{5/2}$ level) at 980 nm excitation; this excitation energy of Yb^{3+} is transferred to the $^2H_{11/2}$, $^4S_{3/2}$, and $^4F_{9/2}$ energy levels of the Er^{3+} ions. Subsequently, the energy is transferred from NYF to Ag_2CrO_4, which excited the Ag_2CrO_4 to generate electron/hole pairs. The generation of $^{\bullet}O_2^-$ and holes can further oxidize the pollutants such as MO, MB, RhB, and tetracyclines (TCs). At the same time, H_2O_2 was obtained from NYF/AC suspension through the electron reduction pathway ($e^- \rightarrow O_2^- \rightarrow H_2O_2$). As shown in Figure 16.5b–e, the degradation rates of MO, MB, RhB, and TCs by 20% NYF/AC were 62%, 87%, 55%, and 57%, respectively.

Norfloxacin (NOR), tetracycline, and other antibiotics used in human clinical and animal husbandry are discharged into natural water in large quantities. Antibiotics have stable chemical structure and complex existing forms, which are difficult to be removed by conventional techniques. Residual antibiotics in the environment have become a new kind of microcontaminants, which increase the generation of resistance genes in bacteria and endanger the balance of the ecosystem and human health. They usually have low optical radiation penetration when entering into deep water or polluted waters (such as rivers or lakes), and the weak UV–vis light in water will inevitably inhibit the photocatalytic degradation of NOR. Li et al. have successfully prepared a BiOBr-based NIR-responsive heterojunction photocatalyst with high luminescence and electron energy transfer by using a facile in situ etching glass–ceramic method (FIEG) to grow BiOBr nanosheets on the surface of SrF_2-Bi_2O_3-B_2O_3/Yb^{3+},Tb^{3+} (SBBF) upconversion glass–ceramic (GC) (Li et al. 2020). The core–shell structure of the GC@BiOBr heterojunction photocatalyst was beneficial to increase the charge transfer and reduce the recombination rate of the excited carriers. It can efficiently collect NIR photons to emit UV and visible light upconversion emissions, which can be used in photocatalytic processes. The degradation rate of NOR reached 56% and 88% after SBBF/BiOBr-10 was exposed to NIR light for 180 and 30 minutes, respectively.

Zhang et al. reported a core@shell nanohybrid material synthesized by two steps. Firstly, $NaGdF_4$:Er^{3+},Yb^{3+} nanoparticles (UCNPs) were prepared by a thermal decomposition method. Then, bismuth ferrite (BFO) shells were assembled around UCNPs by the hydrothermal process. Finally, $NaGdF_4$:Er^{3+},Yb^{3+}@$BiFeO_3$ (UCNP@BFO) nanostructures were formed (Zhang et al. 2017). The enhanced photoactivity of UCNP@BFO structure was attributed to the NIR radiation being converted to visible light by UCNPs and then captured by BFO through the FRET process. Thus, this structure optimized the use of solar energy by efficiently collecting the visible and NIR photons. The photocatalytic activity was investigated under monochromatic NIR light (λ = 980 nm). After 15 hours of reaction, ~55% degradation was observed. The photodegradation results of 4-chlorophenol (4-CP) were similar to those observed with MO.

16.2.5 Others

In addition to common pollutants such as RhB, MB, and MO, upconversion photocatalytic materials have also been studied in the degradation of other pollutants (Table 16.5) (Kwon et al. 2018; Li et al. 2019b; Ullah et al. 2019; Ma et al. 2015). For nitrobenzene (NB), a typical organic waste, Kwon et al. used triblock copolymer P123 as a mesoporous template and carbon source and embedded triblock β-$NaYF_4$:Yb^{3+},Tm^{3+},Gd^{3+} upconversion nanorods into the hybrid membrane of carbon-doped mesoporous TiO_2 (Kwon et al. 2018). The possibility of collecting lost photons from TiO_2 by synchronous carbon doping and UC strategy is reported for the first time. mC-TiO_2-β-$NaYF_4$:Yb^{3+},TM^{3+},Gd^{3+} (mC-TiO_2/UCNPs) has the synergistic effect of photon-trapping TiO_2 structure. The incorporation of UCNPs can improve the light trapping efficiency, reduce the bandgap energy, and increase the visible light absorption of C-integrated TiO_2 structure. The results showed that, compared with mTiO_2 (50%), the photoactivity of mC-TiO_2/UCNPs was significantly increased (83%) during the degradation of NB by UV–vis–NIR irradiation.

Similarly, the novel Au/$NaYF_4$:Yb^{3+},Er^{3+}/WO_3·0.33H_2O-$W_{18}O_{49}$ (A/NYYE/WW) heterojunctions synthesized by Li et al. using hydrothermal and photoreduction methods have excellent degradation effect on NB (Li et al. 2019). The local surface plasmonic resonance (LSPR) effect of Au NPs and $W_{18}O_{49}$ can induce the wide-band optical absorption using NIR light. In addition, the resulting LSPR effect can improve the photoluminescence of UCNPs. High-energy $NaYF_4$:Yb^{3+},Er^{3+} (NYYE) emits visible light, which in turn excites Au NPs and tungsten oxide semiconductors via FRET. The plasmonic coupling between Au NPs and $W_{18}O_{49}$ may promote the generation of "hot electrons" in the plasma. On the other hand, in the process of plasmonic energy transfer, photogenerated electrons are transferred from the semiconductor to the surface of Au NPs, which promotes effective charge separation in the heterostructures (Figure 16.6). Compared with WO_3·0.33H_2O (35%) or NYEE/WW (41%), A/NYYE/WW had the best photocatalytic activity (96%) for NB aqueous solution.

Ullah et al. used a simple and rapid microwave-assisted hydrothermal method to obtain $NaYF_4$:Tm^{3+} hexagonal UCNPs and high-crystalline BV powders with customized morphology, crystal phase, and bandgap energy (2.4–2.93 eV) (Ullah et al. 2019). The emission spectra of $NaYF_4$:Tm^{3+} UCNPs and the absorption spectra of BV (monocline) overlapped well in the region of 500 nm, so it can be reasonably inferred that the NIR to UV/visible upconversion light from UCNPs can be used effectively for photoexcitation of BV. Oxidative h^+ and reductive e^- were produced by photoexcitation. Reduced e^- and oxidized h^+ reacted with molecular O_2 and water to form $^{\bullet}OH$, coupled with h^+ formed in the valence band direct oxidation, resulting in the photodegradation of crystal violet (CV) dyes.

Ma et al. synthesized $NaYF_4$:Yb^{3+},Tm^{3+}@TiO_2/Ag (UC@TiO_2/Ag) core–shell composites using a three-step hydrothermal method (Ma et al. 2015). Yb^{3+}, as

Table 16.5 Degradation of other pollutants by typical UC photocatalysts.

Photocatalysts	Energy-level transitions	Light source	Photocatalyst amount	Aqueous solution	Reaction time	Removal percentage (%)	References
mC-TiO_2-β-$NaYF_4$:Yb^{3+},Tm^{3+},Gd^{3+}	$^1D_2 \rightarrow {}^3F_4$, $^1G_4 \rightarrow {}^3H_6$, $^1G_4 \rightarrow {}^3F_4$, $^3F_3 \rightarrow {}^3H_6$, $^3H_4 \rightarrow {}^3H_6$	A Xe lamp equipped with a 420 nm cutoff filter at 300 W; 980 nm		30 ml of a 10 ppm of NB solution	3 h	57% 55% 83%	Kwon et al. (2018)
Au/$NaYF_4$:Yb^{3+},Er^{3+}/WO_3·0.33H_2O-$W_{18}O_{49}$	$^2H_{11/2} \rightarrow {}^4I_{15/2}$, $^4S_{3/2} \rightarrow {}^4I_{15/2}$, $^4F_{9/2} \rightarrow {}^4I_{15/2}$	A 300 W xenon lamp ($\lambda \geq 420$ nm)	30 mg	50 ml of 20 ppm NB solution	100 min	96%	Li et al. (2019)
$NaYbF_4$:Tm^{3+}/$BiVO_4$	$^1I_6 \rightarrow {}^3H_6$, $^1I_6 \rightarrow {}^3F_4$, $^1D_2 \rightarrow {}^3H_6$, $^1D_2 \rightarrow {}^3F_4$, $^1G_4 \rightarrow {}^3H_6$	NIR	50 mg	25 ml of CV (10 mg/l)	6 h	80%	Ullah et al. (2019)
$NaYF_4$:Yb^{3+},Tm^{3+}@TiO_2/Ag	$^1I_6 \rightarrow {}^3F_4$, $^1D_2 \rightarrow {}^3H_6$, $^1D_2 \rightarrow {}^3F_4$, $^1G_4 \rightarrow {}^3H_6$	UV-filtered Xe lamp	—	50 ml 1×10^{-5} mol/l R6G solution	120 min	96%	Ma et al. (2015)

Source: Based on Kwon et al. 2018; Li et al.2019; Ullah et al. 2019 and Ma et al. 2015.

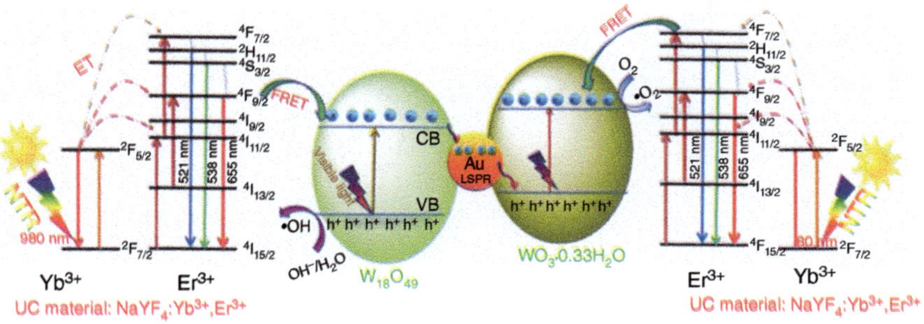

Figure 16.6 The proposed photocatalytic mechanism of the A/NYYE/WW photocatalyst under vis–NIR light irradiation. Source: Li et al. (2019). Reprinted with permission of Royal Society of Chemistry.

a sensitizer, was the main absorber of NIR excitation. Three energy transfers from Yb^{3+} to Tm^{3+} form $^3H_5$, $^3F_2$, and $^1G_4$ energy levels. Finally, the $^1D_2 \rightarrow {}^3F_4$, $^1G_4 \rightarrow {}^3H_6$, $^3P_0 \rightarrow {}^3F_4$, and $^1D_2 \rightarrow {}^3H_6$ transitions produce blue light at 450 and 474 nm and UV light at 349 and 362 nm. The presence of precious metal silver nanoparticles can enhance upconversion luminescence, resulting in stronger UV emission. In addition, Ag nanoparticles can be used as co-catalysts to further improve the photocatalytic performance. When activated by UV light, TiO_2 can produce electrons and holes in the conduction and valence bands. Because the Fermi level of TiO_2 is higher than that of Ag, the photogenerated electrons of TiO_2 can be transferred to Ag, and electrons are removed by molecular oxygen to form superoxide $^{\cdot}O_2^-$ radical anions, while the holes can react with H_2O to form $^{\cdot}OH$. Therefore, the organic molecules present in the solution will react with these oxidants, resulting in oxidative degradation. Electron transfer from TiO_2 to Ag nanoparticles can effectively inhibit the recombination of excited electrons and holes, thus improving the photocatalytic performance of the nanostructures. In the UV–vis absorption experiment, under the full spectrum light of xenon lamp, about 96% of the R6G dye in the solution was degraded by UC@TiO_2/Ag within 120 minutes, while only about 64% of the dye was degraded by pure TiO_2 under the same conditions. About 35% of the dyes were degraded by UC@TiO_2/Ag in 120 minutes under xenon lamp with UV filter, while approximately 8% of the dye was degraded by pure TiO_2.

16.3 Degradation of Inorganic Pollutants

The upconversion materials can also be applied to the photocatalytic treatment of inorganic heavy metal ions in wastewater (Table 16.6). Cr(VI) and chromium salts in sewage are carcinogens, so all compounds containing Cr(VI) are toxic. It does great harm to crops and other organisms (GracePavithra et al. 2019; Rahman and Singh 2019; Xia et al. 2019). Therefore, researchers have carried out a lot of research on the transformation of Cr(VI). At present, the removal technologies of Cr(VI) mainly include adsorption and photocatalysis.

Table 16.6 Degradation of inorganic pollutants by typical UC photocatalysts.

Photocatalysts	Light source	Photocatalyst amount	Aqueous solution	Reaction time	Removal percentage (%)	References
$NaYF_4$:Yb^{3+}, Tm^{3+}@$NaYF_4$@$Zn_{0.5}Cd_{0.5}S$	780–2500 nm 320–3500 nm	5 mg	50 ml of Cr(VI) solution (20 mg/l)	30 min	38.1% 98.7%	Wang et al. (2018)
$NaYF_4$:Yb^{3+}, Tm^{3+}@$NaYF_4$:Yb^{3+}, Er^{3+}@$Zn_xCd_{1-x}S/TiO_2$	Xe lamp (1500 mW/cm^2) equipped with (or without) a UV–vis filter	15 mg	50 ml of aqueous Cr(VI) solution (20 mg/l)	30 min	More than 70% ~99%	Zhao et al. (2018)
X/$NaTaO_3$/Er^{3+}:$YAlO_3$ (X = Ag, Au, and Pt)	Mercury lamp (UV light source); halogen lamp (visible light source); xenon lamp (simulated sunlight source)	50 mg	50 ml of Cr(VI) aqueous solution (1.0 ppm)	120 min	97.0% (pH = 3.0)	Li et al. (2017)
(Yb^{3+},Tm^{3+})-$NaYF_4$/BiOX (X = Cl, Br, I) (Yb^{3+},Er^{3+})-$NaYF_4$/BiOX (X = Cl, Br, I)	980 nm NIR laser		1 : 1 mixed air and nitrogen gas with 1 ppm of NO		6.2% 4.1%	Wu et al. (2017)

Wang et al. developed $NaYF_4:Yb^{3+},Tm^{3+}@NaYF_4@Zn_{0.5}Cd_{0.5}S$ core–shell nanoparticles ($UCNPs@Zn_{0.5}Cd_{0.5}S$ CSN) by a two-step solution process (Wang et al. 2018). After calcination at 400 °C, $NaYF_4:Yb^{3+},Tm^{3+}@NaYF_4@Zn_{0.5}Cd_{0.5}S$ core–shell nanoparticles ($UCNPs@Zn_{0.5}Cd_{0.5}S$ YSN) were obtained. The fluorescence emission of Tm^{3+} was effectively quenched, which proved the effective energy transfer between the UCNPs and $Zn_{0.5}Cd_{0.5}S$ components. The conversion rates of Cr(VI) were 38.1% and 98.7% under simulated sunlight and NIR light, respectively, which showed excellent photocatalytic activity of Cr(VI) reduction. The Cr(VI) conversion rate of $UCNPs@Zn_{0.5}Cd_{0.5}S$ YSN was higher than that of $UCNPs@Zn_{0.5}Cd_{0.5}S$ CSN. It was attributed to the enhanced crystallization degree, which improved the enhanced energy transfer efficiency and separation efficiency of the photogenerated electrons and holes. More importantly, the Cr(VI) conversion rate of $UCNPs@Zn_{0.5}Cd_{0.5}S$ YSN was higher than UCNPs@ZnS YSN under either simulated sunlight or NIR light, indicating that the introduction of CdS can improve the energy transfer efficiency between UCNPs with shells. The rationalization mechanism of Cr(VI) photochemical reduction is as follows:

$$Zn_{0.5}Cd_{0.5}S + h\nu \rightarrow h^+ + e^- \tag{16.1}$$

$$H_2O + h^+ \rightarrow {}^{\cdot}OH + H^+ \tag{16.2}$$

$$Cr_2O_7{}^{2-} + 14H^+ + 6e^- \rightarrow 2Cr^{3+} + 7H_2O \tag{16.3}$$

Similarly, Zhao et al. used simple hydrothermal method to combine $NaYF_4:Yb^{3+}$, $Tm^{3+}@NaYF_4:Yb^{3+},Er^{3+}$ (UCNPs) with $Zn_xCd_{1-x}S$ alloy semiconductor and then adjusted the proportion of Zn and Cd to synthesize $UCNPs@Zn_xCd_{1-x}S/TiO_2$ yolk–shell composite structure (Zhao et al. 2018) (Figure 16.7a). Amorphous TiO_2 binded UCNPs closely to $Zn_xCd_{1-x}S$, which played an important role in the formation of the $UCNPs@Zn_xCd_{1-x}S/TiO_2$ nanoparticles. The preparation process and reaction mechanism are shown in Figure 16.7b. Under a 30 minute irradiation of NIR light using a 1500 mW/cm^2 Xe lamp equipped with a UV–vis filter, the reduction rate of Cr(VI) on 15 mg of $UCNPs@Zn_{0.64}Cd_{0.36}S/TiO_2$ catalyst was over 70%. About 99% of Cr(VI) can be removed under the irradiation of simulated solar sunlight. This was because the NIR light in the xenon lamp was absorbed and converted into UV–vis light, which can be utilized by $Zn_{0.64}Cd_{0.36}S/TiO_2$, thus improving the utilization of sunlight and promoting the enhancement of photocatalytic performance.

In order to expand the light response range of wide-bandgap semiconductor photocatalyst ($NaTaO_3$), Li et al. prepared visible-light-driven photocatalyst $NaTaO_3/Er^{3+}:YAlO_3$ by combining the upconversion luminator ($Er^{3+}:YAlO_3$) with $NaTaO_3$ (Li et al. 2017a). Ag, Au, and Pt co-catalysts were deposited on the surface of $NaTaO_3/Er^{3+}:YAlO_3$, respectively, which could accept photogenerated electrons as active sites. The photogenerated electron (e^-) was transferred from the conduction band of $NaTaO_3$ to Ag, Au, and Pt particles. Because of its high reducibility, Cr(VI)

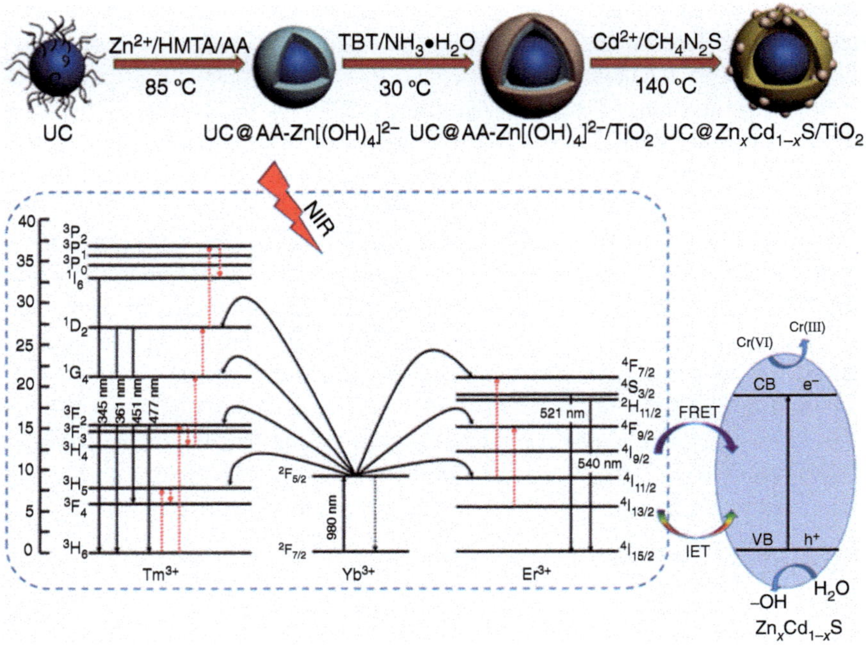

Figure 16.7 (a) Schematic illustration of the synthesis of UCNPs@$Zn_xCd_{1-x}S/TiO_2$ nanocomposites; (b) IET and FRET between UCNPs and $Zn_xCd_{1-x}S$, and the photochemical reduction mechanism of Cr(VI). Source: Zhao et al. (2018). Reprinted with permission of Elsevier.

can be easily converted to Cr(III). At the same time, H_2O was oxidized to O_2 in the valence band of $NaTaO_3$. The process might look something like this:

$$[Er^{3+} : YAlO_3] + h\nu(\text{visible} - \text{light}) \rightarrow [Er^{3+} : YAlO_3]^* \tag{16.4}$$

$$[Er^{3+} : YAlO_3]^* \rightarrow [Er^{3+} : YAlO_3] + h\nu(\text{ultraviolet} - \text{light}) \tag{16.5}$$

$$NaTaO_3 + h\nu(\text{ultraviolet} - \text{light}) \rightarrow e^-(CB) + h^+(VB) \tag{16.6}$$

$$e^-(CB) + X(Ag, Au, Pt) \rightarrow e^-(X(Ag, Au, Pt)) \tag{16.7}$$

$$Cr(VI) + e^-(X(Ag, Au, Pt)) \rightarrow Cr(III) + X(Ag, Au, Pt) \tag{16.8}$$

$$h^+(VB) + H_2O \rightarrow O_2 + H^+ \tag{16.9}$$

The results showed that the upconversion luminescence agent (Er^{3+}:$YAlO_3$) and the co-catalyst (Ag, Au, and Pt) can jointly promote the photocatalytic conversion of Cr(VI) from $NaTaO_3$ by using visible light. When the molar ratio of Er^{3+}:$YAlO_3$ to $NaTaO_3$ was 0.3 : 1.0 and the content of Au was 1.0 wt%, the prepared Au/$NaTaO_3$/Er^{3+}:$YAlO_3$ nanocomposites had the highest photocatalytic Cr(VI) conversion activity.

In addition to typical Cr(VI), nitric oxide (NO) is one of the serious gaseous pollutants in atmospheric environment. It is very harmful to humans that can cause respiratory infections, heart disease, lung cancer, and stroke. Biological filtration, wet washing, photocatalysis, and selective catalytic reduction are commonly used to clean up NO pollution. The UC photocatalytic material is not only widely used in the reduction of Cr(VI), but researchers have found that it also has a good effect in the removal of NO. Wu et al. have prepared the oxygen-rich void BioX (X = Cl, Br, and I) nanoparticles by the precipitation method and coupled upconversion phosphors with representative BiOBr nanoparticles through a simple method (Wu et al. 2017). Pure BiOBr did not react to 980 nm NIR light, resulting in no significant effect on NO removal. However, (Yb^{3+},Tm^{3+})-$NaYF_4$/BiOBr and (Yb^{3+},Er^{3+})-$NaYF_4$/BiOBr can be converted into visible light such as blue light and green light on NIR light excitation, and BiOBr can re-absorb to generate electron transition, thus improving the photocatalytic performance. Under 980 nm NIR laser irradiation, the removal rate was 6.2% and 4.1%, respectively, higher than that of pure BiOBr.

16.4 Photocatalytic Hydrogen Production

Resource shortage and environmental deterioration are two key problems that must be solved in today's society. Replacement of fossil fuels with renewable fuels has become the key to the sustainable development of human society. Hydrogen is a kind of high energy density fuel, and its combustion product is water. It is an ideal clean energy that can replace fossil fuels and achieve sustainable development (Momirlan and Veziroglu 2005; Navarro et al. 2007). In recent years, the research and development of hydrogen energy has attracted extensive attention in the field of energy. Hydrogen energy can be obtained by hydrocarbon pyrolysis, water gas reaction, water electrolysis, and other methods, but it needs a lot of energy supply (Fu et al. 2008; Turner 2004). Using cheap and efficient catalysts to achieve photohydrolysis of water for hydrogen production under sunlight can fundamentally solve the problems of environmental pollution and energy crisis faced by the development of human society, which also reveals the practical feasibility of converting solar energy into chemical energy from the side.

At present, photochemical hydrogen evolution is the most economical and sustainable means of hydrogen production in the hydrogen energy industry, in which the efficient catalyst plays an important role in promoting the process of hydrogen evolution. Therefore, semiconductor materials with good photoelectric response characteristics have attracted more and more attention (Wang et al. 2014; Cao et al. 2020; Zhang and Guo 2013). However, the photocatalytic efficiency is low because of the narrow absorption range of solar spectrum and the easy recombination of photogenerated carriers. Therefore, it is necessary to develop new catalyst substitutes. Photocatalysts with UC effect have attracted extensive attention in recent years because it can convert visible light or NIR light to UV light or visible light, thus improving the activity of photocatalyst and improving the utilization rate of sunlight (Li et al. 2017, 2019; Zhao et al. 2018; Anushree and Sadhana 2018; Gao et al. 2019;

Zhu et al. 2019; Zhang et al. 2018, 2019; Shu et al. 2018; Bao et al. 2019; Shang et al. 2019). The fundamental cause of hydrogen production through photocatalyst with UC effect is the conduction band of semiconductor under sunlight irradiation. The electrons react with water to produce hydrogen only if the conduction band is greater than the H^+/H_2 potential.

Since 1972, Fujishima and Honda discovered the photocatalytic decomposition of water on the semiconductor TiO_2 electrode, multiphase photocatalysis began to develop rapidly. As a wide-bandgap semiconductor, TiO_2 has been widely studied for its low cost, high stability, and good biosafety. Anushree and Sadhana (2018) successfully combined the UC material $NaYF_4$:Yb^{3+}, Er^{3+} (noted as (Yb, Er)NYF) with the TiO_2 semiconductor. The results showed that no hydrogen production was detected on TiO_2 after five hours of irradiation under simulated light, but the hydrogen production reached 6.71 μmol/h after loading the UC material. At the same time, the precious metal Au was used to prepare (Yb,Er)NYF-TiO_2/Au by photodeposition. The surface plasmonic resonance property and the upconversion property of (Yb,Er)NYF can synergically enhance the broadband absorption, generation, and transfer of photoelectric charge. Au as plasma nanoparticles on (Yb,Er)NYF-TiO_2/Au had a variety of solar energy collection effects and showed good catalytic activity. As listed in Table 16.7, the photocatalytic hydrogen production experiments showed that the H_2 production rate of (Yb,Er)NYF-TiO_2/Au was 350 μmol/h under the condition of solar AM 1.5, which was about 3.1 times that of Au/TiO_2 (Degussa P-25). It provided a way for the future design of plasma-enhanced wideband absorption materials.

Metal–organic frameworks (MOFs) are a class of crystalline porous materials with diverse and customizable structures. Li et al. made use of the plasma effect and the upconversion effect to prepare a novel broadband spectrum-responsive MOF composite (β-$NaYF_4$:Yb^{3+},Tm^{3+},Er^{3+}-Pt@MOF/Au, noted as UCNPs-Pt@MOF/Au) for the first time (Li et al. 2017). MOF was mainly responsive to UV light, the plasmonic Au absorbed the visible light, while UCNPs converted NIR light to UV and visible light, which were collected by MOF and Au to generate light collection from UV to NIR light (Figure 16.8a). In addition, the spatial separation of Au and Pt NPs by MOF enabled the catalytic substrate/product to be accessible to the Pt active site, which also contributed to the generation of "bare and clean" Au NPs (Figure 16.8b). By assembling layer by layer, the MOF shells can be grown on UCNPs of controlled thickness. The optimized UCNPs-Pt@MOF/Au composite showed a good H_2 production rate (280 mol/g/h) under simulated solar illumination. The first integration of plasma and upconversion effects in MOF composites greatly expanded light absorption, and Pt further promoted charge separation, which was the first hydrogen-producing photocatalyst to be quite active in all the three light-exposed regions.

Gao et al. combined Pt/CdS with the upconversion component $NaYF_4$:Yb^{3+},Er^{3+} (NYF) to form the Pt/CdS/$NaYF_4$:Yb^{3}, Er^{3+} (Pt/CdS/NYF) photocatalyst (Gao et al. 2019). Under the irradiation of 980 nm laser, Pt/CdS/NYF can upconvert the input NIR light to visible light and excite CdS to produce hydrogen and oxygen. Isotope experiments confirmed that both hydrogen and oxygen were produced from water.

Table 16.7 Summarized application of UC photocatalytic materials in hydrogen production.

Photocatalysts	Energy-level transitions	Light source	Photocatalyst amount	H_2 production activity	Reaction time (h)	References
$NaYF_4$:Yb^{3+},Er^{3+}-TiO_2/Au	$^1I_6 \rightarrow {}^3H_6$, $^2H_{11/2} \rightarrow {}^4I_{15/2}$, $^4S_{3/2} \rightarrow {}^4I_{15/2}$, $^4F_{9/2} \rightarrow {}^4I_{15/2}$	UV cutoff (400–1000 nm) visible cutoff (>400 nm, <800 nm) IR cutoff (200–700 nm)	500 mg	6.71 μmol/h 130.19 μmol/h 81.5 μmol/h	6	Anushree and Sadhana (2018)
β-$NaYF_4$:Yb^{3+},Tm^{3+}, Er^{3+}-Pt@MOF/Au	$^1I_6 \rightarrow {}^3H_6$, $^1I_6 \rightarrow {}^3F_4$, $^1D_2 \rightarrow {}^3H_6$, $^1G_4 \rightarrow {}^3H_6$, $^1D_2 \rightarrow {}^3H_4$, $^2H_{11/2} \rightarrow {}^4I_{15/2}$, $^4S_{3/2} \rightarrow {}^4I_{15/2}$	420–800 nm; 980 nm; simulated solar light	420–800 nm or simulated solar light: 10 mg NIR light: 2 mg	99 μmol/g/h 34 μmol/g/h 280 μmol/g/h		Li et al. (2017)
Pt/CdS/$NaYF_4$:Yb^3,Er^{3+}	$^2H_{11/2} \rightarrow {}^4I_{15/2}$, $^4S_{3/2} \rightarrow {}^4I_{15/2}$, $^4F_{9/2} \rightarrow {}^4I_{15/2}$	980 nm, 2.8 W	50 mg	3.38 μmol/g/h	1	Gao et al. (2019)
YF_3:Yb^{3+},Er^{3+}/ZnS	$^2H_{11/2} \rightarrow {}^4I_{15/2}$, $^4S_{3/2} \rightarrow {}^4I_{15/2}$, $^4F_{9/2} \rightarrow {}^4I_{15/2}$	<365 nm 420–760 nm 980 nm	50 mg	191.3 μmol/g/h 127.9 μmol/g/h 54.2 μmol/g/h	5	Li et al. (2019)
$NaYF_4$:Yb^{3+},Tm^{3+},Gd^{3+}/C_3N_4	$^1I_6 \rightarrow {}^3F_4$, $^1D_2 \rightarrow {}^3H_6$, $^1D_2 \rightarrow {}^3F_4$, $^1G_4 \rightarrow {}^3H_6$, $^1G_4 \rightarrow {}^3F_4$	980 nm	40 mg	311.6 μmol/g	3	Zhu et al. (2019)
β-$NaYF_4$:Yb^{3+}, Tm^{3+}@Ag_3PO_4@BP	$^1I_6 \rightarrow {}^3H_6$, $^1I_6 \rightarrow {}^3F_4$, $^1D_2 \rightarrow {}^3H_6$, $^1D_2 \rightarrow {}^3F_4$, $^1G_4 \rightarrow {}^3H_6$, $^1D_2 \rightarrow {}^3H_4$, $^1G_4 \rightarrow {}^3F_4$, $^3F_3 \rightarrow {}^3H_6$, $^3H_4 \rightarrow {}^3H_6$	980 nm, 10 W	15 mg	145.93 μmol/g/h	4	Zhang et al. (2019)
β-Tm^{3+}, Yb^{3+}:$NaYF_4$:MoS_2/Ta_2O_5				Shows a significant increase	5	Shu et al. (2018)

(continued)

Table 16.7 (Continued)

Photocatalysts	Energy-level transitions	Light source	Photocatalyst amount	H_2 production activity	Reaction time (h)	References
$NaYF_4$:Yb^{3+},Er^{3+}@$W_{18}O_{49}$	$^2H_{11/2} \rightarrow {}^4I_{15/2}$, $^4S_{3/2} \rightarrow {}^4I_{15/2}$, $^4F_{9/2} \rightarrow {}^4I_{15/2}$	980 nm, 2 W/cm^2	2 mg	35 times higher than that of BH_3NH_3 hydrolysis	1	Zhang et al. (2018)
$NaYF_4$:Yb^{3+},Er^{3+},Nd^{3+}/$W_{18}O_{49}$	$^2H_{11/2} \rightarrow {}^4I_{15/2}$, $^4S_{3/2} \rightarrow {}^4I_{15/2}$, $^4F_{9/2} \rightarrow {}^4I_{15/2}$, $^4F_{3/2} \rightarrow {}^4I_{9/2}$	980 nm laser diode	2 mg	About 2.5 μmol	1	Bao et al. (2019)
Er^{3+}-Yb^{3+}:TiO_2/MoO_{3-x}	$^2H_{11/2} \rightarrow {}^4I_{15/2}$, $^4S_{3/2} \rightarrow {}^4I_{15/2}$, $^4F_{9/2} \rightarrow {}^4I_{15/2}$	980 nm, 7 W/cm^2	2 mg	6.7 μmol/h	1	Shang et al. (2019)
$NaLuF_4$:Gd^{3+},Yb^{3+}, Tm^{3+}@$NaLuF_4$:Gd^{3+}, Yb^{3+}@SiO_2@Ag/g-C_3N_4	$^1G_4 \rightarrow {}^3H_6$, $^1D_2 \rightarrow {}^3F_4$, $^1D_2 \rightarrow {}^3H_6$, $^1I_6 \rightarrow {}^3F_4$, $^6P_{7/2} \rightarrow {}^8S_{7/2}$	Simulated solar light	20 mg	873 μmol/g/h	5	Zhao et al. (2018)

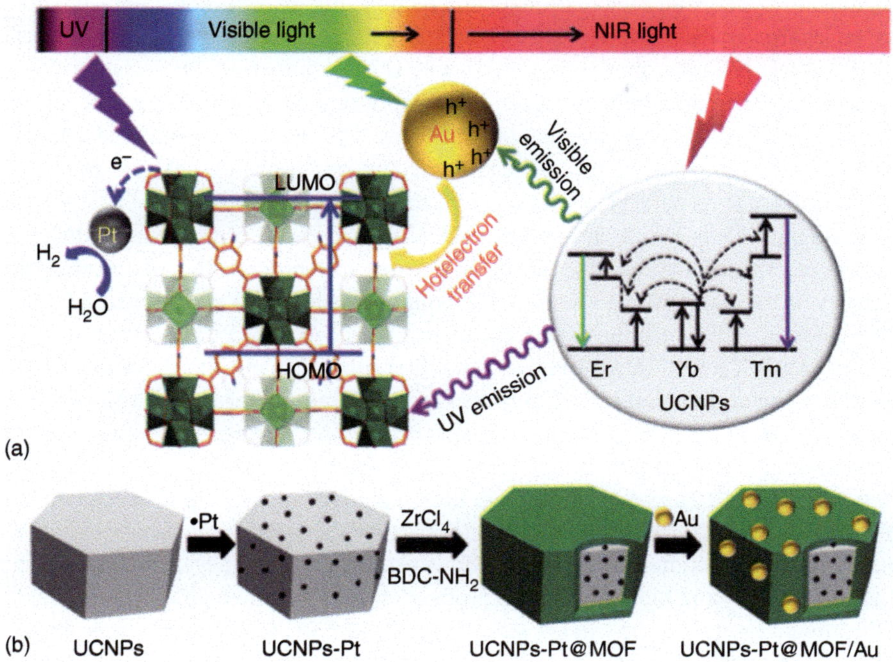

Figure 16.8 (a) The light absorption of each component in the MOF composites and involved mechanism for photocatalytic hydrogen production; (b) schematic illustration of the synthetic process for the UCNPs-Pt@MOF/Au composites. Source: Li et al. (2017). Reprinted with permission of John Wiley and Sons.

The highest hydrogen evolution rate of 3.38 μmol/g/h had been achieved. The apparent quantum efficiency (AQE) under NIR light was 0.008%. These photocatalysts used for hydrogen production use precious metals or alloys as co-catalysts, which to some extent inhibits large-scale utilization.

Li et al. also studied the H_2 production effect of the developed YF_3:Yb^{3+},Er^{3+}/ZnS (YYE/ZnS) heterojunction catalyst. YYE was beneficial for hydrogen production on ZnS semiconductors, and $^{\cdot}O_2^-$ and $^{\cdot}OH$ radicals were the main active substance (Li et al. 2019). According to the experimental results, the H_2 production rate of ZnS was very low (179.2 μmol/g/h). YYE showed little sign of hydrogen production, while the H_2 production rate of the YYE/ZnS catalyst (372.3 μmol/g/h) was twice that of ZnS.

Zhu et al. used a one-dimensional semiconductor material with visible light absorption, C_3N_4 nanotubes (C_3N_4 NTs). $NaYF_4$:Yb^{3+},Tm^{3+},Gd^{3+} (NYFG) and $NaYF_4$:Yb^{3+},Tm^{3+} (NYF) upconversion nanoparticles (UCNPs) were modified on C_3N_4 NTs to construct heterojunction structures (Zhu et al. 2019). The particle size of UCNPs was about 20 nm, and they were uniformly distributed on C_3N_4 NTs. In the ternary system of Yb^{3+}-Tm^{3+}-Gd^{3+}, the high level activated Tm^{3+} ions can transfer energy to Gd^{3+}, resulting in the excitation of Gd^{3+}. Compared with NYF, NYFG can provide higher UV and visible light emission, and bright fluorescence emission can be absorbed by C_3N_4 NTs. Therefore, NYFG and C_3N_4 NTs had more significant synergistic effect, stronger interaction, higher emission intensity, and faster charge transfer. Under 980 nm laser irradiation, the H_2 generation rate of NYFG/C_3N_4 NT heterojunction with 15 wt% UCNPs was the highest, which was 311.6 mol/g. The apparent quantum efficiency was 0.80‰, which was about 1.4 times that of NYF/C_3N_4 NTs nanocomposites. This novel nanomaterial design optimized the upconversion of NIR photons and provided new insights into the broad-spectrum-responsive materials for photocatalytic research.

Although UC materials can broaden the absorption edge of the semiconductor, the rapid recombination of electron/hole pairs still hinders the photocatalytic activity of the photocatalytic material. Zhang et al. successfully synthesized black phosphorus (BP)-decorated three-dimensional β-$NaYF_4$:Yb^{3+},Tm^{3+}@Ag_3PO_4 (NYF@Ag_3PO_4@BP) micro-rods for the first time (Zhang et al. 2019). This Z-scheme heterojunction can quickly transfer electrons and inhibit the recombination of electron–hole pairs and the photocorrosion of Ag^+. The experimental results showed that the hydrogen yield of this catalyst was up to 145.93 μmol/g/h, which was more than 10.2 and 6.1 times higher than that of BP and NYF@BP. As shown in Figure 16.9, Ag_3PO_4 had no photocatalytic effect on hydrogen production because of its positive conduction band (0.45 eV). However, BP had more negative conduction band (−0.18 eV) and higher hydrogen production activity. When UC was combined with them, energy can be transferred to Ag_3PO_4 and BP, resulting in a significant increase in hydrogen production. Yb^{3+} can be transformed from the ground state ($^2F_{7/2}$ level) to the excited state ($^2F_{5/2}$ level) at 980 nm excitation, and the energy generated in this process can be transferred to the activators Tm^{3+}: $^3H_5$, $^3F_2$, $^1G_4$, $^1D_2$, and $^3P_2$. Subsequently, the higher levels of the Tm^{3+} ions were transferred to lower levels because of a radioactive transition. Ag_3PO_4 and BP absorbed energy and produced photogenerated electron/hole pairs. The electrons of Ag_3PO_4 can quickly recombine with the photoinduced holes on BP, and the electrons of BP can make full use of its band-edge potential for the reduction reaction to produce H_2.

The low absorption of light makes it impossible for it to provide more energy to excite electrons. For the problem of limited light absorption, the material with UC effect is the best choice to modify the wide-bandgap semiconductor. Shu et al. prepared Ta_2O_5 particles by a simple Pechini-type sol–gel method, and the co-catalyst MoS_2 was deposited on the surface of the β-Tm^{3+},Yb^{3+}:$NaYF_4$/Ta_2O_5 nanocomposite particles (Shu et al. 2018). The results showed that the photocatalytic activity of Ta_2O_5 for H_2 evolution can be significantly improved because of the addition of upconversion luminescence agents (β-Tm^{3+},Yb^{3+}:$NaYF_4$) and the co-catalyst (MoS_2). The highest photocatalytic activity of β-Tm^{3+}, Yb^{3+}:$NaYF_4$/MoS_2-Ta_2O_5 catalysts with a 40 : 100 mass ratio, which was heat treated at 450 °C for 60 minutes under NIR light irradiation, was observed.

In addition to introducing materials that can improve the UV efficiency, LSPR effects are often used to improve the UV efficiency of NIR photons to visible light. Zhang et al. synthesized $NaYF_4$:Yb^{3+}, Er^{3+}/$W_{18}O_{49}$ nanowires (NWs) heterostructure via self-assembly of $NaYF_4$:Yb^{3+}, Er^{3+} NPs onto the non-metallic plasmonic layer of $W_{18}O_{49}$ NWs grown on an fluorine doped tin oxide (FTO) glass substrate (Zhang et al. 2018). The wide LSPR absorption band of $W_{18}O_{49}$ NWs overlaps with the excitation band and emission band of $NaYF_4$:Yb^{3+},Er^{3+} NPs, which can

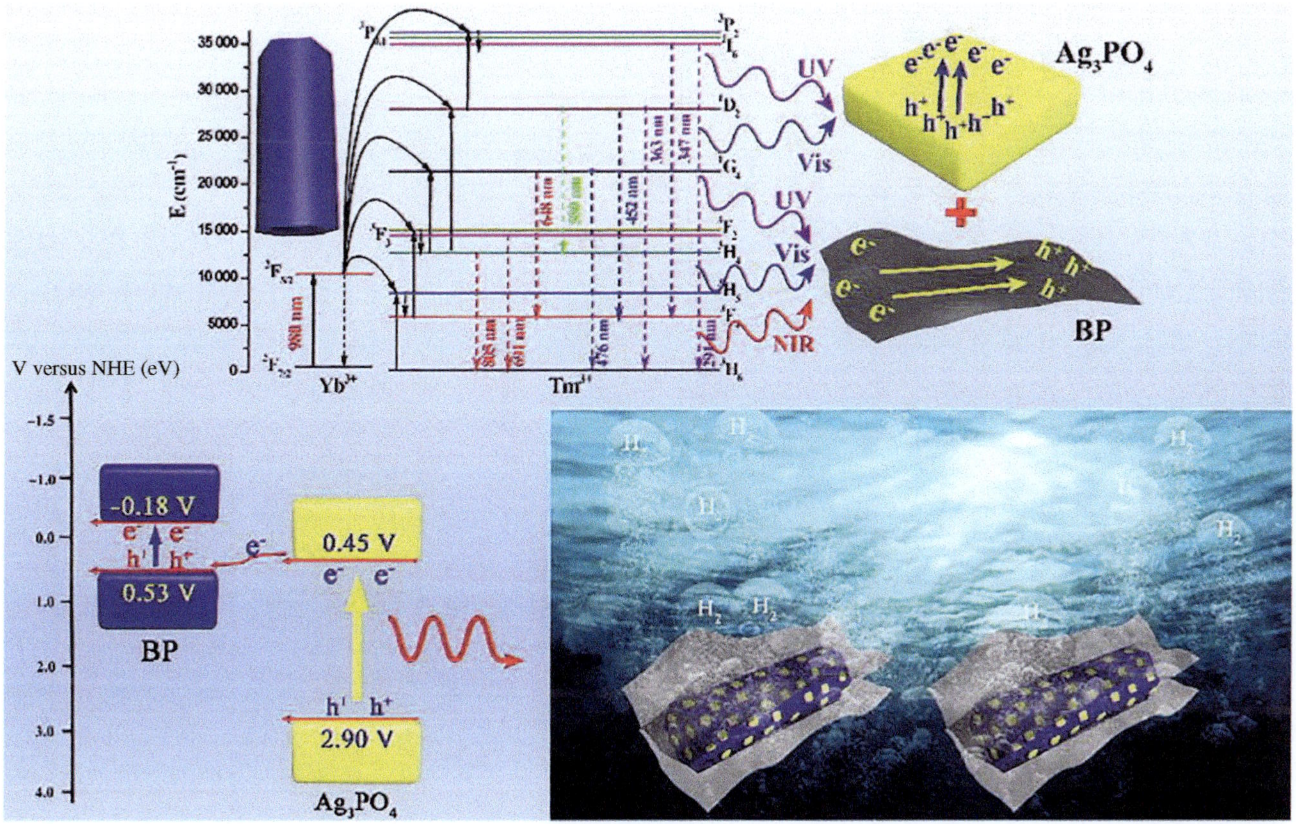

Figure 16.9 Possible mechanisms for photocatalytic hydrogen evolution of NYF@Ag_3PO_4@BP composites. Source: Zhang et al. (2019). Reprinted with permission of Elsevier.

induce the NIR plasmonic energy upconversion process. According to theoretical calculations, the NIR excitation LSPR of $W_{18}O_{49}$ NWs was believed to be the main reason for the enhancement of $NaYF_4$:Yb^{3+}, Er^{3+} NPs. Meanwhile, the plasmonic enhanced upconversion luminescence can be partially absorbed by $W_{18}O_{49}$ NWs to re-excite its high-energy LSPR, resulting in enhanced upconversion luminescence selectivity of $NaYF_4$:Yb^{3+},Er^{3+}/$W_{18}O_{49}$ heterostructures. Based on the plasmonic energy transfer process, the catalytic efficiency of $NaYF_4$:Yb^{3+},Er^{3+}@$W_{18}O_{49}$ NWs quasi-core/shell structure was about 35 times higher than the H_2 production of single ammonia borane (BH_3NH_3).

Although the heterojunction $NaYF_4$:Yb^{3+},Er^{3+}/$W_{18}O_{49}$ promotes H_2 generation by hydrolyzing BH_3NH_3 under NIR light irradiation and improved the H_2 production rate, the photothermal effect from the heterojunction will affect the luminescence efficiency of UC material and reduce the photocatalytic performance. The energy transfer from Er^{3+} to Nd^{3+} of $NaYF_4$:Yb^{3+},Er^{3+},Nd^{3+}/$W_{18}O_{49}$ heterostructures constructed by Bao et al. can use photothermal effect to improve the photocatalytic activity (Bao et al. 2019). As shown in Figure 16.10, the overlapping of

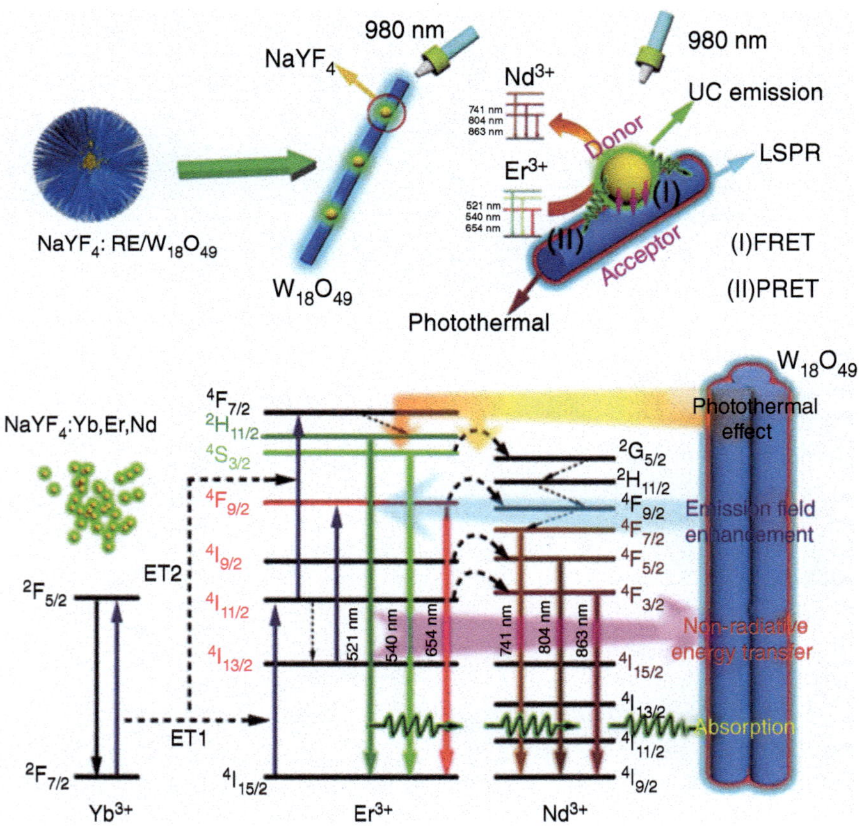

Figure 16.10 Schematic of the interaction between the LSPR effect of $W_{18}O_{49}$ NWs and the energy-transfer UC process of $NaYF_4$:Yb^{3+},Er^{3+},Nd^{3+} nanoparticles. Source: Bao et al. (2019). Reprinted with permission of Royal Society of Chemistry.

LSPR of $W_{18}O_{49}$ and the emission spectra of $NaYF_4$:Yb^{3+},Er^{3+},Nd^{3+} were conducive to the energy transfer from $NaYF_4$:Yb^{3+},Er^{3+},Nd^{3+} to $W_{18}O_{49}$ through non-radiative processes. Then, the localization of plasmonic resonance energy transfer (PRET) at the excitation and emission sites of $NaYF_4$:Yb^{3+},Er^{3+},Nd^{3+} by the LSPR of $W_{18}O_{49}$ can improve the intensity of UC and ultimately improve the catalytic H_2 evolution. Compared with the $NaYF_4$:Yb^{3+},Er^{3+}/$W_{18}O_{49}$ heterostructure, the amount of H_2 evolution of $NaYF_4$:Yb^{3+},Er^{3+},Nd^{3+}/$W_{18}O_{49}$ heterostructure increased about 1.5 times. It effectively solved the problem that the emission intensity of UC was reduced due to the photothermal effect, which led to the weakening of photocatalytic activity. It provides an effective method for the study of catalytic activity in the NIR field.

Shang et al. also used LSPR effect to improve the luminescence intensity of UC materials (Shang et al. 2019). First, a nanosemiconductor heterostructure (NSH) composed of Er^{3+},Yb^{3+}:TiO_2 nanosheets (NTs) and plasmonic MoO_{3-x} NTs was prepared by a two-step solvothermal method. Plasmonic MoO_{3-x} NTs can amplify NIR energy and transmit it to Er^{3+},Yb^{3+}:TiO_2, which enhanced the fluorescence intensity of Er^{3+},Yb^{3+}:TiO_2 by 182 times. Under the excitation of 980 nm laser, the H_2 production rate of the catalyst was 1.5 and 61 times higher than that of MoO_{3-x} NTs and BH_3NH_3, respectively. Under simulated sunlight (AM 1.5), the H_2 production rate was 2.5 and 52 times higher than that of MoO_{3-x} NTs and BH_3NH_3, respectively. It was a proposed method that can provide an efficient strategy for improving LSPR-driven UC luminescence and the efficiency of H_2 production.

16.5 Conclusion

In summary, we focus on the design progress of photocatalysts with UC effect. In particular, many examples were selected and discussed to illustrate the energy conversion and synthesis strategies of UCNP-based composite nanostructures for enhanced photocatalytic degradation of pollutants and hydrogen production. It was believed that upconversion nanoparticle-mediated NIR energy conversion has wide and important applications in energy conversion, catalysis, environment, energy, and other fields.

References

Anushree, A.C. and Sadhana, S.R. (2018). Synergistic plasmonic and upconversion effect of the (Yb,Er)NYFTiO$_2$/Au composite for photocatalytic hydrogen generation. *The Journal of Physical Chemistry C* 122: 26307–26314. https://doi.org/10.1021/acs.jpcc.8b05480.

Anwer, H., Mahmood, A., Lee, J. et al. (2019). Photocatalysts for degradation of dyes in industrial effluents: opportunities and challenges. *Nano Research* 12 (5): 955–972. https://doi.org/10.1007/s12274-019-2287-0.

Auzel, F. (2004). Upconversion and anti-Stokes processes with f and d ions in solid. *Chemical Reviews* 104: 139–173. https://doi.org/10.1002/chin.200416212.

Bai, L., Jiang, W., Gao, C. et al. (2016). Facet engineered interface design of $NaYF_4$:Yb,Tm upconversion nanocrystals on BiOCl nanoplates for enhanced near-infrared photocatalysis. *Nanoscale* 8: 19014–19024. https://doi.org/10.1039/C6NR05720A.

Balaji, R., Kumar, S., Reddy, K. et al. (2017). Near-infrared driven photocatalytic performance of lanthanide-doped $NaYF_4$@CdS core-shell nanostructures with enhanced upconversion properties. *Journal of Alloys and Compounds* 724: 481–491. https://doi.org/10.1016/j.jallcom.2017.07.050.

Bao, Y., Zhang, Z., Cao, B. et al. (2019). Energy transfer from Er to Nd ions by the thermal effect and promotion of the photocatalysis of the $NaYF_4$:Yb,Er,Nd/$W_{18}O_{49}$ heterostructure. *Nanoscale* 11: 7433–7439. https://doi.org/10.1039/C9NR00409B.

Bushuyev, O.S., De Luna, P., Dinh, C.T. et al. (2018). What should we make with CO_2 and how can we make it? *Joule* 2 (5): 825–832. https://doi.org/10.1016/j.joule.2017.09.003.

Cao, S., Piao, L., and Chen, X.B. (2020). Emerging photocatalysts for hydrogen evolution. *Trends in Chemistry* 2 (1): 57–70. https://doi.org/10.1016/j.trechm.2019.06.009.

Chatti, M., Adusumalli, V., Ganguli, S. et al. (2016). Near-infrared light triggered superior photocatalytic activity from MoS_2-$NaYF_4$:Yb^{3+}/Er^{3+} nanocomposites. *Dalton Transactions* 45: 12384–12392. https://doi.org/10.1039/C6DT02548J.

Chen, Z., Chen, T., Gong, W. et al. (2013). Effect of Li^+ ions doping on microstructure and upconversion emission of $Y_2Ti_2O_7$:Er^{3+}/Yb^{3+} nanophosphors synthesized via a sol-gel method. *Journal of the American Ceramic Society* 96 (6): 1857–1862. https://doi.org/10.1111/jace.12271.

Chen, Y., Mishra, S., Ledoux, G. et al. (2014). Direct synthesis of hexagonal $NaGdF_4$ nanocrystals from a single-source precursor: upconverting $NaGdF_4$:Yb^{3+},Tm^{3+} and its composites with TiO_2 for near-IR-driven photocatalysis. *Chemistry - An Asian Journal* 9 (9): 2415–2421. https://doi.org/10.1002/chin.201539014.

Chen, E., Yin, W., Bai, S. et al. (2015). Synthesis of vis/NIR-driven hybrid photocatalysts by electrostatic assembly of $NaYF_4$:Yb,Tm nanocrystals on g-C_3N_4 nanosheets. *Materials Letters* 146: 87–90. https://doi.org/10.1016/j.matlet.2015.02.010.

Chen, X., Jin, L., Kong, W. et al. (2016). Confining energy migration in upconversion nanoparticles towards deep ultraviolet lasing. *Nature Communications* 7: 10304. https://doi.org/10.1038/ncomms10304.

Chen, S., Li, R., Li, X. et al. (2018a). Electrospinning: An enabling nanotechnology platform for drug delivery and regenerative medicine. *Advanced Drug Delivery Reviews* 132: 188–213. https://doi.org/10.1016/j.addr.2018.05.001.

Chen, Z., Fu, M., Huang, X. et al. (2018b). Magnetic infrared responsive photocatalyst: fabrication, characterization, and photocatalytic performance of β-$NaYF_4$:Yb^{3+},Tm^{3+}/TiO_2/Fe_3O_4@SiO_2 composite. *Research on Chemical Intermediates* 44: 6369–6385. https://doi.org/10.1007/s11164-018-3495-9.

Cheng, Q., Sui, J., and Cai, W. (2012). Enhanced upconversion emission in Yb^{3+} and Er^{3+} codoped $NaGdF_4$ nanocrystals by introducing Li^+ ions. *Nanoscale* 4 (3): 779–784. https://doi.org/10.1039/C1NR11365H.

Cheng, E., Zhou, S., Li, M. et al. (2017). Synthesis of g-C_3N_4-based $NaYF_4$:Yb,Tm@TiO_2 ternary composite with enhanced Vis/NIR-driven photocatalytic activities. *Applied Surface Science* 410: 383–392. https://doi.org/10.1016/j.apsusc.2017.03.052.

Cui, C., Tou, M., Li, M. et al. (2016). Heterogeneous semiconductor shells sequentially coated on upconversion nanoplates for NIR-light enhanced photocatalysis. *Inorganic Chemistry* 56: 2328–2336. https://doi.org/10.1021/acs.inorgchem.6b03079.

DaCosta, M.V., Doughan, S., Han, Y. et al. (2014). Lanthanide upconversion nanoparticles and applications in bioassays and bioimaging: a review. *Analytica Chimica Acta* 832: 1–33. https://doi.org/10.1016/j.aca.2014.04.030.

Dong, N., Pedroni, M., Piccinelli, F. et al. (2011). NIR-to-NIR two-photon excited CaF_2:Tm^{3+},Yb^{3+} nanoparticles: multifunctional nanoprobes for highly penetrating fluorescence bio-imaging. *ACS Nano* 5: 8665–8671. https://doi.org/10.1021/nn202490m.

Dong, S., Xu, J., Jia, T. et al. (2019). Upconversion-mediated $ZnFe_2O_4$ nanoplatform for NIR-enhanced chemodynamic and photodynamic therapy. *Chemical Science* 10 (15): 4259–4271. https://doi.org/10.1039/c9sc00387h.

Du, P., Luo, L., Park, H.K. et al. (2016). Citric-assisted sol-gel based Er^{3+}/Yb^{3+}-codoped $Na_{0.5}Gd_{0.5}MoO_4$: A novel highly-efficient infrared-to-visible upconversion material for optical temperature sensors and optical heaters. *Chemical Engineering Journal* 306: 840–848. https://doi.org/10.1016/j.cej.2016.08.007.

Duo, S., Zhang, J., Zhang, H. et al. (2016). Synthesis of β-$NaYF_4$:Yb^{3+},Tm^{3+}@TiO_2 and β-$NaYF_4$:Yb^{3+},Tm^{3+}@TiO_2@Au nanocomposites and effective upconversion-driven photocatalytic properties. *Optical Materials* 62: 240–249. https://doi.org/10.1016/j.optmat.2016.10.005.

Fu, X., Long, J., Wang, X. et al. (2008). Photocatalytic reforming of biomass: a systematic study of hydrogen evolution from glucose solution. *International Journal of Hydrogen Energy* 33 (22): 6484–6491. https://doi.org/10.1016/j.ijhydene.2008.07.068.

Gao, W., Tian, B., Zhang, W. et al. (2019). NIR light driven catalytic hydrogen generation over semiconductor photocatalyst coupling up-conversion component. *Applied Catalysis B: Environmental* 257 (15): 117908. https://doi.org/10.1016/j.apcatb.2019.117908.

GracePavithra, K., Jaikumar, V., Kumar, P.S. et al. (2019). A review on cleaner strategies for chromium industrial wastewater: Present research and future perspective. *Journal of Cleaner Production* 228: 580–593. https://doi.org/10.1016/j.jclepro.2019.04.117.

Guo, X., Song, W., Chen, C. et al. (2013). Near-infrared photocatalysis of β-$NaYF_4$:Yb^{3+},Tm^{3+}@ZnO composites. *Physical Chemistry Chemical Physics* 15: 14681–14688. https://doi.org/10.1039/C3CP52248B.

Guo, X., Di, W., Chen, C. et al. (2014). Enhanced near-infrared photocatalysis of $NaYF_4$:Yb,Tm/CdS/TiO_2 composites. *Dalton Transactions* 43: 1048–1054. https://doi.org/10.1039/c3dt52288a.

He, L., Ni, Q., Mu, J. et al. (2020). Solvent-assisted self-assembly of a metal-organic framework based biocatalyst for cascade reaction driven photodynamic therapy. *Journal of the American Chemical Society* 142 (14): 6822–6832. https://doi.org/10.1021/jacs.0c02497.

Huang, S., Zhu, N., Lou, Z. et al. (2014). Near-infrared photocatalysts of $BiVO_4/CaF_2:Er^{3+},Tm^{3+},Yb^{3+}$ with enhanced upconversion properties. *Nanoscale* 6: 1362–1368. https://doi.org/10.1039/C3NR05266D.

Huang, M., Yuan, B., Dai, L. et al. (2015). Toward NIR driven photocatalyst: fabrication, characterization, and photocatalytic activity of β-$NaYF_4:Yb^{3+},Tm^{3+}/g\text{-}C_3N_4$ nanocomposite. *Journal of Colloid and Interface Science* 460: 264–272. https://doi.org/10.1016/j.jcis.2015.08.063.

Huang, H., Li, H., Wang, Z. et al. (2019). Efficient near-infrared photocatalysts based on $NaYF_4:Yb^{3+},Tm^{3+}@NaYF_4:Yb^{3+},Nd^{3+}@TiO_2$ core@shell nanoparticles. *Chemical Engineering Journal* 361: 1089–1097. https://doi.org/10.1016/j.cej.2018.12.174.

Jin, G., He, R., Liu, Q. et al. (2019). Near-infrared light-regulated cancer theranostic nanoplatform based on aggregation-induced emission luminogen encapsulated upconversion nanoparticles. *Theranostics* 9 (1): 246–264. https://doi.org/10.7150/thno.30174.

Kwon, H., Mota, F.M., Chuang, K. et al. (2018). Enhancing solar light-driven photocatalytic activity of mesoporous carbon-TiO_2 hybrid films via upconversion coupling. *ACS Sustainable Chemistry & Engineering* 6: 1310–1317. https://doi.org/10.1021/acssuschemeng.7b03658.

Lee, J.S. and Kim, Y.J. (2011). The effects of preparation conditions on the structural and up-conversion properties of $NaYF_4:Yb^{3+},Er^{3+}$ nano powders. *Optical Materials* 33 (7): 1111–1115. https://doi.org/10.1016/j.optmat.2010.10.011.

Li, C., Wang, F., Zhu, J. et al. (2010). $NaYF_4$:Yb,Tm/CdS composite as a novel near-infrared-driven photocatalyst. *Applied Catalysis B: Environmental* 100: 433–439. https://doi.org/10.1016/j.apcatb.2010.08.017.

Li, Z., Li, C., Mei, Y. et al. (2013). Synthesis of rhombic hierarchical YF_3 nanocrystals and their use as upconversion photocatalysts after TiO_2 coating. *Nanoscale* 5 (7): 3030–3036. https://doi.org/10.1039/C3NR34018J.

Li, X., Yu, J., Jaroniec, M. et al. (2016a). Hierarchical photocatalysts. *Chemical Society Reviews* 45 (9): 2603–2636. https://doi.org/10.1039/C5CS00838G.

Li, Y., Dong, L., Huang, C. et al. (2016b). Decoration of upconversion nanoparticles@mSiO_2 core-shell nanostructures with CdS nanocrystals for excellent infrared light triggered photocatalysis. *RSC Advances* 6: 54241–54248. https://doi.org/10.1039/C6RA09261F.

Li, B., Wang, Y., Tian, F. et al. (2017a). Preparation of new visible-light driven nanocomposite photocatalysts, $X/NaTaO_3/Er^{3+}:YAlO_3$ (X = Ag, Au and Pt), for photocatalytic conversion of Cr(VI). *Journal of Industrial and Engineering Chemistry* 54: 398–407. https://doi.org/10.1016/j.jiec.2017.06.020.

Li, D., Yu, S., and Jiang, H. (2017b). From UV to near-infrared light-responsive metal-organic framework composites: plasmon and upconversion enhanced photocatalysis. *Advanced Materials* 30 (27): 1707377. https://doi.org/10.1002/adma.201707377.

Li, X., Yang, K., Yu, C. et al. (2019a). Advances towards the utilization of Vis-NIR light energy by coating $YF_3:Yb^{3+},Er^{3+}$ over ZnS microspheres triggering hydrogen production and pollutants disposal. *Journal of Materials Chemistry C* 7: 8053–8062. https://doi.org/10.1039/C9TC02068C.

Li, X., Yang, K., Yu, C. et al. (2019b). Broadband photocatalysis using a Z-scheme heterojunction of Au/$NaYF_4$:Yb,Er/$WO_3 \cdot 0.33H_2O$-$W_{18}O_{49}$ via a synergetic strategy of upconversion function and plasmonic effect. *Inorganic Chemistry Frontiers* 6: 3158–3167. https://doi.org/10.1039/C9QI00823C.

Li, G., Huang, S., Zhu, N. et al. (2020). Near-infrared responsive upconversion glass-ceramic@BiOBr heterojunction for enhanced photodegradation performances of norfloxacin. *Journal of Hazardous Materials* 403: 123981. https://doi.org/10.1016/j.jhazmat.2020.123981.

Liu, Q., Sun, Y., Yang, T. et al. (2011). Sub-10 nm hexagonal lanthanide-doped $NaLuF_4$ upconversion nanocrystals for sensitive bioimaging in vivo. *Journal of the American Chemical Society* 133 (43): 17122–17125. https://doi.org/10.1021/ja207078s.

Liu, Q., Yang, T., Feng, W. et al. (2012). Blue-emissive upconversion nanoparticles for low-power-excited bioimaging in vivo. *Journal of the American Chemical Society* 134 (11): 5390–5397. https://doi.org/10.1021/ja3003638.

Liu, Y., Meng, X., Bu, W. et al. (2019a). Upconversion-based photodynamic cancer therapy. *Coordination Chemistry Reviews* 379: 82–98. https://doi.org/10.1016/j.ccr.2017.09.006.

Liu, Z., Hao, J., Wang, Y. et al. (2019b). Decorating Ag_3PO_4 nanodots on mesoporous silica-functionalized $NaYF_4$:Yb,Tm@$NaLuF_4$ for efficient sunlight-driven photocatalysis: synergy of broad spectrum absorption and pollutant adsorption-enrichment. *Inorganic Chemistry Frontiers* 6: 3529–3538. https://doi.org/10.1039/C9QI01003C.

Ma, Y., Liu, H., Han, Z. et al. (2015). Non-ultraviolet photocatalytic kinetics of $NaYF_4$:Yb,Tm@TiO_2/Ag core@comby shell nanostructures. *Journal of Materials Chemistry A* 3: 14642–14650. https://doi.org/10.1039/c5ta03143e.

Mi, C., Tian, Z., Cao, C. et al. (2011). Novel microwave-assisted solvothermal synthesis of $NaYF_4$:Yb,Er upconversion nanoparticles and their application in cancer cell imaging. *Langmuir* 27 (23): 14632–14637. https://doi.org/10.1021/la204015m.

Momirlan, M. and Veziroglu, T.N. (2005). The properties of hydrogen as fuel tomorrow in sustainable energy system for a cleaner planet. *International Journal of Hydrogen Energy* 30 (7): 795–802. https://doi.org/10.1016/j.ijhydene.2004.10.011.

Na, H., Woo, K., Lim, K. et al. (2013). Rational morphology control of beta-$NaYF_4$:Yb,Er/Tm upconversion nanophosphors using a ligand, an additive, and lanthanide doping. *Nanoscale* 5: 4242–4251. https://doi.org/10.1039/C3NR00080J.

Navarro, R.M., Pena, M.A., and Fierro, J.L.G. (2007). Hydrogen production reactions from carbon feedstocks: fossil fuels and biomass. *Chemical Reviews* 107 (10): 3952–3991. https://doi.org/10.1021/cr0501994.

Qin, W., Zhang, D., Zhao, D. et al. (2010). Near-infrared photocatalysis based on YF_3:Yb^{3+},Tm^{3+}/TiO_2 core/shell nanoparticles. *Chemical Communications* 46 (13): 2304–2306. https://doi.org/10.1039/b924052g.

Qiu, X., Zhou, Q., Zhu, X. et al. (2020). Ratiometric upconversion nanothermometry with dual emission at the same wavelength decoded via a time-resolved technique. *Nature Communications* 11 (1): 4. https://doi.org/10.1038/s41467-019-13796-w.

Rahman, Z. and Singh, V.P. (2019). The relative impact of toxic heavy metals (THMs) (arsenic (As), cadmium (Cd), chromium (Cr)(VI), mercury (Hg), and lead (Pb)) on the total environment: an overview. *Environmental Monitoring and Assessment* 191 (7): 419. https://doi.org/10.1007/s10661-019-7528-7.

Rao, L., Bu, L., Cai, B. et al. (2016). Cancer cell membrane-coated upconversion nanoprobes for highly specific tumor imaging. *Advanced Materials* 28 (18): 3460-+. https://doi.org/10.1002/adma.201506086.

Ren, H., Koshy, P., Chen, W. et al. (2017). Photocatalytic materials and technologies for air purification. *Journal of Hazardous Materials* 325: 340–366. https://doi.org/10.1016/j.jhazmat.2016.08.072.

Renero-Lecuna, C., Martin-Rodriguez, R., Valiente, R. et al. (2011). Origin of the high upconversion green luminescence efficiency in beta-$NaYF_4$:2%Er^{3+},20%Yb^{3+}. *Chemistry of Materials* 23 (15): 3442–3448. https://doi.org/10.1021/cm2004227.

Rochat, J., Demenge, P., and Rerat, J.C. (1978). Toxicologic study of a fluorescent tracer: rhodamine B. *Toxicology European Research* 1 (1): 23–26.

Runowski, M., Stopikowska, N., Szeremeta, D. et al. (2019). Upconverting lanthanide fluoride core@shell nanorods for luminescent thermometry in the first and second biological windows: beta-$NaYF_4$:Yb^{3+}-Er^{3+}@SiO_2 temperature sensor. *ACS Applied Materials & Interfaces* 11 (14): 13389–13396. https://doi.org/10.1021/acsami.9b00445.

Runowski, M., Wozny, P., Lis, S. et al. (2020). Optical vacuum sensor based on lanthanide upconversion-luminescence thermometry as a tool for ultralow pressure sensing. *Advanced Materials Technologies* 5 (4): 1901091. https://doi.org/10.1002/admt.201901091.

Schafer, H., Ptacek, P., Zerzouf, O. et al. (2008). Synthesis and optical properties of KYF_4/Yb,Er nanocrystals, and their surface modification with undoped KYF_4. *Advanced Functional Materials* 18: 2913–2918. https://doi.org/10.1002/adfm.200800368.

Shang, J., Xu, X., Liu, K. et al. (2019). LSPR-driven upconversion enhancement and photocatalytic H_2 evolution for Er-Yb:TiO_2/MoO_{3-x} nano-semiconductor heterostructure. *Ceramics International* 45: 16625–16630. https://doi.org/10.1016/j.ceramint.2019.05.203.

Shao, Y., Liu, B., Di, Z. et al. (2020). Engineering of upconverted metal-organic frameworks for near-infrared light-triggered combinational photodynamic/chemo-/immunotherapy against hypoxic tumors. *Journal of the American Chemical Society* 142 (8): 3939–3946. https://doi.org/10.1021/jacs.9b12788.

Shu, X., Zhao, C., Zhang, X. et al. (2018). Highly efficient near-infrared light photocatalytic hydrogen evolution over MoS_2 supported Ta_2O_5 combined with an up-conversion luminescence agent (β-Tm^{3+},Yb^{3+}:$NaYF_4$/MoS_2-Ta_2O_5 nanocomposite). *New Journal of Chemistry* 42: 6134–6143. https://doi.org/10.1039/C7NJ04993E.

Spasiano, D., Marotta, R., Malato, S. et al. (2015). Solar photocatalysis: materials, reactors, some commercial, and pre-industrialized applications. A comprehensive approach. *Applied Catalysis B: Environmental* 170: 90–123. https://doi.org/10.1016/j.apcatb.2014.12.050.

Su, W., Zheng, M., Li, L. et al. (2014). Directly coat TiO_2 on hydrophobic $NaYF_4$:Yb,Tm nanoplates and regulate their photocatalytic activities with the core size. *Journal of Materials Chemistry A* 2: 13486–13491. https://doi.org/10.1039/C4TA02756F.

Tang, Y., Di, W., Zhai, X. et al. (2013). NIR-responsive photocatalytic activity and mechanism of $NaYF_4$:Yb,Tm@TiO_2 core-shell nanoparticles. *ACS Catalysis* 3: 405–412. https://doi.org/10.1021/cs300808r.

Tian, Y., Tian, Y., Huang, P. et al. (2016). Effect of Yb^{3+} concentration on upconversion luminescence and temperature sensing behavior in Yb^{3+}/Er^{3+} co-doped $YNbO_4$ nanoparticles prepared via molten salt route. *Chemical Engineering Journal* 297: 26–34. https://doi.org/10.1016/j.cej.2016.03.149.

Tian, Q., Yao, W., Wu, W. et al. (2019). NIR light-activated upconversion semiconductor photocatalysts. *Nanoscale Horizons* 4 (1): 10–25. https://doi.org/10.1039/C8NH00154E.

Tou, M., Mei, Y., Bai, S. et al. (2016). Depositing CdS nanoclusters on carbon-modified $NaYF_4$:Yb,Tm upconversion nanocrystals for NIR light enhanced photocatalysis. *Nanoscale* 8: 553–562. https://doi.org/10.1039/C5NR06806A.

Tou, M., Luo, Z., Bai, S. et al. (2017). Sequential coating upconversion $NaYF_4$:Yb,Tm nanocrystals with SiO_2 and ZnO layers for NIR-driven photocatalytic and antibacterial applications. *Materials Science and Engineering C* 70: 1141–1148. https://doi.org/10.1016/j.msec.2016.03.038.

Turner, J.A. (2004). Sustainable hydrogen production. *Science* 305 (5686): 972–974. https://doi.org/10.1126/science.1103197.

Ullah, S., Ferreira-Neto, E.P., Hazra, C. et al. (2019). Broad spectrum photocatalytic system based on $BiVO_4$ and $NaYbF_4$:Tm^{3+} upconversion particles for environmental remediation under UV-vis-NIR illumination. *Applied Catalysis B: Environmental* 243: 121–135. https://doi.org/10.1016/j.apcatb.2018.09.091.

Wang, F. and Liu, X. (2009). Recent advances in the chemistry of lanthanide-doped upconversion nanocrystals. *Chemical Society Reviews* 38 (4): 976–989. https://doi.org/10.1002/chin.200925210.

Wang, W., Huang, W., Ni, Y. et al. (2013a). Graphene supported beta $NaYF_4$:Yb^{3+},Tm^{3+} and N doped P25 nanocomposite as an advanced NIR and sunlight driven upconversion photocatalyst. *Applied Surface Science* 282: 832–837. https://doi.org/10.1016/j.apsusc.2013.06.066.

Wang, W., Ding, M., Lu, C. et al. (2013b). A study on upconversion UV-vis-NIR responsive photocatalytic activity and mechanisms of hexagonal phase $NaYF_4$:Yb^{3+},Tm^{3+}@TiO_2 core-shell structured photocatalyst. *Applied Catalysis B: Environmental* 144: 379–385. https://doi.org/10.1016/j.apcatb.2013.07.035.

Wang, C., Song, K., Feng, Y. et al. (2014a). Preparation of $NaLuF_4$:Gd,Yb,Tm-TiO_2 nanocomposite with high catalytic activity for solar light assisted photocatalytic degradation of dyes and wastewater. *RSC Advances* 4 (74): 39118–39125. https://doi.org/10.1039/C4RA05575F.

Wang, W., Huang, W., Ni, Y. et al. (2014b). Different upconversion properties of β-$NaYF_4$:Yb^{3+},Tm^{3+}/Er^{3+} in affecting the near-infrared-driven photocatalytic activity of high-reactive TiO_2. *ACS Applied Materials & Interfaces* 6: 340–348. https://doi.org/10.1021/am404389g.

Wang, H., Zhang, L., Chen, Z. et al. (2014c). Semiconductor heterojunction photocatalysts: design, construction, and photocatalytic performances. *Chemical Society Reviews* 43 (15): 5234–5244. https://doi.org/10.1039/C4CS00126E.

Wang, W., Li, Y., Kang, Z. et al. (2015). A NIR-driven photocatalyst based on $NaYF_4$:Yb,Tm@TiO_2 core-shell structure supported on reduced graphene oxide. *Applied Catalysis B: Environmental* 182: 184–192. https://doi.org/10.1016/j.apcatb.2015.09.022.

Wang, W., Zhang, F., Zhang, C. et al. (2017a). Fabrication of zinc oxide composite microfibers for near-infrared-light-mediated photocatalysis. *ChemCatChem* 9: 3611–3617. https://doi.org/10.1002/cctc.201700781.

Wang, N., Zhang, F., Zhang, C. et al. (2017b). TiO_2 composite nanotubes embedded with CdS and upconversion nanoparticles for near infrared light driven photocatalysis. *Chinese Journal of Catalysis* 38: 1851–1859. https://doi.org/10.1016/S1872-2067(17)62929-2.

Wang, W., Huang, C., Zhang, C. et al. (2018a). Controlled synthesis of upconverting nanoparticles/$Zn_xCd_{1-x}S$ yolk-shell nanoparticles for efficient photocatalysis driven by NIR light. *Applied Catalysis B: Environmental* 224: 854–862. https://doi.org/10.1016/j.apcatb.2017.11.037.

Wang, W., Dong, W., Huang, C. et al. (2018b, 2018). UCNPs@$Zn_{0.5}Cd_{0.5}S$ core-shell and yolk-shell nanostructures: Selective synthesis, characterization, and near-infrared-mediated photocatalytic reduction of Cr(VI). *Journal of Nanomaterials* 1293847. https://doi.org/10.1155/2018/1293847.

Wang, M., Song, J., Zhou, F. et al. (2019). NIR-triggered phototherapy and immunotherapy via an antigen-capturing nanoplatform for metastatic cancer treatment. *Advanced Science* 6 (10): 1802157. https://doi.org/10.1002/advs.201802157.

Wang, J., Wang, H., Zuo, S. et al. (2020). Synergistic effects of lanthanide surface adhesion and photon-upconversion for enhanced near-infrared responsive photodegradation of organic contaminants in wastewater. *Environmental Science: Nano* 7: 3333–3342. https://doi.org/10.1039/D0EN00670J.

Wen, P., Zong, M., Linhardt, R.J. et al. (2017). Electrospinning: A novel nano-encapsulation approach for bioactive compounds. *Trends in Food Science & Technology* 70: 56–68. https://doi.org/10.1016/j.tifs.2017.10.009.

Wu, J., Song, Y., Han, B. et al. (2015). Synthesis and characterization of UV upconversion material Y_2SiO_5:Pr^{3+},Li^+/TiO_2 with enhanced the photocatalytic properties under a xenon lamp. *RSC Advances* 5: 49356–49362. https://doi.org/10.1039/C5RA06416C.

Wu, X., Zhang, K., Zhang, G. et al. (2017). Facile preparation of BiOX (X = Cl, Br, I) nanoparticles and up-conversion phosphors/BiOBr composites for efficient degradation of NO gas: oxygen vacancy effect and near infrared light responsive mechanism. *Chemical Engineering Journal* 325: 59–70. https://doi.org/10.1016/j.cej.2017.05.044.

Xia, S., Song, Z., Jeyakumar, P. et al. (2019). A critical review on bioremediation technologies for Cr(VI)-contaminated soils and wastewater. *Critical Reviews in Environmental Science and Technology* 49 (12): 1027–1078. https://doi.org/10.1080/10643389.2018.1564526.

Xu, D., Lian, Z., Fu, M. et al. (2013). Advanced near-infrared-driven photocatalyst: Fabrication, characterization, and photocatalytic performance of β-$NaYF_4$:Yb^{3+},Tm^{3+}@TiO_2 core@shell microcrystals. *Applied Catalysis B: Environmental* 142-143: 377–386. https://doi.org/10.1016/j.apcatb.2013.05.062.

Xu, Z., Quintanilla, M., Vetrone, F. et al. (2015). Harvesting lost photons: plasmon and upconversion enhanced broadband photocatalytic activity in core@shell microspheres based on lanthanide-doped $NaYF_4$, TiO_2, and Au. *Advanced Functional Materials* 25: 2950–2960. https://doi.org/10.1002/adfm.201500810.

Xu, J., Xu, L., Wang, C. et al. (2017). Near-infrared-triggered photodynamic therapy with multitasking upconversion nanoparticles in combination with checkpoint blockade for immunotherapy of colorectal cancer. *ACS Nano* 11 (5): 4463–4474. https://doi.org/10.1021/acsnano.7b00715.

Xu, J., Han, W., Yang, P. et al. (2018). Tumor microenvironment-responsive mesoporous MnO_2-coated upconversion nanoplatform for self-enhanced tumor theranostics. *Advanced Functional Materials* 28 (36): 1803804. https://doi.org/10.1002/adfm.201803804.

Xu, J., Gulzar, A., Yang, P. et al. (2019a). Recent advances in near-infrared emitting lanthanide-doped nanoconstructs: mechanism, design and application for bioimaging. *Coordination Chemistry Reviews* 382: 104–134. https://doi.org/10.1016/j.ccr.2018.11.014.

Xu, X., Sun, Y., Zhang, Q. et al. (2019b). Synthesis and photocatalytic activity of plasmon-enhanced core-shell upconversion luminescent photocatalytic $Ag@SiO_2@YF_3:Ho^{3+}@TiO_2$ nanocomposites. *Optical Materials* 94: 444–453. https://doi.org/10.1016/j.optmat.2019.05.038.

Xue, J., Wu, T., Dai, Y. et al. (2019). Electrospinning and electrospun nanofibers: methods, materials, and applications. *Chemical Reviews* 119 (8): 5298–5415. https://doi.org/10.1021/acs.chemrev.8b00593.

Yang, W., Li, X., Chi, D. et al. (2014). Lanthanide-doped upconversion materials: emerging applications for photovoltaics and photocatalysis. *Nanotechnology* 25 (48): 482001. https://doi.org/10.1088/0957-4484/25/48/482001.

Yang, M., Gao, M., Hong, M. et al. (2018). Visible-to-NIR photon harvesting: progressive engineering of catalysts for solar-powered environmental purification and fuel production. *Advanced Materials* 30 (47): 1802894. https://doi.org/10.1002/adma.201802894.

Yao, C., Wang, W., Wang, P. et al. (2018). Near-infrared upconversion mesoporous cerium oxide hollow biophotocatalyst for concurrent pH-/H_2O_2-responsive O_2-evolving synergetic cancer therapy. *Advanced Materials* 30 (7): 1704833. https://doi.org/10.1002/adma.201704833.

Yuan, P., Lee, Y., Gnanasammandhan, M.K. et al. (2012). Plasmon enhanced upconversion luminescence of $NaYF_4$:Yb,Er@SiO_2@Ag core-shell nanocomposites for cell imaging. *Nanoscale* 4 (16): 5132–5137. https://doi.org/10.1039/C2NR31241G.

Zhang, K. and Guo, L. (2013). Metal sulphide semiconductors for photocatalytic hydrogen production. *Catalysis Science & Technology* 3 (7): 1672–1690. https://doi.org/10.1039/C3CY00018D.

Zhang, F., Zhang, C., Peng, H. et al. (2016a). Near-infrared photocatalytic upconversion nanoparticles/TiO_2 nanofibers assembled in large scale by electrospinning. *Particle & Particle Systems Characterization* 33: 248–253. https://doi.org/10.1002/ppsc.201600010.

Zhang, F., Hao, L., Wang, Y. et al. (2016b). Hydrothermal-assisted crystallization for the synthesis of upconversion nanoparticles/CdS/TiO_2 composite nanofibers by electrospinning. *CrystEngComm* 18: 6013–6018. https://doi.org/10.1039/C6CE00987E.

Zhang, F., Wang, W., Cong, H. et al. (2017a). Facile synthesis of upconverting nanoparticles/zinc oxide core-shell nanostructures with large lattice mismatch for infrared triggered photocatalysis. *Particle & Particle Systems Characterization* 34: 1600222. https://doi.org/10.1002/ppsc.201600222.

Zhang, Q., Deng, J., Xu, Z. et al. (2017b). High-efficiency broadband C_3N_4 photocatalysts: synergistic effects from upconversion and plasmons. *ACS Catalysis* 7: 6225–6234. https://doi.org/10.1021/acscatal.7b02013.

Zhang, J., Huang, Y., Jin, L. et al. (2017c). Efficient upconverting multiferroic core@shell photocatalysts: visible-to-near-infrared photon harvesting. *ACS Applied Material Interfaces* 9: 8142–8150. https://doi.org/10.1021/acsami.7b00158.

Zhang, Y., Wang, L., Ma, X. et al. (2018a). Up-conversion nanoparticles sensitized inverse opal photonic crystals enable efficient water purification under NIR irradiation. *Applied Surface Science* 435: 799–808. https://doi.org/10.1016/j.apsusc.2017.11.195.

Zhang, J., Qiao, B., Liang, Z. et al. (2018b). Near-infrared light-induced photocatalysis of $NaYF_4$:Yb,Tm@Cu_2O core-shell nanocomposites. *Optical Materials* 84: 89–93. https://doi.org/10.1016/j.optmat.2018.06.053.

Zhang, Z., Liu, Y., Fang, Y. et al. (2018c). Near-infrared-plasmonic energy upconversion in a nonmetallic heterostructure for efficient H_2 evolution from ammonia borane. *Advanced Science* 5: 1800748. https://doi.org/10.1002/advs.201800748.

Zhang, Q., Yang, F., Xu, Z. et al. (2019a). Are lanthanide-doped upconversion materials good candidates for photocatalysis? *Nanoscale Horizons* 4 (3): 579–591. https://doi.org/10.1039/C8NH00373D.

Zhang, J., Zhao, S., Xu, Z. et al. (2019b). Near-infrared light-driven photocatalytic $NaYF_4$:Yb,Tm@ZnO core/shell nanomaterials and their performance. *RSC Advances* 9: 3688–3692. https://doi.org/10.1039/C8RA07861K.

Zhang, Y., Wang, L., Park, S.H. et al. (2019c). Single near-infrared-laser driven Z-scheme photocatalytic H_2 evolution on upconversion material@Ag_3PO_4@black phosphorus. *Chemical Engineering Journal* 375: 121967. https://doi.org/10.1016/j.cej.2019.121967.

Zhang, K., Zhou, M., Yu, C. et al. (2020). Enhanced visible-NIR-driven photocatalytic activities over $NaYF_4$:Yb^{3+},Er^{3+}/Ag_2CrO_4 composite. *Journal of Materials Science* 55: 10435–10452. https://doi.org/10.1007/s10853-020-04723-2.

Zhang, K., Zhou, M., Yu, C. et al. (2021). High value-added fluorescence upconversion agents-assisted nano-semiconductors for efficient wide spectral response photocatalysis: Exerting energy transfer effect and applications. *Journal of Rare Earths* 39 (3): 243–260. https://doi.org/10.1016/j.jre.2020.03.009.

Zhao, F., Khaing, K.K., Yin, D. et al. (2018a). Large enhanced photocatalytic activity of g-C_3N_4 by fabrication of a nanocomposite with introducing upconversion nanocrystal and Ag nanoparticles. *RSC Advances* 8: 42308–42321. https://doi.org/10.1039/C8RA07901C.

Zhao, M., Wang, W., Huang, C. et al. (2018b). Facile synthesis of UCNPs/$Zn_xCd_{1-x}S$ nanocomposites excited by ear-infrared light for photochemical reduction and removal of Cr(VI). *Chinese Journal of Catalysis* 39: 1240–1248. https://doi.org/10.1016/S1872-2067(18)63061-X.

Zhou, S., Deng, K., Wei, X. et al. (2013). Upconversion luminescence of $NaYF_4$:Yb^{3+},Er^{3+} for temperature sensing. *Optics Communications* 291: 138–142. https://doi.org/10.1016/j.optcom.2012.11.005.

Zhou, J., Liu, Q., Feng, W. et al. (2015). Upconversion luminescent materials: advances and applications. *Chemical Reviews* 115 (1): 395–465. https://doi.org/10.1021/cr400478f.

Zhou, S., Zhang, Q., Zhao, D. et al. (2018). Synthesis, properties and mechanism of photodegradation of core-shell structured upconversion luminescent $NaYF_4$:Yb^{3+},Er^{3+}@BiOCl. *Applied Organometallic Chemistry* 32 (4): e4230. https://doi.org/10.1002/aoc.4230.

Zhu, X., Su, Q., Feng, W. et al. (2006). Anti-Stokes shift luminescent materials for bio-applications. *Chemical Society Reviews* 46 (4): 1025–1039. https://doi.org/10.1039/C6CS00415F.

Zhu, Y., Zheng, X., Lu, Y. et al. (2019). Efficient upconverting carbon nitride nanotubes for near-infrared-driven photocatalytic hydrogen production. *Nanoscale* 11: 20274–20283. https://doi.org/10.1039/C9NR05276C.

17

Upconverting Nanoparticles in the Detection of Fungicides and Plant Viruses

Vishab Kesarwani and Vineet Kumar Rai

Indian Institute of Technology (Indian School of Mines), Laser and Spectroscopy Laboratory, Department of Physics, Dhanbad 826004, Jharkhand, India

17.1 Introduction

The agricultural sector plays a crucial role in the GDP growth, especially for the underdeveloped and developing nations. Developing economies are considered as dual economies having a traditional agricultural sector and a highly productive modern capitalist sector. Despite being considered as a low-productivity sector, several studies have proved the importance of this sector in the economic growth of a country (Dethier and Effenberger 2012). Johnston and Mellor have explicitly established the importance of agriculture in the economic development of a country (Johnston and Mellor 1961). According to their study, agriculture plays an important role in economic development via production and consumption linkages. Based on the study, it was further concluded that the progress strategy of a country should be based on agriculture, and the high productivity agriculture sector will lead to industrialization, which in turn decrease the poverty (Singer 1979). The green revolution has been very helpful in achieving high agricultural productivity, but it also had its own drawbacks. One of the most major setbacks is the increment in crop homogeneity, i.e. growing only a few high yielding varieties of crops. Because of increased crop homogeneity, the seeds get prone to diseases and pests. As a result, the use of different types of pesticides has been increased to protect those few varieties that have an adverse effect on the environment, soil, and human health.

The fungicide is a class of pesticides that are used to treat fungal infections of crops. Fungal infections are supposed to be harmful to the growth of high-quality crops. Although the use of fungicides has promoted high-quality yield of crops by reducing the crop losses due to fungi, a prolonged and heavy dosage of the same is very much toxic. Several researchers have investigated in detail the detrimental effects of fungicides on human health, environment, and aquatic ecosystem. Not only this but some of the antifungal agents have direct and indirect effects on soil microorganisms. For example, glucopyranosyl antibiotic is very harmful for the soil bacteria as it obstructs the synthesis of amino acids (Brauer et al. 2019; Zubrod et al. 2019).

Upconverting Nanoparticles: From Fundamentals to Applications, First Edition. Edited by Vineet K. Rai.

Thus, the use of fungicides in agriculture needs to be monitored with great sincerity, so that a healthy ecosystem and high agricultural productivity could be maintained without affecting the beneficial microorganisms.

Plant viruses are also a major foe of crops that hinder their productivity. The viruses are parasitic organisms that interact with hosts to alter their behavior, resulting in enhanced transmission of diseases. Plant virology (study of plant viruses) actually started in the late nineteenth century, when Dutch microbiologist Martinus Beijerinck and Russian researcher Dmitri Iwanowski independently discovered the agent causing mosaic disease in tobacco, which has been later named as tobacco mosaic virus (TMV) (Gergerich and Dolja 2006). According to the article published by Lefeuvre et al., there are about 31 virus families (Figure 17.1). Several excellent reviews are available in the literature describing about their interactions with hosts, lifestyle, spread, genes, and crop yield losses due to different types of viral diseases (Roossinck 2010; Pallas and Garcia 2011; Lefeuvre et al. 2019).

Thus, proper monitoring of fungicides and viral plant diseases by choosing appropriate detection methods or techniques will be very much helpful in establishing a high-productivity agricultural sector, which will further boost the economy of a nation. Although the traditional detection methods of fungicides and plant viruses are reliable and accurate, but because of the ever-increasing demand for high and qualitative agricultural productivity, those techniques are not very much suitable because they require labor, specific reagents, and elaborative sample preparation,

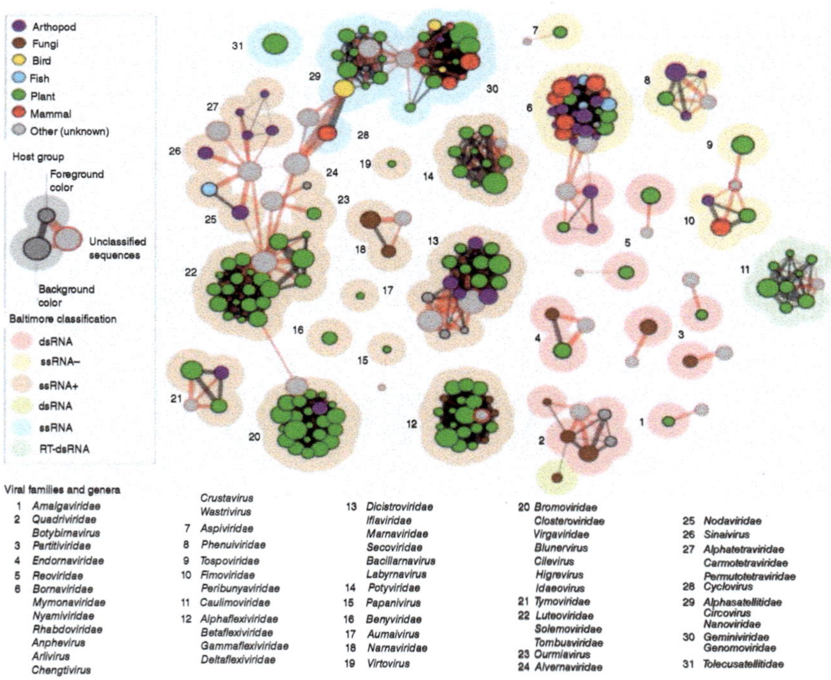

Figure 17.1 Different families of plant virus species. Source: Lefeuvre et al. (2019). Reprinted with permission of Springer Nature BV.

which make them time-consuming, expensive, and hence impractical to be used as a preliminary tool for processing a large number of samples (Sankaran et al. 2010). The current time demands are to explore the detection techniques which are cost-effective, user-friendly, specific, rapid, nondestructive, and highly sensitive. Optical detection techniques involving upconverting nanoparticles (UCNPs) can meet up these requirements (Kalyani et al. 2021).

Ever since its discovery, optical biosensing via frequency upconversion in lanthanide-doped nanomaterials has been a hot topic of research. This type of sensing is free from the drawbacks that come across in conventional methods such as background interference of complex materials, target damaging, spectral overlap of multi-detection signals, etc. (Chen et al. 2016; Wang et al. 2016; Fan et al. 2016; Zhu et al. 2017). Various sensing methods based on fluorescence via UCNPs such as Forster resonance energy transfer (FRET)/luminescence resonance energy transfer (LRET), inner filter effect (IFE), photoinduced electron transfer (PET) etc., have been developed so far. However, the main problem that limits its application is the low quantum yield of UCNPs. The techniques applied to enhance the quantum yield of UCNPs have been covered elsewhere in the book. The present chapter covers the application of UCNPs in visual detection of fungicides and plant viruses. The relevant mechanisms and strategies involved in the detection have been introduced followed by some reported works and future challenges.

17.2 Visual Detection of Fungicides

As discussed above, fluorescence-based sensing/detection can be done through a variety of mechanisms. Therefore, it is important for the readers to have a proper understanding of the detection mechanisms involved in the visual detection of fungicides. Hence, a brief literature about the widely used mechanisms is provided below.

17.2.1 Detection Mechanisms

17.2.1.1 Forster Resonance Energy Transfer (FRET)

FRET is a nonradiative energy transfer phenomenon from an excited donor fluorophore to a nearby acceptor fluorophore via dipole–dipole interaction. When rare-earth-doped materials are used as donors, then this process is known as luminescence resonance energy transfer (LRET) (Hu et al. 2016). A typical energy-level diagram "known as Jablonski diagram" explaining the FRET mechanism is shown in Figure 17.2.

As the definition signifies, in FRET/LRET, resonant energy transfer takes place between the donor and acceptor; thus, choosing a suitable donor and acceptor is very important for an efficient energy transfer. Traditionally, organic molecules were used as donors because these molecules can be prepared very easily. However, the organic molecules have certain drawbacks such as weak signals, short fluorescence lifetime, and low chemical stability that hinder an efficient FRET phenomenon (Shi et al. 2015). On the other hand, using UCNPs as donors are advantageous because of their strong luminescence properties, high chemical

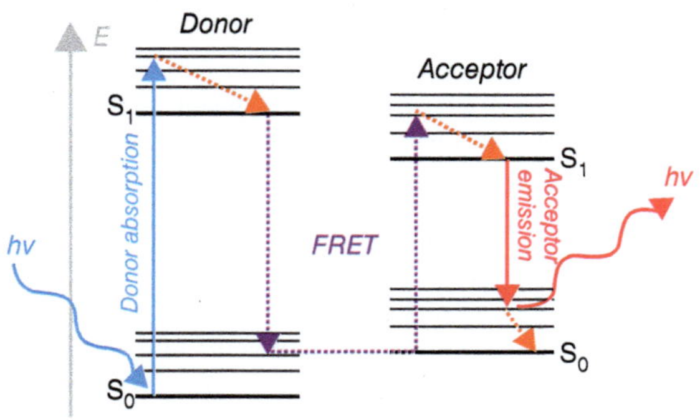

Figure 17.2 Jablonski diagram of FRET. Source: Hochreiter et al. (2015).

stability, negligible autofluorescence, better signal-to-noise ratio, large anti-Stokes shifts, and large fluorescence lifetimes (Wang et al. 2016; Zhang et al. 2018).

Talking about the energy acceptors, generally, those ions or molecules are suitable whose absorption spectra overlaps with the emission spectra of donors, and it must be in close proximity of the donor (typically <10 nm). In principle, the energy transfer efficiency (η) of FRET is given as (Hochreiter et al. 2015; Zhang et al. 2018)

$$\eta = \frac{1}{1 + \left(\frac{R}{R_0}\right)^6} \tag{17.1}$$

where "R_0" is the Forster distance at which the energy transfer efficiency is 50% and "R" is the donor–acceptor distance. From Eq. (17.1), it is evident that the donor–acceptor distance plays a significant role in efficient energy transfer upconversion. However, in rare-earth-doped UCNPs, the distance between every donor and acceptor ion at the surface of UCNPs is different from each other. Hence, the particle size also decides the efficiency of FRET. Another factor that affects the energy transfer efficiency of FRET/LRET is the quenching efficiency of the acceptor. That is why, several types of energy acceptors such as organic molecules/dyes, carbon nanomaterials, and metal and semiconductor NPs are now being examined by the researchers worldwide. Based on FRET, the following detection strategies/modes are applied for constructing the sensing assays (Wang et al. 2016; Su et al. 2017):

- Quenching or recovering the luminescence intensity by changing the distance between the donor (UCNPs) and acceptor via addition of analytes. This is called as fluorescence turn-on mode (Figure 17.3).
- Quenching or recovering the luminescence intensity by changing the absorption spectra of the acceptor via addition of analytes. This is called as fluorescence turn-off mode (Figure 17.4).
- The ratiometric mode in which the analyte concentration is determined by the emission ratios at two different wavelengths.

17.2.1.2 Inner Filter Effect (IFE)

In fluorescence spectroscopy, the attenuation of the traveling excitation light by the sample filled in the cuvette is called the inner filter effect. The energy-level diagram

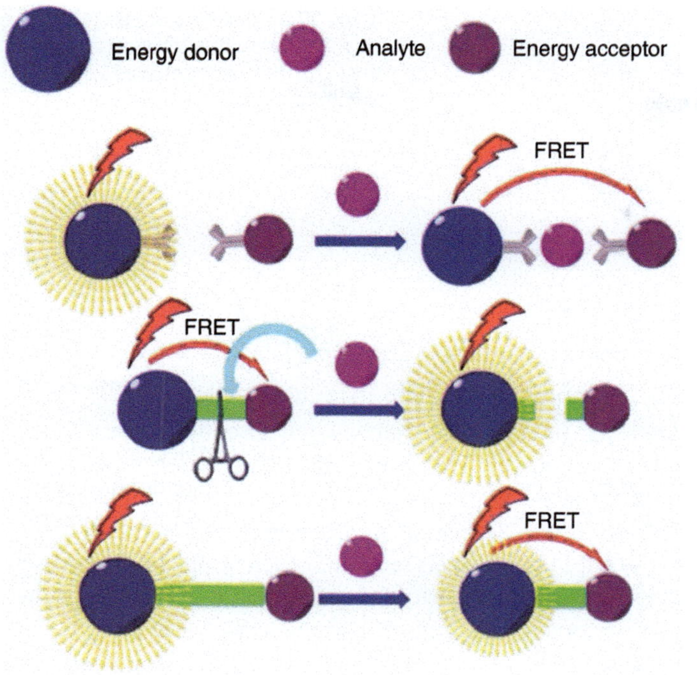

Figure 17.3 The distance between donor and acceptor can be changed by the amount of analytes, so that the FRET process can be activated or deactivated (top). Donors and acceptors are brought close to each other (middle). The analytes break the linkage between donors and acceptors, thus inhibiting the FRET process (bottom). Source: Wang et al. (2016). Reprinted with permission of Royal Society of Chemistry.

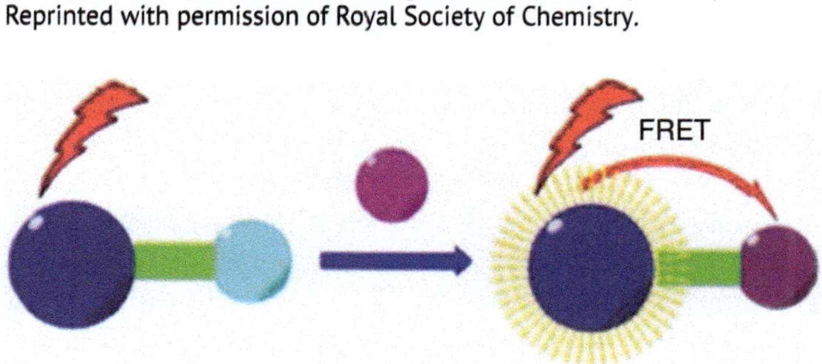

Figure 17.4 Structures as well as the absorption spectra of acceptors are changed by analytes. Source: Wang et al. (2016). Reprinted with permission of Royal Society of Chemistry.

describing this effect is shown in Figure 17.5. At first, this phenomenon has been observed by Stokes in quinine (Chen et al. 2018a; Zhang et al. 2019). He noticed that the blue fluorescence of concentrated quinine comes only from its surface when irradiated by ultraviolet (UV) light. It could happen because the concentration of the quinine sample was relatively high. Hence, most of the UV radiation has been absorbed by the sample surface. After passing through quinine, the excitation light was feeble, so it could not excite it to produce blue fluorescence. IFE is rarely observed

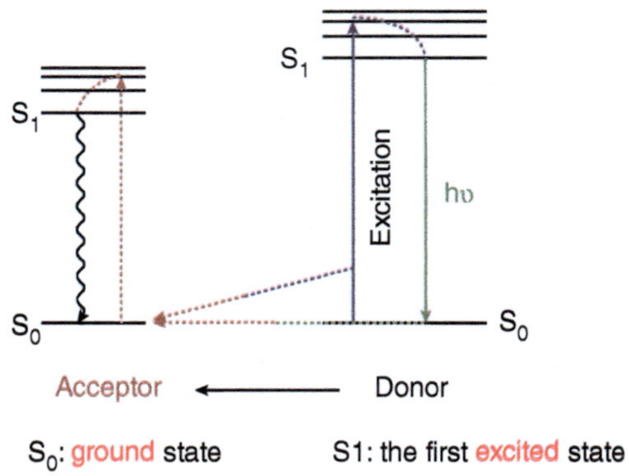

Figure 17.5 Energy-level diagram of the inner filter effect. Source: Zhang et al. (2019). Reprinted with permission of Elsevier.

in dilute (low optical density) samples. The phenomenon described above is more precisely known as the primary inner filter effect (pIFE). The reabsorption of the emitted light by the absorber/quencher is known as the secondary inner filter effect (sIFE) (Panigrahi and Mishra 2019). Because of the absorption of excitation or emission radiation, the fluorescence intensity gets reduced, which results in a nonlinear relationship between the fluorescence intensity and concentration of optically active ions. A linear relationship between them indicates the absence of an IFE.

The IFE-based sensing enjoys all the advantages of optical sensing discussed above. However, in addition to those, it has one more advantage. The IFE process is more flexible because, unlike FRET/LRET, its efficiency does not depend on the donor–acceptor distance. Also, in IFE-based biosensing, the acceptor absorbs the excited-state energy and attenuates the energy reaching to the donor.

Like FRET/LRET, the IFE-based sensing system can be constructed in three different working modes, turn-on, turn-off, and ratiometric. Furthermore, it requires two basic units, the first one is the donor/absorber and the second one is the analyte-independent acceptor/fluorescer. Organic dyes such as rhodamine, quantum dots of different element groups, rare-earth-doped UCNPs, nanoclusters, carbon dots, and graphene quantum dots are used as suitable acceptors in IFE-based sensing systems (Chen et al. 2018a). For absorbers, four categories of materials are being used because their absorbing capacity and selectivity directly influence the performance of the sensing system. Those four categories suitable for absorbers are nanomaterials (usually AuNPs and AgNPs), catalysts, chromoionophore, and the analyte itself (Chen et al. 2018a).

For better sensing performance, the following additional requirements should be kept in mind (Chen et al. 2018a):

- Like FRET, a proper overlap between the absorption spectrum of the absorber and the emission/excitation spectrum of the donor should exist. The efficiency of IFE-based sensing depends on the extent of overlap between the two (Figure 17.6).
- For quantitative detection of analytes, the absorption of the acceptor/absorber should exhibit a sensitive and selective response with respect to the analyte concentration.
- The fluorescence of the donor should be independent of the analyte.

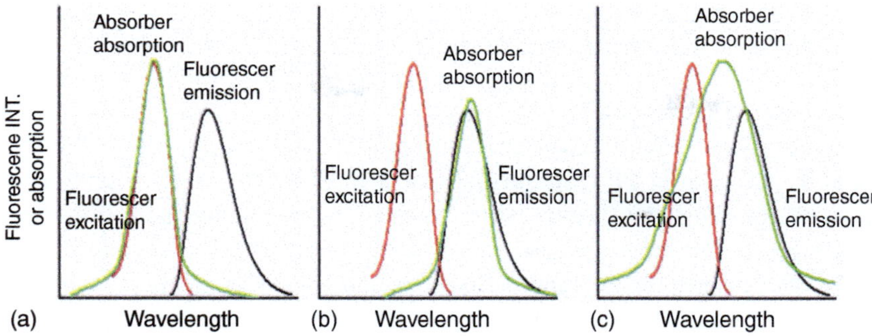

Figure 17.6 Conditions for IFE-based sensing, absorption spectrum of the absorber overlaps with (a) excitation spectrum, (b) emission spectrum, and (c) both excitation and emission spectrum of the acceptor/fluorescer. Source: Chen et al. (2018a). Reprinted with permission of Elsevier.

- The absorption and fluorescence properties of the sensing units should be independent of the external factors.
- The absorber should not act as a fluorescence quencher, i.e. the donor and absorber must be in the charge repulsion condition.

17.2.1.3 Photoinduced Electron Transfer (PET)

PET is a phenomenon in which electron is transferred from the excited level of the donor to the excited level of the acceptor (shown in Figure 17.7a). This type of electron transfer mainly takes place because of a photochemical reaction when the donors and acceptors are close to each other. The phenomenon of electron transfer is very much significant for most of the fields such as solar cells, electronic devices, biosensing, etc. PET sensors also operate on the same strategy where the addition of an analyte interrupts the PET taking place between the donor and the acceptor. This interruption results in fluorescence enhancement, fluorescence quenching, or

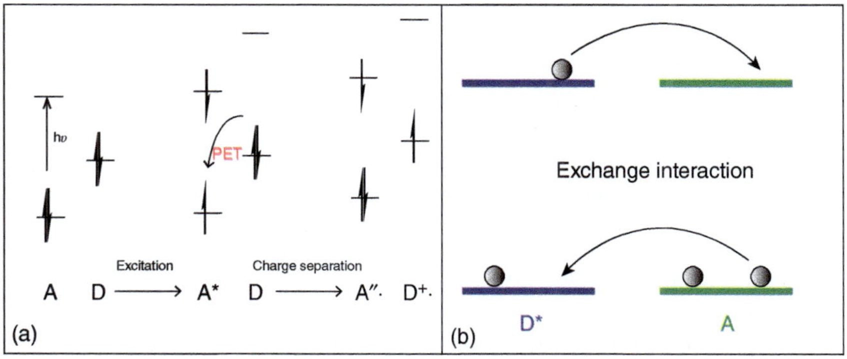

Figure 17.7 Energy-level diagram of (a) photoinduced electron transfer. Source: Escudero (2016). Reprinted with permission of American Chemical Society. (b) Electron exchange. Source: Strieth-Kalthoff et al. (2018). Reprinted with permission of Royal Society of Chemistry.

change in the fluorescence intensity depending on the nature of the donor and the acceptor (Escudero 2016; Zhang et al. 2020).

17.2.1.4 Electron Exchange (EE)

It is a short energy transfer mechanism (typically 1 nm) theoretically proposed by D. L. Dexter in 1953 (Dexter 1953; Strieth-Kalthoff et al. 2018). Hence, it is also known as Dexter energy transfer or Dexter electron transfer. In this phenomenon, an excited electron is transferred from the donor to the acceptor via a nonradiative path (see Figure 17.7b). The electron exchange (EE) occurs if the wave functions of the donor and acceptor overlap with each other. In general, pesticides are electron deficient and hence act as electron acceptors in this energy transfer mechanism (Edgell et al. 1993; Likhtenshtein 1996).

17.2.2 Significant Works on Upconversion-Based Fungicide Detection

The detection strategies described in Figures 17.3 and 17.4 are applied to construct several detection platforms such as test papers, lateral flow strips, microarrays, etc., to ensure rapid and cost-effective detection of the target materials. Some of the recent works related to the visual detection of fungicides using UCNPs and FRET/LRET mechanism are summarized in this chapter.

A smartphone-based paper sensor (limit of detection [LOD] 0.1 μM) to detect thiram, a fungicide that can cause severe headaches, dizziness, fatigue, nausea, diarrhea, and other gastrointestinal complaints if inhaled, has been reported. First of all, $NaYF_4$:Yb/Tm@PAA-Cu nanoprobe has been developed (Mei et al. 2016). After that, the test paper has been fabricated by ultrasonic agitation of a filter paper immersed in the aqueous solution of the upconverting nanoprobe and then drying it at 50° C. The smartphone-based sensing system (shown in Figure 17.8a) was fabricated by using 3D printing technology. The constructed system mainly consists of three parts, a holder for 980 nm mini laser, a dark cavity having an optical fiber, and a chamber for placing the as-prepared test paper and a phone holder having an optical window with a diameter of 1 cm to contact with the phone camera. Now, keeping the laser power at 180 mW, the changes in the UC luminescence intensity of the test paper on adding different amounts of thiram was observed and captured with the phone camera. The observed UC intensity was quantified by a self-written Android program installed in the smartphone. It was found that the graph of luminescence intensity or quenching efficiency versus the logarithm of thiram concentration follows a linear relation (Figure 17.8b,c) expressed by the following equation:

$$\text{Quenching efficiency (\%)} = 30.9094 + 10.746 \times \log[\text{thiram}/1 \times 10^{-5}\ \text{M}] \tag{17.2}$$

The variation in luminescence intensity of the test paper with respect to the amount of thiram is shown in Figure 17.8d. Finally, the reliability and usefulness of the sensing system are investigated in real samples. For this measurement, different amounts of thiram have been added to apple juice. The quenching efficiency of the recorded spectrum has been determined before and after the addition of thiram.

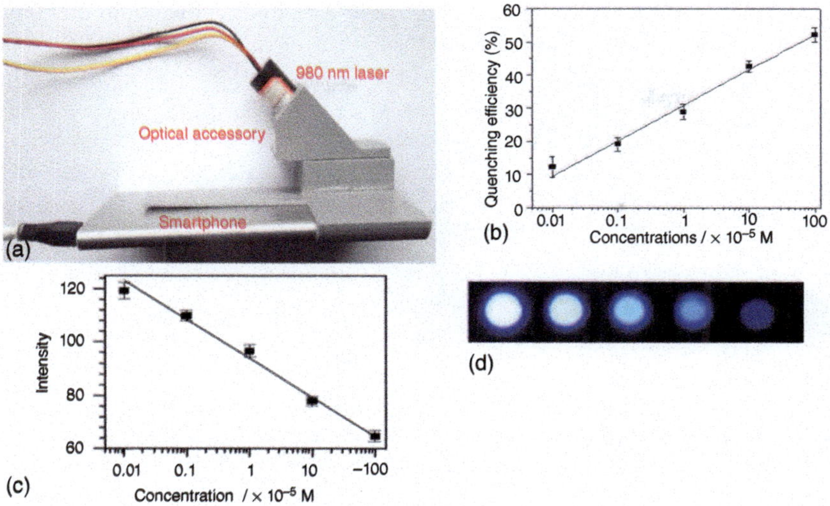

Figure 17.8 (a) Image of the fabricated sensing system. (b) Graph of quenching efficiency versus log of thiram concentration. (c) Graph of UC intensity versus log of thiram concentration. (d) Quenching of UC intensity of the test paper on adding different amounts of thiram (0–100 μM). Source: Mei et al. (2016). Reprinted with permission of Elsevier.

These values have then been used in Eq. (17.2) to calculate the concentrations of thiram in the solution. The obtained results were satisfactory.

A similar smartphone-based detection system for thiram has been reported (Sun et al. 2019). In this work, white-light-emitting (Tm/Tb/Eu/Yb:$NaGdF_4$) UCNPs modified with lead (II)-dithizone complex [$Pb(DZ)_2$] has been prepared. The UC emission spectra of the WL-UCNPs show three prominent peaks at ~475 nm (blue), ~545 nm (green), and ~650 nm (red). However, after the addition of lead(II)-dithizone complex (energy acceptor), the blue and green intensity has been decreased because of FRET. The addition of different amounts of thiram dissociates the $Pb(DZ)_2$ complex, and a new complex Pb-thiram is formed that ruins the energy transfer pathway from $Pb(DZ)_2$ complex to WL-UCNPs because of which the blue and green intensities get recovered. The authors have used two types of detection schemes for the detection of thiram. The first one is the calorimetric detection scheme (Figure 17.9a) and the second one is the smartphone-based detection scheme (Figure 17.9b). For full details of the work, the readers can refer to the original article (Sun et al. 2019).

The works described above for the detection of fungicide thiram used organic molecules as acceptors. Owing to its advantages such as strong molar extinction coefficient, high specific surface area, and good biocompatibility, the biosensing applications of MnO_2 nanosheets have also been explored broadly by many researchers (Yao et al. 2017, 2019). The detection of carbendazim (CBZ) fungicide via the LRET mechanism between UCNPs and MnO_2 nanosheets, the latter acting as the acceptor, has been reported (Ouyang et al. 2021). At first, they prepared poly-ethermine (PEI)-modified UCNPs along with MnO_2 nanosheets as donor–acceptor pairs. After

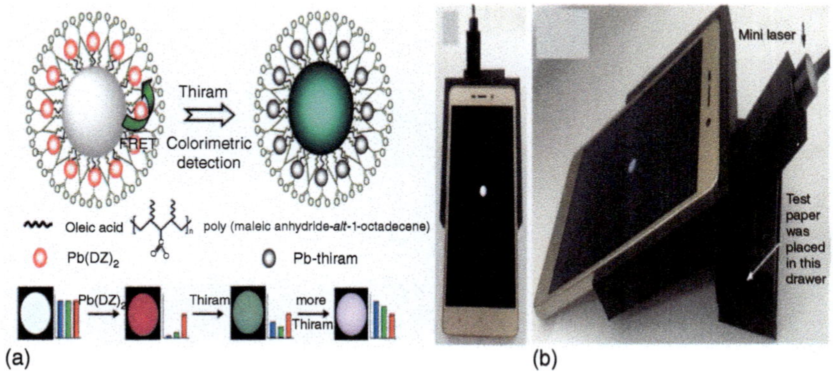

Figure 17.9 (a) Schematic diagram of calorimetric detection of thiram via WL-UCNPs. (b) Smartphone-based sensing system for the detection of thiram via WL-UCNP-modified test paper. Source: Sun et al. (2019). Reprinted with permission of Springer.

that, the classical glutaraldehyde method was employed to conjugate the aptamer. An aptamer is a kind of a single-stranded DNA or RNA sequence enriched by in vitro systematic evolution of ligand by exponential enrichment (SELEX) to enhance its specificity, affinity, and selectivity toward desired biomolecules and UCNPs (Jin et al. 2017). The aptamer can bind itself with MnO_2 nanosheets via van der Waals forces, which in turn decrease the fluorescence intensity of the UCNPs. After the addition of CBZ, the fluorescence intensity has been recovered because of the attenuation in the physisorption of aptamer on the MnO_2 surface (Figure 17.10a). The sensitivity of the aptasensor has been calculated by plotting the intensity variation of 544 nm band against the logarithm of CBZ concentration (Figure 17.10b). The graph is fitted linearly according to the equation,

$$y = 493.37x + 638.39 \tag{17.3}$$

and the value of LOD comes out to be 0.05 ng/mL.

The treatment of CBZ fungicide via near-infrared (NIR)-driven photocatalytic action of $NaYF_4$:Er/Yb@CdS core–shell UCNPs has been reported (Balaji et al. 2017). Photocatalysis is an eco-friendly method of resolving the problems of environmental pollutants. The core–shell nanoparticles have been prepared by a two-step wet chemical method. On performing the photocatalytic degradation, 64% degradation of CBZ was achieved after six hours of NIR irradiation, and the reaction kinetics was found to be of pseudo first-order (Balaji et al. 2017). Several metal ions such as Cu^{2+}, Zn^{2+}, Mg^{2+}, Fe^{3+} etc., are very effective against fungal infections and are widely used as fungicides (Ahmed et al. 2017). One such metal ion is Ag^{+}, which is always considered as precious and safe for human beings. However, using Ag^{+} more than the permitted value (<0.05 ppm, according to World Health Organization) can have serious environmental and human health issues (Liu et al. 2012; Yang et al. 2013; Zhao et al. 2013). Thus, in order to control and cure the negative effects of Ag^{+}, several techniques such as ion-selective electrodes (ISEs), voltammetry, UV–vis, fluorescence-based methods, etc., have been applied by many researchers (Wygladacz et al. 2005; Achenbach et al. 2005; Zhang et al. 2006, 2015;

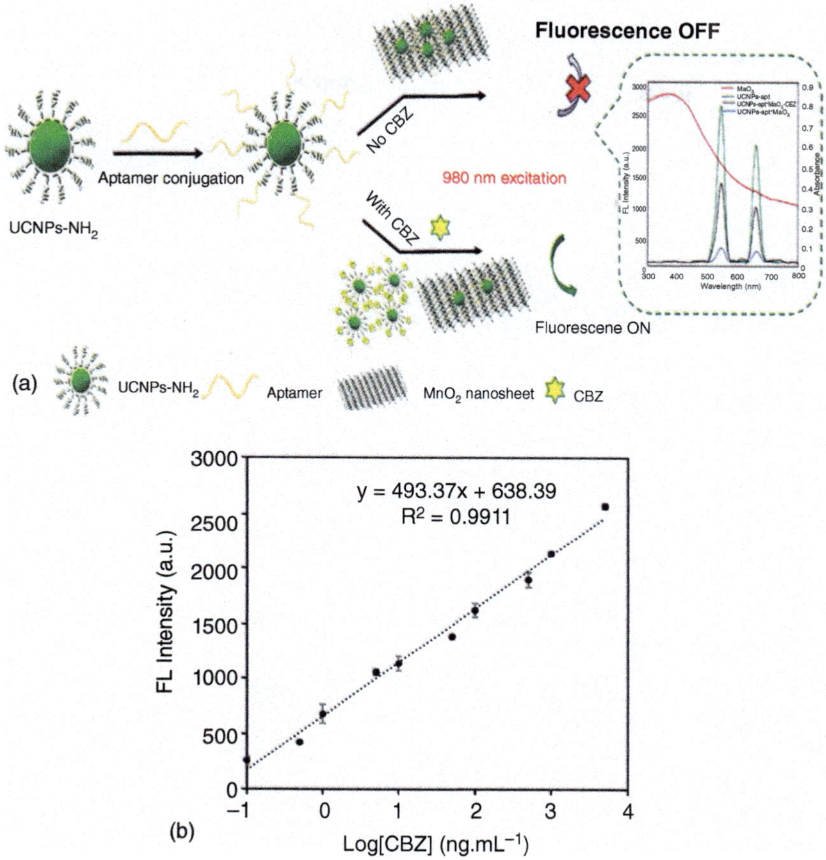

Figure 17.10 (a) Schematic representation of fluorescence-based aptasensor for CBZ detection. (b) Standard curve of 544 nm at different CBZ concentrations. Source: Ouyang et al. (2021). Reprinted with permission of Elsevier.

Shervedani and Babadi 2006; Mohadesi and Taher 2007; Liu et al. 2012; Gan et al. 2012; Sung and Wu 2014; Lv et al. 2014). Among these, the fluorescence-based optical methods again proved to be advantageous in many respects such as detection speed, sensitivity, and efficiency.

One such study on Ag^+ ion detection using UCNPs has been reported by Q. Long et al. The mechanism involved in this detection was the IFE. Polyethylenimine-modified $NaYF_4$:Yb^{3+},Tm^{3+} UCNPs have been prepared by a solvothermal method. The silver ion can oxidize *o*-phenylenediamine (OPD), which induces a yellow color and an absorption peak at ~450 nm. This oxidized OPD can quench the UC luminescence intensity via IFE, and the quenching is proportional to the concentration of the Ag^+ ions. In this study, the LOD for the Ag^+ ions comes out to be ~33 nM, and the feasibility of detection was examined in tap water. The detailed scheme of detection is given in Figure 17.11 (Long et al. 2017). The UCNPs using the same host material ($NaYF_4$) doped with Gd, Ho, and Yb has been applied for the Fe^{3+} detection in water (Chen et al. 2018b). The prepared UCNPs showed emission peaks at ~546,

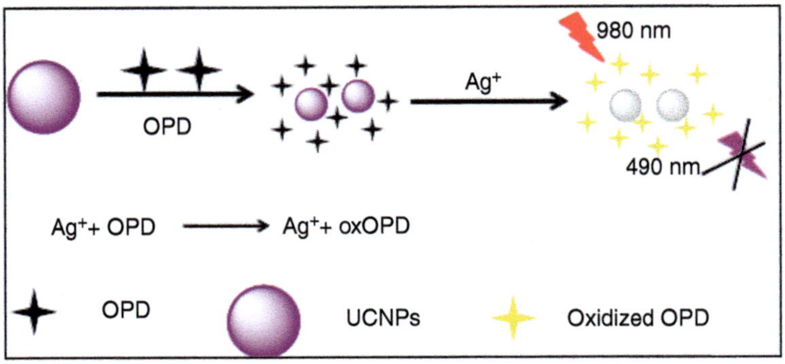

Figure 17.11 Scheme for IFE-based Ag^+ detection. Source: Long et al. (2017). Reprinted with permission of Springer.

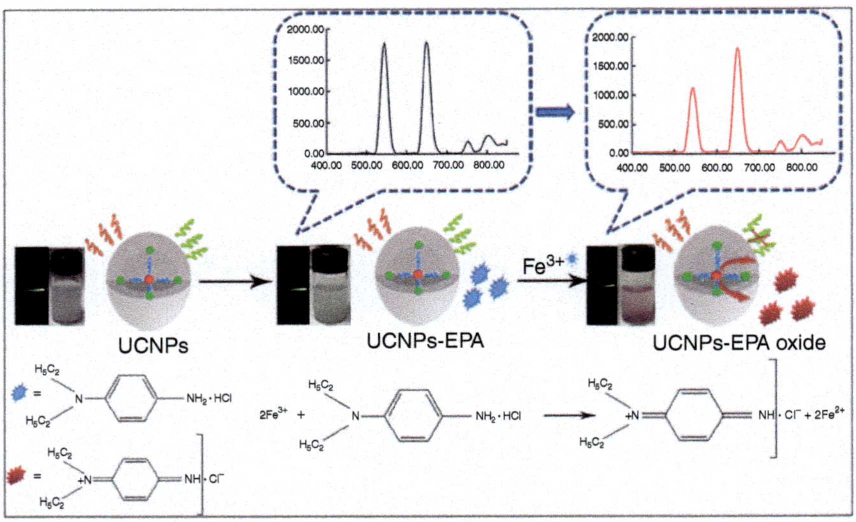

Figure 17.12 IFE-based detection of the Fe^{3+} ions. Source: Chen et al. (2018b). Reprinted with permission of Springer.

~657, ~758, and ~812 nm upon 980 nm excitation. The addition of Fe^{3+} ions led to the oxidation of *N*,*N*-diethyl-*p*-phenylenediamine monohydrochloride (EPA) acting as an acceptor, which further reduces the fluorescence intensity of ~546 nm peak (Figure 17.12) because of IFE. Based on this observation, the concentration of the Fe^{3+} ions was quantified by the fluorescence intensity ratio of ~546 and ~758 nm peaks (Chen et al. 2018b).

The UCNPs have shown its potential not only in the detection of fungicides but also in the detection of other pesticides, pathogenic bacteria, mytotoxins, antibiotics, several other metal ions and other food contaminants. A proper discussion of all the methods and materials applied for the detection is beyond the scope of this book. However, several groups of researchers have published excellent reviews related to detection of these food contaminants (Zhang et al. 2018; Yan et al. 2018; Samsidar et al. 2018; Pehlivan et al. 2019; Peltomaa et al. 2020; Levine 2021; Kalyani et al. 2021).

17.3 Detection of Plant Viruses

The field of science that makes the use of nanotechnology to deal with the diagnosis, detection, and management of plant viral diseases is known as nanophytovirology. Among several kinds of plant diseases, those caused by plant viruses are the most dangerous and hard to manage. The reasons behind this problem are their dynamic genome diversity, genome structure, transmission by a variety of vectors (including human), rapid evolution, and broad adaptation potential to the environmental conditions (Goldbach 1998; Lauring et al. 2013; French and Holmes 2020). As compared to the traditional methods, modified NPs of size ranging from 1 to 100 nm such as nanometals, metallic oxides, quantum dots, carbon dots, UCNPs, etc., have proven to be advantageous because of their properties such as better surface-to-volume ratio, smaller size, high tunability and reactivity, etc., which results in better detection properties such as low inputs, enhanced efficiency, and low ecological toxicity (Elmer and White 2018; Fu et al. 2020). The upcoming subsection will provide a brief overview of various mechanisms through which the NPs interact with plant viruses.

17.3.1 Plant Virus Detection/Management Strategies

For the management and detection of plant viruses, mainly four types of strategies are used. All those are summarized in Figure 17.13, and a brief description is provided below.

17.3.1.1 Direct Interactions

In this strategy, the NPs are injected inside the plant species (before the viral infection), which prohibits the duplication of virus in them and hence systematically prevents the virus spread. Several studies have been published, which proves the effectiveness of this strategy. The antiviral activity of SiO_2 and ZnO nanoparticles against TMV has been explored by the researchers (Cai et al. 2019). They suggested that direct interaction of the NPs with viral capsid proteins results in structural disruption that led to TMV aggregation and rapid deactivation of viral particles. A similar result has been obtained for the turnip mosaic virus (TuMV) interacting with the Fe_2O_3 NPs (Hao et al. 2018). Adeel et al. observed that the insertion of carbon-based NPs (carbon nanotubes and C_{60}) in healthy *Nicotiana benthamiana* prevents the replication and spread of TMV (Adeel et al. 2021). Despite the promising results, there is a lack of robustness in several studies because the provided explanations were usually hypothetical. Also, the evolution of infection with time and effective treatment regimes have been left completely unexplored.

17.3.1.2 Indirect Interactions

This strategy focuses on the development and growth of plant species by developing plant defense responses or via enhanced photosynthesis. Several researchers have reported that a precise and well-timed dose of NPs can enhance the secretion of plant hormones such as salicylic acid, jasmonic acid, etc., which is beneficial for the

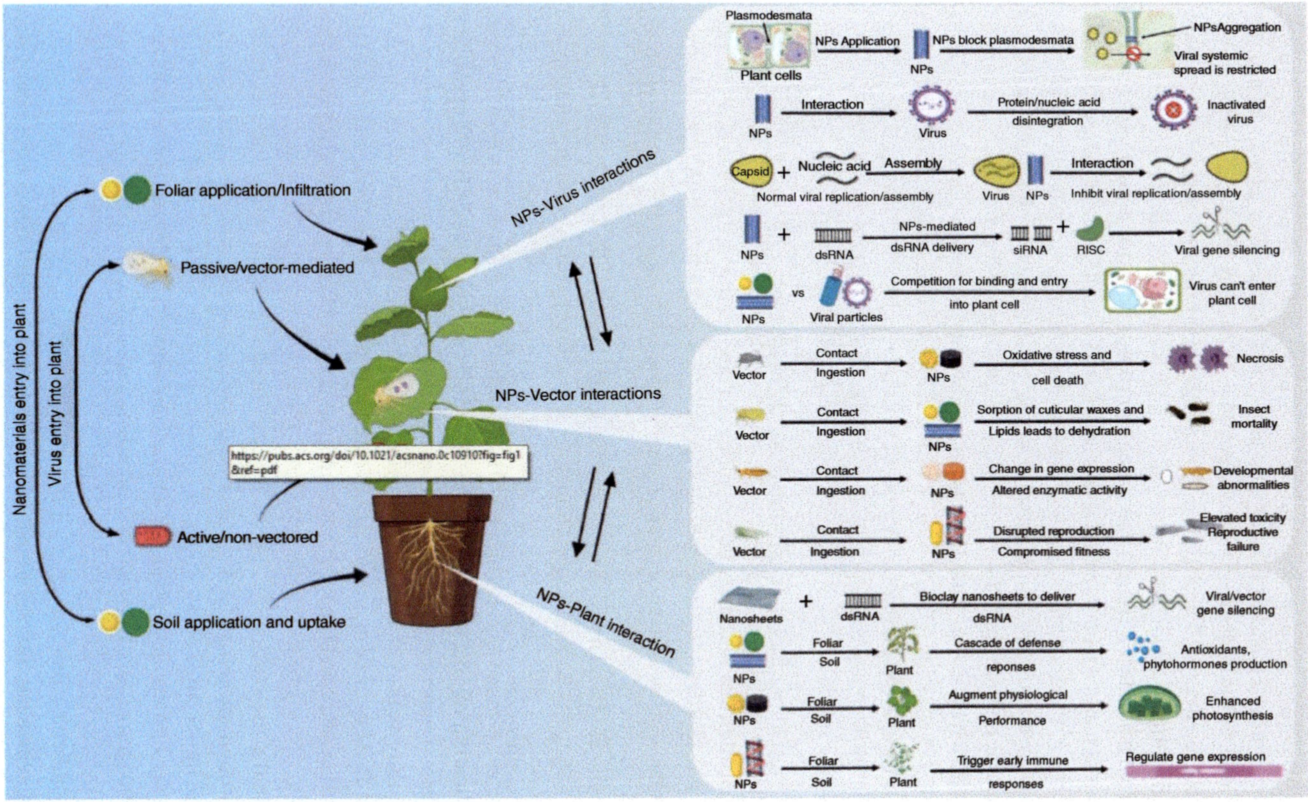

Figure 17.13 Interaction mechanisms between NPs and plant virus vector systems. Source: Farooq et al. (2021). Reprinted with permission of American Chemical Society.

growth and development of plant species (Elmer and White 2018; Hao et al. 2018; Cai et al. 2019; Fu et al. 2020; Adeel et al. 2021). However, the amount of dosing and the number of doses necessary for developing long time defense needs to be explored more.

17.3.1.3 NPs as Biosensors for Virus Detection

The biosensing ability of NPs is one of the most favorite topics among researchers as far as the applications of NPs are concerned. Until now, a number of biosensing probes have been developed for the detection of human and animal viruses using quantum dots, AuNPs, ZnNPs, AgNPs, graphene, and UCNPs (Zhang et al. 2013; Mokhtarzadeh et al. 2017). However, such sensing systems for plant viruses have not been developed much yet. Section 17.3.2 will present some of the developments in the field of plant virus sensing systems, which make use of the ideas applied for developing human and animal virus sensing probes/systems.

17.3.1.4 RNAi Process for Antiviral Protection

RNA interference (RNAi), a biological process that suppresses the effect of pathogens via double-stranded (ds) RNA molecules, is also an effective way to combat with the plant viral diseases. In this strategy, NPs serve as delivery agents, which supply dsRNA to target plants to trigger the RNAi process. Mitter et al. have demonstrated two different works based on this process. In the first work, protection against pepper mild mottle virus (PMMoV) and cucumber mosaic virus (CMV) has been achieved via mechanical inoculation (Mitter et al. 2017). In the next work, the same team has utilized this approach to protect two plants (*Nicotiana benthamiana* and *Vigna unguiculata*) from vectored-mediated virus transmission (Worrall et al. 2019).

17.3.2 Significant Works on Plant Virus Detection Based on UCNPs

The application of NPs in plant viral diseases is an emerging field, and most of the mechanisms regarding this are yet to unfold. To the best of our knowledge, the maiden study regarding the antiviral activity of NPs has been published by Aref et al. (2012). They have demonstrated the in vitro activity of AuNPs against the barley yellow mosaic virus (BaYMV). Because of the action of AuNPs, there occurred a complete dissociation of viral particles. However, there are few drawbacks in this study. Firstly, the efficacy of AuNPs against the BaYMV virus in living plant host was not known properly. Secondly, the absence of ionic or bulk comparisons due to which the importance of particle size of NPs cannot be accessed.

For biosensing applications, quantum dots are preferred over organic dyes and fluorescent proteins because they have qualities such as narrow emission spectra and broad absorption spectra that allow simultaneous excitation of multiple colors and size-tunable emissions. However, quantum dots are usually excited by blue or UV radiation, which can induce autofluorescence from biological tissues, which in turn decreases the sensitivity of detection (Ehlert et al. 2008). The UCNPs get excited from NIR sources that keep them free from the problem of autofluorescence (Wang et al. 2009). Although the UCNPs have several additional advantages as compared to that of other fluorescent materials used for biosensing applications, but until now, the plant virus detection application of UCNPs has not been explored much. There are certain aspects related to the implementation of NPs (especially UCNPs) in plant virus detection, which needs to be investigated in detail. Those aspects have been discussed later on in this chapter. Before that, a brief description of some works regarding the involvement of UCNPs in plant virus detection has been provided. Some works on human and animal virus detection via UCNPs have also been cited because that field is much more developed and the readers may use the knowledge gathered there for the detection of plant viruses.

Mingzhe Zhang and his group synthesized antibody-labeled UCNPs coupled with immunomagnetic separation for detecting tomato ringspot virus (ToRSV), bean pod mottle virus (BPMV), and arabis mosaic virus (ArMV). First of all, Fe_3O_4/SiO_2 magnetic NPs of size ~100 nm and coated with different antibodies have been prepared to capture the target viruses. After that, antibody-conjugated UCNPs/SiO_2 were added to form sandwich complexes (shown in Figure 17.14). The UCNPs have been used as fluorescent signal probes. The measurement was performed using a 980 nm diode laser excitation source. This method removes the problem of autofluorescence as well as facilitates the separation of samples by magnetic forces, thereby showing its potential for detecting plant virus in environmental safety and other fields. Readers can consult with the reference (Zhang et al. 2013) for full details of the work. A DNA-based detection of tomato yellow leaf curl virus (TYLCV) utilizing AuNPs has been reported by A. Razmi et al. (2019).

Lateral flow immunoassays based on the Er^{3+}/Yb^{3+}:$NaYF_4$ core@shell nanoparticles modified with citric acid has been fabricated by Fengping Hou and his team for the detection of foot mouth disease virus (FMDV) (Hou et al. 2020). Recently, the UCNP@mSiO_2-NH_2 nonviral DNA delivery system has been prepared by the co-precipitation method. The flow cytometry result shows that the prepared system has good transfection efficiency (Wang et al. 2020).

Excellent reviews are available, which describe some of the biosensors fabricated till date using quantum dots, graphene, antibodies, UCNPs, etc., for the detection of plant viruses (Hong and Lee 2018; Kumar and Arora 2020; Farooq et al. 2021). Table 17.1 provides a summary of some works.

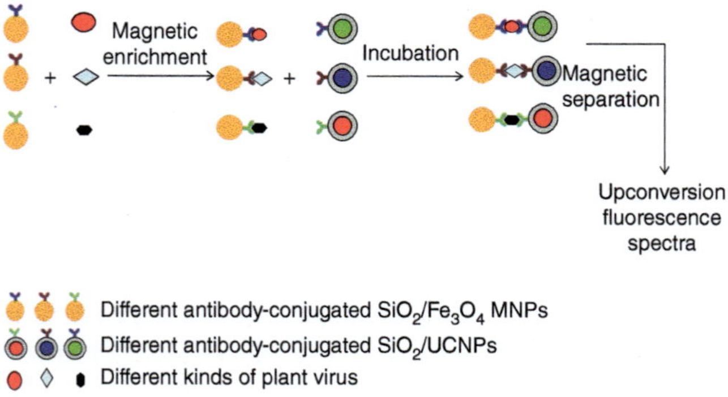

Figure 17.14 Schematic diagram of plant virus detection using antibody-conjugated UCNPs. Source: Zhang et al. (2013).

Table 17.1 Summary of some works performed for plant virus biosensing.

Virus	Nano- and biomolecular conjugate	Limit of detection	Mode of detection
PVY	Monoclonar antibody	0.31 mg/ml	SPR
MCMV	Anti-MCMV antibody	1 ng/ml	SPR
BSMV	Specific oligonucleotide from RNA	14.7 ng/ml	SPR
ASPB	DNA aptamer from coat protein	500 mg/ml	SPR
CPMV	Monoclonal antibody	16 mg/ml	SPR
CPMV	Surface-immobilized CPMV CdSc-ZnS core	—	FRET
CaMV	23 mer derived from CaMV 35S PbS nanoparticle	4.38×10^{-12} mol/L	FRET
CTV	Antibody to CTV coat protein-InP	2 nM for antibody	Electrochemical
CTV	CTV-CP antibody-CdTe	220 ng/ml	Electrochemical
CTV	AuNPs-CTV-CP/Ads-CTV-CP antibody, AuNP/QD	130 ng/ml	FRET
CTV	CTV-CP antibody-CdTe	198 ng/ml	FRET
GVA	Grapevine virus A type proteins ZnO films	1 pg/ml–10 ng/ml	FRET
ToRSV, BPMV, and ArMV	Antibody, Fe_2O_4/SiO_2 MNPs, and SiO_2/UCNPs (upconversion nanoparticles)	100 ng/ml	FRET

Source: Hong and Lee (2018).

17.4 Future Challenges Regarding NP-Based Fungicide and Plant Virus Detection

From the above discussion, one can know that fungicides and plant viruses are major threats to the economy in the agricultural sector. Although the chemical fungicides protect the crops from fungal infections, their heavy dosage and prolonged use can have detrimental effects on crop quality, soil fertility, and human health. On the other hand, viruses, whether plant, animal, or human, are the most dangerous species as it is not easy to tackle viral infections because they tend to adapt and change according to environmental conditions. The ongoing pandemic of COVID-19 is a current example of this tendency of viruses.

As far as the field of optical fungicide detection involving UCNPs (or NPs) is concerned, a lot of research has been performed already using different types of principles, and it has been found that all those sensing mechanisms provide promising results such as high sensitivity, better stability, and accuracy. Even though upconversion-based optical sensors have proved to be potential candidates for fungicide detection, the main problem is their practicality as most of the sensors are still in the phase of laboratory trials and verifying the proof of concept. The fluorescence-based fungicide/pesticide detection systems mainly consist of two units: the recognition unit and the nanoparticles. For the commercialization of such sensing devices, the following points must be taken into consideration (Yan et al. 2018):

- Stability of the recognition unit, i.e. the influence of external conditions such as temperature and pH on it.
- Synthesis of functionalized NPs (or UCNPs) with narrow size distribution, as uneven distribution of NPs will lead to inaccurate detection.

The case of plant virus detection involving NPs is somewhat more complicated, the reason being obvious. For plant virus detection, the synthesis route of NPs or UCNPs must be chosen properly because the synthesis route greatly influences the size, morphology, structure, geometry, dispersity, stability, and other physiochemical properties of NPs (Stankic et al. 2016). The size and charge of NPs influence their entry into host cells and their interactions with biomolecules, whereas their surface charge and shape determine their cellular uptake efficiency (Mody et al. 2010; Shang et al. 2014). The timing and interval of nanoparticle application and dosage are also very crucial factors for controlling and detecting plant viral diseases. It was observed that promoting the growth and development of plants with the help of NPs through biostimulatory effects would be more beneficial rather than treating them after the viral infection (Cai et al. 2019; Adeel et al. 2021). Thus, more attention should be paid to the proper development and growth of plants than to the treatment and detection of viral diseases with the help of NPs. Furthermore, the influence of environmental conditions on the efficacy of the nanoscale antiviral approach should also be examined in detail because to date such studies are very few.

References

Achenbach, J.C., Nutiu, R., and Li, Y. (2005). Structure-switching allosteric deoxyribozymes. *Analytica Chimica Acta* 534 (1): 41–51. https://doi.org/10.1016/j.aca.2004.03.080.

Adeel, M., Farooq, T., White, J.C. et al. (2021). Carbon-based nanomaterials suppress tobacco mosaic virus (TMV) infection and induce resistance in *Nicotiana benthamiana*. *Journal of Hazardous Materials* 404: 124167. https://doi.org/10.1016/j.jhazmat.2020.124167.

Ahmed, F.S., Nagwa, M.S., Nermine, N.A., and Ayah, M. (2017). Fungicidal activity of some metallic ions, fungicides and essential oils for preventing biodeterioration of old manuscripts. *Journal of Advances in Microbiology* 3 (2): 1–14. https://doi.org/10.9734/jamb/2017/33853.

Aref, N., Alkubaisi, N., Marraiki, N., and Hindi, A. (2012). Multi-functional effects of gold nano-particles inducing plant virus resistance crops. In: *Proceedings of the 5th Annual World Congress of Industrial Biotechnology-2012*, vol. 25. Xi'an: BIT Congress.

Balaji, R., Kumar, S., Reddy, K.L. et al. (2017). Near-infrared driven photocatalytic performance of lanthanide-doped $NaYF_4$@CdS core-shell nanostructures with enhanced upconversion properties. *Journal of Alloys and Compounds* 724: 481–491. https://doi.org/10.1016/j.jallcom.2017.07.050.

Brauer, V.S., Rezende, C.P., Pessoni, A.M. et al. (2019). Antifungal agents in agriculture: friends and foes of public health. *Biomolecules* 9 (10): 521–542. https://doi.org/10.3390/biom9100521.

Cai, L., Liu, C., Fan, G. et al. (2019). Preventing viral disease by ZnO NPs through directly deactivating TMV and activating plant immunity in *Nicotiana benthamiana*. *Environmental Science: Nano* 6 (12): 3653–3669. https://doi.org/10.1039/c9en00850k.

Chen, C., Li, C., and Shi, Z. (2016). Current advances in lanthanide-doped upconversion nanostructures for detection and bioapplication. *Advanced Science* 3 (10): 1600029. https://doi.org/10.1002/advs.201600029.

Chen, S., Yu, Y.L., and Wang, J.H. (2018a). Inner filter effect-based fluorescent sensing systems: a review. *Analytica Chimica Acta* 999: 13–26. https://doi.org/10.1016/j.aca.2017.10.026.

Chen, M., Kutsanedzie, F.Y., Cheng, W. et al. (2018b). A nanosystem composed of upconversion nanoparticles and *N*,*N*-diethyl-*p*-phenylenediamine for fluorimetric determination of ferric ion. *Microchimica Acta* 185 (8): 1–8. https://doi.org/10.1007/s00604-018-2902-7.

Dethier, J.J. and Effenberger, A. (2012). Agriculture and development: a brief review of the literature. *Economic Systems* 36 (2): 175–205. https://doi.org/10.1016/j.ecosys.2011.09.003.

Dexter, D.L. (1953). A theory of sensitized luminescence in solids. *The Journal of Chemical Physics* 21 (5): 836–850. https://doi.org/10.1063/1.1699044.

Edgell, K.W., Erb, E.J., Wesselman, R.J., and Longbottom, J.E., and Collaborators(1993). Gas chromatographic/electron capture detection method for determination of chlorinated acids in water: collaborative study. *Journal of AOAC International* 76 (5): 1098–1112. https://doi.org/10.1093/jaoac/76.5.1098.

Ehlert, O., Thomann, R., Darbandi, M., and Nann, T. (2008). A four-color colloidal multiplexing nanoparticle system. *ACS Nano* 2 (1): 120–124. https://doi.org/10.1021/nn7002458.

Elmer, W. and White, J.C. (2018). The future of nanotechnology in plant pathology. *Annual Review of Phytopathology* 56 (1): 111–133. https://doi.org/10.1146/annurev-phyto-080417-050108.

Escudero, D. (2016). Revising intramolecular photoinduced electron transfer (PET) from first-principles. *Accounts of Chemical Research* 49 (9): 1816–1824. https://doi.org/10.1021/acs.accounts.6b00299.

Fan, W., Bu, W., and Shi, J. (2016). On the latest three-stage development of nanomedicines based on upconversion nanoparticles. *Advanced Materials* 28 (21): 3987–4011. https://doi.org/10.1002/adma.201505678.

Farooq, T., Adeel, M., He, Z. et al. (2021). Nanotechnology and plant viruses: an emerging disease management approach for resistant pathogens. *ACS Nano* 15 (4): 6030–6037. https://doi.org/10.1021/acsnano.0c10910.

French, R.K. and Holmes, E.C. (2020). An ecosystems perspective on virus evolution and emergence. *Trends in Microbiology* 28 (3): 165–175. https://doi.org/10.1016/j.tim.2019.10.010.

Fu, L., Wang, Z., Dhankher, O.P., and Xing, B. (2020). Nanotechnology as a new sustainable approach for controlling crop diseases and increasing agricultural production. *Journal of Experimental Botany* 71 (2): 507–519. https://doi.org/10.1093/jxb/erz314.

Gan, T.T., Zhang, Y.J., Zhao, N.J. et al. (2012). Hydrothermal synthetic mercaptopropionic acid stabled CdTe quantum dots as fluorescent probes for detection of Ag^+. *Spectrochimica Acta Part A: Molecular and Biomolecular Spectroscopy* 99: 62–68. https://doi.org/10.1016/j.saa.2012.09.005.

Gergerich, R.C. and Dolja, V.V. (2006). Introduction to plant viruses, the invisible foe. *The Plant Health Instructor* 478. https://doi.org/10.1094/phi-i-2006-0414-01.

Goldbach, R. (1998). All you wanted to know about plant virus control. *Trends in Plant Science* 3 (12): 490. https://doi.org/10.1016/s1360-1385(98)01341-7.

Hao, Y., Yuan, W., Ma, C. et al. (2018). Engineered nanomaterials suppress turnip mosaic virus infection in tobacco (*Nicotiana benthamiana*). *Environmental Science: Nano* 5 (7): 1685–1693. https://doi.org/10.1039/c8en00014j.

Hochreiter, B., Pardo-Garcia, A., and Schmid, J.A. (2015). Fluorescent proteins as genetically encoded FRET biosensors in life sciences. *Sensors* 15 (10): 26281–26314. https://doi.org/10.3390/s151026281.

Hong, S. and Lee, C. (2018). The current status and future outlook of quantum dot-based biosensors for plant virus detection. *The Plant Pathology Journal* 34 (2): 85–92. https://doi.org/10.5423/ppj.rw.08.2017.0184.

Hou, F., Liu, H., Zhang, Y. et al. (2020). Upconversion nanoparticles-labelled immunochromatographic assay for quantitative biosensing. *New Journal of Chemistry* 44 (36): 15498–15506. https://doi.org/10.1039/d0nj03156a.

Hu, S., Cao, H., Wu, X. et al. (2016). Upconversion luminescence and magnetic turning of $NaLuF_4$:Yb^{3+}/Tm^{3+}/Gd^{3+} nanoparticles and their application for detecting acriflavine. *Journal of Nanomaterials* 2016: 1–9. https://doi.org/10.1155/2016/2463479.

Jin, B., Wang, S., Lin, M. et al. (2017). Upconversion nanoparticles based FRET aptasensor for rapid and ultrasensitive bacteria detection. *Biosensors and Bioelectronics* 90: 525–533. https://doi.org/10.1016/j.bios.2016.10.029.

Johnston, B.F. and Mellor, J.W. (1961). The role of agriculture in economic development. *The American Economic Review* 51 (4): 566–593.

Kalyani, N., Goel, S., and Jaiswal, S. (2021). On-site sensing of pesticides using point-of-care biosensors: a review. *Environmental Chemistry Letters* 19: 345–354. https://doi.org/10.1007/s10311-020-01070-1.

Kumar, V. and Arora, K. (2020). Trends in nano-inspired biosensors for plants. *Materials Science for Energy Technologies* 3: 255–273. https://doi.org/10.1016/j.mset.2019.10.004.

Lauring, A.S., Frydman, J., and Andino, R. (2013). The role of mutational robustness in RNA virus evolution. *Nature Reviews Microbiology* 11 (5): 327–336. https://doi.org/10.1038/nrmicro3003.

Lefeuvre, P., Martin, D.P., Elena, S.F. et al. (2019). Evolution and ecology of plant viruses. *Nature Reviews Microbiology* 17 (10): 632–644. https://doi.org/10.1038/s41579-019-0232-3.

Levine, M. (2021). Fluorescence-based sensing of pesticides using supramolecular chemistry. *Frontiers in Chemistry* 9: 27. https://doi.org/10.3389/fchem.2021.616815.

Likhtenshtein, G.I. (1996). Role of orbital overlap and local dynamics in long-distance electron transfer in photosynthetic reaction centres and model systems. *Journal of Photochemistry and Photobiology A: Chemistry* 96 (1–3): 79–92. https://doi.org/10.1016/1010-6030(96)04305-5.

Liu, S., Tian, J., Wang, L., and Sun, X. (2012). Highly sensitive and selective colorimetric detection of Ag (I) ion using 3,3′,5,5′,-tetramethylbenzidine (TMB) as an indicator. *Sensors and Actuators B: Chemical* 165 (1): 44–47. https://doi.org/10.1016/j.snb.2012.02.002.

Long, Q., Wen, Y., Li, H. et al. (2017). A novel fluorescent biosensor for detection of silver ions based on upconversion nanoparticles. *Journal of Fluorescence* 27 (1): 205–211. https://doi.org/10.1007/s10895-016-1947-4.

Lv, Y., Zhu, L., Liu, H. et al. (2014). Single-fluorophore-based fluorescent probes enable dual-channel detection of Ag^+ and Hg^{2+} with high selectivity and sensitivity. *Analytica Chimica Acta* 839: 74–82. https://doi.org/10.1016/j.aca.2014.06.010.

Mei, Q., Jing, H., Li, Y. et al. (2016). Smartphone based visual and quantitative assays on upconversional paper sensor. *Biosensors and Bioelectronics* 75: 427–432. https://doi.org/10.1016/j.bios.2015.08.054.

Mitter, N., Worrall, E.A., Robinson, K.E. et al. (2017). Clay nanosheets for topical delivery of RNAi for sustained protection against plant viruses. *Nature Plants* 3 (2): 1–10. https://doi.org/10.1038/nplants.2016.207.

Mody, V.V., Siwale, R., Singh, A., and Mody, H.R. (2010). Introduction to metallic nanoparticles. *Journal of Pharmacy and Bioallied Sciences* 2 (4): 282. https://doi.org/10.4103/0975-7406.72127.

Mohadesi, A. and Taher, M.A. (2007). Stripping voltammetric determination of silver(I) at carbon paste electrode modified with 3-amino-2-mercapto quinazolin-4(3H)-one. *Talanta* 71 (2): 615–619. https://doi.org/10.1016/j.talanta.2006.05.001.

Mokhtarzadeh, A., Eivazzadeh-Keihan, R., Pashazadeh, P. et al. (2017). Nanomaterial-based biosensors for detection of pathogenic virus. *TrAC Trends in Analytical Chemistry* 97: 445–457. https://doi.org/10.1016/j.trac.2017.10.005.

Ouyang, Q., Wang, L., Ahmad, W. et al. (2021). A highly sensitive detection of carbendazim pesticide in food based on the upconversion-MnO_2 luminescent resonance energy transfer biosensor. *Food Chemistry* 349: 129157. https://doi.org/10.1016/j.foodchem.2021.129157.

Pallas, V. and Garcia, J.A. (2011). How do plant viruses induce disease? interactions and interface with host components. *Journal of General Virology* 92 (12): 2691–2705. https://doi.org/10.1099/vir.0.034603-0.

Panigrahi, S.K. and Mishra, A.K. (2019). Inner filter effect in fluorescence spectroscopy: as a problem and as a solution. *Journal of Photochemistry and Photobiology C: Photochemistry Reviews* 41: 100318. https://doi.org/10.1016/j.jphotochemrev.2019.100318.

Pehlivan, Z.S., Torabfam, M., Kurt, H. et al. (2019). Aptamer and nanomaterial based FRET biosensors: a review on recent advances (2014–2019). *Microchimica Acta* 186 (8): 1–22. https://doi.org/10.1007/s00604-019-3659-3.

Peltomaa, R., Benito-Peña, E., Gorris, H.H., and Moreno-Bondi, M.C. (2020). Biosensing based on upconversion nanoparticles for food quality and safety applications. *Analyst* 146: 13–32. https://doi.org/10.1039/d0an01883j.

Razmi, A., Golestanipour, A., Nikkhah, M. et al. (2019). Localized surface plasmon resonance biosensing of tomato yellow leaf curl virus. *Journal of Virological Methods* 267: 1–7. https://doi.org/10.1016/j.jviromet.2019.02.004.

Roossinck, M.J. (2010). Lifestyles of plant viruses. *Philosophical Transactions of the Royal Society B* 365 (1548): 1899–1905. https://doi.org/10.1098/rstb.2010.0057.

Samsidar, A., Siddiquee, S., and Shaarani, S.M. (2018). A review of extraction, analytical and advanced methods for determination of pesticides in environment and foodstuffs. *Trends in Food Science and Technology* 71: 188–201. https://doi.org/10.1016/j.tifs.2017.11.011.

Sankaran, S., Mishra, A., Ehsani, R. et al. (2010). A review of advanced techniques for detecting plant diseases. *Computers and Electronics in Agriculture* 72 (1): 1–13. https://doi.org/10.1016/j.compag.2010.02.007.

Shang, L., Nienhaus, K., and Nienhaus, G.U. (2014). Engineered nanoparticles interacting with cells: size matters. *Journal of Nanobiotechnology* 12 (1): 1–11. https://doi.org/10.1186/1477-3155-12-5.

Shervedani, R.K. and Babadi, M.K. (2006). Application of 2-mercaptobenzothiazole self-assembled monolayer on polycrystalline gold electrode as a nanosensor for determination of Ag(I). *Talanta* 69 (3): 741–746. https://doi.org/10.1016/j.talanta.2005.11.009.

Shi, J., Tian, F., Lyu, J., and Yang, M. (2015). Nanoparticle based fluorescence resonance energy transfer (FRET) for biosensing applications. *Journal of Materials Chemistry B* 3 (35): 6989–7005. https://doi.org/10.1039/c5tb00885a.

Singer, H. (1979). Policy implications of the lima target. *Industry and Development* 3: 17–22.

Stankic, S., Suman, S., Haque, F., and Vidic, J. (2016). Pure and multi metal oxide nanoparticles: synthesis, antibacterial and cytotoxic properties. *Journal of Nanobiotechnology* 14 (1): 1–20. https://doi.org/10.1186/s12951-016-0225-6.

Strieth-Kalthoff, F., James, M.J., Teders, M. et al. (2018). Energy transfer catalysis mediated by visible light: principles, applications, directions. *Chemical Society Reviews* 47 (19): 7190–7202. https://doi.org/10.1039/C8CS00054A.

Su, Q., Feng, W., Yang, D., and Li, F. (2017). Resonance energy transfer in upconversion nanoplatforms for selective biodetection. *Accounts of Chemical Research* 50 (1): 32–40. https://doi.org/10.1021/acs.accounts.6b00382.

Sun, H., Mei, Q., Shikha, S. et al. (2019). White-light emissive upconversion nanoparticles for visual and colorimetric determination of the pesticide thiram. *Microchimica Acta* 186 (2): 1–9. https://doi.org/10.1007/s00604-019-3231-1.

Sung, Y.M. and Wu, S.P. (2014). Highly selective and sensitive colorimetric detection of Ag(I) using *N*-1-(2-mercaptoethyl) adenine functionalized gold nanoparticles. *Sensors and Actuators B: Chemical* 197: 172–176. https://doi.org/10.1016/j.snb.2014.02.044.

Wang, M., Mi, C., Zhang, Y. et al. (2009). NIR-responsive silica-coated $NaYbF_4$: Er/Tm/Ho upconversion fluorescent nanoparticles with tunable emission colors and their applications in immunolabeling and fluorescent imaging of cancer cells. *The Journal of Physical Chemistry C* 113 (44): 19021–19027. https://doi.org/10.1021/jp906394z.

Wang, C., Li, X., and Zhang, F. (2016). Bioapplications and biotechnologies of upconversion nanoparticle-based nanosensors. *Analyst* 141 (12): 3601–3620. https://doi.org/10.1039/c6an00150e.

Wang, J.H., Chen, H.Y., Chuang, C.C., and Chen, J.C. (2020). Study of near-infrared light-induced excitation of upconversion nanoparticles as a vector for non-viral DNA delivery. *RSC Advances* 10 (67): 41013–41021. https://doi.org/10.1039/d0ra05385f.

Worrall, E.A., Bravo-Cazar, A., Nilon, A.T. et al. (2019). Exogenous application of RNAi-inducing double-stranded RNA inhibits aphid-mediated transmission of a plant virus. *Frontiers in Plant Science* 10: 265. https://doi.org/10.3389/fpls.2019.00265.

Wygladacz, K., Radu, A., Xu, C. et al. (2005). Fiber-optic microsensor array based on fluorescent bulk optode microspheres for the trace analysis of silver ions. *Analytical Chemistry* 77 (15): 4706–4712. https://doi.org/10.1021/ac050856s.

Yan, X., Li, H., and Su, X. (2018). Review of optical sensors for pesticides. *TrAC Trends in Analytical Chemistry* 103: 1–20. https://doi.org/10.1016/j.trac.2018.03.004.

Yang, H., Liu, X., Fei, R., and Hu, Y. (2013). Sensitive and selective detection of Ag^+ in aqueous solutions using Fe_3O_4@Au nanoparticles as smart electrochemical nanosensors. *Talanta* 116: 548–553. https://doi.org/10.1016/j.talanta.2013.07.041.

Yao, C., Wang, J., Zheng, A. et al. (2017). A fluorescence sensing platform with the MnO_2 nanosheets as an effective oxidant for glutathione detection. *Sensors and Actuators B: Chemical* 252: 30–36. https://doi.org/10.1016/j.snb.2017.05.136.

Yao, T., Liu, A., Liu, Y. et al. (2019). Ratiometric fluorescence sensor for organophosphorus pesticide detection based on opposite responses of two fluorescence reagents to MnO_2 nanosheets. *Biosensors and Bioelectronics* 145: 111705. https://doi.org/10.1016/j.bios.2019.111705.

Zhang, X.B., Han, Z.X., Fang, Z.H. et al. (2006). 5,10,15-Tris(pentafluorophenyl) corrole as highly selective neutral carrier for a silver ion-sensitive electrode. *Analytica Chimica Acta* 562 (2): 210–215. https://doi.org/10.1016/j.aca.2006.01.056.

Zhang, M., Chen, W., Chen, X. et al. (2013). Multiplex immunoassays of plant viruses based on functionalized upconversion nanoparticles coupled with immunomagnetic separation. *Journal of Nanomaterials* 2013: 1–8. https://doi.org/10.1155/2013/317437.

Zhang, Y., Jiang, H., and Wang, X. (2015). Cytidine-stabilized gold nanocluster as a fluorescence turn-on and turn-off probe for dual functional detection of Ag^+ and Hg^{2+}. *Analytica Chimica Acta* 870: 1–7. https://doi.org/10.1016/j.aca.2015.01.016.

Zhang, Z., Shikha, S., Liu, J. et al. (2018). Upconversion nanoprobes: recent advances in sensing applications. *Analytical Chemistry* 91 (1): 548–568. https://doi.org/10.1021/acs.analchem.8b04049.

Zhang, J., Zhou, R., Tang, D. et al. (2019). Optically-active nanocrystals for inner filter effect-based fluorescence sensing: achieving better spectral overlap. *TrAC Trends in Analytical Chemistry* 110: 183–190. https://doi.org/10.1016/j.trac.2018.11.002.

Zhang, Z., Feng, Q., Yang, M., and Tang, Y. (2020). A ratiometric fluorescent biosensor based on conjugated polymers for sensitive detection of nitroreductase and hypoxia diagnosis in tumor cells. *Sensors and Actuators B: Chemical* 318: 128257. https://doi.org/10.1016/j.snb.2020.128257.

Zhao, J., Fan, Q., Zhu, S. et al. (2013). Ultra-sensitive detection of Ag^+ ions based on Ag^+-assisted isothermal exponential degradation reaction. *Biosensors and Bioelectronics* 39 (1): 183–186. https://doi.org/10.1016/j.bios.2012.07.073.

Zhu, X., Su, Q., Feng, W., and Li, F. (2017). Anti-Stokes shift luminescent materials for bio-applications. *Chemical Society Reviews* 46 (4): 1025–1039. https://doi.org/10.1039/c6cs00415f.

Zubrod, J.P., Bundschuh, M., Arts, G. et al. (2019). Fungicides: an overlooked pesticide class. *Environmental Science and Technology* 53 (7): 3347–3365. https://doi.org/10.1021/acs.est.8b04392.

18

Upconversion Nanoparticles in Biological Applications

Poulami Mukherjee and Sumanta Kumar Sahu

Indian Institute of Technology (Indian School of Mines), Department of Chemistry, Dhanbad 826004, India

18.1 Introduction

Over the past decade, nanomaterials have gathered attention in different biomedical applications such as targeted drug delivery, bioimaging, biosensors, etc. Those nanomaterials that can envisage the biological process and picturize the interaction between nanoparticles and biological molecules at the cellular level are highly appreciated. They usually possess unique optical, magnetic, or acoustic properties. Therefore, designing multifunctional nanoparticles is growing as a versatile tool in biomedical research such as bioimaging and guided therapy. Various nanomaterials such as metal complex, semiconductor nanocrystals, surface modified silica, organic dyes, and inorganic quantum dots have been used in biomedical applications. These nanomaterials suffer from some limitations such as (i) low tissue penetration because of the excitation and emission wavelength fall in the ultraviolet (UV) range, (ii) toxicity of heavy metal, (iii) normal cell death due to exposure to UV light continuously, and (iv) autofluorescence and low signal to the background ratio. Taking care of the problems stated due to UV excitations, near-infrared (NIR) excitation was considered for the investigation of nanoparticles. Since the late 1990s, upconversion nanoparticles (UCNPs) have been introduced as a new class of luminescent nanomaterials. After that, UCNPs have been widely used in biomedical research because of their remarkable properties. Presently, lanthanide (Ln)-doped upconversion nanocrystals, dye-sensitized UCNPs, and triplet–triplet annihilation (TTA)-based UCNPs are extensively used in biomedical fields such as biosensing, diagnosis, imaging, and cancers treatment (Muhr et al. 2014; Zhao et al. 2011; Ji et al. 2011; Ye et al. 2016). In this chapter, the potential of lanthanide-based UCNPs in biomedical applications has been discussed. Unlike other luminescent materials, UCNPs absorb low-energy NIR light and emit high-energy UV–visible light via photon upconversion. The photon upconversion procedure is carried out under low power excitation, generally continuous wave (CW) laser. The absorption of UCNPs in the NIR region enables deep tissue penetration, which is highly

Upconverting Nanoparticles: From Fundamentals to Applications, First Edition. Edited by Vineet K. Rai.

needed in therapy-based treatment. Moreover, UCNPs show less autofluorescence, as biological samples are optically transparent in this wavelength region and cause low phototoxicity compared to nanomaterials excited in the UV–vis region (DaCosta et al. 2014). They show a sharp emission peak, unlike other fluorescent nanomaterials.

Further, UCNPs possess a longer lifetime than other conventional fluorescent nanomaterials, which makes them more applicable in bioimaging, biosensing, and theranostics. They have excellent photostability and do not suffer from photobleaching, which makes them useful in a biological system by providing continuous luminescence for prolonged exposure to radiation. There are more advantages of using UCNPs such as (i) they can be easily functionalized by biomolecules, (ii) they exhibit minimum non-specific adsorption, (iii) they are easily soluble and dispersible over a wide pH range, and (iv) perturbation of a native molecule in the biological environment by the UCNPs and the coated group is negligible. This chapter includes the fabrication of UCNPs and surface modifications with various inorganic and organic ligands or polymers and their applications in bioimaging, drug delivery, and cancer therapy.

18.2 Upconversion Nanoparticles in Bioimaging

Molecular imaging via photoluminescence (PL) spectroscopy is a vital part of biomedical research. It has opted as a crucial method in medical diagnosis, DNA sequencing, screening, and imaging-guided therapy. It is helpful in a broad area of biological samples, including cells, tissues of a live object, and in vitro and in vivo imaging using PL spectroscopy. Therefore, PL imaging has become an essential tool to visualize biological molecules with subcellular resolution. However, conventional PL imaging agents have limited their abilities to be used in theranostics because of their inappropriate optical properties. They are generally organic fluorophore, metal complex, fluorescent protein, and quantum dots. These materials usually show Stokes shift and excited in the UV–vis range, which has low tissue penetration and low signal-to-background ratio (autofluorescence and light scattering). As the "optical transparency window" of biological tissues lies in the NIR range 700–1100 nm, the UCNPs possess deeper light penetration along with lower autofluorescence, phototoxicity, and photodamage. Thus, UCNPs have emerged as promising bioimaging nanomaterials.

18.2.1 Cell Imaging

In 2008, Chatterjee et al. first reported polyethyleneimine (PEI)-coated UCNPs attached with folic acid (FA) for targeted imaging of human HT29 adenocarcinoma cells and human OVCAR3 ovarian carcinoma cells (Chatterjee et al. 2008). The material shows strong UC emission without significant background autofluorescence when excited at 980 nm. UCNPs have become a reliable imaging probe for their high efficiency of photon upconversion and are used in various cases of cell imaging

(Nam et al. 2011). Wu et al. synthesized core–shell $NaGdF_4:Er^{3+},Yb^{3+}/NaGdF_4$ for multimodal imaging. The upconversion (UC) luminescence was non- photobleaching and non-blinking. Therefore, high cellular uptake of SK-BR-3 cells was found without background imaging. The nanoprobe showed T_1-weighted positive magnetic resonance imaging (MRI) contrast images (Wu et al. 2009). Various cell lines such as HeLa cells (Bai et al. 2008; Cao et al. 2010), HepG2 cells (Das et al. 2010), ovarian cancer cells (Boyer et al. 2010), and Huh7 liver cancer cells (Akiyama et al. 2010) were used for imaging by non-functionalized UCNPs. Further, biomolecular recognition ligands are introduced with UCNPs for targeted imaging of cancer cells. A variety of targeting moieties have been reported. For instance, folic acid is used to target folate receptor, which is overexpressed on different cancer cells such as human HT29 adenocarcinoma cells, human colon cancer cells, and HeLa cells (Xiong et al. 2009; Chatterjee et al. 2008). Folic acid-functionalized UCNPs were prepared through the reaction of 6-aminohexanoic acid-coated $NaYF_4:Yb^{3+}/Er^{3+}$ UCNPs with the carboxylic acid-activated folic acid through the EDC chemistry. The arginyl glycyl aspartic acid (RGD) peptide is attached for $\alpha v\beta 3$ antigen receptor on U87MG cells (human glioblastoma) (Zako et al. 2009), and the rabbit anti-CEA8 antibodies are utilized for carcinoembryonic antigen targeted on HeLa cells (Wang et al. 2009). Thus, attaching targeting ligands with UCNPs lead to enhanced cellular uptake and opens up new opportunities for diagnosis. The imaging of live organisms by using UCNPs was initiated by Lim et al., where the $Y_2O_3:Yb^{3+}/Er^{3+}$ nanoparticle was fabricated for nematode *Caenorhabditis elegans* worms imaging (Lim et al. 2006). When excited at 980 nm, it is clearly visualized that the nanoparticles were well distributed in the intestine of the organism. UCNPs are also used for in vivo trafficking, which is important for the investigation of cell transplants (Idris et al. 2009). Lymphatic node imaging is also a very crucial diagnostic procedure for cancer therapy. This was first exhibited by Kobayashi et al., where researchers have shown that the mouse lymph nodes can be visualized by green and NIR emitting UCNPs (Kobayashi et al. 2009). Further, Wang et al. synthesized three types of UCNPs for multiplexed lymphangiography (Wang et al. 2011b, p. 6145). Multicolor imaging in vivo is also picturized using UCNPs with lanthanide dopant without autofluorescence (Tian et al. 2010). Then, Yu et al. fabricated $NaYF_4:Yb^{3+}/Er^{3+}/La^{3+}$ for in vivo multicolor imaging with deep tissue sensitivity (Yu et al. 2010). Vascular imaging has drawn great attention as it offers important information about vessel configuration and abnormalities. Hilderbrand et al. focused on that topic and reported polyethylene glycol (PEG)-coated $Y_2O_3:Yb^{3+}/Er^{3+}$ for nude mice vascular imaging (Hilderbrand et al. 2009). The polymer is used to minimize the bonding with inexact tissues and extend the particle's circulation half-life in the blood. In the case of in vivo deep tissue imaging, the UC emission wavelength in the UV–vis range can limit the efficacy as biomolecules may absorb it. However, the concept of NIRin-NIRout UCNPs can rule out the problem and provide high-contrast in vivo small animal imaging. This is because the optical transparency window for the biological system lies between 700 and 100 nm. Tm^{3+} ions show this type of two-photon emission between 750 and 850 nm because of the $^3H_4$–$^3H_6$ transition upon excitation at 980 nm. The first report on NIRin–NIRout UC PL was by using

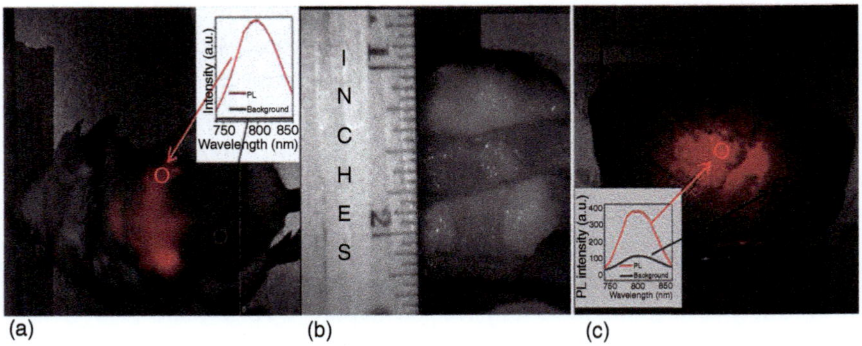

Figure 18.1 (a) Whole-body images of a BALB/c mouse injected via tail vein with the hyaluronic acid coated α-(NaYbF4:0.5% Tm^{3+})/CaF_2 core/shell UCNPs. (b) Bright-field image of the pork tissue (side view), showing the imaging depth and (c) UC PL image of the cuvette containing α-($NaYbF_4$:0.5% Tm^{3+})/CaF_2 core/shell UCNPs covered with a pork tissue. The insets in (a) and (c) show the spectra of the NIR UC PL showing emission at 800 nm. Source: Chen et al. (2012a). Reprinted with permission of American Chemical Society.

$NaYF_4$:Yb^{3+}/Tm^{3+} UCNPs (Nyk et al. 2008). For further improvement, researchers have modified the Tm^{3+}-doped UCNPs (Figure 18.1) (Chen et al. 2012a).

18.2.2 Multimodal Imaging

To overcome the limitation of the single imaging approach, researchers have expanded imaging applications and introduced multimodal imaging concepts, which combine several different imaging processes. UCNPs and their nanocomposites by introducing other functionalities have drawn special attention as multimodal imaging probes for in vivo studies. These kinds of imaging include radiological modalities such as MRI, positron emission tomography (PET), and computed tomography (CT). They can provide anatomic, molecular, physiologic, and genomic information to diagnose the disease more precisely and get knowledge about the treatment response.

For several years, lanthanide-based UCNPs have been used for dual-mode UC PL and MRI. Among them, Gd^{3+} is used as a potential MRI contrast agent. It possesses seven 4f unpaired electrons responsible for changing the spin–lattice relaxation time of proton coordinated or nearby waters in T_1-weighted MRI. The other types of contrast agents are to change the spin–spin relaxation time in T_2 MRI. Gd^{3+} ions can be incorporated into the UCNPs in two ways: (i) co-doping of Yb, Er, and Tm ions in the Gd^{3+}-based host matrix and (ii) the formulation of core–shell nanomaterials where the core shows UC PL and the cell consists of Gd^{3+}-based matrix for MRI. Park et al. first reported PEG-phospholipid-functionalized $NaGdF_4$:Yb/Er nanoparticles for dual-mode imaging of SK-BR-3 cells (Park et al. 2009). Then, other researchers focused on synthesizing nanomaterials based on the first strategy such as $NaYF_4$:Tm^{3+}/Yb^{3+}/Gd^{3+} (Zhou et al. 2010; Johnson et al. 2011) and $NaGdF_4$:Yb^{3+}/$Er^{3+}$$Tm^{3+}$@$SiO_2$-Au@PEG5000 (Xing et al. 2012). On the other hand, nanocomposites built up on the second method include $NaYF_4$:Yb^{3+}/Er^{3+}@$NaGdF_4$, $NaYbF_4$:Tm^{3+}@$NaGdF_4$, $NaYF_4$:Tm^{3+}/Yb^{3+}/Gd^{3+}@mSiO_2 (Liu et al. 2012d), and

$NaYF_4$@SiDTTA-Gd^{3+} (Li et al. 2009). Incorporation of Mn^{2+} into the UCNP lattice is another challenging way to synthesize UC PL and MRI contrast agents. Fe_3O_4 nanoparticles are superparamagnetic in nature. Because of its large magnetic moment, Fe_3O_4 nanoparticles can be combined with UCNPs to act as T_2 MRI contrast agents (Cheng et al. 2011). In this work, authors have synthesized $NaYF_4$:Yb^{3+}/Er^{3+}@Fe_3O_4@Au via a simple layer-by-layer strategy. This multifunctional nanoparticle shows optical and T_2-weighted MRI (Figure 18.2)

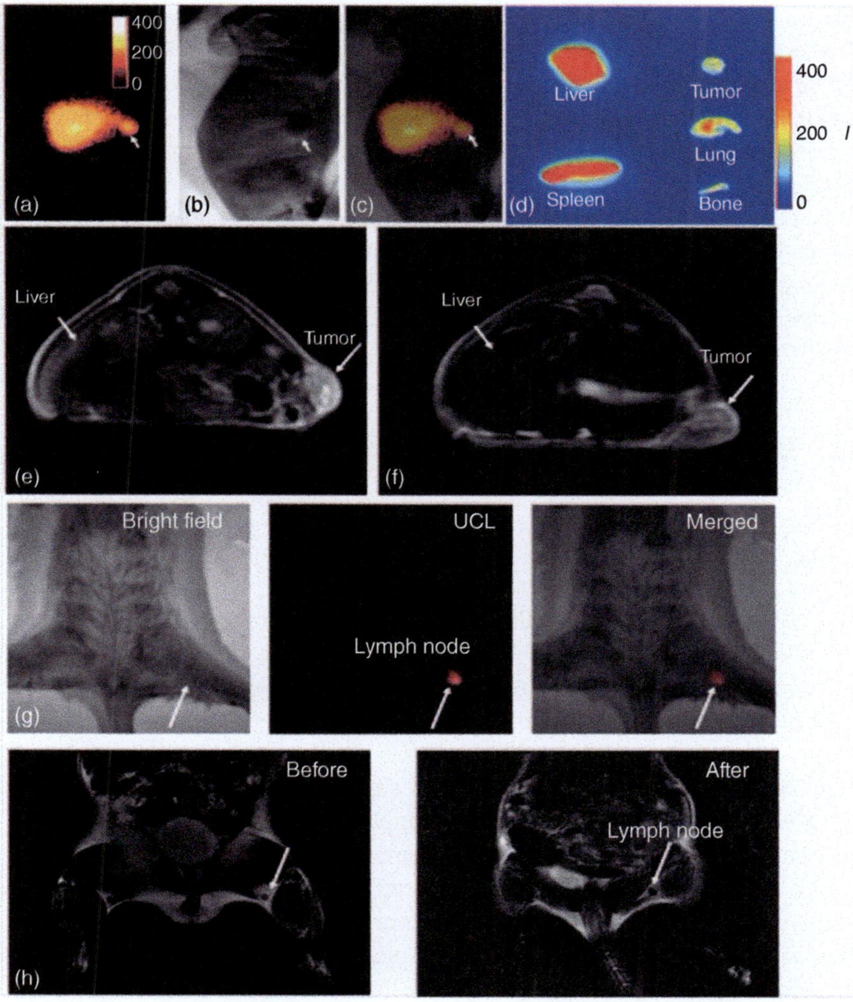

Figure 18.2 (a) The UC PL, (b) bright-field, and (c) merged images of a KB tumor-bearing mouse after 1 hour intravenous injection of PEGylated multifunctional nanoparticles. (d) Ex vivo upconversion photoluminescence (UCPL) imaging showing the accumulation of multifunctional nanoparticles in the liver, spleen, tumor, bone, and lung of the injected mouse after 24 hours of injection. (e and f) T_2-weighted images of KB-tumor bearing nude mice with and without injection of multifunctional nanoparticles. (g) Multimodal UC PL and (h) T2MR imaging for in vivo lymphangiography mapping using multifunctional nanoparticles. Source: Cheng et al. (2011). Reprinted with permission of Wiley.

(Cheng et al. 2011). X-ray CT is an imaging procedure where tomographic images of a particular portion of the body are generated by using computer-processed X-rays. These images are beneficial in biomedical research to diagnose and monitor therapeutic effects as lanthanides possess high atomic number and K-edge values within the X-ray spectrum, thus providing strong X-ray attenuation and showing enhanced CT. Thus, the multipurpose use of UCNPs as UC PL and CT enhancers has become a new aspect in biomedical research. He et al. synthesized $NaGdF_4$ UCNPs as UC PL and CT contrast agents (He et al. 2011). Then, researchers have developed different UCNPs for excellent CT imaging (Liu et al. 2012a; Zhang et al. 2011).

PET is a nuclear medicine imaging technique that produces three-dimensional images in a body by detecting pairs of γ-rays emitted indirectly by a positron-emitting radionuclide in the imaging contrast agents. UC, PL, and PET in a single platform are attracting biomedical applications as they provide more anatomical details of living organisms. Usage of ^{18}F atoms in PET becomes more popular in clinical applications for whole-body imaging. As ^{18}F has a short life, quick coupling occurs into the probe. For example, Sum et al. reported ^{18}F-labeled UCNPs by immersion of UCNPs with ^{18}F-containing solution for five minutes. It has been observed that the ^{18}F-labeled system can be used for in vivo PET imaging and UC (Sun et al. 2011).

18.3 Upconversion Nanoparticles in Drug Delivery

A smart drug delivery system (DDS) is essential to enhance the stability, solubility, targeting ability, biodistribution, and pharmacokinetics of drugs. UCNPs have become a promising candidate in drug delivery as they possess unique optical properties and high optical tunability. The small particle size of UCNPs facilitates endocytosis, which results in higher efficacy in tumor treatment. Moreover, the large surface area of UCNPs causes prolonged drug retention. Because of the tunable functionality of UCNP surface, attaching targeting ligand or other functional moieties can be easily possible to get the significant therapeutic effect and multifunctional usage. Thus, they can overcome the limitation of conventional imaging materials and serve as versatile nanoplatforms in drug delivery. Various UCNPs have been used widely for disease diagnosis and treatment.

UCNPs are synthesized for drug loading, drug targeting to the diseased site, and releasing the drug and transporting to the tumor cells to destroy them. Wang et al. synthesized PEI-modified UCNPs and encapsulated folic acid as a targeting ligand for camptothecin (CPT)/doxorubicin (DOX) delivery and targeted treatment (Wang et al. 2010). Bao et al. synthesized a nanotransport system where UCNPs and DOX were encapsulated in gel nanoparticles decorated with PEI and dimethyl maleic anhydride (DMMA). The nanocomposite shows prolonged drug circulation, more possibility of the drug arriving at the tumor site because of the enhanced permeability and retention (EPR) effect and rapid drug release (Figure 18.3) (Bao et al. 2019).

There are mainly two types of approaches to construct a UCNP-based DDS: (i) core–shell and (ii) hollow sphere with the mesoporous surface. In the first

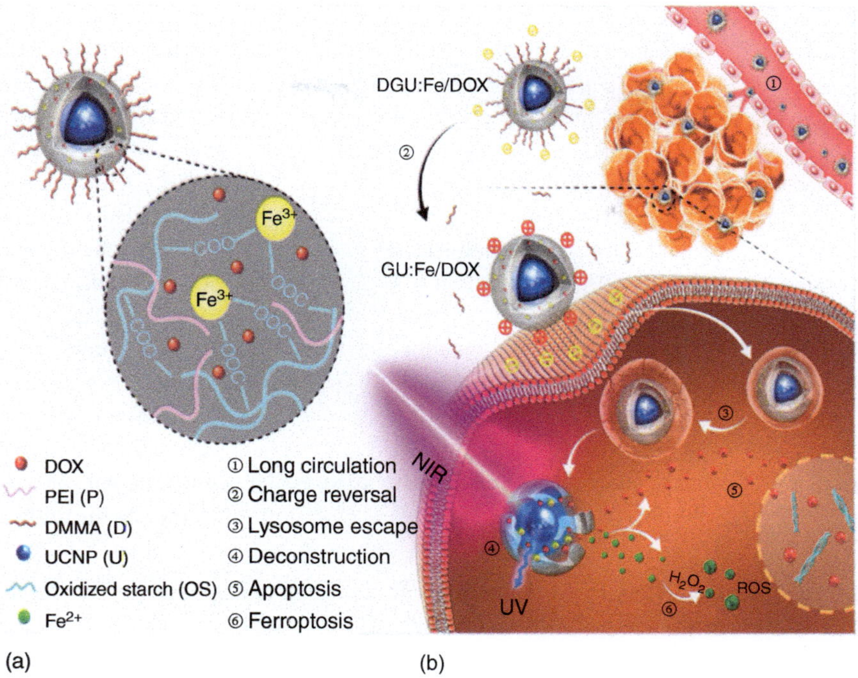

Figure 18.3 Schematic representation of UCNP-based drug delivery system: (A) DOX loaded the Fe^{3+} cross-linked structure of the polymer nanolongan carrier containing UCNPs as the core further modified with PEI and DMMA. (B) Corresponding anti-cancer mechanism for nanolongan. ① Passive accumulation of dimethylmaleic anhydride decorated gel upconversion (DGU):Fe/DOX nanolongan with a long circulation and improved EPR effect. ② pH-activated (pH 6.8) surface shift of negative DGU:Fe/DOX to positive gel upconversion (GU):Fe/DOX nanolongan at tumor site for enhanced cellular uptake. ③ Lysosome escape of GU:Fe/DOX nanolongan by proton sponge effect. ④ NIR-responsive deconstruction of nanolongan under the action of UCNPs. ⑤ Apoptosis of released DOX in the nucleus. ⑥ Ferroptosis of ROS by the Fenton reaction of reduced Fe^{2+} with tumor cellular H_2O_2 at the cytoplasm. Source: Bao et al. (2019). Reprinted with permission of American Chemical Society.

method, the hydrophobic–hydrophobic interaction occurs between the hydrophobic ligand on the UCNP surface and hydrophobic drugs. Thus, hydrophobic drugs are loaded into the core of the hydrophobic pockets on the UCNP surface. For example, the amphiphilic polymer was coated on oleic acid (OA)-capped UCNPs ($NaYF_4$:Yb/Er) to create a hydrophobic pocket and DOX was loaded into that. The system shows pH-responsive, controlled drug release (Figure 18.4) (Wang et al. 2011a, p. 1110). In another work, a nanocomposite was synthesized by conjugating $NaYF_4$:Yb/Er UCNPs with iron oxide, which shows imaging-guided and magnetic targeted drug delivery (Xu et al. 2011a).

Anti-cancer drugs are deposited in the pores of the material coated with UCNPs such as mesoporous silica shells. In this case, the shell material possesses a high surface area and large pore diameter, enabling a high amount of drug accommodation. For example, mesoporous silica-coated β-$NaYF_4$:Yb^{3+}/Er^{3+} UCNPs

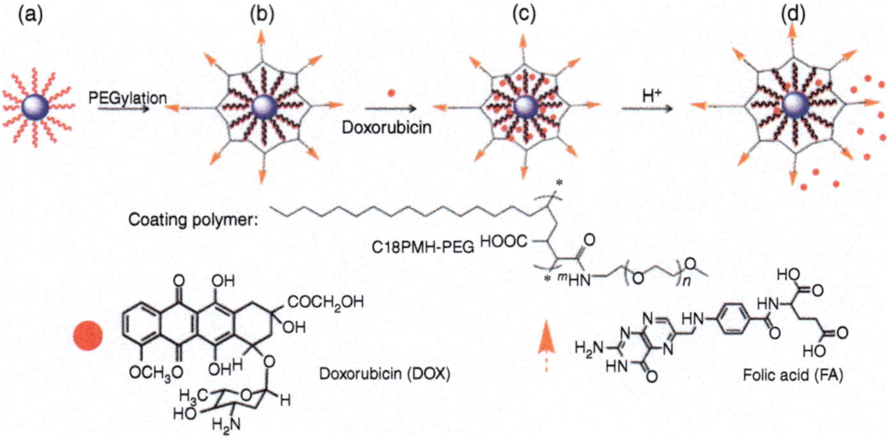

Figure 18.4 Schematic representation of a polymer-coated UCNP-based drug delivery system exhibiting the loading and releasing DOX from UCNPs. Source: Wang et al. (2011a, p. 1110). Reprinted with permission of Elsevier.

were synthesized for ibuprofen delivery via an electrospinning process (Hou et al. 2011). In the second method, hollow UCNPs with a mesoporous shell are used to load the drug. A hollow structure allows sufficient drug loading with satisfactory PL imaging. For example, a monodisperse core–shell Yb(OH)-CO_3@$YbPO_4$:Er^{3+} hollow sphere was developed for DOX loading (Xu et al. 2011b). It is observed that DOX is shuttled into cells by a core/shell hollow sphere carrier. Then, it is released inside the cells after endocytosis. In another study, Yb(OH)-CO_3@$YbPO_4$:Er^{3+} hollow spheres have been developed for DOX delivery (Dong et al. 2013).

18.3.1 Different Types of Surface Modification

18.3.1.1 Polymer Coating

UCNPs have low water dispersibility and short-term colloidal stability, which diminishes the effectiveness of the DDS. To overcome this, UCNPs have been coated with various polymers such as polyvinylpyrrolidone (PVP), PEI, and PEG. Besides providing good hydrophilicity, these polymers also render more biocompatibility and different surface charge as well as good stability. For example, researchers have fabricated hydrophilic PEG-functionalized UCNPs for DOX delivery and imaging as well as MRI. The Yb- and Er-doped $NaYF_4$ core and $NaGdF_4$ shell were decorated and then coated with PEG to enhance hydrophilicity via covalent bonding. The nanocomposite is further conjugated with the trans-activator of transcription (TAT) peptide for nuclear targeting and nuclear transport process. The system was used to improve the efficiency of DOX delivery to the HeLa cells, as it can deliver DOX molecules directly to the HeLa cell (Liu et al. 2012b).

18.3.1.2 Silica Coating

Researchers have paid attention to silica-coated UCNPs as they have good biocompatibility, excellent stability, low cost, high loading capability, and tunable surface functionality. Various DDS consists of mesoporous silica-coated UCNPs because of their remarkable advantages. For example, mesoporous silica was coated on the surface of $\beta NaYF_4:Yb^{3+},Er^{3+}@\beta NaGdF_4:Yb^{3+}$ UCNPs, and core–shell nanostructure was formed followed by further surface modification with PEG. The nanocomposite was utilized in drug delivery, MRI, and in vitro cytotoxicity studies. When the cells were excited at 980 nm, green emission was exhibited, and the cellular uptake of $mSiO_2$–UCNPs–PEG was noticeable (Li et al. 2013). In another research, Yang et al. reported a uniform-sized monodispersed UCNP-based DDS. The mesoporous silica nanocomposite was synthesized with a LaF_3:Yb,Er UCNPs core and a mesoporous silica shell. The nanocomposite shows prominent green UC luminescence when excited with 980 nm laser light. It was used as a potential DDS by using an ibuprofen model drug (Yang et al. 2010).

18.3.1.3 Metal Oxide-Coated UCNPs

Metal oxide-coated UCNPs are used for high tumor targeting, high drug loading, and multifunctional uses. Different metal oxides such as iron oxide (Fe_3O_4), titanium oxide (TiO_2), cerium oxide (CeO_2), zirconium oxide (ZrO_2), copper oxide (CuO), manganese oxide (MnO) nanoparticles are widely used in the biomedical field. Dual-core–shell nanomaterial (Fe_3O_4–$NaYF_4@TiO_2$) was reported for DOX delivery (Shen et al. 2014). TiO_2 acted as a sonosensitizer for sonodynamic therapy (SDT), and $NaYF_4$ nanoparticles (NPs) acted as UC luminescence agents. When the nanocomposite was modified with hyaluronic acid, the cellular uptake increased significantly. Thus, hyaluronic acid modified nanoparticle shows considerable tumor targeting.

18.3.1.4 Functionalization of UCNPs

UCNPs are functionalized with various molecules to increase drug loading or conjugation with particular cancer biomarkers, and β-cyclodextrin (β-CD) is one of them because it can internalize guest molecules by acting as a host. It possesses high solubility with less toxicity. It can render UCNPs good water dispersibility when conjugated with them. A novel β-CD derivative, i.e. 6-phosphate-6-deoxy-β-cyclodextrin, was functionalized on the surface of UCNPs, which shows less cytotoxicity (Ma et al. 2014).

18.3.1.5 Metal–Organic Framework Coating

Metal–organic frameworks are the classes of porous materials that have drawn special attention because of their high surface area. Nanoscale metal–organic frameworks (NMOFs) are widely used with UCNPs as a smart multifunctional DDS. They are used as a unique drug carrier because of their outstanding properties such as high pore volume, large surface area, good biodegradability, tunable functionality, and less cytotoxicity. To date, various UCNP NMOFs have been studied as DDSs. Among them, UIO-66, ZIF-8, IRMOF3, MIL-101, MIL-100, MIL-53-Fe, and MIL-88B are rampantly used. Chowdhuri et al. synthesized folic acid encapsulated core–shell $NaYF_4$:Yb/Er@ZIF-8/FA for tumor targeting and pH-responsive fluorouracil (5-FU) delivery (Chowdhuri et al. 2016). Mukherjee et al. developed magnetic nanocomposites $NaGdF_4$:Yb/Er@MIL-53(Fe)/FA in a facile method for optical imaging and targeted delivery of DOX. The nanocomposites can be used as an MRI contrast agent (Mukherjee et al. 2019).

18.3.2 Drug Release

Drug release is also a very important part of the construction of the DDS. After loading the drug and transporting it to the cell, it should be released to the diseased cell without destroying the normal cells. Therefore, the formulation strategy of the DDS is very crucial in the case of controlled drug release.

18.3.2.1 NIR-Triggered Drug Delivery System

The NIR-triggered DDSs increase the tissue penetration depth and reduce the tissue autofluorescence. Liu et al. reported a DOX-loaded $NaYF_4$:Tm,Yb@$NaYF_4$@SiO_2-azo for NIR-triggered anti-cancer drug release (Liu et al. 2013). The UCNPs emit photons in the UV–vis region under NIR light excitation, which creates rotation–inversion movement of azobenzene that propels the release of the drug. Further, He et al. reported DOX-loaded β-phase $NaYF_4$:Tm,Yb@$NaYF_4$@SiO_2 graphed with the ruthenium complex. Here, the ruthenium complex acts as a photoactive molecular value by absorbing the blue light emitted from UCNPs, facilitating the cleavage of the Ru complex and releasing DOX (He et al. 2015).

18.3.2.2 pH and Thermoresponsive Drug Release

Extracellular microenvironment of tumor tissues has low pH than normal cells. Therefore, they are more acidic in nature than normal cells and blood. Utilizing this phenomenon, researchers have developed a pH-responsive drug release system to overcome normal cell death during anti-cancer drug treatment. Chen et al. developed a pH-responsive DDS eccentric-(concentric-UCNPs@SiO_2)@PAA (polyacrylic acid) and eccentric UCNPs@PAA@SiO_2 nanocarriers. The system exhibits ultra-high drug storage capacity and sensitive pH-responsive properties (Chen et al. 2013). Carboxyl methyl chitosan, methyl acrylic acid (MAA), can act as pH-responsive DDSs. Moreover, the diseased tissues show comparatively high temperatures than the normal tissues. Some polymers can shrink at high temperatures. Dai et al. developed a novel stimuli-responsive smart DDS where $NaYF_4$:Yb/Er UCNPs are coated with hydrogel polymer shell: P(NIPAM-*co*-MAA) where PNIPAM (poly(*N*-isopropylacrylamide)) is a thermoresponsive polymer and MAA is a pH-responsive polymer. Thus, when an anti-cancer drug DOX release study is performed, the system shows both pH and thermoresponsive drug release (Dai et al. 2012).

18.4 Upconversion in Photodynamic Therapy

Photodynamic therapy is an emerging technique for diagnosis and treatment, which involves the administration of photo-triggered chemical substances known as photosensitizers (PSs). Photodynamic therapy involves mainly three components: light source, the PS, and oxygen within the tissues of the disease site. In this process, the PS absorbs the light and becomes activated to a short-lived excited singlet state and then undergoes intersystem crossing by spin flipping and thereby producing a long-lived excited triplet state. Then, in the presence of triplet molecules, especially molecular oxygen, the PS triggers a number of photochemical reactions and relaxes to the ground state. There are two types of photochemical reactions: type I and type II reaction. In type I reaction, the triplet-state PS directly interacts with intracellular molecules such as protein, nucleic acid, and lipids and generates radicals that again react with O_2 or H_2O to produce oxygenated products such as hydroxyl radicals ($OH\cdot$), superoxide anion radicals ($O_2\cdot^-$), and hydrogen peroxides (H_2O_2). In type II reaction, the energy of excited PS is transferred to the triplet ground-state molecular oxygen and thereby produces singlet-state oxygen ($^1O_2$). The reactive oxygen species (ROS) formed in photodynamic therapy (PDT) is responsible for tumor cell death and therapeutic effect. Most of the work on PS demonstrates through type II mechanism and $^1O_2$ plays a predominant role in tumor tissue destruction. There are two factors for the effectiveness of photodynamic therapy. In the first one, PS should accumulate at the tumor region along with the confined illumination of the light source. The other factor is the competence of $^1O_2$ production and the availability of molecular oxygen in the neighboring tissues. For selectivity in the tumor cells, amphiphilicity of PS is an important property as it requires some hydrophobicity when travels to the target tissues. Then, the PS needs to bind to the tumor cells for which some lipophilicity is necessary (Lucky et al. 2015). Furthermore, the PS must have negligible toxicity, capacity to absorb light, the high quantum yield of the triplet state, and a high lifetime in the triplet state. PSs are divided as first-generation, second-generation, third-generation, and fourth-generation PS. The first generation of PS is basically porphyrin-based PSs, which were synthesized in the year 1970s and 1980s. Porphyrins have two ranges of absorption. The 380–500 nm region is called the Soret band and the 500–750 nm region is called Q-bands. The second-generation PS includes porphyrinoid compounds consisting of porphyrin or porphyrin-based macrocyclic molecules such as chlorins, phthalocyanines, bacteriochlorins, texaphyrins along with their metalated derivatives, and nonporphyrinoid compounds such as anthraquinones, phenothiazines, and xanthenes. These PS possess absorption maxima in longer wavelengths, usually at greater than 630 nm. Therefore, they provide deeper penetration of tissues, high extinction coefficient, and produce a higher amount of $^1O_2$. In the third-generation PS, PSs are conjugated with several biomolecules such as an antibody, peptide, and somatostatin analog to gain better selectivity in tumor targeting. The fourth-generation PS includes the synthesis of nanoformulations. Incorporation of nanoparticles as a carrier of PS is done either by encapsulating the PS to the nanoparticle system or by physically adsorbed or covalently attached to the nanoparticle surface. They can overcome the previously synthesized PSs and provide selective delivery of PS to the tumor tissues and light absorption of PS to the longer wavelength.

Most of the PS used in PDT are activated by visible or UV light. The therapeutic effect of PDT for tumors has been hampered by the limitation of tissue penetration depth of visible light. Moreover, the majority of tissue chromophore shows strong absorption in the visible region. Thus, the clinical application of PDT is hindered, causing incomplete treatment of tumor and relapse of tumor. The biological tissues have negligible light absorption in the NIR region, ranging from 700 to 1100 nm, which is called the window of optical transparency and is ideal for imaging as well as phototherapy. UCNPs are capable of converting NIR light to visible light, which can excite the PS by energy transfer.

The idea of UCNPs in PDT was first incorporated by Zhang et al. They coated a thin layer of silica on $NaYF_4$:Yb^{3+}/Er^{3+} UCNPs with further doping merocyanine-540 as PS. Then, the nanostructure was conjugated with tumor-targeting antibody for targeted PDT under NIR light. It is revealed that when the NIR light is irradiated upon PS-impregnated UCNPs, the viability of Michigan Cancer Foundation-7 (MCF-7)/AZ bladder cancer cells can be significantly decreased (Zhang et al. 2007).

The PS incorporation in the silica shell is extensively practiced in the case of UCNP-based PDT systems because the layer of silica is stable and resistant to pH-dependent or enzymatic degradation (Zhang et al. 2007; Chen et al. 2012b). Moreover, the pore size of the silica shell can be tuned easily so that PS sizes can be adjusted. Further, Qian et al. reported mesoporous silica-coated $NaYF_4$:Yb^{3+},Er^{3+} for zinc(II)-phthalocyanine (ZnPc) PS loading. ZnPc was loaded into the silica layer via physical adsorption. ZnPc-encapsulated mesoporous silica is stable and produces singlet oxygen ROS when illuminated by 980 nm (Figure 18.5) (Qian et al. 2009).

The covalently attached PS provides more stability as they are linked by chemical bonding. The advantage of this method is that the PS need not be released from the system to exert a cell-killing effect. Liu et al. described photodynamic therapy by covalently attaching the carboxyl group of hyaluronic acid functionalized PS Rose Bengal (RB) with an amino group on the surface of UCNPs to form an amide bond (Liu et al. 2012c). The PS loading efficiency and energy transfer from nanoparticles to PS become improved. UC PL imaging and in vitro PDT treatment were observed in a single UCNPs-PS system (Liu et al. 2012c). PDT treatment has also effectively worked in the case of in vivo. Wang et al. synthesized a PDT system where PEGylated amphiphilic polymer-coated β-$NaYF_4$:Yb^{3+}/Er^{3+} nanoparticles were non-covalently attached with the porphyrin derivative chlorine 6 (Ce6) (Wang et al. 2011b, p. 6145). When NIR light is irradiated, highly efficient PDT treatment of tumors and tumor cell death was observed. The nanocomplex causes 70% tumor elimination without

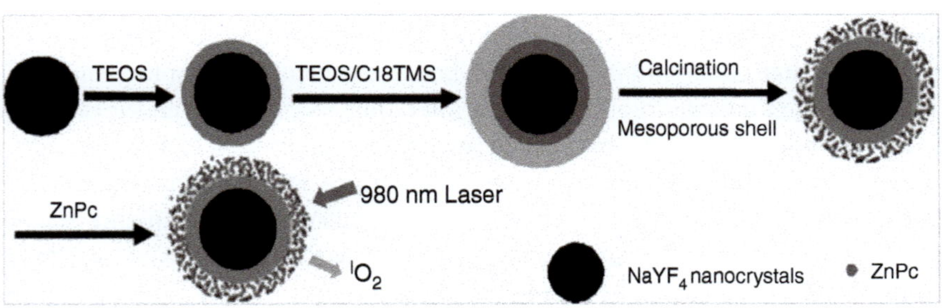

Figure 18.5 Schematic representation of ZnPC PS-loaded mesoporous silica shell-loaded UCNPs for NIR-induced photodynamic therapy. Source: Qian et al. (2009). Reprinted with permission of Wiley.

re-growth within two months. It is also shown that UCNP-Ce6 shows greater penetration depth when irradiated by NIR than only Ce6 by visible light. Further, the UCNP nanocomplex is gradually clear without showing significant toxicity. Then, the tumor selectivity has been achieved by attaching a targeting ligand, which helps to accumulate UCNP-based nanocomplex to the tumor site via receptor recognition. Researchers have synthesized mesoporous silica-coated UCNPs with a tumor-targeting ligand that allows selective accumulation in tumors and provides high loading of PS (Qian et al. 2009). Further, the therapeutic effect of single PS can be enhanced by incorporating two different kinds of PSs to absorb both green and red emission peaks of UCNPs. Idris et al. developed mesoporous silica-coated UCNPs for ZnPc and merocyanine 540 (MC540) loading. The absorbance peak of two PS overlaps with the two UC PL emission peaks. Thus, dual PS-loaded UCNPs render greater PDT efficiency. When the ZnPc/MC540 co-loaded UCNPs conjugated with folic acid (FA), the targeted PDT efficiency of melanoma tumor is observed. It is revealed that in vivo delayed tumor growth occurs after injecting the nanocomplex, followed by NIR irradiation, which was the first evidence of the study of UCNP-based tumor-specific PDT treatment (Idris et al. 2012).

18.4.1 Surface Modification of UCNPs for PDT

The surface chemistry of UCNPs establishes the stability, solubility, pharmacokinetic property, and ability to adjust targeting ligand and PS. After the synthesis of hydrophobic UCNPs, they should be imparted some hydrophilicity to utilize in the biomedical field. This is performed by modification of the core by a stabilizer or surface coating. The UCNP surface can be coated by PEG, PEI, PEG block copolymers, bovine serum albumin (BSA), or chitosan.

PEI coating is advantageous because it can be synthesized in the one-pot hydrothermal method. Moreover, the synthesized amine-functionalized UCNPs can be easily covalently conjugated with other functional groups via EDC (1-ethyl-3(3-(dimethylamino)propyl-carbodiimide hydrochloride, NHS (*N*-Hydroxysulfosuccinimide) chemistry. Chatterjee et al. synthesized PEI-modified ZnPC-loaded $NaYF_4$ nanocrystal for HT29 human colonic adenocarcinoma cell treatment. The physical adsorption of non-polar ZnPC on the highly polar UCNPs leads to 97% PS loading. Moreover, the folic acid conjugation renders selective targeting of HT29 cells via folate receptor (overexpressed on that cancer cells)-mediated endocytosis. In vitro tumor destruction occurs when the nanocluster is excited by 980 nm NIR light for five minutes (Chatterjee and Yong 2008). However, as PEI is a toxic polymer, PEGylation is widely used in biomedical research instead of PEI. It can improve the water dispersibility of UCNPs as well as frequently removed from the immune cells. Ungun et al. demonstrated thiamine pyrophosphate (TPP)-encapsulated PEGylated $NaYF_4$:Yb^{3+},Er^{3+} UCNPs, which produced a higher amount of singlet oxygen. The PEG coating enhanced the oxygen penetration and diffused the produced singlet oxygen and progressed the biocompatibility (Ungun et al. 2009). There is one limitation of using PEGylation: it takes multiple steps to synthesize. Therefore, the PEG block copolymers PEG-*b*-PCL (polycaprolactone) and PEG-*b*-PLA (polylactic acid) have been introduced. Further, Wang et al. developed PEG-grafted poly(maleic anhydride-*alt*-1-octadecene) (C18PMH-PEG) amphiphilic polymer with imparting UCNPs brilliant water solubility (Wang et al. 2011b, p. 6145). In another research, the UCNPs were coated with PEG chains and charge-reversible polymer consisting of dimethylmaleic acid (DMMA) groups via the layer-by-layer (LbL) technique. In this process, the electrostatic interaction occurs where the outer PEG-coated surface is negatively charged at normal physiological pH 7.4 and switched to a positively charged bare surface at 6.8. Thus, the system helped to increase tumor-homing properties (Wang et al. 2013). Chitosan is tremendously used in biomedical research and also for wrapping of UCNPs among all biodegradable polymers. As a positively charged non-antigenic and hydrophilic polymer, chitosan has worth using in the biological applications, as it can accumulate within the negatively charged tumor cell membrane via electrostatic interaction. Further, the amphiphilic nature of chitosan is favorable to anchor various hydrophobic PSs such as pyropheophorbid-a (PPA), Ce6, and different targeting moieties.

Cui et al. reported chitosan-wrapped ZnPc PS-loaded $NaYF_4:Yb^{3+},Er^{3+}$ UCNPs via a hydrophobic interaction, which happens to be one of the enhanced PS payloads of 10.8%. It is revealed that when the nanocomposite is irradiated at 980 nm with 0.6 W power for 10 minutes in vitro, PDT was observed with 80% of MCF-7 breast cancer cell killing after 24 hour incubation. The cell viability of chitosan-wrapped UCNPs was 80% (Cui et al. 2012). Modification of UCNP surface with silica coating is very much stable as silica is resistant to enzymatic degradation. Moreover, the loading of PS into the silica shell is extensively practiced in photodynamic therapy because of their tunable pore size. Silica modification often causes particle agglomeration in physiological conditions. Therefore, mesoporous silica coating was advantageous over simple amorphous silica coating, which favors the hydrophilic PS loading. Researchers have synthesized the UCNPs coated with mesoporous silica, loaded with ZnPC PS, and studied in vitro PDT treatment of MB49-prostate-specific antigen (PSA) cells. About 100 μg/ml concentration of nanocomposites caused 70% cell destruction upon exposure to 980 nm NIR laser light with power 0.5 W for five minutes (Qian et al. 2009).

Idris et al. coated the surface of UCNPs $NaYF_4$:Yb/Tm with photocatalyst TiO_2 to achieve a uniform core–shell structure. Upon 980 nm excitation, TiO_2 produced more than one type of ROS such as superoxide anion, hydroxyl radical, hydrogen peroxide, etc. In vitro study confirmed the ability of generated ROS to destruct the cancer cells (Idris et al. 2014).

Multifunctional nanocarriers were synthesized by introducing a NMOF for the UCNP-based PDT system. NMOFs are a new class of nanomaterials that consist of metal ions and multiple organic linkers. They possess functional tunability, highly adjustable structure, good biocompatibility, and high porosity. The NMOFs in the nanoplatform play varieties of important roles for PDT. They can not only impart the stability or biocompatibility but also hold PS and encapsulate drug molecules. Sometimes, porphyrin-based metal organic frameworks (MOFs) can themselves act as a PS and can also play a pivotal role in acting as an O_2 reservoir.

Shi et al. reported a multifunctional nanoplatform based on UCNPs and meso-tetra (4-carboxyphenyl) porphine (TCPP)-MOF covered with ultra-small TiO_2 nanoparticles named as (upconversion metal organic framework [UMOF]-TiO_2). When the nanoplatform was illuminated with 980 nm laser light, the emitted ultraviolet and visible light from UCNPs can stimulate MOFs and TiO_2 to produce different kinds of cytotoxic ROS with a cause of in vitro and in vivo cell death (Shi et al. 2020). Xie et al. synthesized a pH-responsive DDS for tumor hypoxia and chemo-photodynamic therapy. $NaYF_4$:Yb/Er@$NaYbF_4$:Nd@$NaGdF_4$ UCNPs were prepared, which show UC luminescence and MRI. Then, the UCNPs were coated with mesoporous silica, which loaded the PS RB. The nanocomposite was further modified with the zeolite imidazole framework (ZIF-90), which performed as an O_2 reservoir and releases O_2 in the presence of tumor environment, which enhances tumor hypoxia and increases the PDT efficiency. Folic acid and anti-cancer drug DOX were further conjugated covalently on the surface of nanocomposites for targeted drug delivery (Xie et al. 2018).

18.5 Photothermal Therapy

Photothermal therapy is a minimally invasive treatment that includes photoabsorbers, which generate heat after light absorption to kill cancer cells. The non-invasiveness and spatial specificity made photothermal therapy (PTT) an advantageous and alternative therapeutic platform over surgery, chemotherapy, or other procedures. Nanomaterials that have high optical absorption have been successfully used in PTT. Gold nanostructure, palladium nanosheet, carbon nanomaterials such as grapheme, carbon nanotubes, or organic nanoparticles with conductive polymers are generally used as photothermal agents. UCNPs cannot convert light into heat directly as lanthanide ions have low extinction coefficients. However, they can be conjugated with plasmonic nanoparticles with strong extinction coefficients to achieve PTT ablation of cancer cells. UCNPs absorb photon energy and dissipate the absorbed energy to the surrounding solution or transfer energy to the PTT reagent, leading to local overheating followed by protein denaturation and cell killing. PTT reagents show high absorbance in the tissue-transparent NIR optical window. PTT reagents can also be designed for targeting elements. Thus, local irradiation at the diseased site causes dual selective treatment. Therefore, PTT causes good specificity, high efficacy of cancer cell removal, and low incidental side effect.

Various multifunctional nanoparticles have been reported with PL, magnetic properties, and photothermal effect. Superparamagnetic iron oxides are widely used in MRI because of their special magnetic properties and good biocompatibility. They can be manipulated by an external magnetic field. Taking the idea from this, Cheng et al. developed ultra-small superparamagnetic Fe_3O_4 nanoparticles coated with 200 nm-sized $NaYF_4$:Yb^{3+}/Er^{3+} UCNPs. The nanocomposite was further coated with a thin gold shell via seed-induced reduction growth. The superparamagnetic iron oxide nanoparticles (IONPs) helped to afford magnetic properties as well as reduced the fluorescence quenching effect of gold nanostructure to UCNPs. Thus, UCNP@IONP@AuMFNPs were successfully used for magnetically targeted and dual-mode imaging-guided PTT (Cheng et al. 2011). Because of the good plasmonic resonance performance of silver, it can be used in PTT treatment. Dong et al. synthesized silver-coated β-$NaYF_4$:Yb/Er nanoparticles and utilized them in PTT experiments. PTT was performed with HepG2 cells and BCap-37 cells. Upon exposure to 980 nm NIR light, the optimum mortality approaches 95% after reaching the power density to 1.5 W/cm^2, which is lower than the reported gold nanorods and gold nanoshells (Dong et al. 2011). Then, researchers have tried to draw attention to cancer cell-targeted PTT. Those multifunctional nanoparticles were further conjugated with folic acid and the final nanoprobe was used for upconversion luminescence (UCL), MR imaging, and carcinoma KB cell targeting and magnetically guided PTT. The PEGylated nanoprobe was used for in vitro dual-modal UCL/MR imaging and for targeted PTT of HeLa cells as well as in vivo dual-mode UCL/MR imaging of KB tumor-bearing mice. Upon exposure to the 808 nm NIR light, tumors of the mice were completely eliminated after multifunctional nanoprobe injection and magnetic tumor targeting as the synthesized Au-shelled UCNP-based nanoprobes have strong NIR optical absorption (Cheng et al. 2012). Further, carbon-coated upconversion-based nanocomposites $NaLuF_4$:Yb,Er@$NaLuF_4$@Carbon (csUCNP@C) with diameter 77 nm were prepared by Zhu et al. It is revealed that the carbon shell was a brilliant photothermal agent and can heat the nanocomposites when irradiated by 730 nm laser light. The synthesized nanocomposites destroyed the HeLa cells selectively without killing the adjacent cells (Zhu et al. 2016). Huang et al. have synthesized core–shell nanoplates $NaYF_4$:Yb/Er@polypyrrole (PPy). The nanocomposites exhibited efficient PTT for in vitro and in vivo HeLa cells as well as showed CT imaging, UCL imaging, and infrared thermal imaging (Huang et al. 2016). $LiYF_4$:Yb^{3+}/Tm^{3+}@SiO_2 UCNPs were coated with a layer of chitosan hydrogel via cross-linking with a photocleavable cross-linker. The material can act as an excellent nanotransducer. The nanocomposite was also a good drug carrier (Jalani et al. 2016). Ultrasmall plasmonic CuS nanoparticles were decorated on silica-coated UCNPs. These nanocomposites offered effective PTT thermal ablation and increased localized radiation dose, which improve radiation therapy (Xiao et al. 2013). Cai et al. reported a smart nanocarrier $NaGdF_4$:Yb/Er@$NaGdF_4$ coated with mesoporous silica and further conjugated with gold nanoparticles DOX delivery, NIR-controlled drug release, and synergistic photothermal therapy (Figure 18.6). Under the 980 nm irradiation, it has been observed that due to the high photothermal efficiency of Au NPs, the release of DOX was enhanced. The reason behind the thermal effect is that green emission of UCNPs when excited by NIR light was absorbed by Au NPs via fluorescence resonance energy transfer (FRET). Thus, the nanoclusters demonstrated a superior synergistic therapy in cancer cells that combined chemotherapy and photothermal therapy in a single platform compared to a particular therapy method (Cai et al. 2017).

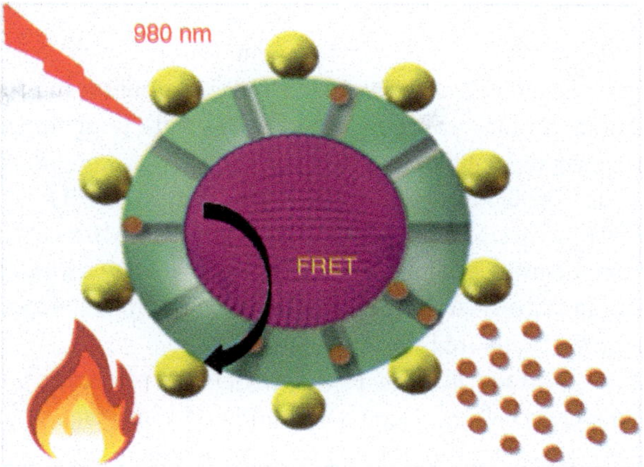

Figure 18.6 Photothermal therapy and NIR-controlled drug release of nanocomposites (the emulsion-based self-assembled 3D upconversion nanoparticle clusters of ($NaGdF_4$:Yb/Er@$NaGdF_4$) EBSUCNPs@SiO_2–DOX@Au. Source: Cai et al. (2017). Reprinted with permission of American Chemical Society.

References

Akiyama, H., Tokuzen, K., Otsuka, H. et al. (2010). Application of liposome-encapsulated ceramic phosphors for cancer cell imaging under near infrared excitation. *Journal of Physics: Conference Series* 232 (1): 012001. IOP Publishing. https://doi.org/10.1088/1742-6596/232/1/012001.

Bai, Y., Yang, K., Wang, Y. et al. (2008). Enhancement of the upconversion photoluminescence intensity in Li+ and Er^{3+} codoped Y_2O_3 nanocrystals. *Optics Communications* 281 (10): 29302932. https://doi.org/10.1016/j.optcom.2008.01.014.

Bao, W., Liu, X., Lv, Y. et al. (2019). Nanolongan with multiple on-demand conversions for ferroptosis–apoptosis combined anti-cancer therapy. *ACS Nano* 13 (1): 260–273. https://doi.org/10.1021/acsnano.8b05602.

Boyer, J.C., Manseau, M.P., Murray, J.I., and Van Veggel, F.C. (2010). Surface modification of upconverting $NaYF_4$ nanoparticles with PEG–phosphate ligands for NIR (800 nm) biolabeling within the biological window. *Langmuir* 26 (2): 1157–1164. https://doi.org/10.1021/la902260j.

Cai, H., Shen, T., Kirillov, A.M. et al. (2017). Self-assembled upconversion nanoparticle clusters for NIR-controlled drug release and synergistic therapy after conjugation with gold nanoparticles. *Inorganic Chemistry* 56 (9): 5295–5304. https://doi.org/10.1021/acs.inorgchem.7b00380.

Cao, T., Yang, T., Gao, Y. et al. (2010). Water-soluble $NaYF_4$:Yb/Er upconversion nanophosphors: synthesis, characteristics and application in bioimaging. *Inorganic Chemistry Communications* 13 (3): 392–394. https://doi.org/10.1016/j.inoche.2009.12.031.

Chatterjee, D.K. and Yong, Z. (2008). Upconverting nanoparticles as nanotransducers for photodynamic therapy in cancer cells. *Nanomedicine (London)* 3 (1): 73–82. https://doi.org/10.2217/17435889.3.1.73.

Chatterjee, D.K., Rufaihah, A.J., and Zhang, Y. (2008). Upconversion fluorescence imaging of cells and small animals using lanthanide doped nanocrystals. *Biomaterials* 29 (7): 937–943. https://doi.org/10.1016/j.biomaterials.2007.10.051.

Chen, G.Y., Shen, J., Ohulchanskyy, T.Y. et al. (2012a). (R-$NaYbF_4$:Tm^{3+})/CaF_2 core/shell nanoparticles with efficient near-infrared to near-infrared upconversion for high-contrast deep tissue bioimaging. *ACS Nano* 6: 828. https://doi.org/10.1021/nn302972r.

Chen, F., Zhang, S., Bu, W. et al. (2012b). A uniform sub-50 nm-sized magnetic/upconversion fluorescent bimodal imaging agent capable of generating singlet oxygen by using a 980 nm laser. *Chemistry A European Journal* 18 (23): 7082–7090. https://doi.org/10.1002/chem.201103611.

Chen, L., Li, L., Zhang, L. et al. (2013). Designed fabrication of unique eccentric mesoporous silica nanocluster-based core–shell nanostructures for pH-responsive drug delivery. *ACS Applied Materials & Interfaces* 5 (15): 7282–7290. https://doi.org/10.1021/am401627u.

Cheng, L., Yang, K., Li, Y. et al. (2011). Facile preparation of multifunctional upconversion nanoprobes for multimodal imaging and dual-targeted photothermal therapy. *Angewandte Chemie International Edition* 50: 7385–7390. https://doi.org/10.1002/ange.201101447.

Cheng, L., Yang, K., Li, Y. et al. (2012). Multifunctional nanoparticles for upconversion luminescence/MR multimodal imaging and magnetically targeted photothermal therapy. *Biomaterials* 33 (7): 2215–2222. https://doi.org/10.1016/j.biomaterials.2011.11.069.

Chowdhuri, A.R., Laha, D., Pal, S. et al. (2016). One-pot synthesis of folic acid encapsulated upconversion nanoscale metal organic frameworks for targeting, imaging and pH responsive drug release. *Dalton Transactions* 45 (45): 18120–18132. https://doi.org/10.1039/C6DT03237K.

Cui, S., Chen, H., Zhu, H. et al. (2012). Amphiphilic chitosan modified upconversion nanoparticles for in vivo photodynamic therapy induced by near-infrared light. *Journal of Materials Chemistry* 22 (11): 4861–4873. https://doi.org/10.1039/C2JM16112E.

DaCosta, M.V., Doughan, S., Han, Y., and Krull, U.J. (2014). Lanthanide upconversion nanoparticles and applications in bioassays and bioimaging: a review. *Analytica Chimica Acta* 832: 1–33. https://doi.org/10.1016/j.aca.2014.04.030.

Dai, Y., Ma, P.A., Cheng, Z. et al. (2012). Up-conversion cell imaging and pH-induced thermally controlled drug release from $NaYF_4$:Yb^{3+}/Er^{3+}@ hydrogel core–shell hybrid microspheres. *ACS Nano* 6 (4): 3327–3338. https://doi.org/10.1021/nn300303q.

Das, G.K., Chan, P.P.Y., Teo, A. et al. (2010). *In vitro* cytotoxicity evaluation of biomedical nanoparticles and their extracts. *Journal of Biomedical Materials Research, Part A* 93A: 337. https://onlinelibrary.wiley.com/doi/10.1002/jbm.a.32533.

Dong, B., Xu, S., Sun, J. et al. (2011). Multifunctional $NaYF_4$:Yb^{3+},Er^{3+}@Ag core/shell nanocomposites: integration of upconversion imaging and photothermal therapy. *Journal of Materials Chemistry* 21 (17): 6193–6200. https://doi.org/10.1039/C0JM04498A.

Dong, L., An, D., Gong, M. et al. (2013). PEGylated upconverting luminescent hollow nanospheres for drug delivery and in vivo imaging. *Small* 9 (19): 3235–3241. https://doi.org/10.1002/smll.201300433.

He, M., Huang, P., Zhang, C.L. et al. (2011). Dual phase-controlled synthesis of uniform lanthanide-doped $NaGdF_4$ upconversion nanocrystals via an OA/ionic liquid two-phase system for in vivo dual-modality imaging. *Advanced Functional Materials* 21: 4470–4477. https://doi.org/10.1002/adfm.201101040.

He, S., Krippes, K., Ritz, S. et al. (2015). Ultralow-intensity near-infrared light induces drug delivery by upconverting nanoparticles. *Chemical Communications* 51 (2): 431–434. https://doi.org/10.1039/C4CC07489K.

Hilderbrand, S.A., Shao, F., Salthouse, C. et al. (2009). Upconverting luminescent nanomaterials: application to in vivo bioimaging. *Chemical Communications* 28: 4188–4190. https://doi.org/10.1039/B905927J.

Hou, Z., Li, C., Ma, P. et al. (2011). Electrospinning preparation and drug-delivery properties of an up-conversion luminescent porous $NaYF_4$:Yb^{3+},Er^{3+}@silica fiber nanocomposite. *Advanced Functional Materials* 21 (12): 2356–2365. https://doi.org/10.1002/adfm.201100193.

Huang, X., Li, B., Peng, C. et al. (2016). $NaYF_4$:Yb/Er@PPy core–shell nanoplates: an imaging-guided multimodal platform for photothermal therapy of cancers. *Nanoscale* 8 (2): 1040–1048. https://doi.org/10.1039/C5NR06394A.

Idris, N.M., Li, Z., Ye, L. et al. (2009). Tracking transplanted cells in live animal using upconversion fluorescent nanoparticles. *Biomaterials* 30 (28): 5104–5113. https://doi.org/10.1016/j.biomaterials.2009.05.062.

Idris, N.M., Gnanasammandhan, M.K., Zhang, J. et al. (2012). In vivo photodynamic therapy using upconversion nanoparticles as remote controlled nanotransducers. *Nature Medicine* 18 (10): 1580–1585. http://www.nature.com/doifinder/10.1038/nm.2933 https://doi.org/10.1038/nm.2933

Idris, N.M., Lucky, S.S., Li, Z. et al. (2014). Photoactivation of core–shell titania coated upconversion nanoparticles and their effect on cell death. *Journal of Materials Chemistry B* 2 (40): 7017–7026. https://doi.org/10.1039/C4TB01169D.

Jalani, G., Naccache, R., Rosenzweig, D.H. et al. (2016). Photocleavable hydrogel-coated upconverting nanoparticles: a multifunctional theranostic platform for NIR imaging and on-demand macromolecular delivery. *Journal of the American Chemical Society* 138 (3): 1078–1083. https://doi.org/10.1021/jacs.5b12357.

Ji, S., Wu, W., Wu, W. et al. (2011). Ruthenium(II) polyimine complexes with a long-lived 3IL excited state or a 3MLCT/3IL equilibrium: efficient triplet sensitizers for low-power Upconversion. *Angewandte Chemie* 123 (7): 1664–1667. https://doi.org/10.1002/ange.201006192.

Johnson, N.J., Oakden, W., Stanisz, G.J. et al. (2011). Size-tunable, ultrasmall NaGdF4 nanoparticles: insights into their T_1 MRI contrast enhancement. *Chemistry of Materials* 23 (16): 3714–3722. https://doi.org/10.1021/cm201297x.

Kobayashi, H., Kosaka, N., Ogawa, M. et al. (2009). In vivo multiple color lymphatic imaging using upconverting nanocrystals. *Journal of Materials Chemistry* 19 (36): 6481–6484. https://doi.org/10.1039/B910512C.

Li, Z., Zhang, Y., Shuter, B., and Muhammad Idris, N. (2009). Hybrid lanthanide nanoparticles with paramagnetic shell coated on upconversion fluorescent nanocrystals. *Langmuir* 25 (20): 12015–12018. https://doi.org/10.1021/la903113u.

Li, C., Yang, D., Ma, P.A. et al. (2013). Multifunctional upconversion mesoporous silica nanostructures for dual modal imaging and in vivo drug delivery. *Small* 9 (24): 4150–4159. https://doi.org/10.1002/smll.201301093.

Lim, S.F., Riehn, R., Ryu, W.S. et al. (2006). In vivo and scanning electron microscopy imaging of upconverting nanophosphors in *Caenorhabditis elegans*. *Nano Letters* 6 (2): 169–174. https://doi.org/10.1021/nl0519175.

Liu, Y., Ai, K., and Lu, L. (2012a). Nanoparticulate X-ray computed tomography contrast agents: from design validation to in vivo applications. *Accounts of Chemical Research* 45 (10): 1817–1827. https://doi.org/10.1021/ar300150c.

Liu, J.N., Bu, W., Pan, L.M. et al. (2012b). Simultaneous nuclear imaging and intranuclear drug delivery by nuclear-targeted multifunctional upconversion nanoprobes. *Biomaterials* 33 (29): 7282–7290. https://doi.org/10.1016/j.biomaterials.2012.06.035.

Liu, K., Liu, X., Zeng, Q. et al. (2012c). Covalently assembled NIR nanoplatform for simultaneous fluorescence imaging and photodynamic therapy of cancer cells. *ACS Nano* 6 (5): 4054–4062. https://doi.org/10.1021/nn300436b.

Liu, J., Bu, W., Zhang, S. et al. (2012d). Controlled synthesis of uniform and monodisperse upconversion core/mesoporous silica shell nanocomposites for bimodal imaging. *Chemistry A European Journal* 18 (8): 2335–2341. https://doi.org/10.1002/chem.201102599.

Liu, J., Bu, W., Pan, L., and Shi, J. (2013). NIR-triggered anti-cancer drug delivery by upconverting nanoparticles with integrated azobenzene-modified mesoporous silica. *Angewandte Chemie International Edition* 125 (16): 4471–4475. https://doi.org/10.1002/ange.201300183.

Lucky, S.S., Soo, K.C., and Zhang, Y. (2015). Nanoparticles in photodynamic therapy. *Chemical Reviews* 115 (4): 1990–2042. https://doi.org/10.1021/cr5004198.

Ma, C., Bian, T., Yang, S. et al. (2014). Fabrication of versatile cyclodextrin-functionalized upconversion luminescence nanoplatform for biomedical imaging. *Analytical Chemistry* 86 (13): 6508–6515. https://doi.org/10.1021/ac5010103.

Muhr, V., Wilhelm, S., Hirsch, T., and Wolfbeis, O.S. (2014). Upconversion nanoparticles: from hydrophobic to hydrophilic surfaces. *Accounts of Chemical Research* 47 (12): 3481–3493. https://doi.org/10.1021/ar500253g.

Mukherjee, P., Kumar, A., Bhamidipati, K. et al. (2019). Facile strategy to synthesize magnetic upconversion nanoscale metal–organic framework composites for theranostics application. *ACS Applied Bio Materials* 3 (2): 869–880. https://doi.org/10.1021/acsabm.9b00949.

Nam, S.H., Bae, Y.M., Park, Y.I. et al. (2011). Long-term real-time tracking of lanthanide ion doped upconverting nanoparticles in living cells. *Angewandte Chemie International Edition* 50 (27): 6093–6097. https://doi.org/10.1002/anie.201007979.

Nyk, M., Kumar, R., Ohulchanskyy, T.Y. et al. (2008). High contrast in vitro and in vivo photoluminescence bioimaging using near infrared to near infrared up-conversion in Tm^{3+} and Yb^{3+} doped fluoride nanophosphors. *Nano Letters* 8 (11): 3834–3838. https://doi.org/10.1021/nl802223f.

Park, Y.I., Kim, J.H., Lee, K.T. et al. (2009). Nonblinking and nonbleaching upconverting nanoparticles as an optical imaging nanoprobe and T_1 magnetic resonance imaging contrast agent. *Advanced Materials* 21 (44): 4467–4471. https://doi.org/10.1002/adma.200901356.

Qian, H.S., Guo, H.C., Ho, P.C.L. et al. (2009). Mesoporous-silica-coated up-conversion fluorescent nanoparticles for photodynamic therapy. *Small* 5 (20): 2285–2290. https://doi.org/10.1002/smll.200900692.

Shen, S., Guo, X., Wu, L. et al. (2014). Dual-core@ shell-structured Fe_3O_4–$NaYF_4$@TiO_2 nanocomposites as a magnetic targeting drug carrier for bioimaging and combined chemo-sonodynamic therapy. *Journal of Materials Chemistry B* 2 (35): 5775–5784. https://doi.org/10.1039/C4TB00841C.

Shi, Z., Zhang, K., Zada, S. et al. (2020). Upconversion nanoparticle-induced multimode photodynamic therapy based on a metal–organic framework/titanium dioxide nanocomposite. *ACS Applied Materials & Interfaces* 12 (11): 12600–12608. https://doi.org/10.1021/acsami.0c01467.

Sun, Y., Yu, M., Liang, S. et al. (2011). Fluorine-18 labeled rare-earth nanoparticles for positron emission tomography (PET) imaging of sentinel lymph node. *Biomaterials* 32 (11): 2999–3007. https://doi.org/10.1016/j.biomaterials.2011.01.011.

Tian, Z., Chen, G., Li, X. et al. (2010). Autofluorescence-free in vivo multicolor imaging using upconversion fluoride nanocrystals. *Lasers in Medical Science* 25 (4): 479–484. https://doi:10.1007/s10103-009-0663-6.

Ungun, B., Prud'homme, R.K., Budijono, S.J. et al. (2009). Nanofabricated upconversion nanoparticles for photodynamic therapy. *Optics Express* 17 (1): 80–86. https://doi.org/10.1364/OE.17.000080.

Wang, M., Mi, C., Zhang, Y. et al. (2009). NIR-responsive silica-coated NaYbF4:Er/Tm/Ho upconversion fluorescent nanoparticles with tunable emission colors and their applications in immunolabeling and fluorescent imaging of cancer cells. *The Journal of Physical Chemistry C* 113 (44): 19021–19027. https://doi.org/10.1021/jp906394z.

Wang, F., Banerjee, D., Liu, Y. et al. (2010). Upconversion nanoparticles in biological labeling, imaging, and therapy. *Analyst* 135 (8): 1839–1854. https://doi.org/10.1039/C0AN00144A.

Wang, C., Cheng, L., and Liu, Z. (2011a). Drug delivery with upconversion nanoparticles for multifunctional targeted cancer cell imaging and therapy. *Biomaterials* 32 (26): 1110–1120. https://doi.org/10.1016/j.biomaterials.2010.09.069.

Wang, C., Tao, H., Cheng, L., and Liu, Z. (2011b). Near-infrared light induced in vivo photodynamic therapy of cancer based on upconversion nanoparticles. *Biomaterials* 32 (26): 6145–6154. https://doi.org/10.1016/j.biomaterials.2011.05.007.

Wang, C., Cheng, L., Liu, Y. et al. (2013). Imaging-guided pH-sensitive photodynamic therapy using charge reversible upconversion nanoparticles under near-infrared light. *Advanced Functional Materials* 23 (24): 3077–3086. https://doi.org/10.1002/adfm.201202992.

Wu, S., Han, G., Milliron, D.J. et al. (2009). Non-blinking and photostable upconverted luminescence from single lanthanide-doped nanocrystals. *Proceedings of the National Academy of Sciences* 106 (27): 10917–10921. https://doi.org/10.1073/pnas.0904792106.

Xiao, Q., Zheng, X., Bu, W. et al. (2013). A core/satellite multifunctional nanotheranostic for in vivo imaging and tumor eradication by radiation/photothermal synergistic therapy. *Journal of the American Chemical Society* 135 (35): 13041–13048. https://pubs.acs.org/doi/10.1021/ja404985w.

Xie, Z., Cai, X., Sun, C. et al. (2018). O2-loaded pH-responsive multifunctional nanodrug carrier for overcoming hypoxia and highly efficient chemo-photodynamic cancer therapy. *Chemistry of Materials* 31 (2): 483–490. https://doi.org/10.1021/acs.chemmater.8b04321.

Xing, H., Bu, W., Zhang, S. et al. (2012). Multifunctional nanoprobes for upconversion fluorescence, MR and CT trimodal imaging. *Biomaterials* 33 (4): 1079–1089. https://doi.org/10.1016/j.biomaterials.2011.10.039.

Xiong, L.Q., Chen, Z.G., Yu, M.X. et al. (2009). Synthesis, characterization, and in vivo targeted imaging of amine-functionalized rare-earth up-converting nanophosphors. *Biomaterials* 30 (29): 5592–5600. https://doi.org/10.1016/j.biomaterials.2009.06.015.

Xu, H., Cheng, L., Wang, C. et al. (2011a). Polymer encapsulated upconversion nanoparticle/iron oxide nanocomposites for multimodal imaging and magnetic targeted drug delivery. *Biomaterials* 32 (35): 9364–9373. https://doi.org/10.1016/j.biomaterials.2011.08.053.

Xu, Z., Li, C., Hou, Z. et al. (2011b). Monodisperse core–shell structured up-conversion Yb(OH) CO_3@$YbPO_4$:Er^{3+} hollow spheres as drug carriers. *Biomaterials* 32 (17): 4161–4173. https://doi.org/10.1016/j.biomaterials.2011.02.026.

Yang, Y., Qu, Y., Zhao, J. et al. (2010). Fabrication of and drug delivery by an upconversion emission nanocomposite with monodisperse LaF_3:Yb,Er core/mesoporous silica shell structure. *European Journal of Inorganic Chemistry* 33: 5195–5199. https://doi.org/10.1002/ejic.201000778.

Ye, C., Zhou, L., Wang, X., and Liang, Z. (2016). Photon upconversion: from two-photon absorption (TPA) to triplet–triplet annihilation (TTA). *Physical Chemistry Chemical Physics* 18 (16): 10818–10835. https://doi.org/10.1039/C5CP07296D.

Yu, X., Li, M., Xie, M. et al. (2010). Dopant-controlled synthesis of water-soluble hexagonal $NaYF_4$ nanorods with efficient upconversion fluorescence for multicolor bioimaging. *Nano Research* 3 (1): 51–60. https://doi:10.1007/s12274-010-1008-2.

Zako, T., Nagata, H., Terada, N. et al. (2009). Cyclic RGD peptide-labeled upconversion nanophosphors for tumor cell-targeted imaging. *Biochemical and Biophysical Research Communications* 381 (1): 54–58. https://doi.org/10.1016/j.bbrc.2009.02.004.

Zhang, P., Steelant, W., Kumar, M., and Scholfield, M. (2007). Versatile photosensitizers for photodynamic therapy at infrared excitation. *Journal of the American Chemical Society* 129 (15): 4526–4527. https://doi.org/10.1021/ja0700707.

Zhang, G., Liu, Y., Yuan, Q. et al. (2011). Dual modal in vivo imaging using upconversion luminescence and enhanced computed tomography properties. *Nanoscale* 3 (10): 4365–4371. https://doi.org/10.1039/C1NR10736D.

Zhao, J., Ji, S., and Guo, H. (2011). Triplet–triplet annihilation based upconversion: from triplet sensitizers and triplet acceptors to upconversion quantum yields. *RSC Advances* 1 (6): 937–950. https://doi.org/10.1039/C1RA00469G.

Zhou, J., Sun, Y., Du, X. et al. (2010). Dual-modality in vivo imaging using rare-earth nanocrystals with near-infrared to near-infrared (NIR-to-NIR) upconversion luminescence and magnetic resonance properties. *Biomaterials* 31 (12): 3287–3295. https://doi.org/10.1016/j.biomaterials.2010.01.040.

Zhu, X., Feng, W., Chang, J. et al. (2016). Temperature-feedback upconversion nanocomposite for accurate photothermal therapy at facile temperature. *Nature Communications* 7 (1): 1–10. https://doi.org/10.1038/ncomms10437.

Index

Upconverting Nanoparticles: From Fundamentals to Applications, First Edition. Edited by Vineet K. Rai.

q

r

s

t

u